Concepts in Gene Therapy

Concepts in Gene Therapy

Editors
Michael Strauss · John A. Barranger

W DE G

Walter de Gruyter · Berlin · New York 1997

Editors

Prof. Dr. Michael Strauss
Humboldt-University at
Max-Delbrück-Center for
Molecular Medicine
Robert-Rössle-Str. 10
D-13125 Berlin
Germany

Danish Cancer Society
Division of Cancer Biology
Strandboulevarden 49
DK-2100 Copenhagen
Denmark

Dr. John A. Barranger
Department of Human Genetics
University of Pittsburgh
Pittsburgh, PA 15261
USA

Cover illustration

In vivo expression of transgenic β-galactosidase activity in dog oral mucosal epithelium transfected with p-CMVβ-gal DNA using the Accell gene gun method (courtesy of A. L. Rakhmilevich and N.-S. Yang, see contribution pp. 107)

With 64 figures and 22 tables.

Library of Congress Cataloging-in-Publication Data

Concepts in gene therapy / editors, Michael Strauss, John A. Barranger.
p. cm.
Includes bibliographical references and index.
ISBN 3 11 014984 2 (alk. paper)
1. Gene therapy. I. Strauss, Michael. 1950– . II. Barranger, John A.
RB155.8.C66 1997
616'.042–dc21 97-31282
CIP

Die Deutsche Bibliothek – Cataloging-in-Publication Data

Concepts in gene therapy / ed. Michael Strauss ; John A. Barranger. – Berlin ; New York : de Gruyter, 1997
ISBN 3-11-014984-2

♾ Printed on acid-free paper which falls within the guidelines of the ANSI to ensure permanence and durability.

Converting and typesetting by: Knipp Medien und Kommunikation, Dortmund. – Printing: Karl Gerike GmbH, Berlin. – Binding: Heinz Stein, Berlin. – Cover Design: Hansbernd Lindemann, Berlin.
Printed in Germany.

Preface

Recombinant DNA technology and molecular biology have paved the way for the study of strategies for gene therapy. The first clinical trial was carried out in 1990. Despite these advances, progress remains uncertain in terms of clinical success. It was predictable that therapy would not be instant. Considerable basic and clinical research will need to be done to advance the approach. In this regard, a variety of aproaches has already been shown to work in animal models. Moreover, the purpose of the first clinical phase I trials was to demonstrate the safety of the gene transfer procedures rather than efficacy. It is obvious that moving from the laboratory to human patients is more than just upscaling of the procedures It requires consideration of many additional parameters. This book adresses these questions and provides guidance to achieving success.

Gene therapy will emerge as an important tool in medical therapy. We have invited many leading scientists in gene therapy research to contribute their ideas about developments in their special field of interest. The concept of the book is to provide thc reader trained in medicine or natural sciences as well as students with basic information about both the current state of the art and the most likely future developments. By discussing impediments and problems related to gene therapy, the book should help to inspire specialists and newcomers in the field to think about alternatives and more successful approaches.

The book is divided into six sections reflecting the major aspects of gene therapy research emphasis. Gene transfer vectors are clearly the bottle-neck for many kinds of gene therapy and, therefore, they are discussed first. Retroviral vectors which have been used in the vast majority of applications so far are introduced by Salmons and Günzburg with an emphasis on future developments including circumvention of complement-dependent inactivation, targeting of resting cells and achievement of tissue specificity.

Adenoviral vectors are introduced by Yeh, Perricaudet and colleagues. These vectors are efficient tools for gene transfer in vivo, but suffer from the problem of causing immune responses. The authors present alternative ways of getting past this problem. Another vector system with some special advantages is derived from herpes viruses. Goins, Glorioso and colleagues give a concise overview of the status of this vector type which shows great promise for gene delivery to the central nervous system. Adeno-associated viruses vectors are presented by McCarthy and Samulski. After the problems associated with these vectors are solved, these viruses may become the favorable vectors for certain applications. Viral vectors are by far the most efficient tools for gene transfer, but artificial vectors would have some advantages. Sandig and Strauss discuss various properties of viruses which could be useful for the development of future artificial vectors including mechanisms for cellular uptake, nuclear transport, integration or replication. One promising method of gene transfer in vitro and in vivo which is not based on viruses is targeted gene delivery by artificial particles designed for specific receptor binding. The pioneers of the method, Wu and Wu, summarize recent results obtained with this system. Finally, gene gun technology is introduced by Rakhmilevich and Yang as an alternative way for delivery of DNA to various tissues.

In the second section, the four major techniques for inactivation of gene function are introduced. Willnow and Herz describe the gene knockout technology and its application to the generation of animal models for genetic diseases. They focus on models for diseases of the lipid metabolism. Grifman, Soreq and coworkers discuss in detail the application of antisense oligonucleotides for therapeutic purposes and describe several model systems for therapy of neurodegenerative diseases. Heidenreich and Eckstein give an overview to the development and application of synthetic ribozymes which may be superior to simple antisense oligonucleotides with regard to the degree of inhibition of the target gene function if the issues of stability and toxicity can be solved. At the end of this section, Milich and Sullenger describe progress towards expression of ribozymes based on natural hammerhead or hairpin structures. The strategy of correcting gene defects by trans-splicing as an alternative to gene replacement is also discussed.

The third section deals with the requirements of various cellular target systems for gene transfer. Baum, Ostertag and colleagues describe the basics of gene transfer and transgene expression in hematopoietic cells, give a detailed background to the biology of this complex system and discuss not only the fields of application but also the biological and technical hurdles in achieving therapeutic levels of gene transfer and expression. Cichon and Strauss discuss aspects of the liver as a target for gene therapy focusing mainly on liver-specific gene delivery as well as problems which have to be solved, in particular the immunostimulatory activity of adenoviral vectors which appears to be most severe in the liver compared to other tissues. Liu, Barranger and coworkers discuss the requirements for gene therapy by *ex vivo* gene transfer to transplantable cells and argue for primary myoblasts as a suitable target cell for gene therapy of Gaucher disease. This target cell may also be important for other diseases where high level and long-lasting secretion of a gene product is required. Bohl and Heard introduce the history and development of artificial tissues, discuss secretion of therapeutic proteins from neo-organs and give good reasons why this strategy may be the first successful one in treatment of genetic diseases where a secreted gene product is required.

Some of the most important genetic disease which may be curable by gene therapy in the future are discussed in the fourth section. Coutelle presents the various basic and applied clinical aspects and problems related to the treatment of cystic fibrosis including results of the first clinical trials. Among the points to consider for future developments the strategy of fetal gene transfer is discussed. Dickson introduces the concepts for gene therapy for dystrophin deficiency and discusses the alternative ways of gene delivery. Gotthardt and Schuster elaborate on the genetic and biochemical background of familial hypercholesterolemia and delineate the alternative forms of treatment. This chapter also discusses the implications of the first clinical trial for the treatment of this disease. Barranger and coworkers introduce lysosomal storage disorders and strategies for their treatment.

From the large number of potential targets among non-genetic and infectious diseases we have selected two representative ones which are dealt with in the fifth section. Ghivizzani, Robbins and colleagues describe a strategy for gene therapy of arthritis. This non-curative approach may be of great importance as a model for treatment of disease-related pain. Böhnlein discusses the plethora of strategies for gene therapy for HIV-1 disease. This includes in particular intracellular immunization.

The sixth and final section is devoted to cancer. Culver introduces the suicide strategy and presents results from the various clinical applications. Chapters by Cayeux, Dörken and Blankenstein and separately by Schmidt discuss the topic of tumor cell vaccination using genetically modified cells. Whereas the first of the two contributions focuses at cooperative effects between B7.1 and IL-4 or IL-7, the second one illuminates the dosage problem. In the last article, Strauss, Brand and Sandig deal with tumor suppressor genes as therapeutic tools in the context of adenoviruses. In this chapter, the authors argue for the development of combinatorial approaches which target the major deregulated genes of the cell cycle machinery.

It is the hope of the editors that this book not only fills a gap in the existing spectrum of books dealing with gene therapy but will also stimulate further thoughts about the various strategical, technical and logistical aspects of gene therapy. The potential of gene therapy is great, but we are still at the very beginning. New ideas and technologies are required before gene therapy can establish itself as a clinical tool. It is a field where basic researchers and clinicians have to collaborate to make the idea work. We hope that students and colleagues are inspired by some of the ideas or strategies discussed in this book

Michael Strauss
John A. Barranger

Summer 1997

Contents

II Methods for Inactivation or Restoration of Gene Function

I Methods of Gene Delivery

1. Retroviral Vectors

Brian Salmons and Walter H. Günzburg

1.1 Introduction

Retroviruses are unique in that they carry two copies of a single stranded RNA genome in the virus particle but, upon infection of a target cell, this RNA is converted into a double stranded DNA form and integrated into the target cell genome. The synthesis of viral DNA from RNA is performed by a virally transmitted enzyme activity (reverse transcriptase). A second viral enzyme (integrase) is responsible for the integration of this double stranded DNA form into host cell chromosomal DNA. The integrated DNA form is termed a provirus. Integration of the provirus is essentially random with respect to the host cell chromosomal DNA, although there may be some preference for actively transcribed regions. A consequence of integration is that the provirus is stably inherited by all the offspring or daughter cells of the originally infected cell, as if it were a normal cellular gene. It is this property of retroviruses that has made them attractive as the basis of gene transfer vehicles (vectors).

Many different retroviruses have been isolated from a wide range of different species, most commonly from mice, birds and cats but also from others including humans. Often they have been isolated because of their association with tumour formation. Historically these viruses were divided into acutely transforming and slow transforming viruses. The acutely transforming viruses were found to have acquired cellular genes (oncogenes), which normally play important roles in cell cycle control, in place of one or more viral structural gene, as a result of a rare recombination event. The acquired oncogenes became a stable part of the genetic make-up of these acutely transforming viruses and were transfered, along with the viral genome, into the infected target cell. The resulting deregulated expression of these oncogenes, under the control of the viral regulatory elements, contribute to tumorigenesis. Ironically it was this observation that lead to the idea of using retroviruses as vector systems: instead of oncogenes, therapeutic or other genes of interest could be inserted into the retroviral genome using focused molecular biological means. Such genes would then be delivered to the target cell in the same manner as the oncogenes contained within the naturally occuring acutely transforming retroviruses.

Although evidence has accumulated indicating that non-oncogene bearing retroviruses are involved in the tumorigenic process in animals, it is generally accepted that retroviruses, because of the multistep nature of tumorigenesis do not play the overriding role. Hence the name slow-transforming for such non-oncogene bearing retroviruses. The mechanism by which non-oncogene bearing retroviruses are implicated in tumorigenesis involves their essentially random integration into, or in the vicinity of, a cellular gene involved in growth control (proto-oncogene or tumour suppressor gene). Since the chance of a single retroviral integration occuring in such a gene locus is extremely low, multiple successive integrations are required before such an event is even likely to occur. This necessitates a retrovirus capable of integration, production of new retroviral particles, followed by their integration, another phase of virus production, and so on. Normal, non-oncogene carrying retroviruses can follow such a course, however, as will become apparent in the following chapter, recombinant retroviruses, designed to carry therapeutic genes for gene therapy, do so at the expense of viral gene information. Thus the resulting viruses are replication defective, only able to integrate once, and not able to produce any new retroviral particles. The chance of such a retroviral vector mediated single integration event causing a genetic change involved in tumorigenesis is, as already pointed out above, extremely low, and, when weighed against the possible benefits of retrovirally mediated therapeutic gene transfer, negligible.

1.2 General Retroviral Structure and Replication

Retroviruses are enveloped viruses with a diameter of 80-120nm (Fig. 1.1). The envelope is derived from the plasma membrane of the virus producing cell and carries, inserted into it, the viral transmembrane (TM) envelope (Env) protein. Linked to the TM protein, on the outside of the virion, is the viral surface Env protein (SU) which interacts specifically with the viral receptor on the surface of the target cell, mediating entry. The presence or absence of an appropriate receptor on the target cell determines whether the cell can be infected by a particular virus. The receptors used by the retroviruses for entry are normal cellular proteins with normal household functions. To date only a few of these proteins that have an additional function as viral receptors have been identified (1). The most commonly used retrovirus for vector construction is the Murine Leukaemia Virus (MLV). The receptors for two forms of the MLV SU protein have been identified. The ecotropic SU protein interacts with an amino acid transporter protein on the target cell surface (2-3). The active protein is only expressed on murine cells and thus limits the infection spectrum of the ecotropic MLV to cells of murine origin. In contrast, the amphotropic SU protein interacts with a cellular phosphate transporter protein (4-5), extending infection of the amphotropic MLV to non-murine cells. Secondary receptors may be required for completion of the infection event (6).

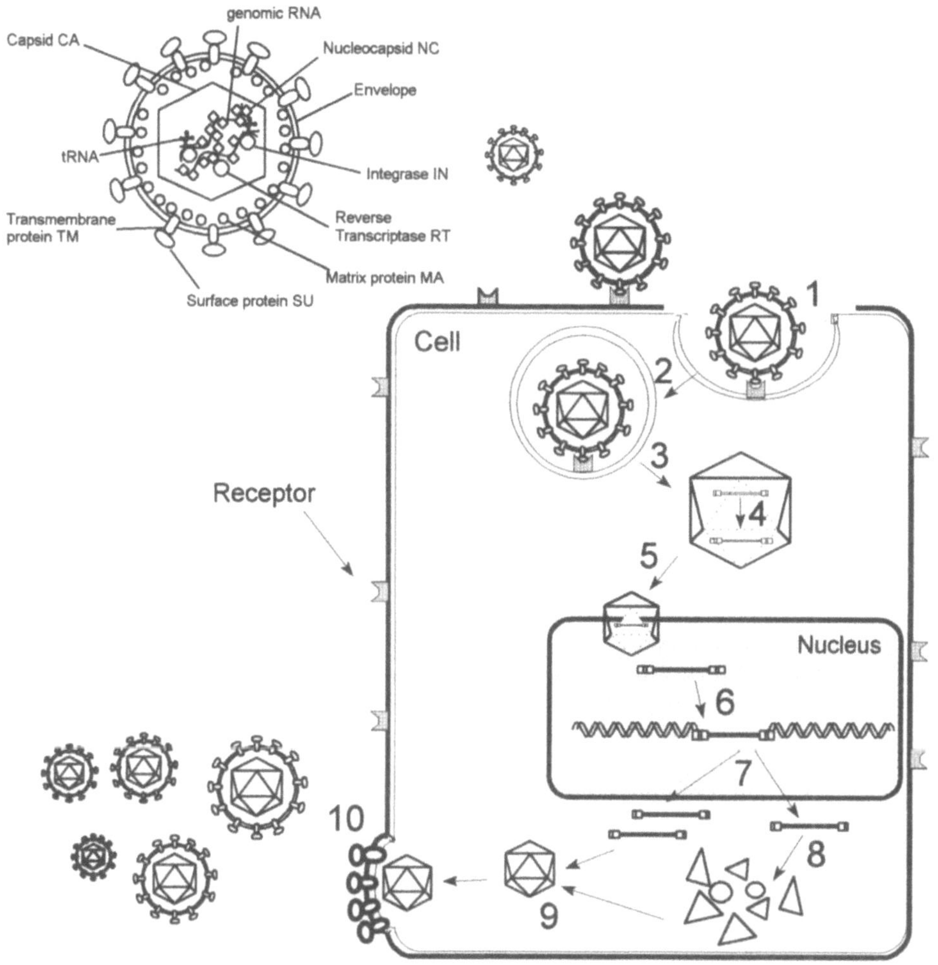

The specific interaction between the SU protein and the cellular receptor is thought to result in changes in the conformation of both the Env protein and the receptor necessary for the entry of the virus into the cell. Two virus uptake mechanisms have been proposed: either cellular uptake by micropinocytosis into vesicles followed by fusion (e.g. ecotropic MLV) or direct extra-cellular fusion of the viral envelope with the cell membrane (7). The best studied example of the latter is Human Immunodeficiency Virus (HIV) in which the SU protein gp120 is shed, thereby unmasking a fusogenic domain carried by the viral TM which leads to fusion of the viral membrane with that of the cell to be infected (6). Viral entry and fusion does not seem to be a prerequisite for later reverse transcription and expression (8), although certain conformational changes may be necessary. After fusion, the viral envelope is lost (uncoating) releasing the viral core or capsid (Gag) into the target cell cytoplasm. The capsid is composed of the Matrix (MA) and capsid (CA) proteins surrounding a

◁ Fig. 1.1: The structure of a retrovirus particle and the infection/replication cycle.
A schematic diagram of a retrovirus particle is shown. Two identical single stranded RNA molecules form the genome of the virus. Each RNA molecule is associated with a cellular tRNA (used after infection as a primer for DNA synthesis) as well as with an enzyme complex consisting of an RNA dependent DNA polymerase (reverse transcriptase or RT) with associated RNase H activity, and integrase. The genomic RNA is also protected by viral nucleocapsid (NC) proteins. This whole RNA-protein nucleocapsid is encased in a capsid or core structure which consists of a complex of virus capsid proteins. The outermost covering of the virus particle consists of a modified host cell plasma membrane into which are inserted the two virus envelope proteins. One of these is a transmembrane protein that links the virus outer layer or envelope to the second, surface or SU, virus envelope protein. The virus envelope is linked to the capsid by the matrix protein.
The infection/replication cycle is also shown. (1) Infection is initiated by an interaction between the virus SU proteins and specific receptors located on the surface of the host cell. (2) This interaction results in uptake and internalisation of the virus particle which then leads to (3) the loss of the virus envelope (uncoating) which occurs by fusion with either the host cell plasma membrane (not shown) or with endosomal vesicle membranes. (4) The liberated viral core undergoes a reorganisation of its structure, and reverse transcription of the virus genomic RNA into a double stranded DNA form takes place, using the tRNA as a primer. (5) The complex then relocates to the nucleus where (6) the double stranded DNA molecule is inserted into the host cell genomic DNA by the viral integrase enzyme. (7) The integrated DNA form of the virus genome, known as the provirus, is then transcribed like any cellular gene by the host cell transcription machinery from the viral promoter. (8) Newly transcribed viral RNAs are then used for translation to virus proteins and as the genomic RNA for virus particles. (9) The virus proteins and RNA assemble into new cores or capsids which interact with regions of the host cell membrane that contain the newly synthesised virus envelope proteins and (10) newly produced virus particles bud out of the infected cell where they undergo the final stages of maturation to form fully infectious virus particles.

nucleoprotein complex consisting of two identical strands of the viral genomic RNA associated with the Gag nucleocapsid (NC) protein. The viral enzymes encoded by the Pol region are also associated with the NC-RNA complex.

Conformational changes in the capsid that occur during or after its liberation into the host cell cytoplasm, activate the Pol encoded enzyme reverse transcriptase (RT), which then synthesises a double stranded DNA copy of the viral RNA. During this process unique sequences at either end of the viral genomic RNA are duplicated and placed at both ends of the newly synthesised DNA, generating a relatively long repeated sequence at each end of the molecule, termed a long terminal repeat (LTR). The double stranded DNA copy, flanked by the two LTRs, is translocated to the nucleus. Nuclear entry requires nuclear membrane breakdown, and thus is limited to actively dividing cells (9-10). One exception to this rule is HIV, which has the ability to infect, albeit less efficiently, non-dividing cells (11-12). This appears to be due to the presence of a nuclear localization signal in the HIV MA protein (13) as well as the HIV accessory gene product Vpr (14) which in a redundant fashion, influence nuclear localization (15).

Once in the nucleus the viral DNA copy is maintained episomally in a linear form before becoming integrated (16-17). During the process of integration, the viral Pol encoded integrase (IN) enzyme recognizes the terminal bases of the LTRs and generates two 2bp long overhanging ends by staggered cleavage. Concurrently, the integrase cuts the cellular DNA in such a way as to leave a 4 to 6bp long overhanging end. The site of cellular DNA cleavage appears to be essentially random, although both active chromatin and sites positioned within the major groove of the DNA appear to be favoured (reviewed in 18). The overhanging viral and cellular ends are ligated even though they may not be complementary. After ligation such non-complementary stretches, as well as any single strand gaps, are repaired by cellular enzymes, leading to the duplication of 4-6bp at the integration site. The result is an integrated DNA form, known as a provirus (since transcription and translation of this molecule gives rise to new progeny virus particles) which is 4bp (two from each end) shorter than the pre-integrative DNA form. One important feature of the integrated provirus, also relevant for gene therapy applications, is that it always is found as a co-linear DNA with the structure LTR-retroviral genes-LTR. This is in sharp contrast to other integrated DNAs after DNA virus or naked DNA integration, which are present in the genome as permutated forms.

After integration, the LTR plays a second important role in that it is the major region of the viral DNA that both regulates transcription and determines the sites of transcription initiation and termination (19). Retroviruses carry at least three genes which encode the viral core proteins (gag gene) forming the inner structure of the virus, the enzymes reverse transcriptase and integrase (pol gene) and the viral envelope proteins (env gene) that interact with cellular receptors thereby initiating the infection event. Transcription is regulated by promoter and enhancer elements within the 5' LTR, but proceeds as for any other cellular gene. Additionally some retroviruses (e.g. HIV) encode trans-regulatory factors, or contain cis-regulatory elements that affect the quantity and nature of the transcribed RNA (reviewed by 20). Initially full length transcripts are generated, a fraction of these are then processed to subgenomic forms resulting from one (e.g. MLV) or multiple (e.g. HIV) splicing event(s). The Gag and Pol proteins are translated as polyprotein precursors from the genomic length transcripts. These precursor proteins are sequentially proteolytically cleaved into the mature Gag (MA;CA;NC) and Pol (RT;IN) proteins, a process that is autocatalytic (a protease domain (PR) is present in the precursor protein) and that is an ongoing process as the virus assembles, buds from the host cell and matures outside of the host cell. Other domains of these precursor molecules, such as the NC region, specifically interact with genomic length RNA destined to become the progeny virus genome (21-22). This ensures that viral RNAs are packaged within the progeny virus, rather than cellular RNAs. This interaction requires one or more specific RNA sequences, known as packaging signals, on the genomic RNA, some of which are obligatory and located at the beginning of the Gag region, just 3' of the Splice Donor (SD) for subgenomic RNA production (see below), ensuring that only full length viral RNAs can be packaged.

The Env proteins (SU and TM) are translated from a spliced subgenomic viral RNA. These proteins are synthesised as a precursor form with the structure NH_2-SU-TM-COOH. This precursor protein carries a rapidly removed signal peptide at the amino-terminus, and thus enters the secretory membrane system of the host cells. During its passage through the secretory system, the precursor becomes modified by a number of glycosylation steps and is finally cleaved into the mature SU and TM proteins by a cellular protease. However both proteins remain associated by disulphide bonds and are finally inserted into the host cell membrane.

The final steps of assembly and release of progeny virions occur at the cell membrane. As a result of the interaction between NC proteins and full length viral RNAs carrying the packaging signal, two RNA molecules become associated with the Gag proteins to form a new core (21-22). In addition the core contains the enzymes RT and IN required for reverse transcription and integration in the target cell, as well as tRNA primers needed to initiate this process. The core buds through the cell membrane in areas where the concentration of inserted viral Env proteins is relatively high. The newly released virions undergo further maturation steps associated with protein cleavage, conformational changes and reorganisation of viral proteins within the core to give the final infectious progeny virions. It should be stressed that the whole process of retroviral infection and virus production, at least for MLV, does not harm the producing cell in any obvious way.

1.3 Retroviral Vector Development

The first retroviral vector systems were most often derived from murine leukaemia virus (MLV). There were (and are still) a number of reasons for choosing MLV as a basis for such gene delivery systems, including that the biology of this retrovirus is particularly well understood, that the MLV genome was among the earliest retroviral genomes molecularly cloned and that these viruses are able to infect cells efficiently in culture. Retrovirus vector systems have been constructed based upon other retroviruses in the meantime, notably human and simian immunodeficiency viruses, bovine leukaemia virus and mouse mammary tumour virus (23), but the titres of virus produced by these systems are often much lower than those based on MLV. Even so, vector systems based on these viruses deserve further research to improve their gene transfer capabilities since they may be useful for gene delivery to certain target cells. HIV based vectors for example may prove invaluable for delivery of therapeutic genes to T-lymphocytes. Retrovirus like elements (retrotransposons) such as VL30 have also been used as a vector system. These elements do not encode virus structural proteins and thus are dependent on MLVs to provide such proteins (24-25). VL30 based vectors may reduce the possibility of producing replication competent virus (see below).

The very first attempts at retroviral vector construction involved a retroviral vector in which heterologous genes had been cloned between the LTRs at the expense of retroviral structural genes. With the exception of Rous sarcoma virus (RSV), retroviruses do not have the capacity to carry additional genetic information. This is presumably due to the inability of viral RNAs longer than the usual genome length to be efficiently packaged into virus particles. It is thus not feasable to insert additional genes (such as therapeutic genes) into retroviral proviruses without removing one or more retroviral genes. However, this renders the retrovirus defective since the products of the gag, pol and env genes are all required for the production of infectious virus. These replication incompetent vectors can be rescued by use of a complete functional retrovirus able to supply, in trans, all retroviral proteins necessary for the life cycle. In this original scenario, both the normal complete full length genomic RNA from the rescuing retrovirus, as well as the full length RNA carrying the heterologous gene sequence from the retroviral vector, could be packaged into progeny virions (Fig. 1.2). It soon became apparent that the normal full length RNAs had a tremendous packaging advantage over those RNAs carrying heterologous sequences, resulting in only minimal yields of recombinant virus. Most of the virus recovered was the rescuing retrovirus. Now we know that this was at least in part due to the presence of facultative packaging signals (see below) throughout the retroviral genome. Substitution of retroviral sequences with heterologous genes always leaves the recombinant retrovirus at a packaging disadvantage compared to the normal standard virus. The only way around this problem was to create a system in which the normal virus could not compete for packaging.

One way to circumvent this problem is to establish cell lines that provide the necessary Gag, Pol and Env viral proteins in trans. The RNA that encodes these proteins should however not be available for packaging into virions. Such retroviruses were found to exist naturally and molecular analysis of their genome revealed that a sequence located between the SD for env transcripts and the AUG iniation codon for Gag protein synthesis was lacking, but that this deletion did not affect protein synthesis or virus production (26). This implies that the RNA sequences required for packaging are located in this region. This finding is supported by the observations that (1) replication competent retroviruses usually package full length genomic RNA and only very rarely subgenomic transcripts suggesting that the signals required for packaging are located within the sequences that are not shared by these two transcripts and (2) acutely transforming retroviruses have lost most of the structural gene regions but yet are still packagable. In fact, it appears that multiple signals may be involved in packaging, including sequences located in the U5 region (27) but by far the most dominant of these is the Ψ region located at the 5' end of the gag gene (28).

These thoughts lead to the first real retroviral vector systems, which exploited the previously observed packaging signal mutation, preventing the packaging of such mutated RNA. Two components were constructed: (1) the vector which basically consisted of the flanking LTRs (essential for reverse transcription and integration) and the packaging signal Ψ allowing the rest of the space (up to a maximum of the natural retroviral genomic length) to be used for the cloning of heterologous genes; and (2) the packaging cell carrying a complete retrovirus with a mutated Ψ signal (Fig. 1.3) such as Ψ2 producing ecotropic virus (28) or Ψ-am producing amphotropic virus (29). Upon transfection of the retroviral vector into the packaging cell line, the retroviral vector RNA, because it carried the functional Ψ, became packaged into the retroviral proteins produced by the packaging construct. The disadvantage of this kind of vector system was that a single recombination event between the packaging

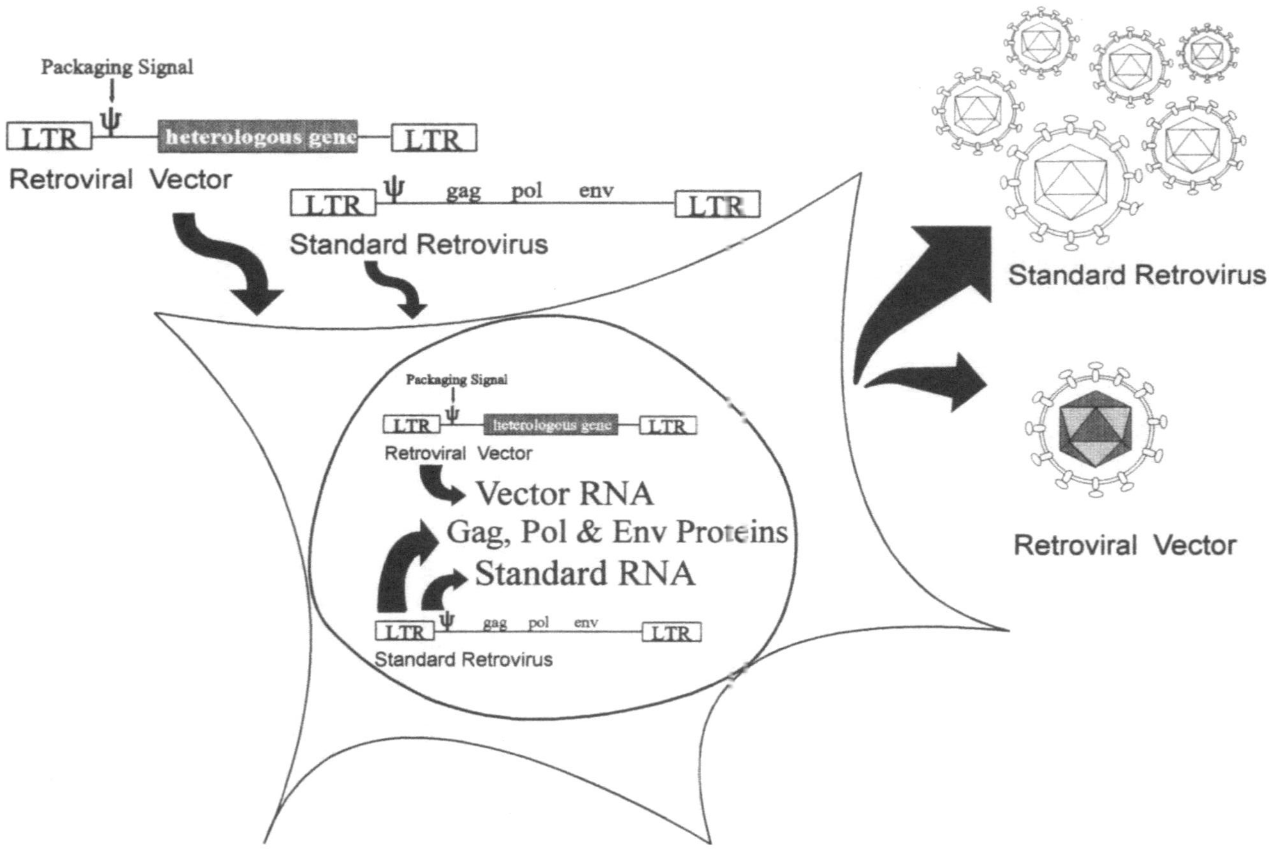

Fig. 1.2: Retroviral vector virus produced with help from standard retrovirus.
A cell line is produced that contains a retroviral vector construct with the structure LTR-heterologous gene-LTR. The necessary virus proteins for virus particle production are supplied by a standard, replication competent virus, usually the same virus from which the vector is derived. Both the vector RNA and the standard virus genomic RNA are packaged into newly synthesised virus particles since both carry the packaging signal Ψ. However, because of faculatitive packaging signals present on the standard virus genomic RNA but absent from the vector RNA, almost all of the virus particles produced are of the standard type.

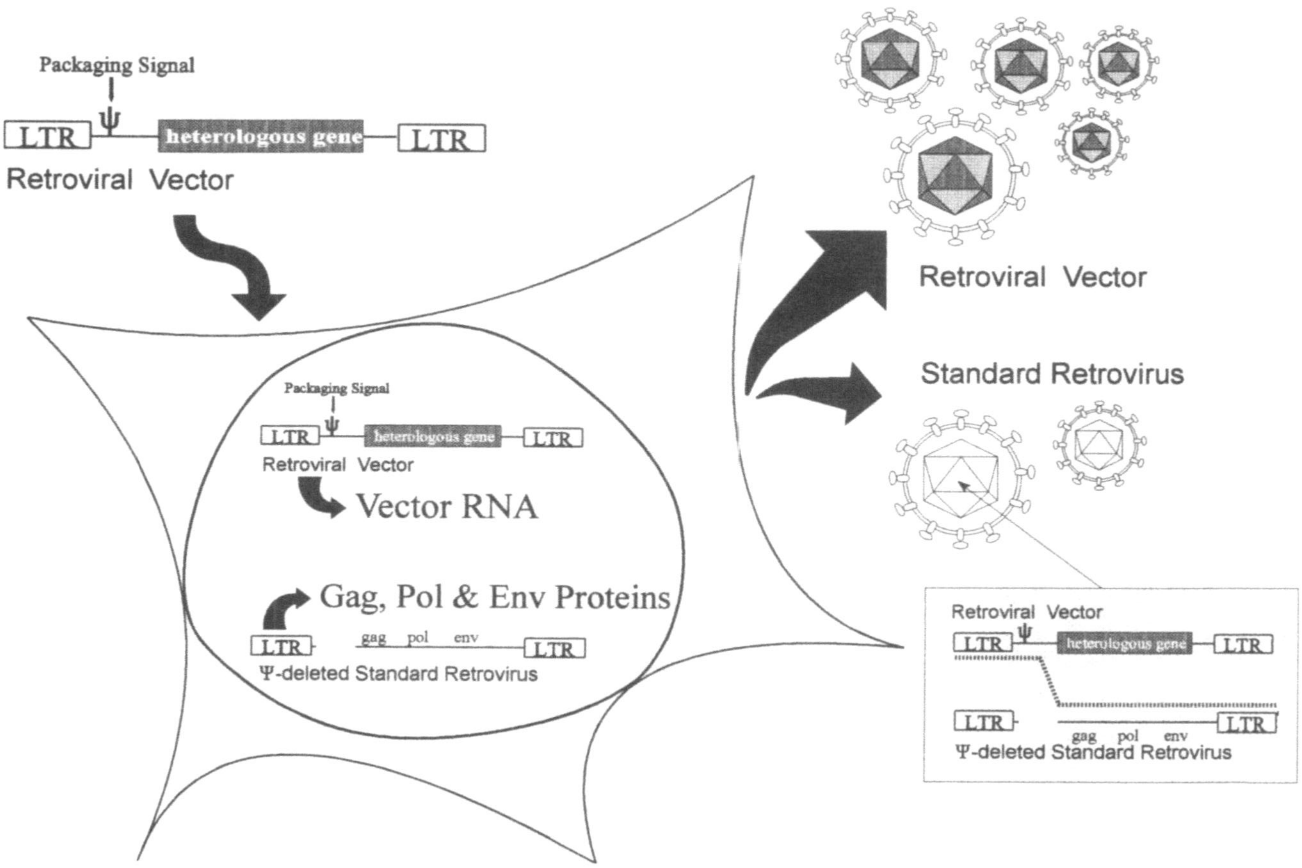

Fig. 1.3: Retroviral vector produced from first generation packaging cells.
The vector construct is introduced into a packaging cell line that carries a retroviral provirus which no longer contains the packaging signal **Ψ**. This construct gives rise to normal levels of viral structural proteins and enzymes but the RNA that is used for the translation of these proteins is poorly able to be packaged since it lacks the **Ψ** signal. Initially almost all virus particles produced from this system carry the retroviral vector genome. However, one recombination event between the vector and the packaging construct can lead to standard, replication competent, virus production.

construct and the vector construct would recreate the normal standard virus. Given the length of homology shared by the two constructs this would be predicted to occur fairly frequently as indeed was observed in practice (30-31). Not only does the generation of such replication competent standard virus reduce considerably the effective titres of recombinant virus, but additionally it poses a safety concern.

The retroviral vector construct, which was packaged in the packaging cell line, produced as recombinant virus, and then able to infect one target cell, was effectively immobilized in the target cell, unable to get out by forming new virus particles due to the lack of retroviral proteins required for virus assembly in the target cell. This meant that each retroviral vector could only infect one cell, integrate at one site, and hence that the chance of such an integration affecting a cellular gene involved in growth control was negligible. In contrast, the standard retrovirus was by no means immobilized in the target cell. It carried with itself all the genetic information needed to make new virions, and hence could amplify itself and infect target cells ad infinitum. Here the chance of one of these exponentially increasing number of retroviral integration events eventually affecting an important cellular gene was very real. The production of such Replication Competent Virus (RCV) is unacceptable for gene therapy applications and thus efforts were made to improve safety in a second generation of vector system.

Second generation retroviral vector systems were designed to include additional alterations thus requiring at least two recombination events to take place before a replication competent virus could be produced (Fig. 1.4). The packaging cell line PA317, producing amphotropic virus, carries a retroviral construct with a deletion in the 5' LTR, a mutated Ψ signal, and a SV40 polyadenylation signal in place of the 3'LTR (32). Unfortunately these vectors have still been shown to undergo recombination experimentally (33). When used in combination with a retroviral vector that carries an artificially inserted stop codon in the Gag reading frame instead of the Gag AUG, any recombinants generated however will not be able to express their Gag and Pol proteins and thus virus assembly and release will be inhibited (34-36).

Currently most retroviral vector systems are based upon the third generation concept in which the Gag and Pol gene products are expressed from one construct and the Env products from another in the packaging cell line (Fig. 1.5). In these safe vector systems at least three recombination events are required for production of RCV. The best known examples of these third generation packaging cell lines are the GP+E-86 line producing ecotropic virus and the GP+envAm12 producing amphotropic virus (37-38). The observation that direct repeats are unstable in retroviral vectors, often leading to the loss of sequences located between the repeats has also been exploited to prevent the generation of RCV. SNV based retroviral vectors have been constructed in which the packaging sequences were flanked by direct repeats. Upon infection of target cells, it could be demonstrated that the majority of the vector proviruses had deleted the packaging sequences, presumably due to template switching during reverse transcription (39).

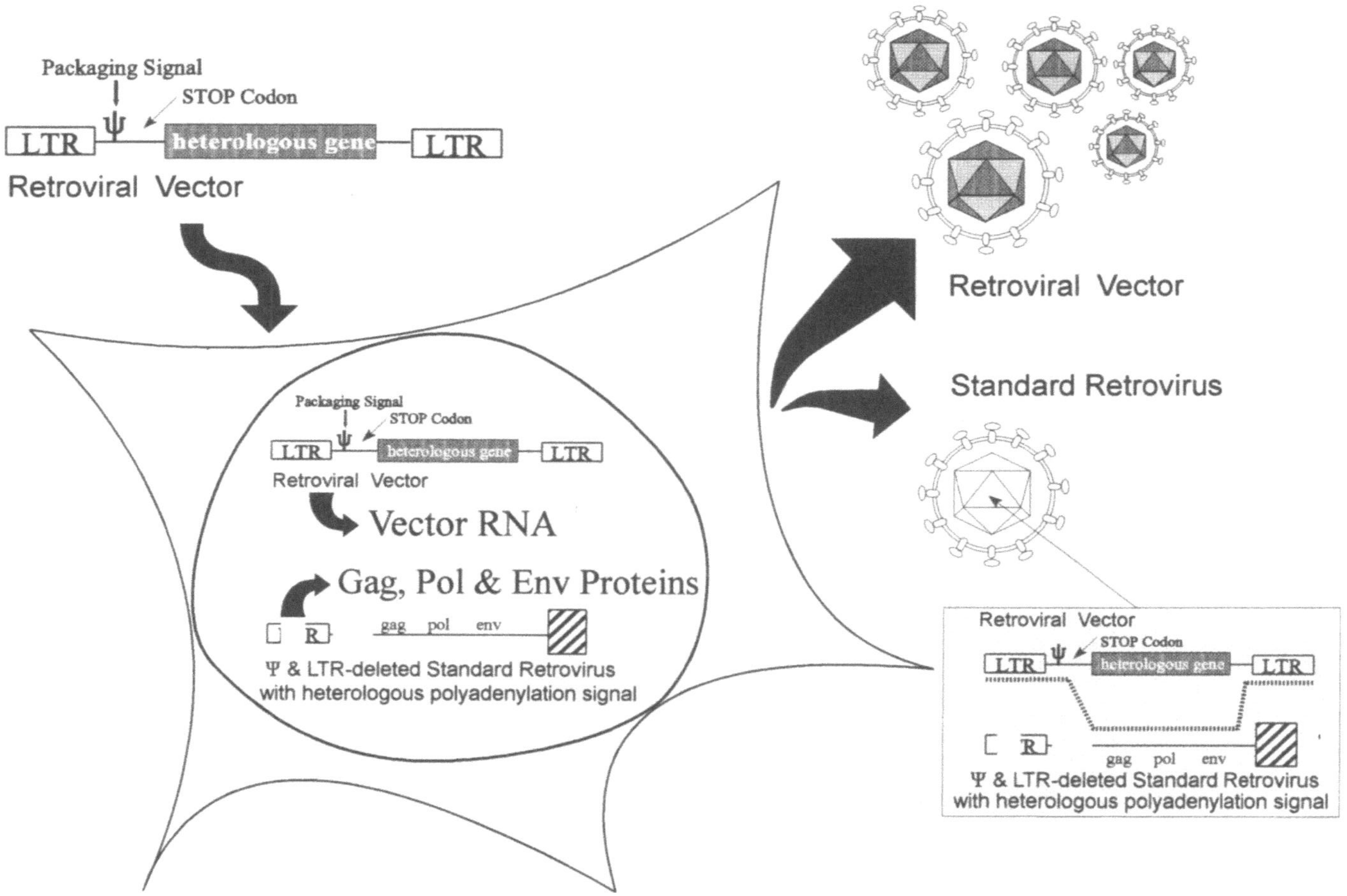

Fig. 1.4: Retroviral vector produced from second generation packaging cells.
The vector construct is introduced into a packaging cell line that carries a modified retroviral provirus lacking part of the 5'LTR and the packaging signal, and in which the 3'LTR has been replaced by heterologous polyadenylation signals (shaded box). Two recombination events between the vector and the packaging construct must occur before standard, replication competent, virus production can occur.

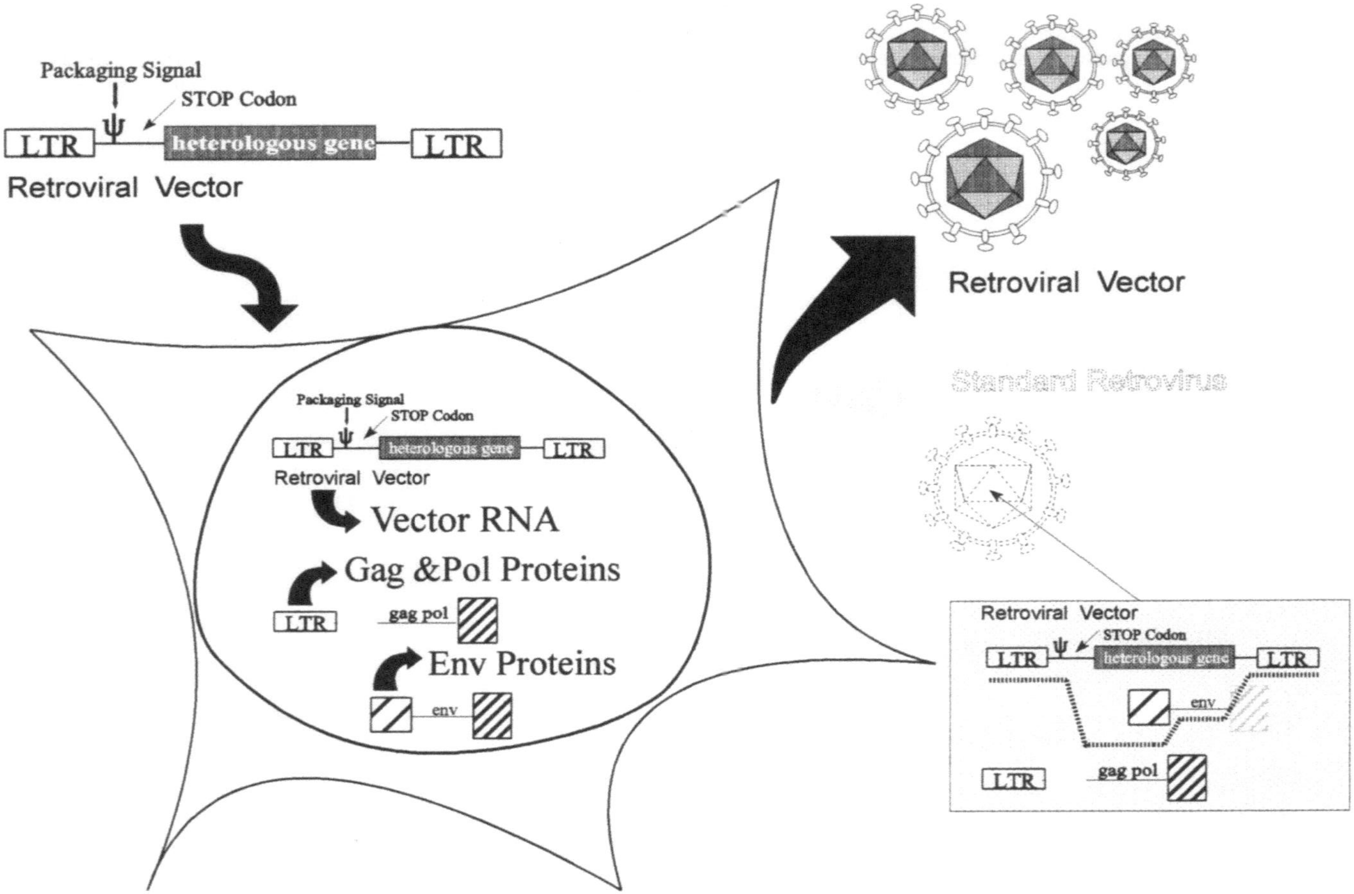

Fig. 1.5: Retroviral vector produced from third generation packaging cells
The vector construct is introduced into a packaging cell line that carries two independent constructs, one of which expresses the gag and pol gene products and a second that is used for expression of the Env proteins. At least three recombination events between the vector and the two packaging constructs are required for the production of standard, replication competent, virus.

The heterologous genes carried in retroviral vectors can be expressed in a number of different ways (Fig. 1.6). The retroviral promoter within the LTR can drive heterologous gene expression when such genes are cloned into the position formerly occupied by Gag. Indeed, in many applications where it is advantageous to transfer and express two genes, for instance a therapeutic gene and a marker gene, the second gene can be inserted into the Env position and expressed of the subgenomic viral RNA. Unfortunately the retroviral promoter is not particularly powerful and also suffers the disadvantage of being shut-down or silenced after a variable period in many cell types (40-41). Methylation and host factors have been identified as culprits for this shut-off, and efforts have been made to alter retroviral LTR and downstream sequences to prevent transcriptional silencing in certain cell types (reviewed in 42).

Heterologous promoters have also been used to express heterologous genes (Fig. 1.6) either within the body of the retroviral vector in syn or anti orientation, or within the LTR region in double copy (DC) vectors (43). DC vectors utilise a unique feature of the retroviral life cycle, the reverse transcription of viral genomic RNA into a double stranded form. During this process sequences from the 5' end of the RNA are duplicated and placed additionally at the 3' end of the DNA, whilst sequences from the 3' end of the RNA are copied onto the 5' end of the DNA. The process generates the identical LTR structures which flank the viral genome. Inclusion of heterologous promoter-gene cassettes into the 3' sequence which becomes duplicated after reverse transcription ensures that the promoter-gene is present twice (i.e. in double copy) in the target cell.

A similar strategy has been utilized to create a series of vectors which carry heterologous promoters or enhancer elements within their LTRs, so as to transcribe viral RNA from the normal start site, but under the control of the heterologous promoter. If this promoter is a tissue specific or conditional promoter, regulated transcription can be achieved (Fig. 1.6). Retroviral vectors carrying heterologous regulatory elements in place of (44) or in addition to (45) the retroviral regulatory elements, have been used to direct expression of genes carried by the retroviral vector to specific cell types. The entire promoter – enhancer region can be removed from the retroviral vector and substituted with heterologous promoters from genes that are expressed in a tissue specific manner (46-47). These vectors, called ProCon vectors since they undergo promoter conversion, may be safer vectors since they entirely lack viral promoter sequences thereby reducing the frequency of recombination with viral sequences in the producer or target cell. Further it has yet to be shown that a promoter from a cellular gene can activate or inactivate cellular genes in the context of a retrovirus. The strategy of tissue specific targeting, either at the level of expression or at the level of infection will be of paramount importance for future in vivo gene therapy protocols (reviewed in 48).

The use of internal ribosome entry sites (IRES) has been pioneered to allow two or more genes to be expressed from the same transcript expressed from a single promoter (49-50). Such vectors are reported to give higher titre, permit the insertion of larger heterologous gene segments, and show more stable expression of trans-

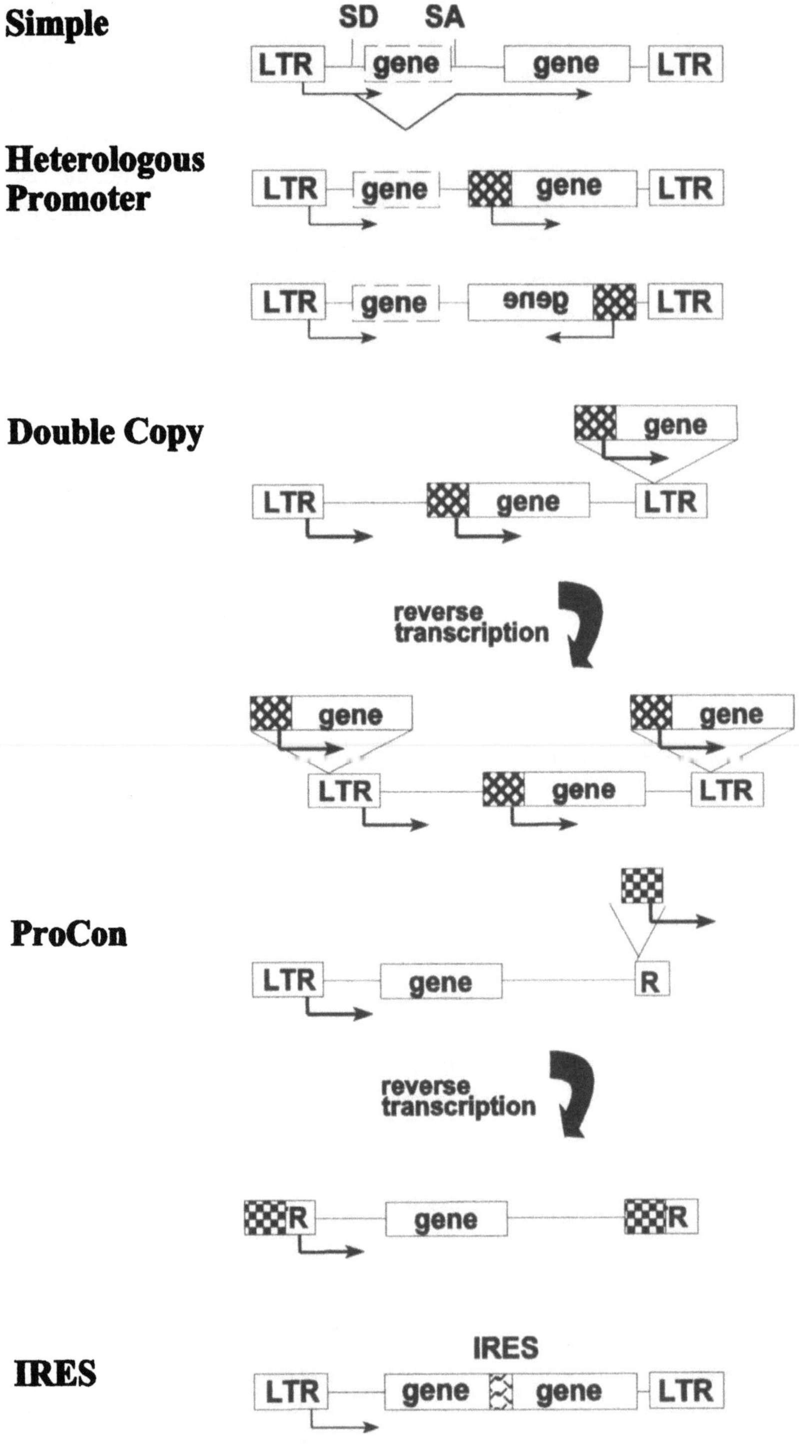
Simple
SD
SA
LTR
gene
gene
LTR
Heterologous Promoter
LTR
gene
gene
LTR
LTR
gene
gene
LTR
Double Copy
gene
LTR
gene
LTR
reverse transcription
gene
gene
LTR
gene
LTR
ProCon
LTR
gene
R
reverse transcription
R
gene
R
IRES
IRES
LTR
gene
gene
LTR

◁ Fig. 1.6: Classes of retroviral vector constructs.
A number of different types of retroviral vectors are shown. The simple type carry one or two heterologous genes (dotted box and closed box) in place of the gag, pol and env genes. Both of the genes are under the transcriptional control of the MLV promoter contained within the retroviral long terminal repeat (LTR). The first gene is expressed from full length transcripts (analogous to the genomic gag-pol transcript) and the second gene is expressed as a spliced transcript (analogous to the env transcript). Alternatively, the second gene can be expressed from an internal heterologous promoter either in the sense or antisense orientation. Double copy (DC) retroviral vectors carry a gene, often linked to a promoter, in the 3'LTR of the vector. As a consequence of infection and reverse transcription this heterologous expression unit becomes duplicated so that a second copy is present in the 5'LTR.
ProCon retroviral vectors carry a heterologous promoter (often from a gene that is expressed in one or a few cell types for targeting purposes) in place of the entire viral enhancer/promoter (U3) normally present in 3'LTR. After infection and reverse transcription, the heterologous promoter becomes duplicated and placed in the 5' LTR as the only promoter that directs expression of the retrovirally carried gene. IRES or dicistronic vectors carry two genes expressed from the MLV promoter as one transcript. Both genes are translated from the transcript, the translation of the second gene being facilititated by an internal ribosome entry site (IRES).

fered genes compared to two-gene, two promoter vectors (see above). They may also help to overcome the reported interference between multiple promoters present in the same retroviral vector (40, 51-52). A combination of the dicistronic IRES type vector with the ProCon vector would remove any possibility of retroviral promoter interference, whilst at the same time ensuring adequate expression of multiple IRES seperated genes from the heterologous promoter inserted in the LTR.

1.4 Use of Retroviral Vectors for Gene Therapy

Basically two approaches have been taken to gene therapy: ex vivo and in vivo. Ex vivo involves the removal of cells from a patient, their genetic modification in the laboratory, and then their subsequent reintroduction into the patient. The most interesting target cells for such ex vivo approaches are haematopoetic cells – easily won and readily reinfused. However most of the applications of haematopoetic cell modification are for inborn disease, and thus require long term correction. This illustrates one of the first problems associated with the use of retroviral vectors: titre. The haematopoetic progenitor cells that need to be corrected in order to guarantee long term correction only represent a very small percentage of the total haematopoetic cells. $CD34^+$ progenitor cells can be enriched using fast cell sorting techniques. However very high efficiencies of retroviral transfer are required to be able to transduce enough stem cells to bring about long lasting clinical benefit. Many strategies including concentration (53-54), spin down infection, temperature shift to 32° C (54), phosphate starvation (55), treatment of vector producing cells with sodium butyrate (56-58) and the use of modified or mixed infections have been tried in an effort to increase titre from the usual 10^6 colony forming units (cfu)/ml to a reported maximum of about 10^9 cfu/ml. Of these methods perhaps the most interesting is the modified/mixed approach.

The infection of a retrovirus is dependent on the presence of the appropriate receptor on the target cell. These are often present in limiting amounts only on certain cells at certain times and thus govern the maximum titre effectively attainable. In contrast the receptor for another enveloped virus, Vesicular Stomatitis Virus (VSV), is abundantly present on many different cell types. Experiments in which the surface protein of VSV has been incorporated into retroviral particles (a laboratory enhanced process based on a naturally occuring phenomenon known as pseudotyping (6)) has resulted in titres of $>10^9$ cfu/ml (59).

The use of VSV surface protein/receptor interaction not only increases the titre of vector produced, but, since the VSV receptor is pleiotropically expressed, also broadens the range of cell types infectable by the retroviral vector. A second strategy that has been utilized to broaden host range in the laboratory is the mixed infection of target cells with retroviral vectors and defective adenoviruses (60). The adenoviral receptor, although not identified to date, is abundantly expressed. It appears that the retrovirus can hitchhike a lift with the adenovirus during the latter's uptake. Unfortunately the titres of this system are still low, making it difficult to envisage its practical use, but this mechanism may be exploitable in combination with other viruses to increase both host range and titre in the future.

Recent reports however (61-63) suggest that even in the face of the current problems with low titre it may be feasible to use retroviral vectors to establish long term correction in haematopoetic cells. These investigators were able to use natural selection for the function of the transfered gene to maintain corrected haematopoetic cells expressing the Adenosine Deaminase (ADA) gene. Expression of this gene protects lymphocytes against the destructive effects of the ADA substrate. In its absence the immune system cannot develop properly. Ex vivo correction of haematopeotic cells with a functional ADA gene copy resulted in long term (up to three years has so far been measured) repopulation by ADA expressing lymphocytes.

In vivo gene therapy involves the direct delivery of the vector into the patient followed by some sort of targeting procedure to ensure that the therapeutic effects of the transfered genes are limited to the correct tissue or organ. Again high titres are a pre-requisite for efficient in vivo infection although many of the approaches discussed above to artificially raise titre (e.g. spin down infection or infection at 32° C) will not be feasible in the in vivo situation. One way round this problem is to implant immunologically shielded virus producing cells directly in the patient, thereby ensuring a long term continuous supply of vector (Saller, R.M.; Stange, J.; Mitzner, S.; Heinzmann, U.; B.S. and W.H.G. submitted).

A second problem associated with the use of retroviruses for gene therapy is the restriction of these vectors to the infection of proliferating cells. With the exception of HIV, nuclear membrane breakdown is a prerequisite for the movement of the viral core into the nucleus. HIV-based vectors may offer an alternative to classic MLV-based retroviral vectors for the infection of non-dividing cells, since these carry a number of redundant nuclear localization signals that actively transport the nucleoprotein complex into the nucleus (reviewed in 15). It has recently been shown that it is feasible to pseudotype such HIV vectors into MLV virions, allowing high efficiency infection of quiescent cells (64, 65, 66). Assuming that the psychological problems associated with the use of HIV based vectors can be overcome, such strategies may open the door to the use of retroviral vectors for long term transduction of differentiated cells. Other strategies, involving the use of subcomponents, such as HIV Gag and Vpr proteins, possibly linked to proteins that can push a quiescent cell into one round of cell division, used in combination with conventional retroviral systems may also allow infection of quiescent cells. In the meantime this specific property MLV-based of retroviral vectors to preferentially infect rapidly proliferating cells, often cited as a disadvantage, can be utilized to target rapidly proliferating cells, such as tumour cells, in a background of quiescent, end-differentiated normal cells.

Virus infection is naturally cleared by the immune system. Retroviruses are directly inactivated by human serum complement (67). Retroviral vectors suffer the same fate in vivo. Inactivation is non-lytic and depends both on the retrovirus and on the cells from which the virus has been produced. MLV produced from mouse cells is much more readily inactivated than MLV produced from human or mink cells (68). Thus it should be possible to use combinations of virus and packaging cells that show high resistance to complement inactivation. HIV-1 and Human T-cell Leukaemia Virus (HTLV) are not inactivated by human serum complement (69-70). The construction of retroviral vector systems based on such complex viruses is difficult, however it may be possible to use the surface proteins from HIV or HTLV in an analagous way to that already described for the VSV surface proteins, and pseudotype conventional retroviral vectors into these complement resistent envelopes (71). Alternatively the administration of monoclonal antibodies that block specific components of the complement pathway may also prove useful (72). Despite these problems of complement inactivation, local, rather than systemic administration of vector, for instance by direct intra-tumoral injection, should be successful (73-74).

1.5 Retroviral Vectors in Perspective

The majority of clinical gene therapy trials involve the use of retroviruses as gene transfer vehicles. This is in part due to the fact that retroviruses have been more intensively studied, and thus more is known about them, than any other virus, especially in the last decade as a result of the emergence of HIV as a causative agent for Acquired Immune Deficiency Syndrome (AIDS). Additionally current retroviral vector design benefits from over 15 years work on improving safety and efficiency. Although retroviral vectors have some disadvantages: relatively low titre, infection only of dividing cells and complement inactivation, these problems can be overcome as discussed above. In contrast, retroviral vectors possess many real advantages for long term correction of inborn gene defects. These include stable, long term gene transfer, currently one of the best gene transfer capacities (7.5kb) and currently the safest packaging/vector systems.

Whilst adenovirus based vectors can be produced at higher titre and infect both quiescent and proliferating cells, they maintain their genome, and any included therapeutic genes, episomally. This means that this information becomes lost with time, making the use of these vectors really only attractive for the treatment of acquired disease such as cancer. Adenovirus vector mediated therapy of inborn errors necessitates repeated treatment and this is connected with immune problems (75) and inflammation at high titre (76-77). Future improvements in adenoviral vector design, and in using vectors based on different adenoviruses for multiple treatments may resolve these problems. Future improvements in adenovirus packaging systems should also allow an increase in the transfer capabilities of these vectors above that of retroviruses.

Adeno-associated vectors are still in their infancy. The low cloning capacity (4kb) and the lack of adequate packaging cell lines limit their current use. The advantages and disadvantages of current viral and non-viral transfer systems have been extensively reviewed elsewhere (48, 78, 79).

1.6 Conclusions

Retroviral vectors will continue to be improved and form the backbone of current gene therapy protocols. Problems associated with insertional activation/inactivation of cellular genes can be, and are being successfully overcome by using new, safe, third generation vector systems. Risk can never be totally removed, but then any medical procedure brings with it a certain risk. These risks must be weighed up against the possible benefit to the patient. In the case of cancer therapy, all current chemo- or radiation therapies are highly toxic to the patient. In comparison gene therapy with retroviral vectors is benign. Our concerns must now lie with the efficacy of the vector systems and of the transfered therapeutic principles, rather than with converting a vector related negligible risk to an unachievable non-existent risk.

References

1. Weiss, R. A. and Tailor, C. S. (1995). Retrovirus Receptors. Cell *82*, 531-533.
2. Kim, J. W., Closs, E. I., Albritton, L. M., and Cunningham, J. M. (1991).Transport of cationic amino acids by the mouse ecotropic retrovirus receptor. Nature 352, 725-728.
3. Wang, H., Kavanaugh, M. P., North, R. A., and Kabat, D. (1991). Cell-surface receptor for ecotropic murine retroviruses is a basic amino-acid transporter. Nature 352, 729-731.
4. Miller, D. G. and Miller, A. D. (1994). A family of retroviruses that utilize related phosphate transporters for cell entry. J. Virol. *68*, 8270-8276.

5. van Zeijl, M., Johann, S. V., Closs, E., Cunningham, J., Eddy, R., Shows, T., and O'Hara, B. (1994). A human amphotropic retrovirus receptor is a second member of the gibbon ape leukemia virus receptor family. Proc. Natl. Acad. Sci. USA *91,* 1168-1172.
6. Weiss, R. A. (1993). Cellular receptors and viral glycoproteins involved in retrovirus entry. In: The Retroviridae, Vol. 2, J. A. Levy, ed. (New York, USA: Plenum Press), pp 1-108.
7. Ragheb, J. A. and Anderson, W. F. (1994). pH-independent murine leukemia virus ecotropic envelope-mediated cell fusion: implications for the role of the R peptide and p12E TM in viral entry. J. Virol. *68,* 3220-3231.
8. Janka, I., Günzburg, W. H., Brem, G., Erfle, V., and Salmons, B. (1993). Sub-retroviral particles as gene transfer vectors that by-pass retrovirus receptor interaction restrictions. Biochem. Mol. Biol. Int. *31*, 349-357.
9. Roe, T., Reynolds, T. C., Yu, G., and Brown, P. O. (1993). Integration of murine leukaemia virus DNA depends on mitosis. EMBO J. *12*, 2099-2108.
10. Lewis, P. F. and Emerman, M. (1994). Passage through mitosis is required for oncoretroviruses but not for the human immunodeficiency virus. J.Virol. *68*, 510-516.
11. Lewis, P., Hensel, M., and Emerman, M. (1992). Human immunodeficiency virus infection of cells arrested in the cell cycle. The EMBO J. *11*, 3053-3058.
12. Dobrescu, D., Ursea, B., Pope, M., Asch, A. S., and Posnett, D. N. (1995). Enhanced HIV-I replication in Vß12 T cells due to human cytomegalovirus in monocytes: evidence for a putative herpesvirus superantigen . Cell *82*, 753-763.
13. Bukrinsky, M. I., Haggerty, S., Dempsey, M. P., Sharova, N., Adzhubel, A., Spitz, L., Lewis, P., Goldfarb, D., Emerman, M., and Stevenson, M. (1993). A nuclear localization signal within HIV-1 matrix protein that governs infection of non-dividing cells. Nature 365, 666-669.
14. Heinzinger, N. K., Bukrinsky, M. I., Haggerty, S. A., Ragland, A. M., Kewalramani, V., Lee, M.-E., Gendelman, H. E., Ratner, L., Stevenson, M., and Emerman, M. (1994). The Vpr protein of human immunodeficiency virus type 1 influences nuclear localization of viral nucleic acids in nondividing host cells. Proc. Natl. Acad. Sci. USA *91*, 7311-7315.
15. Stevenson, M. (1996). Portals of entry: uncovering HIV nuclear transport pathways. Trends in Cell Biology 6, 9-15.
16. Fujiwara, T. and Mizuuchi, K. (1988). Retroviral DNA integration: structure of an integration intermediate. Cell *54*, 497-504.
17. Brown, P. O., Bowerman, B., Varmus, H. E., and Bishop, J. M. (1989). Retroviral integration structure of the initial covalent product and its precursor, and a role for the viral IN protein. Proc. Natl. Acad. Sci. (USA) *86*, 2525-2529.
18. Goff, S. P. (1992). Genetics of retroviral integration. Ann. Rev. Genet. *26*, 527-544.
19. Majors, J. (1990). The structure and function of retroviral long terminal repeats. Curr. Topics Microbiol. Immunol. *157*, 49-92.
20. Cullen, B. R. (1991). Human immunodeficiency virus as a prototypic complex retrovirus. J.Virol. *65*, 1053-1056.
21. Zhang, Y. Q. and Barklis, E. (1995). Nucleocapsid protein effects on the specificity of retrovirus RNA encapsidation. J. Virol. *69*, 5716-5722.
22. Berkowitz, R. D., Hammarskjold, M.-L., Helga-Maria, C., Rekosh, D., and Goff, S. P. (1995). 5' regions of HIV-I RNAs are not sufficient for encapsidation: implications for the HIV-1 packaging signal. Virology *212*, 718-723.
23. Salmons, B. and Gunzburg, W. H. (1993). Targeting of retroviral vectors for gene therapy. Human Gene Therapy *4*, 129-141.
24. Chakraborty, A. K., Zink, M. A., and Hodgson, C. P. (1995). Expression of VL30 vectors in human cells that are targets for gene therapy. Biochem. Biophys. Res. Commun. *209*, 677-683.
25. French, N. S. and Norton, J. D. (1995). Construction of a retroviral vector incorporating mouse VL30 retrotransposon-derived, transcriptional regulatory sequences . Anal. Biochem. *228*, 354-355.
26. Shank, P. R. and Linial, M. (1980). Avian oncovirus mutant (SE21Q1b) deficient genomic RNA: characterization of a deletion in the provirus. J.Virol. *36*, 450-456.
27. Murphy, J. E. and Goff, S. P. (1989). Construction and analysis of deletion mutations in the U5 region of Moloney murine leukemia virus: effects on RNA packaging and reverse transcription. J. Virol. *63*, 319-327.
28. Mann, R., Mulligan, R. C., and Baltimore, D. (1983). Construction of a retrovirus packaging mutant and its use to produce helper-free defective retrovirus. Cell *33*, 153-159.

29. Cone, R. D. and Mulligan, R. C. (1984). High-efficiency gene transfer into mammalian cells: generation of helper free recombinant retrovirus with broad mammalian host range. Proc. Natl. Acad. Sci. USA *81*, 6349-6353.
30. Miller, A. D., Trauber, D. R., and Buttimore, C. (1986). Factors involved in production of helper virus-free retrovirus vectors. Somatic Cell. Mol. Genet. *12*, 175-183.
31. Muenchau, D. D., Freeman, S. M., Cornetta, K., Zwiebel, J. A., and Anderson, W. F. (1990). Analysis of retroviral packaging lines for generation of replication-competent virus. Virol. *176*, 262-265.
32. Miller, A. D. and Buttimore, C. (1986). Redesign of retrovirus packaging cell lines to avoid recombination leading to helper virus production. Mol. Cell. Biol. *6*, 2895-2902.
33. Bosselman, R. A., Hsu, R.-Y., Bruszewski, J., Hu, S., Martin, F., and Nicolson, M. (1987). Replication-defective chimeric helper proviruses and factors affecting generation of competent virus: expression of moloney murine leukaemia virus structural genes via the metallothionein promoter. Mol. Cell. Biol. *7*, 1797-1806.
34. Bender, M. A., Palmer, T. D., Gelinas, R. E., and Miller, A. D. (1987). Evidence that the packaging signal of moloney murine leukaemia virus extends into the gag region. J. Virol. *61*, 1639-1646.
35. Morgenstern, J. P. and Land, H. (1990). Advanced mammalian gene transfer: high titre retroviral vectors with multiple drug selection markers and a complementary helper-free packaging cell line. Nucl. Acids Res. *18*, 3587-3596.
36. Scarpa, M., Cournoyer, D., Muzny, D. M., Moore, K. A., Belmont, J. W., and Caskey, C. T. (1991). Characterization of recombinant helper retroviruses from Moloney-based vectors in ecotropic and amphotropic packaging cell lines. Virology *80*, 849-852.
37. Markowitz, D., Goff, S., and Bank, A. (1988). Construction and use of a safe and efficient amphotropic packaging cell line. Virol. *167*, 400-406.
38. Markowitz, D., Goff, S., and Bank, A. (1988).A safe packaging line for gene transfer: separating viral genes on two different plasmids. J. Virol. *62*, 1120-1124.
39. Julias, J. G., Hash, D., and Pathak, V. K. (1995). E- vectors: Development of novel self-inactivating and self-activating retroviral vectors for safer gene therapy. J.Virol. *69*, 6839-6846.
40. Xu, L., Yee, J. K., Wolff, J. A., and Friedmann, T. (1989). Factors affecting long-term stability of Moloney murine leukemia virus-based vectors. Virology *171*, 331 341.
41. Richards, C. A. and Huber, B. E. (1993). Generation of a transgenic model for retrovirus-mediated gene therapy for hepatocellular carcinoma is thwarted by the lack of transgene expression. Human Gene Therapy *4*, 143-150.
42. Naviaux, R. K. and Verma, I. M. (1992). Retroviral vectors for persistent exprssion in vivo. Current Biology *3*, 540-547.
43. Hantzopoulos, P. A., Sullenger, B. A., Ungers, G., and Gilboa, E. (1989). Improved gene expression upon transfer of the adenosine deaminase minigene outside the transcriptional unit of a retroviral vector. Proc. Natl. Acad. Sci. USA *86*, 3519-3523 .
44. Vile, R. G., Diaz, R. M., Miller, N., Mitchell, S., Tuszyanski, A., and Russell, S. J. (1995). Tissue-specific gene-expression from Mo-MLV retroviral vectors with hybrid LTRs containing the murine tyrosinase enhancer promoter. Virology *214*, 307-313.
45. Ferrari, G., Salvatori, G., Rossi, C., Cossu, G., and Mavilio, F. (1995). A retroviral vector containing a muscle-specific enhancer drives gene expression only in differentiated muscle fibers. Human Gene Therapy *6*, 733-742.
46. Günzburg, W. H., Saller, R., and Salmons, B. (1995). Retroviral vectors directed to predefined cell types for gene therapy. Biologicals *23*, 5-12.
47. Salmons, B., Saller, R. M., Baumann, J., and Günzburg, W. H. (1995). Construction of retroviral vectors for targeted delivery and expression of therapeutic genes. Leukemia *9*, Suppl. 1, 53-60.
48. Günzburg, W. H. and Salmons, B. (1996). Development of retroviral vectors as safe, targeted gene delivery systems. J. Molecular Medicine *74*, 171-182.
49. Levine, F., Yee, J. K., and Friedmann, T. (1991). Efficient gene expression in mammalian cells from a dicistronic transcriptional unit in an improved retroviral vector. Gene *108*, 167-174.
50. Günzburg, W. H., Salmons, B., Zimmermann, B., Mueller, M., Erfle, V., and Brem, G. (1991). A mammary-specific promoter directs expression of growth hormone not only to the mammary gland, but also to Bergman glia cells in transgenic mice. Mol. Endocinol. *5*, 123-133.
51. Li, M., Hantzopoulos, P. A., Banerjee, S. C., Zhao, S. C., Scweitzer, B. I., Gilboa, E., and Bertino, J. R. (1992). Comparison of the expression of a mutant dihydrofolate reductase under control of different internal promoters in retroviral vectors. Human Gene Therapy *3*, 381-390.

52. McLachlin, J. R., Mittereder, N., Daucher, M. B., Kadan, M., and Eglitis, M. A. (1993). Factors affecting retroviral vector function and structural integrity. Virology *195*, 1-5.
53. Paul, R. W., Morris, D., Hess, B., Dunn, J., and Overall, R. W. (1993). Increased viral titer through concentration of viral harvests from retroviral packaging lines. Human Gene Therapy *4*, 609-615.
54. Kotani, H., Newton, P. B., Zhang, S., Chiang, Y. L., Otto, E., Weaver, L., Blaese, R. M., Anderson, W. F., and McGarrity, G. J. (1994). Improved methods of retroviral vector transduction and production for gene therapy. Human Gene Therapy *5*, 19-28.
55. Bunnell, B. A., Muul, L. M., Donahue, R. E., Blaese, R. M., and Morgan, R. A. (1995). High-efficiency retroviral-mediated gene transfer into human and nonhuman primate peripheral blood lymphocytes . Proc. Natl. Acad. Sci. USA *92*, 7739-7743.
56. Olsen, J. C. and Sechelski, J. (1995). Use of sodium-butyrate to enhance production of retroviral vectors expressing CFTR cDNA. Human Gene Therapy *6*, 1195-1202.
57. Pages, J. C., Loux, N., Farge, D., Briand, P., and Weber, A. (1995). Activation of Moloney murine leukemia-virus LTR enhances the titer of recombinant retrovirus in Psi-crip packaging cells. Gene Therapy *2*, 547-551.
58. Soneoka, Y., Cannon, P. M., Ramsdale, E. E., Griffiths, J. C., Romano, G., Kingsman, S. M., and Kingsman, A. J. (1995). A transient three-plasmid expression system for the production of high titer retroviral vectors. Nucl. Acids Res. *23*, 628-633.
59. Burns, J. C., Friedmann, T., Driever, W., Burrascano, M., and Yee, J. K. (1993). Vesicular stomatitis virus G glycoprotein pseudotyped retroviral vectors: Concentration to very high titer and efficient gene transfer into mammalian and nonmammalian cells. Proc. Natl. Acad. Sci. USA *90*, 8033-8037.
60. Adams, R. M., Wang, M., Steffen, D., and Ledley, F. D. (1995). Infection by retroviral vectors outside of their host-range in the presence of replication-defective adenovirus. J. Virol. *69*, 1887-1894.
61. Bordignon, C., Notarangelo, N., Nobili, N., Ferrari, G., Casorati, G., Panina, P., Mazzolari, E., Maggioni, D., Rossi, C., Servida, P., Ugazio, A. G., and Mavilio, F. (1995). Gene therapy in peripheral blood lymphocytes and bone marrow for ADA- immunodeficient patients; gene transfer to hematopoietic progenitor cells using a retrovirus vector. Science *270*, 470-475.
62. Blaese, R. M., Culver, K. W., Miller, A. D., Carter, C. S., Fleisher, T., Clerici, M., Shearer, G., Chang, L., Chiang, Y., Tolstoshev, P., Greenblatt, J. J., Rosenberg, S. A., Klein, H., Berger, M., Mullen, C. A., Ramsey, W. J., Muul, L., Morgan, R. A., and Anderson, W. F. (1995). T-lymphocyte-directed gene therapy for ADA- SCID: initial trial results after 4 years; severe combined immunodeficiency therapy by adenosine-deaminase gene transfer using a retrovirus vector. Science *270*, 475-480.
63. Kohn, D. B., Weinberg, K. I., Nolta, J. A., Heiss, L. N., Lenarsky, C., Crooks, G. M., Hanley, M. E., Annett, G., Brooks, J. S., Elkhoureiy, A., Lawrence, K., Wells, S., Moen, R. C., Bastian, J., Williamsherman, D. E., Elder, M., Wara, D., Bowen, T., Hershfield, M. S., Mullen, C. A., Blaese, R. M., and Parkman, R. (1995).Engraftment of gene-modified umbilical cord blood cells in neonates with adenosine-deaminase deficiency; human severe combined immunodeficiency gene therapy using hematopoietic stem cells transduced with a retrovirus vector. Nat.Med. *1*, 1017-1023.
64. Page, K. A., Landau, N. R., and Littman, D. R., (1990) Construction and use of a human immunodeficiency virus vector for analysis of virus infectivity. J. Virol. *64*, 5270-5276.
65. Naldini, L., Blömer, U., Gallay, P., Ory, D., Mulligan, R., Gage, F. H., Verma, I. M., and Trono, D., (1996) In vivo delivery and stable transduction of non-dividing cells by a lentiviral vector. Science *272*, 263-267.
66. Reiser, J., Harmison, G., Kluepfel-Strahl, S., Brady, R. O., Karlsson, S., and Schubert, M. (1996) Transduction of non-dividing cells using pseudotyped defective high-titer HIV-type 1 particles. Proc. Natl. Acad. Sci. USA *93*, 15266-15271.
67. Welsh, R. M., Cooper, N. R., Jensen, F. C., and Oldstone, M. B. A. (1975). Human serum lyses RNA tumour viruses. Nature *257*, 612-614.
68. Takeuchi, Y., Cosset, F.-L. C., Lachmann, P. J., Okada, H., Weiss, R. A., and Collins, M. K. L. (1994). Type C retrovirus inactivation by human complement is determined by both the viral genome and the producer cell. J. Virol. *68*, 8001-8007.
69. Banapour, B., Sernatinger, J., and Levy, J. A. (1986). The AIDS-associated retrovirus is not sensitive to lysis or inactivation by human serum. Virology *152*, 268-271.
70. Hoshino, H., Tanaka, H., Miwa, M., and Okada, H. (1984). Human T-cell leukaemia virus is not lysed by human serum. Nature *310*, 324-325.
71. Wilson, C., Reitz, M. S., Okayama, H., and Eiden, M. V. (1989). Formation of infectious hybrid virions with gibbon ape leukemia virus and human T cell leukemia virus retroviral envelope gly-

coproteins and the GAG proteins and Pol proteins of Moloney murine leukemia virus. J. Virol. *63*, 2374-2378.
72. Rother, R. P., Squinto, S. P., Mason, J. M., and Rollins, S. A. (1995). Protection of retroviral vector particles in human blood through complement inhibition. Human Gene Therapy *6*, 429-435.
73. Oldfield, E. H., Ram, Z., Culver, K. W., Blaese, R. M., Devroom, H. L., and Anderson, W. F. (1993). Gene therapy for the treatment of brain tumors using intra-tumoral transduction with the thymidine-kinase gene and intravenous ganciclovir; using mouse retro virus vector sensitizes tumor to ganciclovir. Hum.Gene Ther. *4*, 39-69.
74. Culver, K.W. (1994). Gene therapy: a handbook for physicians (New York, USA: Mary Ann Liebert Inc.).
75. Yei, S., Mittereder, N., Tank, K., O Sullivan, C., and Trapnell, B. C. (1994). Adenovirus-mediated gene transfer for cystic fibrosis: quantitative evaluation of repeated in vivo vector administration to the lung. Gene Therapy *1*, 192-200.
76. Knowles, M. R., Hohneker, K. W., Zhou, Z., Olsen, J. C., Noah, T. L., Hu, P. C., Leigh, M. W., Engelhardt, J. F., Edwards, L. J., Jones, K. R., Grossman, M., Wilson, J. M., Johnson, L. G., and Boucher, R. C. (1995). A controlled study of adenoviral-vector-mediated gene transfer in the nasal epithelium of patients with cystic fibrosis; cystic fibrosis transmembrane conductance regulator gene transfer efficacy and safety for application in gene therapy. N. Engl. J. Med. *333*, 823-831.
77. Crystal, R. G., McElvaney, N. G., Rosenfeld, M. A., Chu, C.-S., Mastrangeli, A., Hay, J. G., Brody, S. L., Jaffe, H. A., Eissa, N. T., and Danel, C. (1994). Administration of an adenovirus containing the human CFTR cDNA to the respiratory tract of individuals with cystic fibrosis. Nature Genetics *8*, 42-51.
78. Hodgson, C. P. (1995). The vector void in gene therapy. Bio/Technology *13*, 222-225.
79. Günzburg, W.H. and Salmons, B. (1995). Virus vector design in gene therapy. Mol. Med. Today *1*, 410-417.

Acknowledgements

This work was supported in part by the EU Biotechnology grant BIO4-95-0100 and by the FORGEN programme of the Bayerische Forschungsstiftung.

2. Adenoviral Vectors

Patrice Yeh, Jean-François Dedieu, Emmanuelle Vigne,
Cécile Orsini and Michel Perricaudet

In the last decade, defective recombinant adenoviruses have emerged as potent vehicles for direct in vivo gene delivery in a number of adherent somatic and tumor cells. Since 1993, they are being evaluated in phase I clinical trials that target cancer or hereditary diseases. Preclinical and clinical data indicate that humoral and cellular immunity to the virus and the virus-infected cells represent recurring issues that must be answered for safe and effective adenovirus-based therapies. The exhaustive analysis of the host response to E1-deleted vectors is indeed required. We will review here the basic features of the virus biology, together with the rationale underlying current anti-inflammatory strategies that are being developed for gene transfer, and hopefully therapy, with a clinical grade vector.

2.1 Adenoviruses

Adenoviruses (Ad) have been isolated from man to bird. 47 human serotypes have been identified which are classified into 6 subgroups (A to F). They are mostly associated with acute respiratory infections, epidemic conjunctivitis and infantile gastroenteritis (36). Certain human serotypes are tumorigenic in newborn rodents, but attempts to establish them as a causative agent in human cancer have been uniformly negative. In fact, several serotypes have been used with an excellent safety profile as live enteric vaccines (62). This is the case for serotypes 2 (Ad2) and 5 (Ad5), two closely related, non tumorigenic, adenoviruses belonging to subgroup C. The biology of these viruses has been extensively studied, especially with regard to their oncogenic potential and the various cellular regulation mechanisms they target to refocus gene expression for their own "purpose" (for a review see (63). The nucleotide sequence of Ad5 is very close to that of Ad2 (14).

Subgroup C adenoviruses are small (70 nm in diameter) non enveloped DNA viruses which induce a cytopathic infectious cycle in a large variety of dividing or quiescent human cells. The viral particle exhibits an icosahedric symmetry. Its tridimensional structure has recently been reconstituted from X-ray and cryoelectron micrographs (69, 70). The virion is composed of an outer shell (capsid) surrounding a viral core containing a 36 kb linear, but supercoiled, chromosome in close association with basic virus-encoded proteins (pV, pVII and mu). These proteins have been reported to fold the viral chromosome in a highly organized structure which could participate in the timing of gene expression following its entry into the nucleus (80). The capsid is composed of 252 subunits, 240 hexons (referred to as pII) and 12 pentons (pIII and pIV). Proteins pIIIa, pVI, pVIII, pIX, and a 23K protease are also included. The Ad2 or Ad5 chromosome is characterized by the presence of inverted terminal repeats (ITR) of 103 nucleotides at its extremities, in covalent association with a set of virally encoded proteins (the 55 kDa terminal protein) which are crucial for nuclear matrix attachement, efficient transcription and the initiation of viral DNA replication (61). A cis-acting packaging signal (Ψ) composed of repeated motifs is located at the immediate vicinity of the left ITR (30).

The infectious cycle is completed within 20 to 24 hours in HeLa cells (63). It leads to cell death by attrition and the release of 10^3 infectious particles per cell. The virion uptake proceeds through a sequential two-steps interaction of the penton with different cellular receptors. The penton is a multimeric complex composed of the so-called penton base and a trimer of a typical protruding tail-head structure, the fiber protein (pIV), which mediates the initial binding of the virus to a yet unidentified cell surface receptor. A member of the integrin class of cell adhesion molecules (the vitronectin receptor $\alpha_V\beta_3$ or $\alpha_V\beta_5$), is subsequentely recruited through an interaction with the RGD motif of the penton base, leading to endocytosis of the virus through chlathrin-coated vesicles. In the early endosome compartment, there is a progressive uncoating of the particle and the release in the cytoplasm of a nucleoparticle containing the viral genome and proteins required for intracellular trafficking and nuclear entrance (32). Once inside the nucleus the viral chromosome remains extrachromosomal and transcription is initiated.

2.2 Biology of the Virus

The virus genetic information is extremely compacted, both strands of the DNA molecule are coding, and nearly all transcription units are alternatively spliced into a complex pattern of mRNAs. The viral transcription units are functionally referred to as early (E1A, E1B, E2, E3 and E4) and late (LTU), depending upon their timely expression relative to the onset of viral DNA replication. In HeLa cells, the switch from the early, regulatory, phase to the late phase of infection occurs between 6 and 8 hours post-infection. The pIX and IVa2 transcripts do not belong to either class and are usually referred to as "delayed early" (Fig. 2.1A).

Region E1A, the so-called immediate early region, is first transcribed following infection of quiescent cells. Its major products, the 289R and 243R regulatory proteins, are required in trans for transcriptional activation of all viral units during the early phase of infection (Fig. 2.2A). In particular, these multifunctional proteins tar-

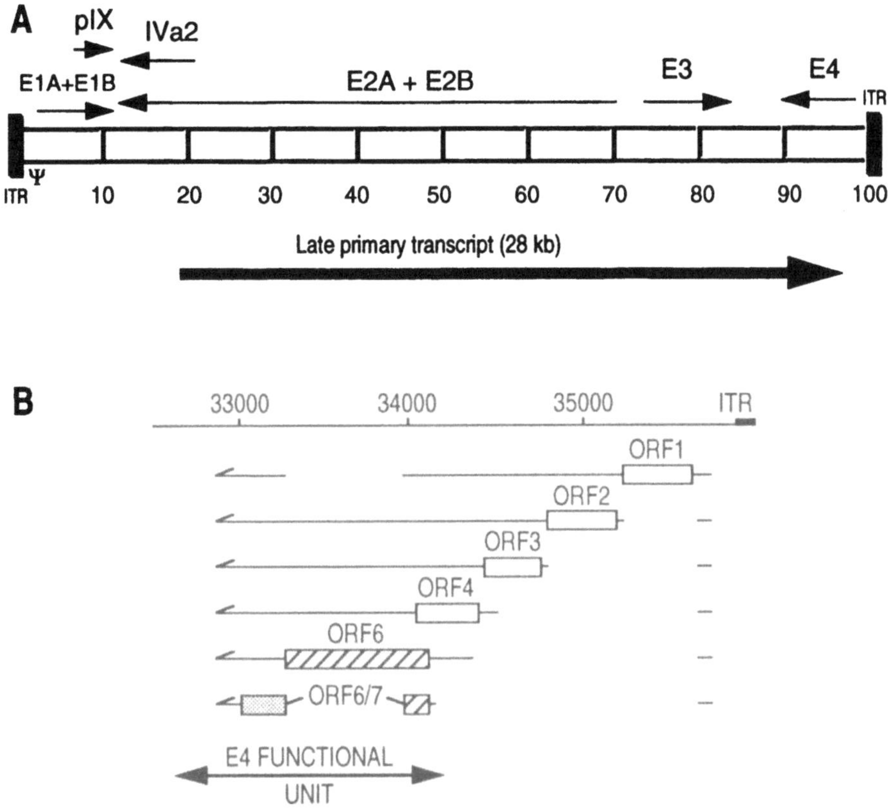

Fig. 2.1: Organization of the Ad5 genome and the E4 region.
Shematic representation of the viral genome (panel A). By convention, the 36 kb viral chromosome is divided into 100 map units. The main transcription units include 5 early (regulatory) regions (E1A, E1B, E2, E3 and E4), 2 delayed early (pIX and IVa2), and one late transcription unit (LTU). In the late phase of infection, transcription from the major late promoter (MLP) generates a large primary transcript which is extensively processed into approximately 20 mRNAs mostly encoding the virion structural components. The inverted terminal repeats (ITR) at both ends of the genome serve as origins of replication. The virus-associated RNAs have been omitted for clarity.
Genetic organization of the E4 region (panel B). The most abundant mRNAs encoding the ORF1, ORF2, ORF3, ORF4, ORF6 and ORF6/7 gene products are represented (19). The sequence encompassing ORF1, ORF2, ORF3, and ORF4 is dispensable for virus growth (34). Therefore, the distal (ORF6+ORF7) moiety of the locus constitutes the E4 functional unit integrated in the IGRP2 cell line (87).

get cell cycle regulators, including E2F-pRB complexes, kinases-associated cyclins, and p300, leading to cell cycle progression into the S-phase (for a review see (63)), a conflictual signal driving quiescent cells towards apoptosis. E1B encodes two regulatory proteins (referred to as 19K and 55K) which antagonize this cellular response at different levels (18, 56, 89). Indeed, the ability of the E1A and E1B proteins to deregulate cell cycle control constitutes one of the most useful in vitro model to study the molecular mechanisms underlying programmed cell death or immortalization. The E1B 55K also binds to one of the E4 regulatory protein (ORF6) in the late phase of infection, a complex which facilitates export of the viral mRNAs from the nucleus to the cytoplasm (7). Region E2 encodes a set of proteins directly involved in viral DNA replication, including the 72K DNA binding protein (DBP, the E2A-encoded product), as well as a 140 kDa DNA polymerase and a 80 kDa precursor of the terminal protein both encoded by E2B. In addition to its requirement for viral DNA replication, the DBP is also involved in the indirect transactivation of the late transcription unit (LTU) (13, 91). E3 is a complex region which is dispensable for viral growth in vitro. It encodes immunomodulatory proteins which interfere with the host immunosurveillance. For example, the E3 19 kDa glycoprotein (gp19K) binds and sequesters certain alleles of MHC class I molecules within the endoplasmic reticulum, thereby substracting the infected cell from identification and destruction by memory lymphocytes (10, 25). The E3 14.7 kDa protein, and a complex between the 10.4 kDa and 14.5 kDa E3 proteins also protect the infected cells from TNF-induced cytolysis (29). Region E3 has also been reported to downregulate viral gene expression, presumably at the level of translation (90). The E4 region encodes at least 6 different proteins which are generated by alternative splicing throughout the infectious cycle (19). Among them, the ORF3 and ORF6 proteins are particularly important because they encode a redundant function required for virus growth (8, 37), obviously because they somehow regulate the overall process of post-transcriptional processing of the late primary transcript (Fig. 2.2B) and/or the nuclear export of the corresponding mRNAs (34, 51).

The LTU encodes a large primary transcript of ca. 28 kb. It is processed into approximately 20 mRNAs that can be grouped into five families (L1 to L5), each of which consisting of a set of alternatively spliced molecules with identical 3' ends depending on the utilization of 5 distinct polyadenylation signals (Fig. 2.2B). Most late mRNAs exhibit in their 5' end a common 201 nucleotides leader sequence which is generated by the junction of three short exons. Sometimes, an internal exon is also included which encodes a 13.6 kDa protein of unknown function (67). The late mRNAs encode most of the virion structural proteins, together with non structural proteins which are mostly involved in the assembly and maturation of the virions (for a review see (63)).

2.3 E1-deleted Vectors

Recombinant adenoviruses do not propagate in most human cells because they lack the E1A and E1B regulatory genes. Large deletions which do not impair virus growth in vitro have also been engineered in the E3 (5) and E4 regions (1, 34). If combined, these deletions would allow a cloning capacity of approximately 9 kb. Even so, 80% of the viral genome would be retained within the recombinant backbone, a coding

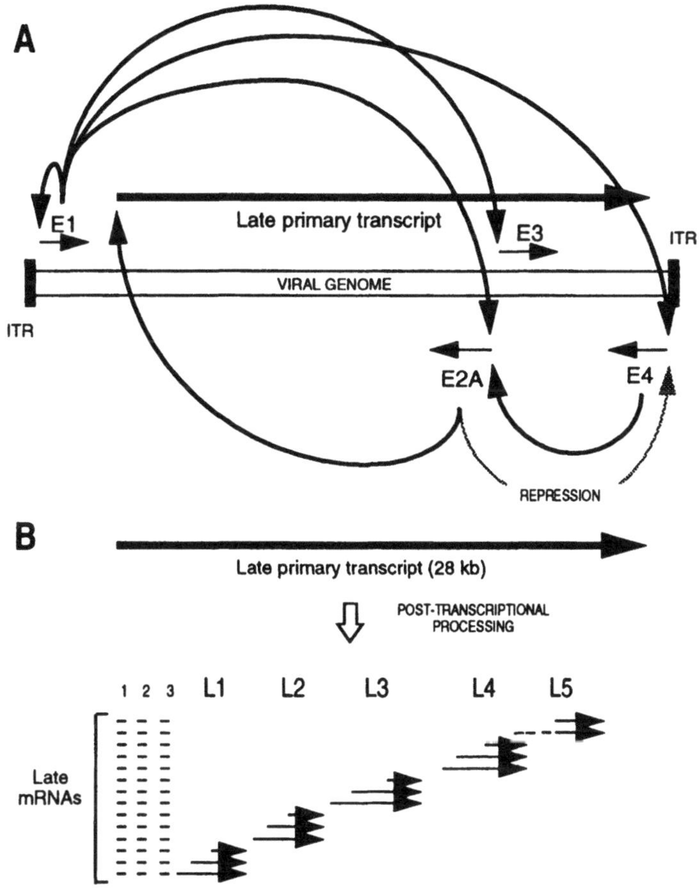

Fig. 2.2: Expression of the viral genome.
Transactivation of the viral promoters (panel A). All early viral promoters are transactivated by the products of the E1A immediate early region. The 72K E2A gene product transativates the MLP while repressing the E4 promoter (13, 50). The E4 ORF6/7 protein potentiates the E2F-mediated transactivation function of the E2A early promoter (2, 6, 58). The IVa2-mediated transactivation of the major late promoter (73) is not represented.
Post-transcriptional processing of the large primary transcript of the LTU (panel B). 1, 2 and 3 refer to the components of the tripartite leader present in all late mRNAs. E4 defective deletants are unable to accumulate the corresponding mRNAs in the cytoplasm.

potential for more than 20 proteins, most of them being generated following extensive post-transcriptional processing of the late primary transcript (Fig. 2.2B).

The technology to construct and recover E1-deleted recombinant viruses relies on the 293 cell line, a human embryonic kidney cell line in which a left terminal fragment of the Ad5 chromosome, including the complete E1 region and more than 0.5 kb extending into the IVa2 gene (E. Vigne, unpublished), has been integrated within the cellular genome (31). Although insertion within E3 is possible, the transgene is usually inserted in place of El in the recombinant backbone. The recombinant virus is usually recovered from 293 cells following a homologous recombination between a restricted (i.e., non infectious) viral genome and an overlapping plasmid corresponding to the left end of the desired recombinant. It can be isolated by the ability of 293-derived monolayers to form lytic plaques upon addition of an agar overlay. Following progressive amplification on 293 cells, the recombinant can be concentrated by step and isopycnic CsCl gradient centrifugations. Viral titers can be as high as 10^{11} plaque-forming unit (pfu)/ml, and the concentrated virus is stable when stored at -70°C in 10% glycerol. This efficient process is not satisfactory for the recovery of a clinical grade recombinant virus because i) viral plaque purification is not a true cloning process, and ii) reincorporation of the E1A and E1B genes arise upon homologous recombination between the viral genomes and the cellular El flanking sequences (47). In fact, viral stocks are systematically contaminated with replication-competent adenoviruses (RCA). Human embryonic retinoblasts immortalized with Ad5 El sequences have been recently used for the construction and efficient propagation of E1-deleted recombinants (24).

Because the viral genome is quite large, its in vitro manipulation is always tedious. Recently, a yeast artificial chromosome containing a complete adenoviral genome has been constructed, from which infectious viral DNA can be recovered using established yeast genetic tools (41). A bacterial-based genetic system has also been developed in our laboratory which allows the introduction of any modification anywhere in the backbone by using homologous recombination. This procedure is particularly useful as it allows a true cloning of the recombinant genome in Escherichia coli, before the corresponding virus be recovered after its transfection in a suitable packaging cell line (J. Crouzet, manuscript in preparation).

2.4 Intrinsic Properties of E1-deleted Vectors

2.4.1 Residual (Leaky) Expression of the Viral Genome in vitro

Infection of non-permissive cells with an E1-deleted virus is not innocuous, especially at high moi (46). It has been previously reported that human primary nasal polyp epithelial cells actually allowed limited viral DNA replication following infection with an E1-deleted recombinant. Furthermore, early (E4) and late (L5) transcription were also detected when established cell lines were infected (59). Our in vitro studies indeed confirm that most human cell lines replicate and express the recombinant genome to some extent. For example, a 100-fold increase in the copy number of viral genomes occurs between 3 and 4 days following infection of Hep3B cells in conditions allowing lacZ transduction in 80% of the cells (Fig. 2.3), while the figure is lower (10-fold increase) when a similar proportion of KB cells are in-

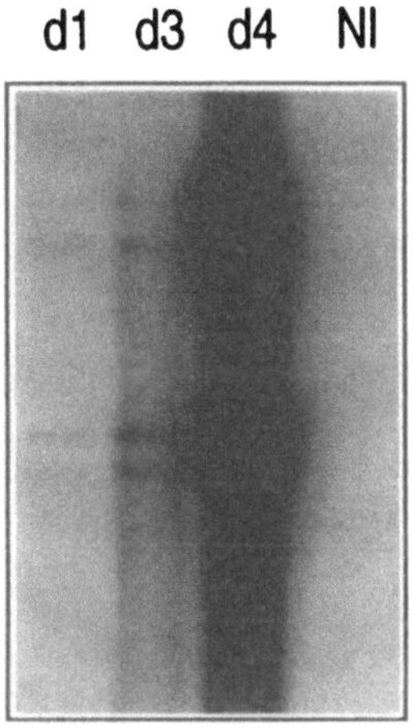

Fig. 2.3: Viral DNA replication in Hep3B cells infected with an E1-deleted virus.
Hep3B cells were infected with virus RSVbGal at a multiplicity of infection of 20 tdu/cell (i.e., the quantity of virus which transduces 80% of the cells). The DNA was extracted by the Hirt procedure and digested by NdeI at 1 (lane d1), 3 (lane d3), or 4 (lane d4) days post-infection. NI refers to non infected cell extracts. Following electrophoresis, the fragments were transferred onto Hybond-N membranes (Amersham) and hybridized with a radiolabeled probe corresponding to HinIII-restricted RSVbGal DNA. Comparative dilution experiments indicate that there is a 100-fold increase in the number of recombinant genomes between 3 and 4 days following infection.

fected (E. Vigne, unpublished result). Significant transcription of E2A is also apparent in many cell lines, including KB (N. Hanania, personal communication), and Hep3B (JF. Dedieu, unpublished result). Identical conclusions were reached for late mRNAs. For example, Northern blot analysis demonstrates that Hep3B cells infected with an E1-deleted recombinant express significant levels of the L3 pVI- and L3 pII (hexon)-encoding RNAs (Fig. 2.5). This also apply for the L5 pIV (fiber)-encoding RNA. Accordingly, the synthesis of the fiber protein can be demonstrated by FACS analysis with specific anti-fiber antibodies (MAB805, Chemicon) in a wide variety of infected cells, including KB and A549 (E. Vigne and M. Maratrat, unpublished results).

In vitro transcription from the E4 region has been demonstrated when HeLa or CF-PAC human cell lines are infected with an E1-deleted recombinant (59). To further document the activity of the E4 promoter, we have constructed an E1-deleted virus expressing the lacZ-reporter gene from the E4 promoter in its original environment. Infection experiments with this reporter virus supported a significant expression of the E4 promoter in vitro (e.g., in KB, HeLa or HepG2 cells), and in vivo (N. Hanania, unpublished results). Because previous studies have demonstrated that the E4 promoter is strongly stimulated by E1A products in vitro (26), these observations support the existence of endogenous activities substituting for particular E1A-transactivating functions (hereafter referred to as E1A-like) in certain cellular backgrounds (see below). In that respect, expression of the E4 locus by endogenous

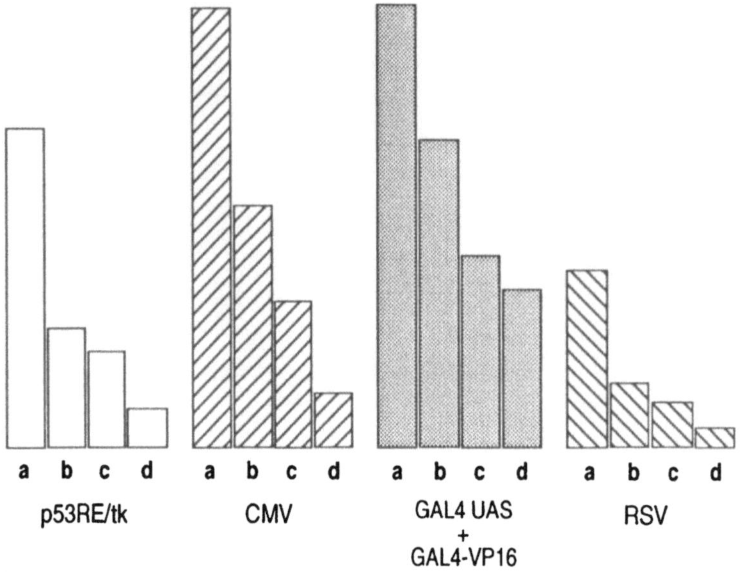

Fig. 2.4: (ORF6+ORF7) influences gene expression
NIH 3T3 cells were transfected with 100 ng of a chloramphenicol acetyl transferase (CAT) expression cassette, in the absence (lanes a) or in the presence of 100 ng (lanes b), 250 ng (lanes c), or 750 ng (lanes d) of a CMV-driven (ORF6+ORF7) expression cassette. A minimal promoter with p53 (p53RE/tk) or GAL4 (GAL4 UAS) responsive elements, the CMV or the RSV promoters were used to drive the synthesis of the CAT-encoding mRNAs. In the case of the GAL4 UAS-driven CAT expression cassette, the GAL4-VP16 transactivator was also transfected. Results are expressed as relative units following quantification of CAT activity with a Phosphoimager.

E1A-like activities is of special concern because E4 encodes pleiotropic proteins regulating cell cycle regulators, together with the cell machinery for post-transcriptional processing and/or nuclear export of the mRNAs. Furthermore, E4 is also involved in host cell protein synthesis shutoff (33). That E4 expression may be detrimental to cell viability is supported by the observation that dexamethasone induction of the distal (ORF6+ORF7) part of the E4 region (encoding the ORF6+ORF6/7 proteins) interferes with cell viability (87). In addition, introduction of a CMV-driven (ORF6+ORF7) expression cassette results in a dose dependent, non specific, inhibition of the cell machinery in Saos, H358, and NIH3T3 cells (L. Bracco, personal communication, see also Fig. 2.4).

2.4.2 In vivo Behaviour

Gene transfer with human-derived E1-deleted recombinants is not restricted to human cells: it has been documented in many non-permissive (chicken, mice, dogs, rats, rabbits, baboons...), as well as permissive (cotton rats) species (53). In fact, Ad2 or Ad5 endocytosis may actually not solely rely on the vitronectin receptor, but perhaps on other (highly conserved) cell surface molecules (3). In any case, local or systemic delivery of the recombinants results in the transduction of a wide variety of adherent, tumor or somatic, cells from many organs (liver, lung, heart, brain, blood vessels, skeletic muscle cells...).

Adenovirus-mediated gene transfer (AMGT) is indeed remarkably efficient. For example, almost all hepatocytes can be transduced following the delivery of 10^{10} particles of a lacZ-expressing E1-deleted virus in the portal vein of adult mice (46). AMGT is also intrinsically immunogenic. Specifically, a number of preclinical and clinical studies have reported a local, dose-dependent, inflammatory response in the target organs of immunocompetent hosts (35, 42, 65). It has also been demonstrated that transgene expression is substantially extended in immunodeprived or immunodeficient adults (22, 83). This is also true when the virus is administered to newborn animals (55, 71), to immunopriviledged organs (4), or to human xenografts implanted within immunodeficient hosts (21). These observations outline the central importance of the immune system in the clearance of the infected cells, very likely because specific vector-encoded antigens presented at the cell surface are recognized by memory cytotoxic lymphocytes (CTL) (20, 83).

Expression of the viral genome actually occurs in vivo as evidenced by immunodetection of early (DBP) and late (hexon) viral proteins following gene transfer into a number of non-permissive hosts including mice (22, 83), rats (11), and baboons (28). It is admitted that synthesis of virus-encoded antigens in vivo triggers the destruction of the infected cells following their presentation to immune cellular effectors by MHC class I molecules. Epitopes derived from certain transgene also participate in this destructive process, as demonstrated in DBA-2 mice for the lacZ-gene product of E. coli (45). The relative importance of viral or transgene-derived MHC class I-restricted epitopes remains largely unknown, in part because AMGT has been studied in strains of inbred mice with different abilities to present and recognize foreign antigens. In an attempt to clarify this point, we have monitored the persistance of infected hepatocytes following intravenous (i.v.) administration of virus AdRSV$_\beta$Gal, an E1E3-deleted virus expressing a nuclear-targeted β-galactosidase from the RSV LTR (72), in a lacZ-transgenic (Tg) mice (15). Interestingly, this model animal is immunotolerant for the E. coli enzyme but it does not express the enzyme in the liver (JM. Caillaud, P. Delaere, and P. Opolon, personal communication), making it possible to monitor the persistance of transduced hepatocytes infected with a lacZ-expressing reporter adenovirus. After ten consecutive backcrosses with C57BL/6 mice, the persistance of infected hepatocytes from the Tg-infected animal was compared to that of C57BL/6-infected controls. Three days after the systemic administration of 2.5 10^9 lacZ-transducing units (tdu) of virus AdRSV$_\beta$Gal (i.e., the dose required to transduce 2.5 10^9 W162 cells), both strains of mice expressed the enzyme in 55 to 70 % of their hepatocytes. This figure dropped very rapidly in the non-transgenic mice as half of the infected hepatocytes were lost 10 days post-infection. In contrast, more than 6 weeks were required to achieve the same result in the lacZ-transgenic animals. These observations strongly support that an important component of the host immune response is targeted towards the transgene itself, at least in the C57BL/6 genetic background. Interestingly, although the extent of the inflammatory infiltration was significantly reduced in the liver of the Tg animals 2 weeks after infection, it was still noticeable around the lacZ-expressing hepatocytes (JM. Caillaud and P. Opolon, personal communication), a result which also suggests the additional involvement of virus-derived antigens in the CTL-mediated clearance of the infected cells (J. F. Dedieu et al., manuscript in preparation).

2.5 Doubly-defective Vectors

2.5.1 Rationale

Endogenous factors substituting for E1A for its ability to transactive the viral promoters may overcome the block in transcription imposed by the E1 deletion in vivo. Indeed, viral and cellular E1A-like activities have been documented in vitro in different cell types. For example, HeLa cells are somehow permissive following infection with an E1A-deleted adenovirus at high multiplicity of infection (64). An E1A-like activity of cellular origin has also been described in undifferentiated mouse embryonal carcinoma cells by vertue of its ability to transactivate the E2A and E3 E1A-inducible promoters (6, 38). Potential cellular candidates with an E1A-like activity include c-myc and a transcription factor induced by IL-6 (52, 68). The latter observation is puzzling in light of the elevated levels of this major inflammatory cytokine reported in a cystic fibrosis patient administred with an E1-deleted recombinant expressing the CFTR gene (16). It is actually possible that the host inflammatory response to AMGT is indeed initiated by leaky viral expression in a small subset of infected cells (i.e., those exhibiting an E1A-like factor or those infected with an RCA), before being generalized to all recipient cells through an intercellular immune signalling cascade involving IL-6. In that respect, safety and efficacy which may be regarded as two different issues associated with AMGT are probably closely linked.

2.5.2 The E2 Approach

To design safer and more potent vectors, the ts125 thermosensitive mutation has been included in the backbone of an E1-deleted adenovirus expressing the lacZ reporter gene. This point mutation, located within the gene encoding the 72K DBP (E2A), severely impairs the initiation of viral DNA replication at 39°C. As such, the switch from early to late gene expression does not occur and the doubly defective virus must be grown in 293 cells at the permissive temperature of 32°C (22).

Mice hepatocytes infected with this recombinant virus persisted for a longer period of time than those infected with an E1-deleted control virus. The persistance of lacZ expression was however below than that observed in nude animals. This observation suggests that the immune clearance of the infected cells was not totally circomvented, a situation which could result from partial inactivation of the DBP mutant protein at 37 °C, and/or some level of MHC class I-restricted presentation of viral or β-galactosidase-derived epitopes to activated cytotoxic lymphocytes (CTL). Importantly, the fact that the inflammatory response could be reduced as compared to that induced with an E1-deleted recombinant demonstrates that crippling viral gene expression (safety issue) can increase transgene expression in vivo (efficacy issue). A longer persistance of gene expression and a reduced inflammation, together with diminished levels of hexon proteins were also observed when the virus was administred into the lungs of cotton rats or baboons (20, 28).

2.5.3 The E4 Approach

We and others have approached the silencing of viral gene expression by the additional deletion of the E4 region in the backbone of E1-deleted vectors. This regulatory region has been targeted because it is required for late protein synthesis, virus particle assembly, and ultimately virus growth (23, 33, 78). E4 is also involved in the transactivation of the E2A promoter (2, 6, 58), and somehow interferes with the replication of viral DNA during infection (9, 33, 77, 78). The deletion of this important regulatory function from the virus genome implies that it must be provided in trans during virus amplification, either by a helper virus (48, 74), or within a transcomplementing (packaging) cell line (44, 76, 79, 87).

We have previously reported the characterization of 293-derived cell lines expressing either the entire E4 locus (IGRP4 cells), or only its distal (ORF6+ORF7) moiety (IGRP2 cells, see Fig. 2.2B) from a dexamethasone-inducible promoter. Functional analysis has demonstrated that these cell lines allowed the construction of several doubly-defective (E3+) deletants expressing the lacZ gene from the RSV LTR (87). In vitro infection experiments demonstrate that the additional deletion of the E4 region drastically impairs viral DNA replication in different cell lines, including KB and Hep3B (E. Vigne, unpublished result). The levels of early (E2A) and late (e.g., L3 or L5) mRNAs were also reduced in these cells (Fig. 2.5). The systemic administration of similar doses of the E1E4-doubly deleted viruses, or an E1-deleted otherwise isogenic virus in mice was shown to transduce a similar number of hepatocytes following infection (i.e., between 50 and 80% of lacZ-expressing cells with doses ranging from 2 to 5 10^9 tdu/animal; N. Aubailly and JM. Caillaud, personal communication). These viruses are currently being compared in a variety of in vivo gene transfer models, including the C57BL/6 lacZ-transgenic mice, to document their behavior in terms of persistance, and the host response to them.

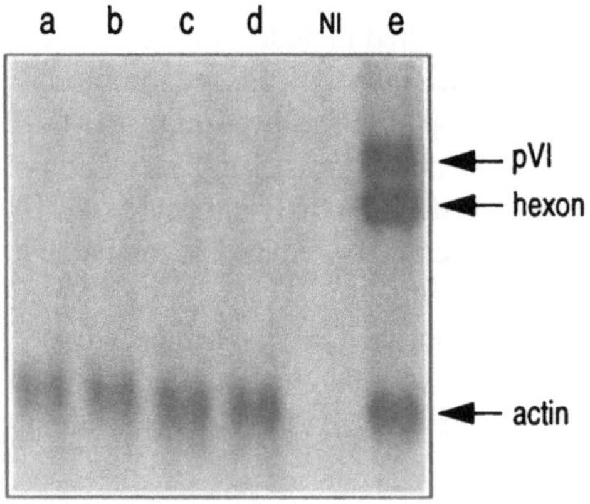

Fig. 2.5: In vitro behaviour of E1- or E1E4-deleted recombinant adenoviruses.
Northern blot analysis of Hep3B-infected cells. The polyA+ RNAs were extracted 2 days following infection at a moi of 20 tdu/cell. In all cases, 80% of the cells were checked and were found positive for X-GAL staining (infection control). The RNA was transferred onto Hybond-N membranes and hybridized with an antisens radiolabeled riboprobe common to the L3 pVI- and hexon-encoding mRNAs, as well as with an actin specific probe (loading control). Hep3B cells were infected with virus Ad$_{\beta}$Gal/dl1004 (lane a), Ad$_{\beta}$Gal/dl1007 (lane b), Ad$_{\beta}$Gal/dl1011 (lane c), Ad$_{\beta}$Gal/dl1014 (lane d), or with virus RSV$_{\beta}$Gal(E3+) which is an E4+ otherwise isogenic control virus (87). NI refers to non infected cell extracts.

2.6 From Gene Transfer to Therapy

Safe and efficient virus-mediated gene therapy requires a comprehensive understanding of the vector's behaviour in vitro and in vivo. Previous studies have demonstrated that the host response to most viral infection typically includes an early phase of inflammation in which macrophages, polynuclear neutrophils and granular cytotoxic lymphocytes (natural killer cells) rapidly infiltrate the infected organ. This (innate) response is characterized by the induction of elevated levels of inflammatory cytokines including interferons (IFN), IL-6 and IL-1. It is followed by a delayed, adaptive, infiltration by immune effectors mostly composed of activated lymphocytes. A similar biphasic inflammatory response has also been reported following Ad5 infection of permissive, as well as non-permissive animal models (27, 49, 54). As reported by several groups, this also apply when a defective, E1-deleted, recombinant is administred to a non-permissive mouse model (JM. Caillaud and P. Opolon, personal communication).

The immediate response to AMGT is dose dependent and most likely directed against the viral input (i.e., in the absence of viral gene expression), outlining the intrinsic inflammatory nature of recombinant adenoviruses. Exogenous antigens are presented by MHC class II to T-helper cells and this interaction subsequentely promotes B cells to produce specific neutralizing antibodies. Recombinant adenoviruses also comply to this rule as they elicit a humoral immune response that prevents gene transfer upon a second, intravenous or intratracheal, administration (17, 42, 43, 66, 82, 88). Adenovirus driven-T-helper cells activation may actually trigger a broader response, perhaps through secretion of IFN-γ by the Th1 subset of CD4 lymphocytes, upregulating MHC class I expression and enhancing the destruction of the infected cells (86). The importance of IFN-γ in this destructive process is supported by the observation that AMGT increases the level of class I heavy-chain RNA in the liver of C57BL/6 mice, but not in IFN-γ-depleted, or knockout, mice (86). Furthermore, AMGT rapidly (i.e., in the absence of protein synthesis) induces transcription of a number of IFN-responsive genes (57). Taken together, these observations predict that more defective versions of the vectors may still cause a substantial inflammation in vivo, a prerequesite for mounting an efficient adaptive humoral and cellular immune response. In that regard, specific immunosuppresive drugs may prove useful to overcome the inefficiency of primary or repeated gene transfer (17, 20, 40, 75, 84).

A complementary approach to this problem relies on the optimization of transgene expression so therapeutic levels could be achieved with a reduced dose of the recombinant virus. For example, inclusion of genomic sequences in a factor IX expression cassette resulted in a dramatic increase of the protein levels in the plasma of immunocompetent mice, together with a reduced hepatotoxicity and a sustained expression of the transgene (39). This strategy however requires an extended cloning capacity for large transgenes (e.g., dystrophin). In that regard, the possibility to engineer infectious genomes by homologous recombination in yeast or in bacteria is particularly relevant because consecutive rounds of modifications (deletions) can be easily introduced with no requirement for any intermediate recombinant virus. As an example, we have engineered a ΔE1ΔE3ΔE4 genome in E. coli and recovered high titers of the corresponding virus (i.e., a virus with a cloning capacity of 9 kb) following its transfection in IGRP2 cells (J. Crouzet, manuscript in preparation). Rescue of pIX expression (12), together with the deletion of additional viral genes (60) also extend the theoritical cloning capacity of the vectors.

Current E1-deleted recombinants express vector-encoded, potentially antigenic, proteins in vivo and elicit a MHC class I-restricted CTL-mediated destruction of the infected cells. Recent studies support this model as infected cells from $_{\beta}2$-microglobulin (81) or perforin (86) knockout mice were not efficiently destructed despite an important lymphocytic infiltration of the target organ was noted (86). Again, it should be stressed that the epitopes which are normally presented and recognized by activated CTLs remain unknown. The nature of all the cellular effectors of this destructive process is also under investigation and it has been suggested that CD4 lymphocytes actually could act as primary CTL effectors, especially when CD8 effectors are depleted (85). At present, it seems more and more likely that sophisticated (attenuated) adenoviral vectors, together with specific immunosuppressive drugs will set the future basis of safer and better gene transfer protocols.

Acknowledgements

We thank all members of the laboratory for technical assistance, Karim Benihoud for critical reading and suggestions, and Nathalie Aubailly, Laurent Bracco, Jean-Michel Caillaud, Joel Crouzet, Pia Delaère, Patrice Denèfle, Jean-Marc Guillaume, Nicole Hanania, Michel Maratrat and Paule Opolon for helpful discussions or sharing unpublished data. The authors also wish to acknowledge Rhône-Poulenc and the French Ministry of Research and Industry for financial support (BioAvenir).

References

1. Armentano, D., Sookdeo, C. C., Hehir, K. M., Gregory, R. J., St. George, J. A., Prince, G. A., Wadsworth, S. C., and Smith, A. E. (1995). Characterization of an adenovirus gene transfer vector containing an E4 deletion. Hum. Gene Ther. *6*, 1343-1353.
2. Babiss, L. E. (1989). The cellular transcription factor E2f requires viral E1A and E4 gene products for increased DNA-binding activity and functions to stimulate adenovirus E2A gene expression. J. Virol. *63*, 2709-2717.
3. Bai, M., Campisi, L., and Freimuth, P. (1994). Vitronectin receptor antibodies inhibit infection of HeLa and A549 cells by adenovirus type 12 but not by adenovirus type 2. J. Virol. *68*, 5925-5932.
4. Bennett, J., Wilson, J., Sun, D., Forbes, B., and Maguire, A. (1994). Adenovirus vector-mediated in vivo gene transfer into adult murine retina. Invest. Ophtalmol. Visual Sci. *35*, 2535-2542.
5. Bett, A. J., Prevec, L., and Graham, F. L. (1993). Packaging capacity and stability of human adenovirus type 5 vectors. J. Virol. *67*, 5911-5921.
6. Boeuf, H., Reimund, B., Jansen-Durr, P., and Kédinger, C. (1990). Differential activation of the E2F transcription factor by the adenovirus E1a and EIV products in F9 cells. Proc. Natl. Acad. Sci. U.S.A. *87*, 1782-1786.
7. Bridge, E. and Ketner, G. (1990). Interaction of adenoviral E4 and E1b products in late gene expression. Virology. *174*, 345-353.
8. Bridge, E. and Ketner, G. (1989). Redundant control of adenovirus late gene expression by early region 4. J. Virol. *63*, 631-638.
9. Bridge, E., Medghalchi, S., Ubol, S., Leesong, M., and Ketner, G. (1993). Adenovirus early region 4 and viral DNA synthesis. Virology. *193*, 794-801.
10. Burget, H. G. and Kvist, S. (1987). The E3/19K protein of adenovirus type 2 binds to the domains of histocompatibility antigens required for CTL recognition. EMBO J. *6*, 2019-2026.
11. Byrnes, A. P., Rusby, J. E., Wood, M. J. A., and Charlton, H. M. (1995). Adenovirus gene transfer causes inflammation in the brain. Neurosciences. *66*, 1015-1024.
12. Caravokyri, C. and Leppard, K. N. (1995). Constitutive episomal expression of polypeptide IX (pIX) in a 293-based cell line complements the deficiency of pIX mutant adenovirus type 5. J. Virol. *69*, 6627-6633.
13. Chang, L.-S. and Shenk, T. (1990). The adenovirus DNA-binding protein stimulates the rate of transcription directed by adenovirus and adeno-associated virus promoters. J. Virol. *6*, 2103-2109.
14. Chroboczek, J., Bieder, F., and Jacrot, B. (1992). The sequence of adenovirus type 5 and its comparison with the genome of adenovirus type 2. Virology. *186*, 280-285.
15. Cohen-Tannoudji, M., Morello, D., and Babinet, C. (1992). Unexpected position-dependent expression of H-2 and b2-microglobulin/lacZ transgenes. Mol. Reprod. Dev. *33*, 149-159.
16. Crystal, R. G., McElvaney, N. G., Rosenfeld, M. A., Chu, C., Mastrangeli, A., Hay, J. G., Brody, S. L., Jaffe, H. A., Eissa, N. T., and Danel, C. (1994). Administration of an adenovirus containing the human CFTR cDNA to the respiratory tract of individuals with cystic fibrosis. Nature Genet. *8*, 42-51.

17. Daï, Y., Schwarz, H. M., Gu, D., Zhang, W.-W., Sarvetnick, N., and Verma, I. M. (1995). Cellular and humoral immune responses to adenoviral vectors containing factor IX gene, tolerization of factor IX and vector antigens allows for long-term expression. Proc. Natl. Acad. Sci. USA. *92*, 1401-1405.
18. Debbas, M. and White, E. (1993). Wild-type p53 mediates apoptosis by E1A, which is inhibited by EIB. Genes Dev. *7*, 546-554.
19. Dix, I. and Leppard, K. N. (1993). Regulated splicing of adenovirus type 5 E4 transcripts and regulated cytoplasmic accumulation of E4 mRNA. J. Virol. *67*, 3226-3231.
20. Engelhardt, J. F., Litzky, L., and Wilson, J. M. (1994b). Prolonged transgene expression in cotton rat lung with recombinant adenoviruses defective in E2a. Hum. Gene Ther. *5*, 1217-1229.
21. Engelhardt, J. F., Yang, Y., Stratford-Perricaudet, L. D., Allen, E. D., Kozarsky, K., Perricaudet, M., Yankaskas, J. R., and Wilson, J. M. (1993). Direct gene transfer of human CFTR into human bronchial epithelia of xenografts with E1-deleted adenoviruses. Nature Genet. *4*, 27-34.
22. Engelhardt, J. F., Ye, X., Doranz, B., and Wilson, J. M. (1994a). Ablation of E2A in recombinant adenoviruses improves transgene persistence and decreases inflammatory response in mouse liver. Proc. Natl. Acad. Sci. USA. *91*, 6196-6200.
23. Falgout, B. and Ketner, G. (1987). Adenovirus early region 4 is required for efficient virus particle assembly. J. Virol. *61*, 3759-3768.
24. Fallaux, F. J., Kranenburg, O., Cramer, S. J., Houweling, A., van Ormondt, H., Hoeben, R. C., and van der Eb, A. J. (1996). Characterization of 911, a new helper cell line for the titration and propagation of early region 1-deleted adenoviral vectors. Hum. Gene Ther. *7*, 215-222.
25. Flomenberg, P., Szmulewicz, J., Gutierrez, H., and Lupatkin, H. (1992). Role of the adenovirus E3-19K conserved region in binding major histocompatibility complex class I molecules. J. Virol. *66*, 4778-4783.
26. Gilardi, P. and Perricaudet, M. (1984). The E4 transcriptional unit of Ad2, far upstream sequences are required for its transactivation by E1A. Nucl. Acids Res. *12*, 7877-7888.
27. Ginsberg, H. S., Moldawer, L.L., Sehgal, P. B., Redington, M., Kilian, P. L., Chanock, R. M., and Prince, G. A. (1991). A mouse model for investigating the molecular pathogenesis of adenovirus pneumonia. Proc. Natl. Acad. Sci. USA. *88*, 1651-1655.
28. Goldman, M. J., Litzky, L. A., Engelhardt, J. F., and Wilson, J. M. (1995). Transfer of the CFTR gene to the lung of nonhuman primates with E1-deleted, E2a-defective recombinant adenoviruses, a preclinical toxicology study. Hum. Gene Ther. *6*, 839-851.
29. Gooding, L. R., Ranheim, T. S., Tollefson, A. E., Brady, H. A., and Wold, W. S. M. (1991). The 10,400- and 14,500- dalton proteins encoded by region E3 of adenovirus function together to protect many but not all mouse cell lines against lysis by tumor necrosis factor. J. Virol. *65*, 4114-4123.
30. Grable, M. and Hearing, P. (1992). cis and trans requirements for the selective packaging of adenovirus type 5 DNA. J. Virol. *66*, 723-731.
31. Graham, F. L., Smiley, J., Russel, W. C., and Nairn, R. (1977). Characteristics of a human cell line transformed by DNA from human adenovirus type 5. J. Gen. Virol. *36*, 59-72.
32. Greber, U. F., Willetts, M., Webster, P., and Helenius, A. (1993). Stepwise dismantling of adenovirus 2 during entry into cells. Cell. *75*, 477-486.
33. Halbert, D. N., Cutt, J. R., and Shenk, T. (1985). Adenovirus early region 4 encodes functions required for efficient DNA replication, late gene expression, and host cell shutoff. J. Virol. *56*, 250-257.
34. Hemstrom, C., Nordqvist, K., Pettersson, U., and Virtanen, A. (1988). Gene product of region E4 of adenovirus type 5 modulates accumulation of certain viral polypeptides. J. Virol. *62*, 3258-3264.
35. Herz, J. and Gerard, R. D. (1993). Adenovirus-mediated transfer of low density lipoprotein receptor gene acutely accelarates cholesterol clearance in normal mice. Proc. Natl. Acad. Sci. USA. *90*, 2812-2816.
36. Horwitz, M. S. (1996). Adenoviruses. In: Virology. B.N. Fields, D.M. Knipe, P.M. Howley et al. (Eds.) Raven Publishers: Philadelphia; 2149-2171.
37. Huang, M. M. and Hearing, P. (1989). Adenovirus early region 4 encodes two gene products with redundant effects in lytic infection. J. Virol. *63*, 2605-2615.
38. Imperiale, M. J., Kao, H.-T., Feldman, L. T., Nevins, J. R., and Strickland, S. (1984). Common control of the heat shock gene and early adenovirus genes, evidence for a cellular E1A-like activity. Mol. Cell. Biol. *4*, 867-874.
39. Kaleko, M., Kayda, D., Sakhuja, K., Mehaffey, M., and McClelland, A. (1995). Genomic sequences increase adenoviral vector-mediated factor IX expression 1900 fold and enable sustained expression in vivo. J. Cell. Biochem. *Suppl. 21A*, p. 366 (abstract).

40. Kay, M. A., Holterman, A. X., Meuse, L., Gown, A., Ochs, H. D., Linsley, P. S., and Wilson, C. B. (1995). Long-term hepatic adenovirus-mediated gene expression in mice following CTLA4Ig administration. Nature Genet. *2*, 191-197.
41. Ketner, G., Spencer, F., Tugendreich, S., Connelly, C., and Hieter, P. (1994). Efficient manipulation of the human adenovirus genome as an infectious yeast artificial chromosome clone. Proc. Natl. Acad. Sci. USA. *91*, 6186-6190.
42. Knowles, M. R., Hohnecker, K. W., Zhou, Z., Olsen, J. C., Noah, T. L., Hu, P. C., Leigh, M. W., Engelhardt, J. F., Edwards, L. J., Jones, K. R., Grossman, M., Wilson, J. M., Johnson, L. G., and Boucher, R. C. (1995). A controlled study of adenoviral-vector-mediated gene transfer in the nasal epithelium of patients with cystic fibrosis. N. Engl. J. Med. *333*, 871-873.
43. Kozarsky, K. F. and X. et al. (1994). In vivo correction of LDL receptor deficiency in the Watanabe heritable hyperlipidemic rabbit with recombinant adenoviruses. J. Biol. Chem. *269*, 13695-13702.
44. Krougliak, V. and Graham, F. L. (1995). Development of cell lines capable of complementing E1, E4, and protein IX defective adenovirus type 5 mutants. Hum. Gene Ther. *6*, 1575-1586.
45. Lee, M. G., Abina, M. A., Haddada, H., and Perricaudet, M. (1995). The constitutive expression of the immunomodulatory gp19k protein in E1-, E3- adenoviral vectors strongly reduces the host cytotoxic T cell response against the vector. Gene Ther. *2*, 256-262.
46. Li, Q., Kay, M. A., Finegold, M., Stratford-Perricaudet, L. D., and Woo, S. L. C. (1993). Assessment of recombinant adenoviral vectors for hepatic gene therapy. Hum. Gene Ther. *4*, 403-409.
47. Lochmuller, H., Jani, A., Huard, J., Prescott, S., Simoneau, M., Massie, B., Karpati, G., and Acsadi, G. (1994). Emergence of early region 1-containing replication-competent adenovirus in stocks of replication-defective adenovirus recombinants (DE1+DE3) during multiple passages in 293 cells. Hum. Gene Ther. *5*, 1485-1491.
48. Mitani, K., Graham, F. L., Caskey, C. T., and Kochanek, S. (1995). Rescue, propagation, and partial purification of a helper virus-dependent adenovirus vector. Proc. Natl. Acad. Sci. USA. *92*, 3854-3858.
49. Mullbacher, A., Bellet, A. J. D., and Hla, R. T. (1989). The murine cellular immune response to adenovirus type 5. Immunol. Cell Biol. *67*, 31-39.
50. Nevins, J. R. and Winkler, J. J. (1980). Regulation of early adenovirus transcription, a protein product of early region 2 specifically represses region 4 transcription. Proc. Natl. Acad. Sci. USA. *77*, 1893-1897.
51. Nordqvist, K., Ohman, K., and Akusjarvi, G. (1994). Human adenovirus encodes two proteins which have opposite effects on accumulation of alternatively spliced mRNAs. Mol. Cell. Biol. *14*, 437-445.
52. Onclercq, R., Gilardi, P., Lavenu, A., and Cremisi, C. (1988). c-myc products trans-activate the adenovirus E4 promoter in EC stem cells by using the same target sequence as E1A products. J. Virol. *62*, 4533-4537.
53. Pacini, D. L., Dubovi, E. J., and Clyde Jr, W. A. (1984). A new animal model for human respiratory tract disease due to adenoviruses. J. Infect. Dis. *150*, 92-97.
54. Prince, G. A., Porter, D. A., Jenson, A. B., Horswood, R. L., Chanock, R. M., and Ginsberg, H. S. (1993). The pathogenenesis of type 5 adenovirus in the cotton rat (Sigmodon Hispidus). J. Virol. *67*, 101-111.
55. Ragot, T., Vincent, N., Chafey, P., Vigne, E., Gilgenkrantz, H., Couton, D., Cartaud, J., Briand, P., Kaplan, J.-C., Perricaudet, M., and Kahn, A. (1993). Efficient adenovirus-mediated transfer of a human minidystrophin gene to skeletal muscle of mdx mice. Nature. *361*, 647-650.
56. Rao, L., Debbas, M., Sabbatini, P., Hockenbery, D., Korsmeyer, S., and White, E. (1992). The adenovirus E1A proteins induce apoptosis, which is inhibited by the E1B 19-kDa and Bcl-2 proteins. Proc. Natl. Acad. Sci. USA. *89*, 7742-7746.
57. Reich, N., Pine, R., Levy, D., and Darnell, J. E. (1988). Transcription of interferon-stimulated genes is induced by adenovirus particles but is suppressed by E1A gene products. J. Virol. *62*, 114-119.
58. Reichel, R., Neill, S. D., Kovesdi, I., Simon, M. C., Raychaudhuri, P., and Nevins, J. R. (1989). The adenovirus E4 gene, in addition to the E1A gene, is important for trans-activation of E2 transcription and for E2F activation. J. Virol. *63*, 3643-3650.
59. Rich, D. P., Couture, L. A., Cardoza, L. M., Guiggio, V. M., Armentano, D., Espino, P. C., Hehir, K., Welsh, M. J., Smith, A. E., and Gregory, R. J. (1993). Development and analysis of recombinant adenoviruses for gene therapy of cystic fibrosis. Hum. Gene Ther. *4*, 461-476.
60. Schaack, J., Guo, X., Yew-Wai HO, W., Karlok, M., Chen, C., and Ornelles, D. (1995). Adenovirus type 5 precursor terminal protein-expressing 293 and HeLa cell lines. J. Virol. *69*, 4079-4085.

61. Schaack, J., Ho, W., Freimuth, P., and Shenk, T. (1990). Adenovirus terminal protein mediates both nuclear matrix association and efficient transcription of adenovirus DNA. Genes Dev. *4*, 1197-1208.
62. Scharwtz, T. Z., Togo, Y., and Hornick, R. B. (1974). Clinical evaluation of live, oral types 1, 2 and 5 adenovirus vaccines. Amer. Rev. Respiratory Disease. *109*, 233-238.
63. Shenk, T. (1996). Adenoviridae, the viruses and their replication. In: Virology. B.N. Fields, D.M. Knipe, P.M. Howley et al. (Eds). Raven Publishers: Philadelphia, 2111-2148.
64. Shenk, T., Jones, N., Colby, W., and Fowlkes, D. (1979). Functional analysis of adenovirus-5 host-range deletion mutants defective for transformation of rat embryo cells. C.S.H. Symp. Quant. Biol. *44*, 367-375.
65. Simon, R. H., Engelhardt, J. F., Yang, Y., Zepeda, M., Weber-Pendleton, S., Grossman, M., and Wilson, J. M. (1993). Adenovirus-mediated transfer of the CFTR gene to lung of nonhuman primates, toxicity study. Hum. Gene Ther. *4*, 771-780.
66. Smith, T. A. G. and X. et al. (1993). Adenovirus mediated expression of therapeutic plasma levels of human factor IX in mice. Gene Ther. *5*, 397-402.
67. Soloway, P. D. and Shenk, T. (1990). The adenovirus type 5 i-leader open reading frame functions in cis to reduce the half-life of L1 mRNAs. J. Virol. *64*, 551-558.
68. Spergel, J. M., Hsu, W., Akira, S., Thimmappaya, B., Kishimoto, T., and Chen-Kiang, S. (1992). NF-IL6, a member of the C/EBP family, regulates E1A-responsive promoters in the absence of E1A. J. Virol. *66*, 1021-1030.
69. Stewart, P. L., Burnett, R. M., Cyrklaff, M., and Fuller, S. D. (1991). Image reconstruction reveals the complex molecular organization of adenovirus. Cell. *67*, 145-154.
70. Stewart, P. L., Fuller, S. D., and Burnett, R. M. (1993). Difference imaging of adenovirus, bridging the resolution gap between X-ray crystallography and electron micrography. EMBO J. *12*, 2589-2599.
71. Stratford-Perricaudet, L. D., Levrero, M., Chasse, J.-F., and Perricaudet, M. (1990). Evaluation of the transfer and expression in mice of an enzyme-encoding gene using a human adenovirus vector. Hum. Gene Ther. *1*, 241-256.
72. Stratford-Perricaudet, L. D., Makeh, I., Perricaudet, M., and Briand, P. (1992). Widespread long-term gene transfer to mouse skeletal muscles and heart via an adenovirus vector. J. Clin. Invest. *90*, 626-630.
73. Tribouley, C., Lutz, P., Staub, A., and Kedinger, C. (1994). The product of the adenovirus intermediate gene IVa2 is a transcriptional activator of the major late promoter. J. Virol. *68*, 4450-4457.
74. Vigne, E., Dedieu, J.-F., Orsini, C., Latta, M., Klonjkowski, B., Prost, E., Lakich, M. M., Kremer, E. J., Denèfle, P., Perricaudet, M., and Yeh, P. (1996). Improvement of adenoviral vectors for human gene therapy. In: Genetic Approaches to Noncommunicable Diseases. K. Berg, V. Boulyjenkov V. and Y. Christen (Eds). Springer-Verlag: Berlin, Heidelberg, 113-131.
75. Vilquin, J.-T., Guérette, B., Kinoshita, I., Roy, B., Goulet, M., Gravel, C., Roy, R., and Tremblay, J. P. (1995). FK506 immunosuppression to control the immune reactions triggered by first-generation adenovirus-mediated gene transfer. Hum. Gene Ther. *6*, 1391-1401.
76. Wang, Q. and Finer, M. H. (1995). A packaging cell line for propagation of recombinant adenovirus vector containing two lethal gene-region deletions. Gene Ther. *2*, 775-783.
77. Weiden, M. D. and Ginsberg, H. S. (1994). Deletion of the E4 region of the genome produces adenovirus DNA concatemers. Proc. Natl. Acad. Sci. USA. *91*, 153-157.
78. Weinberg, D. H. and Ketner, G. (1986). Adenoviral early region 4 is required for efficient viral DNA replication and for late gene expression. J. Virol. *57*, 833-838.
79. Weinberg, D. H. and Ketner, G. (1983). A cell line that supports the growth of a defective early region 4 deletion mutant of human adenovirus type 2. Proc. Natl. Acad. Sci. U.S.A. *80*, 5383-5386.
80. Wong, M.-L. and Hsu, M.-T. (1989). Linear adenovirus DNA is organized into supercoiled domains in virus particles. Nucl. Acids Res. *17*, 3535-3550.
81. Yang, Y., Ertl, H. C. J., and Wilson, J. M. (1994b). MHC class I-restricted cytotoxic T lymphocytes to viral antigens destroy hepatocytes in mice infected with E1-deleted recombinant adenoviruses. Immunity. *1*, 433-442.
82. Yang, Y., Li, Q., Ertl, H. C. J., and Wilson, J. M. (1995a). Cellular and humoral immune responses to viral antigens create barriers to lung-directed gene therapy with recombinant adenoviruses. J. Virol. *69*, 2004-2015.

83. Yang, Y., Nunes, F. A., Berencsi, K., Furth, E. E., Gonczol, E., and Wilson, J. M. (1994a). Cellular immunity to viral antigens limits E1-deleted adenoviruses for gene therapy. Proc. Natl. Acad. Sci. USA. *91*, 4407-4411.
84. Yang, Y., Trinchieri, G., and Wilson, J. M. (1995b). Recombinant IL-12 prevents formation of blocking IgA antibodies to recombinant adenovirus and allows repeated gene therapy to mouse lung. Nature Genet. *1*, 890-893.
85. Yang, Y. and Wilson, J. M. (1995c). Clearance of adenovirus-infected hepatocytes by MHC class I-restricted CD4+ CTLs in vivo. J. Immunol. *155*, 2564-2570.
86. Yang, Y., Xiang, Z., Ertl, H. C. J., and Wilson, J. M. (1995d). Upregulation of class I major histocompatibility complex antigens by interferon g is necessary for T-cell-mediated elimination of recombinant adenovirus-infected hepatocytes in vivo. Proc. Natl. Acad. Sci. USA. *92*, 7257-7261.
87. Yeh, P., Dedieu, J.-F., Orsini, C., Vigne, E., Denèfle, P., and Perricaudet, M. (1995). Efficient dual transcomplementation of adenovirus E1 and E4 regions from a 293-derived cell line expressing a minimal E4 functional unit. J. Virol. *70*, 559-565.
88. Yei, S., Mittereder, N., Tank, K., O'Sullivan, C., and Trapnell, B. C. (1994). Adenovirus-mediated gene transfer for cystic fibrosis, quantitative evaluation of repeated in vivo vector administration to the lung. Gene Ther. *1*, 192-200.
89. Yew, P. R. and Berk, A. J. (1992). Inhibition of p53 transactivation required for transformation by adenovirus early 1B protein. Nature. *357*, 82-85.
90. Zhang, X., Bellet, A. J. D., Tha Hla, R., Voss, T., Mullbacher, A., and Braithwaite, A. W. (1994). Down-regulation of human adenovirus E1a by E3 gene products, evidence for translational control of E1a by E3 14.5K and/or E3 10.4K products. J. Gen. Virol. *75*, 1943-1951.
91. Zijderveld, D. C., d'Adda di Fagagna, F., Giacca, M., Marc Timmers, H. T., and van der Vliet, P. (1994). Stimulation of the adenovirus major late promoter in vitro by transcription factor USF is enhanced by the adenovirus DNA binding protein. J. Virol. *68*, 8288-8295.

3. Herpes Simplex Virus Mediated Gene Transfer to Neurons

W. F. Goins, T. Oligino, D. Krisky, P. Marconi, P. L. Poliani, R. Ramakrishnan, D. J. Fink, and J. C. Glorioso

3.1 Introduction

The diagnosis and treatment of human disease using genetically-based approaches has rapidly launched the age of genetic medicine. The gene has now been conceptualized as a pharmaceutical drug of profound therapeutic implications but with as yet unrealized potential. A fundamental problem remains, namely efficient methods for gene transfer particularly to tissues *in vivo*. A variety of gene delivery vehicles have been explored each of which has both advantages and disadvantages but natural gene transfer agents such as viruses remain by far the most efficient and proven vectors. Most endogenous viruses are so well adapted for permanent gene transfer that they are extremely difficult to eliminate from the body and many go unnoticed after the initial encounter which usually does not lead to life threatening disease in otherwise healthy individuals. Many of these viruses remain transcriptionally active in the host even producing immunogenic proteins and thus when needed persistent viruses have acquired mechanisms that permit escape from immune surveillance and destruction of virus infected cells. The human herpesviruses are an important family of viruses which become established in various tissues for the life of the host and with some notable exceptions do not cause subsequent disease unless the host becomes immunologically compromised. However, the modification of these viruses to reduce pathogenicity and increase safety often results in the loss of viral activities which are required for efficient gene delivery and life long association with the host. Thus the challenge is to understand the fundamental biology of these viruses and their adaptations to particular cells and tissues and based on these insights develop the means to render these sophisticated organisms harmless yet effective for targeted gene transfer and appropriate gene expression.

Many hundreds of patients have now participated in phase one gene therapy clincial trials where this methodology could be evaluated for safety and to some extent efficacy. Thus far, attempts at gene therapy for a variety of diseases incuding cancer, metabolic abnormalities, cardiovascular, autoimmune and infectious diseases, has not produced clear evidence that future success is imminent primarily because effective methods for gene delivery and sustainable gene expression are not yet available. Initial studies suggested that one or several vectors might be adapted for use in a variety of applications. However it has now become apparent that the delivery and expression of transgenes will require individualized approaches tailored to the specific disease and target tissue. Some gene therapy applications will require short term transgene expression while others will demand that the transgene be expressed very long term. Although delivery of single genes should be adequate in many cases, other applications (e.g. cancer and autoimmune disease) will likely require the delivery of multiple transgenes.

For the central and peripheral nervous system, where neurons are post-mitotic and intricate cell-cell relationships and regional specialization complicate *ex vivo* gene transfer approaches, direct gene therapy of the nervous system offers considerable appeal. Herpes simplex virus (HSV) has many features which make it an attractive vehicle for neurons. The virus has evolved to persist in neurons life-long in a latent state without genome integration into host chromosomes and without interference with host cell function. The viral lytic genes are silenced by a powerful but poorly understood mechanism involving the immediate early (IE) gene products (1) and a specialized region of the genome remains available for transcription during latency. This combination of features makes HSV potentially ideal for delivery of transgenes to neurons *in vivo* with subsequent long term expression of a transgene using the natural latency promoter system. In addition, the viral genome is readily manipulated as there are few spliced genes and approximately half of the genome is dispensible for virus replication in cell culture providing ample space for incorporation of transgenes of considerable size and number. The virus has a broad host range, it can infect single cells efficiently requiring only single infectious particles for gene transfer and it can be propagated to high titer. Despite these attractive features, even replication defective HSV mutants are cytotoxic for primary neuronal cells in culture raising the possibility that such vectors may not be safe *in vivo* and the activity of the latency promoter is weak in brain neurons and may therefore be unsuitable for some applications. Moreover, other viral and non-viral promoters that might be regulated in a manner to produce appropriate levels of transgene expression become rapidly silenced in the vector (1). This review will describe progress both in alleviating vector cytotoxicity and in releasing promoters for expression in brain.

3.2 HSV Genome and the Virus Life Cycle

The HSV genome is a linear double-stranded DNA molecule 152 Kb in length and containing more than 80 individual genes (2), composed of two unique sequences, the unique long (U_L) and unique short (U_S) segments each flanked by inverted repeat elements (TR_L and IR_L; IR_S and TR_S) that contain both transcriptional control and latency genes and packaging sequences (Fig. 3.1). The viral DNA is packaged into a capsid surrounded by a protein matrix (tegument) comprised of a number of viral functions including VP16, a protein which combines with cellular transcription factors (Oct-1:HCF) to activate HSV immediate early (IE) gene expression on infection (3-7), and UL41 (vhs), a protein which acts to shut-off host protein synthesis (8-10). The entire particle is enclosed in an envelope in which its component glycoproteins mediate viral attachment and fusion of the virus envelope with the cell surface membrane thereby delivering the capsid into the cytoplasm (11, 12).

Viral gene expression in lytic infection occurs in a highly ordered temporal (13) sequence (Fig. 3.2). Viral immediate early (IE) genes are expressed on entry of the viral genome into the nucleus and IE (or a) gene transcription does not require de novo viral protein synthesis. The five IE gene products are (infected cell proteins) ICP0, ICP4, ICP22, ICP27, and ICP47. With the exception of ICP47, the central role of the IE gene products is to transactivate the expression of the early (E or β) genes and their products which in turn are principally responsible for viral DNA replication. In the presence of IE and E gene products and following viral DNA replication, the HSV late (L or γ) genes are expressed, producing the majority of structural proteins required for viral particle formation (capsid, tegument, and envelope glycoproteins).

In natural infections, HSV replicates in epithelial cells of the skin or mucous membranse from which the virus invades sensory neurons and is carried by retrograde axonal transport to the cell body (14) where the viral DNA is released from a modified capsid penton into the nucleus through a pore in the nuclear membrane (11). In sensory neurons the virus may either initiate the replication cycle or alternatively enter a latent state in which quiescent genomes persist as intranuclear episomal elements for the life of the host. Biologically, latency is defined as the presence of quiescent genomes which under the influence of an appropriate stimulus can be reactivated to produce lytic virus. Completely replication defective virus recombinants can be constructed which retain the ability to persist in a molecular state indistinguishable from latency without the threat of viral growth *in vivo* or reactivation from the latent state. Although first generation vectors deleted for the ICP4 IE gene transiently express very low levels of E gene products and no L viral functions (Fig. 3.2), they still remain cytotoxic for cells in culture and only transiently express transgenes both *in vitro* and *in vivo*.

3.3 HSV Vector Cytotoxicity

It is presumed that the cytotoxicity associated with the first generation HSV vectors must result from the expression of HSV gene products, since UV-inactivated wild-type virus is not toxic to cells even at high multiplicities (15). Viruses deleted for the IE essential ICP4 gene (16, 17) do not produce infectious virus particles in standard cell cultures yet remain capable of producing cytotoxic effects such as cytoplasmic blebbing, host cell DNA fragmentation, and chromosome aberrations (18).

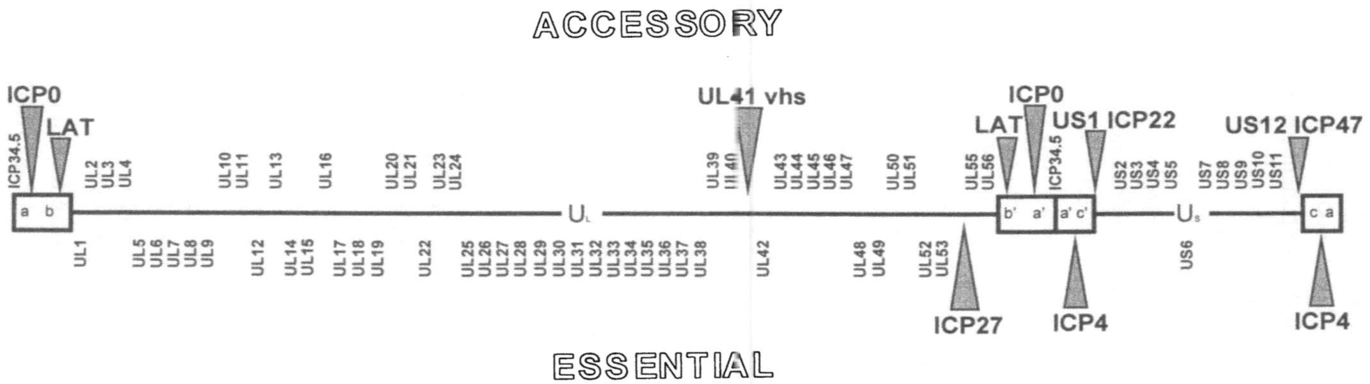

Fig. 3.1: HSV-1 Genome Organization.
Schematic representation of the HSV genome, showing the unique long (U_L) and unique short (U_S) segments, each bounded by inverted repeat (IR) elements. The location of the essential genes which are required for viral replication *in vitro*, and the non-essential or accessory genes, which may be deleted without affecting replication *in vitro*, are indicated. IE gene and other loci targeted for genetic manipulation are highlighted.

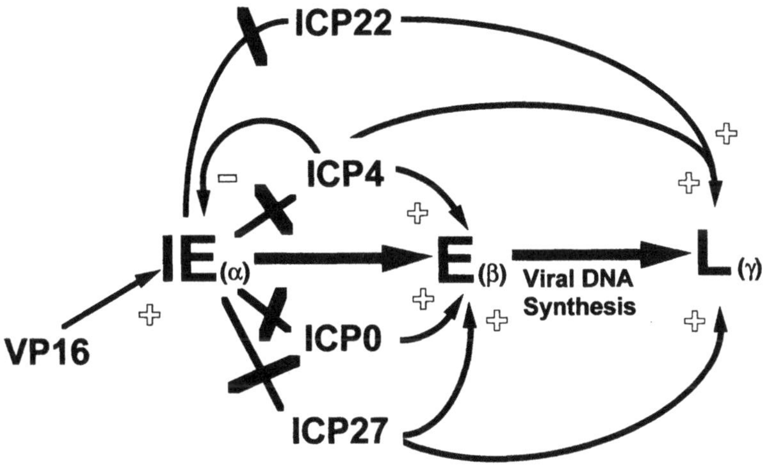

Fig. 3.2: HSV Regulatory Cascade and Vector Design.
The role of ICP4, ICP22, ICP27, ICP0 and VP16 in regulating the temporal cascade of HSV lytic gene expression is shown. Defective viruses deleted for ICP4 are blocked at the immediate early stage, while deletion of the other genes either blocked replication or reduced virus yield.

Even though deletion of ICP4 aborts the expression of both E and L genes, the other four IE gene products (ICP0, ICP27, ICP22 and ICP47) (Figs. 3.1 and 3.2) and ICP6 which is expressed at both IE and E times (19) are overexpressed in the absence of ICP4 (16, 20). Disruption of IE gene expression by interferon also reduces toxicity (18). Deletion of any of the other IE genes alone, or the virion host shut-off (vhs) function which is present within the virus particle, failed to reduce toxicity (18). However, ICP4, ICP0, ICP27 and ICP22 are all toxic to cells in stable transfection assays (21), suggesting that deletion of these genes in combination may be required to reduce toxicity. ICP6 is not toxic in the same transient assays (21) and deletion of ICP47 along with ICP4 did not reduce the toxicity associated with the ICP4 mutants (21). Mutation of the IE transactivating gene product VP16 in the background of an ICP4 deletion mutant resulted in reduced cytotoxicity (21), probably due to the overall reduction of the expression of the other IE genes due to the loss of VP16 activation of the IE gene promoters. Together these data suggest that ICP0, ICP27 and ICP22 may all be responsible for the toxicity found with the first generation ICP4 mutant vectors, and deletion of these genes along with UL41 (vhs) in a single virus will substantially reduce toxicity to a level approaching that observed with UV-irradiated virus.

We and others have endeavored to delete the other IE gene viral functions which contribute to the residual cytotoxicity of the first generation replication defective vectors. To propagate vectors deleted for multiple IE genes, it was necessary to generate a cell line to complement the essential IE gene functions. In order to create a ICP4/ICP27 complementing cell line we transfected Vero cells with a plasmid containing the coding region for ICP4, ICP27 and an SV2-neo cassette for selection.

The coding region for ICP4 and ICP27 were engineered so as not to contain overlapping sequences present within the deletion viruses, thereby avoiding homologous reecombination and rescue of the mutant virus. The 7b ICP4/27 cell line was then used to generate the new multiple IE gene deletion viruses. To date, we have deleted all five targeted cytotoxic genes (ICP4, ICP0, ICP27, ICP22 and UL41) in various combinations (22, 23). The ICP4/ICP22 double deletion mutant DHZ.1 was 9-15 fold less cytotoxic than the first generation single IE gene deletion mutant SHZ.1 ($ICP4^-$) in Vero cells in culture (22). Two multiple deletion viruses THZ.2 ($ICP4^-$/$ICP22^-$/$ICP27^-$) and TOZ.1 ($ICP4^-$/$ICP22^-$/$ICP27^-$/$UL41^-$) were significantly less toxic (55-80% cell survial vs 18-30%) than cells infected with the SHZ.1 single gene deletion mutant. Double (22, 24) and triple (22, 25) deletion mutants in which multiple IE genes were deleted displayed reduced levels of toxicity in Vero cells in culture.

We were also interested in testing the cytotoxicity associated with the battery of mutant viruses in primary cortical neurons. Cortical neuronal cultures isolated from 16 day embryonic rat forebrains were infected (MOI of 1) with the first generation single gene mutant SHZ.1 ($ICP4^-$), and the triple IE gene deletion mutants THZ.2 ($ICP4^-$/$ICP22^-$/$ICP27^-$) or TOZ.1 ($ICP4^-$/$ICP22^-$/$ICP27^-$/$UL41^-$). The difference in cytotoxicity associated with these mutants was observable at 24 hours post-infection; the SHZ.1-infected neurons were essentially destroyed while the THZ.2 and TOZ.1 infected cultures showed no visible cytopathic effect. Cultures infected with THZ.2 and TOZ.1 displayed normal cellular morphology for at least 14 days after infection. In a effort to quantitate the cytotoxicity of these vectors cell survival was determined by the fluorescent alamar Blue method (26). The results of this assay indicated that there was no significant loss of cells from 24 hours to 14 days after infection with THZ.2 and TOZ.1. Analysis of the transcripts produced from the mutant viruses by RT-PCR at 1 and 2 weeks post-infection demonstrated that the viruses had established latency in the primary cultures as defined by LAT expression in the absence of detectable early gene expression (gB transcript). Morever, transgene expression persisted out to at least 7 days with the multiple gene deletion vectors and out to 4 weeks in brain following stereotactic inoculation of the vectors, in contrast to that observed with the first generation single gene deletion virus SHZ.1 (1). Together, these findings are encouraging for the use of non-cytotoxic to transfer genes to neurons and potentaily other target tissues.

The observation that an HSV mutant deleted for multiple functional IE genes also has important implications for the use of HSV vectors for gene transfer to brain and other tissues. The ability of defective viral mutants to establish life-long latency at least in animal models with the expression of LAT as well as genes under control of the viral IE or foreign gene promoters suggests that a useful vector background has been identified. It remains to be determined to what extent other foreign promoters will continue to be active in this mutant vector background and whether they will be responsive to transcriptional signals from the cell.

3.4 Latent Infection and the Latency Promoter System

In latently infected neurons only a single region of the HSV genome, the inverted repeat sequences flanking U_L (TR_L and IR_L) just downstream of (and overlapping) the gene coding for ICP0, remains transcriptionally active producing a family of latency-associated transcripts (LATs) (Fig. 3.3). In molecular terms, latency is defined as the expression of LATs in the absence of detectable lytic gene expression. The most abundant 2 Kb LAT is also readily detected late in lytic infection while, smaller LATs (1.5 and 1.45 Kb) are only detected in latently infected neurons (27-30). LAT is not capped or polyadenylated (27) but rather is a very stable intron spliced from a large 8.77 Kb poly A^+ RNA (31) that appears to be very unstable and difficult to detect in latency. We have recently confirmed that LAT is an intron derived from splicing during latency and that both the internal and external splice donor and acceptor sites are active in sensory ganglia (Alvira and Glorioso, unpublished). LAT has also recently been reported to be in a nonlinear form consistent with a lariat (32). No viral protein has been detected in latently infected cells although there are 16 ORFs in the LAT region of the viral genome and it is possible that a LAT-encoded protein is expressed transiently during the establishment of latency. Although no function has yet been ascribed to the LATs, evidence is accumulating that deletion of the LAT gene affects both the efficiency of the establishment of latency and virus reactivation and may determine the site of reactivation from latency (33-35).

The promoter elements and structural features of the viral DNA in the LAT locus which contribute to the unique activity of LAT during latency has been the object of much research and considerable progress has been made in defining the LAT promoter elements (36-40) (Fig. 3.3). The specific features of the LAT promoter which allow it to escape promoter silencing during latency in neurons are not yet fully characterized. Two latency active promoters (LAPs), LAP1 and LAP2, have been identified upstream of the LAT coding sequence (36, 37, 39-42) (Fig. 3.2). LAP1 contains a TATA box (40, 43, 44) with upstream control elements such as CAAT (41), USF1 (42), CRE (40, 43, 44) and Sp1, YY1 and Brn-2 sites. In addition, there is an enhancer region at the start of transcription that plays a role in up-regulating LAP1 basal activity (40) and also contains an ICP4 binding site that down-regulates LAT expression (45-47). We have recently shown that the TATA box, USF-1, Cre and the putative Brn site all contribute to LAP1 activity (40) (Soares and Glorioso unpublished) *in vivo*.

We have identified a second TATA-less latency-active promoter, LAP2, 5' proximal to the major 2 Kb LAT (38-40, 48). The LAP2 promoter contains putative modulatory sequences including a C/T rich element present in many TATA-less housekeeping gene promoters (49-51), and a polyT stretch of 23 thymidines also found in putative regulatory sequences of higher eukaryotic genes (52-55), both of which are important for LAP2 function *in vitro*. The C/T rich element is bound by cellular

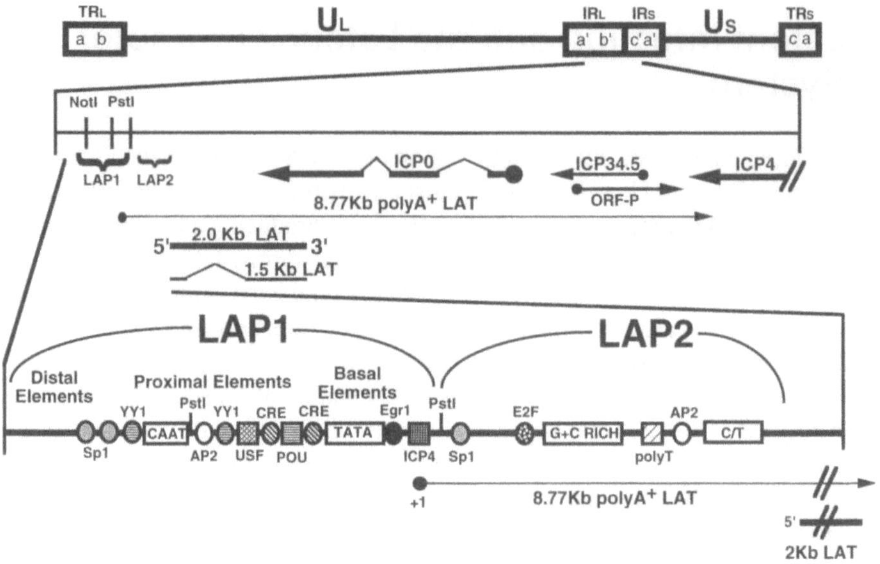

Fig. 3.3: HSV Latency Gene Expression.
The location of the various HSV latency-associated transcripts (LATs) relative to the ICP0, ICP34.5, ORF-P and ICP4 lytic gene RNAs within the prototype genome is depicted. An enlargement of the latency active promoter (LAP) regions LAP1 and LAP2 shows the position of potentially relevant cis-acting elements in LAP1 such as the CRE, USF and TATA box as well as the polyT stretch and C/T element in LAP2.

factors Sp1 and one additional protein NSEP-1 (56), which has been shown to bind the c-myc promoter as well as the EGF-R promoter (57). The polyT stretch interacts with a HMG I(Y) like protein. HMG I(Y) has no intrinsic transcriptional activation activity but rather has been shown to transactivate promoters through protein-protein interactions with known transcription factors such as ATF2 and NFkB (58), in which it facilitates binding of these factors to their respective DNA binding sites and alters the structure of the bound DNA (59). Using an S1 sensitivity assay we mapped an S1 sensitive site to a region between the polyT element and the C/T rich sequence of LAP2. S1 nuclease sensitive sites often correspond to *in vivo* DNaseI sensitive sites (60, 61). DNaseI sensitive sites have been seen in 5' regions of active transcription (60, 62), and are involved in nucleosome exclusion (63, 64). These data suggest that one function of LAP2 may be to disrupt chromatin structure by nucleosome exclusion. The importance of these elements in LAT expression during latency *in vivo* awaits further testing.

To demonstate the ability of LAP to drive foreign gene expression during latency we have engineered the recombinant virus SLZ:LAT (65) with the lacZ reporter gene inserted downstream of LAP2 promoter in the native LAT locus. This recombinant has been shown by RT-PCR to expresses lacZ in the CNS and PNS at 28 days post-infection. The ability of LAP2 to function as a movable genetic element was also examined by introducing either a LAP2-lacZ gene cassette into the ectopic gC locus of the virus (39) (Fig. 3.4A). LAP2 driven expression of either gene was detected long-term (150 days post-infection) in mouse trigeminal ganglion by X-gal staining (Fig. 3.4B) or immunohistochemistry. This recombinant also proved capable of expressing lacZ as detected by RT-PCR in the CNS (Fig. 3.4C) even though the X-gal reaction product was undectable. Together, the data demonstrated that the LAP is functional long-term even in the CNS, although due to the weak levels of transgene expression, further modification of the LAP may be necessary for the production of normal physiologic levels of the therapeutic gene product required for treatment of certain diseases.

3.5 Quantitative Studies of the Establishment of HSV Latency in PNS and CNS

In order to precisely define the number of HSV genomes and the amount of LAT RNA, following infection with a ribonucleotide reductase deleted (RR^-) virus (19), we developed competitive quantitative (CQ)-PCR and CQRT-PCR as modifications of previously published techniques (66, 67). These experiments demonstrated that the RR mutant virus established latency in rat brain with persistence of the viral genome and production of LAT molecules and that once latency was established, the number of viral genomes and LAT RNA molecules remain constant over 8 wks PI (68).

Following corneal scarification with the same virus, both gB mRNA and LAT RNA were found in the TG at 2 days, but only LAT RNA at 7 and 56 days, consistent with the establishment of viral latency at those time points. CQRT-PCR studies determined that the quantitative aspects of latency in the TG were similar to that in the CNS, except that in the TG there were approximately 50 LATs per genome compared to 15/genome in the CNS (69). Using in situ hybridization with a digoxigenin-labeled riboprobe specific for LAT we determined that there were 1 to 2 LAT positive cells per 6 micron section of TG during viral latency at 7 and 56 days PI, or a total of 10-20 LAT-positive cells per ganglion (68), consistent with previous reports (35, 70, 71). However, in contrast to in situ hybridization (ISH) detection of LAT RNA which showed a maximum of 5 LAT-positive cells per section, in situ (IS) PCR detection of viral DNA revealed a maximum of 120 cells per section (68), or approximately 10 times the number of LAT-positive cells (69).

In order to determine whether HSV genome-containing cells expressed low levels of LAT that were not detectable by in situ hybridization, we modified the ISPCR protocol to detect LATs by ISRT-PCR. The number of LAT-positive cells detected by ISRT-PCR was 10 to 75 per section at 1 week and 6 to 54 per section at 8 weeks, numbers that are comparable to the total genome containing cells, and much greater than the number of cells detected by ISH in sections from the same ganglia (68). These results suggest that virtually all the HSV genome-containing cells express

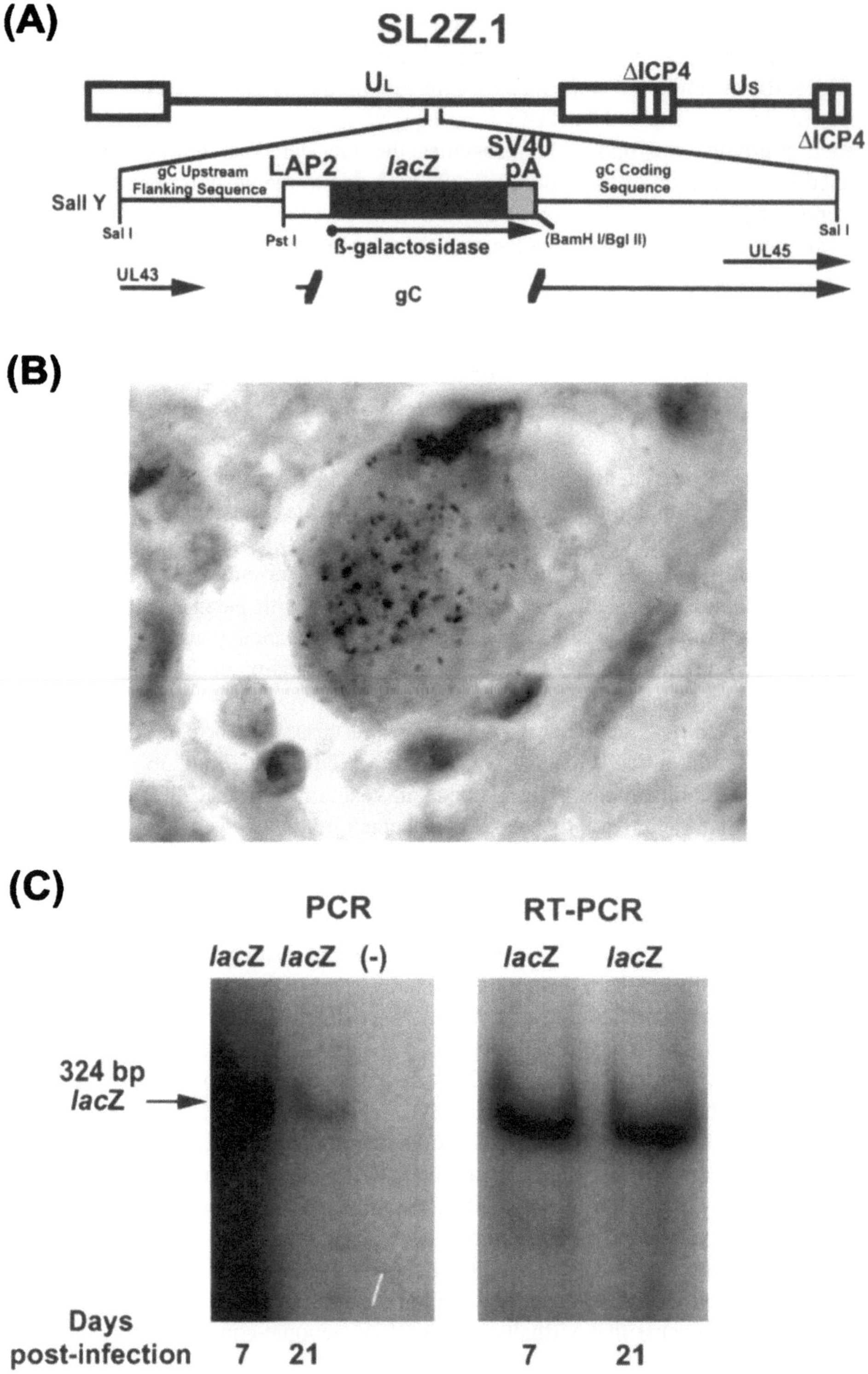
(A)
SL2Z.1
UL
ΔICP4
Us
ΔICP4
SV40
pA
gC Upstream
Flanking Sequence
LAP2
lacZ
gC Coding
Sequence
Sall Y
Sal I
Pst I
ß-galactosidase
(BamH I/Bgl II)
Sal I
UL45
UL43
gC
(B)
(C)
PCR
RT-PCR
lacZ lacZ (-)
lacZ lacZ
324 bp
lacZ
Days
post-infection
7 21
7 21

◁ Fig. 3.4: Replication Defective Vector Mediated Expression in PNS and CNS using the HSV-1 LAP.

(A) Schematic representation of the SL2Z:gC recombinant in which the LAP2-lacZ reporter gene cassette has been inserted into the gC locus of d120 (39).

(B) LAP2 is capable of driving long-term transgene in the mouse PNS as determined by X-gal staining sections from mouse trigeminal ganglia at 150 days post infection.

(C) After stereotactic inoculation of SL2Z:gC into rat hippocampus, reporter gene expression was detected by RT-PCR at 1 and 3 weeks post infection, whereas viral lytic gene promoters as well as the HCMV IE promoter in SHZ.1 ICP4 IE gene deletion muatnt were silent. These results demonstrate that LAP2 can function during latency from a singly deleted mutant to drive transgene expression.

LATs, albeit the majority at low levels again suggesting that expression of transgenes from the HSV-based vectors will require novel approaches.

3.6 Development of Transactivation Systems to Regulate Gene Expression

Because it is possible that the level of transgene expression produced by a promoter construct in an HSV vector may be insufficient to provide the desired biological effect, we have therefore created recombinant viruses in which transgene expression is regulated using targeted bipartite or tripartite transactivation systems based on recombinant transactivators. These systems allow for the specific activation of targeted genes in either a constitutive or regulatable fashion.

3.6.1 Constitutive Activation of Transgene Expression Using the GAL4:VP16 Transactivation System

GAL4:VP16 is a potent transcriptional activator containing the DNA binding domain of the yeast transcriptional regulatory protein GAl4 and the acidic activating domain of the HSV protein VP16 (72). The fusion protein has been demonstrated to strongly activate promoters to which it is targeted by the placement of the 17 bp GAL4 binding site (72-74). Furthermore, minimal GAL4 promoters show no activity in mammalian cells making this an attractive system for the specific activation of genes from the virus genome. In order to determine if transgene expression could be specifically activated from the virus genome using the GAL4:VP16 transactivation system, we created a double recombinant virus which contained both the GAL4:VP16 transactivator and a reporter gene (lacZ) under control of a GAL4 sensitive minimal promoter (75). We were able to demonstrate that this double recombinant virus produced β-galactosidase both *in vitro* and *in vivo* demonstrating the ability of the GAL:VP16 transactivator to stimulate a minimal GAL4 binding site promoter in the viral genome.

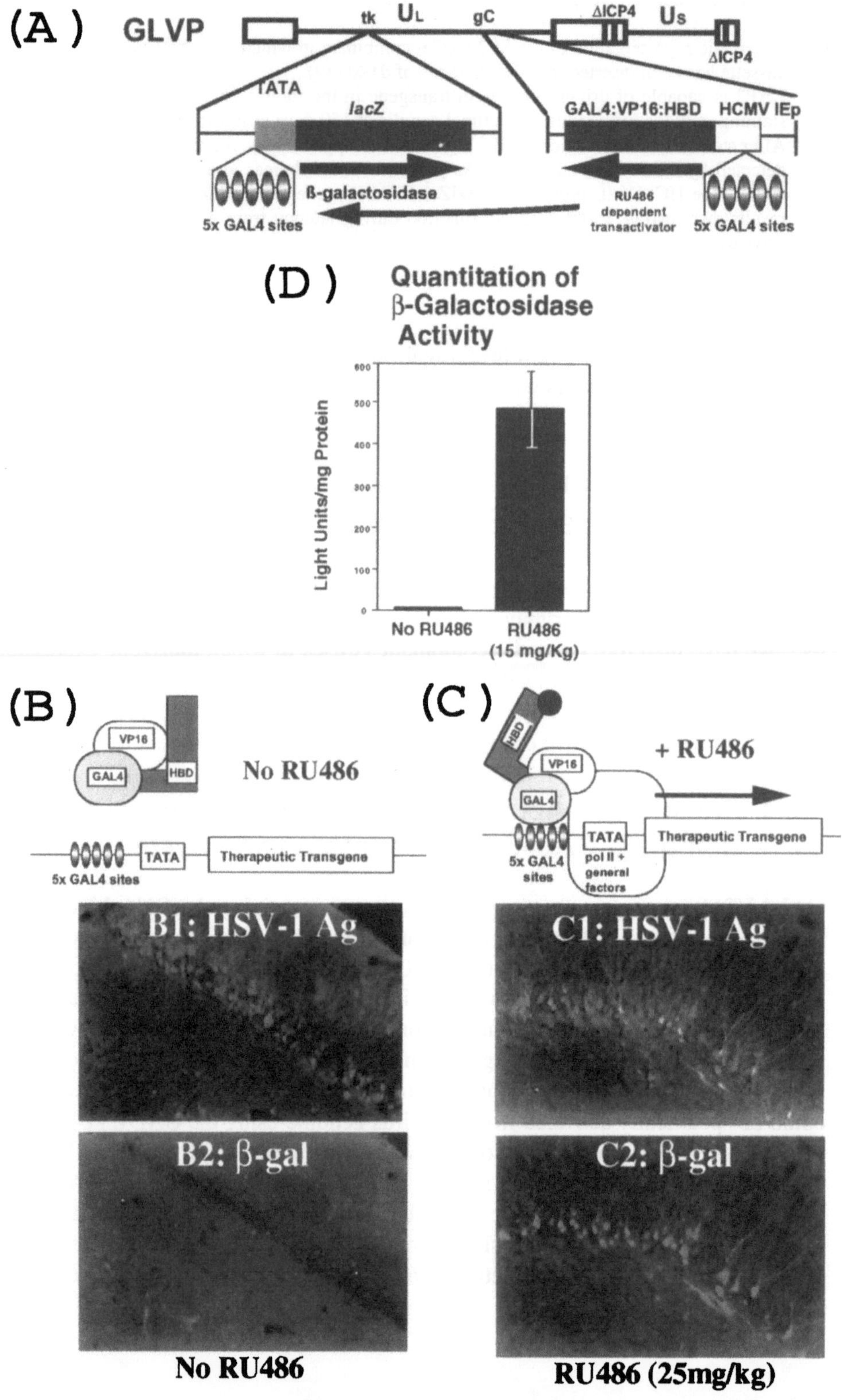
(A)
GLVP
tk
UL
gC
ΔICP4
Us
ΔICP4
TATA
lacZ
GAL4:VP16:HBD
HCMV IEp
ß-galactosidase
RU486 dependent transactivator
5x GAL4 sites
5x GAL4 sites
(D)
Quantitation of β-Galactosidase Activity
Light Units/mg Protein
0
100
200
300
400
500
600
No RU486
RU486 (15 mg/Kg)
(B)
VP16
GAL4
HBD
No RU486
5x GAL4 sites
TATA
Therapeutic Transgene
(C)
HBD
VP16
GAL4
+ RU486
5x GAL4 sites
TATA
pol II + general factors
Therapeutic Transgene
B1: HSV-1 Ag
B2: β-gal
C1: HSV-1 Ag
C2: β-gal
No RU486
RU486 (25mg/kg)

◁ Fig. 3.5: *In vivo* Regulated Transgene Activation from HSV Vectors by Recombinant Transactivator.

(A) The vector GLVP contains the inducible transactivator GAL4:VP16:HBD which can be activated by RU486 and a promoter-reporter cassette that is responsive to activation by the transactivator.

(B) In the absence of the drug, virus can be detected by immunostaining for HSV antigens, (B1) but no β-galactosidase (B2) can be seen in the same region.

(C) Following administration of RU486 (25mg/Kg) the tripartite transactivator binds to the GAL4 binding sites and abundant β-galactosidase expression is seen by fluorescent immunostaining (C2).

(D) Quantitation of β-galactosidase transgene expression in the presence of RU486 displays activation of the transgene promoter by the inducible transactivator.

3.6.2 Drug Inducible Transgene Activation from the Viral Genome

To allow for targeted and regulatable transgene expression from the viral genome the transactivation system described above was modified such that the constitutive transactivator was replaced with one whose activity could be regulated by drug treatment. The drug inducible transactivator differs from the constitutive molecule in that it contains a mutated form of the hormone binding domain (HBD) of the progesterone receptor, which fails to bind progesterone but is activated by binding of the steroid analog RU486, fused to the GAL4:VP16 sequences. This modification renders the trimeric molecule unable to bind to and activate transcription from GAL4 sensitive promoters. In the presence of the RU486, which binds to the HBD and causes a conformational change in the trimeric molecule, the recombinant transactivator can now bind to the GAL4 elements and thereby activate transcription. We have demonstrated that this drug inducible system functions in the context of the viral genome both in culture and in rat brain using a double recombinant virus which contains both the drug inducible transactivator expressed under the transcriptional control of the HCMV IEp and a lacZ reporter gene under control of a minimal GAL4 promoter (Fig. 3.5). This double recombinant virus was injected into the hippocampus of rats and the rats subsequently treated with RU486 (25mg/kg ip) or saline. Immunohistochemistry and β-galactosidase solution assays indicated that in that in the absence of RU486 extremely low levels of β-gal was produced from the virus. However, treatment with RU486 resulted in strong activation of lacZ expression (200 fold induction) in neurons of the hippocampus (Fig. 3.5) (76). This system can therefore be used to specifically activate expression a single or multiple genes at various time points.

3.7 Summary

In natural infection HSV invades neurons of sensory ganglia where it can replicate without apparent neuronal cell destruction and establish a life-long presence as an extrachromosomal element in a quiescent state referred to as latency. During latency,

gene expression is limited to several intranuclear RNA species, the latency associated transcripts (LATs), that do not appear to perturb nerve cell function. HSV can be reactivated from latency where active virus infection can again cause clinical signs of disease and virus spread to other hosts by direct contact with lesions. Virus mutants engineered to be defective for both growth and reactivation from latency nonetheless persist kn an episomal state indistinguishable from latency with repression of lytic genes and expression of LATs. This suggests a paradigm for construction of HSV gene vectors in which replication and pathogenic functions are deleted from the vector genome and the latency promoter system used to drive transgene expression from the defective virus platform. Progress in engineering HSV gene vectors for nervous system applications are discussed emphasizing progress in removing viral functions which cause host cell necrosis and development of promoter systems which provide for appropriate level and duration of transgene expression *in vivo*.

References

1. Ramakrishnan, R., Krisky, D., Marconi, P., Poliani, P. L., Laquerre, S., Schmidt, M., Fink, D. J., and Glorioso, J. C. (1996b). HSV immediate early proteins participate in promoter silencing during the establishment of latency. Science, (submitted).
2. Roizman, B. and Sears, A. (1996). Herpes simplex viruses and their replication. In: Fields Virology, 2, B. N. Fields, D. M. Knipe, P. M. Howley, R. M. Chanock, M. S. Hirsch, J. L. Melnick, T. P. Monath, and Roizman, B.(eds.), (Philadelphia, PA: Lippincott-Raven), pp. 2231-2295.
3. Batterson, W. and Roizman, B. (1983). Characterization of the herpes simplex virion-associated factor responsible for the induction of alpha-genes. J. Virol. *46*, 371-377.
4. Campbell, M. E. M., Palfeyman, J. W., and Perston, C. M. (1984). Identification of herpes simplex virus DNA sequences which encode a trans-acting polypeptide responsible for stimulation of immediate early transcription. J. Mol. Biol. *180*, 1-19.
5. Kristie, T. and Roizman, B. (1986). DNA-binding site of major regulatory protein alpha 4 specifically associated with the promoter-regulatory domains of alpha genes of herpes simplex virus type 1. Proc. Natl. Acad. Sci. USA *83*, 4700-4704.
6. Werstuck, G. and Capone, J. (1993). An unusual cellular factor potentiates protein-DNA complex assembly Oct-1 and Vmw65. J. Biol. Chem. *268*, 1272-1278.
7. Wilson, A., LaMarco, K., Peterson, M., and Herr, W. (1993). The VP16 accessory protein HCF is a family of polypeptides processed from a large precursor protein. Cell *74*, 115-125.
8. Kwong, A. D., Kruper, J. A., and Frenkel, N. (1988). Herpes simplex virus virion host shutoff function. J. Virol. *62*, 912-921.
9. Oroskar, A. and Read, G. (1989). Control of mRNA stability by the virion host shutoff function of herpes simplex virus. J. Virol. *63*, 1897-1906.
10. Read, G. S. and Frenkel, N. (1983). Herpes simplex virus mutants defective in the virion-associated shutoff of host polypeptide synthesis and exhibiting abnormal synthesis of a (immediate early) viral polypeptides. J. Virol. *46*, 498-512.
11. Newcomb, W. and Brown, J. (1994). Induced extrusion of DNA from the capsid of herpes simplex virus type 1. J. Virol. *68*, 443-440.
12. Spear, P. (1993). Membrane fusion induced by herpes simplex virus. In: Viral Fusion Mechanisms, J. Bentz (ed.) (Boca Raton: CRC Press), pp. 201-232.
13. Honess, R. and Roizman, B. (1974). Regulation of herpes simplex virus macromolecular synthesis. I. Cascade regulation of the synthesis of three groups of viral proteins. J. Virol. *14*, 8-19.
14. Stevens, J. G. (1989). Human herpesviruses: a consideration of the latent state. Microbiol. Rev. *53*, 318-332.

15. Leiden, J., Frenkel, N., and Rapp, F. (1980). Identification of the herpes simplex virus DNA sequences present in six herpes simplex virus thymidine kinase-transformed mouse cell lines. J. Virol. *33*, 272-85.
16. DeLuca, N. A., McCarthy, A. M., and Schaffer, P. A. (1985a). Isolation and characterization of deletion mutants of herpes simplex virus type 1 in the gene encoding immediate-early regulatory protein ICP4. J. Virol. *56*, 558-570.
17. Paterson, T. and Everett, R. D. (1990). A prominent serine-rich region in Vmw175, the major regulatory of herpes simplex virus type 1, is not essential for virus growth in tissue culture. J. Gen. Virol. *71*, 1775-1783.
18. Johnson, P., Miyanohara, A., Levine, F., Cahill, T., and Friedmann, T. (1992). Cytotoxicity of a replication-defective mutant herpes simplex virus type 1. J. Virol. *66*, 2952-2965.
19. Desai, P., Ramakrishnan, R., Lin, Z. W., Ozak, B., Glorioso, J. C., and Levine, M. (1993). The RR1 gene of herpes simplex virus type 1 is uniquely transactivated by ICP0 during infection. J. Virol. *67*, 6125-6135.
20. Dixon, R. A. F. and Schaffer, P. A. (1980). Fine-structure mapping and functional analysis of temperature-sensitive mutants in the gene encoding the herpes simplex virus type 1 immediate early protein VP175. J. Virol. *36*, 189-203.
21. Johnson, P., Wang, M., and Friedman, T. (1994). Improved cell survival by the reduction of immediate-early gene expression in replication-defective mutants of herpes simplex virus type 1 but not by mutation of the viron host shutoff function. J. Virol. *68*, 6347-6362.
22. Krisky, D., Marconi, P., Oligino, T., Ramakrishnan, R., Fink, D. J., and Glorioso, J. C. (1996b). Herpes simplex virus multiple immediate early gene deletion mutants display reduced toxicity *in vitro. (submitted).*
23. Marconi, P., Krisky, D., Oligino, T., Poliani, P. L., Ramakrishnan, R., Goins, W. F., Fink, D. J., and Glorioso, J. C. (1996). Replication-defective HSV vectors for gene transfer *in vivo*. Proc. Natl. Acad. Sci. USA *93*, 11319-11320.
24. Samaniego, L., Webb, A., and DeLuca, N. (1995). Functional interaction between herpes simplex virus immediate-early proteins during infection. gene expression as a consequence of ICP27 and different domains of ICP4. J. Virol. *69*, 5705-5715.
25. Wu, N., Watkins, S. C., Schaffer, P. A., DeLuca, N. A. (1996). Prolonged gene expression and cell survival after infection by a herpes simplex virus mutant defective in immediate-early genes encoding ICP4, ICP27, and ICP22. J. Virol. *70*, 6358-6368.
26. Page, B., Page, M., and Noel, C. (1993). A new fluorometric assay for cytotoxicity measurements *in vitro*. Int. J. Oncology *3*, 473-476.
27. Devi-Rao, G. B., Goddart, S. A., Hecht, L. M., Rochford, R., Rice, M. K., and Wagner, E. K. (1991). Relationship between polyadenylated and nonpolyadenylated, HSV type 1 latency-associated transcripts. J. Gen. Virol. *65*, 2179-2190.
28. Spivack, J. and Fraser, N. (1988). Expression of herpes simplex virus type 1 latency-associated transcripts in trigeminal ganglia of mice during acture infection and reactivation of latent infection. J. Virol. *62*, 1479-1485.
29. Stevens, J. G., Wagner, E. K., Devi-Rao, G. B., Cook, M. L., and Feldman, L. T. (1987). RNA complementary to a herpesviruses α gene mRNA is prominent in latently infected neurons. Science *255*, 1056-1059.
30. Wagner, E. K., Flanagan, W. M., Devi-Rao, G. B., Zhang, Y. F., Hill, J. M., Anderson, K. P., and Stevens, J. G. (1988b). The herpes simplex virus latency-associated transcript is spliced during the latent phase of infection. J. Virol. *62*, 4577-4585.
31. Farrell, M. J., Dobson, A. T., Feldman, L. T. (1991). Herpes simplex virus latency-associated transcript is a stable intron. Proc. Natl. Acad. Sci. USA *88*, 790-794.
32. Wu, T.-T., Su, Y.-H., Block, T. M., and Taylor, J. M. (1996). Evidence that two latency-associated transcripts of herpes simplex virus type 1 are non-linear. J. Virol. *70*, 5962-5967.
33. Bloom, D., Hill, J., Devi-Rao, G., Wagner, E., Feldman, L., and Stevens, J. (1996). A 348-base-pair region in the latency-associated transcript facilitates hereps simplex virus type 1 reactivation. J. Virol *70*, 2449-2459.
34. Leib, D. A., Nadeau, K. C., Rundle, S. A., Schaffer, P. A. (1991). Promoter of the latency-associated transcripts of herpes simplex virus type 1 contains a functional cAMP-response element: role of the latency-associated transcripts and cAMP in reactivation of viral latency. Proc. Natl. Acad. Sci. USA *88*, 48-52.

35. Sawtell, N. M. and Thompson, R. L. (1992). Herpes simplex virus type 1 latency-associated transcription unit promotes anatomical site-dependent establishment and reactviation from latency. J. Virol. *66*, 2157-2169.
36. Chen, X., Schmidt, M. C., Goins, W. F., and Glorioso, J. C. (1995). Two herpes simplex virus type-1 latency active promoters differ in their contribution to latency-associated transcript expression during lytic and latent infection. J. Virol. *69*, 7899-7908.
37. Dobson, A. T., Sederati, F., Devi-Rao, G., Flanagan, W. M., Farrell, M. J., Stevens, J. G., Wagner, E. K., and Feldman, L. T. (1989). Identification of the latency-associated transcript promoter by expression of rabbit β-globin mRNA in mouse sensory nerve ganglia latently infected with a recombinant herpes simplex virus. J. Virol. *63*, 3844-3851.
38. French, S., Goins, W., Montesanti, D., Kinniburgh, A., Schmidt, M., and Glorioso, J. (1996b). *In vitro* studies of the TATA-less latency active promoter (LAP2) of herpes simplex virus type-1 identify essential promoter elements which form non-B-DNA structures. J. Gen. Virol. (submitted).
39. Goins, W. F., Sternberg, L. R., Croen, K. D., Krause, P. R., Hendricks, R. L., Fink, D. J., Straus, S. E., Levine, M., and Glorioso, J. C. (1994). A novel latency-active promoter is contained within the herpes simplex virus type 1 UL flanking repeats. J. Virol. *68*, 2239-2252.
40. Soares, M. K., Hwang, D.-Y., Schmidt, M. C., Fink, D. J., and Glorioso, J. C. (1996). Cis-acting elements involved in transcriptional regulation of the herpes simplex virus type-1 latency-associated promoter 1 (LAP1) *in vitro* and *in vivo*. J. Virol. *70*, 5384-5394.
41. Batchelor, A. H., O'Hare, P. O. (1990). Regulation and cell-type-specific activity of a promoter located upstream of the latency-associated transcript of herpes simplex virus type 1. J. Virol. *64*, 3269-3279.
42. Zwaagstra, J. C., Ghiasi, H., Nesburn, A. B., and Wechsler, S. L. (1991). Identification of a major regulatory sequence in the latency-associated transcript (LAT) promoter of herpes simplex virus type 1 (HSV-1). Virology *182*, 287-297.
43. Ackland-Berglund, C., Davido, D., and Leib, D.A. (1995). The role of the cAMP-response element and TATA box in expression of the herpes simplex virus type 1 latency-associated transcripts. Virology *210*, 141-151.
44. Rader, K. A., Ackland Berglund, C. E., Miller, J. K., Pepose, J. S., Leib, D. A. (1993). In vivo characterization of site-directed mutations in the promoter of the herpes simplex virus type 1 latency-associated transcripts.
45. Batchelor, A. H., Wilcox, K. W., and O'Hare, P. (1994). Binding and repression of the latency-associated promoter of herpes simplex virus by the immediate early 175K protein. J. Gen. Virol. *75*, 753-767.
46. Farrell, M. J., Margolis, T. P., Gomes, W. A., and Feldman, L.T. (1994). Effect of the transcription start region of the herpes simplex virus type 1 latency-associated transcript promoter on expression of productively infected neurons *in vivo*. J. Virol. *68*, 5337-5343.
47. Rivera-Gonzalez, R., Imbalzano, A. N., Gu, B., and DeLuca, N.A. (1994). The role of ICP4 repressor activity in the temporal regulation of the IE-3 and latency-associated transcript promoter during herpes simplex virus type-1 infection. Virology *202*, 550-564.
48. French, S. W., Schmidt, M. A., and Glorioso, J. C. (1996b). Involvement of a high-mobility-group protein in the transcriptional activity of herpes simplex virus latency-active promoter 2. Mol. Cell. Biol. *16*, 5393-5399.
49. Hoffman, E. K., Trusko, S. P., Freeman, N. A., and George, D. L. (1987). Structural and functional characterization of the promoter region of the mouse c-Ki-ras gene. Mol. Cell. Biol. *7*, 2592-2596.
50. Johnson, A. C., Jinno, Y., and Merlino, G. T. (1988). Modulation of epidermal growth factor receptor proto-oncogene transcription by a promoter site sensitive to S1 nuclease. Mol. Cell. Biol. *8*, 4174-4184.
51. Kasai, Y., Chen, H., and Flint, S. J. (1992). Anatomy of an unusual RNA polymerase II promoter containing a downstream TATA element. Mol. Cell. Biol. *12*, 2884-2897.
52. Jorcyk, C. L., Watson, D. K., Mavrothalassitis, G. J., Papas, T. S. (1991). The human ETS1 gene: genomic structure, promoter characterization and alternative splicing. Oncogene *6*, 523-532.
53. Kim, R. Y., Lietman, T., Piatigorsky, J., and Wistow, G. J. (1991). Structure and expression of the duck α-enolase/t-crystallin-encoding gene. Gene *103*, 193-200.
54. Koller, E., Hayman, A. R., Trueb, B. (1991). The promoter of the chicken α2(VI) collagen gene has features characteristic of housekeeping genes and of proto-oncogenes. Nucleic Acids Res. *19*, 485-491.

55. Tomarev, S.I., Zinovieva, R. D., and Piatigorsky, J. (1992). Characterization of squid crystallin genes. J. Biol. Chem. *267*, 8604-8612.
56. Kolluri, R. and Kinniburgh, A. J. (1991). Full length cDNA sequence encoding a nuclease-sensitive element DNA binding protein. Nucleic Acids Res. *19*, 4771.
57. Kolluri, R., Torrey, T. A., and Kinniburgh, A. J. (1992). A CT promoter element binding protein: definition of a double-strand and a novel single-strand DNA binding motif. Nucleic Acids Res. *20*, 111-116.
58. Thanos, D. and Maniatis, T. (1992). The high mobility group protein HMG I(Y) is required for NF-kB-dependent virus induction of the human IFN-β gene. Cell *71*, 777-789.
59. Falvo, J. V., Thanos, D., and Maniatis, T. (1995). Reversal of intrinsic DNA bends in the IFN-β gene enhancer by transcription factors and the architectural protein HMG I(Y). Cell *83*, 1101-1111.
60. Larsen, A. and Weintraub, H. (1982). An altered DNA conformation detected by S1 nuclease occurs at specific regions in active chick globin chromatin. Cell *29*, 609-622.
61. McKeon, C., Schmidt, A., and Crombrugghe, B. (1984). A sequence conserved in both the chicken and mouse α2(I) collagen promoter contains sites sensitive to S1 nuclease. J. Biol. Chem. *259*, 6636-6640.
62. Elgin, S.C.R. (1981). DNAase I-hypersensitive sites of chromatin. Cell *27*, 413-415.
63. Almer, A., Rudolph, H., Hinnen, A., and Horz, W. (1986). Removal of positioned nucleosomes from the yeast PHO5 promoter upon PHO5 induction releases additional upstream activating DNA elements. EMBO J. *5*, 2689-2696.
64. Pina, B., Bruggemeier, U., Beato, M. (1990). Nucleosome positioning modulates accessibility of regulatory proteins to the mouse mammary tumor virus promoter. Cell *60*, 719-731.
65. Chen, X., Schmidt, M., Goins, W., Fink, D., Hendricks, R., and Glorioso, J. (1996). HSV-1 vector-mediated lacZ gene transfer to the nervous system: long-term gene expression using the viral latency-active promoter, Gene Ther. (submitted).
66. Becker-Andre, M. and Hahlbrook, K. (1989). Absolute mRNA quantitation using the polymerase cahin reaction (PCR): a novel approach by a PCR aided transcript totration assay (PATTY). Nucl. Acids Res. *17*, 9437-9446.
67. Gilliland, G. S., Perrin, S., and Bunn, H. F. (1990). Competitive PCR for quantitation of mRNA. In: PCR protocols: a guide to methods and applications., M. A. Innis, J. J. Sninsky and T. J. White (eds.), (New York: Academic Press), pp. 60-69.
68. Ramakrishnan, R., Fink, D. J., Guihua, J., Desai, P., Glorioso, J.C., and Levine, M. (1994b). Competitive quantitative polymerase chain reaction (PCR) analysis of herpes simplex virus type 1 DNA and LAT RNA in latently infected cells of the rat brain. J. Virol. *68*, 1864-1870.
69. Ramakrishnan, R., Levine, M., and Fink, D. (1994a). PCR-based analysis of herpes simplex virus type 1 latency in the rat trigeminal ganglion established with a ribonucleotide reductase-deficient mutant. J. Virol. *68*, 7083-7091.
70. Coen, D. M., Kosz-Venchak, M., Jacobson, J. G., Leib, D. A., Bogard, C. L., Schaffer, P. A., Tyler, K. L., and Knipe, D. M. (1989). Thymidine kinase-negative herpes simplex virus mutants establish latency in mouse trigeminal ganglia but do not reactivate. Proc. Natl. Acad. Sci. USA *86*, 4736-4740.
71. Jacobson, J., Leib, D., Goldstein, D., Bogard, C., Schaffer, P., Weller, S., and Coen, D. (1989). A herpes simplex virus ribonucleotide reductase deletion mutant is defective for productive, acute, and reactivatible infections for mice and for replication in mouse cells. Virology *173*, 276-283.
72. Sadowski, I., Ma, J., Triezenberg, S., Ptashne, M. (1988). GAL4/VP16 is an unusually potent transcriptional activator. Nature *335*, 563-564.
73. Carey, M., Leatherwood, J., and Ptashne, M. (1990). A potent GAL4 derivative activates transcription at a distance *in vitro*. Science *247*, 710-712.
74. Chasman, D. I., Leatherwood, M., Carey, M., Ptashne, M., and Kornberg, R. D. (1989). Activation of yeast polymerase II transcription by herpesvirus VP16 and GAL4 derivative *in vitro*. Mol. Cell. Biol. *9*, 4746-4749.
75. Oligino, T., Poliani, P. L., Marconi, P., Bender, M. A., Schmidt, M. C., Fink, D. J., and Glorioso, J.C. (1996a) *In vivo* transgene activation from an HSV-based gene vector by GAL4:VP16. Gene Ther. *3*, 892-899.
76. Oligino, T., Poliani, P. L., Wang, Y., Tsai, S.Y., O'Malley, B. W., and Glorioso, J. C. (1996b). Drug inducible transgene expression in brain using a herpes simplex virus vector. J. Virol. (submitted).

4. Adeno-Associated Viral Vectors

D. M. McCarty and R. J. Samulski

4.1 Introduction

Adeno-associated virus (AAV) based gene therapy research has been driven by our desire to employ the unique biology and life cycle of this human parvovirus. Few organisms exhibit as many natural features which seem to lend themselves to this purpose. The foremost property of this virus which suits it to applications in gene therapy is the tendency to establish latent infections by integrating into the chromosomal DNA of non-permissive cells. Because permissivity generally depends on the presence of a co-infecting helper virus, this suggests that any uninfected cell could potentially serve as host for latent AAV. The second prominent feature which distinguishes AAV from many other potential viral vectors is its lack of pathogenicity. Ostensibly, a virus which has never been associated with clinical disease poses relatively little risk when adapted as a vector. Other attractive characteristics are its stability, genetic simplicity, and ease of genetic manipulation. While each of these points remains valid, some obstacles to the application of recombinant AAV (rAAV) vectors have recently come to light. It is not surprising, given the cryptic nature of this virus, that new insights into its biology have surfaced only after extensive research with the rAAV vectors which are so much more easily detected. In this review we will discuss recent observations dealing with rAAV mediated transduction as well as some of the potential solutions that currently are being explored.

4.2 Recombinant AAV Vectors

Genetic manipulation of AAV became possible when the viral genome was cloned into bacterial plasmids (1, 2). Because these constructs contained intact palindromic terminal repeats (TR), which are the AAV origins of replication, it was possible to rescue the virus by transfecting the plasmid DNA into cells infected with adenovirus (Ad). This allowed mutations to be made in the parental plasmid which subsequently could be used to generate mutant virus by transfection into permissive cells (3-6). In general, rAAV vectors are still made in the same way. The two AAV specific genes, *rep* and *cap* are deleted, leaving only the TR sequences of the parent virus (7). A heterologous gene is cloned between the TRs and becomes part of the viral replicon when the plasmid is co-transfected with a second plasmid expressing the *rep* and *cap* genes. Single-stranded progeny DNA molecules from the rAAV construct are then encapsidated as long as they are within approximately 5% of the size of the wild type (wt) AAV genome (8). While relatively simple to execute, this procedure is inefficient and difficult to scale upward due to the need to transfect cells. A great deal of effort is being made to establish producer cell lines which would eliminate this step (9-11). Also, incremental improvements in rAAV titers are being achieved through variations on the trasfection procedure (11-14). However, the availability of even limited quantities of rAAV has allowed a great deal of progress to be made in understanding transduction by these vectors.

A source of confusion in the field of rAAV vector research has been the lack of standard terminology to describe the infectious dose used in each experiment. This has arisen in part because it is not as straight-forward to quantify AAV as with most viruses. The requirement for superinfection with a helper virus precludes the use of plaque assays to measure the infectious titer. It is therefore necessary to detect either the viral gene products or the replicating genome in order to infer the titer.

The most broadly relevent technique for titering rAAV vectors is the replication center assay. This assay relies upon the detection of replicated vector DNA in single cells co-infected with Ad and wtAAV (15, 16). While this assay is technically demanding, it does measure two parameters that are common to all rAAV based vectors, i.e. the presence of intact terminal repeats and proper encapsidation into an infectious particle. Because the TRs are the only *cis* active sequences required for integration, this also serves to measure the number of potential long-term transducing units (17). As long as a standard cell type is used in the replication center assay, it yields a value that can be compared to any rAAV construct or to the wild type virus.

One infectious unit as measured in the replication center assay typically corresponds to approximately 20-100 physical virus particles (18). The relationship between particles and infectious units will differ, of course, between different cell types and tissues. We have adopted 293 cells as a standard for measuring infectious rAAV vectors but similar titers are obtained from HeLa cells. While the infectious center assay provides a consistent standard for the comparison of various vector titers, it does not necessarily predict a third value, which is the transducing unit titer. Again, this is greatly influenced by cell type as well as promoter activity and specificity. In 293 cells infected with rAAV vectors expressing the bacterial lacZ gene under the control of the cytomegalovirus immediate early promoter (CMV), the ratio of infectious titer to transducing units is approximately 10:1. Therefore, the ratio of physical virus particles to transducing units is approximately 200-1000:1.

For the purpose of this review, we will attempt to frame our discussions in terms of infectious particles per cell. It is arguably more straight-forward to report rAAV titers in physical virus particles, which do not change from one cell type or infection condition to another. However, with the aim of elucidating the mechanisms of rAAV transduction, we feel that focusing on the number of potentially active genomes entering the nucleus may be more meaningful. In some of the specific experiments discussed, we have estimated the ratios of infectious units to cells from reported values in physical particles or in transducing units, using our own experience as a guide.

4.3 Barriers to rAAV Transduction

The ability of rAAV to efficiently transduce cells has been determined largely through experiments *in vitro* using immortalized cell lines. The relevance of these results to the transduction of primary and non-dividing cells *in vitro* and *in vivo* recently has come into question. Observations which first raised these concerns stemmed from discrepancies in the titer of rAAV virus stocks before and after purification on CsCl density gradients or heating for 30 min. at 56 °C. Both of these procedures are intended to remove or inactivate contaminating adenovirus. It was soon realized that the titer of rAAV was not actually changing, but that the ability to detect rAAV infection as a transduction event was enhanced in the presence of Ad co-infection. The original rAAV titer could be fully restored by adding back the adenovirus. The helper effects that Ad normally provides for wtAAV were not expected to contribute to the expression of the recombinant viruses because the vectors were not capable of replication and the transgenes were not driven by either Ad or AAV promoters. This previously unrecognized Ad helper function was observed with many heterologous promoters which were not expected to respond to Ad specific transcription factors.

There were several possible steps where Ad could have exerted such an effect:

1) participation in the uptake of the virus or the intracellular release of the rAAV genome from the capsid.
2) conversion of the single stranded rAAV virion DNA to the duplex form which could serve as template for transcription.
3) stimulation of transcription from the transgene via an enhancer effect from the AAV TR. The evidence presented below suggests that it is the synthesis of the second strand of the rAAV genome that is limiting for the transduction of normal cells.

Adenovirus infection is a complex process involving several discreet steps as a prelude to viral gene expression. Among these are: adsorption and entry, endosomal disruption, and transport to the nucleus. The idea that Ad had a role in AAV adsorption, penetration, or translocation to the nucleus had been previously dismissed based on early efforts to define the Ad helper effect for wtAAV (19). Similarly, the novel helper effect that Ad provides for rAAV vector transduction did not correlate with any of these steps. The induction could still be observed when cells were infected with rAAV six hours prior to superinfection with Ad, making co-adsorption and penetration an unlikely possibility (20). Further, there was no requirement for an intact Ad virion. Transfected Ad DNA was sufficient to induce high levels of transduction in rAAV infected cells. This suggested that an Ad specific or Ad induced gene product was mediating the enhancement of transduction.

A series of well characterized Ad mutants was used to determine which gene product (s) was responsible for the helper effect (20, 21). Mutants in E1, VAI, E2, or E4 were co-infected with rAAV and the transduction rates were compared. Only the E4 deletion mutants failed to enhance transduction. Additional mutants were used to map this effect to E4 open reading frame 6 (ORF 6). The possibility that other interactions with the Ad genome were essential for the transduction helper effect was eliminated by using a transfected plasmid expressing only the E4 ORF 6 gene product. Interestingly, the E4 ORF 3 gene product, which is redundant with ORF 6 for some activities associated with lytic Ad infection (22, 23), could not provide the helper effect for rAAV vectors.

The identification of the factor which mediated the enhancement of rAAV transduction allowed the partial characterization of the mechanism behind this effect. The idea that E4 ORF 6 might promote uncoating of the virion was eliminated by comparing the DNase sensitivity of rAAV genomes in cells tranfected with plasmids expressing E4 ORF 6 versus cells transfected with a control plasmid (Summerford and Samulski, unpublished). These studies demonstrated that most of the virion DNA was DNase sensitive at 24 hr post-infection regardless of the expression of the Ad protein. No differences in virion released DNAs were detected between E4 expressing cells and controls.

The possibility remained that the E4 protein was directly or indirectly modulating transcription of the rAAV transgene through *cis* activities associated with the TR. The TRs do have an enhancer-like activity which is dependent upon the *trans* effects of the AAV Rep protein (24, 25). However, it was difficult to reconcile this notion with the magnitude of the effect. Up to 1,000-fold increases were observed in the number of cells transduced. Further, rAAV transgenes did not show any such response to Ad infection when they were in the proviral state or when transfected as plasmid DNA. In the latter two situations, the transgene would be in the same position relative to the TRs as in the infecting virion DNA.

The observations were consistent with an induction of second strand synthesis by Ad or the Ad E4 ORF 6 gene product. Evidence supporting this conclusion was found in experiments correlating the induction of transgene expression with the conversion of the single-stranded virion DNA to duplex (20, 21). This was observed as the appearance of dimer size molecules when DNA from rAAV infected cells was analyzed on denaturing agarose gels. The dimer length molecules comprised duplex monomers which were covalently linked at one end via the priming of DNA synthesis from the hairpinned TR of single-stranded virion DNA. The mechanism for inducing synthesis of the rAAV second strand is not known. Other functions that have been associated with E4 ORF 6 are:

1) binding to the cellular tumor suppresser protein, p53, blocking its transcriptional transactivation activity (26);
2) enhancing the splicing of leader sequences of specific Ad late mRNAs (22, 27);
3) prevention of concatemer formation during Ad DNA replication (23);
4) binding to E1b to facilitate Ad mRNA transport (28, 29);
5) antagonist to E4 ORF 4 in the regulation of Ad DNA replication (30).

There is still some uncertainty as to whether products of Ad E1 are necessary for enhancement of rAAV transduction. Fisher et al. have suggested that E1 at least augments the effect based on two observations (21). First, Ad mutants that express E4, but are deleted in E1, enhance transduction in 293 cells (which constituitively express E1) but not in HeLa cells. However, this can easily be explained on the basis of the transactivation effect of E1 on the E4 promoter. Second, in a 293 cell line expressing E4 under an inducible metallothionein promoter, rAAV transduction is enhanced in a Zn^{2+} concentration dependent manner. In contrast, in a similar HeLa cell line the enhancement does not reach proportionate levels. It would appear however, that the primary effect on rAAV transduction is mediated by E4 ORF 6 while E1 probably is an indirect effector through activation of E4 expression or reporter gene expression.

While the induction of rAAV second-strand synthesis by the E4 ORF 6 gene product might well represent an adaptation evolved by wtAAV to further exploit its Ad helper, the phenomenon is clearly not unique to Ad infected cells. A broad spectrum of cell culture treatments recently have been found to facilitate the same step. These include hydroxyurea, UV and X-ray irradiation, some topoisomerase inhibitors, and heat shock (20, 31). Most of these treatments, with the possible exception of heat shock, result in DNA damage. It is a reasonable hypothesis that the induction of DNA damage repair synthesis is the mechanism allowing the conversion of single-strand rAAV virion DNA to double-strand transcriptionally active templates in a manner analogous to filling a gapped region of chromosomal DNA. However, there have not been any direct demonstrations of repair enzymes such as DNA Pol ε physically associating with rAAV DNA. Experiments with mutant cell lines lacking p53 and pRB suggest that these factors are not necessary for the induction of second-strand synthesis (Ferrari and Samulski, unpublished). This argues against the possibility that apoptotic pathways mediated by these factors lead to enhanced transduction. However, it should be noted that while this cell line is deficient in p53 and Rb, many other genetic alterations have also occurred which may obscure the definitive answer to this question. Analysis of rAAV transduction in primary mouse embryo cells deficient in p53 or Rb would resolve this issue.

The observation that cytotoxic treatments could induce rAAV second strand synthesis, and hence transduction, was reminiscent of earlier experiments wherein similar treatments were shown to induce helper virus independent replication of wtAAV (16, 32, 33). In these cases, however, replication of viral DNA was dependent upon UV or hydroxyurea treatment whether the genome was presented as single-stranded virion DNA or as duplex plasmid. This indicated that the changes that took place in the cells following DNA damage contributed to more than simply overcoming the barrier to duplex formation.

Yalkinoglu et al. made the more remarkable observation that plasmids containing a single copy of the AAV TR were replicated in cells treated with genotoxic agents (34). The replication origin on these molecules was mapped to the TR using cell-free extracts from carcinogen induced cells. This suggests that the AAV TR can specifically recruit replication factors under conditions of cell stress. This provides a very plausible mechanism for the induction of second-strand synthesis in the rAAV vectors. The conditions which enhance rAAV transduction might induce the same factors to be recruited by the palindromic TR, which is already in a duplex hairpin conformation in the single-stranded virion DNA. Further, it suggests that AAV has made multiple adaptations to exploit cellular responses to environmental factors. This aspect of AAV biology has not been extensively explored as yet but will certainly have a great impact on the future use of rAAV vectors for gene therapy.

4.4 Transduction of Non-Dividing Cells

Several of the recent reports of enhancement of rAAV transduction have stemmed from attempts to characterize the apparent disparity in transduction rates between transformed cells and non-dividing cells. Russell et al. investigated the ability of rAAV vectors to transduce stationary versus dividing primary human fibroblasts (35). When a *neo* reporter construct was used, little difference in transduction rates were noted between cells in stationary and rapidly dividing cultures. However, transduction rates differed by greater than 20-fold when an alkaline phosphatase (AP) reporter was used such that the cultures did not require selection (and cell division) to measure rAAV gene expression. Further, when transduced cells in a stationary culture were evaluated for the incorporation of ^{3}H thymidine as a marker for passage through S phase, it was found that 90% of the transduced cells had replicated nuclear DNA. This indicated that, although, rAAV vectors were capable of transducing cells in the absence of cell division, the efficiency was greater in cells which were undergoing DNA synthesis.

The discrepancy between *neo* and AP transduction in non-dividing cells was postulated to be caused by the persistence of single-stranded virion DNA in the nuclei of infected cells. Stationary human fibroblast cells retained approximately 10% of the original infectious dose of virion DNA after twelve days in culture. When the cells were passaged for G-418 selection and allowed to divide, these were available to be converted into duplex DNA for expression of the transgene and the observed G-418 resistance.

Differences in susceptibility to rAAV transduction have also been noted between primary and transformed cells. Halbert et al.compared primary human epithelial cells to the epithelial cell line, IB3, which is immortalized by an Ad/SV40 chimera (36). The transformed cells were transduced at a 60-fold higher frequency than the primary cells even though both cultures were actively dividing. This effect might have derived in part through the expression of the Ad E4 region in the transformed cell line. However, similar effects (up to 15-fold) were observed in cells transformed with SV40 large T-Ag or the HPV E6, E7 genes. Importantly, these effects were not manifested until the oncogenically transformed cells had been passaged greater than five times. This indicated that it was not the oncogene which directly mediated the enhancement of transduction with rAAV vectors. Rather, it more likely can be attributed to the manifold changes in endogenous gene expression which are the hallmark of cellular transformation.

While it is not yet clear why there are differences in transduction rates between primary and immortalized cells, the evidence for transduction in the absence of cell division is overwhelming. Recombinant AAV vectors have been used to transduce adult animal neural and muscle tissue which are clearly not dividing (37, 38). Further, experiments designed to correlate rAAV transduction with any specific phase of the cell cycle have indicated that passage through S-phase is neither necessary for transduction nor sufficient to ensure that vector gene expression will take place in a rAAV infected cell (Ferrari, Snyder, and Samulski, manuscript in preparation).

4.5 Episomal Expression

It is becoming evident that the primary rate limiting step for rAAV transduction in many cell types is conversion of the single-stranded virion DNA to duplex. The new double-stranded molecule is then either a substrate for gene expression or for integration into the host chromosome. Estimates of the integration efficiency of rAAV vectors have ranged from 0.4% to 80%, depending on the presence of the AAV Rep protein and the method of detecting transduction. In those cases where non-selectable markers were used, there was compelling evidence that a significant fraction, or even a preponderance, of the observed transgene expression was from episomal templates (39-41). Further perspective on the nature of rAAV transduction may be gained from previous studies with wtAAV.

It has been well established that wtAAV displays a strong preference for integration into a specific region of chromosome 19. This faculty is dependent upon the expression of AAV Rep protein (42, 43). The chromosome 19 integration site contains a sequence known to bind to Rep protein as well as a putative site for the single-strand endonucleolytic activity of Rep (44-47). Cleavage at this site might therefore be the initiating event for wtAAV integration. Whether or not this is the case, the dependence of targeted integration on the Rep protein *in trans* suggests that some virion DNA was converted to duplex prior to integration. Thus, it is clear that an episomal wtAAV genome is capable of gene expression and it is likely that this capability is shared by rAAV.

The status of the AAV genome in infected cells was evaluated in several recent studies. Flotte and colleagues looked for rAAV specific DNA in either the chromosomal or episomal fractions of cells transduced with a LacZ expressing vector (40). While episomal copies of the genome were readily detected by Southern blotting, no rAAV specific sequences were observed in association with the cellular genomic DNA. This suggested that integration is a relatively uncommon fate for rAAV molecules.

The electrophoretic mobility of the episomal vector was consistent with a linear molecule 8 kb in length, which was approximately twice the expected size of the double-stranded monomer vector genome. The authors proposed that this represented either a duplex dimer or a circular form of the vector DNA molecule. While the duplex dimer form is normally observed during AAV DNA replication, it is not clear how such a molecule would be derived in the absence of the AAV *rep* gene products. The synthesis of the complementary strand of single-stranded virion AAV initiates from one of the hairpinned terminal repeat sequences (48-52). Termination of DNA synthesis either occurs at the 5' end of the opposite TR or, the 5' end may be displaced and synthesis continued to extend and copy the TR. Either termination will yield a monomer duplex product which would not be expected to reinitiate DNA replication in the absence of the active Rep protein (53, 54).

The anomalous mobility of the episomal rAAV genomes might represent a circular form of AAV DNA. Circularized AAV DNA could arise either by annealing of the complementary TR sequences (55, 56) or less likely, by recombination between the TRs of one or more AAV molecules [compelling arguments against the joining of multiple rAAV DNA molecules have been advanced by N. Muzyczka and co-workers (17)] . The ability of a single copy of the TR to promote *rep*-independent replication of plasmids in response to genotoxic stress (as described above) would provide a mechanism for amplifying the DNA containing these structures and can lead to the accumulation of multimeric or concatenated circular molecules (34). The detailed characterization of the episomal structures observed in rAAV transduced cells will have an tremendous impact on the use of these vectors in terms of their ability to stably transduce cells and, potentially, on the understanding of the mechanism of vector integration and persistence.

4.6 Adeno-Associated Virus Integration

The pertinent question for the application of AAV as a gene therapy vector is with what frequency do the episomes integrate into chromosomal DNA. The potential of AAV as an *integrating vector* and an indication of its relative efficiency were first illustrated by Berns and colleagues (57). Populations of Detroit 6 cells were infected with wtAAV at doses ranging from 0.0175 to 175 MOI. The pools of cells were then serially passaged and periodically assayed for the expression of AAV proteins following challenge with adenovirus. Cells infected at MOI 175 were capable of rescue of AAV for at least 47 passages. Those infected at doses of MOI 1.75 and 0.0175 lost the ability to rescue AAV after 34 and 8 passages respectively. This suggested that either episomal copies of AAV were lost by dilution or degradation or that cells harboring integrated AAV were diluted out of the culture. In either case, it must be concluded that a only a small number of cells actually contained integrated provirus when infected at an MOI of 1.75.

The cells which were infected at MOI 175 were further characterized at the clonal level. At passage 39 post-infection, the cells were plated at limiting dilutions to obtain single cell clones. When each randomly selected clonal cell line was subsequently challenged with Ad, 18 of 63 (29%) were found to contain rescuable AAV. Since each cell should have initially contained an average of 175 genomes, this suggested that each infectious virus had a 0.17% chance of establishing a latent infection.

Similar experiments were performed by Laughlin et al., who infected KB cells at MOI 200 and were able to demonstrate latent AAV in approximately 10% of cloned cells from the infected culture (58). Handa et al. found that approximately 2% of KB cells carried latent AAV after infection at MOI 500 (59). Laughlin et al. went on to show that the AAV infection had no effect on cell viability upon replating. These experiments made it clear that wtAAV could establish latency in a significant proportion of a cell population but only when infecting at high multiplicity.

4.7 Recombinant AAV Vector Integration

The early studies with rAAV vectors containing marker genes utilized vectors which retained the AAV *rep* gene. These consistently yielded similarly low rates of integration. Hermonat and Muzyczka infected D6 cells with 0.1 infectious units per cell of rAAV expressing the *neo* gene (60). This vector yielded stable $G418^R$ colonies representing 0.04 percent of the cell culture, suggesting that 0.4% of infecting genomes had integrated. This was remarkably similar to the number previously observed with wtAAV at high multiplicities of infection. However, when the rAAV vector was titrated to higher multiplicities, the number of transduced cells rose to only 3% of the culture at MOI 1000 (10% if G418 selection was delayed). Similar results were reported by Tratschin et al. using a *rep*$^+$ *neo* vector on 293 and HeLa cells (61). This suggested that either only a subpopulation of cells was susceptible to transduction or that high multiplicity infections with the Rep$^+$ vector or with contaminating wtAAV interfered with transduction. However, Tratschin et al. performed some of their experiments using a vector preparation comprising 80% wtAAV and found similar transduction rates, thus arguing against interference from the wt virus. The basis of the apparent plateau in rAAV *neo* transduction has not as yet been resolved. It is not likely to be a limitation in the number of cells which can be infected with the virus since populations of 293 and HeLa cells are easily transduced to near 100% levels using rAAV-LacZ vectors.

McLaughlin et al. were the first to construct and characterize a *rep*$^-$ rAAV vector containing only the TRs and polyadenylation signal from the parent virus (17).

Comparison of this *neo* expressing vector against the previously characterized *rep*$^+$ vector at low MOI revealed that the transduction frequencies were far greater in the absence of the *rep* gene. Using the *rep*$^-$ vector, approximately 80% of the infectious genomes yielded G418^R colonies at an MOI of 0.03. The *rep*$^+$ vector again yielded a 0.4% transduction rate in parallel experiments at the same multiplicity of infection. Similarly high transduction rates (>50% of infectious genomes were integrated) were observed by Samulski et al. using *neo* expressing *rep*$^-$ rAAV vectors (7). More recently, Russel and co-workers (35) demonstrated *neo* transduction rates of approximately 30% in primary human fibroblast cultures (deduced from the approximate equivalence of 600 physical particles to three transducing units).

It has been suggested that the difference in transduction rate between *rep*$^+$ and *rep*$^-$ vectors stems from Rep mediated repression of the heterologous promoters from which the vector genes are transcribed . Samulski et al. were able to generate rAAV vector stocks free of contaminating wt virus and then titrated wtAAV against a constant dose of the recombinant (7). This resulted in a slight increase in transduction rate as the ratios of wt to vector approached 1:1 and a decreased rate at a ratio of 10:1. This suggests that the role of *rep* in AAV integration is not simple and, in the light of more recent observations on its effect on specific integration into human chromosome 19, may point to alternate integration pathways for *rep*$^+$ and *rep*$^-$ vectors.

In contrast to the above reports of efficient *neo* transduction, recent experiments in our laboratory have suggested significantly lower integration rates from rAAV vectors carrying the bacterial *lacZ* gene. When plates of HeLa cells were infected in parallel and evaluated for transient versus stable transduction, approximately 1% of the total transduced cell population appeared to contain integrated vector (Ferrari and Samulski, manuscript in preparation). While the sensitivity of transgene detection may account for these differences, many other explanations are possible. Selection with G418 may mimic a stress response that facilitates viral integration as mentioned above. The issue of the rate of integration versus the rate of transient transduction with rAAV vectors lacking the *rep* gene deserves a great deal more attention and will be a primary focus of research in the use of these vectors.

4.8 Mechanism of AAV Integration

The mechanism of integration is not well understood for either the wtAAV or recombinant virus. Because the AAV Rep protein has been implicated as an active factor in wtAAV integration, we must consider the possibility that rAAV is integrated by a different mechanism, relying solely upon cellular factors. The most pronounced difference in the rep$^+$ versus *rep*$^-$ integration is the chromosome 19 site preference in *rep*$^+$ vectors (42, 43). Data from *in vitro* reconstituted systems demonstrate that Rep protein can mediate the attachment of AAV DNA to the chromosome 19 integration site (44). Further, cell-free *in vitro* replication systems suggest that Rep can initiate DNA replication by enzymatic cleavage of a specific site in the chromosomal sequence, which is analogous to the cleavage site within the AAV terminal repeat (62). These observations, combined with analyses of large numbers of AAV integration products (45, 63, 64), have lead to the recent proposal of a model for AAV integration (65). Berns and colleagues postulate that a Rep-AAV protein-DNA complex catalyses a single-strand nick at the integration site which serves as primer for a DNA polymerase. During ensuing elongation, the polymerase complex is transferred from the chromosomal template to the viral template, creating the first covalent link between the two sequences. A subsequent transfer at a downstream site would complete the integration.

While the model outlined above is consistent with observations of viral integration products, it does not explain the integration of rAAV vectors which lack the *rep* gene. There is data suggesting that Rep is associated with the AAV virion, which could serve as a catalyst for integration (66). However, it is difficult to reconcile this possibility with the kinetics of Rep-DNA complex formation, which suggest that multiple protein molecules are required to join two DNA molecules together (67). It is therefore more likely that cellular enzymes are interacting with the AAV TR to promote integration independently of Rep.

While there are clear differences between wtAAV and rAAV vector integration, most notably, chromosome 19 specificity, there are recognizable similarities (Yang and Samulski, manuscript in preparation). Most rAAV junctions contain deletions of variable length at the ends of the TR sequences, as do the wtAAV junctions. Also, microhomologies of two to five base pairs between TR and chromosomal sequences are usually evident at the break points. These observations suggest that the two kinds of integration events are mediated by the same process. The role of Rep may be to potentiate site-specific integration and perhaps to inhibit non-specific integration.

4.9 Transduction *in vivo*

Much of the recent advance in the understanding of rAAV vectors has come about through the need to resolve the incongruities between *in vitro* and *in vivo* observations. Although several recent studies have shown great promise in terms of duration of transgene expression *in vivo*, there has been a shortfall in the efficiency of transduction which was unexpected based on previous results *in vitro*.

One of the first experiments to demonstrate the utility of rAAV vectors *in vivo* was aimed at transduction of brain tissue in rats. Kaplitt et al. injected rAAV-CMV-LacZ vectors into various regions of adult rat brains and subsequently observed expression by histochemical staining (68). Beta-galactosidase activity was detected in a relatively high proportion of cells (estimated to approach 10%) in the regions surrounding the 2 microliter injections. Positive cells were still detected up to 3 months post injection. The authors subsequently injected a rAAV vector coding for tyrosine hydroxylase (TH), a gene which was potentially therapeutic for Parkinson's disease. Using an immunohistochemical assay for vector expression, they were able to show that TH was expressed primarily in neurons and a few glial cells. However, the number of positive cells diminished over the 2 months following injection. In spite of this, the vector gene expression did appear to result in behavioral improvements in a rat model for Parkinson's disease. A double blinded study in non-human primates has confirmed the results from the rat model. Histochemical, enzymatic, and behavioral changes were confirmed in monkeys that received rAAV carrying TH when compared to rAAV LacZ controls (During, PNAS, submitted).

This study demonstrated the ability of rAAV vectors to transduce non-dividing and highly differentiated cells. The use of local injection of the vector provided a distinct advantage in the subsequent analysis of transduction over protocols relying on systemic inoculation. This approach is well suited for rAAV vector research due to the limitations still encountered in growing large quantities of virus stocks.

A more systematic survey of rAAV transduction of various neuronal tissues was recently conducted by McCown et al. (37). Vectors expressing *lacZ* under the control of the CMV promoter were injected into either adult or neonatal rats. The authors noted differential expression of the vector depending on the site of injection and the age of the animal. While initial transduction rates were high in some tissues such as hippocampus, the duration of transgene expression was relatively short. Other tissues exhibited a more stable number of transduced cells whether they initially were transduced efficiently, as was the inferior colliculus, or not (olfactory tubercle). In all areas, large multipolar neurons were the predominant transduced cell type. However, *lacZ* positive astrocytes were occasionally identified. Whether vector expression in these cells was inefficient or vector virion was not taken up remains to be determined. In addition, whether the loss of vector expression was the result of degradation of the DNA template or inactivation of the promoter remains an open ended question. Importantly however, there was no indication that the transduced cells were being cleared via an immune response. In no instance was there any evidence of neurotoxicity associated with vector administration. While illustrating that rAAV does not behave uniformly in all regions of the brain, this research does support the potential for long term transduction of specific regions using currently available vectors.

Stable transduction of muscle tissue with a rAAV vector has been reported recently and the experiments have been carried beyond the one year mark (38). The vector was delivered by local injection to muscles in the hind leg of 3-week old mice. Mice were sacrificed at various times post-injection and muscle tissue was sectioned and stained for expression of the *LacZ* marker. The average number of positive staining muscle fibers was substantially greater than in mice treated with an Ad-*LacZ* vector and did not change in mice examined from four days post-injection to 19 months. Similar results were obtained with adult mice. This suggested that the majority of transduction events (detected on day four) went on to form stable integrants or episomes.

Southern blotting of total cellular DNA from the transduced muscle tissue at various time points allowed some analysis of the fate of the rAAV vector DNA. Uncut DNA yielded a high molecular weight (greater than 12 Kb) smear of *LacZ* hybridization signal. Digestion with a single-cut restriction enzyme yielded several products including: a monomeric vector length fragment, a diffuse background hybridization from approximately 3 to 12 Kb, and a fragment consistent with the product of a tail to tail concatemer of the vector. The momomeric digestion product suggested that the proviral DNA was composed of a tandem array of vector sequences. In contrast to previous observations (40) there was no evidence of dimer size molecules. The vector specific sequences were linked to high molecular weight DNA which may have been chromosomal or, alternatively, large episomes consisting of tandem arrays or concatemeric circles. The persistence of a background of hybridization after digestion with the single-cut enzyme is consistent with some or all of the vector specific sequences being linked to chromosomal DNA.

While a great many questions remain to be answered regarding the efficiency, stability, and the very nature of rAAV transduction *in vivo*, these experiments suggest that rAAV vectors may be well suited to use in non-dividing, fully differentiated cell types. In addition to neural and muscle tissues, rAAV vectors have recently been used to transduce retinal and cochlear cells (69, 70). Further, the tyrosine hydroxylase vector experiments outlined above for the treatment of Parkinson's disease are a compelling demonstration of the potential utility of expressing therapeutic genes in these tissues.

4.10 Enhancement of rAAV Transduction *in vivo*

As previously discussed, an important factor in the efficiency of transduction *in vitro* is the ability of the infecting rAAV genome to generate its complementary strand. It is reasonable to expect that a similar mechanism with similar rate limitations will exist *in vivo*. In two recent reports, investigators have taken the principles learned through enhancement of rAAV transduction *in vitro* and applied them to animal models. Alexander et al. used gamma radiation treatment to overcome the transduction barrier (71). They injected an alkaline phosphatase (AP) expressing vector into the brains of rats that had received localized cranial irradiation or into mock treated animals. The sites chosen for vector injection were the hippocampus, the ventricular system, and the scalp intradermis.

The gamma radiation treatment substantially increased the number of transduced cells observed following injection into the lateral ventricle. Non-irradiated animals had undetectable levels of AP expression while irradiated animals displayed numerous transduced cells in the pia-arachnoid and choroid epithelium. The pattern of transduced cells appeared to follow the path of cerebral spinal fluid flow through the side of the brain which was injected. This was in contrast to the observations of McCown et al. who noted little transduction of ventricular epithelium beyond the local injection site. However, this may be explained by the increase in sensitivity gained by the radiation mediated enhancement, or simply to differences in injection procedures. Transduction was not observed in the contralateral side of the brain, which was not injected with vector. The stability of vector expression was evaluated at three months post-injection. In most animals, barely detectable levels of AP activity revealed that the increase in the initial transduction did not necessarily lead to increases in the duration of transgene expression. This is consistent with data from our laboratory indicating that UV mediated enhancement of rAAV transduction does not lead to linear increases in stable transduction in HeLa cells (Ferrari and Samulski, manuscript in preparation).

As in the ventricular system, the gamma irradiation treatment substantially increased the number of transduced cells following intradermal scalp injection. Transduction was confined to the striated muscle cells at this site. The duration of transgene expression was not determined in this tissue.

In the hippocampus, AP expression was readily detectable 12 days post-injection, as had been noted previously (68). Surprisingly, the gamma irradiation treatment did not enhance the rate of transduction in this tissue. By six weeks post-injection, the level of expression had dropped 10-fold in some animals and had become undetectable in others. No differences were observed in the duration of expression between irradiated and non-irradiated animals. Again, it was not ascertained whether the loss of transgene expression was due to instability of the vector or a reduction in promoter activity.

The authors went on to show that a rAAV vector expressing a secreted form of AP could transduce epithelial cells of the ventricular system and that its gene product could be detected in the cerebral spinal fluid. While promising, it should be noted that the doses of radiation used in these experiments would preclude use in clinical applications.

The previously discussed report from Fisher et al., characterizing the effect of Ad E4 on rAAV transduction, also included two experiments *in vivo* (21). In the first, 10^{11} particles of rAAV-*lacZ* were used to transduce liver tissue in mice following infection with adenovirus. The pre-treatment with adenovirus, at a multiplicity sufficient to infect approximately 25% of the liver hepatocytes, allowed rAAV mediated transduction of 10 to 25% of the liver. Without Ad co-infection, or with mutants lacking efficient E4 expression, the transduction rate was less than 0.01%. This suggested that all of the hepatocytes had taken up the rAAV vector and only required a secondary event to trigger expression from the virion DNA. This was further demonstrated by showing that an rAd vector expressing alkaline phosphatase could be used to induce *lacZ* expression from rAAV and that the two markers co-localized in most transduced hepatocytes. Similar results were obtained by instillation of a mixture of rAAV and Ad in mouse lungs.

Although Ad co-infection does not constitute a workable gene therapy protocol, these experiments clearly show that rAAV vectors can be taken up and expressed by potentially useful cell types efficiently. Because a broad range of treatments can induce gene expression from these vectors *in vitro*, it is likely that suitable non-invasive and safe treatments will be found to produce this effect in animals and humans.

4.11 Conclusions

We are in the early stages of understanding and fully exploiting the potential of rAAV gene delivery vectors. We are still faced with some conflicting information as to the transduction and integration efficiencies of these vectors and the uncertainty spans a range between less than 1% and greater than 80%, depending, in part, upon the method of detection and the presence of the AAV *rep* gene. This situation largely reflects our limited understanding of the AAV integration mechanism either in the presence or absence of Rep protein. Therefore, two of the most pressing problems in the use of rAAV vectors will be to better define the mechanism of integration and to understand the nature and limitations of persistent expression in the absence of integration.

Related to these questions is the problem of increasing the efficiency of transduction using various enhancing agents, many of which have genotoxic activity. Although there are inherent risks in administering these treatments, the potential gain is great and the use of some of these agents has been warranted and amply justified in therapies for cancer. It remains to adapt these treatments, which have proven effective *in vitro* for increasing rAAV transduction rates greater than 100-fold, to acceptable treatments which can be applied to clinical situations.

While we probably have not yet achieved the greatest potential that rAAV has to offer toward transduction efficiency, many of the reports discussed above begin to reveal the value of these vectors even without further enhancement. These results hold great promise for long term gene delivery using rAAV *in vivo* and suggest that systematic testing of tissue should be carried out to determine the true range of rAAV vectors *in vivo* before efforts to enhance transduction are applied.

References

1. Laughlin, C. A., Tratschin, J.-D., Coon, H., and Carter, B. J. (1983). Cloning of infectious adeno-associated virus genomes in bacterial plasmids. Gene, *23*, 65-73.
2. Samulski, R. J., Berns, K. I., Tan, M., and Muzyczka, N. (1982). Cloning of adeno-associated virus into pBR322: rescue of intact virus from the recombinant plasmid in human cells. Proc Natl Acad Sci USA, *79*, 2077-2081.
3. Hermonat, P. L., Labow, M. A., Wright, R., Berns, K. I., and Muzyczka, N. (1984). Genetics of adeno-associated virus: isolation and preliminary characterization of adeno-associated virus type 2 mutants. J Virol, *51*, 329-333.
4. Tratschin, J.-D., Miller, I. L., and Carter, B. J. (1984). Genetic analysis of adeno-associated virus: properties of deletion mutants constructed *in vitro* and evidence for an adeno-associated virus replication function. J Virol, *51*, 611-619.
5. Tratschin, J.-D., Tal, J., and Carter, B. J. (1986). Negative and positive regulation in trans of gene expression from adeno-associated virus vectors in mammalian cells by a viral Rep gene product. Mol Cell Biol, *6*, 2884-2894.
6. Labow, M. A., Hermonat, P. L., and Berns, K. I. (1986). Positive and negative autoregulation or the adeno- associated virus type 2 genome. J Virol, *60*, 251-258.
7. Samulski, R. J., Chang, L.-S., and Shenk, T. (1989). Helper-free stocks of recombinant adeno-associated viruses: normal integration does not require viral gene expression. J Virol, *63*, 3822-3828.

8. Muzyczka, N. (1992). Use of adeno-associated virus as a general transduction vector for mammalian cells. Curr Top Microbiol Immunol, *158,* 97-129.
9. Holscher, C., et al. (1994). Cell lines inducibly expressing the adeno-associated virus (AAV) *rep* gene: requirements for productive replication of rep-negative AAV mutants. J Virol, *68(11)*, 7169-77.
10. Trempe, J. P. and Yang, Q. (1993). Characterization of a cell line that expresses the AAV replication proteins (abstract). In: Fifth Parvovirus Workshop. Crystal River, FL.
11. Clark, K. R., Voulgaropoulou, F., Fraley, D. M., and Johnson, P. R. (1995). Cell lines for the production of recombinant adeno-associated virus. Human Gene Therapy, *6(10)*, 1329-41.
12. Tamayose, T., Yukihiko, H., and Shimada, T. (1996). A New Strategy for Large-Scale Preparation of High-Titer Recombinant Adeno-Associated Virus Vectors by Using Packaging Cell Lines and Sulfonated Cellulose Column Chromatography. Human Gene Therapy, *7*, 507-513.
13. Flotte, T. R. et al. (1995). An improved system for packaging recombinant adeno-associated virus vectors capable of *in vivo* transduction. Gene Therapy, *2(1)*, 29-37.
14. Mamounas, M., Leavitt, M., Yu, M., and Wong-Staal, F. (1995). Increased titer of recombinant AAV vectors by gene transfer with adenovirus coupled to DNA-polylysine complexes. Gene Therapy, *2(6)*, 429-32.
15. Snyder, R. O., Xiao, X., and Samulski, R. J. (1996). Production of recombinant adeno-associated virus vectors. In: Current protocols in Human Genetics, N. Dracopoli et al. (eds.). John Wiley & Sons Ltd.: New York, 12.1.1-12.2.23.
16. Yakobson, B., Koch, T., and Winocour, E. (1987). Replication of adeno-associated virus in synchronized cells without the addition of a helper virus. J Virol, *61(4)*, 972-81.
17. McLaughlin, S. K., Collis, P., Hermonat, P.,L., and Muzyczka, N. (1988). Adeno-associated virus general transduction vectors: analysis of proviral structures. J Virol, *62(6)*, 1963-73.
18. de la Maza, L. M. and Carter, B. J. (1980). Heavy and light particles of adeno-associated virus. J Virol, *33,* 1129-1137.
19. Rose, J. A. and Koczot, F. J. (1972). Adenovirus-associated virus multiplication: VII. Helper requirement for viral deoxyribonucleic acid and ribonucleic acid synthesis. J Virol, *10(1)*, 1-8.
20. Ferrari, F. K., Samulski, T., Shenk, T., and Samulski, R. J. (1996). Second-Strand Synthesis is a Rate Limiting Step for Efficient Transduction by Recombinant Adeno-Associated Virus Vectors. Journal of Virology, *70(5)*.
21. Fisher, K. J. et al. (1996). Transduction with recombinant adeno-associated virus for gene therapy is limited by leading-strand synthesis. Journal of Virology, *70(1)*, 520-32.
22. Ohman, K., Nordqvist, K., and Akusjarvi, G. (1993). Two adenovirus proteins with redundant activities in virus growth facilitates tripartite leader mRNA accumulation. Virology, *194(1)*, 50-8.
23. Weiden, M. D. and Ginsberg, H. S. (1994). Deletion of the E4 region of the genome produces adenovirus DNA concatemers. Proceedings of the National Academy of Sciences of the United States of America, *91(1)*, 153-7.
24. McCarty, D. M., Christensen, M., and Muzyczka, N. (1991). Sequences required for the coordinate induction of the AAV p19 and p40 promoters by the Rep protein. J Virol, *65,* 2936-2945.
25. Beaton, A., Palumbo, P., and Berns, K. I. (1989). Expression from the adeno-associated virus p5 and pl9 promoters is negatively regulated in trans by the *rep* protein. J Virol, *63,* 4450-4454.
26. Dobner, T., Horikoshi, N., Rubenwolf, S., and Shenk, T. (1996). Blockage by Adenovirus E4orf6 of Transcriptional Activation by the p53 Tumor Suppressor. Science, *272,* 1470-1473.
27. Nordqvist, K., Ohman, K., and Akusjarvi, G. (1994). Human adenovirus encodes two proteins which have opposite effects on accumulation of alternatively spliced mRNAs. Molecular & Cellular Biology, *14(1)*, 437-45.
28. Samulski, R. J. and Shenk, T. (1988). Adenovirus ElB 55-M, polypeptide facilitates timely cytoplasmic accumulation of adeno-associated virus mRNAs. J Virol, *62,* 206-210.
29. Sarnow, P. et al. (1984). Adenovirus early region 1B 58,000-dalton tumor antigen is physically associated with an early region 4 25,000-dalton protein in productively infected cells. Journal of Virology, *49(3)*, 692-700.
30. Bridge, E., Medghalchi, S., Ubol, S., Leesong, M., and Ketner, G. (1993). Adenovirus early region 4 and viral DNA synthesis. Virology, *193(2)*, 794-801.
31. Russell, D. W., Alexander, I. A., and Miller, A. D. (1995). DNA synthesis and topoisomerase inhibitors increase transduction by adeno-associated virus vectors. Proc. Natl. Acad. Sci. USA, *92,* 5719-5723.

32. Yalkinoglu, A. O., Heilbronn, R., Burkle, A., Schlehofer, J. R., and zur-Hausen, H. (1988). DNA amplification of adeno-associated virus as a response to cellular genotoxic stress. Cancer Res, *48(11)*, 3123-9.
33. Yakobson, B., Hrynko, T. A., Peak, M. J., and Winocour, E. (1989). Replication of adeno-associated virus in cells irradiated with UV light at 254 nm. J Virol, 63*(3)*, 1023-30.
34. Yalkinoglu, A. O., Zentgraf, H., and Hubscher, U. (1991). Origin of adeno-associated virus DNA replication is a target of carcinogen-inducible DNA amplification. Journal of Virology, *65(6)*, 3175-84.
35. Russell, D. W., Miller, A. D., and Alexander, I. E. (1994). Adeno-associated virus vectors preferentially transduce cells in S phase. Proc Natl Acad Sci USA, *91(19)*, 8915-9.
36. Halbert, C. L., Alexander, I. E., Wolgamot, G. M., and Miller, A. D. (1995). Adeno-associated virus vectors transduce primary cells much less efficiently than immortalized cells. J Virol, *69(3)*, 1473-9.
37. McCown, T. J., Xiao, X., Li, J., Breeze, G. R., and Samulski, R. J. (1996). Differential and Persistent Expression Patterns of CNS Gene Transfer by an Adeno-Associeated Virus (AAV) Vector. Brain Research, *713*, 99-107.
38. Xiao, X., Li, J., and Samulski, R. J. (1996). Efficient Long-Term Gene Transfer into Muscle Tissue of Immunocompetent Mice by Adeno-Associated Virus Vector. J. Virol., *70(11)*, 8098-8108.
39. Bertran, J. et al. (1996). Recombinant Adeno-Associated Virus-Mediated High-Efficiency, Transient Expression of the Murine Cationic Amino Acid Transporter (Ecotropic Retroviral Receptor) Permits Stable Transduction of Human HeLa Cells by Ecotropic Retrovirual Vectors. J. Virol., *70(10)*, 6759-6766.
40. Flotte, T. R., Afione, S. A., and Zeitlin, P. L. (1994). Adeno-associated virus vector gene expression occurs in nondividing cells in the absence of vector DNA integration. Am J Respir Cell Mol Biol, *11(5)*, 517-21.
41. Goodman, S. et al. (1994). Recombinant adeno-associated virus-mediated gene transfer into hematopoietic progenitor cells. Blood, *84(5)*, 1492-500.
42. Kearns, W. G. et al. (1996). Recombinant adeno-associated virus (AAV-CFTR) vectors do not integrate in a site-specific fashion in an immortalized epithelial cell line. Gene Therapy, *3*, 748-755.
43. Walsh, C. E. et al. (1992). Regulated high level expression of a human gamma-globin gene introduced into erythroid cells by an adeno-associated virus vector. Proc Natl Acad Sci USA, *89(15)*, 7257-61.
44. Weitzman, M. D., Kyostio, S. R., Kotin, R. M., and Owens, R. A. (1994). Adeno-associated virus (AAV) Rep proteins mediate complex formation between AAV DNA and its integration site in human DNA. Proc Natl Acad Sci U S A, *91(13)*, 5808-12.
45. Kotin, R. M., Linden, R. M., and Berns, K. I. (1992). Characterization of a preferred site on human chromosome 19q for integration of adeno-associated virus DNA by non-homologous recombination. Embo J, *11(13)*, 5071-8.
46. McCarty, D. M. et al. (1994). Identification of linear DNA sequences that specifically bind the adeno-associated virus Rep protein. J Virol, *68(8)*, 4988-97.
47. Snyder, R. O. et al. (1993). Features of the adeno-associated virus origin involved in substrate recognition by the viral Rep protein. J Virol, *67(10)*, 6096-104.
48. Muzyczka, N. (1991). In vitro replication of adeno-associated virus DNA. Seminars in VIROLOGY, 2, 281-290.
49. Lusby, E., Fife, K. H., and Berns, K. I. (1980). Nucleotide sequence of the inverted terminal repetition in adeno-associated virus DNA. J Virol, *34*, 402-409.
50. Hauswirth, W. W. and Berns, K. I. (1979). Adeno-associated virus DNA replication. Virology, *93*, 57-68.
51. Hauswirth, W. W. and Berns, K. I. (1977). Origin and termination of adeno-associated virus DNA replication. Virology, *78*, 488-499.
52. Straus, S. E., Sebring, E. D., and Rose, J. A. (1976). Concatemers of alternating plus and minus strands are intermediates in adenovirus-associated virus DNA synthesis. Proceedings of the National Academy of Sciences USA, *73*, 742-746.
53. Ward, P. and Berns, K. I. (1995). Minimum origin requirements for linear duplex AAV DNA replication *in vitro*. Virology, *209(2)*, 692-5.
54. Ni, T. H., Zhou, X., McCarty, D. M., Zolotukhin, I., and Muzyczka, N. (1994). In vitro replication of adeno-associated virus DNA. J Virol, *68(2)*, 1128-38.

55. Koczot, F. J., Carter, B. J., Garon, C. F., and Rose, J. A. (1973). Self-complementarity of terminal sequences within plus or minus strands of adenovirus-associated virus DNA. Proceedings of the National Academy of Sciences of the United States of America, *70(1)*, 215-9.
56. Berns, K. I. and Kelly, T. J. J. (1974). Visualization of the inverted terminal repetition in adeno-associated virus DNA. J Mol Biol, *82,* 267-271.
57. Berns, K. I., Pinkerton, T. C., Thomas, G. F., and Hoggan, M. D. (1975). Detection of adeno-associated virus (AAV)-specific nucleotide sequences in DNA isolated from latently infected Detroit 6 cells. Virology, *68,* 556-560.
58. Laughlin, C. A., Cardellichio, C. B., and Coon, H. C. (1986). Latent infection of KB cells with adeno-associated virus type 2. J Virol, *60(2)*, 515-24.
59. Handa, H., Shiroki, K., and Shimojo, H. (1977). Establishment and characterization of KB cell lines latently infected with adeno-associated virus type 1. Virology, *82,* 84-92.
60. Hermonat, P. L. and Muzyczka, N. (1984). Use of adeno-associated virus as a mammalian DNA cloning vector: transduction of neomycin resistance into mammalian tissue culture cells. Proc Natl Acad Sci USA, *81,* 6466-6470.
61. Tratschin, J. D., Miller, I. L., Smith, M. G., and Carter, B. J. (1985). Adeno-associated virus vector for high-frequency integration, expression, and rescue of genes in mammalian cells. Mol Cell Biol, *5(11)*, 3251-60.
62. Urcelay E., Ward, P., Wiener, S. M., Safer, B., and Kotin, R. M. (1995). Asymmetric replication *in vitro* from a human sequence element is dependent on adeno-associated virus Rep protein. Journal of Virology, *69(4)*, 2038-46.
63. Giraud, C., Winocour, E., and Berns, K. I. (1994). Site-specific integration by adeno-associated virus is directed by a cellular DNA sequence. Proc Natl Acad Sci USA, *91(21)*, 10039-43.
64. Kotin, R. M. and Berns, K. I. (1989). Organization of adeno-associated virus DNA in latently infected Detroit 6 cells. Virology, *170,* 460-467.
65. Linden, R. M., Winocour, E., and Berns, K. I. (1996). The recombination signals for adeno-associated virus site-specific integration. P.N.A.S, *93,* 7966-7972.
66. Prasad, K. M. and Trempe, J. P. (1995). The adeno-associated virus Rep78 protein is covalently linked to viral DNA in a preformed virion. Virology, *214(2)*, 360-70.
67. McCarty, D. M., Ryan, J. H., Zolotukhin, S., Zhou, X., and Muzyczka, N. (1994). Interaction of the adeno-associated virus Rep protein with a sequence within the A palindrome of the viral terminal repeat. J Virol, *68(8)*, 4998-5006.
68. Kaplitt, M. G. et al. (1994). Long-term gene expression and phenotypic correction using adeno-associated virus vectors in the mammalian brain. Nature Genetics, *8*, 148-154.
69. Zolotukhin S., Potter, M., Hauswirth, W. W., Guy, J., and Muzyczka, N. (1996). A "humanized" green fluorescent protein cDNA adapted for high-level expression in mammalian cells. Journal of Virology, *70(7)*, 4646-54.
70. Lalwani, A. K., Walsh, B. J., Reilly, P. G., Muzyczka, N., and Mhatre, A. N. (1996). Development of *in vivo* gene therapy for hearing disorders: introduction of adeno-associated virus into the cochlea of the guinea pig. Gene Therapy, *3*, 588-592.
71. Alexander, I. E., Russel, D. W., Spence, A. M., and Miller, A. D. (1996). Effects of Gamma Irradiation on the Transduction of Dividing and Nondividing Cells in Brain and Muscle of Rats by Adeno-Associated Virus Vectors. Human Gene Therapy, *7*, 841-850.

5. Virus Functions for Artificial Vectors

Volker Sandig and Michael Strauss

5.1 Introduction

Gene transfer into mammalian cells by physical, chemical or biological methods has a history of almost two decades. The possibility of manipulating certain functions of a cell or of the whole organism at a molecular level by expression of a newly introduced gene was the basis for rapid development of a new strategy for treatment of diseases- Gene therapy. The essence of gene therapy is to deliver genetic information to a specific target tissue by an efficient, safe and nontoxic way. Although over hundred clinical protocols in gene therapy have already been approved and carried out, the technology is still in its infancy. The development of suitable vectors for *in vivo* gene transfer remains the most challenging task. In contrast to classical pharmacotherapy, very large molecules have to be delivered to the majority of cells of a certain tissue. Since even a single DNA molecule of useful size (10Kb) in a soluble form has an extension of 200nm, diffusion rate and mechanical access are barriers against effectiveness. Viruses are natural vehicles for gene transfer into mammalian cells. They have evolved specific mechanisms for efficient condensation and packaging of their genomes as well as for cell attachment, penetration, maintenance and replication. Therefore, viruses play a major role in the first developmental phase of gene therapy. However, viral life cycles are mostly incompatible with the aim of gene therapy. Extensive modifications of viral structures and functions would be required. For several applications, it would be desirable to combine features of different viruses.

This chapter will discuss viral features with respect to their benefit for current and future vectors for gene therapy.

5.2 Currently Used Viral Vectors

5.2.1 Retroviruses

This vector type has undergone a period of more than 10 years of development leading to well established strategies for vector design and production. Therefore retroviruses have been chosen as vectors for most of the recent clinical trials for *ex vivo* gene therapy (for review see Vile and Russell, 1995).

The vector essentially contains the gene of interest in a viral capsule. Viral coding sequences have almost completely been removed from the vector. Only viral LTRs and the packaging signal are required in *cis* for genome synthesis, encapsidation, reverse transcription and integration are flanking the foreign gene. The viral proteins *gag, pol* and *env* are derived from C-type oncoviruses and are provided in *trans* by appropriate helper cell lines. Minimal sequence overlap between vector and helper sequences and separation of the viral genes within the helper cell line reduce the chance for generation of replication competent viruses.

The most important feature of retroviruses is integration of the provirus into the host genome resulting in transmission of the transferred gene to all the progeny of an infected cell. The advantage of integration has to be paid by the possibility of insertional mutagenesis. However, careful experiments and theoretical models have shown that this risk is rather low provided no replication competent virus is present in the vector preparation. Further development of retroviral vectors is aiming at stabilisation against concentration procedures (Miyanohara et al., 1994), altered host spectrum, and infection of resting cells. Other subfamilies (Lentivirinae, Spumavirinae) of retroviruses are studied extensively for this reason (Naldini et al., 1996; Russell and Miller, 1996).

5.2.2 Adenoviruses

Adenoviruses are becoming increasingly important as vectors for *in vivo* gene therapy. The reason for this is that adenoviruses have reproducibly demonstrated the highest efficiency of gene transfer. The first generation of recombinant viruses most widely used today in preclinical studies contains 90 % of the 36kb genome. Deletion of the early region E1 which contains the main transactivator prevents productive infection and allows, together with a deletion in the early region E3, for insertion of foreign sequences of up to 8 kb (Bett et al., 1994). Whereas E3 is dispensable for virus production, E1gene functions have to be provided by an appropriate helper cell line. However, even in the absence of the transactivator E1 some replication at high multiplicity of infection can take place and genes of the E2, E4 and late regions are expressed. As a result, a strong cellular and humoral immune responses are induced against the virus infected cells leading to expression shut-off or destruction of these cells (Yang et al., 1994). Additional deletion of viral genes is rather complicated because of large overlapping open reading frames distributed on both strands of the genome, extensive splicing and the toxicity of many viral gene products when expressed constitutively in helper cells. Nevertheless, new vectors lacking the E4 region or the E2A region have been designed (Armentano et al., 1995; Yeh et al.,1996; Gorziglia et al., 1996). Whether these viruses indeed allow for prolonged gene expression *in vivo* remains to be demonstrated. The design of vectors containing only the viral terminal repeats, the packaging signal and the gene of interest but missing all viral genes is also feasible (Kochanek et al., 1995; Fisher et al., 1996). Propagation of these viruses requires support by a helper virus which has to be physically separated from the therapeutic virus or disabled for packaging (Kochanek et al., 1996; our unpublished results).

5.2.3 Adeno-Associated Virus

Advantages of AAV as a gene transfer vector are lack of pathogenicity and tumorogenicity, high frequency of integration into the host genome and their broad host range (Rolling et al.,, 1995). For replication, the virus requires coinfection with a helper virus. Although herpes virus and SV40 can also fulfil the helper function, adenoviruses are best suited. They provide activators of gene expression (E1, E4 regions), the E2A ssDNA-binding protein, the VA RNAs and the general environment for virus replication. The structural proteins and the 4 rep proteins also required for replication are encoded by AAV itself. For insertion of foreign genes parts or all of the AAV genes have to be deleted. A minimal AAV vector contains only a packaging signal and the terminal repeats and allows for insertion of up to 4.7kb. The AAV genes are provided by transient plasmid transfection, via a recombinant adenovirus (Thraser et al., 1995) or by a stably transfected cell line. The latter is difficult to generate since rep proteins suppress the growth of immortalised cell lines (Flotte et al., 1995a; Clark et al., 1995). Whereas wild-type AAV integrates specifically into human chromosome 19 (Kotin et al., 1994), recombimnant viruses are missing this important feature (Flotte at al., 1995b). Minimal AAV vectors do not produce viral proteins in the target cell and superinfection of the cells with a second vector seems to be possible. If the production of high titer stocks containing more than $1x10^9$ infectious particles/ml be simplified and cross contamination with wild type virus can be avoided, these vectors have good chances to be used for stable *in vivo* gene transfer.

5.2.4 Herpes Viruses

This virus family is of increasing interest as a basis of gene therapy vectors because they naturally establish themselves in the latent state in the infected cells without integrating into the host genome. Whereas herpes simplex virus which is the origin of most advanced herpes-based vectors is maintained only in quiescent neuronal cells, Epstein-Barr virus has the potential for gene delivery to stem cells and their differentiated progeny. Importantly, latent HSV infection does not alter the cell metabolism. Moreover, the virus carries a promoter system uniquely capable of escaping repression that shuts off the expression of HSV lytic genes. Three alternative strategies for development of herpes viral vectors are imaginable: (i) Even in replication competent vectors, 50% of the 150kb genome is dispensible leaving enough space for foreign sequences in the 150kb genome. Deletion of individual genes can substantially attenuate the virus and reduce the neurovirulence typical for HSV (Fink et al., 1992). However, recombination with viruses present in the patient could reverse attenuation. (ii) Replication-defective vectors typically lack one essential immediate early regulatory gene e.g. VP16, the viral transactivator. These viruses cannot be reactivated but the latency system is preserved (Glorioso et al., 1995). (iii) Amplicon vectors contain only signals required in *cis* for replication and packaging. Based on the rolling circle mechanism for HSV replication which is supported by an helper virus vector concatamers are packaged and delivered to the target cell (Späte and Frenkel, 1985). The benefit of such amplicons is reduced by the presence of helper particles in the vector preparation (Frenkel et al., 1994). HSV would be the ultimate vector for gene transfer to the CNS if the strong primary inflammatory response to the introduced virus could be eliminated.

5.3 Viral Mechanisms for Cellular Attachment

Attachment to cellular receptors is the first step for a viral particle as a stable chemical agent to enter a life cycle. Effective cell binding which is followed by penetration and uncoating to a large extent determines the effectivity of virus reproduction. Evolution has refined this process.

There is broad diversity in strategies used for cell binding. Almost any cell surface molecule, regardless of its structure and function, may facilitate virus entry. Simple and abundant structures like heparan sulfate (HSV, CMV), lipids or sialic acid on glycoproteins (influenza virus, polyoma virus, Sendai virus) can serve as cellular attachment devices as well as single membrane proteins such as integrins (picornaviruses), signalling receptors (vaccinia) and immunoglobuline like molecules (HIV, SFV). For review see (Patterson and Oxford, 1986).

Most of these receptors are broadly distributed among cell types. Thus, preferential infection of one tissue is primarily the result of entry into or distribution throughout the organism and intracellular restrictions then that of exclusive receptor expression. Therefore, the use of viral capsids alone to determining cell type restriction is rather limited.

Moreover, cell tropism is often modified and infectivity is increased by binding of soluble factors which can themselves interact with cell surfaces. Enhancement by virus specific antibodies in combination with or without complement has been described for hantavirus (Jao et al., 1992), influenca virus (Ochiai et al., 1992) and HIV (Joault et al., 1989; Lund et al., 1995). Polymerised albumin and IL6 stimulate cell attachment of Hepatitis B virus (Colucci et al., 1987).This modification does not necessarily abolish cell type specificity. HIV and HBV show a strong preference for CD4+ T-cells or hepatocytes respectively and the envelopes could be used to direct a vector to these cells (Hofmann et al., 1995).

Different members of one virus family often exploit completely different receptors for cell entry. In the case of enveloped viruses, the surface protein of one virus can easily be substituted by that of another virus thereby enhancing virus stability or changing cell tropism (Gall et al., 1996; Krasnykh et al., 1996; Kiem et al., 1995). One of the striking features of viruses is the presence of multiple receptor ligands on a single virus particle based on capsomere structure of the virus coat (Wickham et al., 1990). Provided the receptor concentration and density is adequate, multiple interactions lead to enhanced binding, support the concentration of coated pits or destabilises the cellular membrane. By this mechanism, cellular entry of virus particles is facilitated. Approaches for receptor-mediated gene delivery try to mimic this viral feature: Receptor ligands are chemically linked to a polycation such as polylysine. After complexation with the DNA to be delivered, an artificial viral core is formed surrounded by binding sites. In this respect virally explored cell surface molecules such as the integrins (Heart et al., 1995), CD4 or the EGF receptor (Christiano et al., 1996; Shimizu et al., 1996) are equally attractive targets as other widespread receptors like those for transferin (Wagner et al., 1990), surfactant protein A (Ross et al., 1996), asialoglycoprotein (Findeis et al., 1994) and insulin (Huckett et al., 1990).

Receptors which recycle between the cell surface and the endosome compartment seem to be advantageous for uptake of the particle. However these artificial systems still lack behind the efficiency of viral infection. Complexes exceeding the size of natural viruses considerably have to be taken up by the phagocytotic pathway which is well developed only in specialised phagocytes. In contrast, pinocytosis of fluid and small particles takes place in all nucleated cells at high speed.

A real breakthrough in DNA uptake was achieved using new protocols that are supposed to allow for formation of very small complexes with a size of about 10nm (Perales et.al., 1994; Findeis et al., 1994). Using such complexes, reasonable *in vivo* gene transfer to the liver seems to be feasible (see also Wu and Wu; Chapter 6).

This gene targeting technique will undergo a rapid development in the near future directed towards search for alternative DNA complexing agents which would also facilitate intracellular transport such as nonlinear cascade polymers (Kukowska-Latallo et al., 1996) and small peptide or monosacharide ligands instead of large proteins. This would further reduce immunogencity of the vector particles and facilitate repeated application of the vector. (Erbacher et al., 1995; Remy et al., 1995)

5.4 Viral Entry into the Cell

After binding to its receptor, the virus enters the cell. To this end, many viruses explore physiological host cell functions such as receptor-mediated endocytosis. However, after inclusion into the endosome the viral nucleic acid is still separated from the cytoplasm by one (nonenveloped viruses) or even two membranes (enveloped viruses). Clever mechanisms for endosomal release protect the virus from lysosomal destruction. Since the therapeutic DNA of "artificial viruses" enters the cell via the same pathway, these mechanisms are of substantial importance for the effectivity of gene transfer.

For interaction with the endosomal membrane, hydrophobic domains have to be exposed, triggered by conformational changes which are induced through receptor binding, cleavage by host proteases or the low pH in the endosomal compartment. The adenovirus capsid mediates rupture of the endosome when the inner pH is decreased releasing even cointernalised material into the cytoplasm (Greber et al., 1993). Polio- and rhinoviruses deliver their RNA into the cytoplasm through a tight pore (Prcla et al., 1995). Pore formation via fusion of the viral and endosomal membranes has also been described for enveloped viruses, e.g. Semliki forest virus, vesicular stomatitis virus and influenza virus (Patterson and Oxford, 1986)

The adenoviral mechanism seemed to be well suited for application in receptor-mediated gene transfer, because a size limit for the released material does not exist (Prcla et al., 1995). One only has to make sure, that the DNA complex reaches the endosome together with the virus. Indeed, adenovirus particles linked to a transferrin-polylysine DNA complex by a streptavidin-biotin bridge increased gene transfer efficiency by 2-3 orders of magnitude (Wagner et al., 1992) Similar strategies have been described by others (Christiano et al.,1993; Fisher and Wilson, 1994) Since viral transcription is not required and even disadvantageous for gene transfer empty capsids were tested in the same way (Fisher and Wilson, 1994). These attempts were not successful because incompletely processed viral proteins failed to induce membrane rupture. However, inactivation of the viral DNA by UV or psoralen treatment did not influence transfer efficiency (Cotten et al., 1992). The complex vector is able to transduce almost 100% of the cells in vitro and has therapeutic potential e.g. for *ex vivo* transduction of cancer cells (Zatloukal et al., 1995) Despite the revolutionary results in vitro this vector type is not effective *in vivo*. Major drawbacks are the large size and the addition of immunogenic components to the otherwise immunologically inert vector.

Surprisingly, small endosome disruptive amphipatic peptides derived from influenza hamagglutinin HA2 or the N-terminus from rhinovirus VP1 can substitute for adenovirus in transfection complexes (Plank et al., 1994; Zauner et al., 1995). In the viral context, the influenza hamagglutinin peptide is activated in a multistep process including receptor binding as well as proteolysis and exposure to low pH. Moreover, for its activity the peptide must be orientated at a specific angle to the membrane. When the chronological order of these steps is interrupted in an experimental setting, the virus is denatured (White et al., 1994). Whereas the rhinovirus pore normally restricts the size of the released molecule to 10kD, a separate short fusion peptide from hemagglutinin or rhinovirus surface protein mediates the release of larger components. It destabilises the membrane by formings a helices with hydrophobic amino acids directed to one surface of the helix and several acidic amino acids mediating pH sensitivity. However, multiple interactions with the membrane by well positioned peptides are required for effective membrane rupture This may be achieved by introducing the peptides into self-assembling structures or rigid polymer structures (Cotten and Wagner, 1993; Haensler and Szoka, 1993) To our knowledge, peptides for endosomal breakage have not yet been applied to *in vivo* gene transfer. A better understanding of virus entry itself and efforts in protein design are required to generate DNA-peptide complexes which exhibit extreme surface charges and exert their fusion potential only inside the target cell.

Another principle of cell entry may also be of interest for gene transfer: Several enveloped (Sendai virus, Rous sarcoma virus, HIV, Hepatitis A) and nonenveloped (SV40, Polyoma) viruses exploit an pH-independent mechanism for cell entry. Membrane fusion /penetration of these viruses can take place intracellulary as well as at the plasma membrane. It is unclear whether the process must be activated by specific signals. Sendai virus (the hemagglutinating virus of Japan, HVJ) can even fuse efficiently with liposomes within a wide range of pH values without the requirement of proteins in the membrane. This feature was adopted to create fusion-active liposomes loaded with DNA. Neutral liposomes containing DNA and DNA-binding proteins are produced by reverse phase evaporation and fused with previously inactivated Sendai virus. These HVJ liposomes liberate their content directly into the cytoplasm and the chromosomal protein HMG1 help the DNA to reach the nucleus. The HVJ-liposome complex is one of the few vectors described so far which mediates efficient *in vivo* gene transfer. Targets are the liver (Kato et al., 1991), the vascular wall (von der Leyen et al., 1995), the kidney (Tomita et al., 1992) and the brain (Kato et al., 1994). Obviously, there is no tissue specificity of gene delivery. Unless the liposomes are infused into specific vessels e.g. the portal vein, the hepatic artery, the coronary artery or a segment of the aorta via a balloon catheter most of the material ends up in the reticulo-endotelial system. Since Sendai virus is able to induce syncitia formation, one would expect this unwanted effect to occur the artificial vector too. However, adverse side effects of gene transfer have not been found and the HVJ liposomes are claimed to be safe since no HVJ associated pathogenicity is known in humans. Nevertheless, it would be desirable to reduce the viral part of the vector only to a single protein or peptide.

Cell entry appears to be the weakest point on the way towards foreign gene expression, not only for artificial vectors but also for many viruses (Marsh and Helenius 1989). New findings in this field of virus research will have a strong impact on the future progress in the gene therapy technology.

5.5 Transport to the Nucleus

Once a DNA virus is cleared of the cellular membrane system and localises free in the cytoplasma it moves directly to the nucleus where its genome gets activated unless it codes for its own transcription machinery (pox viruses). The same is basically true for DNA-based artificial vectors.

Movement across the cytoplasm does not pose a serious problem. The involvement of mictrotubules in directed transport has been reported for several groups of viruses (Miles et al., 1980; Charlton and Volkman, 1993). The nuclear membrane functions as a real barrier separating the nuclear compartment from the cytoplasm. Free passage through the nuclear pores is only allowed for proteins with a molecular weight of less than 20-40kD whereas selective transport of larger proteins requires specific nuclear localisation signals (NLS) (for review see Boulikas, T., 1993). The transport of nucleic acids through the pore (transport of mRNA into the cytoplasm and that of snRNA in both directions) is facilitated by the association with proteins (Fischer et al., 1993 and 1994).

Viruses have adopted to the cells policy and exhibit nuclear localisation signals on proteins required inside the nucleus. Moreover, several viral proteins such as the terminal protein and the core protein V from adenovirus type 5, the HBV core protein and the influenza nucleoprotein which carry a NLS seem to mediate passage of the viral genome through the nuclear pore (Zhao and Padmanabhan, 1988; Eckhardt et al., 1991; O'Neil, R. E. et al., 1995). Onco-retroviruses devoid of such signals require mitosis for genome entry into the host nucleus. Infection by these viruses, therefore, is limited to dividing cells. This is the main disadvantage of the currently used retroviral vectors because cell division is an rare event in most of the target tissues for gene therapy (liver, brain and muscle). In contrast, HIV and other members of the lentivirus family are able to infect stationary terminally differentiated cells. In the case of HIV, three proteins promote nuclear entry of the preintegration complex. The Matrix protein and the integrase contain typical NLS and can interact with caryopherins whereas vpr acts via an alternative pathway. (Bugrinsky et al., 1992; Heinzinger et al., 1994; Gallay et al., 1996). This feature makes lentiviruses extremely attractive for vector design. Very recently hybrid vectors have been designed containing the HIV core proteins including MA and Vpr and the surface protein from Murine leucemia virus or vesicular stomatitis virus which are less discriminating with regard to the cell type they can infect (Naldini et al., 1996). Effective and stable gene transfer to nondividing adult rat neurons has been achieved with this new vector. Once a packaging system allowing for higher titers is constructed, the vector would be extremely useful also for gene transfer to resting totipotent hematopoetic stem cells.

In nonviral vectors, nuclear proteins with strong DNA binding activity can take over the function of the virus core. The chromosomal protein HMG 1 is able to condense and bend DNA, thereby forming particles of about 50 nm which are by themselves able to transfect cells (Böttger et al., 1988, our observations). Since HMG1 is normally located between nucleosomes in actively transcribed regions of the genome, it may not only facilitate nuclear entry but also transiently stabilise the transferred DNA inside the nucleus and enhance its access to transcription factors (Bianchi et al., 1989; Travers et al., 1994; Varga-Weisz et al., 1994). Unfortunately, the HMG1-DNA particles dissociate under conditions leading to dilution. Therefore it would be wise to wrap the complex with a membrane. This has been done by HJV liposomes. In this context, HMG1 as well as histones and nucleoplasmin enhanced transcription of the transferred gene by 5-10fold (Kaneda et al., 1995).

Taking into account all the different functions required to transfer the foreign gene to its destination a modular system should be designed. However, even a single molecule may fulfil all requirements. The polyoma virus major capsid protein associates into particle-like structures, which bind and incorporate DNA independently from specific packaging signals thereby protecting DNA from degradation, they bind the receptor, cause penetration of the cell membrane and mediate entry into the nucleus by a strong NLS. When short genes are encapsidated in these particles by a diffusion procedure and the particles are incubated with rodent or human cells, relatively efficient transient or stable gene transfer is feasible (Forstova et al., 1995). The use of polyoma capsids is restricted by the size of DNA which can be packaged (2kb). More effective condensation of the DNA may increase this limit towards the size of the virus genome (5kb) which would be sufficient for some applications.

Basic physiological processes such as endocytosis and intracellular traffic, the mechanisms underlying some of the first steps in viral infection, are highly conserved among eukaryotic cells. For instance, nuclear localisation signals which are functional in mammalian cells are also able to direct a protein to an insect cell nucleus (Forstova et al., 1993). In contrast, transcriptional control elements are often restricted to a single cell type of a multicellular organism and largely differ between species. This phenomenon may have an important implication: Viruses from evolutionary distant hosts could provide a save capsule to transfer a foreign gene into human cells, embedded into a completely silent genome. The gene of interest would only be active inside the target cell due to the specificity of the promoter. This is of considerable advantage, because viral gene expression is the main factor in induction of a CTL response against the transfected cell.

We have constructed recombinant baculoviruses containing reporter genes driven by the CMV or RSV promoters (Hofmann et al., 1995). Surprisingly, these vectors not only enter mammalian cells efficiently, they also direct strong expression of a reporter gene. A clear-cut cell type specificity could be observed. Expression levels in human hepatoma cell lines and primary human hepatocytes exceed those obtained in other cells by more than one order of magnitude and transduction efficiency in hepatocytes can reach 100%. Some aspects of the infection mechanism in insect cells are known. Cell entry occurs via adsorptive endocytosis, the major envelope protein, gp64, mediates pH-dependent fusion of the endosomal and viral membranes (Volkmann et al., 1986) and the nucleocapsid transport to the cell nucleus is associated with actin cable formation (Charlton et al., 1993). Since weak bases such as chloroquine which interfere with pH decrease in endosomal maturation also inhibit baculovirus infection of hepatocytes, similar pathways are likely to be used in the two different cellular systems.

5.6 Stability of Transferred Genes

For most applications, stable or at least long term persistence of the introduced gene is the ultimate goal. One would expect that naked DNA which is not organized in nucleosomes, not embedded within the nuclear matrix and physically separated from chromosomes will be rapidly lost. This DNA should be unable to replicate along with chromosomal DNA during S phase and to segregate during mitosis due to the lack of a centromer. Even in nondividing cells, the foreign DNA may be recognized by the cellular repair machinery and, to a certain degree subjected to degradation. Nevertheless, expression of a nonintegrated gene for even one month or longer has been found after gene transfer to the liver by polylysine complexes (Wu et al., 1995). Condensation by histone and nonhistone proteins such as HMG1 may be one way to protect the foreign DNA.

Whereas viruses undergoing only a lytic cycle of replication normally lack mechanisms which stabilise the genome, viruses causing latent infection provide several alternative solutions to this problem: (i) integration into host chromosomes, (ii) extrachromosomal stabilisation in nondividing cells without replication and (iii) establishment of an extrachromosomal replicon.

5.6.1 Integration into the Host Genome

Chromosomal localisation is the most secure way to distribute the transferred gene to 100% of the daughter cells in proliferating cell populations. The interest in both retroviruses and adeno-accociated viruses as vectors pays tribute to this feature. Considerable progress has been made in the understanding of the underlying mechanisms. In the case of retroviruses, integration involves a set of co-ordinated cleavage and joining reactions carried out by a single viral protein, the integrase. It is encoded by the gag-pol gene and separated by selective proteolysis from the reverse transcriptase. The ends of the linear proviral DNA produced by reverse transcription are modified resulting in 3'recessed ends, strand brakes are introduced into the target DNA, the ends are joined and the single stranded gaps are repaired. Integration takes place at random sites preferentially in actively transcribed DNA regions (Shih, C. et al., 1988; Mooslehner, K. et al., 1990) This process is highly effective because the viral life cycle completely depends upon integration. Purified viral integrase as well as the recombinant protein expressed in bacteria or insect cells can perform the multistep reaction in vitro (Bushman, F. D. et al., 1990; Vincent, K. A. et al., 1993). Moreover, a linearized plasmid carrying U3 and U5 regions of the virus at both ends can be used instead of the authentical donor DNA. The integration event appears to be most efficient if the ends are precleaved by the restriction enzyme NdeI that already produces the 3'recessed ends, normally generated by the integrase (Katz et al., 1990). Preferential integration at a preselected site could be achieved by fusing the integrase to a sequence specific DNA binding protein (Bushman, 1994; Goulaouic et al., 1996). Since binding of the integrase protein to its substrate is extremely strong (Ellison and Brown, 1994) it seems possible that such a complex could be delivered by a liposome based vector. This would eliminate restrictions in the size of DNA to be integrated and would make the use of retroviral packaging cell lines unnecessary. To our knowledge, such systems have not been described so far. However, attempts have been made to use the bacterial recombinase recA for homologous integration (Zarling, D., pers. communication). Mammalian recombinases which have been isolated recently could be more useful in this respect. They could provide a chance to increase the recombiantion rate and make a gene replacement strategy feasible.

Replication and integration processes of adeno-associated virus are much less understood. The site specificity of integration (chromosome 19q13.3) (Kotin et al., 1992) observed for wild type AAV was originally considered as a big advantage because it would reduce the chance of detrimental insertional mutagenesis. However, recombinant viruses, especially those lacking the rep proteins, appear to integrate at random (Walsh et al., 1993) The rep68 and rep78 proteins provided within the viral capsid were suggested to be obligatory for the AAV integration. They bind to the terminal repeats and the chromosomal target site and provide helicase and endonuclease activities (Chiorini et al., 1994). Unfortunately, adenovirus exhibits its helper function not only in the cell line where the AAV vector is produced, but also in the target cell. Some of the gene products (E1, E4) enhance integration drastically, inducing the appearance of a double-stranded replicative form of the vector (Fisher et al., 1996). The most important feature of AAV vectors is their ability to transduce quiescent cells. The advantage of integration may also be reduced by genetic instability of the integrated vector (Giraud, C. et al., 1994).

5.6.2 Extrachromosomal Persistence without Replication

Another way to achieve long term persistence of a viral genome is its association with proteins in nondividing cells. The latent state of herpes simplex virus does not require viral protein synthesis. Neither integration nor replication of the circular double-stranded genome occur and the latency related transcript is the only viral RNA found. Nevertheless, reporter genes can be expressed from herpes simplex virus vectors for several months in the brain and the liver (Bloom et al., 1995; Lu, B. et al., 1995).

For adenoviruses a typical latent state has not been described. However, infected B lymphocytes can carry the virus and survive over longer periods due to extremely low expression of the E1 genes. The behaviour of recombinant viruses in target cells very much resembles that state. In those viruses, the E1 region has been replaced by foreign genes preventing lytic infection outside of the appropriate helper cell line. Infection takes place and the virus genome persists extrachromosomally as an linear molecule without or with low level replication. The ends of the DNA are protected by covalent linkage of the terminal protein (TP). Moreover, its unprocessed form, the preterminal protein, attaches the genome to the chromosomal matrix (Fredman et al.,

1993). Since processing of the terminal protein occur before virus assembly, at least a first round of replication seems necessary for stabilisation.

If immunological extinction of the infected cells can be prevented, long-lasting expression from adenoviral vectors could be feasible in resting cells. This stabilisation strategy is not generally applicable.

5.6.3 Extrachromosomal Replication

Extrachromosomal stability requires controlled DNA replication, even in slowly proliferating cells. The best studied example for this pathway is latent infection of human B cells in vitro with Epstein-Barr virus. These cells become immortalised lymphoblasts carrying the 173kb EBV genome as a supercoiled replicating plasmid. Whereas as many as 11 viral genes may be required for immortalisation (Kempkes et al., 1995), EBNA 1 is the only EBV encoded protein able to support DNA replication. It binds specifically to two regions (dyad symmetry region and family repeats) within the latent origin of replication of the virus, oriP (Hearing et al., 1992; Harrison et al., 1994). This interaction directs the cellular replication machinery towards the origin (Gahn and Schildkraut, 1989; Su et al., 1991). In addition, EBNA 1 is necessary for retention of the plasmids carrying the oriP inside the nucleus. EBNA1 binding renders the plasmids to minichromosomes, which are associated with metaphase chromosomes and distributed between daughter cells (Middleton and Sugden, 1994). Already in the 1980th the EBNA 1 gene and oriP have been transferred to bacterial plasmids. These shuttle vectors between bacteria and mammalian cells are extremely useful for a wide variety of applications such as long term gene expression for biotechnological purposes, gene isolation and mutation analysis (Jalanko et al., 1988; Peterson and Legerski, 1991). The replicon is under strong control: EBV based plasmids replicate once per cell cycle and are, therefore, not amplified (Yates and Guan, 1991).

Since relatively stable gene transfer is achieved in all cells reached by the vector, rearrangements or deletions are unlikely to occur, and very large DNA fragments can be stabilised (Sun et al., 1994), the latent replicon should also be of considerable interest for gene therapy. It depends mainly on the safety of the components whether application of this system for gene therapy could be approved. It is supported by a number of observations: (i) Neither expression of EBNA 1 nor the presence of multiple plasmid copies interfered with a set of hepatocyte specific functions *in vivo* (Lutfalla et al., 1989). (ii) Although EBNA 1 is one of the few viral antigens expressed in Burkitt lymphoma and other EBV induced tumors it seems to have no immortalising or transforming potential (Gross et al., 1986). (iii) 90% of the adult human population carries EBV in an asymptomatic state in normal B cells where EBNA1 is also expressed (Miller, 1990).

Recently, the gly-ala amino acid repeats of EBNA-1 have been implicated in inhibiting antigen processing and MHC1 restricted presentation (Levitskaya et al.,

1995). Hence, a mini EBV system may overcome immune surveillance and allow for efficient *in vivo*-persistence.

The EBV based vector seems to be particularly suitable for gene transfer to B cells. The approach has been used by Banerjee and coauthors to transfer the FA gene to B lymphoblasts from an Franconi aneamia patient (Banerjee et al., 1995). In this experiment, not only the latent replicon but an infectious nontransforming EBV-minivirus was used. However, in the absence of any selection pressure, the FA cDNA was not really stable within this rapidly growing cell population. In this respect, the transfer of EBV vectors carrying selectable genes into more slowly cycling cell populations such as hepatocytes, where the replicon is also functionally active (Lutfalla, G. et al., 1989, our own observation), would have better chances for long term stabilisation of the vector within the target cell.

The interaction between the origin and EBNA 1 not only mediates replication, it is also responsible for co-localisation of the episome with chromosomes during mitosis (nuclear retention). Whereas the EBVorigin can be replaced by cellular replication origens (Krysan et al., 1993; Wohlgemuth et al., 1996), the second function is of even more importance because transfer of the centromer function to a vector would be much more complicated.

The replicon of the bovine papillomavirus-1 (BPV-1) may provide similar functions (Ohe et al., 1995). Genes required for replication (E1 and E2) have been separated from transforming functions (E6 and E7) and recently a well defined minimal replicon has been established that is stably maintained in mouse and human cells at 30 copies/cell (Ohe et al., 1995; Piirsoo et al., 1996).

Replication capability of the vector is not only a prerequisite for its persistence in dividing cells, it also provides the basis for higher expression levels. Polynucleotide based immunisation strategies require a high level expression at least for several days in cells presenting the antigen. For safety reasons, integration into the host genome should be ruled out.

SV40 based vectors generating 100.000 DNA copies/cell has been extensively used to increase gene expression in biotechnology (Gething and Sambrook, 1981; Mellon et al., 1981; DiMaio et al., 1982). Nevertheless, they cannot be applied in a therapeutic setting due to the oncogenic potential of the T-antigen required for vector replication.

Alphaviruses provide a simple and safe replication mechanism based on RNA. Since the cell does not provide the reverse transcriptase function, integration of such an RNA vector is highly unlikely. The alphavirus genome, a plus strand RNA molecule (a typical mRNA), is translated immediately after cell entry to generate the viral replicase. Nothing else than this enzyme and replication signals at both ends of the RNA are required for RNA amplification, which takes place in the cytoplasm of the cell. The intermediate minus strand RNA serves as a template for the synthesis of full-length genomic RNA and a subgenomic transcript coding for the structural proteins (Schlesinger and Schlesinger, 1990).

Two members of the alphavirus family, Sindbis and Semliki forest virus, have been used for vector design. The gene of interest is cloned into a plasmid which serves as a template for *in vitro* transcription by phage polymerases (Rice et al., 1987). The transcript which codes for the foreign gene controlled by the subgenomic RNA promoter and the replicase gene can be applied as a vaccine. After the RNA has entered the cytoplasm, active replication starts resulting in 10^5 RNA molecules per cell. Efficient translation signals lead to high level production of the foreign protein reaching up to 3% of the cell protein (Xiong et al., 1989). Compared to nonreplicative RNA or DNA, elevated protein levels and sustained expression has been found *in vivo* (Johanning et al., 1995) This mode of amplification is of special importance for overexpression of therapeutic ribozymes.

5.7 Conclusion

We have outlined and discussed some features of animal viruses which are of particular interest for vector development. Despite the great potential of pure viral vectors for gene delivery their future application in gene therapy will be limited, mainly due to their immunogenicity. However, a number of viral features appear to function also in the context of a heterologous or artificial vector system and would be of great advantage for such vectors. Thus, the development of modular artificial vectors will make increasing use of the unique properties of viruses. We believe that there will be no single vector type useful for all purposes. On the contrary, every particular application requires its specially tailored vector, not only because of the particular therapeutic gene but also due to other demands including level and duration of gene expression as well as the percentage of cured cells necessary to obtain a therapeutic effect. The plethora of viral mechanisms for cellular entry, integration or replication open up a broad perspective for the development of a multitude of more or less virus-like artificial vectors. It is almost impossible to make a prediction as to the outcome of such a development before several of these imaginable new types of vectors have been tested. We are looking forward to exiting new developments in this field in the near future.

References

Armentano, D., Sookdeo, C. C., Hehir, K. M., Gregory, R. J., St. George, J. A., Prince, G. A., Wadsworth, S. C., and Smith, A. E. (1995). Characterization of an adenovirus gene transfer vector containing an E4 deletion. Hum. Gene Ther. *6*, 1343-1353.

Banerjee, S., Livanos, E. and Vos, J. M. (1995). Therapeutic gene delivery in human B-lymphoblastoid cells by engineered non-transforming infectious Epstein-Barr virus. Nat. Med. *1*, 1303-1308.

Bett, A. J., Haddara, W., Prevec, L., and Graham, F. L. (1994). An efficient and flexible system for construction of adenovirus vectors with insertions or deletions in early regions 1 and 3. Proc. Natl. Acad. Sci. U. S. A. *91*, 8802-8806.

Bianchi, M. E., Beltrame, M., and Paonessa, G. (1989). Specific recognition of cruciform DNA by nuclear protein HMG1. Science *243*, 1056-1059.

Bloom, D. C., Maidment, N. T., Tan, A., Dissette, V. B., Feldman, L. T., and Stevens, J. G. (1995). Long-term expression of a reporter gene from latent herpes simplex virus in the rat hippocampus. Brain Res. Mol. Brain Res. *31*, 48-60.

Bottger, M., Vogel, F., Platzer, M., Kiessling, U., Grade, K., and Strauss, M. (1988). Condensation of vector DNA by the chromosomal protein HMG1 results in efficient transfection. Biochim. Biophys. Acta *950*, 221-228.

Boulikas, T. (1993). Nuclear localization signals (NLS). Crit. Rev. Eukaryot. Gene Expr. *3*, 193-227.

Bukrinsky, M. I., Sharova, N., Dempsey, M. P., Stanwick, T. L., Bukrinskaya, A. G., Haggerty, S., and Stevenson, M. (1992). Active nuclear import of human immunodeficiency virus type 1 preintegration complexes. Proc. Natl. Acad. Sci. U. S. A. *89*, 6580-6584.

Bushman, F. D. (1994). Tethering human immunodeficiency virus 1 integrase to a DNA site directs integration to nearby sequences. Proc. Natl. Acad. Sci. U. S. A. *91*, 9233-9237.

Bushman, F. D. and Craigie, R. (1991). Activities of human immunodeficiency virus (HIV) integration protein in vitro: specific cleavage and integration of HIV DNA. Proc. Natl. Acad. Sci. U. S. A. *88*, 1339-1343.

Charlton, C. A. and Volkman, L. E. (1993). Penetration of Autographa californica nuclear polyhedrosis virus nucleocapsids into IPLB Sf 21 cells induces actin cable formation. Virology *197*, 245-254.

Chiorini, J. A., Wiener, S. M., Owens, R. A., Kyostio, S. R., Kotin, R. M., and Safer, B. (1994). Sequence requirements for stable binding and function of Rep68 on the adeno-associated virus type 2 inverted terminal repeats. J. Virol. *68*, 7448-7457.

Clark, K. R., Voulgaropoulou, F., Fraley, D. M., and Johnson, P. R. (1995). Cell lines for the production of recombinant adeno-associated virus. Hum. Gene Ther. *6*, 1329-1341.

Colucci, G., Beazer, Y., and Waksal, S. D. (1987). Interactions between hepatitis B virus and polymeric human serum albumin. II. Development of syngeneic monoclonal anti-anti-idiotypes which mimic hepatitis B surface antigen in the induction of immune responsiveness. Eur. J. Immunol. *17*, 371-374.

Cotten, M., Wagner, E., Zatloukal, K., Phillips, S., Curiel, D. T., and Birnstiel, M. L. (1992). High-efficiency receptor-mediated delivery of small and large (48 kilobase gene constructs using the endosome-disruption activity of defective or chemically inactivated adenovirus particles. Proc. Natl. Acad. Sci. U. S. A. *89*, 6094-6098.

Cristiano, R. J. and Roth, J. A. (1996). Epidermal growth factor mediated DNA delivery into lung cancer cells via epidermal growth factor receptor. Cancer Gene Therapy. *3*, 4-10.

DiMaio, D., Treisman, R., and Maniatis, T. (1982). Bovine papillomavirus vector that propagates as a plasmid in both mouse and bacterial cells. Proc. Natl. Acad. Sci. U. S. A. *79*, 4030-4034.

Eckhardt, S. G., Milich, D. R., and McLachlan, A. (1991). Hepatitis B virus core antigen has two nuclear localization sequences in the arginine-rich carboxyl terminus. J. Virol. *65*, 575-582.

Ellison, V. and Brown, P. O. (1994). A stable complex between integrase and viral DNA ends mediates human immunodeficiency virus integration in vitro. Proc. Natl. Acad. Sci. U. S. A. *91*, 7316-7320.

Erbacher, P., Roche, A. C., Monsigny, M., and Midoux, P. (1995). Glycosylated polylysine DNA complexes: Gene transfer efficiency in relation with the size and the sugar substitution level of glycosylated polylysines and with the plasmid size. Bioconjugate Chemistry. *6*, 401-410.

Findeis, M. A., Wu, C. H., and Wu, G. Y. (1994). Ligand-based carrier systems for delivery of DNA to hepatocytes. Methods Enzymol. *247*, 341-351.

Fink, D. J., Sternberg, L. R., Weber, P. C., Mata, M., Goins, W. F., and Glorioso, J. C. (1992). In vivo expression of beta-galactosidase in hippocampal neurons by HSV-mediated gene transfer. Hum. Gene Ther. *3*, 11-19.

Fischer, U., Sumpter, V., Sekine, M., Satoh, T., Luhrmann, R (1993). Nucleo-cytoplasmic transport of U snRNPs: definition of a nuclear location signal in the Sm core domain that binds a transport receptor independently of the m3G cap. EMBO J. *12*, 573-83

Fischer, U., Heinrich, J., van-Zee, K., Fanning,E., Luhrmann, R. (1994). Nuclear transport of U1 snRNP in somatic cells: differences in signal requirement compared with Xenopus laevis oocytes. J. Cell Biol.; *125* 971-80.

Fisher, K. J., Choi, H., Burda, J., Chen, S. J., and Wilson, J. M. (1996). Recombinant adenovirus deleted of all viral genes for gene therapy of cystic fibrosis. Virology *217*, 11-22.

Fisher, K. J., Gao, G. P., Weitzman, M. D., DeMatteo, R., Burda, J. F., and Wilson, J. M. (1996). Transduction with recombinant adeno-associated virus for gene therapy is limited by leading-strand synthesis. J. Virol. *70*, 520-532.

Fisher, K. J. and Wilson, J. M. (1994). Biochemical and functional analysis of an adenovirus-based ligand complex for gene transfer. Biochem. J. *299*, 49-58.

Flotte, T. R., Barraza Ortiz, X., Solow, R., Afione, S. A., Carter, B. J., and Guggino, W. B. (1995). An improved system for packaging recombinant adeno-associated virus vectors capable of in vivo transduction. Gene Ther. *2*, 29-37.

Flotte, T. R. and Carter, B. J. (1995). Adeno-associated virus vectors for gene therapy. Gene Ther. *2*, 357-362.

Forstova, J., Krauzewicz, N., Sandig, V., Elliott, J., Palkova, Z., Strauss, M., and Griffin, B. E. (1995). Polyoma virus pseudocapsids as efficient carriers of heterologous DNA into mammalian cells. Hum. Gene Ther. *6*, 297-306.

Forstova, J., Krauzewicz, N., Wallace, S., Street, A. J., Dilworth, S. M., Beard, S., and Griffin, B. E. (1993). Cooperation of structural proteins during late events in the life cycle of polyomavirus. J. Virol. *67*, 1405-1413.

Fredman, J. N. and Engler, J. A. (1993). Adenovirus precursor to terminal protein interacts with the nuclear matrix in vivo and in vitro. J. Virol. *67*, 3384-3395.

Frenkel, N., Singer, O., and Kwong, A. D. (1994). Minireview: the herpes simplex virus amplicon – a versatile defective virus vector. Gene Ther. *1 Suppl 1*, S40-6.

Gahn, T. A. and Schildkraut, C. L. (1989). The Epstein-Barr virus origin of plasmid replication, oriP, contains both the initiation and termination sites of DNA replication. Cell *58*, 527-535.

Gallay, P., Stitt, V., Mundy, C., Oettinger, M., and Trono, D. (1996). Role of the karyopherin pathway in human immunodeficiency virus type 1 nuclear import. J. Virol. *70*, 1027-1032.

Gall, J., Kass Eisler, A., Leinwand, L., and Falck Pedersen, E. (1996). Adenovirus type 5 and 7 capsid chimera: fiber replacement alters receptor tropism without affecting primary immune neutralization epitopes. J. Virol. *70*, 2116-2123.

Gething, M. J. and Sambrook, J. (1981). Cell-surface expression of influenza haemagglutinin from a cloned DNA copy of the RNA gene. Nature *293*, 620-625.

Giraud, C., Winocour, E., and Berns, K. I. (1994). Site-specific integration by adeno-associated virus is directed by a cellular DNA sequence. Proc. Natl. Acad. Sci. U. S. A. *91*, 10039-10043.

Glorioso, J. C., Bender, M. A., Goins, W. F., Fink, D. J., and DeLuca, N. (1995). HSV as a gene transfer vector for the nervous system. Mol. Biotechnol. *4*, 87-99.

Gorziglia, M. I., Kadan, M. J., Yei, S., Lim, J., Lee, G. M., Luthra, R., and Trapnell, B. C. (1996). Elimination of both E1 and E2A from adenovirus vectors further improves prospects for in vivo human gene therapy. J. Virol. *70*, 4173-4178.

Goulaouic, H. and Chow, S. A. (1996). Directed integration of viral DNA mediated by fusion proteins consisting of human immunodeficiency virus type 1 integrase and Escherichia coli LexA protein. J. Virol. *70*, 37-46.

Greber, U. F., Willetts, M., Webster, P., and Helenius, A. (1993). Stepwise dismantling of adenovirus 2 during entry into cells. Cell *75*, 477-486.

Gross, T. G., Sakai, K., and Volsky, D. J. (1986). Transfer of the Epstein-Barr virus (EBV) DNA fragment coding for EBNA-1, the putative transforming antigen of EBV, into normal human lymphocytes: gene expression without cell transformation. Biochem. Biophys. Res. Commun. *134*, 1260-1268.

Harrison, S., Fisenne, K., and Hearing, J. (1994). Sequence requirements of the Epstein-Barr virus latent origin of DNA replication. J. Virol. *68*, 1913-1925.

Hart, S. L., Harbottle, R. P., Cooper, R., Miller, A., Williamson, R., and Coutelle, C. (1995). Gene delivery and expression mediated by an integrin-binding peptide. Gene Ther. *2*, 552-554.

Hearing, J., Mulhaupt, Y., and Harper, S. (1992). Interaction of Epstein-Barr virus nuclear antigen 1 with the viral latent origin of replication. J. Virol. *66*, 694-705.

Heinzinger, N. K., Bukinsky, M. I., Haggerty, S. A., Ragland, A. M., Kewalramani, V., Lee, M. A., Gendelman, H. E., Ratner, L., Stevenson, M., and Emerman, M. (1994). The Vpr protein of human immunodeficiency virus type 1 influences nuclear localization of viral nucleic acids in nondividing host cells. Proc. Natl. Acad. Sci. U. S. A. *91*, 7311-7315.

Hofmann, C., Sandig, V., Jennings, G., Rudolph, M., Schlag, P., and Strauss, M. (1995). Efficient gene transfer into human hepatocytes by baculovirus vectors. Proc. Natl. Acad. Sci. U. S. A. *92*, 10099-10103.

Huckett, B., Ariatti, M., and Hawtrey, A. O. (1990). Evidence for targeted gene transfer by receptor-mediated endocytosis. Stable expression following insulin-directed entry of NEO into HepG2 cells. Biochem. Pharmacol. *40*, 253-263.

Jalanko, A., Kallio, A., Ruohonen Lehto, M., Soderlund, H., and Ulmanen, I. (1988). An EBV-based mammalian cell expression vector for efficient expression of cloned coding sequences. Biochim. Biophys. Acta *949*, 206-212.

Johanning, F. W., Conry, R. M., Lobuglio, A. F., Wright, M., Sumerel, L. A., Pike, M. J., and Curiel, D. T. (1995). A sindbis virus mRNA polynucleotid vector achieves prolonged and high level heterologous gene expression in vivo. Nucleic. Acids. Research. *23*, 1495-1501.

Jouault, T., Chapuis, F., Olivier, R., Parravicini, C., Bahraoui, E., and Gluckman, J.C. (1989). HIV infection of monocytic cells: role of antibody-mediated virus binding to Fc-gamma receptors. AIDS *3*, 125-133.

Kaneda, Y., Morishita, R., and Tomita, N. (1995). Increased expression of DNA cointroduced with nuclear protein in adult rat liver. J. Mol. Med. *73*, 289-297.

Kato, K., Yoneda, Y., Okada, Y., Kiyama, H., and Shiosaka, S. (1994). Gene transfer and the expression of a foreign gene in vivo in post-mitotic neurons of the adult rat brain using the hemagglutinating virus of the Japan-liposome method. Brain Res. Mol. Brain Res. *25*, 359-363.

Katz, R. A., Merkel, G., Kulkosky, J., Leis, J., and Skalka, A. M. (1990). The avian retroviral IN protein is both necessary and sufficient for integrative recombination in vitro. Cell *63*, 87-95.

Kempkes, B., Pich, D., Zeidler, R., and Hammerschmidt, W. (1995). Immortalization of human primary B lymphocytes in vitro with DNA. Proc. Natl. Acad. Sci. U. S. A. *92*, 5875-5879.

Kiem, H. P., von Kalle, C., Schuening, F., and Storb, R. (1995). Gene therapy and bone marrow transplantation [see comments]. Curr. Opin. Oncol. *7*, 107-114.

Kochanek, S., Clemens, P. R., Mitani, K., Chen, H. H., Chan, S., and Caskey, C. T. (1996). A new adenoviral vector: replacement of all viral coding sequences with 28kb of DNA independently expressing full length dystrophin and ß galactosidase. Proc. Natl. Acad. Sci. U. S. A . *93*, 5731-5736.

Kotin, R. M., Linden, R. M., and Berns, K. I. (1992). Characterization of a preferred site on human chromosome 19q for integration of adeno-associated virus DNA by non-homologous recombination. EMBO J. *11*, 5071-5078.

Kukowska Latallo, J. F., Bielinska, A. U., Johnson, J., Spindler, R., Tomalia, D. A., and Baker, J. R. (1996). Efficient transfer of genetic material into mammalian cells using Starburst polyamidoamine dendrimers. Proc. Natl. Acad. Sci. U. S. A. *93*, 4897-4902.

Krasnykh, V. N., Mikheeva, G. V., Doulas, J. T., and Curiel, D. T. (1996). Generation of recombinant adenovirus vectors with modified fibers for altering viral tropism. J.Virol. *70* 6839-6846

Lu, B., Gupta, S., and Federoff, H. (1995). Ex vivo hepatic gene transfer in mouse using a defective herpes simplex virus-1 vector. Hepatology *21*, 752-759.

Lund, O., Hansen, J., Soorensen, A. M., Mosekilde, E., Nielsen, J. O., and Hansen, J. E. (1995). Increased adhesion as a mechanism of antibody-dependent and antibody-independent complement-mediated enhancement of human immunodeficiency virus infection. J. Virol. *69*, 2393-2400.

Lutfalla, G., Armbruster, L., Dequin, S., and Bertolotti, R. (1989). Construction of an EBNA-producing line of well-differentiated human hepatoma cells and of appropriate Epstein-Barr virus-based shuttle vectors. Gene *76*, 27-39.

Mellon, P., Parker, V., Gluzman, Y., and Maniatis, T. (1981). Identification of DNA sequences required for transcription of the human alpha 1-globin gene in a new SV40 host-vector system. Cell *27*, 279-288.

Middleton, T. and Sugden, B. (1994). Retention of plasmid DNA in mammalian cells is enhanced by binding of the Epstein-Barr virus replication protein EBNA1. J. Virol. *68*, 4067-4071.

Miles, B. D., Luftig, R. B., Weatherbee, J. A., Weihing, R. R., and Weber, J. (1980). Quantitation of the interaction between adenovirus types 2 and 5 and microtubules inside infected cells. Virology *105*, 265-269.

Miller, G. (1990) Epstein-Barr virus. In Fields Virology. Fields, B. N., Kniepe, M. D. eds. Ravens Press, New York pp 1921-1958.

Mitani, K., Graham, F. L., Caskey, C. T., and Kochanek, S. (1995). Rescue, propagation, and partial purification of a helper virus-dependent adenovirus vector. Proc. Natl. Acad. Sci. U. S. A. *92*, 3854-3858.

Mooslehner, K., Karls, U., and Harbers, K. (1990). Retroviral integration sites in transgenic Mov mice frequently map in the vicinity of transcribed DNA regions. J. Virol. *64*, 3056-3058.

Naldini, L., Blomer, U., Gallay, P., Ory, D., Mulligan, R., Gage, F. H., Verma, I. M., and Trono, D. (1996). In vivo gene delivery and stable transduction of nondividing cells by a lentiviral vector [see comments]. Science *272*, 263-267.

Naldini, L., Blomer, U., Gallay, P., Ory, D., Mulligan, R., Gage, F. H., Verma, I. M., and Trono, D. (1996). In vivo gene delivery and stable transduction of nondividing cells by a lentiviral vector [see comments]. Science *272*, 263-267.

O'Neill, R. E., Jaskunas, R., Blobel, G., Palese, P., and Moroianu, J. (1995). Nuclear import of influenza virus RNA can be mediated by viral nucleoprotein and transport factors required for protein import. J. Biol. Chem. *270*, 22701-22704.

Ochiai, H., Kurokawa, M., Matsui, S., Yamamoto, T., Kuroki, Y., Kishimoto, C., and Shiraki, K. (1992). Infection enhancement of influenza A NWS virus in primary murine macrophages by anti-hemagglutinin monoclonal antibody. J. Med. Virol. *36*, 217-221.

Patterson, S. and Oxford, J. S. (1986). Early interactions between animal viruses and the host cell: relevance to viral vaccines. Vaccine *4*, 79-90.

Perales, J. C., Ferkol, T., Molas, M., and Hanson, R. W. (1994). An evaluation of receptor mediated gene transfer using synthetic DNA ligand complexes. European Jurnal of Biochemistry *226*, 255-266.

Peterson, C. and Legerski, R. (1991). High-frequency transformation of human repair-deficient cell lines by an Epstein-Barr virus-based cDNA expression vector. Gene *107*, 279-284.

Prcla, E., Plank, C., Wagner, E., Blaas, D., and Fuchs, R. (1995). Virus mediated release of endosomal content in vitro: different behavior of adenovirus and rhinovirus serotype 2. J. Cell Biology. *131*, 111-123.

Remy, J. S., Kichler, A., Mordvimov, V., Schuber, F. and Behr, J. P. (1995). Targeted gene transfer into hepatoma cells with lipopolyamine condensed DNA particles presenting galactose ligands: A stage toward artificial viruses. Proc. Natl. Acad. Sci. U. S. A. *92*, 1744-1748.

Rice, C. M., Levis, R., Strauss, J. H., and Huang, H. V. (1987). Production of infectious RNA transcripts from Sindbis virus cDNA clones: mapping of lethal mutations, rescue of a temperature-sensitive marker, and in vitro mutagenesis to generate defined mutants. J. Virol. *61*, 3809-3819.

Robinson, W. E., Jr., Montefiori, D. C., and Mitchell, W. M. (1990). Complement-mediated antibody-dependent enhancement of HIV-1 infection requires CD4 and complement receptors. Virology *175*, 600-604.

Rolling, F. and Samulski, R. J. (1995). AAV as a viral vector for human gene therapy. Generation of recombinant virus. Mol. Biotechnol. *3*, 9-15.

Ross, G. F., Morris, R. E., Ciraolo, G., Huelsman, K., Bruno, M., Whitsett, J. A., Baatz, J. E., and Korfhagen, T. R. (1995). Surfactant protein A-polylysine conjugates for delivery of DNA to airway cells in culture. Hum. Gene Ther. *6*, 31-40.

Russell, D. W. and Miller, A. D. (1996). Foamy virus vectors. J. Virol. *70*, 217-222.

Schlesinger, S. and Schlesinger, M. J. (1990) Replication of Togaviridae and Flavaviridae. In: Fields Virology. Fields, B. N., Kniepe, M. D. eds. New York: Ravens Press, pp 697-711.

Shih, C. C., Stoye, J. P., and Coffin, J. M. (1988). Highly preferred targets for retrovirus integration. Cell *53*, 531-537.

Shimizu, N., Chen, J. B., Gamou, S., and Takayanagi, A. (1996). Immunogene approach toward cancer therapy using erytrocyte growth factor receptor mediated gene delivery. Cancer Gene Therapy. *3*, 113-120.

Spaete, R. R. and Frenkel, N. (1985). The herpes simplex virus amplicon: analyses of cis-acting replication functions. Proc. Natl. Acad. Sci. U. S. A. *82*, 694-698.

Su, W., Middleton, T., Sugden, B., and Echols, H. (1991). DNA looping between the origin of replication of Epstein-Barr virus and its enhancer site: stabilization of an origin complex with Epstein-Barr nuclear antigen 1. Proc. Natl. Acad. Sci. U. S. A. *88*, 10870-10874.

Tomita, N., Higaki, J., Morishita, R., Kato, K., Mikami, H., Kaneda, Y., and Ogihara, T. (1992). Direct in vivo gene introduction into rat kidney. Biochem. Biophys. Res. Commun. *186*, 129-134.

Travers, A. A., Ner, S. S., and Churchill, M. E. (1994). DNA chaperones: a solution to a persistence problem? Cell *77*, 167-169.

Varga Weisz, P., van Holde, K., and Zlatanova, J. (1994). Competition between linker histones and HMG1 for binding to four-way junction DNA: implications for transcription. Biochem. Biophys. Res. Commun. *203*, 1904-1911.

Vile, R. G. and Russell, S. J. (1995). retroviruses as vectors. .British Medical Bulletin. *51*, 12-30.

Vincent, K. A., Ellison, V., Chow, S. A., and Brown, P. O. (1993). Characterization of human immunodeficiency virus type 1 integrase expressed in Escherichia coli and analysis of variants with amino-terminal mutations. J. Virol. *67*, 425-437.

Volkman, L. E. (1986). The 64K envelope protein of budded Autographa californica nuclear polyhedrosis virus. Curr. Top. Microbiol. Immunol. *131*, 103-118.

von der Leyen, H. E., Gibbons, G. H., Morishita, R., Lewis, N. P., Zhang, L., Nakajima, M., Kaneda, Y., Cooke, J. P. and Dzau, V. J. (1995). Gene therapy inhibiting neointimal vascular lesion: in vivo transfer of endothelial cell nitric oxide synthase gene. Proc. Natl. Acad. Sci. U. S. A. *92*, 1137-1141.

Wagner, E., Zatloukal, K., Cotten, M., Kirlappos, H., Mechtler, K., Curiel, D. T., and Birnstiel, M. L. (1992). Coupling of adenovirus to transferrin-polylysine/DNA complexes greatly enhances receptor-mediated gene delivery and expression of transfected genes. Proc. Natl. Acad. Sci. U. S. A. *89*, 6099-6103.

Wagner, E., Zenke, M., Cotten, M., Beug, H., and Birnstiel, M. L. (1990). Transferrin-polycation conjugates as carriers for DNA uptake into cells. Proc. Natl. Acad. Sci. U. S. A. *87*, 3410-3414.

Walsh, C. E., Liu, J. M., Miller, J. L., Nienhuis, A. W., and Samulski, R. J. (1993). Gene therapy for human hemoglobinopathies. Proc. Soc. Exp. Biol. Med. *204*, 289-300.

Watts, C. and Marsh, M. (1992). Endocytosis: what goes in and how? J. Cell Sci. *103*, 1-8.

Wickham, T. J., Granados, R. R., Wood, H. A., Hammer, D. A., and Shuler, M. L. (1990). General analysis of receptor-mediated viral attachment to cell surfaces. Biophys. J. *58*, 1501-1516.

White J. M. (1994). Fusion of influenza virus in endosomes: role of the hemaglutinin. In: Cellular receptors for animal viruses. Wimmer, E. ed. Could Spring Habor Laboratory Press.

Wohlgemuth, J. G. Kang, S. H., Bulboaca, G. H., Nawotka, K. A., and Calos, M. P. (1996). Long-term gene expression from auronomously replicating vectors in mammalian cells. Gene Therapy *3*, 503-512.

Wu, G. W., Chowdhury, J. R., Bommineni, V. R., Basu, S. K., Wu, C. H., and Chowdhury, N. R. (1995). Fate of DNA targeted to hepatocytes by asialoglycoproteinpolylysine conjugates. Proc. Assoc. Am. Physicians. *107*, 211-217.

Xiong, C., Levis, R., Shen, P., Schlesinger, S., Rice, C. M., and Huang, H. V. (1989). Sindbis virus: an efficient, broad host range vector for gene expression in animal cells. Science *243*, 1188-1191.

Yang, Y., Nunes, F. A., Berencsi, K., Furth, E. E., Gonczol, E., and Wilson, J. M. (1994). Cellular immunity to viral antigens limits E1-deleted adenoviruses for gene therapy. Proc. Natl. Acad. Sci. U. S. A. *91*, 4407-11X.

Yao, J. S., Kariwa, H., Takashima, I., Yoshimatsu, K., Arikawa, J., and Hashimoto, N. (1992). Antibody-dependent enhancement of hantavirus infection in macrophage cell lines. Arch. Virol. *122*, 107-118.

Yates, J. L. and Guan, N. (1991). Epstein-Barr virus-derived plasmids replicate only once per cell cycle and are not amplified after entry into cells. J. Virol. *65*, 483-488.

Yeh, P., Dedieu, J. F., Orsini, C., Vigne, E., Denefle, P., and Perricaudet, M. (1996). Efficient dual transcomplementation of adenovirus E1 and E4 regions from a 293-derived cell line expressing a minimal E4 functional unit. J. Virol. *70*, 559-565.

Zhao, L. J. and Padmanabhan, R. (1991). Three basic regions in adenovirus DNA polymerase interact differentially depending on the protein context to function as bipartite nuclear localization signals. New Biol. *3*, 1074-1088.

6. Targeted Gene Delivery and Expression in Hepatocytes

George Y. Wu and Catherine H. Wu

Parenchymal liver cells are the only cells of the adult mammalian body that possess of large numbers of high affinity cell-surface receptors that can recognize galactose-terminal (asialo-) glycoproteins (1, 2). Binding of an asialoglycoprotein by the receptor leads to internalization of the ligand-receptor complex in membrane-bound vesicles, endosomes. Subsequently, acidification of the endosomes results in segregation of the ligand from the receptor. The latter is usually transported to lysosomes where degradation occurs (3).

6.1 A DNA Carrier System Targetable to Hepatocytes

Asialoglycoproteins have been used previously to deliver a variety of biologically active compounds to liver cells (4, 5). In these studies, the agents were coupled to asialoglycoproteins by covalent links. In 1987, targeted delivery of DNA to liver cells via asialoglycoprotein receptors was demonstrated. This was accomplished by designing a protein conjugate consisting of two components: 1) a cell targeting component – an asialoglycoprotein ligand covalently bound to 2) a DNA binding component consisting of a polymer containing multiple positive charges (polylysine). In the original description, a natural glycoprotein was converted into an asialoglycoprotein. However, a number of investigators have subsequently used synthetic asialoglycoproteins (6, 7, 8, 9).

6.2 Targeted Gene Delivery *in Vitro*

To test this carrier system in vitro, a plasmid carrying the chloramphenicol acetyltransferase gene (CAT) driven by an SV-40 viral promoter was added to the conjugate to form a soluble complex (10). The complex was transfected into an asialoglycoprotein (AsG) receptor (+) cell line, and an AsG receptor (-) cell line.

After incubation, no CAT activity could be demonstrated in AsG receptor (-) cells or in AsG receptor (+) cells treated with separate components of the complex. However, CAT activity was detected in complex-treated AsG receptor (+) cells. CAT gene expression was completely inhibited by the addition of an excess of asialoglycoprotein to compete with the complex for receptor uptake (6). This indicated that cellular recognition of the complex was directed by the asialoglycoprotein component.

6.3 Targeted Gene Delivery *in Vivo*

To determine whether an asialoglycoprotein-based DNA delivery system could target DNA to hepatocytes in vivo, ^{32}P-labeled DNA alone or as a complex with AsOR-PL conjugate, were injected intravenously into groups of rats. After 10 min, in rats treated with DNA alone, most of the radioactivity, 55%, was found still circulating in the blood. Seventeen percent of the injected counts were detected in the liver. In contrast, in rats injected with labeled DNA in the form of a complex, 85% of the counts were detected in the liver while only 5% of the radioactivity remained in the blood. This organ distribution of the complex was similar to that of ^{125}I-asialoorosomucoid carrier alone suggesting that the complex retained its ability to be recognized by AsG receptors in vivo (11).

To determine whether any of this targeted DNA remained intact after uptake by the liver in vivo, complexed DNA was injected intravenously, and after 24 hours, liver DNA was extracted and analyzed by dot-blot using a ^{32}P-labeled CAT cDNA probe. CAT DNA sequences were detected in virtually all parenchymal cells in transfected liver, but not in control liver DNA (7). Some Kupffer cell uptake was also noted, but was minimal compared to that of parenchymal cells. CAT activity was detected only in livers from rats treated with complexed DNA, but not in kidney, spleen, or lungs (7).

To determine whether genes driven by *natural* mammalian regulatory elements could be similarly delivered and expressed in hepatocytes, a plasmid containing the CAT gene driven by albumin promoter/enhancer regions was used. The complex injected intravenously into adult rats resulted in transient CAT gene expression that reached a peak level of 10 units/g liver was reached after 24 hours, but was undetectable by 96 hours after injection (12).

6.4 Transient Gene Expression in Animal Models of Inherited Metabolic Disorders

In collaboration with James Wilson, a plasmid carrying a full-length cDNA for the human LDL receptor was prepared. The plasmid was complexed with the AsG-PL conjugate and injected intravenously into Watanabe rabbits to test the feasibility of targeting the gene for the low density lipoprotein (LDL) receptor in a model for familial hypercholesterolemia (13). Livers removed 10 min after injection were demonstrated to contain approximately 1000 copies of plasmid per cell by Southern blots. However, these levels declined progressively with time, and by 48 hours plasmid DNA was less than 0.1 copies/cell. To study the transcriptional activity of the recombinant gene, total RNA was extracted and analyzed by RNase protection assay. Exogenous LDL receptor mRNA was detected at 4 hours, reached a peak at 24 hours and decreased to undetectable levels by 72 hours after transfection. Maximal levels of recombinant LDL receptor mRNA occurred at 24 hours were estimated to be 2-4% of normal endogenous levels. The metabolic effects of hepatocyte-directed gene transfer in WHHL rabbits injected with complexed LDL receptor gene or CAT gene were analyzed by measuring changes in total serum cholesterol. After administration of the LDL receptor gene complex, there was a rapid, but transient decline in serum cholesterol that lasted 6 days. The drop in cholesterol levels was maximal at 2 days post-injection and was 25% to 30% of pretreatment values (9). Treatment with the CAT gene complex had no effect on serum cholesterol.

6.5 Prolongation of Targeted Gene Expression

Efforts to prolong the observed transient expression were based on stimulation of hepatocyte replication to enhance the probability of integration of the foreign gene into the genome of foreign cells. In normal adult liver, there are few dividing hepatocytes. However, hepatocytes can be induced to replicate in response to injury resulting in regeneration, e.g. partial hepatectomy (14). To determine if gene expression could be prolonged in hepatocytes stimulated to replicate by partial hepatectomy, rats were injected with complexed DNA. Thirty min later, two-thirds partial hepatectomies were performed. In these rats, CAT activity was detected for at least 11 weeks post-transfection (12).

The use of the DNA delivery system combined with hepatocyte replication to correct a metabolic dysfunction in genetically defective animals was studied in Nagase analbuminemic rats (15). A plasmid containing the gene for the human serum albumin driven by mouse albumin enhancer-rat albumin promoter sequences was constructed and complexed with the targetable conjugate. The resulting soluble complex was injected intravenously into Nagase rats and two-thirds partial hepatectomies were performed. Two weeks after injection, densitometric quantitation revealed an average copy number of 1000 copies of the plasmid/diploid genome most of which was found to be in plasmid form. Using an RNase protection assay, the level of human albumin mRNA was estimated to be between 0.01% and 0.1% of the normal level of rat albumin in normal rats. The presence of circulating human albumin was determined by Western blot analysis that showed no detectable albumin at 24 hrs, but appearance of human albumin in the serum within 48 hours. Human serum albumin was first detectable at a level of 0.05 μg/ml, but increased in concentration to a peak level of 34 μg/ml by 2 weeks post-injection. This level remained stable through 4 weeks. No antibodies directed against the human albumin were detected during the 4 weeks of the study (12).

Perales et al. have used a galactose derivative of polylysine to form a condensed DNA complex in the presence of high salt. Such complexes were found to result in high levels and prolonged expression (up to at least 140 days) of a secreted marker gene product, human factor IX, in mice without partial hepatectomy or any agent to prolong expression (7).

6.6 The Mechanism of Persistence of Targeted Gene Expression Achieved by Partial Hepatectomy

In an attempt to define the mechanism(s) responsible for persistent gene expression following partial hepatectomy, the molecular state of the retained, liver-associated transgenes was characterized. Southern blot analysis of DNA from liver tissues harvested about 40 min after in vivo gene transfer and partial hepatectomy contained high levels of transgene DNA (1,000-10,000 copies/cell). The predominant form of this DNA appeared to be episomal based on analyses of uncut DNA, or DNA as determined by exposure to a single cutter endonuclease. The existence of episomal DNA in liver was confirmed in experiments in which intact plasmid was rescued from total hepatocyte DNA by transformation of bacteria. The plasmid DNA recovered from the livers retained a bacterial pattern of methylation suggesting that the plasmid had not replicated in the eukaryotic cell. These results are consistent with the hypothesis that the majority of transgene sequences are retained as stabilized plasmids. The specific form of DNA that is transcriptionally active was not identified in these studies. This represents a new mechanism for retaining foreign DNA in eukaryotic cells in vivo and has implications both for the development of somatic gene therapies and the pathogenesis of viral diseases (16).

In collaboration with J. Roy Chowdhury, to determine the mechanism of long-term persistence and expression after partial hepatectomy, the intracellular location of the persisting DNA was examined as a function of time. A plasmid expressing bacterial chloramphenicol acetyl transferase (CAT) was targeted into the liver of rats in vivo by receptor-mediated endocytosis. The internalized DNA was measured by Southern blot analysis. Twenty minutes after administration, 80-85% of the plasmid appeared in the liver, 80% of which was within hepatocytes, representing 12,000 to 18,000 copies per hepatocyte. In sham-operated rats, the transgene concentration decreased to 8-12% and 2-4% of the initial levels in 4 and 24 hr, respectively, and became undetectable at 7 days. In rats subjected to 66% hepatectomy 20 min after DNA administration, approximately 20%, 9% and 7% of the plasmid in the residual liver persisted at 4 hr, 24 hr and 7 days, respectively. Liver homogenates were centrifuged at 750 X g for 20 min. The pellet was subfractionated into nuclear and membranous fractions; of these, the internalized plasmid was present mainly in the membranous fraction. Plasmid in the 750 X g membrane fraction was DNase sensitive and became undetectable in 24 hr. The plasmid in isolated nuclei was undetectable by Southern blot, but was demonstrated by polymerase chain reaction (PCR). The 750 X g supernatant was subfractionated on a Percoll gradient. Twenty minutes after administration, the internalized plasmid was distributed bimodally in a plasma membrane/endosome-enriched fraction and a lysosome-enriched fraction. In sham-operated controls, the plasmid in all fractions was degraded by 24 hr. In the 66% hepatectomy group, plasmid in the lysosomal fraction became undetectable in 24 hr. In contrast, the plasmid in the plasma membrane/endosome fraction was DNase resistant and persisted undegraded in a transfection-competent form throughout the experiment (7 days), indicating that cytoplasmic vesicles are the main site of persistence of the endocytosed DNA (17).

6.7 Incorporation of Endosomolytic Agents

To enhance the levels of targeted gene expression, adenovirus was incorporated into the DNA carrier system. In initial experiments by Cristiano et al., polylysine was coupled to adenovirus by random attachment of polycation to the virus, resulting in enhanced expression in vitro (18). However, in order to retain specific targeting to hepatocytes, the inherent infectivity of the virus had to be blocked. To accomplish this, a different coupling strategy had to be employed based on the fact that normal infectivity of adenovirus for host tissue involves recognition of sites on the fibers of the virus and receptor on the surface of target cells. Furthermore, it is known that only the virus fibers contain carbohydrate. We hypothesized that if carbohydrate residues could be specifically activated, the recognition sites on the fibers of the virus might be obscured. Wild-type virus and replication defective dl312 adenovirus were treated and finally coupled to an asialoorosomucoid-polylysine conjugate. To assess possible changes in infection specificity, modified viruses were exposed to various cells in culture and the viability determined by microscopically counting cells capable of trypan blue exclusion as a function of time, see Table 6.1.

Table 6.1 shows that modified wild-type virus had greatly decreased infectivity toward normally susceptible HeLa S3 [asialoglycoprotein receptor (-)] and SK Hep1 [asialoglycoprotein receptor (-)] cells leaving 91% and 86% viable, respectively. However, with Huh 7 [asialoglycoprotein receptor (+)] cells, modified virus retained its infectivity leaving only 19% of cells viable under identical conditions.

To assess foreign gene expression, modified virus was complexed to a plasmid, pSVHBV, containing the gene for hepatitis B surface antigen, as a marker. Huh 7, receptor (+), cells treated with modified wild-type, and modified replication-defective dl312 virus complexed to DNA raised antigen levels by approximately 13- and 30-fold, respectively, compared to asialoglycoprotein-polylysine DNA complex alone, see Table 6.2.

Table 6.1 Effect of Modification of Adenovirus on Infection Specificity+

Addition	HeLa S3*	% Control	Huh 7*	% Control	SK Hep1*	% Control
None	1.1 ± .10	100	1.4 ± .13	100	1.4 ± .10	100
Wild type	.11 ± .01	10	.43 ± .15	31	.20 ± .02	14
Modified wild-type	1.0 ±.05	91	.026 ± .02	19	1.2 ± .30	86
Modified wild-type + AsOR**	1.0 ± .05	91	1.3 ± .02	93	1.4 ± .30	100
Wild-type + AsOR	.10 ± .05	9	.37 ± .03	27	.23 ± .20	16
dl312	1.2 ± .20	109	.98 ± .15	70	1.2 ± .11	86
Modified dl312	1.0 ± .11	91	1.3 ± .10	93	1.2 ± .11	86
Modified dl312 + AsOR	1.0 ± .10	91	1.4 ± .12	100	1.3 ± .12	93

+ reproduced with permission from J. Biol. Chem. (1994). *269*, 11542-11546.
* number of viable cells (x 10^{-6}) determined by trypan blue exclusion.
** AsOR, asialoorosomucoid

Table 6.2: Effect of Modified Virus on Targeted Gene Expression+

Addition	HBV surface antigen production (pg/106 cells/24 hrs)
Untreated control cells	0
AsOR-PL-DNA complex alone	5.4 ± 1.1
Complex + wild-type virus	14 ± 2.5
Complex + dl312 virus	15 ± 5.5
Modified wild-type virus-DNA complex	70 ± 28
Modified dl312 virus-DNA complex	160 ± 15
Modified dl312 virus-DNA complex + 1000-fold AsOR	4.6 ± 1.0
Modified dl312 virus-DNA complex + dl312 virus (1:100)	150 ± 25

+ reproduced with permission from J. Biol. Chem. (1994). *269*, 11542-11546.

Competition with a large excess of an asialoglycoprotein blocked the enhancement by more than 95%. The transfection efficiency of modified virus was found to be 50-fold higher than the complex without virus. Yet, specificity was retained exclusively for asialoglycoprotein receptor-bearing cells (19).

6.8 Targeted Delivery of Antisense DNA

We considered the possibility that single-stranded DNA in the form of antisense oligomers could be targeted in the same manner as had been demonstrated for double-stranded DNA. Because of the specificity of the action of antisense DNA in the inhibition of the synthesis of specific proteins based on hybridization of antisense to target mRNA, it was conceivable that receptor-mediated delivery of antisense DNA specific for hepatitis B viral (HBV) mRNA sequences could inhibit viral gene expression in target cells. To test this hypothesis, a 21-mer oligo DNA sequence complementary to the polyadenylation signal for human hepatitis B virus (HBV) was complexed to a soluble DNA-carrier system. A cell line, HepG2 (2.2.15) that possesses asialoglycoprotein receptors (20) and which is permanently transfected with hepatitis B virus was exposed to complexed antisense DNA or controls. In the presence of complexed antisense DNA, the concentration of hepatitis B surface antigen in medium was decreased by **80%** by day 1, and by greater than **95%** through the 6th day compared to untreated cells. There was *no* significant increase in surface antigen concentration in the presence of complexed antisense DNA after the first day of exposure. This inhibition was blocked by competition with an excess of free asialoglycoprotein. Protein secretion from cells was not affected, and could not account for the decrease in HBV surface antigen concentration in the medium after exposure to antisense DNA. Also, total protein synthesis remained unchanged by exposure to complexed antisense sequences under identical conditions. Finally, HBV DNA in the medium and cell layers after 24 hrs exposure to complexed antisense sequences was 80% lower than in controls. Exposure of cells to a random 21-mer oligo DNA sequence under identical conditions failed to alter HBV surface antigen concentration or HBV DNA in medium or cells (21).

6.9 Conclusions

We conclude that an asialoglycoprotein-based DNA carrier system can target DNA to highly selective receptors on parenchymal hepatocytes and result in new gene expression to alter an abnormal cellular phenotype, or to introduce antisense nucleotides to suppress endogenous gene expression.

Acknowledgments

The secretarial assistance of Mrs. Rosemary Pavlick is gratefully acknowledged. This work was supported in part by grants from: NIH DK 42182 (GYW) and TargeTech, Inc./Immune Response Corp. (CHW). (GYW and CHW hold equity in the Immune Response Corp.)

References

1. Ashwell G., Morell A., The role of surface carbohydrates in the hepatic recognition and transport of circulating glycoproteins. Adv. Enzymol. 1974, 41, 99-128.
2. Stockert R.J., Morell A.G., Ashwell G., Targeted Diagnosis and Therapy Using Specific Receptors and Ligands. In: Wu G, Wu C, ed. Liver Diseases. New York: Marcel Dekker, 1991, 41-64.
3. Schwartz A., Trafficking of asialoglycoproteins and the asialoglycoprotein receptor. In: Wu G., Wu C., ed. Liver Diseases. Targeted Diagnosis and Therapy Using Specific Receptors and Ligands. New York: Marcel Dekker, 1991, 3-40.
4. Wu G., Wu C., Rubin M., Acetaminophen hepatotoxicity and targeted rescue: A model for specific chemotherapy for hepatocellular carcinoma. Hepatology, 1985, 5, 5709-5713.
5. Wu G., Wu C., Stockert R., Model for specific rescue of normal hepatocytes during methotrexate treatment of hepatic malignancy. Proc Natl Acad Sci U S A. Proc. Natl. Acad. Sci. USA , 1983, 80, 3078-3080.
6. Midoux P., Mendes C., Legrand A., Raimond J., Mayer R., Monsigny M., Roche A.C., Specific gene transfer mediated by lactosylated poly-L-lysine into hepatoma cells. Nucl. Acids Res., 1993, 21, 871-878.
7. Perales J.C., Ferkol T., Beegan H., Ratnoff O.D., Hanson, R.W., Gene transfer in vivo: sustained expression and regulation of genes introduced into the liver by receptor-targeted uptake. Proc. Natl. Acad. Sci. USA, 1994, 91, 4086-4090.
8. Chen J., Stickles R.J., Daichendt K.A., Galactosylated histone-mediated gene transfer and expression. Hum. Gene. Ther., 1994, 5, 429-435.
9. Plank C., Zatloukal K., Cotten M., Mechtler K., Wagner E., Gene transfer into hepatocytes using asialoglycoprotein receptor mediated endocytosis of DNA complexed with an artificial tetra-antennary galactose ligand. Bioconj. Chem., 1992, 3, 533-539.
10. Wu G., Wu C., Evidence for targeted gene delivery to Hep G2 hepatoma cells *in vitro*. Biochemistry 1988, 27, 887-892.
11. Wu G.Y., Wu C.H., Receptor-mediated gene delivery and expression *in vivo*. J. Biol. Chem., 1988, 263, 14621-14624.
12. Wu C.H., Wilson J.M., Wu G.Y., Targeting genes: Delivery and persistent expression of a foreign gene driven by mammalian regulatory elements *in vivo*. J. Biol. Chem ., 1989, 264, 16985-16987.
13. Wilson J.M., Grossman M., Wu C.H., Chowdhury N.R., Wu G.Y., Chowdhury J.R., Hepatocyte-directed gene transfer in vivo leads to transient improvement of hypercholesterolemia in LDL receptor-deficient rabbits. J. Biol. Chem., 1992, 267, 963-967.
14. Fabrikant J.I., The kinetics of cellular proliferation in regenerating liver. J. Cell Biol., 1968, 36, 551-565.
15. Wu G.Y., Wilson J.M., Shalaby F., Grossman M., Shafritz D.A., Wu C.H., Receptor-mediated gene delivery *in vivo*: Partial correction of genetic analbuminemia in Nagase rats. J. Biol. Chem. ,1991, 266, 14338-14342.
16. Wilson J.M., Grossman M., Cabrera J.A., Wu C.H., Wu, G.Y., A novel mechanism for achieving transgene persistence *in vivo* following somatic gene transfer into hepatocytes. J. Biol. Chem., 1992, 267, 11483-11489 .
17. Roy-Chowdhury N., Wu C.H., Wu G.Y., Yerneni P.C., Bommineni V.R., Roy-Chowdhury J., Fate of DNA targeted to the liver by asialoglycoprotein receptor-mediated endocytosis in vivo: prolonged

persistence in cytoplasmic vesicles after partial hepatectomy. J. Biol. Chem. , 1993, 268, 11265-11271.

18. Cristiano R. J., Smith L. C., Woo, S. L. C., Hepatic gene therapy: adenovirus enhancement of receptor-mediated gene delivery and expression in primary hepatocytes. Proc. Natl. Acad. Sci. USA, 1993, 90, 2122-2126.
19. Wu G.Y., Zhan P.L., Sze L.L., Rosenberg A.R., Wu C.H., Incorporation of adenovirus into a ligand-based DNA carrier system results in retention of the original receptor specificity and enhances targeted gene expression. J. Biol. Chem. 1994, 269, 11542-11546.
20. Sells M.A., Chen M.L., Acs G., Production of hepatitis B virus particles in Hep G2 cells transfected with cloned hepatitis B virus DNA. Proc. Natl. Acad. Sci. USA , 1987, 84, 1005-1009.
21. Wu G.Y., Wu C.H., Specific inhibition of hepatitis B viral gene expression *in vitro* by targeted antisense oligonucleotides. J. Biol. Chem. 1992, 267,12436-12439.

7. Particle-Mediated Gene Delivery System for Cancer Research

Alexander L. Rakhmilevich and Ning-Sun Yang

7.1 Introduction

Most of the current approaches for cancer gene therapy deal with the modification of tumor cells or lymphocytes *ex vivo,* rather than with direct, *in vivo* gene delivery into the tumor or tumor-bearing host. We describe here a developing *in vivo* particle-mediated gene delivery technology, which can lead to regression of established murine tumors followed by the development of immunological memory. Whereas the gene gun technology has also been shown to be effective for a range of *ex vivo* gene transfer systems, this aspect has been reviewed elsewhere (1, 2) and will not be emphasized here.

7.2 Cancer Immunotherapy: Complex Problems and New Strategies

It is generally believed, based on the results from clinical studies, that while immunotherapy strategies for specific cancers using cytokines have yet to show routine efficacy, limited but increasing success rate also holds considerable therapeutic promise (3, 4). Many patients in early clinical trials, however, experienced significant incapacitating side effects, mainly resulting from the systemic administration of physiologically intolerable cytokine dosages (3, 5, 6). To avoid or reduce such toxicity problems, many studies have been performed to evaluate localized, peritumoral cytokine administration, either by a direct, *in situ* protein injection method or by a gene transfer approach. Initial successes of the latter approach in animal models (7, 8) introduced a new strategy for cytokine therapy for cancers. Since most cytokine proteins have a short half-life in the human blood or tissues, many cancer investigators now hypothesize that cytokine gene therapy may provide an alternative method for the induction of high-level, localized cytokine production resulting in antitumor immune activity with greatly reduced side effects.

Experiments with animal models suggest that *in situ* production of certain cytokines by genetically modified tumor or immune cells can effectively induce inflammatory and/or tumor cell-specific immune responses, which inhibit growth or cause regression of transplantable tumors. For example, IL-12, IL-2, IL-4, IFN-γ and GM-CSF have been shown to mediate either T-cell dependent or inflammatory responses which lead to tumor regression (9-12). In many studies, mice that were vaccinated with the cytokine gene-modified tumor cells developed long lasting immunity against the unmodified, parental tumor cells (10-12). Recently, clinical trials have also been initiated to test cytokine gene-modified autologous tumor cells (13-15) or tumor infiltrating lymphocytes (16, 17) to treat patients with advanced cancers.

Most immunologically-based cancer gene therapy approaches are currently formulated on the hypothesis that greater presentation of tumor antigens will enhance *in vivo* T-cell recognition of tumor-specific or -associated peptides. These include the following experimental models: a) cytokine genes transfected into tumor cells (10-12); b) genes for tumor-associated antigens such as MAGE-1 (18) or mutant p53 product (19) transfected into tumor cells; c) "suicide" genes (thymidine kinase) transferred into tumor cells to mediate the direct and, apparently immunologically-mediated, bystander cell-killing effects (20); d) transfection of tumor cells with allogeneic MHC class I antigen genes (21); and e) expression of B7.1 co-stimulatory molecule genes in tumor cells to enhance tumor antigen presentation (22). Encouraging results have been obtained from these and other animal models, suggesting that gene therapy might provide a new modality for immunotherapy. However, if these findings are to offer potential applications to the treatment of human cancer, many basic research and technological barriers must be overcome first (23-25). These include the need to activate systemic immunity, simplify and upgrade gene delivery technology, and increase transgene expression above the levels currently achievable for most experimental systems. Unlike the theoretical requirements for gene therapy of genetic diseases, it has recently been suggested that a high level of transient or short-term (days to weeks) transgene expression, rather than moderate or low levels of stable, long-term (months to life-span) expression, may be more desirable or sufficient for various cancer gene therapy approaches. Furthermore, it has also been postulated that effective delivery of candidate therapeutic genes at high dosages *in situ* to localized, immuno-reactive or -competent tissues, may first mediate or induce a strong, local immune response which later can result in a systemic immunity against the metastatic or recurrent cancer.

As a potential modality of treatment for cancer, specific regimens and dosages of transgenic proteins to be administered via a gene therapy protocol have to be carefully defined in pre-clinical and clinical studies. Before then, a key initial step is the demonstration of efficacy of a candidate gene therapy protocol in experimental animal systems. However, a major distinction between animal model systems and the pre-clinical studies/clinical trials is the nature of the target tissues for gene transfer. Unfortunately, the great majority of the current experimental studies in rodent or other tumor models have employed only immortalized tumor cell lines as the tumor source. It is well known that human tumor cells, obtained from the tissues freshly excised from patients, most often do not behave in tissue culture like the established, immortalized cell culture lines. As a result of this difference, efficacy or efficiency of gene transfer into human tumor or other somatic tissues, *in vivo* or *ex vivo*, were routinely found to be drastically and unpredictably different from those of transplantable rodent tumor models. Therefore, this major difference in gene transfer capability between these two biological systems must be effectively addressed in future gene delivery and expression vector research.

7.3 Current Techniques for *in vivo* Gene Transfer

The clinical delivery of genes capable of initiating and augmenting an antitumor immune response is dependent on defining effective delivery systems and augmenting gene expression in the tumor or targeted normal tissue. The advent of gene delivery biotechnology has allowed the introduction of marker and candidate therapeutic genes directly into the tumor, to adjacent normal tissues, or to various leukocytes, tumor cells, or stromal cells after their removal from the patient. Two technical approaches of gene therapy, *ex vivo* or *in vivo,* are classified according to the gene transfer method utilized. The *ex vivo* approach involves removing tumor or other desirable cell types from the host, genetically modifying them via gene transfer, and reintroducing them back into the host, to deliver specific transgenic products. The *in vivo* approach delivers therapeutic genes directly, sometimes *in situ*, into the tumor-bearing host to transfect the tumor cells or the surrounding normal tissues. It is also possible that a combination of *ex-vivo* and *in vivo* gene transfers may be desirable for certain future cancer gene therapy protocols.

7.3.1 Virus-mediated Gene Transfer

Application of retroviral vectors to gene therapy requires the production of replication-deficient viruses with high-titer infectivity. Newly developed amphotropic retrovirus vectors can confer stable gene transfer to many different cell types, mainly established cell lines. Therefore, retroviral vectors are currently much more effective under *in vitro* conditions than *in vivo*. Furthermore, many tumor cell types in primary culture preparations or fresh *ex vivo* cell explant samples (e.g., prostate and breast cancer cells) are apparently refractory to retroviral transduction. In addition, *ex vivo*-transduced cell populations routinely need to be enriched by efficient culturing for days or often weeks under selective pressure. Recently, Nabel and associates have demonstrated in comparative studies that the particle-mediated gene transfer method is often 10-fold more efficient than retrovirus systems in transduction of human peripheral blood or T cells, for both transient and stable gene transfers (26, 27). Thus, while the retrovirus vector system has been shown to be efficacious for a wide range of immortalized cell culture lines, it also has some inherent problems, including: a) the *ex vivo* manipulation can be very costly and time-consuming; b) the co-transfer or sequential delivery of different genes is technically difficult; and most importantly, c) retroviral gene delivery requires target cell proliferation for gene transfer, thus transfection of slow growing tumor or normal cells is limited (28).

Recently, researchers have successfully employed recombinant adenovirus (AV) as a gene transfer vector for a broad range of mammalian cell systems. Rosenfeld et al. (29, 30) used a recombinant adenovirus vector to transfer a recombinant human gene into lung respiratory epithelium *in vivo*. Recent works by Chen et al. (31), Cotten et al. (32, 33) and others have extended the use of the AV vector to gene therapy of studies in different tumor models. A potential disadvantage of this system, however, is that certain viral antigens are expressed using current AV vector systems, causing acute inflammation or other immunological side effects in test animals (34, 35) and in human clinical trials (36).

7.3.2 Direct Gene Transfer

The limitations of virus-mediated delivery systems have fueled attempts to develop alternative, non-viral means of gene delivery into somatic cells *in vivo*. Other than efficiency, direct gene transfer into somatic cells may also have theoretical advantages compared to viral vectors, including safety, cost-effectiveness, immunogenicity and vector product reliability issues. Although, to our knowledge, there have been no direct long-term comparative studies, many investigators believe that non viral gene transfer methods should have a reduced risk of causing malignancy or acute toxicity, which are potentially associated with viral vectors (37).

Direct gene transfer via intramuscular injection of naked plasmid DNA can deliver genes to non- dividing, "resting" somatic cells (38); however, it can be technically unreliable, and is not readily applicable to non-muscle somatic tissues. Direct intratumor injection of a DNA/liposome complex containing allogenic MHC class I cDNA can result in a significant antitumor effect in mice (21). Limited, but detectable antitumor effect was also reported in early clinical studies in melanoma patients (39). Routine applications of this *in vivo* gene delivery method for solid tumor tissues are being actively investigated by many researchers.

7.4 Particle-Mediated Gene Transfer: Techniques and Potential Application to Gene Therapy

The particle-mediated gene transfer by gene gun technology developed at Agracetus by McCabe, Yang, and colleagues (40), is a new approach for mammalian gene transfer. It uses a shock wave to accelerate DNA-coated gold particles into target cells (1). At submicrogram quantities of DNA per dose for *in vitro*, *ex-vivo*, or *in vivo* gene transfer, the gene gun physically enables intracellular delivery of up to thousands of DNA copies per cell into targeted tissues. Resulting transgene expression levels are often significantly higher than those achieved by other DNA delivery methods, such as lipofection, calcium phosphate precipitation, or electroporation (1, 41, 42). Moreover, it takes only about 10 to 30 seconds to complete a single treatment with the newly developed, hand-held *Accell* helium pulse gene gun (Fig. 7.1A), which greatly increases the convenience, manipulability and potential utility of the *Accell* technology for gene therapy research and for human clinical applications. We have shown that epidermal tissues of mouse, rhesus monkey, rat, dog, and pig are excellent *in vivo* target tissues for this gene therapy approach. Recently, we have further shown that human skin xenografts implanted on SCID mice can also serve as supreme tissue targets for gene gun-mediated *in vivo* gene transfer and expression of HLA genes (43). We thus anticipate that human skin, a fully immunocompetent organ, should function as an excellent target tissue for gene gun-mediated DNA delivery in candidate therapeutic studies and in future clinical trials.

During the past several years we and others have successfully demonstrated that the Accell gene gun technology can be effectively applied to a variety of mammalian somatic cells and tissues (1, 41, 44). In particular, we have actively evaluated the application of the *Accell* gene gun method to *ex-vivo* gene transfer into human prostrate (44), sarcoma and melanoma (45), breast (41) and other cancer cells. Significant advantages of the gene gun technology include the ability to: 1) physically deliver intracellular naked DNA into targeted cells, regardless of the receptors or structural features of the target cell membranes; 2) accurately direct particle delivery to the tissue of interest for *in situ* gene transfer; 3) transfect resting, non-dividing, or differentiated cell types, irrespective of cell lineage (46); and 4) modify the levels and pattern of transgene expression over time via a repeated gene treatment regimen (47). Therefore, the gene gun technique potentially can overcome some of the deficiencies of virus-mediated gene transfer.

7.4.1 Simultaneous Delivery of Multiple Therapeutic Genes

Several groups (48, 49) have demonstrated that the gene gun methods can be effectively employed to deliver multiple reporter or candidate therapeutic genes, at desirable specific molar ratios, into a wide spectrum of mammalian somatic cell, tissue or organ systems. We have shown that the hand-held gene gun devices can be applied efficiently *in situ* with a high degree of versatility to various somatic tissues of a mouse or rat (1). Some of the findings on *in vivo* gene transfer are now being extended to large experimental animals including dogs (Fig. 1B), pigs, and rhesus monkeys. Thus, the technology offers the potential for custom-tailored delivery of multiple therapeutic genes at the locations of primary and metastatic tumors originating in different organs.

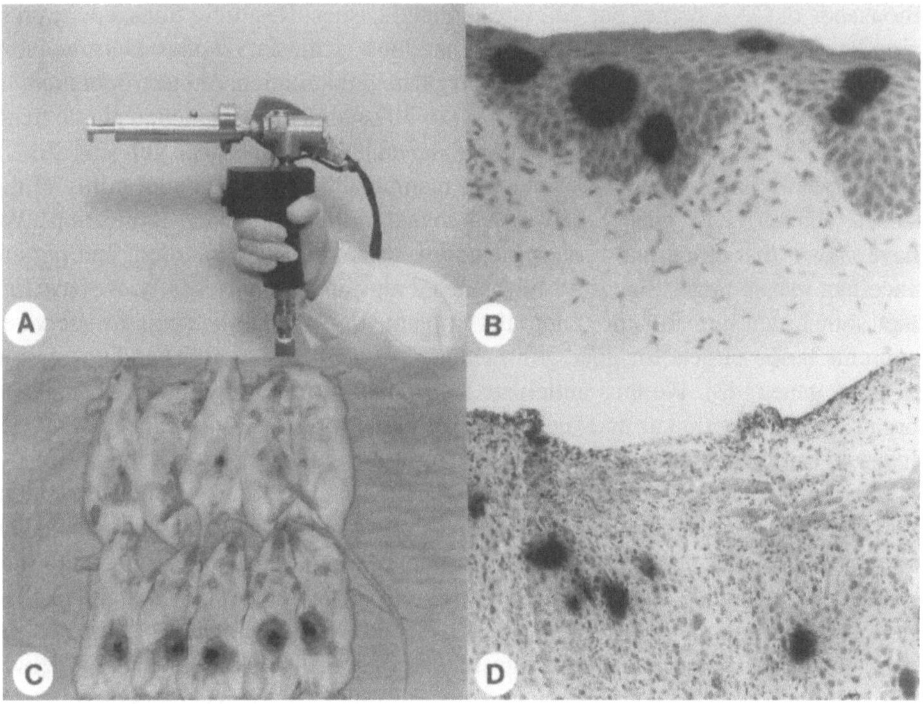

Fig. 7.1.A: General design of the Accell® helium-pulse gene gun (courtesy of Dr. McCabe, Agracetus, Inc.).

Fig. 7.1.B: *In vivo* expression of transgenic β-galactosidase activity in dog oral mucosal epithelium transfected with p-CMVβ-gal DNA using the Accell gene gun method.

Fig. 7.1.C: Antitumor effect of gene gun-mediated, *in vivo* IL-12 gene delivery observed in Renca carcinoma model. Mouse skin overlying and surrounding the established (0.5 cm in diameter) tumor was transfected with IL-12 (upper row) or Luc (lower row) cDNA expression vectors on day 7 after intradermal tumor cell implantation, following by two additional transfections on days 9 and 11. The photograph of the mice was taken on day 20 post tumor cell implantation.

Fig. 7.1.D: Transgenic CMVβ-gal expression in mouse Renca tumors transfected *in situ* using the Accell gene gun. Note that DNA/gold particle penetration reached up to 30 cell layers deep into the subcutaneous tumor mass, when surgically exposed and transfected from the dermal skin flap side.

7.4.2 High Level Expression of Candidate Therapeutic Genes by Local Delivery

For *in vivo* gene transfer into various somatic tissues of rats or mice, we found that epidermis, liver and pancreas tissues are highly susceptible to the particle mediated gene transfer, and exhibit the highest levels of reporter and cytokine gene expression (44, 48). Kidney, spleen and colon tissue express medium levels, while muscle, vascular and heart tissues express the lowest levels. Since efficient transgene expression can be obtained by gene gun transfections in many tissues types *in vivo,* this method may address one of the major difficulties currently associated with the cytokine therapy approach for cancer, and it may also offer promise for a cancer vaccine strategy.

Specifically, we have shown that 50 to 200 ng of transgenic IFN-γ and TNF-α proteins were produced at the gene gun-treated site at 24 hours post transfection of mouse epidermal tissue. We estimate that each target mouse epidermis site contains between 1 to 5 million cells, and approximately 10 to 15 % of treated cells transiently produce transgenic cytokines. This can result in effective secretion of transgenic cytokines into circulation at a level of 30 to 200 pg/ml serum in test mice (50).

7.4.3 *In vivo* Promoter Usage, Long-Term Transgene Expression

Using the *Accell* gene gun method, Cheng et al. (48) compared and analyzed the relative promoter strength of 5 viral and 5 mammalian cellular promoters for *in vivo* expression in skin, liver, muscle and pancreas tissues: a quantitative analysis for *in vivo* promoter activity in five different tissues was not previously technically possible using other available gene transfer methods (48). These investigators also evaluated the level, duration and cellular pattern of reporter gene expression in various somatic tissues, and determined that gene gun transfected dermis and abdominal muscle tissue can achieve long-term expression of a luciferase transgene, occasionally lasting for more than 18 months in test rats (48). Therefore, the particle-mediated gene transfer method can confer long-term *in vivo* transgene expression in specific muscle, or derived cell types, which may allow the sustained therapeutic transgene production in muscle tissues to combat tumor recurrence.

7.5 Antitumor Effects of Gene Gun-Mediated Cytokine Gene Transfer *in vivo*: Transgenic Cytokine Secreted from Normal Skin Cells Results in Tumor Regression

We have recently reported that a gene gun-mediated, *in vivo* cytokine gene therapy protocol can reduce tumor growth in mice (50). Combination treatments with IFN-γ and TNF-α genes, or IFN-γ and IL-2 genes, or treatment with an IL-6 gene alone, starting shortly after implantation of tumor cells, inhibited tumor growth and prolonged the survival of tumor-bearing mice.

Our recent studies have shown that a dramatic, immunologically-mediated antitumor effect can be achieved via gene gun delivery of IL-12 cDNA into the skin tissue overlying various implanted tumors (47). Particle-mediated delivery of the IL-12 gene resulted in suppression of tumor growth or in complete regression of established tumors which were 4-8 mm in diameter at the beginning of the gene therapy protocol (Fig. 7.1C). Only 1 to 4 gene transfer treatments *in vivo* were required to achieve these anti-tumor effects in six experimental tumor models. Notably, local IL-12 gene therapy increased the survival of mice from spontaneous metastases of P815 tumors. The antitumor effect of IL-12 gene therapy was shown to be $CD8^+$ T cell-mediated and led to tumor-specific immunological memory. It is important to emphasize that the IL-12 levels detected at the site of skin transfection were between 1/400 to 1/ 40,000 of the dosage (0.1-10 μg of recombinant IL-12 protein, which caused, following systemic administration, both antitumor effects and toxicity in mice (51-53).

Histological examination of the transfected mouse skin tissue overlying a 7-day intradermal Renca tumor revealed that IL-12 transgene expression was detected mainly in the epidermal cell layers, which is consistent with our previous reports using other cytokine genes (1, 50). These findings strongly suggest that the antitumor effect of this skin transfection protocol is largely the result of production and secretion of transgenic IL-12 by normal epidermal cells in the vicinity of the tumor. We and others have documented (40, 48, 49) that the *Accell* gene gun technology can be readily and effectively applied to *in vivo* gene transfer into skin from drastically different experimental animal species, including rodents, dog, pig and rhesus monkey. Together, the above findings provide the foundation for future application of gene gun technology to research using large animals as experimental systems or in pre-clinical studies for cancer gene therapy. Furthermore, we observed that the 1-3 μm DNA-coated gold particles delivered by the newly developed pulse gene gun can effectively penetrate 20 to 30 epidermal cell layers, and result in expression of β-gal and HLA-B7 genes in human skin engrafted onto SCID/hu mice (43). We therefore suggest that, for human applications, skin and subcutaneous tumors may serve as good *in vivo* target tissues for expression and treatment with candidate therapeutic cytokine genes. Alternatively, tumor cells can be transfected *in situ* on non-resectable tumors or on residual tissues after the surgical exposure of the bulk tumor mass. As an example, we have shown that, when a mouse skin flap containing an implanted tumor was surgically exposed and transfected from the dermal side, transgene expression was obtained in up to 30 cell layers deep into the targeted tumor tissue (Fig. 7.1D).

7.6 Gene Gun-Mediated *ex vivo* Gene Transfer to Clinically-Relevant Tissue Samples

In addition to *in vivo* gene transfer, the gene gun technique has also been shown to be efficient for *ex vivo* gene transfer into a wide range of mammalian somatic tissue or primary culture systems, namely monolayer cultures, cells in suspension, freshly isolated tissue or cell clumps, tissue sections and explants, including human clinical tumor samples. Our laboratories have shown that the gene gun method is up to 100-

fold more efficient than three other direct gene transfer methods for primary cultures of mammary epithelial cells (41), fetal or adult brain cells (54) and oligodendrocytes (42). Woffendin et al. (26) and Fox et al. (27) demonstrated particle-mediated, stable gene transfer to primary human T cells. Gene transfer to primary leukocytes (55) and CD34+ hematopoietic progenitor cells (56) were also achieved. We also obtained data demonstrating that primary human prostate tumors (44) and freshly excised human or rodent mammary gland biopsies prepared as organoids (40, 41) can be effectively transfected by gene gun *ex vivo*.

We recently evaluated gene gun-mediated, *ex vivo* cytokine gene transfer into rodent and human tumor cell lines and to freshly isolated human and canine tumor explants with a single gene gun treatment in cell suspension. The expression of murine GM-CSF in γ-irradiated B16 cells was achieved for up to 13 days at levels > 100 ng/ml/106 cells/24 hr, which is comparable to published results obtained using retroviral vectors. Importantly, it was found that vaccination of mice with weakly immunogenic B16 melanoma cells expressing transgenic GM-CSF could prevent the development of subsequent tumors in 58 to 75% of mice challenged with parental B16 tumors (45). The *ex vivo* gene transfer approach may thus be a component of an integrated cancer gene therapy strategy, in combination with *in vivo* gene transfer systems. Various freshly isolated human and canine tumor explants are now being studied as cancer vaccines for cytokine transgene expression using the gene gun technology (45).

The research grade helium pulse-driven gene gun will be manufactured and distributed by Bio-Rad by late 1996. Agracetus has recently built a facility with Good Manufacturing Procedure (GMP) and Good Laboratory Practice (GLP) qualifications, which will allow the production of clinical grade DNA cartridges for future human studies and potential clinical trials using the *Accell* gene gun technology.

Acknowledgments

We thank Carolyn De Luna for editing the manuscript and Joe Burkholder for providing the photographs.

References

1. Yang, N.S. and Sun, W. (1995). Gene Gun and Other Non-viral Approaches for Cancer Gene Therapy. Nature Med. *1*, 481-483.
2. Yang, N.S., Sun, W. H., and McCabe, D. (1996). Developing particle-mediated gene-transfer technology for research into gene therapy. Mol. Med. Today 2, 476-481.
3. Kolitz, J. and Mertelsmann, R. (1991). The Immunology of Human Cancer with Interleukin 2: Present Status and Future Directions. Cancer Investigation 9, 529-542.
4. Maas, R., Dullens, H., and Otter, W. D. (1993). Interleukin-2 in cancer treatment: disappointing or (still) promising? A review. Cancer Immunol Immunother *36*, 141-148.
5. Lotze, M., Matory, Y., Rayner, A. et al. (1986). Clinical Effects and Toxicity of Interleukin-2 in Patients with Cancer. Cancer 58, 2764.

6. Marshall, E. (1995). Cancer Trial of Interleukin-12 Halted. Science *268*, 1555.
7. Gansbacher, B., Zier, K., Daniels, B., Cronin, K., Bannerji, R., and Gilboa, E. (1990). Interleukin 2 Gene Transfer into Tumor Cells Abrogates Tumorigenicity and Induces Protective Immunity. J Exp. Med. *172*, 1217-1224.
8. Golumbek, P., Lazenby, A., Levitsky, H., et al. (1991). Treatment of Established Renal Cancer by Tumor Cells Engineered to Secrete Interleukin-4. Science *254*, 713-716.
9. Heaton, K.M. and Grimm, E.A. (1993). Cytokine combinations in immunotherapy for solid tumors: a review. Cancer Immunol. Immunotherapy *37*, 213-219.
10. Pardoll, D. (1992). New strategies for active immunotherapy with genetically engineered tumor cells. Current Opinion in Immunol. 4, 619-623.
11. Colombo, M.P., Modesti, A., Parmiani, G., and Forni, G. (1992). Local cytokine availability elicits tumor rejection and systemic immunity through granulocyte-T-lymphocyte cross-talk. Cancer Res. *52*, 4853-4857.
12. Colombo, M. and Forni, G. (1994). Cytokine Gene Transfer in Tumor Inhibition and Tumor Therapy: Where are we now? Immunology Today *15*, 48-51.
13. Rosenberg, S. A. (1992). Immunization of cancer patients using autologous cancer cells modified by insertion of the gene interleukin-2 (IL-2). Human Gene Therapy, *3*, 167-177.
14. Cassileth, P. A. and Podack, E. (1995). Phase I study of transfected cancer cells expressing the interleukin-2 gene product in limited stage small cell lung cancer. Human Gene Therapy, *6*, 369-383.
15. Stingl, G. (1996). Phase I study to the immunotherapy of metastatic malignant melanoma by a cancer vaccine consisting of autologous cancer cells transfected with the human IL-2 gene. Human Gene Therapy, *7*, 551-563.
16. Rosenberg, S. A. (1990). Gene therapy of patients with advanced cancer using tumor-infiltrating lymphocytes transduced with the gene coding for tumor necrosis factor. Human Gene Therapy, *1*, 441-480.
17. Rosenberg, S. A., Aebersold, P., Cornetta, K., Kasid, A., Margan, R. A., Moen, R. A., Karson, E. M., Lotze, M. T., Yang, J. C., Topalian, S. L., Merino, M. J., Culver, K., Miller, A. D., Blaese, R. M., and Anderson, W. F. (1990). Gene transfer into humans – immunotherapy of patients with advanced melanoma using tumor infiltrating lymphocytes modified by retroviral gene transduction. N. Engl. J. Med., *323*, 570-578.
18. Toso, J. F., Oei, C., Oshidari, F., Tartaglia, J., Paoletti, E., Lyerly, H. K., Talib, S., and Weinhold, K. J. (1996). MAGE-1-specific precursor cytotoxic T-lymphocytes present among tumor-infiltrating lymphocytes from a patient with breast cancer: characterization and antigen-specific activation. Cancer Res. *56*, 16-20.
19. Noguchi, Y., Chen, Y.N., and Old, L. (1994). A Mouse Mutant p53 Product Recognized by CD4+ and CD8+ T cells. Proc. Natl. Acad. Sci. USA *91*, 3171-3175.
20. Tiberghien, P. (1994). Use of Suicide Genes in Gene Therapy. J. Leukocyte Biology *56*, 203-209.
21. Plautz, G.E., Yang, Z.Y., Wu, B.Y., Gao, X., Huang, L., and Nabel, G.J. (1993). Immunotherapy of malignancy by in vivo gene transfer into tumors. Proc. Natl. Acad. Sci. USA *90*, 4645-4649.
22. Chen, L., Linsey, P., and Hellström, K.E. (1993). Costimulation of T cells for tumor immunity. Immunol. Today *14*, 483-486.
23. Touchette, N. (1996). Not ready for prime time. Nature Medicine 2, 7-8.
24. Kreeger, K. Y. (1996). Gene therapy trials should be driven by basic science. The Scientist *22*, 10-12.
25. Marshall, E. (1996). Less hype, more biology needed for gene therapy. Science *270*, 1751.
26. Woffendin, C., Yang, Z. Y., Udaykumar, Yang, N. S., Sheehy, M., and Nabel, G. (1994). Nonviral and viral delivery of a human immunodeficiency virus protective gene into primary human T cells. Proc. Natl. Acad. Sci. U.S.A. *91*, 11581, 11585.
27. Fox, B., Woffendin, C., Yang, Z. Y. San, Hong, Ranga, U., Gordon, D., Osterholzer, J., and Nabel, G. (1995). Genetic Modification of Human Peripheral Blood Lymphocytes with a Transdominant Negative Form of Rev: Safety and Toxicity. Human Gene Therapy *6*, 997-1004.
28. Miller, D. G., Adam, M. A., and Miller, A. D. (1990). Gene transfer by retrovirus vector occurs only in cells that are actively replicating at the time of infection. Mol. Cell. Biol. *10*, 4239-4242.
29. Rosenfeld, M. A., Siegfried, W., Yoshimura, K., Yoneyama, K., Fukayama, M., Stier, L.E., Paakko, P. K., Giraldi, P., Stratford-Perricaudet, L. D., Perricaudet, M., Jallat, S., Pavirani, A., Iecocq, J., and Crystal, R. G. (1991). Adenovirus-mediated transfer of recombinant alpha-1-antitrypsin gene to the lung epithelium in vivo. Science *252*, 431-434.

30. Rosenfeld, M. A., Yoshimura, K., Trapnell, B. C., Yoneyama, K., Rosenthal, E. R., Dalemans, W., Fukayama, M., Bargon, J., Stier, L.E., Stratford-Perricaudet, L. D., Perricaudet, M., Guggino, W. B., Pavirani, A., Iecocq, J. P., and Crystal, R. G. (1992). In vivo transfer of the human cystic fibrosis transmembrane conductance regulator gene to the airway epithelium. Cell *68*, 143-155.
31. Chen, S-H., Shine, H. D., Goodman, J. C., Grossman, R.G., and Woo, S.L.C. (1994). Gene therapy for brain tumors: Regression of experimental brain tumors by adenovirus-mediated gene transfer in vivo. Proc. Natl. Acad. Sci. USA *91*, 3054-3057.
32. Cotten, M. Wagner., E., Zatloukal, K., Phillips, S., Curiel, D., and Birnstiel, M.L. (1992). Coupling of adenovirus to transferrin-polylysine/DNA complexes greatly enhances receptor-mediated gene delivery and expression of transfected cells. Proc. Natl. Acad. Sci. USA *89*, 6094-6098.
33. Gao, L., Wagner, E., Cotten, M., Agarwal, S., Harris, C., Rømer, M., Miller, L., Hu, P.C., and Curiel, D. (1993). Direct gene transfer into airway epithelium employing adenovirus-polysine-DNA complexes. Human Gene Therapy *4*, 17-24.
34. Yang, Y., Nunes, F. A., Berencsi, K., Furth, E.E., Gönczöl, E., and Wilson, J. (1994). Cellular immunity to viral antigens limits E1-deleted adenoviruses for gene therapy. Proc. Natl. Acad. Sci. USA *91*, 4407-4411.
35. Adenoviral vectors damage normal arteries: use cautioned. (1996). Gene Therapy Weekly, March 4, 2-3.
36. Noguchi, P. (1994). As an oral presentation "FDA perspectives on gene therapy," CHI symposium: "Gene Therapy, New Techniques and Applications." April 25-27, Bethesda, MD.
37. Fox, J. L. (1993). NIHRAC and FDA ponder gene-therapy risks. Bio/Technology *11*, 28-29.
38. Yang, N.S. (1992). Gene transfer into mammalian somatic cells in vivo, in "CRC Critical Reviews in Biotechnology." Toronto, CRC Press, 335-356.
39. Nabel, G.J., Nabel, E.G., Yang, Z.Y., Fo, B.A., Plautz, G.E., Gao, X., Huang, L. Shu, S., Gordon, D., and Chang, A.E. (1993). Direct gene transfer with DNA-liposome complexes in melanoma: expression, biologic activity, and lack of toxicity in humans. Proc. Natl. Acad. Sci. USA *90*, 11307-11311.
40. Yang, N.S., Burkholder, J., Roberts, B., Martinell, B., and McCabe, D. (1990). In vivo and In vitro gene transfer to mammalian somatic cells by particle bombardment. Proc. Natl. Acad. Sci. USA *87*, 9568-9572.
41. Thompson, T. A., Gould, M. N., Burkholder, J. K., and Yang, N.S. (1993). Transient promoter activity in primary rat mammary epithelial cells evaluated using particle bombardment gene transfer. In Vitro Cell Dev. Biol. *29A*, 165-170.
42. Guo, Z., Jiao, S. S., Cheng, L., Wolff, J., Yang, N. S., and Duncan, I. (1996). Efficient gene transfer and expression in mature rat oligodendrocytes in culture. J. Neuroscience Research, 43, 32-41.
43. Kusaka, S., Turner, J., Burlington, W., and Yang, N. S. (1996). Characteristics of in vivo expression of HLA transgenes in human skin grafts on SCID mouse. Submitted.
44. Cheng, L., Sun, J., Pretlow, T., and Yang, N. S. (1996). CWR22 xenograft as an ex-vivo human tumor model for prostate cancer gene therapy. J. Natl. Cancer Inst., 88, 607-611.
45. Mahvi, D. M., Burkholder, J. K., Turner, J., Malter, J., Sondel P., and Yang, N.S. (1996). Particle-mediated gene transfer of GM-CSF cDNA to tumor cells: Implications for a clinically relevant tumor vaccine. Human Gene Therapy 7, 1535-1543.
46. Rajagopalan, L. E., Burkholder, J. K., Turner, J., Culp, J., Yang, N.-S., and Malter, J. (1995). Targeted mutagenesis of GM-CSF cDNA increases transgenic mRNA stability and protein expression in normal cells. Blood, *86*, 2551-2558.
47. Rakhmilevich, A., Turner, J., Ford, M., McCabe, D., Wenn, S. H., Sondel, P. M., Grota, K., and Yang, N. S. (1996). Gene gun-mediated skin transfection with interleukin 12 gene results in regression of established primary and metastatic murine tumors. Proc. Natl. Acad. Sci. USA, 93, 6291-6296.
48. Cheng, L., Ziegelhoffer, P., and Yang, N.S. (1993). I n vivo promoter activity and transgene expression in mammalian somatic tissues evaluated by using particle bombardment. Proc. Natl. Acad. Sci. USA *90*, 4455-4459.
49. Williams, R.S., Johnson, S.A., Riedy, M., DeVit, M.J., McElligot, S.G., and Sanford, J.C. (1991). Introduction of foreign genes into tissues of living mice by DNA-coated Microprojectiles. Proc. Natl. Acad. Sci. USA *88*, 2726-2730.
50. Sun, W. H., Burkholder, J. K., Sun, J., Decker, J., Turner J. G., Lu, X.G., Pugh, T.D., Ershler, W.B., and Yang, N.S. (1995). In vivo cytokine gene transfer by particle bombardment reduces tumor growth in mice. Proc. Natl. Acad. Sci. USA *87*, 9568-9572.

51. Brunda, M., Luistro, L., Warrier, R., Wright, R., Hubbard, B., Murphy, M., Wolf, S., and Gately, M. (1993). Antitumor and Antimetastatic Activity of Interleukin 12 against Murine Tumors. J Exp. Med. *178*, 1223-1230.
52. Nastala, C., Edington, H., McKinney, T., Tahara, H., Nalesnik, M., Brunda, M., Gately, M., Wolf, S., Schreiber, R., Storkus, W., and Lotze, M. (1994). Recombinant IL-12 Administration Induces Tumor Regression in Association with IFN-γ Production. J Immunology *153*, 1697-1706.
53. Orange, J., Salazar-Mather, T., Opal, S., Spencer, R., Miller, A., McEwen, B., and Biron, C. (1995). Mechanism of Interleukin 12-mediated Toxicities during Experimental Viral Infections: Role of Tumor Necrosis Factor and Glucocorticoids. J. Exp. Med. *181*, 901-914.
54. Jiao, S., Cheng, L., Wolff, J. A., and Yang, N.S. (1993). Particle bombardment-mediated gene transfer and expression in rat brain tissues. Bio/Technology *11*, 497-502.
55. Burkholder, J. K., Decker, J., and Yang, N.S. (1993). Rapid transgene expression in lymphocyte and macrophage primary cultures after particle bombardment-mediated gene transfer. J. Immunol. Meth. *165*, 149-156.
56. Ye, Z. Q., Qiu, P., Burkholder, J. K., Turner, J., Culp, J., Roberts, T., Shahidi, N. T., and Yang, N.S. (1996). Particle-mediated gene transfer into CD34+ hematopoietic progenitor cells from human umbilical cord blood. Submitted.

II Methods for Inactivation or Restoration of Gene Function

8. Gene Inactivation by Homologous Recombination – Animal Models of Plasma Lipoprotein Disorders

Thomas E. Willnow and Joachim Herz

8.1 Introduction

In its 10th edition, McKusick's catalog on Mendelian disorders in man lists almost 6,000 inborn errors, approximately 300 of which have been mapped to individual gene defects (1). Typically, the identification of mutations underlying inheritable metabolic defects has been achieved by biochemical analysis of disease phenotypes which led to the identification of the dysfunctional proteins. Subsequently, alterations in the genetic information coding for these gene products was demonstrated in affected individuals (e.g. abetalipoproteinemia, familial hypercholesterolemia, phenylketonuria, thalassemia). Alternatively, positional cloning has been applied successfully to identify disease genes directly (e.g. Huntington's disease, Duchenne muscular dystrophy, cystic fibrosis). The latter approach is likely to become ever more important as efforts to establish a complete high resolution map of the human genome and to sequence the entire three billion base pairs of genetic information are ongoing (2). The detailed characterization of the human genome will aid in the identification of candidate genes localized to chromosomal regions previously implicated in genetic disorders and will allow the identification of disease gene loci. An in-depth knowledge about the nature of these disease gene loci will be important for early diagnosis and preventive treatment of inherited disorders such as certain forms of cancer, neurodegenerative defects or cardiovascular diseases. In addition, it may provide us with the means to correct inborn errors by somatic gene therapy.

Information about the genetic basis of inherited diseases in man did not only come from analysis of the human genome per se but from the analysis of animal models as well. A number of gene defects have been identified in various animal species that parallel known human inherited diseases. Among these are hemophilia in dogs and the low density lipoprotein (LDL) receptor defect in rabbits. Previously, the isolation of animal models has been solely dependent on either screening spontaneously occurring mutants or on random mutagenesis using carcinogens and retroviral insertions. Recent ground breaking advances in mouse genetics have greatly facilitated this process. Today, we are able to introduce specific alterations into any mouse gene and to study the consequence of the alteration *in vivo*. This has been made possible by technological advances in two areas: 1) the ability to introduce defined changes into the mouse genome using homologous recombination, and 2) the capability to reintroduce genetically modified pluripotent embryonic stem cells into the germ line, generating mutant mouse strains. Application of this technology resulted in a burst of information on the genetic basis of metabolic defects both in mice and humans. For example, the inactivation of specific genes in the mouse directly led to the identification of the genetic defect of previously not characterized human inheritable disorders (e.g. Hirschsprung disease) (3). Furthermore, the reenactment of human monogenic diseases in the mouse has enabled us to study the consequence of single gene defects in detail and to dissect the contribution of both genetic and environmental factors to complex metabolic disorders such as diabetes, obesity or atherosclerosis. Finally, the generation of murine models of human genetic defects has given us the tools to design and test gene therapy approaches.

The following chapter will describe methods of homologous recombination currently employed to introduce defined alterations into the mouse genome *in vivo*. Furthermore, it will describe the application of these methods to generate mouse models of plasma lipoprotein disorders. These models have been most informative for our understanding of pathological processes of the plasma cholesterol metabolism and have provided us with the opportunity to design and test somatic cell gene transfer protocols for treatment of cardiovascular diseases.

8.2 Methods for Gene Targeting in the Mouse

8.2.1 Gene Disruption by Homologous Recombination

Cells have the intrinsic capability to swap sequences between two homologous duplex molecules of DNA. This so called homologous recombination event results in a reciprocal exchange of genetic information so that no nucleotides are lost or added to the participating DNA molecules. Homologous recombination not only takes place between homologous chromosomes but also between chromosomes and extrachromosomal plasmid molecules that have been introduced artifically into the cells. In yeast, where homologous recombination occurs with high frequency, this mechanism has been exploited to integrate plasmid molecules specifically into endogenous genes (gene targeting) (4, 5). The application of gene targeting to modify the genome of established mammalian cell lines has also been shown (6, 7). However, as homologous recombination is a rare event in these cells, the efficiency of recovering cells that have undergone homologous rather than random integration of the introduced plasmid DNA has been low. Only after the development of positive/negative selection schemes to enrich for homologous recombination several thousand fold has gene targeting become a feasible and valuable method for studying the consequence of gene modification in higher eukaryotic cells (8, 9).

The positive/negative selection scheme employed in most laboratories for gene targeting has been developed by Capecchi and coworkers (10). It results in inactivation of the target gene by integration of a positive selection marker into the gene coding region. Initially, a replacement-type targeting vector is constructed that contains part of the target gene, functionally disrupted by the positive selection marker (Fig. 8.1). Typically, this marker is the *neo* gene that confers resistance to the aminoglycoside G^{418} on cells that have taken up the construct. To enrich for integration of the vector into the target gene via homologous recombination, a negative selection step is employed simultaneously, eliminating cell clones that have undergone random integration of the plasmid DNA. This was achieved by Mansour et al. (10) by introducing the thymidine kinase gene of herpes simplex virus (HSV-TK) in the targeting vector. HSV-TK converts inactive nucleoside analogs such as 1-[2-deoxy, 2-fluoro-β-D-arabinofuranosyl] – 5 iodouracil (FIAU) and ganciclovir into chain terminators that are incorporated into the DNA of the target cells. In the vector described in Figure 8.1, the HSV-TK gene flanks the homologous gene region and is lost, when the plasmid is integrated by homologous recombination. In contrast, random integration will frequently result in insertion of the HSV-TK gene into the host cell DNA. These cell clones will be eliminated by selection in FIAU. Disruption of the endogenous target gene by *neo* integration can be verified in individual cell clones using Southern blot analysis (Fig. 8.1).

The replacement-type targeting construct described here is suitable for complete inactivation of a target gene. In many instances this "knock-out" of gene function will be appropriate to test the significance of a gene product by studying a loss-of-function phenotype. For the generation of certain mouse models, however, modification rather then elimination of a gene product might be desirable. For example, some inherited disorders in humans are thought to be caused by a gain-of-function of mutated gene products (e.g. Huntington's disease). Therefore, more elaborate protocols of homologous recombination have been developed that can be used to introduce single base pair changes into a gene locus. Presentation of such strategies is beyond the scope of this chapter. However, they have been described by Bronson and Smithies in a recent review (11).

8.2.2 Generation of Mice from Genetically Modified Embryonic Stem Cells

A gene targeting event in diploid cells will generate cells that carry one wild type and one mutant allele. As most mutations are recessive, homozygosity for a gene defect will be necessary for a mutant phenotype to become apparent. To achieve

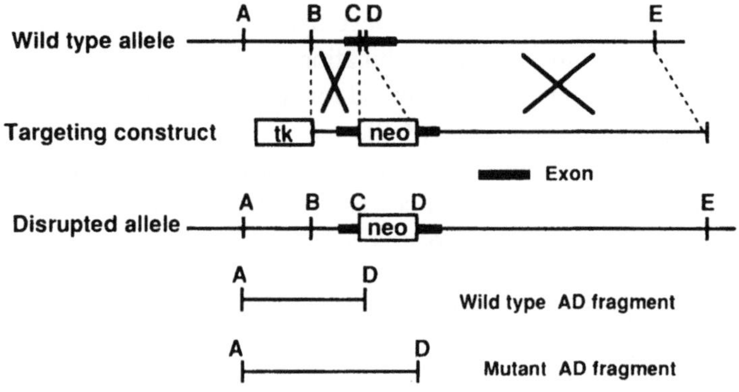

Fig. 8.1: Construction of a replacement-type targeting vector.
The replacement-type targeting construct is assembled from DNA segments BC and DE, which are derived from the murine target gene locus (Wild type allele). The *neo* expression casette (neo) is inserted between fragments BC and DE, physically disrupting exon sequences of the gene region (filled box). The herpes simplex virus thymidine kinase gene (tk) flanks the gene homology region and will be lost in the event of homologous recombination between targeting construct and wild type allele. Integration of the *neo* cassette into the target gene (Disrupted allele) will increase the size of restriction fragment AD. This can be used to differentiate between wild type and mutant gene locus by Southern blot analysis.

homozygosity for the disrupted allele in cells, two rounds of gene targeting have to be performed using two different positive selection markers (e.g. neomycin and hygromycin B resistance) (12). Alternatively, cells that have eliminated the wild type allele by gene conversion of the mutant allele can be selected by increasing the G^{418} concentration in the cell medium (13) or by negative selection of the wild type gene product (14). In most instances however, it will be advantageous to disrupt only one copy of the gene in cells and to establish a mouse strain that carries the mutant allele in its germ line. Mice heterozygous for the mutation can be bred to homozygosity and the effect of the gene inactivation can be studied both *in vivo* and in cultured cells derived from the homozygous mutant mice (15, 16).

The potential to introduce genetic modifications like gene disruptions into mice is based on initial studies by Mintz (17), extended by Evans and Kaufmann (18) and Martin (19). Their experiments demonstrated that embryonic stem cells (ES cells), derived from the inner cell mass of the early mouse embryos, could be propagated and manipulated in culture without loosing their pluripotency. After reintroduction into blastocysts these cells colonized many tissues of the developing embryo including the germ line. This offered the possibility to perform a gene disruption experiment in ES cell cultures and to introduce the genetic alteration into mice by transferring the genetically modified stem cells into mouse embryos.

Currently, two experimental approaches are pursued to introduce modified stem cells into mouse embryos. One possibility is the direct injection of stem cells into the blastocoel cavity of mouse embryos at day 3.5 of gestation using micromanipulation. These blastocysts are subsequently transferred into pseudopregnant foster

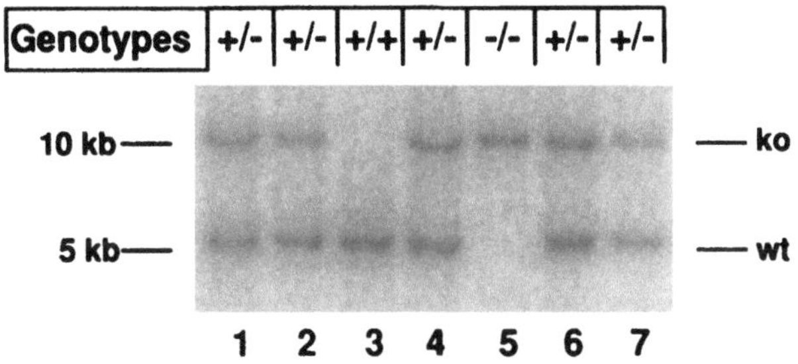

Fig. 8.2: Southern blot analysis of wild type and mutant mice.
DNA was obtained from offspring of mice heterozygous for a disruption of the gp330 gene (78). The DNA was digested with *BamHI* and *Hind*III and subjected to Southern blot analysis using a genomic fragment of the murine gp330 gene locus as hybridization probe. 5 kb and 10 kb restriction fragments indicative of the wild type (wt) and the mutant gene locus (ko) respectively, are indicated. +/+, wild type; +/-, heterozygous; -/- homozygous.

mothers to develop to term (20). Alternatively, the stem cells are co-cultivated with morulae over night, during which time they attach and integrate into the embryonic inner cell mass. The following day, the chimeric blastocysts are transferred into pseudopregnant foster mice (21). The chimeric animals obtained from either procedure are bred to normal mice and their offspring are tested for the presence of the mutant allele that has been transmitted through the germ line. F1 generation animals which are heterozygous for the gene alteration can than be bred to each other to derive mice homozygous for the mutant allele (Fig. 8.2). Whether viable homozygous deficient mice can be obtained depends on the function of the gene inactivated. All possible effects of homozygosity ranging from early embryonic lethality to a wild type phenotype have been reported in the literature (22). Nevertheless, even if homozygosity should result in embryonic lethality, analysis of the developing embryo or of permanent cell lines established from it can be informative.

8.2.3 Conditional Gene Disruption by Site-Specific Recombination

A mutant phenotype, resulting from conventional gene disruption by homologous recombination manifests itself in all cells of an organism and at all stages of ontogenesis. Under certain circumstances however, it might be desirable to induce a gene defect only at distinct stages in development or in a tissue-specific manor. For instance, embryonic lethality caused by disruption of developmental essential genes precludes the analysis of the gene defect in adult animals. Also, a gene product might serve distinct functions in different mouse tissues, which could be studied separately by disrupting the gene in defined cell types only.

To achieve conditional gene inactivation, novel approaches have been developed using site-specific recombination. One approach is based on the unique properties of the Cre recombinase of the bacteriophage P1 which specifically binds to a 34 bp loxP recognition sequence. If two loxP sites are located in the same orientation on a DNA molecule, Cre-mediated recombination will result in excision of the intervening DNA region, leaving a single loxP site behind. In a series of studies, this system has been adapted to achieve inducible and cell-type specific gene disruption in mammalian cells (23-27).

Different vector systems can be applied for Cre-mediated gene disruption. The approach taken by our laboratory is outlined in Figure 8.3A. A replacement-type targeting vector is generated which contains a genomic region of the target gene. This gene region is modified by integration of the *neo* marker and one loxP site in one intron and a second loxP site in an adjacent intron. The modified introns flank an essential exon of the target gene. In addition, the HSV-TK gene is included as a negative selection marker in the construct. Using standard positive/negative selection protocols, this modification is introduced into the ES cell target gene via homologous recombination. ES cells carrying the loxP-modification (Targeted allele, Fig. 8.3A) are injected into blastocysts and the chimeras obtained are bred to derive mice carrying two copies of the loxP-modified allele. These animals are phenotypically normal as integration of *neo* and the loxP sites in the intron sequences most likely does not interfere with the functional expression of the gene. However, induced expression of the Cre recombinase in these mice will result in excision of essential exon sequences flanked by the two loxP sites and in disruption of the gene (Disrupted allele, Fig. 8.3A and B).

Controlling the temporal and spatial expression of the Cre recombinase in loxP-modified mice is key to determining the conditions of the gene inactivation. Several possibilities for temporally controlled expression of the Cre recombinase in loxP-modified cells have been reported. Induction of Cre activity has been achieved by transient transfection of ES cells with a Cre expression construct (27) or by direct injection of recombinant Cre protein into fertilized mouse eggs (28). In animals, inducible gene inactivation was demonstrated by crossbreeding mice homozygous for a loxP-modified gene with transgenic animals expressing the Cre recombinase under an interferon-inducible promoter. Treatment with interferon resulted in complete gene inactivation in liver and spleen, and to a major extent in a variety of other tissues (29). Similarly, spatially restricted gene inactivation was achieved by crossing loxP-modified mice with transgenics carrying the Cre recombinase under tissue-specific promoter elements. In these double transgenic mice, the target gene was inactivated in cell types expressing Cre recombinase but not in other tissues (27). The application of a recombinant adenovirus carrying an expressible Cre gene for somatic gene transfer offers even greater flexibility in controlling Cre expression. Adenovirus particles are able to infect a large variety of cell types *in vivo* when injected locally, allowing tissue specific gene transfer. The onset of gene inactivation is determined by the timing of virus administration. The feasibility of this approach has been shown both in cell culture (30) and in hepatocytes *in vivo* (31).

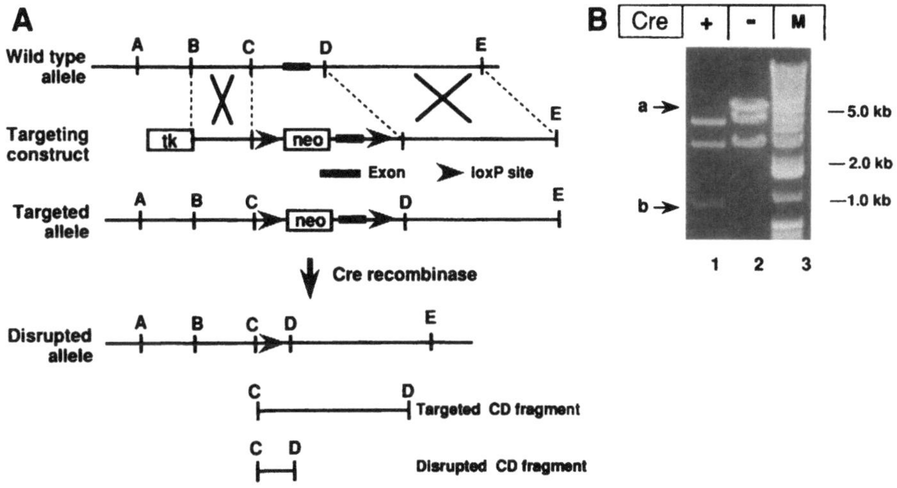

Fig. 8.3: Strategy for Cre-mediated gene inactivation.

(A) Construction of a targeting vector. The replacement-type targeting construct is assembled from DNA segments BC, CD and DE which are derived from the murine target gene locus (Wild type allele). Essential exon sequences in fragment CD (filled box) are flanked by modified introns (solid line) containing the *neo* expression cassette (neo) and two loxP sites (arrow heads). The herpes simplex virus thymidine kinase gene (tk) flanks the homologous gene region. Integration of the targeting vector into the wild type allele does not affect gene expression from the targeted allele. Cre recombinase activity will cause site-specific excision of exon sequences flanked by the loxP sites (Disrupted allele) resulting in size reduction of restriction fragment CD.

(B) Test of Cre-mediated excision of loxP-modified DNA sequences. Normal (-; lane 2) or Cre recombinase-expressing *E. coli* strains (+; lane 1) (kindly provided by L. Parada, Southwestern Medical Center, Dallas, TX) were transformed with a loxP-modified targeting vector for disruption of the murine gp330 gene locus. Subsequently, plasmids were prepared from both strains, digested with *Hind*III and *Sal*I and separated on a 0.8 % agarose gel. The agarose gel was stained with ethidium bromide. Cre-mediated excision of exon sequences on the targeting construct flanked by loxP sites results in convertions of a 5 kb DNA fragment (a; lane 2) into a 0.8 kb fragment (b; lane 1). DNA fragments of the indicated sizes were run as molecular weight marker in lane 3 (M).

Most recently, Cre-mediated recombination has also been used to introduce large deletions (32) or translocations (33, 34) into ES cell chromosomes. In the future, this approach will enable us to generate mouse strains carrying chromosomal aberrations found in human patients. Furthermore, it will be possible to map syntenic regions in the mouse genome by creating deletions implicated in human inheritable diseases.

8.3 Animal Models of Plasma Lipoprotein Disorders

8.3.1 Physiology and Pathophysiology of Lipoprotein Metabolism

The concentration of plasma lipoproteins is determined by complex metabolic pathways involving secretion of nascent lipoproteins, enzymatic modification of the lipoprotein particles in the circulation and receptor-mediated uptake into the liver and peripheral tissues. Production and secretion of triglyceride-rich lipoproteins occurs in two cell types. Hepatocytes produce very low density lipoproteins (VLDL), enterocytes produce chylomicrons that carry dietary lipids and fat soluble vitamins absorbed in the intestine. In the circulation, triglycerides in both lipoprotein particles are hydrolyzed by lipoprotein lipase, which is bound to the surface of capillary endothelial cells. Hydrolysis generates free fatty acids that serve as an energy source for extrahepatic tissues, and converts chylomicrons and VLDL into cholesterol-rich remnant particles. These are designated chylomicron remnants (CR) and VLDL remnants or intermediate density lipoproteins (IDL), respectively. IDL are further metabolized to LDL. As the final step in their metabolism, CR and LDL enter the liver parenchyme, where they are taken up by two hepatic lipoprotein receptors, the LDL receptor and the chylomicron remnant receptor (CR receptor). The latter is identical with the LDL receptor-related protein (LRP) (35-37). Hepatic clearance of LDL is mediated by the LDL receptor which binds apoB-100, the only apoprotein present on LDL. CR, on the other hand, are cleared by the LDL receptor and the CR receptor, recognizing their apoE moieties. ApoB-48, a truncated form of apoB-100 and structural component of CR, does itself not bind to lipoprotein receptors. The plasma concentration of apoB-containing lipoproteins is tightly regulated and constitutes an important risk factor for the development of cardiovascular disease. In general, these lipoprotein particles are considered atherogenic and high plasma concentrations are associated with an increased risk for atherosclerosis and coronary artery disease (CAD) (38).

High density lipoproteins (HDL) in contrast, are considered anti-atherogenic and are associated with reduced incidence of CAD (39). HDL are small, cholesterol-rich lipoproteins produced by peripheral tissues. ApoA-I is the major apoprotein found in these particles. HDL undergo a complicated maturation process in the circulation, ultimately transferring most of their cholesterol content to IDL and LDL. This way, HDL deliver excessive cholesterol from peripheral tissues back to the liver, a process called "reverse cholesterol transport".

Taken together, approximately 17 gene products, apolipoproteins, enzymes and receptors are involved in the metabolism of plasma lipoproteins (40). All of these factors are encoded by single copy genes in the human genome and a substantial number of mutations have been identified in these genes causing disturbances of the

Table 8.1 Mouse models of human monogenic disorders of plasma lipoproteins

Affected gene	Disorder	Mouse model
Apolipoprotein defect		
ApoA-I	A-I deficiency	(42)
	A-I/C-III deficiency	(43)*
ApoB	Fam. hypobetalipoproteinemia	(44, 45)
ApoE	E deficiency	(46, 47)
Enzyme defect		
Lipoprotein lipase	Lipoprotein lipase deficiency	(48)
Hepatic lipase	Hepatic lipase deficiency	(49)
Receptor defect		
LDL receptor	Fam. hypercholesterolemia	(50)

* In this mouse model described by Maeda et al. (43) the apoC-III gene has been disrupted. This results in complete absence of apoC-III protein and reduced intestinal expression of apoA-I and apoA-IV.

lipoprotein metabolism. Some of these defects are listed in Table 8.1. Substantial efforts have been made to understand inheritable defects of the lipoprotein metabolism as they contribute to the development of cardiovascular disease, the major cause of death in western societies (41). In particular, gene targeting has been applied to generate mouse models of known lipoprotein disorders (Table 8.1) and to study both pathological mechanisms and therapeutic approaches. Two of these animal models, the LDL receptor-deficient and the apoE-deficient mouse will be discussed in the following.

8.3.2 The LDL Receptor-Deficient Mouse

Mouse Model for Familial Hypercholesterolemia

The genetic defect of the LDL receptor in familial hypercholesterolemia (FH) results in accumulation of apoB-100 containing IDL and LDL particles in the circulation. Plasma cholesterol concentrations average 300 mg/dl in heterozygous and 700 mg/dl in homozygous patients, well above values found in age-matched normal individuals (180 mg/dl) (51). Pronounced atherosclerosis, xanthomatous accumulation of cholesterol in skin and tendons and premature death from CAD are common in affected individuals.

A mouse model for FH has been generated by targeted disruption of the murine LDL receptor gene (50). As in FH patients, LDL receptor-deficient mice exhibit poor removal of apoB-100-containing particles from the circulation. This results in a two fold increase in plasma cholesterol concentrations, mainly due to LDL accumulation (Fig. 8.4). The rather modest increase in plasma cholesterol levels compared to FH patients can be explained by the lower production rate of LDL in the mouse.

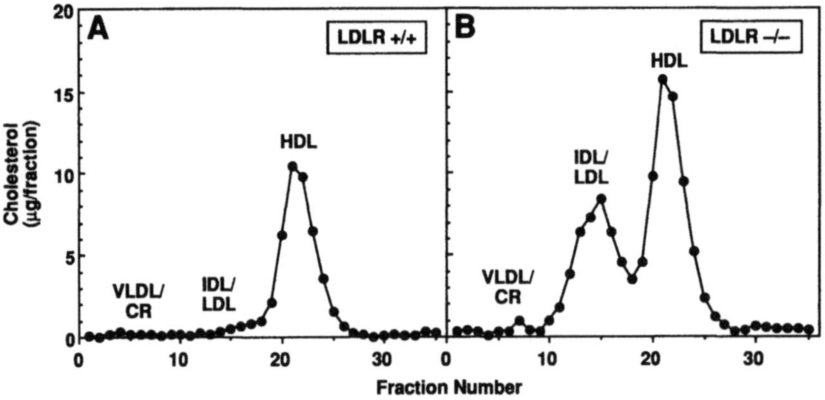

Fig. 8.4: Plasma lipoprotein profile of wild type (A) and LDL receptor-deficient mice (B). Blood was collected by retro-orbital bleeding from mice of the indicated genotypes and 50 µl of plasma analysed by fast performance liquid chromatography (FPLC) on a superose 6 column. The cholesterol content of each fraction was determined spectrofluorimetrically. FPLC separates lipoproteins in the plasma according to size into small HDL, medium sized IDL and LDL and into the large VLDL and CR particles.

The human liver produces VLDL containing apoE and apoB-100. In contrast, in the mouse approximately 70 % of the VLDL produced by hepatocytes contains apoB-48 rather then apoB-100. Only the latter particles are converted to LDL, which accumulate in LDL receptor-deficient organisms. The major portion of the VLDL remnants containing apoB-48 and apoE can still be cleared by the CR receptor which binds apoE. This alternative uptake pathway for apoE-containing lipoproteins also assures the virtually normal clearance of CR in LDL receptor-negative patients and animals (Fig. 8.4) (50, 52-54).

Despite these differences, the LDL receptor-negative mouse proved to be a valuable model for the pathology of FH. Detailed studies have shown that plasma levels in these mice exceed 1,500 mg/dl when fed a diet high in cholesterol, saturated fat and cholic acid (53). This impressive hypercholesterolemia resulted in accumulation of cholesterol-loaded macrophages in the skin and subcutaneous tissues causing massive xanthomatosis, a hallmark of FH. Furthermore, the animals developed atheromatous lesions of the coronary and pulmonary arteries. Development and progression of these lesions followed similar mechanisms as in FH patients (53, 55).

Dissection of Chylomicron Remnant Metabolism

The LDL receptor-deficient mice were also used to analyze whether LRP, a multifunctional 600 kDa cell surface receptor on hepatocytes functions as a receptor for CR (35, 56). Because both LRP and the LDL receptor can bind apoE-containing lipoproteins, mice doubly deficient for both hepatic receptors had to be generated to test this hypothesis in an animal model. Unfortunately, embryos homozygous for a targeted disruption of the LRP gene died *in utero*, precluding the analysis of LRP deficiency in adult mice (57, 58). To circumvent the problem of embryonic lethality, mice were generated that carried a targeted disruption of the receptor-associated protein (RAP), a 39 kDa protein that functions as a chaperone in LRP biosynthesis (37, 59). RAP deficiency resulted in a 70 – 80 % reduction in functional expression of LRP in the livers of $RAP^{-/-}$ mice. The residual LRP activity was sufficient for normal embryonic development, but resulted in significantly impaired clearance of LRP ligands from the circulation. Therefore, these animals constituted a phenotypically hypomorphic model of partial LRP deficiency and could be used to assess the role of LRP in CR metabolism. RAP and LDL receptor doubly deficient mice were generated by breeding $RAP^{-/-}$ and LDL $receptor^{-/-}$ animals and their plasma lipoprotein profiles were compared to that of the single knockout strains (37). RAP-deficient mice had a lipoprotein profile indistinguishable from wild type animals. This was due to the unaffected LDL receptor activity clearing apoB-100 and apoE-containing lipoproteins from the circulation. The LDL receptor-deficient mice accumulated IDL and LDL but not CR, and doubly deficient animals showed a marked increase in both the IDL/ LDL and the CR fraction. These results supported the hypothesis that hepatic clearance of CR is mediated by a dual receptor system, consisting of the LDL receptor and LRP.

Somatic Gene Therapy for Familial Hypercholesterolemia

One out of every 500 individuals is heterozygous for an LDL receptor gene defect, making FH one of the most common inheritable diseases and an important target for somatic gene therapy. LDL receptor-deficient mice were used to test the feasibility of such a therapeutic approach. Experiments published by Stratford-Perricaudet et al. (60) had demonstrated the potential to use recombinant human adenovirus to transfer expressible foreign genes into mouse cells *in vivo*. This system was then applied to deliver a functional copy of the mouse LDL receptor cDNA into hepatocytes of wild type mice. Intravenous injection of the animals with the LDL receptor-virus resulted in quantitative infection of liver cells, whereas other tissues were affected only marginally. Virus-mediated overexpression of the LDL receptor in hepatocytes elicited a marked reduction in total plasma cholesterol concentrations (61). Similarly, the systemic application of the LDL receptor-virus in LDL receptor-deficient mice resulted in the complete reversal of the hypercholesterolemic phenotype, due to selective removal of LDL but not HDL from the circulation (50). Infection of LDL $receptor^{-/-}$ mice with a control virus carrying the β-galactosidase gene of *E. coli* did not induce any changes in the lipoprotein profile. Therapeutic amounts of hepatic LDL receptors were measured four days after virus administration and continued for up to one week, after which expression of the transgene ceased. The disappearance of adenovirus-mediated gene expression in general is caused by loss of virus DNA from the host cell and by reaction of cytotoxic T-cells against virus infected cells (62). Despite the only transient success of this gene transfer experiment, it demonstrated the principal possibility for gene therapy of FH, once technical obstacles of the gene delivery system can be overcome.

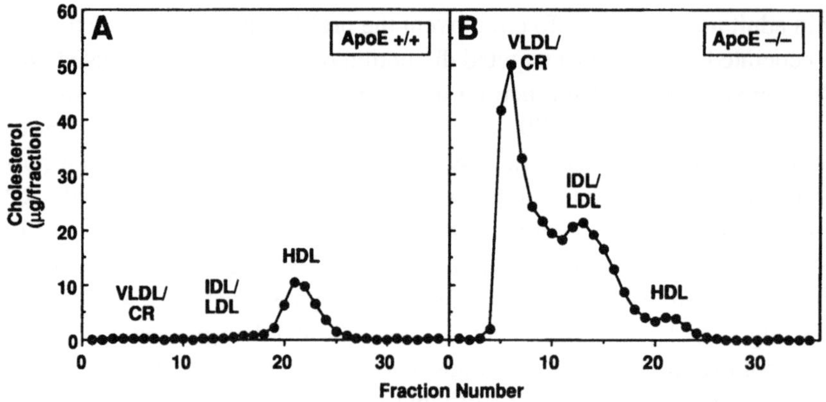

Fig. 8.5: Plasma lipoprotein profile of wild type (A) and apoE-deficient mice (B).
Blood was collected by retro-orbital bleeding from mice of the indicated genotypes and 50 µl of plasma analysed by FPLC on a superose 6 column. The cholesterol content of each fraction was determined spectrofluorimetrically. The fractions containing HDL, IDL and LDL as well as VLDL and CR are indicated.

8.3.3 The apoE-deficient Mouse

Murine model for apoE-deficiency

Apolipoprotein E is a 34 kDa protein found in several lipoprotein particles and is present in three isoforms in the human population. Isoforms apoE3 and apoE4 bind to lipoprotein receptors with high affinity, whereas apoE2 binds only weakly. As a consequence, patients homozygous for the APOE2 gene accumulate IDL and CR in their circulation, a defect referred to as type III hypercholesterolemia or dysbetalipoproteinemia (63). To develop a murine model for functional apoE-deficiency, Plump et al. (46) and Zhang et al. (47) generated mice with a targeted disruption of the apoE gene. Already on normal chow, apoE-deficient mice exhibited dramatic accumulation of CR and IDL in their circulation, apoB-100 mediated clearance of LDL, however, was unaffected (Fig. 8.5). Due to the high plasma cholesterol concentrations reaching 400 to 600 mg/ dl apo$E^{-/-}$ mice developed spontaneous atherosclerosis (46, 47). The value of the apoE-deficient mouse as a small animal model for hypercholesterolemia and atherosclerosis was confirmed in studies addressing the pathology of atheromatous lesion formation and progression on normal and cholesterol-enriched diets (64).

The LDL receptor mediates cellular uptake of lipoproteins by binding of two apolipoproteins, apoE and apoB-100. Because apoE binds with significantly higher affinity to the LDL receptor than apoB-100, it was postulated that apoE could act as a competitive inhibitor of apoB-100 in the circulation. This was supported by the finding that the amount of LDL necessary to achieve half-maximal binding to the LDL receptor was ten fold higher *in vivo* than *in vitro* (65). Comparison of the LDL metabolism in wild type and apoE-deficient mice offered the unique opportunity to test this hypothesis. By measuring the uptake of radiolabeled LDL into livers of $apoE^{-/-}$ and control mice Woollett et al. (66) were able to show that the K_m for hepatic LDL uptake in the apoE-deficient mice was only 13 % of that seen in the controls (44 mg/dl versus 329 mg/dl). As a consequence, the LDL clearance rate in the knockout mice doubled and the concentration of apoB-100 containing particles in the circulation dropped. At the same time, the net rates of cholesterol absorption and cholesterol synthesis in the liver were unchanged. These data convincingly demonstrated that *in vivo*, apoE-containing particles act as competitor of hepatic LDL uptake and increase steady-state levels of LDL in the circulation.

Somatic apoE Gene Transfer

The possibility of reversing the apoE-deficient phenotype by hepatocyte gene transfer was tested by intravenous infection of $apoE^{-/-}$ mice with a recombinant adenovirus containing the human apoE cDNA (67). The successful gene transfer into liver cells resulted in circulating apoE levels reaching up to 600 mg/dl four days after virus administration. These concentrations were sufficient to normalize plasma lipoprotein levels and significantly decrease the extent of aortic lesion formation in these animals compared to control virus infected mice. As already observed for other adenovirus-mediated gene transfer experiments, high apoE expression was only transient and plasma cholesterol concentrations gradually rose to half the level of untreated apoE-deficient mice within 30 days after virus administration (67).

ApoE is predominantly synthesized by parenchymal cells of the liver, but expression has also been shown in other cell types, including macrophages (68). Although experimental evidence for the significance of apoE synthesis by macrophages was missing, this offered the possibility to correct apoE-deficiency by bone marrow transplantation (69, 70). Bone marrow transplantation had been used previously for successful permanent gene therapy and offered an alternative to the transient virus-mediated approaches (reviewed in 71). Four weeks after receiving wild type bone marrow, average plasma cholesterol levels in apoE knockout mice fell to 125 mg/dl, due to specific reduction of VLDL, IDL and LDL concentrations (69). Apparently, apoE secreted by circulating macrophages associated with lipoproteins of hepatic and intestinal origin, enabling their clearance by hepatic lipoprotein receptors. To test the effectiveness of the gene transfer, $apoE^{-/-}$ animals that had received bone marrow transplantation were challenged with a cholesterol-enriched diet. Although circulating apoE levels in these animals were only 12 % of wild type levels, this was sufficient to protect the mice from diet-induced hypercholesterolemia. The increase in plasma cholesterol concentration after three months on the high-fat diet was only moderate and atherosclerotic lesions were at an early stage of development when compared to untreated knockout mice (69).

Anti-Atherogenic Effect of apoA-I

Epidemiological studies show considerable variability in the susceptibility for atherosclerosis in the normal population not affected by known familial deficiencies of the cholesterol metabolism. Apparently, both environmental and genetic factors contribute to the risk for CAD and a number of gene loci affecting atherosclerosis susceptibility in inbred mouse strains have been mapped (reviewed in 72). Most likely, these genetic factors are heterogeneous in nature, complicating the choice of target gene for genetic therapeutic intervention.

Rather than trying to correct specific inborn errors, an approach has been tested to confer increased resistance to the development of atherosclerosis into test animals by increasing their circulating HDL levels. In numerous studies, the inverse correlation of plasma HDL concentrations with risk of atherosclerosis had been established and the production rate of apoA-I seemed to control HDL levels in humans (73). These findings were confirmed in wild type mice infected with an adenovirus carrying the human apoA-I cDNA (74) or transgenic for a copy of the human apoA-I gene (75). In both studies, overexpression of human apoA-I increased HDL levels significantly, resulting in protection from diet-induced atherosclerosis (75).

Subsequently, the same approach was tested in apoE-deficient mice, an animal model for severe hypercholesterolemia. ApoE-deficient mice were bred with animals carrying the human apoA-I transgene (76, 77). The double transgenic animals showed similar elevation of total plasma cholesterol levels as $apoE^{-/-}$ mice. Circulating HDL levels in the transgenic mice however, were increased two fold and this increase correlated with a sixfold reduction in susceptibility for atherosclerosis (76). These experiments suggest the intriguing possibility to treat unrelated genetic causes of increased risk for CAD by modulation of the HDL metabolism using apoA-I gene transfer.

8.4 Conclusions

Still a young discipline in molecular medicine, gene targeting in the mouse has already contributed significantly to our knowledge about genetic factors and their influence on the physiology and pathological abnormalities in humans. Examples described in this chapter have focused on the application of this approach to study plasma lipoprotein disorders. Here, like in many other research areas, the generated mouse models proved useful both in basic and applied medical research. Currently, intense efforts are undertaken to apply gene disruption technology to animal species other than mice. Preferred targets for such gene inactivation approaches are rabbits and rats, because well characterized experimental systems for studying lipoprotein metabolism and hypertension already exist in these species.

References

1. McKusick, V.A. (1992). Mendelian inheritance in man. Baltimore and London: The John Hopkins University Press.

2. Gottesman, M.M. and Collins, F.S. (1994). The role of the human genome project in disease prevention. Preventive Medicine *23*, 591-594.
3. Hosoda, K., Hammer, R.E., Richardson, J.A., Baynash, A.G., Cheung, J.C., Giaid, A., and Yanagisawa, M. (1994). Targeted and natural (piebald-lethal) mutations of endothelin-B receptor gene produce megacolon associated with spotted coat color in mice. Cell *79*, 1267-1276.
4. Hinnen, A., Hicks, J.B., and Fink, G.R. (1978). Transformation of yeast. Proc. Natl. Acad. Sci. USA *75*, 1929-1933.
5. Orr-Weaver, T., Szostak, J.W., and Rothstein, R.J. (1981). Yeast transformation: A model system for the study of recombination. Proc. Natl. Acad. Sci. USA *78*, 6354-6358.
6. Jasin, M. and Berg, P. (1988). Homologous integration in mammalian cells without target gene selection. Genes & Development *2*, 1353-1363.
7. Sedivy, J.M. and Sharp, P.A. (1989). Positive genetic selection for gene disruption in mammalian cells by homologous recombination. Proc. Natl. Acad. Sci. USA *86*, 227-231.
8. Smithies, O. (1993). Animal models of human genetic diseases. Trends Genet. *9*, 112-116.
9. Capecchi, M.R. (1989). Altering the genome by homologous recombination. Science *244*, 1288-1292.
10. Mansour, S.L., Thomas, K.R., and Capecchi, M.R. (1988). Disruption of the proto-oncogene *int*-2 in mouse embryo-derived stem cells: a general strategy for targeting mutations to non-selectable genes. Nature *336*, 348-352.
11. Bronson, S.K. and Smithies, O. (1994). Altering mice by homologous recombination using embryonic stem cells. J. Biol. Chem. *269*, 27155-27158.
12. te Riele, H., Maandag, E.R., Clarke, A., Hooper, M., and Berns, A. (1990). Consecutive inactivation of both alleles of the pim-1 proto-oncogene by homologous recombination in embryonic stem cells. Nature *348*, 649-651.
13. Mortensen, R.M., Conner, D.A., Chao, S., Geisterfer-Lowrance, A.A.T., and Seidman, J.G. (1992). Production of Homozygous Mutant ES Cells with A Single Targeting Construct. Mol. Cell. Biol. *12*, 2391-2395.
14. Willnow, T.E. and Herz, J. (1994). Genetic Deficiency in Low Density Lipoprotein Receptor-related Protein (LRP) Confers Cellular Resistance to Pseudomonas Exotoxin A – Evidence that LRP Is Required for Uptake and Degradation of Multiple Ligands. J. Cell Sci. *107*, 719-726.
15. Pevny, L., Simon, M.C., Robertson, E., Klein, W.H., Tsai, S.-F., D'Agati, V., Orkin, S.H., and Constantini, F. (1991). Erythroid differentiation in chimaeric mice blocked by a targeted mutation in the gene for transcription factor GATA-1. Nature *349*, 257-260.
16. Thomas, J.E., Soriano, P., and Brugge, J.S. (1991). Phosphorylation of c-Src on Tyrosine 527 by another Protein Tyrosine Kinase. Science *254*, 568-570.
17. Mintz, B. and Illmensee, K. (1975). Normal genetically mosaic mice produced from malignant teratocarcinoma cells. Proc. Natl. Acad. Sci. USA *72*, 3585-3589.
18. Evans, M.J. and Kaufmann, M.H. (1981). Establishment in culture of pluripotential cells from mouse embryos. Nature *292*, 154-156.
19. Martin, G.R. (1981). Isolation of a pluripotent cell line from early mouse embryos cultured in medium conditioned by teratocarcinoma stem cells. Proc. Natl. Acad. Sci. USA *78*, 7634-7638.
20. Bradley, A. (1987). Production and analysis of chimeric mice. In: Teratocarcinomas and Embryonic Stem Cells: A Practical Approach. E.J. Robertson, ed. Oxford: IRL Press, 113-151.
21. Wood, S.A., Allen, N.D., Rossant, J., Auerbach, A., and Nagy, A. (1993). Non-injection methods for the production of embryonic stem cell-embryo chimaeras. Nature *365*, 87-89.
22. Shastry, B.S. (1994). More to learn from gene knockouts. Mol. Cell. Biochem. *136*, 171-182.
23. Sauer, B. and Henderson, N. (1988). Site-specific DNA recombination in mammalian cells by the Cre recombinase of bacteriophage P1. Proc. Natl. Acad. Sci. USA *85*, 5166-5170.
24. Lakso, M., Sauer, B., Mosinger, J., Lee, E.J., Manning, R.W., Yu, S.-H., Mulder, K.L., and Westphal, H. (1992). Targeted oncogene activation by site-specific recombination in transgenic mice. Proc. Natl. Acad. Sci. USA *89*, 6232-6236.
25. Orban, P.C., Chui, D., and Marth, J.D. (1992). Tissue- and site-specific DNA recombination in transgenic mice. Proc. Natl. Acad. Sci. USA *89*, 6861-6865.
26. Gu, H., Zou, Y.-R., and Rajewsky, K. (1993). Independent control of immunoglobulin switch recombination at individual switch regions evidenced through Cre-loxP-mediated gene targeting. Cell *73*, 1155-1164.

27. Gu, H., Marth, J.D., Orban, P.C., Mossmann, H., and Rajewski, K. (1994). Deletion of the DNA Polymerase b Gene in T cells Using Tissue-Specific Gene Targeting. Science *265,* 103-106.
28. Araki, K., Araki, M., Miyazaki, J-I., and Vassalli, P. (1995). Site-specific recombination of a transgene in fertilized eggs by transient expression of Cre recombinase. Proc. Natl. Acad. Sci. USA *92,* 160-164.
29. Kühn, R., Schwenk, F., Aguet, M., and Rajewsky, K. (1995). Inducible gene targeting in mice. Science *269,* 1427-1428.
30. Anton, M. and Graham, F.L. (1995). Site-specific recombination mediated by an Adenovirus vector expressing the Cre recombinase protein: a molecular switch for control of gene expression. J. Virol. *69,* 4600-4606.
31. Rohlmann, A., Gotthardt, M., Willnow, T.E., Hammer, R.E. and J. Herz. 1996. Sustained somatic gene inactivation by viral transfer of Cre recombinase. Nature Biotech. 14: 1562-1565.
32. Ramirez-Solis, R., Pentao, L., and Bradley, A. (1995). Chromosome engineering in mice. Nature *378,* 720-724.
33. Smith, A.J.H., de Sousa, M.A., Kwabi-Addo, B., Heppell-Parton, A., Impey, H., and Rabbitts, P. (1995). A site-directed chromosomal translocation induced in embryonic stem cells by Cre-loxP recombination. Nature Genetics *9,* 376-385.
34. van Deursen, J., Fornerod, M., van Rees, B., and Grosveld, G. (1995). Cre-mediated site-specific translocation between nonhomologous mouse chromosomes. Proc. Natl. Acad. Sci. USA *92,* 7376-7380.
35. Herz, J., Hamann, U., Rogne, S., Myklebost, O., Gausepohl, H., and Stanley, K.K. (1988). Surface location and high affinity for calcium of a 500-kd liver membrane protein closely related to the LDL-receptor suggest a physiological role as lipoprotein receptor. EMBO J. *7,* 4119-4127.
36. Willnow, T.E., Sheng, Z., Ishibashi, S., and Herz, J. (1994). Inhibition of Hepatic Chylomicron Remnant Uptake by Gene Transfer of A Receptor Antagonist. Science *264,* 1471-1474.
37. Willnow, T.E., Armstrong, S.A., Hammer, R.E., and Herz, J. (1995). Functional expression of low density lipoprotein receptor-related protein is controlled by receptor-associated protein in vivo. Proc. Natl. Acad. Sci. USA *92,* 4537-4541.
38. Kane, J.P. and Havel, R.J. (1989). Disorders of the biogenesis and secretion of lipoproteins containing the B apolipoproteins. In: The metabolic basis of inherited disease. Scriver, C.R., Beaudet, A.L., Sly, W.S. and D. Valle, eds. New York: McGraw-Hill, 1139-1164.
39. Breslow, J.L. (1989). Familial disorders of high density lipoprotein metabolism. In: The metabolic basis of inherited disease. Scriver, C.R., Beaudet, A.L., Sly, W.S. and D. Valle, eds. New York: McGraw-Hill, 1251-1266.
40. Breslow, J.L. (1994). Lipoprotein metabolism and atherosclerosis susceptibility in transgenic mice. Current Opinion in Lipidology *5,* 175-184.
41. Ross, R. (1986). The pathogenesis of atherosclerosis – An update. N. Engl. J. Med. *314,* 488-500.
42. Williamson, R., Lee, D., Hagaman, J., and Maeda, N. (1992). Marked reduction of high density lipoprotein cholesterol in mice genetically modified to lack apolipoprotein A-I. Proc. Natl. Acad. Sci. USA *89,* 7134-7138.
43. Maeda, N., Li, H., Lee, D., Oliver, P., Quarfordt, S.H., and Osada, J. (1994). Targeted disruption of the apolipoprotein C-III gene in mice results in hypotriglyceridemia and protection from postprandial hypertriglyceridemia. J. Biol. Chem. *269,* 23610-23616.
44. Homanics, G.E., Smith, T.J., Zhang, S.H., Lee, D., Young, S.G., and Maeda, N. (1993). Targeted modification of the apolipoprotein B gene results in hypobetalipoproteinemia and developmental abnormalities in mice. Proc. Natl. Acad. Sci. USA *90,* 2389-2393.
45. Farese, R.V., Ruland, S.L., Flynn, L.M., Stokowski, R.P., and Young, S.G. (1995). Knockout of the mouse apolipoprotein B gene results in embryonic lethality in homozygotes and protection against diet-induced hypercholesterolemia in heterozygotes. Proc. Natl. Acad. Sci. USA *92,* 1774-1778.
46. Plump, A.S., Smith, J.D., Hayek, T., Aalto-Setälä, K., Walsh, A., Verstuyft, J.G., Rubin, E.M., and Breslow, J.L. (1992). Severe Hypercholesterolemia and Atherosclerosis in Apolipoprotein E-Deficient Mice Created by Homologous Recombination in ES Cells. Cell *71,* 343-353.
47. Zhang, S.H., Reddick, R.L., Piedrahita, J.A., and Maeda, N. (1992). Spontaneous hypercholesterolemia and arterial lesions in mice lacking apolipoprotein E. Science *258,* 468-471.
48. Coleman, T., Seip, R.L., Gimble, J.M., Lee, D., Maeda, N., and Semenkovich, C.F. (1995). COOH-terminal disruption of lipoprotein lipase in mice is lethal in homozygotes, but heterozygotes have elevated triglycerides and impaired enzyme activity. J. Biol. Chem. *270,* 12518-12525.

49. Homanics, G.E., de Silva, H.V., Osada, J., Zhang, S.H., Wong, H., Borensztajn, J., and Maeda, N. (1995). Mild dyslipidemia in mice following targeted inactivation of the hepatic lipase gene. J. Biol. Chem. *270*, 2974-2980.
50. Ishibashi, S., Brown, M.S., Goldstein, J.L., Gerard, R.D., Hammer, R.E., and Herz, J. (1993). Hypercholesterolemia. In: LDL Receptor Knockout Mice And Its Reversal By Adenovirus-Mediated Gene Delivery. J. Clin. Invest. 92, 883-893.
51. Goldstein, J.L. and Brown, M.S. (1989). Familial hypercholesterolemia. In: The metabolic basis of inherited disease. Scriver, C.R., Beaudet, A.L., Sly, W.S. and D. Valle, eds. New York: McGraw-Hill, 1215-1250.
52. Kita, T., Goldstein, J.L., Brown, M.S., Watanabe, Y., Hornick, C.A., and Havel, R.J. (1982). Hepatic uptake of chylomicron remnants in WHHL rabbits: A mechanism genetically distinct from the low density lipoprotein receptor. Proc. Natl. Acad. Sci. USA *79*, 3623-3627.
53. Ishibashi, S., Goldstein, J.L., Brown, M.S., Herz, J., and Burns, D.K. (1994). Massive Xanthomatosis and Atherosclerosis in Cholesterol-Fed LDL Receptor-Negative Mice. J. Clin. Invest. *93*, 1885-1893.
54. Rubinsztein, D.C., Cohen, J.C., Berger, G.M., van der Westhuyzen, D.R., Coetzee, G.A., and Gevers, W. (1990). Chylomicron remnant clearance from the plasma is normal in familial hypercholesterolemic homozygotes with defined receptor defects. J. Clin. Invest. *86*, 1306-1312.
55. Tangirala, R.K., Rubin, E.M., and Palinski, W. (1995). Quantitation of atherosclerosis in murine models: correlation between lesions in the aortic origin and in the entire aorta, and differences in the extent of lesions between sexes in LDL receptor-deficient and apolipoprotein E-deficient mice. J. Lipid Res. *36*, 2320-2328.
56. Krieger, M. and Herz, J. (1994). Structures and Functions of Multiligand Lipoprotein Receptors: Macrophage Scavenger Receptors and LDL Receptor-Related Protein (LRP). Ann. Rev. Biochem. *63*, 601-637.
57. Herz, J., Clouthier, D.E., and Hammer, R.E. (1992). LDL Receptor-Related Protein Internalizes and Degrades uPA/PAI-1 Complexes and Is Essential for Embryo Implantation. Cell *71*, 411-421.
58. Herz, J., Clouthier, D.E., and Hammer, R.E. (1993). Correction: LDL Receptor-related Protein Internalizes and Degrades uPA-PAI-1 Complexes and Is Essential for Embryo Implantation. Cell *73*, 428-420.
59. Bu, G., Geuze, H.J., Strous, G.J., and Schwartz, A.L. (1995). 39 kDa receptor-associated protein is an ER resident protein and molecular chaperone for LDL receptor-related protein. EMBO J. *14*, 2269-2280.
60. Stratford-Perricaudet, L.D., Levrero, M., Chasse, J.-F., Perricaudet, M., and Briand, P. (1990). Evaluation of the Transfer and Expression in Mice of an Enzyme-Encoding Gene Using a Human Adenovirus Vector. Hum. Gene Ther. *1*, 241-256.
61. Herz, J. and Gerard, R.D. (1993). Adenovirus-mediated transfer of low density lipoprotein receptor gene acutely accelerates cholesterol clearance in normal mice. Proc. Natl. Acad. Sci. USA *90*, 2812-2816.
62. Dai, Y., Schwarz, E.M., Gu, D., Zhang, W.-W., Sarvetnick, N., and Verma, I.M. (1995). Cellular and humoral immune responses to adenoviral vectors containing factor IX gene: Tolerization of factor IX and vector antigens allows for long-term expression. Proc. Natl. Acad. Sci. USA *92*, 1401-1405.
63. Mahley, R.W. and Rall, S.C. (1989). Type III hyperlipoproteinemia. In: The metabolic basis of inherited disease. Scriver, C.R., Beaudet, A.L., Sly, W.S. and D. Valle, eds. New York: McGraw-Hill, 1195-1213.
64. Zhang, S.H., Reddick, R.L., Burkey, B., and Maeda, N. (1994). Diet-induced atherosclerosis in mice heterozygous and homozygous for apolipoprotein E gene disruption. J. Clin. Invest. *94*, 937-945.
65. Spady, D.K., Meddings, J.B., and Dietschy, J.M. (1986). Kinetic constants for receptor-dependent and receptor-independent low density lipoprotein transport in the tissues of the rat and hamster. J. Clin. Invest. *77*, 1474-1481.
66. Woollett, L.A., Osono, Y., Herz, J., and Dietschy, J.M. (1995). Apolipoprotein E competitively inhibits receptor-dependent low density lipoprotein uptake by the liver but has no effect on cholesterol absorption or synthesis in the mouse. Proc. Natl. Acad. Sci. USA *92*, 12500-12504.
67. Kashyap, V.S., Santamarina-Fojo, S., Brown, D.R., Parrot, C.L., Applebaum-Bowden, D., Meyn, S., Talley, G., Paigen, B., Maeda, N., and Brewer, H.B.J. (1995). Apolipoprotein E deficiency in mice: gene replacement and prevention of atherosclerosis using adenovirus vectors. J. Clin. Invest. *96*, 1612-1620.

68. Basu, S.K., Brown, M.S., Ho, Y.K., Havel, R.J., and Goldstein, J.L. (1981). Mouse macrophages synthesize and secrete a protein resembling apolipoprotein E. Proc. Natl. Acad. Sci. USA *78*, 7545-7549.
69. Linton, M.F., Atkinson, J.B., and Fazio, S. (1995). Prevention of atherosclerosis in apolipoprotein E-deficient mice by bone marrow transplantation. Science *267*, 1034-1037.
70. Boisvert, W.A., Spangenberg, J., and Curtiss, L.K. (1995). Treatment of severe hypercholesterolemia in apolipoprotein E-deficient mice by bone marrow transplantation. J. Clin. Invest. *96*, 1118-1124.
71. Nienhuis, A.W. (1994). Gene transfer into hematopoietic stem cells. Blood cell *20*, 141-148.
72. Paigen, B., Plump, A.S., and Rubin, E.M. (1994). The mouse as a model for human cardiovascular disease and hyperlipidemia. Current Opinion in Lipidology *5*, 258-264.
73. Glueck, C.J., Gartside, P., Fallat, R.W., Sielski, J., and Steiner, P.M. (1976). Longevity syndromes: familial hypobeta and familial hyperalpha lipoproteinemia. J. Lab. Clin. Med. *88*, 941-957.
74. Kopfler, W.P., Willard, M., Betz, T., Willard, J.E., Gerard, R.D., and Meidell, R.S. (1994). Adenovirus-mediated transfer of a gene encoding human apolipoprotein A-I into normal mice increases circulating high-density lipoprotein cholesterol. Circulation *90*, 1319-1327.
75. Rubin, E.M., Krauss, R.M., Spangler, E.A., Verstuyft, J.G., and Clift, S.M. (1991). Inhibition of early atherogenesis in transgenic mice by human apolipoprotein AI. Nature *353*, 265-267.
76. Paszty, C., Maeda, N., Verstuyft, J., and Rubin, E.M. (1994). Apolipoprotein AI transgene corrects apolipoprotein E deficiency-induced atherosclerosis in mice. J. Clin. Invest. *94*, 899-903.
77. Plump, A.S., Scott, C.J., and Breslow, J.L. (1994). Human apolipoprotein A-I gene expression increases high density lipoprotein and suppresses atherosclerosis in the apolipoprotein E-deficient mouse. Proc. Natl. Acad. Sci. USA *91*, 9607-9611.
78. Willnow, T.E., Hilpert, J., Armstrong, S.A., Rohlmann, A., Hammer, R.E., Burns, D.K. and J. Herz. 1996. Defective forebrain development in mice lacking gp330/megalin. Proc. Natl. Acad. Sci. USA 93: 8460-8464.

9. Potential Antisense Oligonucleotide Therapies for Neurodegenerative Diseases

Mirta Grifman, Efrat Lev-Lehman, Dalia Ginzberg, Fritz Eckstein, Haim Zakut and Hermona Soreq

9.1 Introduction

For the first time in the history of neuropharmacological research, it is possible to design short synthetic strands of DNA as effectors of neuromodulatory events or phenomena. Antisense oligodeoxynucleotides (AS-ODNs) are short DNA strands often protected by chemical modification from nucleolytic degradation. They are aimed at suppression of the expression of a target gene or genes or prevention of the enhancement in its expression in cases where such enhancement takes place. They operate by creating stable hybrids with their target sequences in a highly specific and highly selective manner. Once such hybrids are formed, they may remain as such (which temporarily prevents translation) or degrade (which destroys the target mRNA altogether). The well-defined requirements for hybridization (i.e. complementarity and length of the interacting sequences, ionic strength and pH conditions) provide a preliminary option for simple design of such AS-ODN drugs with a high probability that they will efficiently hybridize with their mRNA targets.

Antisense technology lies somewhere between gene and conventional therapies. Like conventional drugs, it involves the administration of a synthetic compound (in this case, an oligodeoxyribonucleotide or an analog thereof; for a chain of 20 nucleotides, this implies a molecular weight of ca. 6,000 g/mol). Also, it is intended to prevent the functioning of a particular protein, similarly to most conventional drugs, and it may be used for short or long periods and can cause transient or long-term side effects like many such drugs. This further means that the development of antisense drugs depends on criteria similar to those familiar for conventional drugs: the effective dose should be as low as possible, the therapeutic window as wide as possible, delivery should be simple and convenient, drug stability high and adverse responses minimal. However, antisense drugs are based on gene sequences and aimed at interaction with such sequences, rather than at binding to particular protein(s). Their capacity to prevent the functioning of the protein products of their target genes is hence indirect. Unlike gene therapy protocols, they must be administered repeatedly; however, their specificity is expected to be far higher than that of protein blockers, thanks to the basic parameters characterizing the hybridization reaction between oligonucleotide chains and the target mRNA. Also, since they interact with mRNAs and not proteins, the number of target molecules they should encounter in a cell is inevitably smaller than the number of molecules to be blocked by protein inhibitors. Therefore, the effective dose of antisense drugs can be lower than that of protein inhibitors. In contrast, the mode of AS-ODNs delivery poses considerable difficulties, and chemical modification(s) are a necessary prerequisite for improving the rather poor stability of unprotected DNA chains. At the same time, administration of an AS-ODN is far simpler than gene therapy and can therefore be available to larger numbers of patients. Lastly, little is yet known about side effects and ways to avoid them.

In vivo suppression of gene expression with AS-ODNs has been demonstrated in various vertebrates including ducks, rats, mice, and pigs by targeting mRNAs encoding diverse proteins such as MYC, cdk2, RAS and NF-κB (1, 2). Direct intracerebroventricular (ICV) administration of AS-ODNs has been shown in several cases to elicit specific mRNA reduction in the brain of rats, resulting in detectable behavioral changes (3, 4, 5, see details below). These experiments prompted intensive research efforts directed toward developing the potential of antisense-based therapies for various diseases of the nervous system and several antisense "drugs" directed toward mRNA transcripts expressed in brain have already reached clinical trials (1, 2). In the following, we discuss the accumulated experience with neurochemically-relevant examples of this concept.

At the chemical level, antisense oligodeoxynucleotides (AS-ODNs) are designed to operate by sequence-specific binding to preselected RNA targets. The underlying physical-chemical principle for such drug-target interaction is hydrogen bonding between the pharmacologic agent and heterocyclic base moieties in its RNA target (6), which conceptually derives from the synthesis of analogs of nucleic acids as model polymer systems (7). These analogs contain the heterocyclic bases of nucleic acids but differ from the natural polymer in the connecting backbone. As an example, polymers derived from 1-vinyl uracil and 1-vinyl adenine were shown to display inhibitory activity against RNA viruses (8). These and other types of synthetic analogs of nucleic acids were described by Halford and Jones nearly 30 years ago as future drugs. The authors argued that as nuclease-resistant uncharged compounds, these analogs might be taken up by cells more readily than unmodified charged oligonucleotides, and would therefore interfere with the function of mRNA in biological systems. A yet more important aspect of these analogs was their longevity. This was first demonstrated for interferon induction by modified double stranded RNA (9). Nuclease-resistant phosphorothioate RNA was then presented as a potentially more effective inducer of interferon by virtue of its longer half-life *in vivo*. This pharmacological advantage has been used since then for blocking various viral sequences (see, for example,10, 11, 12). The same concept is currently being exploited in the design of possible antisense drugs for different uses, including neuropharmacology.

Like all drugs, AS-ODNs should be tested for their dose and time dependence. Thanks to their special mode of functioning through hybridization, they would operate only in those cells expressing their target mRNA transcripts. This leaves their vulnerability to nuclease destruction as a major deficit hampering the development of AS-ODNs into practical drugs, which leads us into the concept of their phosphorothioate modification.

9.2 The Concept of Phosphorothioate Modifications

While the principle of antisense mechanisms has been verified with unmodified oligomers, their therapeutic applications require stabilization for improved half-life. Several types of chemical modification were attempted for this purpose over the years (for reviews, see 1, 2, 13). Of these, the phosphorothioate protection of internucleotide bonds is the one used most frequently.

Phosphorothioate modification of oligodeoxynucleotides has been in existence now for 30 years. It involves the substitution of a non-bridging oxygen atom in normal phosphate backbones in ODNs with a sulfur atom (14, 15, 16). Phosphorothioate oligomers are water soluble, easy to produce by automated procedures and resistant to enzymatic breakdown. Interestingly, cells take up phosphorothioate oligonucleotides more easily than was originally thought. This occurs not only by passive diffusion across the cell membrane but by endocytosis as well.

Once the synthesis of phosphorothioated nucleotides and oligonucleotides was demonstrated, researchers proceeded to demonstrate the use of such oligonucleotides to improve resistance of cells to viral infection in culture (9) and *in vivo* (17). By the mid-1970s, novel solution synthesis methods were exploited to obtain a 13-mer oligodeoxyribonucleotide complementary to a portion of a reiterated terminal sequence in Rous sarcoma virus (RSV). This enabled Zamecnik and Stephenson (18) to assess the influence of such an oligonucleotide on the course of cellular infection by RSV. The researchers also investigated the use of 3' and 5' terminal protection, via source derivatization, as a means of preventing antisense oligomer degradation by exonucleases. This strategy is now widely emulated, using different types of nucleic acid modifications.

Highly efficient, yet uncomplicated automated systems for protein sequencing (19) and oligonucleotide synthesis (20, 21) became widely available at the beginning of the 1980s as part of the advent of genetic engineering and biotechnology. Simple-to-use automated DNA sequencing was successfully marketed in 1987. At about the same time, automated synthetic chemistries and product purification protocols for phosphorothioate oligonucleotide analogs were commercialized. This made available to investigators interested in the antisense approach the necessary technology to more readily conduct their experimental programs.

The main potential value of phosphorothioate oligonucleotides as neurochemically active drugs lies in the specificity of the hybridization reaction. The repertoire of proteins that are expressed in the brain is far higher than in any other tissue, and many of these proteins share sequence homologies with other proteins. This leads to non-specific interactions of numerous drugs with proteins that resemble their intended target. In contrast, once phosphorothioate antisense oligonucleotides are actively taken up by cells, they find their target mRNA chains and selectively associate with them to form highly specific RNA:DNA double-strands. Phosphorothioated oligonucleotides further increase, within these cells, the activity of a particular ribonuclease – RNaseH – which selectively destroys the double stranded RNA chains. Therefore, the nuclease resistance of phosphorothioate AS-ODNs, their functional specificity and their capacity to lead to destruction of their target mRNA make this approach advantageous over the use of protein inhibitors to block expression of protein products of single genes. This would particularly be the case when these protein products, but not nucleotide sequences, resemble homologous molecules.

Three questions which are often asked with regard to the use of AS-ODNs refer to the optimal length of AS-ODNs, to the preferred concentration of such compounds and to the choice of the location of these AS-ODNs within their target mRNA sequence. A recent study (22), demonstrated that AS-ODN-mediated inhibition is strongly dependent on polymer length and concentration but almost independent of location of the target sequence along the mRNA chain. The study employed microinjection into *Xenopus* oocytes of the mRNA for α-amino-3-hydroxy-5-methyl-4-isoxasole propionate receptor and was based on electrophysiological measurement of remaining receptors. The authors compared phosphorothioate AS-ODNs with unmodified, phosphodiester polymers. The choice of a functionally active AS-ODN should hence be an essential initial phase in each study which involves this technology. In both cases they observed length dependence, with half-maximal inhibition decreasing from 9.9 nucleotides for non-protected AS-ODNs to 7.6 for the phosphorothioated ones. Phosphorothioate and phosphodiester AS-ODNs of 12 and 15 nucleotides, respectively, sufficed to block 95% of their target mRNA. However, phosphorothioate protection reduced the AS-ODN concentration needed for half-maximal inhibition from 18 ng/μl to as low as 0.19 ng/μl for 12-mer oligos. This 100-fold difference may make an analytically useful compound into an economically sensible drug, which emphasizes the value of phosphorothioate protection for AS-ODN drug design. Octamer AS-ODNs dispersed throughout the target mRNA, from the initiator AUG to the 3'-untranslated region, reduced by 70% the level of the tested mRNA (22). It should be noted, however, that this is one specific case; with regard to other genes, certain sites on the mRNA may make better targets than others. In another study, 18 different AS-ODNs were designed against various domains in a specific mRNA sequence, but only 3 of those were satisfactorily effective (23). The

Tab. 9.1 Historical milestones in the development of phosphorothioate antisense technology

	Milestone	Example References
1	Synthesis of phosphorothioated nucleotides	14
2	Synthesis of nuclease-resistant nucleic acid analogs	7
3	Induction of interferon *in vitro* and *in vivo* by modified double stranded RNA	9, 17
4	Demonstration of inhibitory activity against RNA viruses	8
5	Suppression of Rous sarcoma virus infection by a 13-mer AS-ODN	18
6	Automation of protein sequencing	19
7	Automation of oligonucleotide synthesis	20, 21
8	Pharmacokinetics and toxicology studies of phosphorothioate AS-ODNs	24
9	Demonstration of sequence-independent protein binding activity of AS-ODNs	24, 79

choice of a functionally active AS-ODN should hence be an essential initial phase in each study which involves this technology.

When administered to live animals, phosphorothioate oligonucleotides were argued to be unevenly distributed in different tissues and organs, with kidney, liver and bone marrow being their main targets (24). At least part of their effects in these tissues may be due to AS-ODN-protein interactions, particularly with nuclear proteins in young, proliferating bone marrow cells (25). Consequently, their main toxicological outcome is paucity of blood platelets, thrombocytopenia (24). However, it is quite clear that what disadvantages the phosphorothioates have or might have are far outweighted by the advantages that this particular modification offers as compared to the potential use of unmodified AS-ODNs or of those carrying other modifications. This is clearly seen by the fact that almost all publications in the antisense field employ this particular modification (for reviews see 1, 2, 26). This development sheds new light on the phosphorothioate invention, the importance of which could not be fully imagined beforehand, and which transported the antisense technology from a basic biomedical science tool to a pharmacological approach of potentially great clinical value. Table 9.1 presents some key historical milestones for the development of phosphorothioate antisense technology.

9.3 False Positive and False Negative Outcome of AS-ODNs

Like many other therapeutic protocols, the use of AS-ODNs may lead to a false positive or a false negative outcome. False positive results will be obtained when the functioning of the target protein is suppressed through mechanisms other than AS-ODN-mRNA hybridization. For example, the host cells may respond to the mere presence of the AS-ODN agent by suppressing production of many proteins, including the intended one. This may imply that the target protein belongs to a susceptible subgroup of gene products whose expression is vulnerable to external stimuli. Alternatively, the selected AS-ODN may exert its effect through non-sequence dependent competition with endogenous host genes or mimic bacterial or viral motifs and therefore elicit responses characteristic of attacks by such entities. Stein and Krieg refer to such false positives in their recent editorial (27). They point out that a positive drug effect is what drug developers aim for; however, such false positives may become a dangerous precedence for subsequent development of other drugs using the antisense principle.

A false negative outcome of AS-ODNs may be yet more frustrating than a false positive, as in this case the AS-ODN agent totally fails to suppress the functioning of the protein translated from its target mRNA. This may happen if the protein in question is very stable, if it is present in the cell in high levels or when total RNA transcription is enhanced as a reaction to the treatment. Worse, yet, it is possible to predict that the suppression effect of certain AS-ODNs will last only very briefly, as it will immediately elicit a selective feedback response leading to increase in the level of its target mRNA. Finally, complex combinations of all of the above possibilities may further obscure the picture, preventing evaluation of the mechanism(s) of response. Also, AS-ODNs targeted to consensus domains in genes that belong to large families may interact with a certain region in many genes (for example, the sequence encoding the membrane spanning domain in a specific family of receptors). Suppression of several receptors may in this case, be wrongly interpreted as general toxicity when in fact it represents a true antisense mechanism, albeit to many sequences.

All this does not preclude the antisense approach, neither does it imply that in using this new technology one encounters more difficulties than when developing conventional drugs. It is conceivable that *any* drug causes complex feedback responses and that false positives and negatives appear for numerous reasons in *any* case of rational drug design. However, as approval of new drugs for human use is increasingly more difficult, one should be aware of the pitfalls that lie ahead, even more so when the drug involved is based on human genome data.

9.4 Drug Delivery

The obvious difficulty in delivery of AS-ODN drugs to the brain stimulated considerable efforts toward solving this problem, and several innovative approaches have been developed which improve the penetration of AS-ODNs into the brain and the targeting of such molecules to specific brain regions. At the mechanical level, high-flow microinfusion at a constant flow rate of 3 μl/min into the gray matter distributes macromolecules into an area with a 1.5 cm radius within 12 hr. (28). This may ascertain that the delivered AS-ODN will be distributed throughout a certain nucleus in the human brain. The concentration of delivered material is rather uniform, which is important for preventing toxicity. A totally different delivery approach involves the use of the avidin-biotin conjugation reaction for improved protection of AS-ODNs against serum nuclease degradation (29). According to this concept the AS-ODN agents are monobiotinylated at their 3'-end. When reacted with avidin, they form tight, nuclease-resistant complexes with 6-fold improved stability over non-conjugated ODNs. A major concern regarding this approach is that the tight complex may prevent the subsequent uptake of these well-protected agents into cells. This additional difficulty may be solved by the use of liposomes for AS-ODN delivery (30). While these phospholipid vesicles were mainly introduced into the field to treat tumors, they may be of value also for delivery into the brain (31). A method which combines the lipophilic advantage of liposomes with the efficiency of direct conjugation is based on chemical conjugation of cholesteryl moieties to the AS-ODN agent (32). When conjugated to the 5'end of a phosphorothioate AS-ODN, the elimination half-life of these agents is shortened (from 55 to 23 hr). The cholesteryl oligonucleotides suppress their target mRNAs in a sequence-specific manner in spite of their reduced bioavailability. However, they are hepatotoxic at concentrations above 1 mg/kg.

The main obstacle for AS-ODN transport into the brain is the blood brain barrier (BBB). This capillary endothelial wall may be penetrated by ICV drug infusion or hyperosmotic treatment (i.e. using mannitol), by using liposomes or by coupling the transported agent to a transport peptide that is known to enter the brain by absorptive or receptor-mediated transcytosis. For example, a monoclonal antibody to the transferrin receptor operates as an efficient BBB transporter (31, 33). When injected intravenously, it concentrates the coupled agent in brain (5-fold over plasma concentration) within 5 hr. The coupling may be chemical or biological (i.e. through avidin-biotin connection). In this case as well, the question arises to what extent such complexes will disintegrate once they reach the inner side of the BBB and enter their target neurons. However, the method certainly increases BBB penetration. Brain oligonucleotide uptake was recently shown to be improved by this method to the extent that 0.1% of the injected dose reach brain within 60 min following intravenous injection (comparable to the brain uptake of morphine) (34).

9.5 Antisense Modulation of Behavioral Phenotypes in Mammals

While several AS-ODN drugs are currently being tested in clinical trials, none of them is yet designed for treating neurodegenerative disease, probably due to the complex and yet unraveled mechanisms underlying the development of such diseases which has prevented the identification of an appropriate target mRNA. However, ample research is being performed which tackles the issue of modulating behavioral phenotypes in mammals, an inevitable phase preceding serious development of rationally designed drugs.

Over the past few years, AS-ODNs were employed for *in vivo* modulation of diverse behavioral phenotypes. Several of those were aimed at preventing anxiety. AS-ODNs which suppress the levels of corticotropin-releasing hormone when ICV infused at three 30 μg doses 12 hr apart, were shown to attenuate social defeat-induced anxiety in rats (4). Direct injection of c-fos AS-ODNs into the amygdala suppressed the c-fos expression induced by anxiety and functioned similarly to antianxiety drugs (35, 36). Infusion of AS-ODNs directed against the V1 vasopressin receptor gene into the septum suppressed social discrimination abilities and anxiety-related behavior in rats (37). Other AS-ODNs had more general effects. An increase in cholinergic neurotransmission occurred in rats injected with AS-ODNs to the 5-ht6 receptor gene; this was accompanied by a phenotype of yawning, stretching and chewing which could be antagonized by atropine in a dose-dependent manner (38). Injection of AS-ODNs targeted at galanin mRNA into the paraventricular nucleus reduced fat ingestion and body weight gain (39), demonstrating a definite function for this intriguing peptide in the mammalian brain. A less direct effect was demonstrated by inhibition of proopiomelanocortin expression. An AS-ODN against β-endorphin mRNA (40) reduced adrenocortiocotropin (ACTH) and β-endorphin concentrations in corticotrophic cell cultures, while not affecting the cellular levels of irrelevant proteins or cell viability and proliferation. When microinfused into the rat hypothalamic arcuate nucleus, this AS-ODN reduced ACTH and β-endorphin levels and suppressed the grooming behavior characteristic of exposure to a novel environment.

Another behaviorally important neuropeptide produced in the arcuate nucleus, neuropeptide Y, controls feeding behavior. Direct anti-neuropeptide Y AS-ODN injection to the arcuate nucleus at daily intervals reduced neuropeptide Y levels by 40% in the arcuate nucleus but not in adjacent nuclei, and suppressed appetite (39). While this finding demonstrates clearly that the effect of this AS-ODN was highly specific, it also indicates the drug distribution problem which may be associated with the use of AS-ODN to suppress expression of genes normally expressed in the brain.

The antisense approach has also been used to explore the behavioral patterns caused by drugs of abuse. One target of such AS-ODNs was the D2 dopamine receptor, thought to control certain behavioral responses to cocaine, particularly the characteristic walking in close circles. Unilateral administration to the rat substantia nigra of AS-ODNs to the mRNA translated into this receptor suppressed the cocaine rotational response of the treated rats while reducing the D2 receptor in the substantia nigra (41) by 40%. This *in vivo* effect was, however, more limited than that observed with the same AS-ODN in cultured retinoblastoma cells, where 1 μM of the AS-ODN suppressed D2 dopamine receptor levels by 57% within 3 days. This study thus sheds light on a problem particular to the use of AS-ODNs in the brain: there is a limited accessibility to nerve cells *in vivo*, even under the cumbersome approach of ICV administration. In another experiment, c-fos AS-ODNs generated apomorphine and amphetamine-induced rotation. To this end, sodium pentobarbital-anaesthesized rats were injected ipsilaterally with an AS-ODN against c-fos and contralaterally with a "sense" ODN into the striatum. When treated 10 hr later with an amphetamine or apomorphine, these rats showed drug-induced rotation toward the AS-ODN-treated side (42). Also, c-fos and Jun-B levels were reduced in striatal neurons at the AS-ODN treated hemisphere, suggesting that the c-fos AS-ODN suppressed the drug-induced behavioral phenotype (43).

The selectivity of behavioral phenotypes associated with the D2 dopamine receptor was examined by *in vivo* administration of a D2 AS-ODN into the striatum of mice with unilateral lesions induced by 6-hydroxydopamine (44). When challenged by acute injections of various agents that cause D2-associated contralateral rotational behavior (i.e. quinpirole) these mice did not rotate. In contrast, injections of the muscarinic cholinergic receptor agonist oxotremorine induced rotational behavior both in control mice and in those treated with the D2 receptor AS-ODN. The reduction in quinpirole-induced rotational behavior was related to the amount and length of time the D2 AS-ODN was given. The effect was visible after 1 day of treatment and almost completely disappeared by 6 days. Yet the effect was reversible within 2 days after cessation of treatment. It was also associated with significant and selective reduction in D2 receptor mRNA within the striatum, suggesting a genuine antisense mechanism. Table 9.2 presents these recent examples for behavioural functions modulated by AS-ODNs.

9.6 AS-ODNs and Human Neuropathology

When considering the use of antisense drugs for treating nervous system diseases, several special arguments should be kept in mind.

First, brain neurons are terminally differentiated cells for which there will be no replacement. This implies that special care should be taken to minimize both sequence-dependent and sequence-independent cytotoxicity of AS-ODNs. Also, the lifetime of a neuron by far exceeds that of cells in other tissues. Therefore, it may be much more difficult to change properties of a neuron than, for example, those of a tumor cell. However, once modified, the affected neuron will survive and not proliferate.

Second, the intricate networks connecting brain neurons predict that suppression of the level of particular mRNA transcripts or their protein products in a particular brain region may affect totally different brain regions due to complex and not-yet-understood signal transduction mechanisms. Another outcome of neuronal complexity is that the feedback responses to AS-ODN-induced changes can hardly be predicted. Therefore, the establishment of AS-ODN therapies for nervous system diseases necessitates careful pre-clinical studies in both cultured neurons and animal models. This, however highlights yet one more difficulty, that of species specificity.

Tab. 9.2 Recent Brain-expressed Targets in AS-ODN Studies

	Protein product of target gene	Modulated behavioral Function	References
1	Corticotropin-releasing hormone	attenuation of social-defeat induced anxiety in rats	4
2	c-fos	suppression of anxiety; generation of apomorphine and amphetamine-induced behavior	35, 36, 42
3	V1 vasopressin receptor	suppression of social discrimination ability and anxiety	37
4	5-ht6 receptor	enhanced cholinergic neurotransmission, yawning, stretching	38
5	galanin	reduced fat ingestion	39
6	XCS-endorphin	suppressed response to novel environment	40
7	neuropeptide Y	suppression of appetite	39
8	D2 dopamine receptor	suppression of cocaine response; suppression of quinpirole response	41, 44

Primate experiments are excluded from many studies because of economic concerns, yet the rodent brain differs from the human in many different aspects. Also, several human diseases of major importance (i.e. Alzheimer's disease, multi-infarct dementia) do not occur in rodents. Finally, due to species-specific differences in codon usage, AS-ODNs studied in rodents may differ from those to be developed later as drugs for human use. This adds to the risk involved in subsequent clinical studies.

Once these conceptual arguments have been met, technical difficulties emerge. As stated above, delivery of AS-ODN drugs to the brain is far more complex than to any other tissue or organ. Drug pharmacokinetics in the brain should also be more complex than in any other tissue, as the rate of removal of AS-ODNs from particular subsets of neurons might depend on transient parameters such as electrophysiological firing rate, synaptic plasticity and circadian rhythm. Yet one more difficulty is late diagnosis: many nervous system diseases are diagnosed in advanced stages, when most of the relevant neurons have already died and the mode of functioning of the remaining ones has been drastically altered from that characteristic of the normal state. Therefore, the ultimate levels of the target mRNA transcripts will be highly variable between patients and with disease progression; however, in most cases it is extremely difficult to define the disease stage, as diagnosis is generally based on subjective criteria and because most nervous system diseases limit the caretaker's capacity to evaluate the patient's mental state. All this is well illustrated for Alzheimer's disease, as is presented in the following.

9.7 The Challenges of Alzheimer's disease

Recent studies unraveled several genes whose protein products are abnormally expressed in the demented brain. However, the primary cause(s) of Alzheimer's disease are not yet known, which complicates the choice of target genes and/or proteins whose functions should be blocked for attenuating the progress of this neurodegenerative disease. One clear fact is that cholinergic neurotransmission is perturbed in the Alzheimer's disease brain. This has been the basis for past and present therapies and it is on this fact that our current report is focused. This, in turn, requires a short introduction.

The Cholinegic Hypothesis
Acetylcholine (ACh) is the agent which mediates neurotransmission in cholinergic brain synapses. Impaired cholinergic neurotransmission may result from pathological or pharmacological insults that alter the balanced interaction between ACh and its receptors (45). Alzheimer's disease (AD) is characterized by a selective loss of ACh-producing neurons (46). The cholinergic hypothesis of AD proposes that loss of cholinergic neurons leads to an ACh deficit which is correlated with progressive degeneration of central cholinergic systems and the accompanying deterioration of cognitive function characteristic of AD patients (46, 47, 48). About 5-10% of cases are genetically determined, but even within this group, there are at least four classes, associated with four different chromosomes. The great majority of AD cases do not have an obvious hereditary component, but most cases of whatever origin, are characterized by microscopic abnormalities in brain tissue, called plaques and tangles. These structures have been for pathologists, the diagnostic feature of AD, and much effort has gone into characterizing the constituent protein, β-amyloid, abnormalities of which lead to the plaques and tangles. The concept that cholinergic imbalance may be etiologically associated with the impaired cognitive performance in AD was supported by studies in aging rats (49 and references therein) and recently strengthened by the finding that transgenic mice which overexpress the enzyme that hydrolyzes ACh, acetylcholinesterase (AChE) in brain neurons display progressive deterioration in spatial learning and memory (50). Altogether, these findings support the notion that an approach aimed at redressing the cholinergic imbalance, e.g. suppression of AChE levels, should be therapeutically beneficial for treating AD patients.

The Pharmacological Approach
A widely accepted approach to the retardation of the AD-related decline of cognitive functions has been to reverse the ACh deficit with cholinesterase inhibitors (48). Normally, AChE serves to restrict duration of the post-synaptic response by hydrolyzing ACh to acetate and choline. In the well-studied vertebrate neuromuscular junction (NMJ), AChE inhibitors have proven effective in prolonging miniature endplate potentials, presumably by extending the lifetime of ACh released into the synap-

tic cleft and thereby increasing the number of receptor-transmitter interactions (45). Tacrine (THA, tetrahydroamino-acridine, Cognex®) – the first FDA-approved AD drug – is a potent AChE inhibitor which relieves cognitive symptoms in 30-50% of mildly to moderately affected AD patients (51, 52). Unfortunately, anti-AChE therapies for AD, including tacrine, are associated with exacerbated cholinergic deficits, hematopoietic irregularities, and hepatotoxicity in some patients (48, 53). Therapeutic response and hypersensitivity to these drug regimens are likely attributable to individually variable interactions of AChE inhibitors with other proteins, especially the homologous, but genetically diverse, serum enzyme butyrylcholinesterase (BuChE;54). Thus, a more selective approach for AD is called for.

A Molecular Biology Approach

When searching for appropriate target genes, the suppression of which could be advantageous for treating AD patients, one remembers the tangle structures that characterize the pathophysiology of this disease. The tangles are composed of disarranged microtubules with abnormally high phosphorylation levels of the microtubule-associated protein tau. When transfected with tau DNA, pheochromocytoma PC12 cells responded to nerve growth factor by extending neurites more effectively than control cells; when transfected with anti-tau DNA, these cells failed to extend neurites in response to nerve growth factor treatment (55). In principle, anti-tau AS-ODNs might hence suppress formation of the tangle structures. Other important hallmarks of AD pathology are the plaque structures which contain precipitated amyloid peptides. When PC12 cells were treated with AS-ODNs to suppress amyloid peptide production, they lost the ability to respond to nerve growth factor by extending neurites (56). However, if nerve growth factor treatment preceded the addition of AS-ODNs, the oligonucleotides had no effect. This study therefore implies that AS-ODN treatment for suppressing amyloid peptide production should be initiated prior to plaque formation, a major difficulty under today's conditions when no prognosis or early diagnosis is available.

Another intriguing property of AD plaques is that they stain positively for AChE activity. AChE has a morphogenic activity in transfected glial cells (57), transiently transgenic *Xenopus* embryos (58) and transgenic mice (59). This suggests that the value of anti-AChE therapies could, at least in part, be related to suppression of this morphogenic activity and hence of plaque growth. Here again, a molecular biology approach may be of benefit. This and the long-recognized low selectivity of anti-AChE inhibitors prompted the search for alternative strategies to block AChE activity *in vivo*. Taking advantage of the divergent nucleotide sequences of the genes which encode AChE and BuChE, we developed the technology for selective inhibition of AChE expression in living cells by use of partially phosphorothioated AS-ODNs (15, 25) targeted against AChEmRNA (60, 61). Thanks to the precise and specific interaction of such AS-ODNs with their mRNA target, we hoped that they would efficiently block biosynthesis of AChE via RNaseH-mediated destruction of the message, or by steric interference with RNA splicing or translation (1). In light of their high specificity and low-dose biological effects, AS-ODNs targeted against AChEmRNA can potentially offer a promising new alternative strategy to cholinomimetic treatment of AD, which require evaluation in cultured neurons as well as in an appropriate mammalian model prior to the initiation of clinical trials. Therefore, and because this is the AS-ODN study with which we are most familiar, we chose to present it in more detail.

9.8 Human Cholinesterase Genes as Potential Targets for Antisense Therapy

Cholinesterases (ChEs) are the enzymes which terminate each nerve impulse by hydrolyzing ACh. There are two distinct ChE genes in all vertebrates. The two human genes, ACHE and BCHE, were cloned, mapped to specific chromosomal sites and expressed in transgenic organisms (reviewed by 62, 63). Interestingly, these two genes are very different in their base composition and nucleotide sequence. This ensures that antisense agents targeted to the RNA product of AChE will not interact with the RNA product of BCHE. The selectivity of AS-ODN agents therefore favors, in this gene family, the antisense approach. In contrast, most inhibitors targeted to one of the ChE proteins, like carbamates and organophosphates, will also interact with the other, since these proteins are 50% identical and >85% homologous (reviewed by 64). As the BCHE gene is frequently subject to mutation, the level and properties of its protein product are highly variable. This situation further complicates the use of ChE inhibitors (54), yet should not interfere with the action and specificity of AS-ODNs targeted at ChE genes.

An important consideration when designing a drug is that its function will not perturb basic biological properties other than that for which it is intended. For example, for neurochemical uses of antisense drugs to suppress the expression of ChEs, one needs to assure sustained neuromuscular communication by selecting AS-ODN doses which will sufficiently inhibit these enzymes in the brain but will not reduce their levels in NMJs that essential amount required for breathing and other motor functions. Furthermore, the use of anti-ChEs in AD is based on the assumption that AChE is primarily involved in the termination of cholinergic neurotransmission. However, as is the case with many other proteins, the biological role of ChEs may include additional functions. For example, accumulating evidence suggests that AChE is also involved in the regulation of nervous system development. To this end, it was shown using several experimental approaches that expression of AChE during the early stages of brain development correlates closely with the major phase of neurite outgrowth (65, 66). When overexpressed in transgenic *Xenopus* tadpoles, the brain and muscle form of the enzyme accumulates in NMJs and enhances their development (58, 67). In the C6 glioma cell line, transfected ACHE DNA induces cytomorphological alterations that lead to process extensions (57). Moreover, it was recently shown that AChE stimulates neurite outgrowth from cultured chick neurons in a manner unrelated to its catalytic activity (68, 69). Therefore, AS-ODNs targeted to the ACHE gene can be expected to affect both neurochemical and morphogenic effects.

There are at least three options for alternative splicing of ACHEmRNA transcripts, with distinct tissue distribution for each of these mRNA subtypes (57, 70) and, possibly, different biological functions. Therefore, one would like to distinguish between ACHEmRNA subtypes involved in specific function(s) and target AS-ODNs toward those domains in the AChEmRNA chain which are required for the function to be perturbed, without interfering with other functions of the AChE protein. To this end, we decided to examine the antisense approach for inhibiting AChE expression in primary murine neuronal cultures. It should be emphasized in this context, that the neuropathological hallmark of AD, the plaque structure, includes abnormally grown processes that cytochemically stain for AChE activity. Therefore, in this case we were interested in suppressing the morphogenic activity of AChE in addition to suppressing some of its catalytic capacity. It so happened that this preclinical study combined several aspects of AS-ODN studies in the use of *in vivo* and *ex vivo* systems and required several different techniques for analyzing the outcome of AS-ODN administration. In the following, we describe these experiments with some details referring to the methods employed and the interpretation of the results.

9.9 Effects of Antisense Oligonucleotides Targeted to Primary Neuron mRNAs

Important considerations in the design of an AS-ODN approach for use with neurons is the uptake mechanism of the AS-ODNs and their potential capacity to block expression of their target genes. The uptake of AS-ODNs directed against neuronal ACh receptors was studied in cultured chick primary neurons (71). Fifteen-mer AS-ODNs added to the culture medium were taken up into the cell bodies in a temperature-dependent, saturable manner (up to 20 μM). Monomeric nucleotides (AMP, ATP) competed effectively with this active uptake process in a manner reminiscent of the endocytosis of AS-ODNs described in non-neuronal cells. The efficiency of uptake depended on the age of the embryos from which neurons were removed but not on the number of days that these neurons were maintained *in vitro*. The functioning of neuronal nicotinic ACh receptors was effectively blocked in these neurons by suppressing expression of the α3 subunits of these receptors (by 80-90%). Electrophysiological analyses demonstrated abnormal properties of remaining receptors, suggesting aberrant assembly of other receptor subunits. This study paved the way for further manipulations of mammalian brain neurons.

Primary Neuron ACHE Antisense Studies

To study the involvement of AChE in mammalian neurite outgrowth and differentiation, we added synthetic 20-mer 3'-terminally phosphorothioated AS-ODNs complementary to two alternative exons in the ACHE gene, to primary neuronal cultures from 14 day-old embryonic (E14) mouse whole brain. Cells grown for 24 hr in serum-free medium on serum-coated dishes were treated with these AS-ODNs or with a control oligonucleotide with the inverted base sequence. Both antisense treatments but not the inverse oligodeoxynucleotide induced the appearance of multilayered cell aggregates and suppressed neurite outgrowth (72). The effect appeared earlier with increasing doses of the AS-ODNs, indicating dose dependence. These oligonucleotide-induced phenotypic changes suggested AChE involvement in neuronal growth and differentiation and supported the development of AS-AChE oligomers as potential drugs to replace mechanism-based AChE inhibitors for AD therapy.

The methodology involved was quite straightforward. Primary mouse neuronal cultures were prepared from embryonic (E14) mouse (Balb/C) whole brains (73). Brains were removed and cells mechanically dissociated and plated in serum-free medium ($2.5X10^6$ cells/ml) in 24-well (1 ml per well) culture dishes coated successively with poly-L-ornithine and culture medium containing 10% fetal calf serum. Cultures grown for 24 hr at 37°C, 5% CO_2 were treated with synthetic 20-mer 3'-terminally phosphorothioated oligonucleotides (25) complementary to either the consensus ACHE exon 2, common to all of the alternative forms of ACHEmRNA (AS-mE2) or the alternative 3' exon 5 (AS-mE5). The inverse sequence of AS-mE2 (inv-mE2) was used as a control. After 24 hr growth, cells were cytochemically examined to determine the remaining levels of ChEs under each treatment as correlated with the morphogenic changes caused by the oligodeoxynucleotide treatments. Figure 9.1 presents the positions of these AS-ODNs along the ACHE gene.

Alternative AS-ODNs with Common Effects

Several parameters may be used to assess the effectiveness of AS-ODN experiments and deduce which was the mechanism(s) through which they exerted their effects. It is commonly assumed that when more than one AS-ODN agent targeted against a certain mRNA species cause similar biological effects, it is likely that it occurred through an antisense mechanism. In the case of the ACHE gene, we had the option to design AS-ODNs against common and alternative regions in the ACHEmRNA transcripts. Common outcome in terms of biological effect would imply that a single function was suppressed whereas different results for each AS-ODN could distinguish between the yet undefined functions of these alternative transcripts.

The mouse ACHE gene which includes 6 exons and 4 introns, gives rise to two alternative mRNAs in mouse primary neuronal cultures:

a) brain and muscle ACHE mRNA which includes exons 1-4 and 6,
b) "readthrough" ACHE mRNA which includes exons 1-4, continued by pseudointron 4, which in certain tissues operates as an exon (70), and exon 5.

AS-mE2 and AS-mE5 were designed to hybridize with specific domains in exon 2 and exon 5, respectively. Therefore AS-mE2 could potentially lead to destruction of both ACHE mRNAs, whereas AS-mE5 could interact with only the readthrough mRNA or with the complete, yet unprocessed, transcript. Cells grown and treated for 24 hr with 0.5 μM of either AS-mE2, AS-mE5 or inv-mE2, were first stained with May-Grunwald and Giemsa stains, following with visible microscopy was performed with a Zeiss inverted microscope.

Untreated and inv-mE2 treated cells formed monolayers of single neurons with thin extensions. This important control experiment demonstrated that the inverse sequence, with identical base composition but no target, was not cytotoxic. Cells treated with either AS-mE2 or AS-mE5 were re-organized in multicellular, multi-layered aggregates connected by a few thick processes. Moreover, the initiation time of the antisense ACHE morphogenic effect was dose-dependent. Thus, cell cultures treated with increasing concentrations (0.1-5.0 μM) of AS-mE2 were monitored for 5 days and the day on which aggregation was first observed was noted. At the highest concentration of oligonucleotide (5.0 μM), cytotoxicity was observed, cells detached from the dishes and died. However, cells treated with 0.1-0.5 μM AS-ODN were morphologically affected but remained viable.

The mouse ACHE gene, controled by the promoter, P, includes 6 exons (light cross-hatching) and 4 introns (dark cross-hatching), and gives rise to two alternative mRNAs in mouse primary neuronal cultures: "brain an muscle" ACHEmRNA includes exons 1-4 and 6, and "readthrough" ACHEmRNA includes exons 1-4, pseudointron 4 (which in certain tissues operates as an exon) and exon 5. The antisense oligonucleotides AS-mE2 and AS-mE5 were designed to hybridize with specific sequences in exons 2 and 5, respectively. Therefore, AS-mE2 can potentially lead to destruction of unprocessed mRNA and both alternatively spliced mRNAs, whereas AS-mE5 can interact only with unprocessed mRNA and the "readthrough" form.

The viability issue is of major concern when vulnerable nerve cells are to bc treated with a potentially cytotoxic agent, and viability tests should probably be included in any preclinical study with AS-ODNs. In our particular case, the viability of neuronal primary cultures which displayed the cytomorphological effect following 24 hr growth in the presence of 0.5 μM AS-mE2 was assayed using a viability/cytotoxicity kit. To this end, we exploited the fact that enzymatic conversion by ubiquitous esterases in viable cells of the permeable, non-fluorescent dye calcein generates an intensely fluorescent green form of the dye. This product is retained within viable cells, producing the green fluorescence (with emission wavelength of about 530 nm). A second dye, ethidium homodimer, was used to identify dead cells, since it penetrates only damaged membranes. In the cell it binds to nucleic acids, producing a bright red fluorescence (>600 nm). Fluorescence microscopy was performed with a Zeiss Axioplan microscope equipped with X40 Achroplan lens, suitable for photographying viable cells under medium, a HC100 camera and a FITC/Texas red 485/578 double excitation filter, at magnification X400. This test proved that the aggregated cells in AS-mE2 treated cultures remained viable.

To demonstrate that the antisense treatement reduces ChE activities in these primary neuronal cultures, neuronal cell cultures were treated for 24 hr with no oligonu-

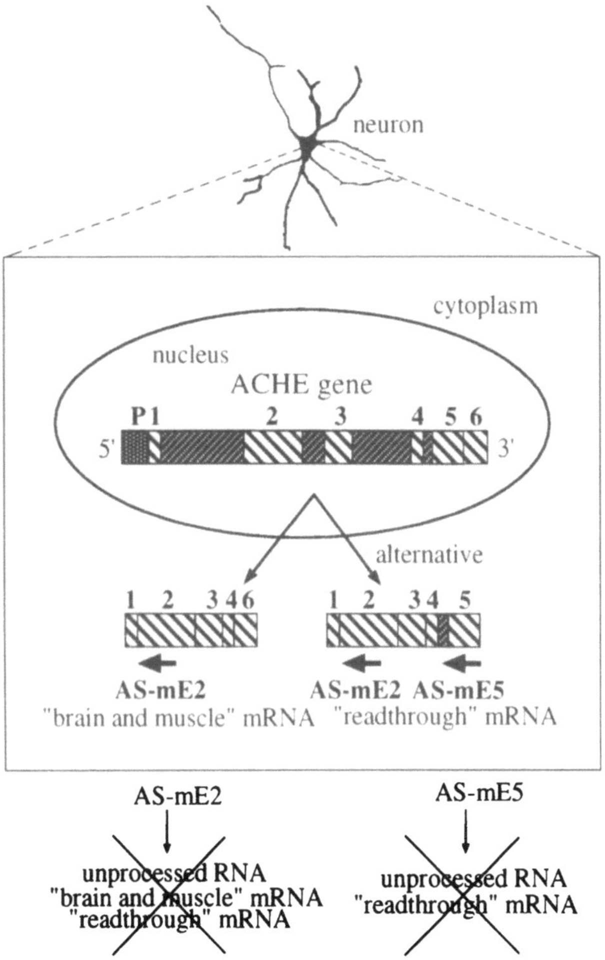

Fig. 9.1: Target sites for antisense oloigonucleotide interactions with ACHEmRNA.

cleotide, or with 0.5 μM of the noted oligomers (AS-mE2 or Inv-mE2). Cells were then stained for AChE activity overnight at 4°C with no prior fixation (59). Stained cells in 20 different microscope fields for each preparation (magnification X1000), were classified by the intensity of staining. Each field contained aproximately 250 neurons. Stained neurons (approxinmately 0.5-5% of total) included:

(1) light brown-stained cells
(2) more intensely stained cells, particularly around the cell body
(3) very intensely, dark brown stained cells, with stain reaching into neuronal extensions.

Within all classes, staining was considerably lower in AS-mE2 treated cells but not in those treated with Inv-mE2, suggesting an antisense mechanism (72).

The antisense suppression of AChE activity was further visualized by electron microscopy. For this purpose, neuronal cell cultures were treated for 24 hr with no oligonucleotide, or with 0.5 μM of the noted oligomers (AS-mE2 or Inv-mE2). Cells were then fixed for 30 min in 4% paraformaldehyde, lightly stained for AChE activity for 4 hr at room temperature and analyzed by transmission electron microscopy as detailed elsewhere (58). Crystals formed by non-enzymatic reaction initiated by the thiocholine product of AChE action on acetylthiocholine appeared in control neurons, but not in neurons treated with AS-mE2. This analysis thus demonstrated the efficacy of the employed AS-ODNs as well as their selectivity. A word of caution is in order. We do not understand the mechanism that connects AChE inhibition to prevention of neurite outgrowth. Antisense suppression of the metabolically unrelated enzyme 5'-nucleotidase also prevents neurite outgrowth (74). Both AChE and 5'-nucleotidase are ectopic enzymes in these cells, and the common effect may be to change, in an unknown manner, the surfaces of the cells. Therefore, the morphogenic changes observed in primary neurons with suppressed AChE activity suggested that suppressing this enzyme's activity may exert morphogenic changes also in other tissues in which AChE is produced, for example in hematopoietic cells. To search for the potential side effects of oligonucleotides for suppression of AChE production in such sites, we employed the antisense approach also to hematopoietic cells and tissues.

9.10 *In vitro* and *in vivo* Tests for Potential Side Effects

In addition to brain neurons and muscle cells, bone marrow cells (red blood cells, lymphocytes, platelet progenitors) from all known vertebrates express ChEs. This raises the concern for side effects of the proposed antisense therapy due to AChE inhibition in hematopoietic cells. Indeed, injection of mice with carbamate ChE inhibitors, alters proliferation of megakaryocytes (platelet progenitors) in rodents (reviewed by 62). In addition, farmers using organophosphorous anti-ChEs as insecticides are at an increased risk of development of leukemia (75). We therefore investigated the therapeutic safety margins of AS-ODNs targeted to the ACHE gene or to the closely related BCHE gene. In previous studies, we found that 25% of leukemic patients carry cholinesterase genes with abnormal copy numbers and structures (76, 77). Altogether, this implied that interference with the expression of blood cell AChE and/or BuChE is associated with enhanced bone marrow proliferation. BuChE is quantitatively more important in the blood (75% of total blood ChEs). Therefore, we considered that the bone marrow damage caused by chemical anti-ChEs could have occurred due to AChE and/or BuChE inhibition. To identify the highest concentration of oligonucleotides which will be therapeutically safe, we have further studied the expression of BuChE in the bone marrow of patients with somatically mutated, aberrantly expressed and amplified ACHE and BCHE genes. This phenomenon is associated with very low platelet counts and abnormal development of megakaryocytes (76), the platelet progenitors from which platelets are formed. We therefore attempted to mimic interference with BCHE gene expression, using phosphorothioated AS-ODNs.

Non Sequence-Dependent Interactions
When 5 µM of a fullly phosphorothioated antisense agent intended to block BCHE gene expression (AS-BCHE) was added to murine bone marrow cells grown in culture in the presence of the cytokine interleukin 3 (IL-3), megakaryocyte progenitor development was severely inhibited (78). When incubated with bone marrow cells, radioactively labeled phosphorthioated oligonucleotides bound to yet unidentified nuclear protein(s) in small, dividing cells. Incubation of bone marrow cellular proteins with these agents also resulted in protein labeling (25). However, this proved to be unrelated to the nucleotide sequence of the employed oligonucleotides – a non-specific interaction which could explain part of the interference with *ex vivo* bone marrow proliferation that was previously observed for AS-BCHE in its totally phosphorothioated form (25). Recent findings (79) strengthen this assumption and attribute part of the megakaryocytopoietic suppression observed with our AS-ODNs as well as with oligomers targeted to other genes and to AS-ODN-protein interactions.

Suppression of platelet production may lead to severe thrombocytopenia, a dangerous toxicological outcome of several other AS-ODNs. For example, this is the main danger involved in rel A antisense therapy, aimed at suppression of production of the P65 subunit of the NFκB transcription factor in tumor cells (24). This phenomenon could have resulted from the antisense effect itself (because developing promegakaryocytes depends on NFκB for their maturation). Alternatively, or additionally, it could reflect binding to and inhibition of essential nuclear proteins in promegakaryocytes, a confirmed property of phosphorothioate ODNs. To prevent, or at least minimize this protein interaction and its consequent cytotoxicity, the following observations were taken into consideration when designing anti-BCHE ODNs:

(1) Phosphorothioate protection is important for RNaseH induction and oligonucleotide stabilization.
(2) This protection may also be cytotoxic and induce part or all of the observed interference with cell development.
(3) Since RNA destruction is primarily initiated at the 3'-end, it can be blocked effectively by interfering with this activity at that end alone.
(4) Blocking only three 3'-terminal phosphorothioates is sufficient to protect the AS-ODN. This partially protected oligomer should be less toxic than, yet equally effective as the fully-blocked oligomer.

Based on these arguments, phosphorothioate protection of AS-BCHE was limited to the three 3'-terminal internucleotidic bonds. When administered in similar doses, thus partially protected AS-ODN reduced production of platelet progenitors as efficiently as their fully protected counterparts (25). However, when incubated with bone marrow proteins, they displayed no binding. Also, their addition to cultures apparently did not interfere with the functioning of genes other than those targeted, and no non-specific reduction in cell proliferation could be observed when irrelevant oligomers were tested. Altogether, this technological improvement demonstrated that antisense inhibition of BCHE mRNA (80) would indeed cause hematopoietic side effects, if used at 5 µM concentrations. To test the validity of this assumption, we moved on to combine cell culture studies with *in vivo* studies.

9.11 Comparative Studies with AS-ODNs for Genes with Closely Related Functions

When mice were injected with sufficient AS-ODNs to achieve a concentration of 5 µM of the oligomer in body fluids blocking BCHE gene expression (AS-BCHE), their platelet progenitors revealed reduced levels of BCHEmRNA (as tested by *in situ* hybridization). This indicated that specific RNA destruction occurred. At the same time, an unrelated gene (i.e. β-actin) remained fully expressed, as tested by reverse transcription followed by PCR amplification (RT-PCR), reflecting low toxicity of these antisense agents. When bone marrow cells from the injected mice were cultured, megakaryocyte colony development was severely inhibited (40% reduction). *In vitro* megakaryocytopoietic cultures incubated with AS-BCHE further displayed a sharp shift in differentiation, from predominantly megakaryocyte to myeloid lineages (81). Thus suppression of BuChE production can cause thrombocytopenia *and* increase myeloid cell counts, two hematopoietic effects which indeed were reported in Alzheimer's disease patients under tacrine treatment (82).

The AS-BCHE studies may be regarded as a necessary precaution, aimed at explaining hematopoietic complications in patients treated with chemical anti-ChEs of relatively limited selectivity. However, the drug(s) we aim to develop is(are) AS-ACHE. Therefore, our goal was to test ACHE AS-ODNs for their effects on hematopoietic development. Intraperitoneal injection of AS-ACHE to block expression of AChE resulted in much more dramatic changes than those observed with AS-BCHE. A single injection of 5 µg/g body weight yielding 5 µM AS-ODN in body fluids, caused drastic reductions in the fractions of bone marrow erythrocytes and lymphocytes at 12 days post-treatment, as well as reciprocal increases in myeloid cells, changes which were almost totally reversed by day 20 (61). The possibility of transcriptional feedback could not be excluded. Moreover, since *in vivo* it is virtually impossible to determine absolute numbers of bone marrow cells, we could not conclude whether erythroid development had been inhibited, if myeloid cell proliferation had been enhanced, or both. To answer these questions, we administered such AS-ACHE oligocnucleotides *ex vivo*, in primary cultures of murine bone marrow cells.

Bone marrow cell cultures have several advantages over *in vivo* analytical studies of antisense effects.

(1) They reveal the absolute number of proliferating stem cells present at plating time, each of which develops into a colony, whereas the other terminally differentiated cells die in culture (essentially, as they would *in vivo* as well). Therefore, they yield important information on the effect of the tested AS-ODNs on hematopoiesis in general.
(2) They are subject to convenient modulation by cytokines. In the presence of interleukin-3 IL-3, both platelet progenitors and myeloid cells will develop, but with added erythropoietin, erythroid cells will develop as well. This enables differential cell counts and examination of the dependence of cell composition on AS-ODN effects and growth factors.
(3) Apart from their importance for evaluation of the safety of AD drugs, the *ex vivo* cultures are a clinically important model system for progenitor cells produced for transplantation. This increasingly popular procedure is currently used for treatment of cancer patients following drastic chemotherapy or irradiation.

When added to erythropoietic cell cultures, 15 μM AS-ACHE caused a dose-dependent increase in colony and cell counts and increased the fraction of myeloid cells at the expense of erythroid cells and megakaryocytes (60). This implies that thrombocytopenia and increased myeloid cell counts can occur under AS-ACHE treatment as well, especially at high concentrations. With IL-3 alone, AS-ACHE decreased colony counts but not cell numbers, and diverted up to 50% of the cells into erythroblasts (which could not develop further for the lack of erythropoietin). Additional tests revealed transient early reduction in ACHEmRNA, followed by 10-fold increase by day 4 post-treatment. This could reflect an antisense mechanism which elicits a feedback response replacing the missing ACHEmRNA molecules and more, and calls for prolonged exposure to AS-ODNs in any therapeutic regimen.

A general change in the pattern of mRNA transcripts demonstrated by using the approach of Differential PCR display (60) supported the possibility of a feedback response. A 10-fold increase in DNA yield and prevention of the DNA fragmentation (83) which appeared in non-treated cultures confirmed the increased cell counts, and the generally healthier appearance of cells under AS-ACHE treatement (fewer vacuoles, larger nuclei, smoother cell surface) strengthened this assumption. As mentioned above, AS-BCHE did not cause such changes, it only reduced the number of megakaryocytes at the expense of meyloid cells (80). Comparison of the effects of AS-ACHE to those of AS-BCHE in erythropoietic cultures thus revealed the wider scope and dominant nature of AS-ACHE over erythroid and megakaryocyte development. However, since all of these effects occurred only at AS-ODN concentrations higher than 5 μM, the hematopoietic studies also confirmed that AS-ACHE should be therapeutically safe at the concentrations effective in neurons (0.1-0.5 μM). While this leaves open the question of body fluids concentration required to reach the desired levels of AS-ODNs in the brain, the safety margin does seem reassuring.

9.12 Discussion

We have presented the design and application of antisense techniques for the treatment of Alzheimer's disease, which illustrates many of the advantages inherent in this approach. We have also been at pains to elaborate on complications we encountered because they may be generally instructive. To sum up the principle concepts we tried to present, several major questions are raised when an antisense drug is to be considered for therapeutic use:

(1) Can it produce the desired biological effect?
(2) Can it reduce the level of the corresponding protein product for prolonged periods?
(3) Does it suppress the level of targeted mRNA?
(4) Does this effect appear both in cultured cells and *in vivo*?
(5) Is the tissue specificity of the relevant AS-ODN sufficient to exclude dangerous side effects?

We have addressed these questions by covering recent publications in the field of molecular brain research and by describing our own research using antisense drugs for the cholinesterase genes. Yet more specifically, we designed two AS-ODNs targeted to the ACHE gene. AS-ACHE ODNs suppressed neurite extension from cultured primary neurons in a dose-dependent manner, suggesting that they might attenuate AD plaque development; they suppressed the level of the AChE protein, although mRNA studies suggested feedback responses that modulate ACHEmRNA levels; and their functions(s) *in vivo* resembled their effects in cultured cells. Also, the sensitivity of neurons to AS-ACHE ODNs exceeded by 50-fold that of hematopoietic cells, providing a potentially safe therapeutic window. While many questions remain to be solved before AS-ACHE ODNs reach clinical trials, their use seems promising.

To test for potential side effects of the proposed drugs, we employed hematopoietic cell cultures. Our findings in these cultures, and in subsequent *in vivo* experiments confirm the suggestion of others of a regulatory role of the ChEs in bone marrow development. The accumulated evidence suggests that AS-ACHE apparently blocks the production of the multi-potential progenitors of both erythroid and megakaryocyte cells. Under such conditions, cultured bone marrow cells can only proliferate or develop into myeloid cells, which indeed they do. This explains complications associated with the use of conventional anti-AChE drugs and emphasizes the advantages of AS-ACHE ODNs as substitutes.

The hypothesis that emerges from the above experiments is that ChEs possibly participate in directing proliferating stem cells toward a differentiated state that eventually will terminate in programmed death. When their expression is blocked, hematopoietic proliferation may be perturbed. The range of AS-ODNs found useful in primary neurons from mice (0.1 to 0.5 μM) will be a starting point for their evaluation in primate pre-clinical trials, to test for efficacy while avoiding hematopoetic side effects. In addition, important implications arise for bone marrow transplantations, since this *ex vivo* procedure includes a phase of cell culture very much like the *in vitro* culture technique detailed above. It would be extremely helpful to amplify the number of proliferating stem cells prior to their reintroduction into the patient – thus shortening considerably the hospitalization time and improving the patient's condition while reducing the volume of cells to be injected. The complication involved in the use of AS-ACHE for neurochemical purposes may thus become a benefit for bone marrow uses.

If reproduced in humans, our procedure further offers the opportunity to improve the proliferative state of bone marrow in patients following chemotherapy or irradiation for any pathology, not only hematopoietic in nature. In addition, this same fact explains certain hematopoietic side effects reported in AD patients under anti-ChE therapy. Special care should therefore be taken to reduce the concentration of AS-ACHE drugs to minimum, to avoid such dangerous side effects.

A general concern as regards to the use of AS-ODNs in brain refers to the impermeability of the BBB, potentially posing a delivery problem. However, the blood brain barrier is known to be disrupted in AD patients (84); the delivery of AS-ACHE can thus be effected with simple i.v. injections. The main remaining difficulty, one which pertains to all other AS-ODN protocols, is that of the feedback response to be expected and its possible long-term effects.

In addition to the value of this therapeutic approach, the studies discussed bear intriguing basic research implications. ChEs are known to be expressed during embryonic development of many other cell types that undergo terminal differentiation (i.e., muscle and bone cells). It would be fascinating to discover whether in those tissues as well, these interesting enzymes are involved in controlling the shift from proliferation to differentiation. Thus, we may expect further basic science spin-offs as antisense technology is developed for clinical use.

Measured in terms of publications and emergence of biotechnology-based new enterprises, the interest and activity in antisense phosphorothioate drugs has grown rapidly during the past 5 years, and has directed much attention to synthesis of oligonucleotide analogs (79, 2). Important, but as yet unresolved issues in antisense technology deal with elucidating the mechanisms of antisense-agent uptake by cells. In addition, refining and understanding of antisense-agent pharmacokinetics and cellular distribution are required. For each drug to be developed, toxicology and efficacy must be evaluated. In addition, the mechanism of action of such drugs is of primary importance, as recent evidence indicates that many of the biomedically-active AS-ODNs operate through protein binding and not necessarily by the antisense mechanism (27). These and many other parameters will determine the future of antisense-based chemotherapies. However, the status of this technology as an innovation of pharmacotherapeutics has already been gained.

Acknowledgements

We thank Dr. D. Patinkin and Ms. R. Timberg (Jerusalem) for their contribution toward these studies, and Drs. D. Glick and S. Seidman for critical evaluation of this manuscript. This research was supported, in part, by the United States Army Medical Research and Development Command (contract DAMD 1797-1-7007, to H.S. and H.Z.), the German Israeli Fund (to H.S. and F.E.), the Israel Ministry of Health (to H.S. and H.Z.) and the E.W.H. Trust, London (to H.Z.).

References

1. Wagner, R. W. (1994). Gene inhibition using antisense oligodeoxynucleotides. Nature *372*, 333-335.
2. Strauss, M. (1996). Antisense Molecules and Ribozymes: Medical Applications. Antisense Technology (in press).
3. McCarthy, M. M., Masters, D. B., Rimvall, K., Schwartz-Giblin, S., and Pfaff, D. W. (1994). Intracerebral administration of antisense deoxynucleotides to GAD65 and GAD67 mRNAs modulate reproductive behavior in the female rat. Brain Res. *636*, 209-220.
4. Skutella, T., Probst, J. C., Jirikowski, G. F., Holsboer, F., and Spanagel, R. (1994). Ventral tegmental area (VTA) injection of tyrosine hydroxylase phosphorothioate antisense oligonucleotide suppresses operant behavior in rats. Neurosci. Lett. *167*, 55-58.
5. Sakai, R. R., He, P. F., Yang, L. Y., Ma, Y. F. Guo, J. J., Reilly, C. N., and Fluharty, S. J. (1994). Intracerebroventicular administration of AT1 receptor antisense oligonucleotides inhibits the behavioral actions of angiotensin II. J. Neurochem. *62*, 2053-2056.
6. Maurizot, J. C., Brahms, J., and Eckstein, F. (1969). Forces involved in the conformational stability of nucleic acids. Nature *222*, 559-561.
7. Halford, M. H. and Jones, A. S. (1968). Synthetic analogues of polynucleotides. Nature *217*, 638-640.
8. Pitha, P. M., Teich, N. M., Lowy, D. R., and Pitha, J. (1973). Inhibition of murine leukemia virus replication by poly (vinyluracil) and poly (vinyladenine). Proc. Natl. Acad. Sci. USA *70*, 1204-1208.
9. De Clercq, E., Eckstein, F., and Merigan, T.C. (1969). Interferon induction increased through chemical modification of a synthetic polyribonucleotide. Science *165*, 1137-1139.
10. Zamecnik, P. C., Goodchild, J., Taguchi, Y., and Sarin P. S. (1986). Inhibition of replication and expression of human T-cell lymphotropic virus type III in cultured cells by exogenous synthetic oligonucleotides complementary to viral RNA. Proc. Natl. Acad. Sci. USA *83*, 4143-4146.
11. Matsukura, M., Shinozuka, K., Zon, G., Mitsuya, R. M., Cohen, J. S., and Broder S. (1987). Phosphorothioate analogs of oligodeoxynucleotide inhibitors of replication and effects on human immunodeficiency virus. Proc. Natl. Acad. Sci. USA *84*, 7706-7710.
12. Agrawal, S., Goodchild, J. Civeira, M. P. Thornton, A. H., Sarin, P. S., and Zamecnik P. C. (1988). Oligodeoxphosphoroamidates and phosphorothioates as inhibitors of human immunodeficiency virus. Proc. Natl. Acad. Sci. USA *85*, 7079-7083.
13. Stein, C. A. and Cheng, Y. C. (1993). Antisense oligonucleotides as therapeutic agents – Is the bullet really magical? Science *261*, 1004-1012.
14. Eckstein, F. (1966). Nucleoside phosphorothioates. J. Am. Chem. Soc. *88*, 4292-4294.
15. Eckstein, F. (1985). Nucleoside phosphorothioates. Annu. Rev. Biochem. *54*, 367-402.
16. Eckstein, F. and Gish, G. (1989). Phosphothioates in molecular biology. Trends Biochem. Sci. *14*, 97-100.
17. De Clercq, E., Eckstein, F., Sternbach, H., and Merigan, T.C. (1970). The antiviral activity of thiophosphate-substituted polyribonucleotides *in vitro* and *in vivo*. Virology *42*, 421-428.
18. Zamecnik, P. C. and Stephenson, M. L. (1978). Inhibition of Rous sarcoma virus replication and cell transformation by a specific oligodeoxynucleotide. Proc. Natl. Acad. Sci. USA *75*, 280-284.
19. Hewick, R. M., Hunkapiller, M. W., Hood, L. E., and Dreyer, W.J. (1981). A gas-liquid solid phase peptide and protein sequenator. J. Biol. Chem. *256*, 7990-7997.

20. Caruthers, M. H. (1985). Gene synthesis machines: DNA chemistry and its uses. Science *230*, 281-285.
21. Alvarado-Urbina, G., Sathe, B. M., Liu, W.-C., Gillen, M. F., Duck, P. D., Bender, R., and Ogilvie, K. K. (1981). Automated synthesis of gene fragments. Science *214*, 270-274.
22. Fakler, B., Herlitze, S., Amthor, B., Zenner, H. P., and Ruppersberg, J. P. (1994). Short antisense oligonucleotide-mediated inhibition is strongly dependent on oligo length and concentration but almost independent of location of the target sequence. J. Biol. Chem. *269*, 16187-16194.
23. Bennett, C. F., Condon, T. P., Grimm, S., Chan, H., and Chiang, M. V. (1994). Inhibition of endothelial cell adhesion molecule expression with antisense oligonucleotides. J. Immunol. *152*, 3530-3540.
24. Sarmiento, U. M., Perez, J. R., Becker, J. M., and Narayan, R. (1994). *In vivo* toxicological effects of rel A antisense phosphorothioates in CD-1 mice. Antisense Res. Develop. *4*, 99-107.
25. Ehrlich, G., Patinkin, D., Ginzberg, D., Zakut, H., Eckstein, F., and Soreq, H. (1994). Use of partially phosphorothioated "Antisense" oligodeoxynucleotides for sequence-dependent modulation of hematopoiesis. Antisense Res. Develop. *4*, 173-183.
26. Stein, C. A. and Narayanan, R. (1994). Antisense oligodeoxynucleotides. Curr. Opin. Oncol. *6*, 587-594.
27. Stein, C. A. and Krieg, A. M. (1995). Problems of interpretation of data derived from *in vitro* and *in vivo* use of antisense oligodeoxynucleotides. Antisense Res. Develop. *4*, 67-69.
28. Morrison, P. F., Laske, D. W., Bobo, H., Oldfield, E. H., Dedrick, R. L. (1994). High-flow microinfusion: tissue penetration and pharmacodynamics. Am. J. Physiol. *266*, R292-305.
29. Boado, R. J. and Pardridge, W. M. (1992). Complete protection of antisense oligonucleotides against serum nuclease degradation by an avidin-biotin system. Bioconjug. Chem. *3*, 519-523.
30. Juliano, R. L. and Akhtar, S. (1992). Liposomes as a drug delivery system for antisense oligonucleotides. Antisense Res. Develop *2*, 165-176.
31. Pardridge, W. M. (1992). Recent development in peptide drug delivery to the brain. Pharmacol. Toxicol. *71*, 3-10.
32. Desjardins, J., Mata, J., Brown, T., Graham, D., Zon, G., and Iversen, P. (1995). Cholesteryl-conjugated phosphorothioate oligodeoxynucleotides modulate CYP2B1 expression *in vivo*. J. Drug Target 2, 477-85.
33. Pardridge, W. M. (1994). Vector-mediated delivery of a polyamide ("peptide") nucleic acid analogue through the blood-brain barrier *in vivo*. Trends Biotechnol. *12*, 239-245.
34. Pardridge, W. M., Boado, R. J., and Kang, Y. S. (1995). Vector-mediated delivery of a polyamide ("peptide") nucelic acid analogue through the blood-brain barrier *in vivo*. Proc. Natl. Acad. Sci., USA. *92*, 5592-5596.
35. Hooper, M. L., Chiasson, B. J., and Robertson, H. A. (1994). Infusion into the brain of an antisense oligonucleotide to the immediate-early gene c-fos suppresses production of fos and produces a behavioral effect. Neurosci. *63*, 917-924.
36. Moller, C., Bing, O., and Heilig, M. (1994). c-fos expression in the amygdala: *in* vivo antisense modulation and role in anxiety. Cell. Molec. Neurobiol. *14*, 415-423.
37. Landgraf, R., Gerstberger, R., Montkowski, A., Probst, J.C., Wotjak, C.T., Holsboer, F., and Engelmann, M. (1995). V1 vasopressin receptor antisense oligodeoxynucleotide into septum reduces vasopressin binding, social discrimination abilities and anxiety-related behavior in rats. J. Neurosci. *15*, 4250-4258.
38. Bourson, A., Borroni, E., Austin, R. H., Monsma, F. J. Jr; and Sleight, A. J. (1995). Determination of the role of the 5-ht6 receptor in the rat brain: a study using antisense oligonucleotides. J. Pharmacol. Exp. Ther. *274*, 173-180.
39. Akabayashi, A., Wahlestedt, C., Alexander, J. T., and Leibowitz, S. F. (1994). Specific inhibition of endogenous neuropeptide Y synthesis in arcuate nucleus by antisense oligonucleotides suppresses feeding behavior and insulin secretion. Molec. Brain Res. *21*, 55-61.
40. Spampinato, S., Canossa, M., Carboni, L., Campana, G., Leanza, G., and Ferri, S. (1994). Inhibition of proopiomelanocortin expression by an oligodeoxynucleotide complementary to beta -endorphin mRNA. Proc. Natl. Acad. Sci. USA. *91*, 8072-8076.
41. Silvia, C. P., King, G. R., Lee, T. H., Xue, Z. Y., Caron, M. G., and Ellinwood, E. H. (1994). Intranigral administration of D2 dopamine receptor antisense oligodeoxynucleotides establishes a role for nigrostriatal D2 autoreceptors in the motor actions of cocaine. Molec. Pharmacol. *46*, 51-57.
42. Dragunow M., Lawlor, P., Chiasson, B., and Robertson, H. (1993). c-fos antisense generates apomorphine and amphetamine-induced rotation. Neuroreport *5*, 305-306.

43. Sommer, N., Melms, A., Weller, M., and Dichgans, J. (1993). Ocular myasthenia gravis. A critical review of clinical and pathological aspects. Doc. Opthalmol. *84*, 309-333.
44. Zhou, L. W., Zhang, S. P., Quin, Z. H., and Weiss, B. (1994). *In vivo* administration of an oligodeoxynucleotide antisense to the D2 dopamine receptor messenger RNA inhibits D2 dopamine receptor-mediated behavior and the expression of D2 dopamine receptors in mouse striatum. J. Pharmacol. Exp. Ther. *268*, 1015-1023.
45. Katz, B. and Miledi, R. (1973). The binding of acetylcholine to receptors and its removal from the synaptic cleft. J. Physiol. Lond. *231*, 549-574.
46. Coyle, J. T., Price, D. L., and DeLong, M. R. (1983). Alzheimer's disease: a disorder of cortical cholinergic innervation. Science *219*, 1184-1189.
47. Katzman, R. (1986). Alzheimer's disease. New Eng. J. Med. *314*, 964-973.
48. Davis, R. E., Emmerling, M. R., Jaen, J. C., Moos, W. H., and Spiegel, K. (1993). Therapeutic intervention in dementia. Crit. Rev. Neurobiol. *7*, 41-83.
49. Kadar, T., Silbermann, M., Weissman, B. A., and Levy, A. (1990). Age-related changes in the cholinergic components within the central nervous system II. Working memory impairement and its relation to hippocampal muscarinic receptors. Mech. Aging Develop. *55*, 139-149.
50. Beeri, R. Andres, C., Lev-Lehman, E., Timberg, R., Huberman, T., Shani, M., and Soreq, H. (1995). Transgenic expression of human acetylcholinesterase induces progressive cognitive deterioration in mice. Curr. Biol. *5*, 1063-1071.
51. Crimson, M. L. (1994). Tacrine: first drug approved for Alzheimer's disease. Ann. Pharmacother. *28*, 744-751.
52. Wagstaff, A. J. and McTavish, Y. (1994). Tacrine. A review of its pharmacodynamic and pharmacokinetic properties, and therapeutic efficacy in Alzheimer's disease. Drugs Aging *4*, 510-540.
53. Knapp, M. J., K, D. S., Solomon, P. R., Pendlebury, W. W., Davis, C. S., and Gracon, S. I. (1994). A 30-week randomized controlled trial of high-dose tacrine in patients with Alzheimer's disease. JAMA *271*, 985-991.
54. Loewenstein-Lichtenstein, Y., Schwarz, M., Glick, D., Norgaard-Pedersen, B., Zakut, H., and Soreq, H. (1995). Genetic predisposition to adverse consequences of anticholinesterases in 'atypical' BCHE carriers. Nature Med. *1*, 1082-1085.
55. Esmaeli-Azad, B., McCarty, J. H., and Feinstein, S. C. (1994). Sense and antisense transfection analysis of tau function: tau influences net microtubule assembly, neurite outgrowth and neuritic stability. J. Cell Sci. *107*, 869-879.
56. Majocha, R. E., Agrawal, S., Tang, J. Y., Humke, E. W., and Marotta, C. A. (1994). Modulation of the PC12 cell response to nerve growth factor by antisense oligonucleotide to amyloid precursor protein. Cell. Mol. Neurobiol. *14*, 425-437.
57. Karpel, R., Sternfeld, M., Ginzberg, D., Guhl, E., Graessman, A., and Soreq H. (1996). Overexpression of alternative human acetylcholinesterase forms modulates extensions in cultured glioma cells. J. Neurochem. *66*, 114-123.
58. Seidman, S., Sternfeld, M., Ben Aziz-Aloya, R., Timberg, R., Kaufer, D., and Soreq, H. (1995). Synaptic versus epidermal accumulation of human acetylcholinesterase is encoded by alternative 3'-terminal exons. Molec. Cell Biol. *14*, 459-473.
59. Andres, C., Beeri, R., Huberman, T., Shani, M., and Soreq, H. (1996). Cholinergic drug resistance and impaired spatial learning in transgenic mice overexpressing human brain acetylcholinesterase. K. Loffelholz, Ed. Prog. Brain. Res. 109, pp. 265-272.
60. Soreq, H., Patinkin, D., Lev-Lehman, E., Ginzberg, D., Grifman, M., Eckstein, F., and Zakut, H. (1994). Antisense oligonucleotide inhibition of acetylcholinesterase gene expression induces progenitor cell expansion and suppresses hematopoietic apoptosis *ex vivo*. Proc. Natl. Acad. Sci USA. *91*, 7907-7911.
61. Lev-Lehman, E., Ginzberg, D., Hornreich, G., Ehrlich, G. Meshorer, A., Eckstein, F., Soreq, H., and Zakut, H. (1994). Antisense inhibition of acetylcholinesterase gene expression causes transient hematopoietic alterations *in vivo*. Gene Therapy *1*, 127-135.
62. Soreq, H. and Zakut, H. (1993). Human Cholinesterases and Anticholinesterases. San Diego Academic Press.
63. Massoulie, J., Pezzementi, L., Bon, S., Krejci, E., and Vallette, F. M. (1994). Molecular and cellular biology of the cholinesterases. Prog. Neurobiol. *41*, 31-91.

64. Schwarz, M., Glick, D., Loewenstein, Y., and Soreq, H. (1995). Engineering of human cholinesterases explains and predicts diverse consequences of administration of various drugs and poisons. Pharmac. Ther. *67*, 283-322.
65. Layer, P. G., Weikert, T., and Willbold, E. (1992). Chicken retinospheroids as developmental and pharmacological *in vitro* models: acetylcholinesterase is regulated by its own and by butyrylcholinesterase activity. Cell Tissue Res. *268*, 409-418.
66. Catalono, S. M., Robertson, R. T., and Killackey, H. P. (1991). Early ingrowth of thalamocortical afferents to the neocortex of the prenatal Rat. Proc. Natl. Acad. Sci. USA *88*, 299-3003.
67. Shapira, M., Seidman, S., Sternfeld, M., Timberg, R., Kaufer, D., Patrick, J. W., and Soreq, H. (1994). Transgenic engineering of neuromuscular junctions in *Xenopus laevis* embryos transiently overexpressing key cholinergic proteins. Proc. Natl. Acad. Sci. USA *91*, 9072-9076.
68. Layer, P. G., Weikert, T., and Alber, R. (1993). Cholinesterases regulate neurite growth of chick nerve cells *in vitro* by means of a non-enzymatic mechanism. Cell Tissue Res. *273*, 219-226.
69. Small, D. H., Reed, G., Whitefield, B., and Nurcombe, V. (1995). Cholinergic regulation of neurite outgrowth from isolated chick sympathetic neurons in culture. J. Neurosci. *15*, 144-151.
70. Karpel, R., Ben Aziz-Aloya R., Sternfeld, M., Ehrlich, G., Ginzberg, D., Tarroni, P., Clementi, F., Zakut, H., and Soreq, H. (1994). Expression of three alternative acetylcholinesterase messenger RNAs in human tumor cell lines of different tissue origins. Exp. Cell Res. *210*, 268-277.
71. Yu, C., Brussaard, A. B., Yang, X., Listerud, M., and Role L.W. (1993). Uptake of antisense oligonucleotides and functional block of acetylcholine receptor subunit expression in primary embryonic neurons. Dev. Genet. *14*, 296-304.
72. Grifman, M., Ginzburg, D., and Soreq, H. (1995). Impairment of neurite extension and apoptosis-dependent DNA fragmentation in primary neuronal cell cultures administered with an ACHE antisense oligonucleotide. J. Neurochem. *65 (Suppl.)*, 82.
73. Weiss, S., Pin, J. P., Sebben, M., Kemp, D.E., Sladeczek, F., Gabrion, J., and Bockaert, J. (1986). Synaptogenesis of cultured striatal neurons in serum free medium: a morphological and biochemical study. Proc. Natl. Acad. Sci. USA *83*, 2238-2242.
74. Zimmermann, H., Heilbronn, A., Braun, N., Carstensen, C., Kegel, B., and Maienschein, V. (1995). Extracellular metabolism of nucleotides in the nervous system. J. Neurochem. *65 (Suppl.)*, 210.
75. Brown, L. M., Blair, A., Gibson, R., Everett, G. D., Cantor, K. P., Shuman, L. M., Burmeister, L. F. Van Lier, S. F., and Dick, F. (1990). Pesticide exposures and other agricultural risk factors for leukemia among men in Iowa and Minnesota. Cancer Res. *50*, 6585-6591.
76. Lapidot-Lifson, Y., Prody, C. A., Ginzberg, D., Meytes, D., Zakut, H., and Soreq, H. (1989). Co amplification of human acetylcholinesterase and butyrylcholinesterase genes in blood cells: Correlation with various leukemias and abnormal megakaryocytopoiesis. Proc. Natl. Acad. Sci. USA *86*, 4715-4719.
77. Zakut, H., Lapidot-Lifson, Y., Beeri, R., Ballin A., and Soreq, H. (1992). *In vivo* gene amplification in non-cancerous cells: Cholinesterase genes and oncogenes amplify in thrombocytopenia associated with lupus erythematosus. Mut. Res. *276*, 275-284.
78. Patinkin, D., Seidman, S., Eckstein, F., Benseler, F., Zakut, H., and Soreq, H. (1990). Manipulations of cholinesterase gene expression modulate murine megakaryocytopoiesis *in vitro*. Molec. Cell Biol. *10*, 6046-6050.
79. Wagner, R. (1995). Toward a broad-based antisense technology. Antisense Res. Develop. *5*, 113-144.
80. Soreq, H., Lev-Lehman, E., Patinkin, D., Grifman, M., Ehrlich, G., Ginzberg, D., Eckstein, F., and Zakut, H. (1995). Antisense oligonucleotides suppressing expression of cholinesterase genes modulate hematopoiesis *in vivo* and *ex vivo*. In: Enzymes of the Cholinesterase Family, D. M. Quinn, A. S. Balasubramanian, B. P. Doctor and P.Taylor, eds. New York: Plenum Press, 1-6.
81. Patinkin, D., Lev-Lehman, E., Zakut, H., Eckstein F., and Soreq, H. (1994). Antisense inhibition of butyrylcholinesterase gene expression predicts adverse hematopoietic consequences to cholinesterase inhibitors. Cell. Molec. Neurobiol. *14*, 459-473.
82. Winker, M.A. (1994). Tacrine for Alzheimer's disease, which patent, what dose? JAMA *271*, 1023-1024.
83. Gavrieli, Y., Sherman, Y., and Ben-Sasson S. (1992). Identification of programmed cll deeath *in situ* via specific labeling of nuclear DNA fragmentation. J. Cell Biol. *119*, 493-501.
84. Harik, S. I. and Kalaria, R. N. (1991). Blood-brain barrier abnormalities in Alzheimer's disease. Ann. N.Y. Acad. Sci. *640*, 47-52.

10. Synthetic Ribozymes: The Hammerhead Ribozyme

Olaf Heidenreich and Fritz Eckstein

10.1 Introduction

Ribozymes can cleave RNA sequences specifically by recognition of their target RNA through Watson-Crick base pair formation to form the substrate-ribozyme complex. Thus, ribozymes have the potential for the sequence specific interference with gene expression. Ribozymes can be introduced into cells by two methods either by endogenous or by exogenous delivery. Endogenous delivery is discussed in the article by B. Sullenger. Our article is concerned with exogenous delivery whereby the ribozyme is prepared, either chemically or by transcription, prior to introduction into the cell. This method resembles the use of antisense oligodeoxynucleotides to inhibit gene expression. The antisense methodology is described in the article by Soreq et al. in this book. Most examples of exogenous delivery of ribozymes so far described require the chemical synthesis of the ribozyme. As the hammerhead ribozyme is the smallest ribozyme it is not surprising that it has been used almost exclusively in this approach. It is for this reason that this review is restricted to this particular ribozyme.

A comprehensive review on different aspects of ribozymes, entitled "RNA Catalysis", can be found in volume 10 of Nucleic Acids & Molecular Biology (1). Applications of ribozymes for the inhibition of gene expression are also discussed and summarized in several recent review articles (2, 3, 4, 5, 6).

10.2 Structure and Function of the Hammerhead Ribozyme

The RNA hammerhead structure was originally detected and described as the RNA motif responsible for the self cleavage of the multimeric RNAs of viroids and virusoids (7, 8, 9). It has also been found in the transcript of the satellite DNA of a newt species (10). The hammerhead structure consists of some 50 nucleotides and is the shortest catalytic RNA found in nature so far. Its secondary structure is com-

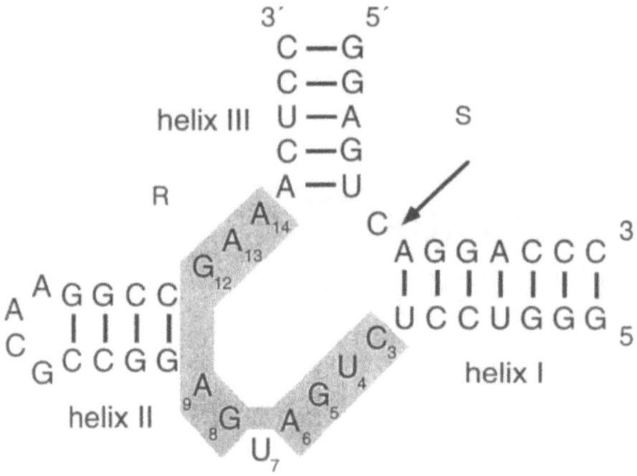

Fig. 10.1: Hammerhead ribozyme.
Twodimensional representation of a hammerhead ribozyme. S, substrate strand; R, ribozyme strand; arrow, position of cleavage. Numbering is according to Hertel et al. (96).

posed of three double stranded helices connected by a core region (Fig. 10.1). Sequence comparisons and extensive mutagenesis studies showed that the core region is very conserved (11, 12). Almost any nucleotide change in this region results in a dramatic loss of cleavage activity. The only exception is position 7 which may be any nucleotide. In contrast to the core region the helical sequences are not conserved. This fact allows targeting of hammerhead ribozymes against many RNA sequences (see below).

Recently, the three dimensional structure of the hammerhead RNA has been solved. Several methods such as X-ray crystallography (13, 14), fluorescence resonance energy transfer (15) and gel electrophoresis techniques (16) have been employed. All approaches suggest a wishbone like structure, where helices → II and → III are stacked with helix → I pointing away. The core region is highly structured and contains a complex hydrogen bond network. Stem II is extended by two non-Watson-Crick GA base pairs. The CUGA sequence from position 3 to 6 in the core forms a sharp turn similar to the uridine turn in tRNAs and has been identified as a potential metal cofactor binding site where the metal ion coordinates to the N^3 of the cytidine at position 3 (14). Another potential site is located at the terminal GC base pair of stem II and is supported by the finding that the orientation of this base pair is conserved (17).

The cleavage mechanism of the hammerhead structure has some analogy with the RNA cleavage by RNases such as RNase A or RNase T1. The hammerhead catalyzes a nucleophilic attack of the 2'-oxygen on the phosphorus (18) resulting in the formation of a nucleoside cyclic 2',3'-phosphate and a 5'-hydroxyl group of the following nucleoside (Fig. 10.2). However, unlike RNases the hammerhead structure does not hydrolyze the cyclic 2',3'-phosphate to the 3'-phosphate. Substrates containing a phosphorothioate at the cleavage site have been used to determine the stereochemical

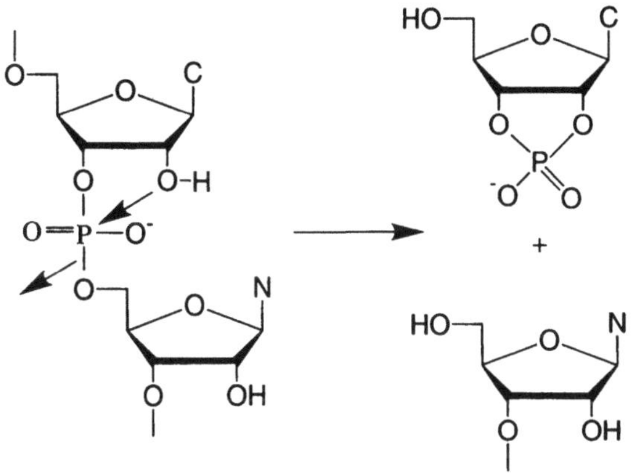

Fig. 10.2: Fig. 10.2: Cleavage mechanism.
The ribozyme catalyzes the nucleophilic attack of the 2'-oxygen on the phosphorus. The resulting cleavage products are a 2',3'-cyclic phosphate and a 5'-hydroxyl group. C, cytosine; N, any nucleotide base.

course of the ribozyme cleavage reaction. In these analogues a non-bridging oxygen is replaced by a sulfur resulting in the existence of two diastereomers (Fig. 10.3; 19). It has been shown with these analogues that the transesterification process proceeds with inversion of configuration of phosphorus indicating an in-line S_N2 mechanism (20, 21). The cleavage is dependent on divalent metal ions such as magnesium, manganese or cobalt (22). However, under physiological conditions, i.e. intracellularly, magnesium is the cofactor for the RNA cleavage. The metal ion presumably plays a dual role, one structural, the other catalytic. In the catalytic role a magnesium hydroxide complex is suggested to deprotonate the 2'-hydroxyl group thereby initiating the nucleophilic attack by the 2'-oxygen (23). The metal cofactor might play a similar role in the hammerhead catalysis as a histidine residue in the catalytic center of the RNase A for the abstraction of the proton from the 2'-OH group. Substitution of the nonbridging pre-Rp oxygen of the scissile phosphodiester linkage by sulfur inhibits the Mg^{2+}-dependent cleavage by a hammerhead ribozyme whereas cleavage in the presence of Mn^{2+} is not affected (22). An Sp-phosphorothioate linkage can be easily cleaved with Mg^{2+} as cofactor (21). Whereas Mn^{2+} coordinates to both oxygen and sulfur, Mg^{2+} strongly prefers oxygen over sulfur. Therefore, these results indicate a coordination of the metal cofactor to the pro-Rp oxygen of the scissile bond in addition to the interaction with the 2'-hydroxyl group.

In nature the hammerhead motif is involved in an intramolecular or in cis cleavage as substrate and catalytic motif are located on the same RNA molecule. However, the catalytic and the substrate part can be separated so that they reside on two different RNAs. Cleavage is then an intermolecular or in trans reaction. Opening of the loops of stem I and III results in such a hammerhead structure where the substrate forms a complex with the catalytic part via Watson Crick hydrogen bonds to form helices I and II (24). After cleavage the products can dissociate and another substrate RNA

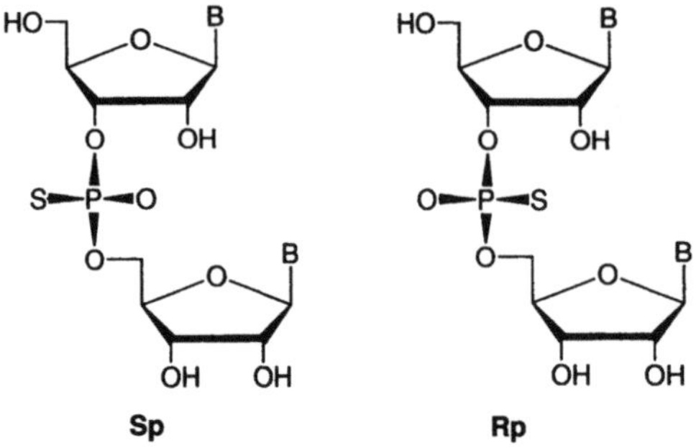

Fig. 10.3: Phosphorothioate diastereomers.

can be bound and cleaved. Therefore, the catalytic part has the potential for multiple turnovers, very much like a protein enzyme. The kinetics of hammerhead ribozyme cleavage have been studied in great detail. Some hammerhead ribozymes reach catalytic efficiencies in the range of 10^6 M^{-1} s^{-1}(25), which reflects the first step in the hybridisation of complementary nucleic acids and approaches a theoretical upper limit (26) making such hammerhead ribozymes optimal enzymes. As the stems of the hammerhead structure are not conserved it can cleave any complementary RNA with a NUH triplet, where N may be any nucleotide and H an A, C or U, but not a G (12). Such triplets are numerous in any mRNA and therefore most RNAs are potential targets for the hammerhead ribozyme.

10.3 The Synthesis of Ribozymes

Oligoribonucleotides can be synthesized by two different approaches: the enzymatic synthesis with RNA polymerases such as T7 RNA polymerase or, like oligodeoxynucleotides, by automated chemical synthesis on a DNA synthesizer. The first ribozymes were synthesized by transcription using T7 RNA polymerase (27). Two different types of templates can be used for the transcription reaction, either a plasmid carrying the T7 RNA polymerase promoter in front of the ribozyme gene or oligonucleotides of 50 to 60 nucleotides in length containing the promoter and the ribozyme gene. Improvements of the original protocolls permit the synthesis of ribozymes in milligram quantities (28). These approaches are relatively simple and can be performed in any laboratory with conventional molecular biology equipment. They also allow the efficient synthesis of ribozymes longer than the hammerhead. One limitation is that chemically modified nucleotides can not be incorporated at a defined single position. Additional procedures such as ligation techniques would be required to achieve this (29). Some modified nucleoside triphosphates such as 2'-amino or 2'-fluoronucleoside triphosphates are efficiently incorporated into RNA by e.g. T7 RNA polymerase whereas others such as 2'-deoxy or 2'-methoxy analogues are no or only poor substrates and are therefore at best only inefficiently incorporated into RNA (30). The sequence context may also affect incorporation efficiencies of analogues and differential incorporation efficacies have been observed during initiation versus the elongation phase of transcription (31).

Because of these limitations considerable efforts have been undertaken to develop efficient methods for the chemical synthesis of ribozymes. As hammerhead ribozymes are short oligoribonucleotides of 35-40 nucleotides in length they can be chemically synthesized by automated procedures with reasonable yield (32; Fig. 10.4). The 2'-hydroxyl group of the ribose moiety poses the major problem for the synthesis and purification of oligoribonucleotides. During the synthesis and deprotection of an oligoribonucleotide this group has to be blocked to prevent the formation of 2'-5'-linkages or strand cleavage but it should be easily removable after completion of the synthesis under mild conditions to avoid RNA degradation. In most methods the t-butyldimethylsilyl (TBDMS) protection group is used for this purpose (33, 34). The synthesis cycle is divided into several steps and consists of the removal of the 5'-protecting DMTr group, the coupling to the next nucleotide unit, capping of the nonreacted 5'-hydroxyl groups and oxidation of the phosphite triester to the phosphate triester internucleotidic linkage. Coupling yield range from 97 to 99% and are similar to those of the oligodeoxynucleotide synthesis. At the end of the synthesis, the protected oligoriboynucleotide is cleaved from the column support. To deprotect the RNA the base and phosphate protection groups are removed by incubation inethanolic ammonia followed by removal of the TBDMS groups with tetrabutylammonium fluoride. Finally, the full length product is separated from shorter fragments by gel electrophoresis or HPLC. A synthesis of a 36 nucleotides long oligoribonucleotide at a 1 µmol scale yields about 100 nmol (1 mg) of full length product.

Automated chemical synthesis permits the selective incorporation of many modifications into the ribozyme to change certain of their properties, such as stability (see below 35, 36). Because of the ease of synthesis the hammerhead ribozyme is so far the only catalytic RNA which has successfully been applied to exogenous delivery approaches.

10.4 What is the Optimal Length of a Ribozyme?

The hammerhead ribozyme cleaves RNA 3' of any NUH triplet. N may be any nucleotide, U is a uridine and H indicates an adenosine, a cytidine or a uridine. A guanosine at the third position prevents cleavage. Longer target RNAs contain several potential cleavage sites but because of factors such as RNA structure or protein binding some sites are more accessible to ribozyme-mediated cleavage than others. It is therefore important to determine the best cleavage site in the target RNA for any cellular ribozyme approach.

As previously mentioned, the hammerhead ribozyme can cleave successively many RNA molecules and can be considered a true enzyme whose kinetic properties may be characterized. Many kinetic studies have been performed under steady

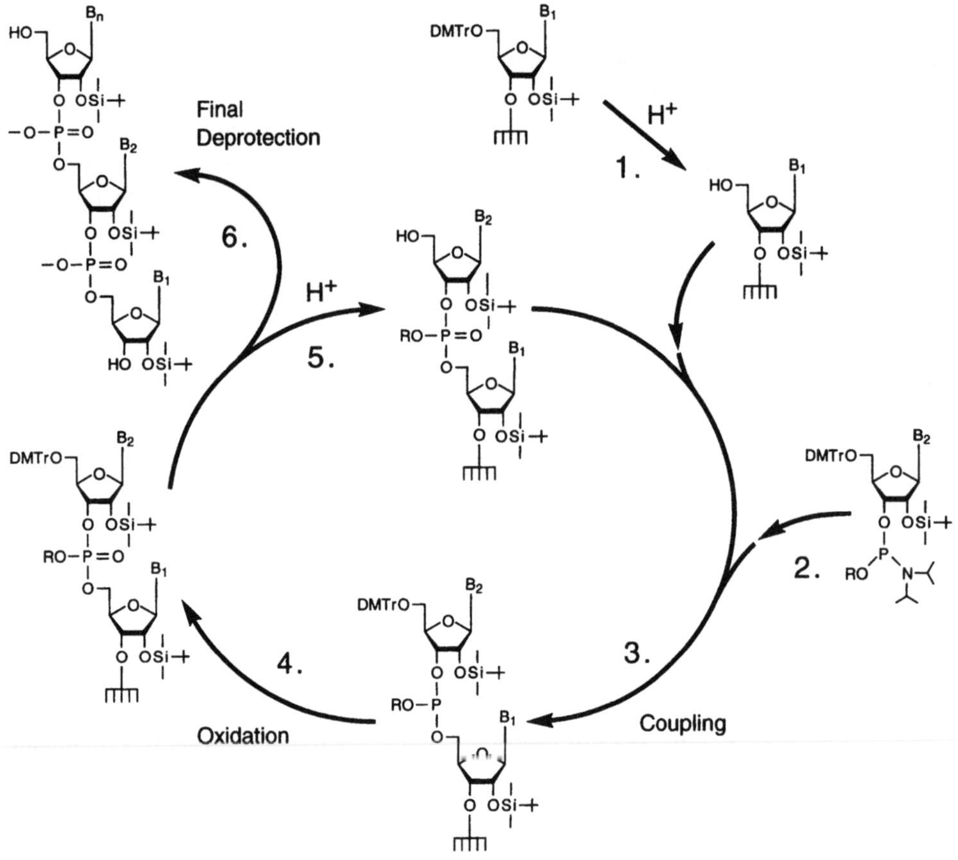

Fig. 10.4: Automated synthesis of oligoribonucleotides.
DMTr, dimethoxy trityl; B, protected nucleotide bases; R, 2-cyanoethyl; 2'-protecting group, tert. butyl dimethyl silyl.
Step 1, removal of DMTr group; 2, addition of nucleoside phosphoramidite; 4, coupling of this phosphoramidite to growing oligonucleotide chain; 5, oxidation of phosphite triester to phosphate ester with iodine; 6, removal of DMTr group for entry into the next coupling cycle; 7, cleavage of the oligoribonucleotide from the polymer support and removal of all phosphate and base protecting groups with ammonia.

state conditions (27, 25). In this case, the substrate is in large excess over the ribozyme with the ribozyme performing multiple turnovers (37). This type of kinetic analysis examines the whole reaction pathway from the binding of the ribozyme to its target RNA to the product release step. Dependent on the sequences and lengths of ribozyme arms and substrate RNA the rate-determining step of the reaction pathway may be any of these steps. In particular, if the ribozyme forms long duplexes with its substrate, the rate-determining step is the release of the cleavage products. Under single turnover conditions where the ribozyme is in excess over the substrate RNA, an average ribozyme molecule cleaves maximally one substrate molecule, and no reaction step after the cleavage step can be analyzed (37). It is important to keep these differences in mind, as it is not clear whether single or multiple turnover conditions are more applicable to the intracellular situation. Despite the fact that most conclusions about ribozyme efficacy or specificity have been drawn from multiple turnover studies, several reports suggest that efficient cleavage of cellular RNA can only be achieved with a considerable ribozyme excess (38, 39, 40).

Generally, the longer the arms of the ribozyme, the more stably it will bind to its complementary RNA. However, the same is true for the complex between ribozyme and cleavage products resulting in very slow product release which slows down turnover (25). Under single turnover conditions only the stability of the ribozyme-substrate complex is important and not the stability of the ribozyme-product complex. Under conditions where the substrate is in excess, however, the ribozyme arms should be shorter to avoid problems associated with product release and to facilitate turnover.

Several results support these considerations. The optimal length of ribozyme arms, for multiple turnover conditions with a short substrate at 37°C, was found to be approximately seven nucleotides (41). Ribozymes of different length were targeted against the same site in a 1000 nucleotides long transcript of the HIV-1 Long Terminal Repeat. Ribozymes with arms I and III severn or eight nucleotides in length cleaved this long transcript under single turnover conditions whereas a ribozyme containing only five nucleotides long arms cleaved this substrate with a 100 fold reduced efficiency (41). However, one must be cautious extrapolating from the ribozyme activity under in vitro conditions to ribozyme efficiency in cell culture.This is exemplified by one study on the inhibition of HIV by ribozymes. Only ribozymes with at least 30 nucleotide long arms caused a significant inhibition of replication in cell culture whereas such ribozymes were very poor catalysts in vitro (42).

There are several limitations to the lenght of ribozyme arms. Firstly, the yield of chemically synthesized ribozymes decreases with increasing length but may not be a limitation for enzymatically synthesized ribozymes. Secondly, longer ribozymes are more likely to contain, in addition to stem II, self complementary sequences which can result in formation of double stranded regions. Such misfolded ribozymes have to unfold before they can bind to their target RNAs. As a result, they cleave RNA more slowly than correctly folded RNAs and possibly less efficiently depending on the stability of such unwanted duplexes (43). Finally, a ribozyme with long arms may not only cleave its perfectly homologous substrate RNA but also related RNA sequences. For example, a hammerhead ribozyme targeted against the mRNA of a mutated dihydrofolate reductase gene cleaved the unmutated RNA as well as the mutated one which contained two mutations in the 20 nucleotides long recognition site (44). Base mismatches between the ribozyme and the target sequence destabilize the ribozyme-substrate complex and often reduce the cleavage efficiency but longer complementary sequences are less affected by mismatches than shorter ones. Therefore, a ribozyme with longer arms is less discriminatory than a ribozyme with shorter arms (45). Mismatches within or next to the cleavage triplet abolish or strongly inhibit the RNA cleavage of a hammerhead ribozyme (46, 47). However, more distant mismatches decrease the cleavage to a much smaller extent.

The results described here demonstrate that an optimal ribozyme for all applications does not exist. The optimal length of the ribozyme arms is dependent on the particular system. A shorter ribozyme is more specific than a longer one which is particularly important for therapeutic approaches directed against oncogenes. Such high specificity may not be necessary or even disadvantageous for the inhibition of viral genes with a high sequence variability such as HIV where longer ribozymes may be more effective in cleaving as many viral sequence variants as possible (48).

10.5 The Influence of RNA Secondary Structure on Ribozyme Catalysis

Hammerhead ribozymes cleave longer RNAs such as mRNAs significantly less efficiently than short oligoribonucleotides. The lower cleavage efficiency is due to secondary structures such as double stranded regions which prevent the binding of the ribozyme to its target sequence. Ribozymes targeted against more accessible single stranded regions cleave RNA more efficiently than those targeted against highly structured regions (41, 49). Therefore, ribozymes should be targeted against single stranded sequences to obtain maximal inhibition of gene expression. How can such accessible RNA regions be identified? A common method to predict single stranded regions is the calculation of secondary structures of RNA with progammes such as RNAfold (50). Indeed, RNA sequences predicted to be single stranded by such programmes have been successfully targeted by ribozyme or antisense approaches in cell culture (38). RNAfold has also been applied to the prediction of ribozyme structures (43, 51). Because intramolecular hybridisation within the ribozyme may reduce its catalytic efficiency an optimal ribozyme should contain single stranded arms for efficient binding to its target sequence. Two reports demonstrated a good correlation between the catalytic efficiencies and the predicted secondary structures of ribozymes (43, 51).

Programmes such as RNAfold calculate the thermodynamically most favoured structures. Protein binding or kinetic processes such as strand displacement are not considered. An alternative to structure calculation for the identification of accessible RNA sequences is in vitro selection. Such methods have the advantage that no knowledge of the structures of the target RNA or the ribozyme are required. Several approaches for such selection have been published. For instance, partial alkaline hydrolysis of a long antisense RNA created a library of RNA molecules of different length which upon incubation with its target RNA resulted in the formation of double stranded hybrids (Fig. 10.5). The hybrids were separated from the nonhybridized single strands and subsequently separated according to length on a denaturing gel. This approach facilitated the identification of short antisense RNAs which inhibited HIV-1

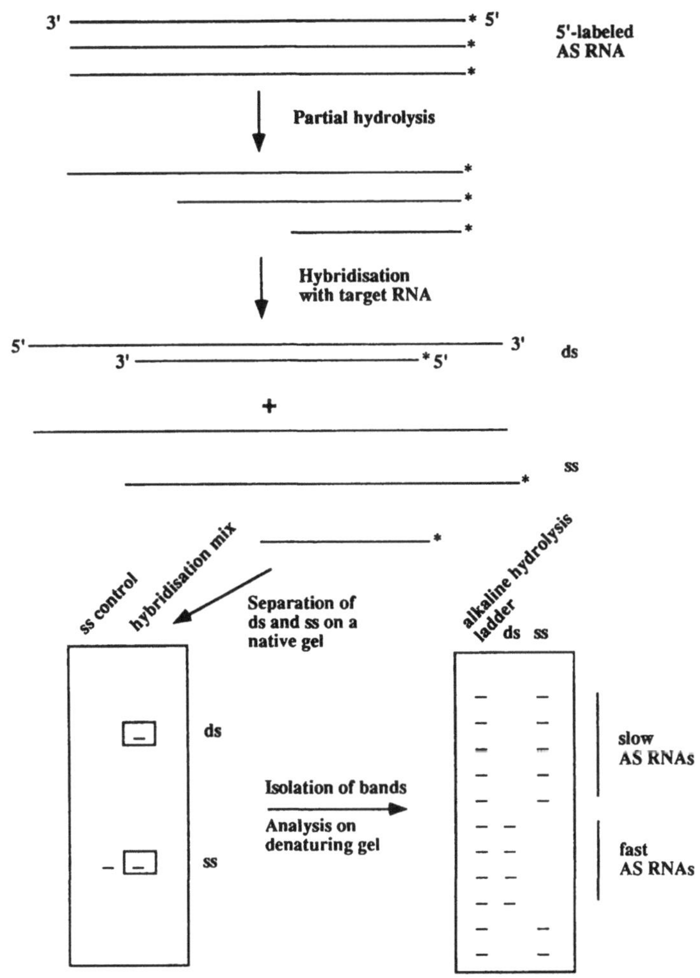

Fig. 10.5: In vitro selection of optimal antisense sequences.
A library of antisense RNAs is created by partial hydrolysis of the 5'-labeled full length RNA. Incubation of the library with the target RNA results in the formation of double stranded species (ds) which are separated from the unhybridised, single stranded species by native gel electrophoresis. The bands corresponding to both species are excised, and the RNAs are extracted and loaded onto a denaturing gel. Finally the sequences of the quickly hybridising species (fast AS RNAs) are determined by comparison with an RNA ladder obtained by partial hydrolysis of the full length AS RNA (according to ref.52).

replication in cell culture five times more efficiently than long antisense RNAs (52). This method, although limited in the present protocoll to hybridising RNAs differing in length at the 3'-end, should also be applicable to ribozymes.

Efficient ribozymes for cleavage of a particular mRNA were also identified by incubation of the target RNA with a ribozyme library with randomized arms (Fig. 10.6; 53). Cleavage products were amplified with the polymerase chain reaction, and the

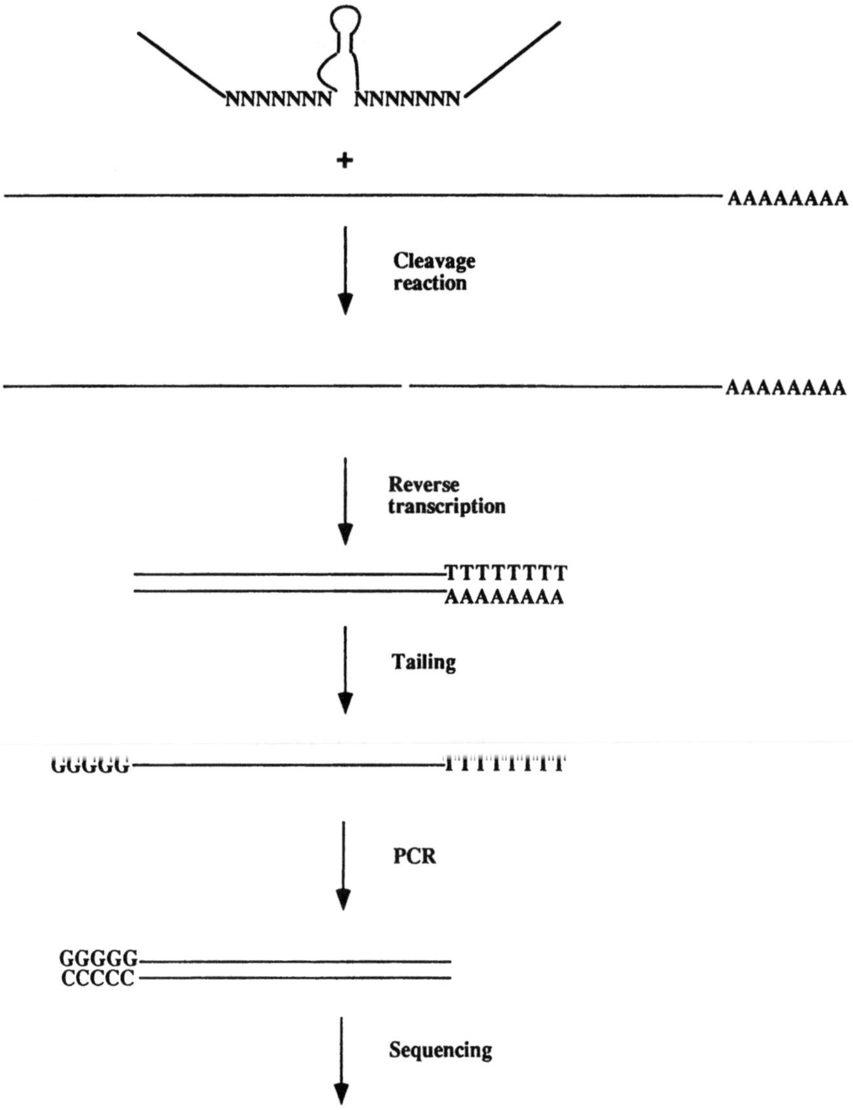

Fig. 10.6: In vitro selection of optimal ribozymes.
A ribozyme library with random sequences in the ribozyme arms is incubated with the target mRNA (cleavage reaction). The fragments are reverse transcribed using an oligo dT primer. An oligo dG tail is added to the 3'-end of the cDNAs followed by PCR amplification. Sequencing of the resulting fragments gives the arm sequences of the optimal ribozymes (according to ref. 53).

cleavage sites were identified by sequencing. As a result ribozymes were identified which efficiently cleaved the mRNA of the human growth hormone in vitro and ex vivo (i.e. in cell culture). A similar approach has been successfully applied to the group I intron (54). These examples demonstrate the power and potential of in vitro selections for the identification of efficient ribozyme or antisense molecules.

10.6 The Influence of RNA-binding Proteins on Ribozyme Catalysis

Most of the ribozyme assays consider only effects of the ribozyme, the target RNA or the metal factor on the RNA cleavage. RNA-binding proteins which play an important role in the intracellular RNA metabolism, are not considered. There are many RNA-binding proteins which are involved in processes such as RNA folding or trafficking. Only the influence of three such RNA-binding proteins on ribozyme catalysis has been investigated in greater detail.

Initially RNA binding proteins were thought to interfere with ribozymes because of possible coverage of the target site. For example, a ribozyme targeted against the U7 RNA cleaved the RNA only in the absence of snRNP particle proteins (55). However, in a later study, ribozyme catalysis was unaffected by the addition of cytosolic or nuclear protein extracts (56). Mor recently though, enhancements in ribozyme-mediated cleavage have been observed by addition of RNA-binding proteins such as the nucleocapsid protein (NC) of HIV-1 or the heterogeneous nuclear ribonucleoprotein A1 (hnRNP A1) (57, 58, 59, 60, 61). This effect is brought about by these proteins by facilitating ribozyme binding to its target sequence and the dissociation of the cleavage products. In addition, it has been shown for NC that it causes dissociation of mismatched ribozyme-RNA complexes and therefore might contribute to an improved specificity of the ribozyme catalysis in vivo (59).

Apparently, the effects discussed here are based on the destabilization of RNA secondary or tertiary structures by such proteins. Partial unfolding of the target RNA facilitates the binding of the ribozyme to its complementary sequence whereas destabilization of the ribozyme-product complex increases the rate of product dissociation. Mismatched double strands are thermodynamically less stable than perfect matches and are more likely to dissociate because of protein -dependent destabilization. Thus, lower concentrations of NC, for example, destabilize such mismatched double strands without affecting the ribozyme-mediated cleavage of the perfectily matched substrate RNA. However, at higher than optimal NC concentrations the cleavage of perfectly complementary RNAs is also inhibited.

RNAs are associated with many RNA-binding proteins inside the cell and it is important to know how they affect ribozyme catalysis. In one case, the cleavage of such endogenous RNA-protein complexes was compared to that of isolated "naked" RNA (40). Nuclei were prepared from HTLV tax transformed cells and incubated with an anti-tax ribozyme. Alternatively, the RNA was isolated from the nuclei prior to addition of the ribozyme. Surprisingly, the ribozyme cleaved the RNA present in the nuclei some 30 fold more efficiently than the isolated RNA. This result supports the view, obtained before with isolated RNA binding proteins, that such proteins are present in the nucleus and that they are capable of supporting ribozyme-mediated cleavage in vivo.

Besides their effects on the ribozyme reaction, proteins may also influence the intracellular stability of ribozymes. It was observed that proteins bound sequence-specifically to a ribozyme which was targeted against tumor necrosis factor-α, and protected it against degradation by intracellular nucleases (62). These ribozyme-protein complexes were still catalytically active suggesting that this interaction not only extends the intracellular ribozyme life span without affecting their catalytic activity.

Taken together, the examples discussed here demonstrate that RNA-binding proteins can influence the ribozyme-mediated RNA cleavage by different modes. While competition for binding sites inhibits the ribozyme, the unfolding of target RNAs or the protection against degradation by some proteins enhances the ribozyme efficacy.

10.7 Colocalisation of Ribozymes and Target RNAs in the Cell

A ribozyme has to reach the cellular compartement where the target RNA is located. This is a particular problem for exogenously delivered antisense oligodeoxynucleotides or ribozymes. Addition of oligonucleotides to the cell culture medium results in their uptake by endocytosis and their localosation in the endosome vesicles (63). Only that fraction of oligonucleotides which escapes from such vesicles into the cytosol or the nucleus can find its target RNA. Approaches to deliver ribozymes exogenously to cells will be discussed in more detail in a following section.

Recently an elegant study examined the effects of colocalisation of ribozyme and substrate RNA (64). Cells were infected with a retroviral vector containing the lacZ gene encoding the β-galactosidase. The corresponding mRNA contained the retroviral packaging sequence and was either translated in the cytosol, or was packaged into retroviral particles. This cell line was superinfected with retroviruses containing an anti-lacZ ribozyme. The ribozyme RNA contained the same packaging sequence as the lacZ mRNA. Interestingly, the ribozyme reduced the titer of infectious lacZ virus but had no effect on the intracellular expression of lacZ. The authors concluded that the ribozyme could cleave the target RNA only during or after it was packaged with the lacZ RNA into retroviral particles thereby destroying the infectious potential of the viral particle. Why the ribozyme could not cleave the unpacked lacZ mRNA remains unclear. These results demonstrate that the colocalisation of ribozyme and target is an important factor for any ribozyme approach.

10.8 Chemically Modified Hammerhead Ribozymes

The low stability of RNA against degradation by nucleases is a considerable obstacle for the exogenous delivery of ribozymes. Therefore, considerable efforts have been undertaken to improve ribozyme stability by chemical modification. For several reasons the amount of chemical modifications in a ribozyme should be kept to a minimum. Several positions of the hammerhead ribozyme cannot be modified without a substantial loss in cleavage activity (2, 65). Thus, a small increase in stability may be paid for by a much lower catalytic efficiency. Excessive modification of oligonucleotides bears the risk of resulting in undesired and unforeseen properties such as has been seen with complete substitution of phosphate by phosphorothioate linkages in oligodeoxynucleotides where an unspecific binding of these oligonucleotide to proteins is sometimes observed (63). Inhibitory effects seen with such oligonucleotides might not be the result of an antisense and sequence-specific effect (66). Such unspecific effects have not been reported for ribozymes yet but the results reported with antisense oligonucleotides should be taken as a sign of caution. In how far any of the modified nucleotides incorporated into ribozymes elicit undesired side effects awaits evaluation.

To avoid unnecessary ribozyme modifications, one should consider the nucleases to be encountered by the ribozyme. If the ribozyme is added directly to serum, it has to be protected against pyrimidine specific endoriboynucleases and a 3'-exonuclease activity (67, 35). In this case, ribozymes should contain 2'-modified pyrimidine nucleosides and 3'-terminal modifications such as phosphorothioate linkages. The amount of nuclease activity in serum differs between different sources. Human serum contains significantly less nuclease activity than fetal calf serum resulting in a longer half-life time for a ribozyme in human than in calf serum (67). If the ribozyme however is delivered to cells complexed to cationic lipids it will be protected against the action of serum nucleases but will encounter mainly 5'- and 3'-exonucleases in the nucleus which should be inhibited by terminal modifications of the ribozyme (68).

Therefore, selected and site-specific incorporation of modifications are necessary to obtain stable and active ribozymes. The automated chemical synthesis of oligoribonucleotides permits not only the site-specific incorporation of one type of modification but also the combination of several different ones.

Phosphorothioate linkages were among the first modifications employed to improve the stability of RNA against nucleases (19). Initially, phosphorothioates were incorporated into polyribonucleotides by enzymatic synthesis where only the Rp diastereomer can be introduced (19; Fig. 10.3). These linkages are now generally introduced into the ribozyme during chemical synthesis by oxidation of the phosphite linkage with sulfurizing agents such as the Beaucage reagent (21) where both diastereomers are formed.Consequently phosphorothioate-containing oligonucleotides are a racemic mixture of diastereomers. This is important, because nucleases often cleave one phosphorothioate diastereomer preferently. A mixture of diastereomers thus ensures that a nuclease with a diastereomeric preference encounters an uncleavable phosphorothioate diastereomer and cannot reach the unmodified core region of the oligoribonucleotide.

Several positions in the core of the hammerhead ribozyme should not contain phosporothioate linkages such as positions 9 and 13 to 15 where their presence results in complete loss of activity (69). Terminal phosphorothioates are tolerated without loss of activity and typically three to five have been used to protect the ribozyme against degradation by exonucleases (35, 36). Exonucleases only cleave 3'-5' phosphate diester internucleotidic linkages. The incorporation of 3'-3'linkages is therefore also a barrier against degradation (36).

Modification of the 2'-hydroxyl group is a common method to increase the resistance of ribozymes against degradation by 2'-hydroxyl-dependent ribonucleases. Substitution of this group by hydrogen, fluorine, amino- and alkoxy groups have been successfully used to prepare stable and active ribozymes (Fig. 10.7). Caution must be taken, however, since the catalytic activity of the ribozyme may be destroyed by conformational and functional changes introduced by these nucleotide analogues. The tendency towards a 3'-endo conformation, which predominates in RNA, increases in the order 2'-amino > 2'-deoxy > 2'-O-methyl > 2'-hydroxy > 2'-fluoro (70, 71). In addition, these modifications have different properties in forming hydrogen bonds. Only the 2'-amino group can act like the 2'-hydroxyl group both as a proton acceptor and donor. In contrast, the 2'-fluoro-, the 2'-O-methyl- and the 2'-O-allyl groups can only accept protons. The 2'-allyl group or the hydrogen of the 2'-deoxy modification cannot serve in either function. Modifications in the binding arms affect the affinity of hammerhead ribozyme to its complementary RNA. For example, 2'-amino groups strongly destabilize nucleic acid duplexes, whereas 2'-fluoro and 2'-methoxy groups have a stabilising effect (72, 73).

The first 2'-modified ribozymes contained 2'-deoxynucleotides at all positions except the seven purine ribonucleosides in the core region (74). Such a ribozyme was resistant against pancreatic ribonuclease A and was 1000-fold more stable in yeast extract than its unmodified counterpart but cleaved RNA 100-fold less efficiently. When only the stems contained 2'-deoxynucleotides and the core remained unmodified, the catalytic activity of such "chimeric" ribozymes was hardly affected. Endoribonucleases present in serum can still degrade these ribozymes but after transfection with cationic lipids they had a threefold improved intracellular stability. Ribozymes containing 2'-deoxynucleotides in the stems in combination with phosphorothioate linkages were 100-fold more stable in serum and were only 15-fold less active than unmodified riboyzmes (67).

2'-O-allylmodified ribozymes with only five unmodified nucleosides in the core region exhibit a much better ratio of increase in stability over loss of catalytic efficiency and cleave RNA with only a 5-fold lower efficiency than unmodified ribozymes (75). However, whereas the latter ribozyme is completely degraded within one minute, 30% of the 2'-O-allylmodified ribozyme remained intact after two hours

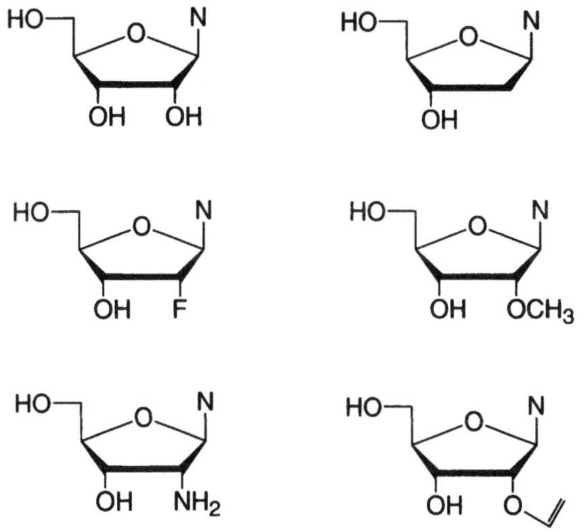

Fig. 10.7: Chemical modifications.
The figure shows from the top left to the bottom right: unmodified ribonucleoside, 2'-deoxynucleoside, 2'-fluoronucleoside, 2'-O-methylnucleoside, 2'-aminonucleoside, 2'-O-allylnucleoside.

of incubation in bovine serum. Thus, this type of ribozyme seems to be suitable for therapeutic applications.

Two other modifications have also been successfully used to synthesize stable and active ribozymes. Replacement of all pyrimidine nucleosides by the corresponding 2'-fluoro or 2'-amino analogues decreased their catalytic efficiencies 7- to 50-fold but increased their stability in serum or cell culture supernatant more than 1000 fold (76). Additional incorporation of terminal phosphorothioate linkages improved their stability further (41). The reduced activity of such ribozymes resulted from the 2'-fluoro modifications at position 4 and 7 in the core. Incorporation of 2'-aminouridines instead of 2'-fluorouridines at these two positions completely restored the activity to that of the unmodified ribozyme (35). Such modified ribozymes remained intact after 24 hours of incubation in concentrated, untreated fetal calf serum. Similar results were obtained for ribozymes with 2'-aminouridines at positions 5 and 7, only five unmodfied purine nucleosides in the core and the rest as 2'-methoxy derivatives (36). A 3'-3'-linkage and 5'-terminal phosphorothioate linkages protected these ribozymes against exonucleases. Such ribozymes had a half-life time in human serum of more than ten days and were as active as unmodified ribozymes. These results demonstrate the possibility to synthesize stable and active ribozymes which is a precondition for any therapeutic application.

10.9 The Exogenous Delivery of Ribozymes

Exogenously added ribozymes or antisense oligonucleotides must pass at least one membrane to find their target RNAs. Because of their negatively charged phosphate groups oligonucleotides cannot penetrate through membranes by passive diffusion but enter the cells by endocytosis. There they are entrapped in endocytic vesicles and only a minor fraction of the oligonucleotides enters the cytoplasm or the cell nucleus (63). Several approaches have been developed to deliver oligonucleotides to these two compartments. One common method uses cationic liposomes such as DOTAP or Transfectam for the exogenous delivery of oligonucleotides and ribozymes (77, 78). This class of lipids contains positively charged head groups which interact with the negatively charged phosphate groups. Cells take up such lipid-oligonucleotide complexes via endocytosis and like the free oligonucleotides the largest part of the complexes remains entrapped in the vesicles (79). However, a larger fraction than of the free oligonucleotides escapes and enters the cytoplasm and the nucleus. This poorly understood process is much more efficient than the delivery without liposomes. However, this approach has several disadvantages. Several lipid formulations such as Lipofectamine or Transfectam are only effective in the absence of serum which limits their suitability in vivo. Others such as DOTAP are less effective than Lipofectamine but the oligonucleotide delivery is not inhibited by serum. In addition, cationic lipids often display severe cytotoxicity at higher concentrations. Despite these problems cationic lipids are the most commonly used method to deliver ribozymes exogenously to cell cultures.

An alternative to cationic lipids is receptor-mediated endocytosis of ribozymes. In analogy to gene therapy protocolls oligonucleotides are complexed to conjugates of polylysine and ligand such as transferrin or folate. The positively charged polylysine moiety complexes to the oligonucleotide, whereas the ligand moiety binds to the cell receptor and triggers the endocytic uptake. This approach is interesting as it may permit cell-type specific delivery of ribozymes. However, the escape of the oligonucleotide from the endocytic vesicle is rather inefficient but the use of adenoviral particles or membrane-fusing peptides may alleviate this bottle neck (80, 81).

Finally, oligonucleotides have been successfully delivered by encapsulation by liposomes and, like polylysine conjugates, they can be coupled to ligands of cellular receptors to allow specific delivery (82). In addition, certain lipid formulations of liposomes enhance the escape from the endosome. Unfortunately, the encapsulation process is very inefficient and because of the pausity of synthetic ribozymes such approaches have not yet been applied to ribozymes.

10.10 Potential Artefacts of Ribozyme Applications

Ribozymes may affect cells not only by their catalytic activity. Like other oligonucleotides they are polyanionic compounds with a sequence-dependent array of many different functional groups such as carbonyl-, amino-, hydroxyl- or phosphate groups. It is therefore not surprising that oligonucleotides including ribozymes interact not only with other nucleic acids but also with proteins. The sequence-specific interaction of an anti-TNF-α ribozyme with stabilizing proteins was already mentioned (62). If a researcher is only interested in a biological effect, he or she may not be concerned about whether it is due to RNA cleavage, to a simple antisense action or to protein-RNA interactions. However, only the understanding of the mechanism of action allows a rational approach to further development of such potentially therapeutic compounds. To be confident the observed effects are a consequence of ribozyme action, a number of controls are required. One control typically consists of using an inactive ribozyme, where one of the core nucleotides has been mutated. For example, changing the guanosine at position 5 to an adenosine abolishes the cleavage activity of the ribozyme without affecting binding to the target sequence. If an inactive ribozyme causes the same effect as the active one, it may simply act as an antisense RNA. In such cases, there is no point in synthesizing a long oligoribonucleotide as a ribozyme when an oligoribonucleotide without the catalytic core has the same effect and can be synthesized more easily. Alternatively, the effect may be the result of sequence-specific binding of the oligoribonucleotide to proteins essential for e.g. cell proliferation. Several studies have linked biological effects of antisense oligonucleotides to such a mechanism (63, 66).

If the effect produced by a ribozyme is caused by the expected interaction and cleavage with the mRNA, the cleavage products should be ideally identified. However, this is generally not possible in a direct manner but requires isolation of the RNA by cell lysis and includes often an amplification procedure such as RT-PCR. Both procedures involve steps which can falsify the analysis. During cell lysis compartments are destroyed which migth have harboured the ribozyme and the substrate RNA separately and which now can interact without the compartmental barrier. Procedures such as DNase treatment or reverse transcription require millimolar concentrations of Mg^{2+} promoting the ribozyme cleavage reaction which might not have occured in the cell because of insufficient concentration of this metal ion there. The intracellular concentration of free, uncomplexed Mg^{2+} is in the range of 200 to 500 µM (83). It is unclear yet if, for example, proteins or other agents permit efficient RNA cleavage at such low concentrations. These complications in identifying that the cleavage reaction has actually occured in the cell, are demonstrated in two articles where it is shown that the target RNA is only cleaved by the ribozyme after cell lysis (84, 85). Thus a combination of controls such as determination not only of RNA but also of protein levels in combination with the inactive ribozyme variant are mandatory to be able to attribute the observed effect to an intracellularly active ribozyme.

10.11 Applications of Ribozymes in Cell Culture and Animal Models

Even though exogenous delivery of ribozymes to cells is a very recent technology several reports describe the successful application of this method. The range of targets range from oncogenes such as bcr/abl to enzymes, to transcription factors and to viruses such as HIV. As discussed below ribozymes have also been tested in a mouse model and were found to be very effective inhibitors of gene expression.

The first report using a preformed ribozyme examined its ability to interfere with the expression of tumor necrosis factor-α (TNA-α). This cytokine is involved in many processes such as inflammation, toxic shock, apoptosis and replication of HIV. A ribozyme which efficiently reduces the production of TNA-α would be a promising tool to control these processes. An anti-TNF ribozyme was synthesized by run-off transcription and delivered to either HL60 cells or PBMC with lipofectine followed by induction of TNA-α (86). The transfection with the ribozyme reduced the amount of TNA-α mRNA and secreted TNA-α by 90%. An antisense RNA targeted against the same sequence was less efficient than the ribozyme showing the importance of the ribozyme activity for the inhibition of expression of this cytokine. The ribozyme contained a 3'-hairpin which improved its intracellular stability compared to a ribozyme without such a structure.

Exogenously delivered ribozymes have a great potential in anticancer therapy. In particular, if a mutation within the oncogenic mRNA can be targeted a high degree of specificity may be achieved. One should keep in mind that only such mutations affect the ribozyme-mediated RNA cleavage which are located within or very close to the cleavage triplet. Several groups examined ribozyme approaches against chronic myelogenic leukemia (CML) which is caused by a reciprocal translocation of chromosomes 9 and 22 resulting in the "Philadelphia" chromosome. In such chromosomes the 5'-part of the bcr gene wich encodes a phosphotransferase is ligated to the 3'-part of the tyrosine kinase abl gene. Two different types of this leukemia exist: the B2A2 type where the second exon of bcr is ligated to the second exon of abl, and the B3A2 type where the third bcr exon is connected to abl.

Two groups targeted ribozymes against the B3A2 type of CML. One group synthesized an unmodified ribozyme by either run-off transcription (77) or chemical synthesis (87), whereas the second group (78) synthesized a modified ribozyme by automated chemical synthesis. The latter ribozyme contained 2'-deoxynucleotides instead of ribonucleotides in its stems. Both groups used cationic lipids for the delivery of the ribozymes to the cells. Both ribozymes inhibited the proliferation of CML cells and reduced the amount of bcr-abl mRNA and the corresponding kinase activity. In contrast, inactive ribozymes or antisense constructs were significantly less efficient emphasizing again the importance of the ribozyme activity for the observed effects. In a somewhat different approach, a ribozyme construct was targeted against the B2A2 type. This type contains three potential cleavage triplets surrounding the breakage point. The group of Reddy (88) synthesized an oligomeric ribozyme which contained three ribozyme units in a row and could cleave all three cleavage triplets. Two different agents for the ribozyme delivery were tested: DOTAP, a cationic lipid, and a polylysine-folate conjugate. Their efficacy was tested by the reduction of B2A2 mRNA as judged by RT-PCR. Interestingly, delivery of the ribozyme with the polylysine-folate conjugate resulted in a 1000-fold reduction of the mRNA level. In contrast, ribozyme delivery with DOTAP caused only a 10-fold reduction. Unfortunately, this study did not include protein data. It is therefore unclear whether the ribozyme-mediated reduction of the bcr/abl mRNA level was really due to intracellular cleavage, or if the ribozyme cleaved its target RNA after cell lysis.

An increasing resistance of cancer cells against cytostatika is a common phenomenon during chemotherapy of cancers. The increased expression of the mdr-1 gene is responsible for this phenomenon. Its product is a glycoprotein which transports many different cytostatika such as vindesin or adriamycin out of the cell. Kiehntopf et al. (89) examined two different anti-mdr ribozymes targeted against the same site. One ribozyme was unmodified, the other contained terminal phosphorothioate linkages combined with 2'-deoxynucleotides in the stems and 2'-fluoropyrimidine nucleosides in the core region (Fig. 10.8a). The ribozymes were delivered to cells with an increased resistance against vindesine and adriamycin. Both ribozymes caused a reduction in resistance of the cells against these cytostatika. In addition, the mdr mRNA, as judged by RT-PCR, and protein levels were diminished. However, the modified ribozyme was somewhat more effective than the unmodified suggesting a higher intracellular stability of the modified ribozyme. Controls such as antisense phosphorothioate oligonucleotides or an inactive ribozyme had no effect on the drug resistance of the cells.Therefore, anti-mdr ribozymes are promising compounds for the reduction of multiple drug resistance and to increase the efficacy of anticancer chemotherapy.

The same group targeted such a modified ribozyme against the transcription factor NF-IL6 (90). This factor is involved in the regulation of several cytokines, and is induced by factors such as TNF-α. Human fibroblasts were transfected several times with the ribozyme complexed to DOTAP. This treatment caused a significant reduction of the NF-IL6 protein levels without affecting NF-κB, another transcription factor which served as control. In addition, the secretion of G-CSF, but not of GM-CSF or IL-6 was impaired. These results are similar to findings obtained with NF-IL6 knock-out mice (91). An inactive control ribozyme did not affect the NF-IL6 amount, whereas an antisense oligodeoxynucleotide caused a similar reduction of NF-IL6 and of the G-CSF secretion as the active ribozyme. This is somewhat surprising, as both the inactive ribozyme and the antisense oligodeoxynucleotide can induce the nuclear RNaseH and, therefore, should have similar effects on NF-IL-6.

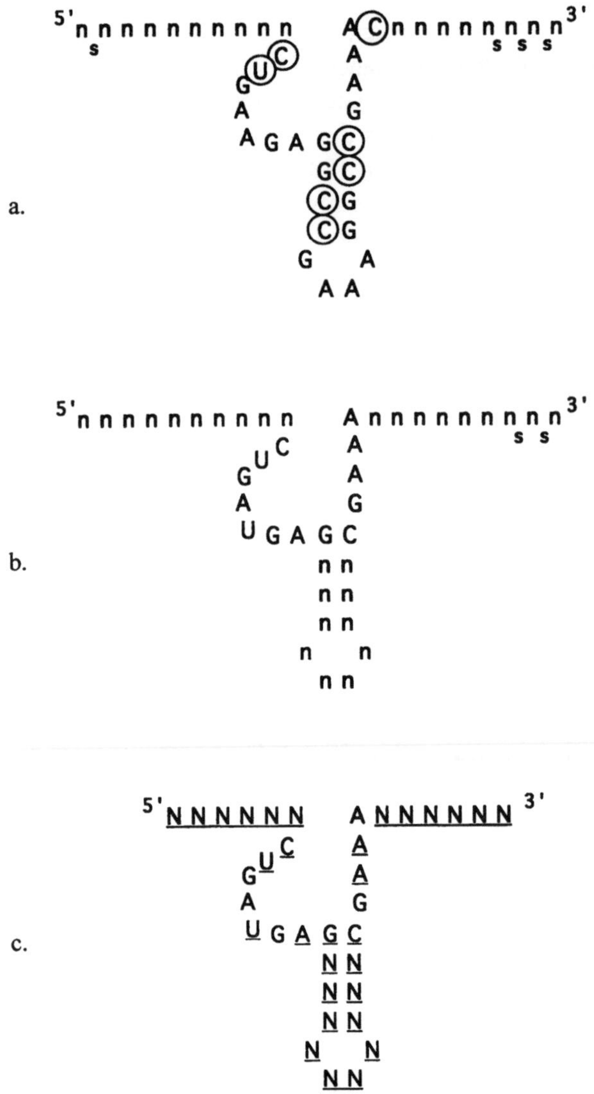

Fig. 10.8: Examples of chemically modified ribozymes.
Upper case letters, unmodified ribonucleosides; lower case letters, 2'-deoxynucleosides; underlined letters, 2'-O-allylnucleosides; circled letters, 2'-fluoropyrimidine nucleosides; s, phosphorothioate linkages. a: Ribozyme used in (89, 90). b: Chimeric Ribozyme used in (92). c: 2'-O-allyl-modified ribozyme used in a mouse model (95).

Unfortunately, no RNA data were included into this study. Thus, other mechanisms than ribozyme or antisense mechanisms cannot be excluded.

Leukocyte-type 12 lypoxygenase is involved in the angiotensin II-induced vascular smooth muscle cell growth resulting in the development of hypertension and artheriosclerosis. Nadler and collegues (92) targeted a chimeric ribozyme with 2'-deoxynucleotides in the stems in addition to 3'-terminal phosphorothioate linkages (Fig. 10.8b), against the lipoxygenase mRNA. The cationic lipid transfectam was used to deliver the ribozyme to vascular smooth muscle cells. RT-PCR revealed a tenfold reduction of lypoxygenase mRNA in ribozyme-transfected cells. In addition, the ribozyme caused a twofold reduction of lypoxygenase protein. The action of the ribozyme seemed to depend on its catalytic activity, as an inactive control showed reduced effects.

Ribozymes are considered to be promising compounds for antiviral therapy. In particular HIV is an attractive target for ribozyme approaches. However, the great genetic variability of HIV complicates the use of ribozymes to inhibit viral replication and spread. Several possible solutions exist to solve this problem. One possibility is to target only highly conserved regions such as some sequences located in the long terminal repeat. Another is the delivery of several ribozymes targeted against different sites in the HIV RNA. Finally, ribozymes with long complementary arms should be less susceptible to point mutations if they are not located around the cleavage triplet. The last approach was chosen by Homan et al. who embedded a ribozyme core into a 413 nucleotide long antisense RNA (93). Such RNAs and the proviral DNA were cotransfected into SW480 cells. Virus produced by these cells was amplified by coculture with MT-2 cells which are very susceptible for HIV infection. Interestingly, the ribozyme inhibited HIV replication some five times more efficiently than the corresponding antisense RNA. A construct containing a catalytically inactive ribozyme core inhibited HIV to a similar degree as the antisense RNA. This result is surprising as the antisense RNA forms a very stable duplex with the complementary viral RNA which should result in an essentially irreversible inhibition of viral translation. The improved efficacy of the ribozyme construct suggests that the intracellular stability of such RNA duplexes is much lower than under in vitro conditions. It implies the presence of helix-destabilizing activities such as RNA helicases or RNA-binding proteins such as hnRNP A1 which cause the dissociation of the RNA doublestrand (94). In such a scenario ribozyme-mediated RNA cleavage would irreversibly inactivate the viral transcript thereby improving the inhibition of HIV.

Recently, ribozymes were tested under real in vivo conditions in a mouse model. Amelogenin is the major translation product during enamel synthesis for mammalian teeth and which is supposed to play a major role in the enamel mineralization. Anti-amelogenin ribozymes were synthesized which contained 2'-O-allylmodified nucleotides (Fig. 10.8c; 95). One ribozyme contained in addition 3'-terminal phosphorothioate linkages to protect it against 3'-exonucleases. The ribozymes were injected close to the developing mandibular molar teeth in newborn mice. Only the amelogenin protein was analyzed by autoradiography of ^{35}S-labeled total protein isolated from teeth buds. Injection of 50 μg of either active ribozyme caused a delay of the amelogenin synthesis. Only 90 hours after the injection of both ribozymes amelogenin synthesis reached control levels. Thus, the additional phosphorothioate linkages did not enhance the inhibitory effect of the ribozyme. In contrast, both an antisense oligonucleotide and a catalytically inactive ribozyme delayed amelogenin synthesis for only 24 hours. A randomized control oligonucleotide had no effect. In addition, scanning electron microscopy revealed regions of the first molars with disturbed enamel with a pronounced hypomineralization but regions with normal enamel were also observed. Second molars were more affected than first molars. These differences can be explained by a key role of the amelogenin in the early phase of enamel development and a less important role in the later stages. This study suggests that chemically modified ribozymes may be very effective in blocking gene expression in vivo. It has the potential as a powerful method to achieve a timed and localized inhibition of gene expression in vivo.

10.12 Perspectives

The therapeutic applications of ribozymes are promising approaches which may result in the development of a new and versatile class of therapeutic agents. However, unlike the impression given in some recent newspaper articles there is still much research to do to obtain such omnipotent "gene shears". One major problem for ribozyme and antisense oligonucleotide approaches is the low efficiency and high toxicity of delivery agents such as cationic liposomes. The development of efficient and specific delivery approaches is an absolute must for such approaches. The future will show if new genereations of cationic liposomes will reach this aim. An exciting alternative is provided by viruses which developed highly efficient mechanisms to deliver their genome to the right cell compartment. Understanding their mechanisms will result in the development of new approaches for the delivery of not only oligonucleotides but also many other classes of drugs.

Much progress has been achieved in stabilizing ribozymes against nucleases. However, it is still unclear how much protection against degradation is really necessary. This question will be increasingly targeted, as more and more cell systems are available to examine the efficacies of differently modified ribozymes.

Despite many problems, the exogenous delivery of preformed ribozymes has made great progress during the last two years. Besides an increasing number of cell culture studies, they are already successfully applied to an animal model. Additional animal studies will be forthcoming. Thus, it seems not to be unlikely that, like antisense oligonucleotides, exogenously delivered ribozymes will enter clinical studies in the not too distant future.

Acknowledgement

Work in the authors' laboratories has been supported by the Deutsche Forschungsgemeinschaft. Critical reading of the manuscript by J. Thomson and P. A. Heaton is gratefully acknowledged.

References

1. Eckstein, F. and Lilley, D. M. J. (eds.). (1996). RNA catalysis. Nucleic Acids and Molecular Biology, vol. 10. Heidelberg: Springer.
2. Bratty, J., Chartrand, P., Ferbeyre, G., and Cedergren, R. (1993). The hammerhead RNA domain, a model ribozyme. Biochim. Biophys. Acta. *1216*, 345-359.
3. Birikh, K. R., Heaton, P. A. and Eckstein, F. (1997). The structure, function and application of the hammerhead ribozyme. Eur. J. Biochem. *245*, 1-16.
4. Sczakiel, G. and Nedbal, W. (1995). The potential of ribozymes as antiviral agents. Trends Microbiol. *3*, 213-217.
5. Whitton, J. L. (1994). Antisense treatment of viral infection. Adv. Virus Res. *44*, 267-303.
6. Kiehntopf, M., Esquivel, E. L., Brach, M. A., and Herrmann, F. (1995). Clinical applications of ribozymes. Lancet. *345*, 1027-1031.
7. Buzayan, J. M., Gerlauch, W. L., and Bruening, G. (1986). Non-enzymatic cleavage and ligation of RNAs complementary to a plant virus satellite RNA. Nature. *323*, 349-353.
8. Prody, G. A., Bakos, J. T., Buzayan, J. M., Schneider, I. R., and Bruening, G. (1986). Autolytic processing of dimeric plant virus satellite RNA. Science. *231*, 1577-1580.
9. Forster, A. C. and Symons, R. H. (1987). Self-cleavage of plus and minus RNAs of a virusoid and a structural model for the active site. Cell. *49*, 211-220.
10. Epstein, L. M. and Gall, J. G. (1987). Self-cleaving transcripts of satellite DNA from the newt. Cell. *48*, 535-543.
11. Bruening, G. (1989). Compilation of self-cleaving sequences from plant virus satellite RNAs and other sources. Methods Enzymol. *180*, 546-558.
12. Ruffner, D. E., Stormo, G. D., and Uhlenbeck, O. C. (1990). Sequence requirements of the hammerhead RNA self-cleavage reaction. Biochemistry. *29*, 10695-10702.
13. Pley, H. W., Flaherty, K. M., and McKay, D. B. (1994). Three-dimensional structure of a hammerhead ribozyme. Nature. *372*, 68-74.
14. Scott, W. G., Finch, J. T., and Klug, A. (1995). The crystal structure of an all-RNA hammerhead ribozyme: a proposed mechanism for RNA catalytic cleavage. Cell. *81*, 991-1002.
15. Tuschl, T., Gohlke, C., Jovin, T. M., Westhof, E., and Eckstein, F. (1994). A three-dimensional model for the hammerhead ribozyme based on fluorescence measurements. Science. *266*, 785-789.
16. Bassi, G. S., Møllegard, N.-E., Murchie, A. I. H., von Kitzing, E., and Lilley, D. M. J. (1995). Ionic interactions and the global conformations of the hammerhead ribozyme. Nature Struct. Biol. *2*, 45-55.
17. Tuschl, T. and Eckstein, F. (1993). Hammerhead ribozymes: Importance of stem-loop II for activity. Proc. Natl. Acad. Sci. USA. *90*, 6991-6994.
18. Symons, R. H. (1992). Small catalytic RNAs. Annu. Rev. Biochem. *61*, 641-671.
19. Eckstein, F. (1985). Nucleoside phosphorothioates. Annu. Rev. Biochem. *54*, 367-402.
20. van Tol, H., Buzayan, J. M., Feldstein, P. A., Eckstein, F., and Bruening, G. (1990). Two autolytic processing reactions of a satellite RNA proceed with inversion of configuration. Nucleic Acids Res. *18*, 1971-1975.
21. Slim, G. and Gait, M. J. (1991). Configurationally defined phosphorothioate-containing oligoribonucleotides in the study of the mechanism of cleavage of hammerhead ribozymes. Nucleic Acids Res. *19*, 1183-1188.
22. Dahm, S. C. and Uhlenbeck, O. C. (1991). Role of divalent metal ions in the hammerhead RNA cleavage reaction. Biochemistry. *30*, 9464-9469.

23. Dahm, S. C., Derrick, W. B., and Uhlenbeck, O. C. (1993). Evidence for the role of solvated metal hydroxide in the hammerhead cleavage mechanism. Biochemistry. *32*, 13040-13045.
24. Haseloff, J. and Gerlach, W. L. (1988). Simple RNA enzymes with new and highly specific endoribonuclease activities. Nature. *334*, 585-591.
25. Hertel, K. J., Herschlag, D., and Uhlenbeck, O. C. (1994). A kinetic and thermodynamic framework for the hammerhead ribozyme reaction. Biochemistry. *33*, 3374-3385.
26. Pörschke, D. and Eigen, M. (1971). Co-operative non-enzymic base recognition. J. Mol. Biol. *62*, 361-381.
27. Uhlenbeck, O. C. (1987). A small catalytic oligoribonucleotide. Nature. *328*, 596-600.
28. Hoffman, L. M. and Johnson, M. G. (1994). Enzymatic synthesis of milligram quantities of ribozymes in small volumes. Biotechniques. *17*, 372-375.
29. Moore, M. J. and Sharp, P. A. (1992). Site -specific modification of pre-mRNA: The 2'-hydroxyl groups at the splice sites. Science *256*, 992-997.
30. Eaton, B. E. and Pieken, W. A. P. (1995). Ribonucleosides and RNA. Ann. Rev. Biochem. *64*, 837-863.
31. Aurup, H., Williams, D. M., and Eckstein, F. (1991). 2'-Fluoro- and 2'-amino-2'-deoxynucleoside-5'-triphosphates as substrates for T7 RNA ppolymerase. Biochemistry. *31*, 9636-9643.
32. Wincott, F., DiRenzo, A., Shaffer, C., Grimm, S., Tracz, D., Workman, C., Sweedler, D., Gonzalez, C., Scaringe, S., and Usman, N. (1995). Synthesis, deprotection, analysis and purification of RNA and ribozymes. Nucleic Acids Res. *23*, 2677-2684.
33. Ogilvie, K. K., Usman, N., Nicoghosian, K., and Cedergren, R. J. (1988). Total chemical synthesis of a 77 nucleotide long RNA sequence having methionine-acceptance activity. Proc. Natl. Acad. Sci. USA. *85*, 5764-5768.
34. Gait, M. J., Pritchard, C., and Slim, G. (1991). Oligoribonucleotide synthesis. In: Oligonucleotides and analogues: A practical approach. Eckstein ed. Oxford: Oxford University Press, 25-48.
35. Heidenreich, O., Benseler, F., Fahrenholz, A., and Eckstein, F. (1994). High activity and stability of hammerhead ribozymes containing 2'-modified pyrimidine nucleosides and phosphorothioates. J. Biol. Chem. *269*, 2131-2138.
36. Beigelman, L., McSwiggen, J. A., Draper, K. G., Gonzalez, C., Jensen, K., Karpeisky, A. M., Modak, A. S., Matulic-Adamic, J., DiRenzo, A. B., Haeberli, P., S. D., Tracz, D., Grimm, S., Wincott, F. E., Thackray, V. G., and Usman, N. (1995). Chemical modification of hammerhead ribozymes. J. Biol. Chem. *270*, 25702-25708.
37. Fersht, A. (1985). Enzyme structure and mechanism. New York: Freeman.
38. L'Huillier, P. J., Davis, S. R., and Bellamy, A. R. (1992). Cytoplasmic delivery of ribozymes leads to efficient reduction in a-lactalbumin mRNA levels in C127I mouse cells. EMBO J. *11*, 4411-4418.
39. Bertrand, E., Pictet, R., and Grange, T. (1994). Can hammerhead ribozymes be efficient tools to inactivate gene function? Nucleic Acids Res. *22*, 293-300.
40. Heidenreich, O., Kang, S.-H., Brown, D. A., Xu, X., Swiderski, P., Rossi, J. J., Eckstein, F., and Nerenberg, M. (1995). Ribozyme-mediated RNA degradation in nuclei suspension. Nucleic Acids Res. *23*, 2223-2228.
41. Heidenreich, O. and Eckstein, F. (1992). Hammerhead ribozyme-mediated cleavage of the long terminal repeat RNA of human immunodeficiency virus type 1. J. Biol. Chem. *267*, 1904-1909.
42. Crisell, P., Thompson, S., and James, W. (1993). Inhibition of HIV-1 replication by ribozymes that show poor activity in vitro. Nucleic Acids Res. *21*, 5251-5255.
43. Denman, R. B. (1993). Using RNAFOLD to predict the activity of small catalytic RNAs. Biotechniques. *15*, 1090-1094.
44. Kobayashi, H., Kim, N., Halatsch, M.-E., and Ohnuma, T. (1994). Specificity of ribozyme designed for mutated DHFR mRNA. Biochem. Pharmacol. *47*, 1607-1613.
45. Herschlag, D. (1991). Implications of ribozyme kinetics for targeting the cleavage of specific RNA molecules in vivo: More isn't always better. Proc. Natl. Acad. Sci. USA. *88*, 6921-6925.
46. Werner, M. and Uhlenbeck, O. C. (1995). The effect of base mismatches in the substrate recognition helices of hammerhead ribozymes on binding and catalysis. Nucleic Acids Res. *23*, 2092-2096.
47. Zoumadakis, M., Neubert, W. J., and Tabler, M. (1994). The influence of imperfectly paired helices I and III on the catalytic of hammerhead ribozymes. Nucleic Acids Res. *22*, 5271-5278.
48. Sczakiel, G. (1994). Antisense strategies for the control of aberrant gene expression. J. Hematother. *3*, 305-313.

49. Xing, Z. and Whitton, J. L. (1992). Ribozymes which cleave arenavirus RNAs: Identification of susceptible target sites and inhibition by target site secondary structure. J. Virol. *66*, 1361-1369.
50. Sczakiel, G., Homann, M., and Rittner, K. (1993). Computer-aided search for effective antisense RNA target sequences of the human immunodeficiency virus type 1. Antisense Res. Dev. *3*, 45-52.
51. Christoffersen, R. E., McSwiggen, J., and Konings, D. (1994). Application of computational technologies to ribozyme biotechnology products. Theochem. J. Mol. Struct. *311*, 273-284.
52. Rittner, K., Burmester, C., and Sczakiel, G. (1993). In vitro selection of fast-hybridizing and effective antisense RNAs directed against the human immunodeficiency virus type 1. Nucleic Acids Res. *21*, 1381-1387.
53. Lieber, A. and Strauss, M. (1995). Selection of efficient cleavage sites in target RNAs by using a ribozyme expression library. Mol. Cell. Biol. *15*, 540-551.
54. Campbell, T. B. and Cech, T. R. (1995). Identification of ribozymes within a ribozyme library that efficiently cleave a long substrate RNA. RNA. *1*, 598-609.
55. Cotten, M., Schaffner, G., and Birnstiel, M. L. (1989). Ribozyme, antisense RNA, and antisense DNA inhibition of U7 small nuclear ribonucleoprotein-mediated histone pre-mRNA processing in vitro. Mol. Cell. Biol. *9*, 4479-4487.
56. Denman, R. B., Purow, B., Rubenstein, R., and Miller, D. L. (1992). Hammerhead ribozyme cleavage of hamster prion pre-mRNA in complex cell-free model systems. Bichem. Biophys. Res. Comm. *186*, 1171-1177.
57. Tsuchihashi, Z., Khosla, M., and Herschlag, D. (1993). Protein enhancement of hammerhead ribozyme catalysis. Science. *262*, 99-102.
58. Bertrand, E. L. and Rossi, J. J. (1994). Facilitation of hammerhead ribozyme catalysis by the nucleocapsid protein of HIV-1 and the heterogeneous nuclear protein A1. EMBO J. *13*, 2904-2912.
59. Herschlag, D., Khosla, M., Tsuchihashi, Z., and Karpel, R. L. (1994). An RNA chaperone activity of non-specific RNA binding proteins in hammerhead ribozyme catalysis. EMBO J. *13*, 2913-2924.
60. Müller, G., Strack, B., Dannull, J., Sproat, B. S., Surovoy, A., Jung, G., and Moelling, K. (1994). Amino acid requirements of the nucleocapsid protein of HIV-1 for increasing catalytic activity of a Ki-ras ribozyme in vitro. J. Mol. Biol. *242*, 422-429.
61. Herschlag, D. (1995). RNA chaperones and the RNA folding problem. J. Biol. Chem. *270*, 20871-20874.
62. Sioud, M. (1994). Interaction between tumour necrosis factor α ribozyme and cellular proteins. J. Mol. Biol. *242*, 619-629.
63. Stein, C. A. and Cheng, Y.-C. (1993). Antisense oligonucleotides as therapeutic agents – Is the bullet really magical? Science. *261*, 1004-1012.
64. Sullenger, B. A. and Cech, T. R. (1993). Tethering ribozymes to a retroviral packaging signal for destruction of viral RNA. Science. *262*, 1566-1569.
65. Heidenreich, O., Pieken, W., and Eckstein, F. (1993). Chemically modified RNA: approaches and applications. FASEB J. *7*, 90-96.
66. Heidenreich, O., Kang, S.-H., Xu, X., and Nerenberg, M. (1995). Application of antisense technology to therapeutics. Mol. Medicine Today. *1*, 128-133.
67. Shimayama, T., Nishikawa, F., Nishikawa, S., and Taira, K. (1993). Nuclease-resistant chimeric ribozymes containing deoxyribonucleotides and phosphorothioate linkages. Nucleic Acids Res. *21*, 2605-2611.
68. Heidenreich, O., Xu, X., Swiderski, P., Rossi, J. J., and Nerenberg, M. (1996). Correlation of activity with stability of chemically modified ribozymes in nuclei suspension. Antisense & Nucleic Acid Drug Devel. *6*, 111-118.
69. Ruffner, D. E. and Uhlenbeck, O. C. (1990). Thiophosphate interference experiments locate phosphates important for the hammerhead RNA self-cleavage reaction. Nucleic Acids Res. *18*, 6025-6029.
70. Guschlbauer, W. and Jankowski, K. (1980). Nucleic Acids Res. *8*, 1421-1433.
71. Saenger, W. (1984). Principles of nucleic acid structure. Heidelberg: Springer.
72. Aurup, H., Tuschl, T., Benseler, F., Ludwig, J., and Eckstein, F. (1994). Oligonucleotide duplexes containing 2'-amino-2'-deoxycytidines: thermal stability and chemical reactivity. Nucleic Acids Res. *22*, 20-24.
73. Monia, B. P., Lesnik, E. A., Gonzalez, C., Lima, W. F., McGee, D., Guinosso, C. J., Kawasaki, A. M., Cook, P. D., and Freier, S. M. (1993). Evaluation of 2'-modified oligonucleotides containing 2'-deoxy gaps as antisense inhibitors of gene expression. J. Biol. Chem. *268*, 14514-14522.

74. Yang, J.-H., Usman, N., Chartrand, P., and Cedergren, R. (1992). Minimum ribonucleotide requirement for catalysis by the RNA hammerhead domain. Biochemistry. *31*, 5005-5009.
75. Paolella, G., Sproat, B. S., and Lamond, A. I. (1992). Nuclease resistant ribozymes with high catalytic activity. EMBO J. *11*, 1913-1919.
76. Pieken, W. A., O. D. B., Benseler, F., Aurup, H., and Eckstein, F. (1991). Kinetic characterization of ribonuclease-resistant 2'-modified hammerhead ribozymes. Science. *253*, 314-317.
77. Lange, W., Cantin, E. M., Finke, J., and Dölken, G. (1993). In vitro and in vivo effects of synthetic ribozymes targeted against BCR/ABL mRNA. Leukemia. *7*, 1786-1794.
78. Snyder, D. S., Wu, Y., Wang, J. L., Rossi, J. J., Swiderski, P., Kaplan, B. E., and Forman, S. J. (1993). Ribozyme-mediated Inhibition of bcr-abl gene expression in a Philadelphia chromosome-positive cell line. Blood. *82*, 600-605.
79. Zabner, J., Fasbender, A. J., Moninger, T., Poellinger, K. A., and Welsh, M. J. (1995). Cellular and molecular barriers to gene transfer by a cationic lipid. J. Biol. Chem. *270*, 18997-19007.
80. Wagner, E., Zatloukal, K., Cotten, M., Kirlappos, H., Mechtler, K., Curiel, D. T., and Birnstiel, M. (1992). Coupling of adenovirus to transferrin-polylysine/DNA complexes greatly enhances receptor-mediated gene delivery and expression of transfected genes. Proc. Natl. Acad. Sci. USA. *89*, 6099-6103.
81. Plank, C., Oberhauser, B., Mechtler, K., Koch, C., and Wagner, E. (1994). The influence of endosome-disruptive peptides on gene transfer using synthetic virus-like gene transfer systems. J. Biol. Chem. *269*, 12918-12924.
82. Zelphati, O., Imbach, J.-L., Signoret, N., Zon, G., Rayner, B., and Leserman, L. (1994). Antisense oligonucleotides in solution or encapsulated in immunoliposomes inhibit replication of HIV-1 by several different mechanisms. Nucleic Acids Res. *22*, 4307-4314.
83. Romani, A. and Scarpa, A. (1992). Regulation of cell magnesium. Arch. Biochem. Biophys. *298*, 1-12.
84. Beck, J. and Nassal, M. (1995). Efficient hammerhead ribozyme-mediated cleavage of the structured hepatitis B virus encapsidation signal in vitro and in cell extracts, but not in intact cells. Nucleic Acids Res. *23*, 4954-4962.
85. Heidenreich, O., Xu, X., and Nerenberg, M. (1996). A hammerhead ribozyme cleaves its target RNA during RNA preparation. Antisense & Nucleic Acid Drug Devel. *6*, 141-144.
86. Sioud, M., Natvig, J. B., and Førre, ø. (1992). Preformed ribozyme destroys tumour necrosis factor mRNA in human cells. J. Mol. Biol. *223*, 831-835.
87. Lange, W., Daskalatis, M., Finke, J., and Dölken, G. (1994). Comparison of different ribozymes for efficient and specific cleavage of BCR/ABL related mRNAs. FEBS Lett. *338*, 175-178.
88. Leopold, L. H., Shore, S. K., Newkirk, T. A., Reddy, R. M. V., and Reddy, E. P. (1995). Multi-unit ribozyme-mediated cleavage of bcr-abl mRNA in myeloid leukemias. Blood. *85*, 2162.
89. Kiehntopf, M., Brach, M. A., Licht, T., Petschauer, S., Karawajew, L., Kirschning, C., and Herrmann, F. (1994). Ribozyme-mediated cleavage of the MDR-1 transcript restores chemosensitivity in previously resistant cancer cells. EMBO J. *13*, 4645-4652.
90. Kiehntopf, M., Herrmann, F., and Brach, M. A. (1995). Functional NF-IL6/CCAAT enhancer-binding protein is required for tumor necrosis factor a-inducible expression of the granulocyte colony-stimulating factor (CSF), but not the granulocyte/macrophage CSF or interleukin 6 gene in human fibroblasts. J. Exp. Med. *181*, 793-798.
91. Tanaka, T., Akira, S., Yoshida, K., Umemoto, M., Yoneda, Y., Shirafuji, N., Fujiwara, H., Suematsu, S., Yoshida, N., and Kishimoto, T. (1995). Targeted disruption of the NF-IL6 gene discloses its essential role in bacteria killing and tumor cytotoxicity by macrophages. Cell. *80*, 353-361.
92. Gu, J.-L., Veerapanane, D., Rossi, J., Natarajan, R., Thomas, L., and Nadler, J. (1995). Ribozyme-mediated inhibition of expression of leukocyte-type 12-lipoxygenase in porcine aortic vascular smooth muscle cells. Circ. Res. *77*, 14-20.
93. Homann, M., Tzortzakaki, S., Rittner, K., Sczakiel, G., and Tabler, M. (1993). Incorporation of the catalytic domain of a hammerhead ribozyme into antisense RNA enhances its inhibitory effect on the replication of human immunodeficiency virus type 1. Nucleic Acids Res. *21*, 2809-2814.
94. Sczakiel, G. and Goody, R. S. (1994). Antisense principle or ribozyme action? Biol. Chem. Hoppe-Seyler. *375*, 745-746.
95. Lyngstradaas, S. P., Risnes, S., Sproat, B. S., Thrane, P. S., and Prydz, H. P. (1995). A synthetic, chemically modified ribozyme eliminates amelogenin, the major translation product in developing mouse enamel in vivo. EMBO J. *14*, 5224-5229.

96. Hertel, K. J., Pardi, A., Uhlenbeck, O. C., Koizumi, M., Ohtsuka, E., Uesugi, S., Cedergren, R., Eckstein, F., Gerlach, W. L., Hodgson, R., and Symons, R. H. (1992). Numbering system for the hammerhead. Nucleic Acids Res. *20*, 3252.

11. Ribozymes as Tools for the Gene Therapist

Lynn Milich and Bruce A. Sullenger

11.1 Introduction

Over the last fifteen years, many studies have revealed that RNA enzymes or ribozymes play important roles in cellular biochemistry and metabolism. The first discovered example of such an RNA catalyst was the self-splicing group I intron from *Tetrahymena thermophilia (T. thermophilia)*, a ciliated protozoan. This now well characterized intron efficiently excises itself from nuclear ribosomal RNA (rRNA) precursors to generate mature rRNAs without the aid of proteins (reviewed in 19, 20, 81). RNase P, a ribozyme found in a variety of cells, functions to produce mature 5' ends of transfer RNAs (tRNAs) (reviewed in 3, 20).

There are also several other "natural" catalytic RNAs that are associated with plant pathogens. The hammerhead and hairpin ribozymes derived from satellite RNA of Tobacco Ringspot Virus (sTRSV) catalyze a self-cleavage reaction that is believed to play a major role in the replication of these RNA molecules *in vivo* (reviewed in 155).

Many of the self-cleaving ribozymes have been engineered into RNA restriction endonucleases that can cleave other RNA molecules in a sequence-specific manner (reviewed in 18). Because of their ability to carry out cleavage in *trans*, the use of ribozymes as inhibitors of viral pathogens and dominant oncogenes has been widely tested and reviewed (117, 130-134, 183). More recently, novel potential therapeutic applications have been proposed for the targeted trans-splicing reaction catalyzed by the group I intron ribozyme from *Tetrahymena* (reviewed in 149). This reaction includes both a cleavage and ligation step and may prove effective in restoring wild type function to mutant cellular transcripts (150). As discussed in section 11.3.2, targeted trans-splicing can also be used to transform viral transcripts into potent antivirals.

The purpose of this chapter is not to provide an exhaustive review of the biochemistry of ribozymes or to catalogue the published results that suggest that ribozymes may become useful tools for the gene therapist. Both of these topics have been extensively dealt with elsewhere (1, 11, 19, 20, 22, 117, 119, 129, 155). In addition, we will leave discussion of the use of synthetic ribozymes and of the various gene transfer and expression systems to the other chapters in this volume. Rather, we will attempt to present a fair and focused account of the potential utility of catalytic RNAs for gene therapy by first presenting an overview of the basic biochemistry of well characterized ribozymes. Then, we will discuss applications of both trans-cleaving and trans-splicing ribozymes for gene therapy. This approach will hopefully enhance the reader's understanding of the potential utility of catalytic RNAs as both gene inhibitors and revisionists.

11.2 Biochemistry of Catalytic RNAs

Currently the most well characterized catalytic RNAs fall into five classes and are most easily distinguished by secondary structure, size and mechanism of cleavage (18). All five types of ribozymes have precisely folded RNA structures (Fig. 11.1) which are required to bring reactive groups into close proximity for catalysis. Group I introns and RNase P are the largest catalytic RNAs typically found to be greater than 200 nucleotides in length. Both cleave RNA and generate products with 3'-hydroxyl and 5'-phosphate termini. Cleavage reactions catalyzed by group I introns occur via transesterification or hydrolysis, while RNase P cleaves its substrates by hydrolysis.

In contrast, hammerhead, hairpin and the Hepatitis Delta Virus (HDV) ribozymes are typically only 30 to 80 nucleotides in length. All three cleave RNA substrates by transesterification to form cleavage products with 2',3' cyclic phosphate and 5'-hydroxyl termini. In their natural forms all of these catalytic RNAs with the exception of RNase P are self cleaving. The group I intron from *Tetrahymena*, the hammerhead, hairpin ribozymes, and HDV have all been altered to facilitate cleavage of exogenous RNA, and to improve their kinetic properties and substrate specificity (18). These and other features which make ribozymes attractive therapeutic agents are presented in more detail below.

11.2.1 The Group I Intron from *Tetrahymena*

The intervening sequence (IVS) present within nuclear precursor rRNAs from *T. thermophilia* is one of the most well characterized catalytic RNAs. This IVS is a member of a growing family of group I introns and is naturally present as a 413-base pair intron in the middle of the 26S rRNA gene found in *Tetrahymena*. The IVS is transcribed as part of the rRNA precursor and excises itself thru a series of cleavage-ligation reactions to form a functional rRNA without the aid of proteins (23, 81). Prior to the discovery of this self-splicing reaction, RNA was thought to be strictly a carrier of genetic information and not able to perform catalysis.

11.2.1.1 The Self-Splicing *Tetrahymena* Intron

The secondary structure of the IVS from *Tetrahymena* was determined by comparative sequence analysis of several group I introns (37, 103). As shown in Figure 11.1A, the phylogenetically conserved secondary structure of group I introns consists of a set of paired regions, P1-P9 (5, 12, 42, 46, 111, 166, 168, 170, 171). Although the crystal structure of a group I intron ribozyme has not been solved, several studies have shown that to adopt the proper three dimensional structure the enzyme must form several tertiary interactions (21, 24, 84, 90, 105, 165). Like proteins, the intron has an interior and an exterior with the catalytic core surrounded by a close-packed layer of RNA helices (90). Several studies have shown that this folded RNA structure participates directly in self-splicing (39, 164, and references below).

The *Tetrahymena* intron excises itself and ligates together its flanking exons by performing two consecutive transesterification reactions (reviewed in 19, 20, 23). The first reaction is initiated by an intron-bound guanosine which serves as a nucleophile and attacks the phosphorous atom at the 5' splice site to form a covalent bond with the first nucleotide at the 5' end of the intron (Fig. 11.2A). The recognition element that defines the exact site of guanosine attack is a non-Watson-Crick G-U base pair that is highly conserved among group I introns (121, 148). The G-U base pair is part of a short duplex called P1. The P1 duplex includes base pairing between the last six nucleotides of the 5' exon and sequences within the intron called the internal guide sequence (IGS) or the 5' exon binding site. In a second transesterification reaction, a free 3'-hydroxyl group at the 3' end of the cleaved 5' exon attacks the phosphorous atom at the 3' splice site resulting in the ligation of the 5' and 3' exons and the excision of the intron. An important step to note in this pathway is the continued P1 base pairing after cleavage at the 5' splice site. This continued base pairing between the intron and 5' exon is necessary to correctly position the 5' exon for the second cleavage step and for ligation with the 3' exon.

Even though it was shown that the excised group I intron maintained its ability to make and break phosphodiester bonds, the RNA was not regenerated in its original form after the self-splicing reaction and therefore was not thought to be a true enzyme (186). However, subsequent alteration of the *Tetrahymena* IVS resulted in a shortened form of the intron that fulfilled the requirements of a true enzyme (185).

11.2.1.2 The Trans-Cleaving *Tetrahymena* Ribozyme

Shortened forms of the IVS from *Tetrahymena* have been derived from a circularized intermediate generated during the self-splicing reaction. These ribozymes do not contain the first 19 to 21 nucleotides of the original excised IVS and when incubated with oligonucleotide substrates can catalyze the cleavage and rejoining of oligonucleotides with multiple turnover (185). In addition, the value of k_{cat}=2 min^{-1} for the shortened form of the intron is within the range of values for protein enzymes, such as EcoRI, that catalyze sequence-specific cleavage of nucleic acids.

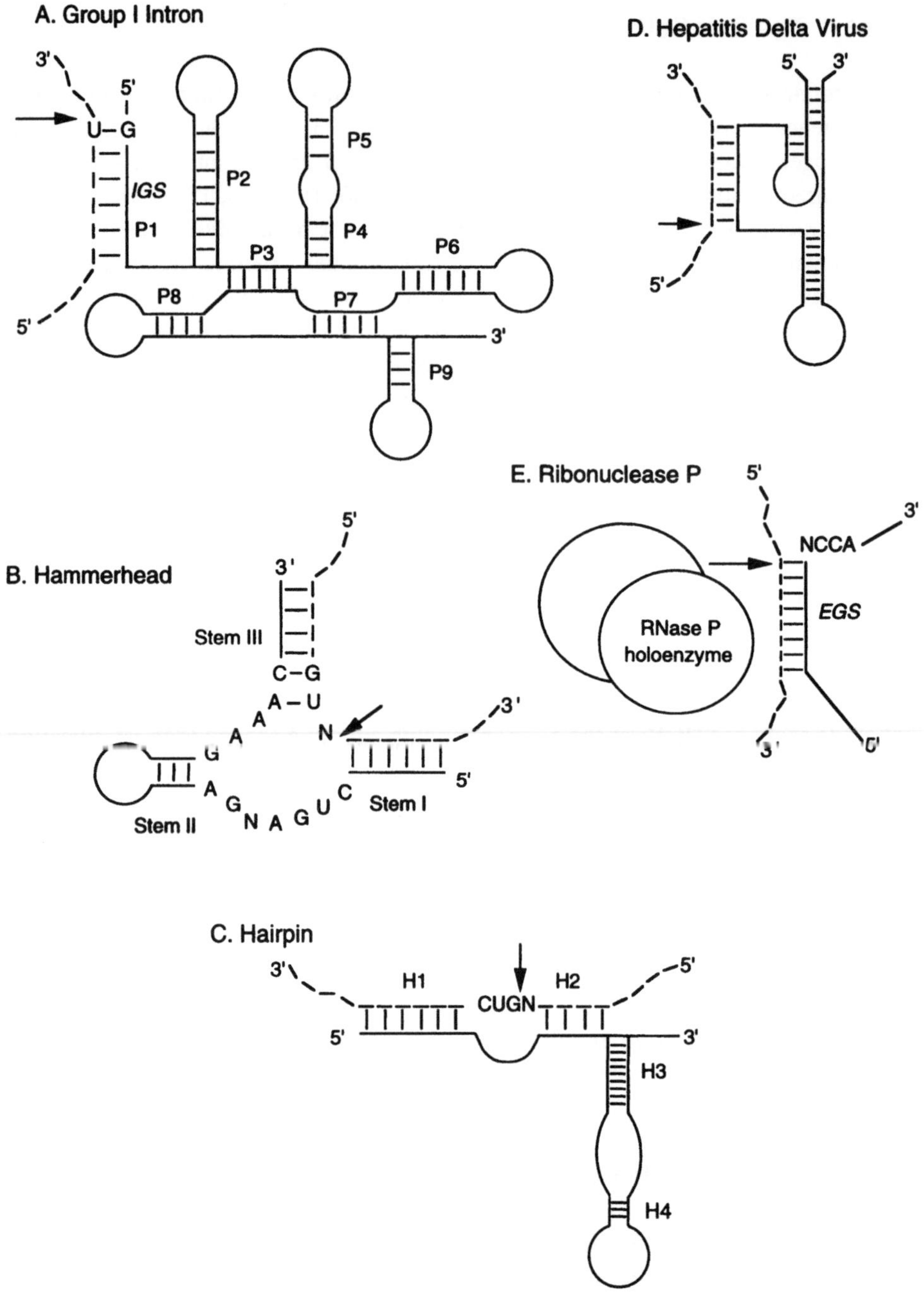

The mechanism of the trans-cleavage reaction is very similar to the first transesterification step in the self-splicing pathway with a few exceptions (22). First, since the 5' exon sequences preceding the IGS have been removed in the shortened form

◁ Fig. 11.1: Secondary structure diagrams for the five classes of catalytic RNAs.
Group I introns, hammerhead, hairpin and Hepatitis Delta Virus (HDV) ribozymes have been engineered to mediate cleavage of exogenous RNA substrates. Ribozymes are indicated by solid lines and RNA substrates by dashed lines. Cleavage sites are indicated by arrows.

(A) Group I ribozyme; P2-P9 represent base paired regions conserved among group I introns. The RNA substrate is shown bound to the internal guide sequence (IGS) within the ribozyme. The substrate is cleaved 3' of the conserved G-U base pair.

(B) Hammerhead ribozyme; shown are two single-stranded regions containing nine highly conserved nucleotides that form the catalytic core. Cleavage of the substrate occurs 3' of the sequence, GUN where N is C, U, or A.

(C) Hairpin ribozyme; two helices, H1 and H2, formed by base pairing between sequences within the ribozyme and substrate form part of the substrate recognition site. Cleavage occurs between the nucleotides 5'NpG3' where N is G, A, U, or C.

(D) HDV ribozyme; the formation of 7 to 8 base pairs between HDV and the substrate 3' of the cleavage site is required for target site recognition. Cleavage occurs with a C, U, or A nucleotide 5' of the cleavage site.

(E) Ribonuclease P; the holoenzyme which consists of an RNA subunit (M1) and protein cofactor is depicted schematically. The external guide sequence (EGS) is shown bound to an RNA target sequence. Cleavage of the substrate occurs at a site adjacent to the RNA duplex and requires the unpaired nucleotides CCA at the 3' end of the EGS.

of the intron, the IGS binds to an exogenous RNA target (Fig. 11.2B). Similar to the mechanism of self-splicing, the site of cleavage depends on the presence of an unconventional G-U base pair followed by base pairing between the IGS and the substrate (2, 38, 184). For the wild type IGS (5"GGAGGG3") binding would occur at nucleotides 5'CCCUCU3"within the target as shown in Figure 11.2B. Cleavage occurs after the reconstructed G-U base pair generating a 5' and 3' cleavage product. However, unlike the self-splicing scenario, the guanosine nucleophile in the trans-cleavage reaction is attached to the 5' end of the 3' cleavage product and not to the ribozyme (compare Fig. 11.2A and B).

As noted, substrate recognition and accurate trans-cleavage require base pairing interaction between the IGS on the ribozyme and the RNA substrate. Substrate specificity can easily be manipulated to target any RNA molecule by altering sequences in the IGS (5, 104, 184). Thus, cleavage can occur at any site in an RNA substrate as long as there is base pairing between the IGS and sequences in the substrate, and a G-U base pair immediately precedes the cleavage site.

Binding of a shortened form of the *Tetrahymena* ribozyme (called L-21 ScaI) to a substrate RNA is 10^3-10^4-fold tighter than would be predicted by base pairing alone (63, 120). Studies on the L-21 ScaI ribozyme, as well as the self-splicing form of the intron, suggest that tertiary interactions contribute to the ribozyme-substrate binding energy. Specific tertiary interactions have been identified which involve several 2'-hydroxyl groups both within the substrate and the intron (8, 76, 122, 147).

This tight binding between the ribozyme and RNA substrate has limited the substrate specificity of the *Tetrahymena* intron (22). Both matched substrates that are only six nucleotides long and substrates that form a single base pair mismatch with the IGS have similar K^S_{on} values despite 10^3-fold weaker binding of the mismatched

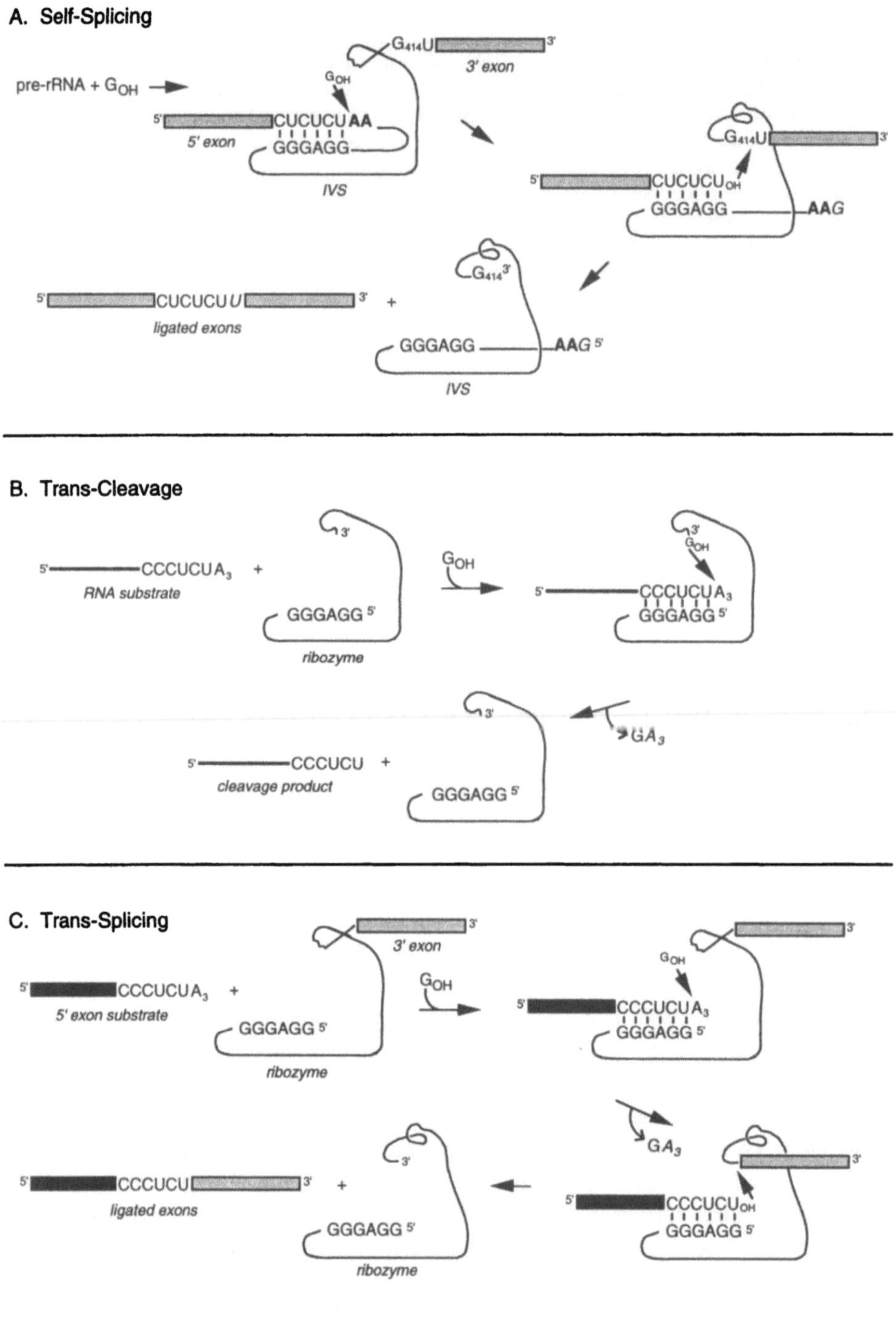

substrate to the ribozyme (64). Thus under conditions of saturating guanosine and 10mM $MgCl_2$, RNA substrates are bound so tightly that the ribozyme cleaves essentially every substrate molecule that binds (discussed further in section 11.2.6).

◁ Fig. 11.2: Cleavage and/or ligation reactions catalyzed by three different forms of the *Tetrahymena* group I intron.

(A) The group I intron is naturally present as an intervening sequence (IVS) within precursor rRNA. With exogenous guanosine acting as a nucleophile the IVS excises itself from the precursor RNA and mediates the ligation of the 5' and 3' exons to generate the mature rRNA.

(B) Shortened forms of the group I intron have been engineered to catalyze sequence-specific endonuclease reactions on exogenous RNA substrates. In trans-cleavage reactions, the guanosine nucleophile is added to the downstream cleavage product (GA_3). Subsequently the 5' cleavage product is released and the *Tetrahymena* ribozyme regenerated to its original form catalyzes multiple cleavage reactions.

(C) The trans-splicing group I ribozyme mediates trans-cleavage as shown in 2B, but also transfers a 3' exon to the 5' cleavage product. Note the wild type IGS (5'GGAGGG3') was used to illustrate both the trans-cleavage and trans-splicing reactions.

Several *in vitro* studies have demonstrated that ribozyme specificity can be improved. These approaches include: i) introduction of small mutations within the *Tetrahymena* ribozyme which decreases its affinity for RNA substrates and results in the release of mismatched substrate RNAs before cleavage occurs (122, 178); ii) lowering the concentration of guanosine which slows the cleavage step so that dissociation of the substrate RNA occurs before cleavage (64); and iii) addition of denaturants such as urea or formamide to the reaction, or lowering the concentration of Mg^{++}. These conditions may slow the chemical step and/or destabilize substrate binding (187).

If the specificity of the group I ribozyme cannot be easily improved, then trans-cleaving ribozymes engineered from group II introns may be better candidates for achieving the levels of substrate specificity that will be required for gene therapy applications. These autocatalytic RNAs which are present in organelles of plants, lower eukaryotes, and prokaryotes, do not appear to form additional tertiary interactions with their substrates (52). Thus, group II introns may be particularly adept at recognizing only the intended RNA sequence in the pool of cellular RNAs.

11.2.1.3 The Trans-Splicing *Tetrahymena* Ribozyme

The self-splicing intron from *Tetrahymena* can also mediate targeted trans-splicing *in vitro* by performing an intermolecular cleavage-ligation reaction (5, 68). In this reaction, the ribozyme with an attached 3' exon recognizes a target RNA (5' exon) by complementary base pairing between the IGS and sequences within the 5' exon (Fig. 11.2C). Like the self-splicing and trans-cleavage reactions, cleavage occurs at a site immediately 3' of a conserved G-U base pair formed between a guanosine nucleotide at the 5' end of the IGS and a uridine nucleotide within the substrate RNA. This results in the release of the sequences downstream of the cleavage site (3' cleavage product). A second nucleophilic attack mediated by the free hydroxyl group at the 3' end of the 5' cleavage product results in ligation of the cleaved 5' exon with the attached 3' exon and release of the free ribozyme.

Like the trans-cleavage reaction, targeted trans-splicing is very versatile and any uridine residue in an RNA molecule can be targeted by simply making the nucleotides in the IGS complementary to the nucleotides which precede an available uridine residue in a targeted 5' exon RNA. Since there are no sequence requirements for the 3' exon in this splicing reaction, essentially any RNA sequence of choice can be spliced onto a targeted 5' exon. Potential therapeutic applications for targeted trans-splicing will be discussed in more detail in section 11.3.2.

11.2.2 The Hammerhead Ribozyme

The hammerhead ribozyme actually represents a catalytic motif that was originally derived by comparing the self-cleavage domains from a number of naturally occurring viroid RNAs and satellite RNAs that replicate in plants (59, 73). The self-cleaving consensus domain consists of a highly conserved catalytic region and three helices, and has been shown to have sequence-specific endoribonuclease activity (Fig. 11.1B) (59). Subsequently, a hammerhead domain of less then 60 contiguous nucleotides was shown to be sufficient for cleavage (14, 48), and two separate oligonucleotides that assembled into a hammerhead structure were shown to mediate a trans-cleavage reaction (162).

In crystallographic studies of the hammerhead ribozyme, the tertiary structure of the hammerhead appears to be "Y shaped" or like that of a "wishbone" (116, 141). Helices I and II are in close juxtaposition while helix III is at the bottom of the molecule. All three are A-form helices.

As shown in Figure 11.1B, the minimal structural requirements for hammerhead catalyzed cleavage include two single-stranded regions that contain nine highly-conserved nucleotide sequences, three helices, and the nucleotides GUN immediately 5' of the cleavage site in the substrate. Results of mutagenesis and kinetic studies suggest that the conserved residues comprise the catalytic core of the ribozyme and are required for cleavage (44, 135). Helices I and III which flank the cleavage site are formed by base pairing with the substrate. This base pairing interaction is extremely important not only because it holds the ribozyme and substrate together but because it precisely positions the ribozyme relative to the cleavage site. The most efficient cleavage has been observed with GUC, GUA, or GUU at the cleavage site although some cleavage also occurs after CUC, UUC and AUC (78). The hammerhead catalyzes transesterification of the 3', 5' phosphodiester bond at the cleavage site which results in the production of RNA with 2', 3' cyclic phosphate and 5'-hydroxyl termini. Although the exact mechanism of the cleavage reaction is not known a divalent metal ion (119) and 2'-hydroxyl groups on specific nucleotides in both the substrate and the ribozyme are required (34, 35, 112, 162).

Much attention has been focused on developing hammerhead ribozymes as therapeutic agents because of their small size, simple secondary structure, and because the helices (I and III) which base pair with the substrate can be designed to target any RNA molecule containing an available cleavage site. In particular, efforts have been directed at optimizing the interaction between the ribozyme and its substrate since both the length (50) and base composition (44) of complementary helices I and III can affect substrate specificity, ribozyme-substrate affinity and turnover.

Other approaches have been used successfully to increase the catalytic activity and substrate specificity of hammerhead ribozymes. Inclusion of a protein derived from the p7 nucleocapsid (NC) protein of HIV-1 in ribozyme reactions *in vitro* was shown to increase the rate of cleavage of an oligonucleotide 10- to 20-fold by increasing the rate of helix association and dissociation (161). Additional studies using HIV-1 NC protein as well as the bacteriophage T4 heteronuclear RNP (hnRNP) A1 protein have shown these proteins similarly enhance ribozyme turnover by increasing the rate of product dissociation (6, 65). Chimeric hammerhead ribozymes that contain DNA in their flanking helices have also been shown to have increased catalytic activity over all-RNA ribozymes as a result of enhanced product dissociation (157) , and formation of a structure between the ribozyme-substrate complex that is more favorable for cleavage (61).

11.2.3 The Hairpin Ribozyme

Like the hammerhead ribozyme, the hairpin ribozyme represents a catalytic motif derived from RNA associated with plant pathogens (155). This small catalytic RNA is found within the negative strand 359-base satellite RNA of Tobacco Ringspot virus (-)sTRSV which is known to mediate self-catalyzed cleavage and ligation reactions as part of its replication pathway (14, 49).

A minimal catalytic domain for this RNA molecule has been identified (45, 58) which consists of a 50-base RNA catalyst that efficiently cleaves an RNA substrate containing 14 bases of satellite RNA sequence (56). Features of secondary structure within this domain, defined from minimum energy RNA folding calculations and supported by mutagenesis studies, include four helices two of which are formed by base pairing between the RNA catalyst and the substrate (H1 and H2) (Fig. 11.1C) (57). These two helices form part of the substrate recognition site and flank a four-base loop within the substrate (5'AGUC3') containing the cleavage site. For the wild type (-)sTRSV cleavage occurs between the nucleotides, 5'ApG3', by transesterification generating a 5' fragment with 2',3' cyclic phosphate termini and a 3' fragment with a 5'-hydroxyl termini (13, 14, 118). Trans-cleavage has also been observed *in vitro* using separate transcripts for the substrate and catalytic region of the hairpin ribozyme with multiple turnover (k_{cat}=2.1 min^{-1}) (56).

The hairpin catalytic motif can be designed to target any number of RNA molecules since there are only two sequence requirements. First, in order to maintain catalytic activity complementary base pairing between the ribozyme and substrate must be maintained. Single base pair mismatches at the ten positions included in the two flanking helices can result in the loss of catalytic activity (57) although single base pair mismatches distal to the cleavage site appear to be tolerated (70). It should also be noted that the composition of base paired substitutions in these helices have a wide range of affects on the kinetic properties of the ribozyme (57) suggesting that base pair substitutions should be optimized for each application.

A second sequence requirement involves the nucleotides that compose the target site. Optimal substrate cleavage occurs with the nucleotides GUC immediately 3' of the cleavage site. The guanosine residue is invariant and is believed to be directly involved in catalysis (27). Catalytic activity has been shown to vary widely when nucleotide substitutions are made at the other three positions.

11.2.4 The Hepatitis Delta Virus (HDV) Ribozyme

HDV is a 1700 nucleotide, covalently closed circular RNA that is associated with Hepatitis B virus in certain patients. It is an animal virus but undergoes autocatalytic self-cleavage as part of its replication cycle similar to the plant pathogenic subviral RNAs, the hammerhead and hairpin ribozymes (reviewed in 155).

Minimal self-cleaving domains have been determined for HDV (82, 174) which were subsequently reduced to include approximately 85 nucleotides from both the genomic (115) and antigenomic RNAs (113). Features of secondary structure were proposed based on nuclease probing and/or site-directed mutagenesis, and were found to be similar for both the genomic and antigenomic self-cleaving sequence elements (113, 127). Included are four stems two of which (stems I and II) generate a tertiary interaction which is a type of pseudoknot (Fig. 11.1D) (113, 127). Stems II, III, and IV appear to be important for stabilizing the catalytic form of the ribozyme (113, 158) while formation of stem I is required for efficient cleavage (113, 158).

The self-cleavage reaction occurs by transesterification and is dependent on divalent cations. Cleavage products have 5'-hydroxyl and 2',3' cyclic phosphate termini. The rates of HDV self-cleavage *in vitro* using either *in vitro*-transcribed HDV RNA (128) or HDV RNA isolated from infected tissue (82) appear to be very slow. However, addition of urea or formamide has been shown to increase the rate of cleavage as much as 50-fold suggesting that these denaturants may be mimicking a viral or cellular RNA binding or unwinding factor which facilitates cleavage *in vivo* (128).

HDV has been engineered to catalyze cleavage reactions in trans with multiple turnover (10, 114). Target site recognition is dependent on the formation of seven to eight base pairs between the target RNA and HDV 3' of the cleavage site. Cleavage occurs with only one nucleotide 5' of the cleavage site (5'N^GG3') and that nucleotide can be C, U, or A but not G (114).

11.2.5 The RNase P Ribozyme

Unlike the other four ribozymes that have been discussed, RNase P is the only catalytic RNA known to act *in vivo* as a true enzyme. A second unique feature of this RNA molecule is that substrate specificity is primarily determined by an external guide sequence (EGS) as opposed to an internal guide sequence which is a natural part of the other ribozymes.

RNase P is a natural component of procaryotic and eukaryotic cells, and its function is to remove the 5' leader sequence from the cells repertoire of precursor tRNAs (reviewed in 1, 20). *In vivo,* the active form of the ribozyme is a ribonucleoprotein (RNP). One of the most well characterized enzymes, RNase P isolated from *Escherichia Coli (E. Coli),* has been shown to consist of a 377-nucleotide RNA subunit (M1) and a 119-amino acid protein (Fig 11.1E). Although it was initially thought that the holoenzyme was required for catalytic activity *in vitro,* M1 RNA preparations from *E. Coli* (54) as well as *in vitro*-transcribed M1 RNA (53) were shown to cleave tRNA precursors at high Mg^{++} concentrations with multiple turnover. Cleavage occurs by hydrolysis to generate 5'-phosphate and 3'-hydroxyl termini. While the RNA subunit of RNase P alone is catalytic, the protein co-factor allows the reaction to proceed efficiently under physiologically-relevant conditions and also facilitates cleavage of a broader substrate range (53).

With its natural pre-tRNA substrate, RNase P removes the 5' leader sequences from the end of the tRNA molecule. Cleavage occurs specifically and accurately 5' of the first nucleotide that is present in the mature tRNA even though there is a small degree of sequence conservation among different tRNAs. Thus the three-dimensional conformation of the pre-tRNA substrates appears to be the structure recognized by this ribozyme. This notion was supported by mutagenesis studies which showed that disruption of tRNA folding decreased the rate of cleavage (101). Subsequently, it was shown that the full tertiary structure of tRNA is not required for cleavage by RNase P (47). What is required is a short RNA fragment, the EGS, that is complementary to the substrate and which has an unpaired CCA sequence at its 3'end (Fig. 11.1E). In the natural pre tRNA substrate the EGS is part of the aminoacyl acceptor stem of the pre tRNA. However, RNase P RNA will cleave single-stranded 5' leader sequences adjacent to any double stranded RNA duplex as long as the unpaired CCA nucleotides are present at the 3' end of the EGS (47). Thus it is conceivable that endogenous RNase P could be targeted to cleave any RNA with the use of an appropriately designed EGS.

11.2.6 Kinetic Analysis: A Pathway to Improve Ribozyme Catalysis

If catalytic RNAs are to become useful therapeutic agents, they probably will have to be able to destroy the vast majority of their target transcripts inside of a cell. Thus, ribozymes that can cleave target RNAs with multiple turnover *in vivo* would be expected to be more effective gene inhibitors than ribozymes that do not cycle. A thorough examination of the *in vitro* kinetic properties of the group I intron from *Tetrahymena* highlights how understanding a ribozyme's reaction pathway can facilitate the design of improved ribozymes for therapeutic applications.

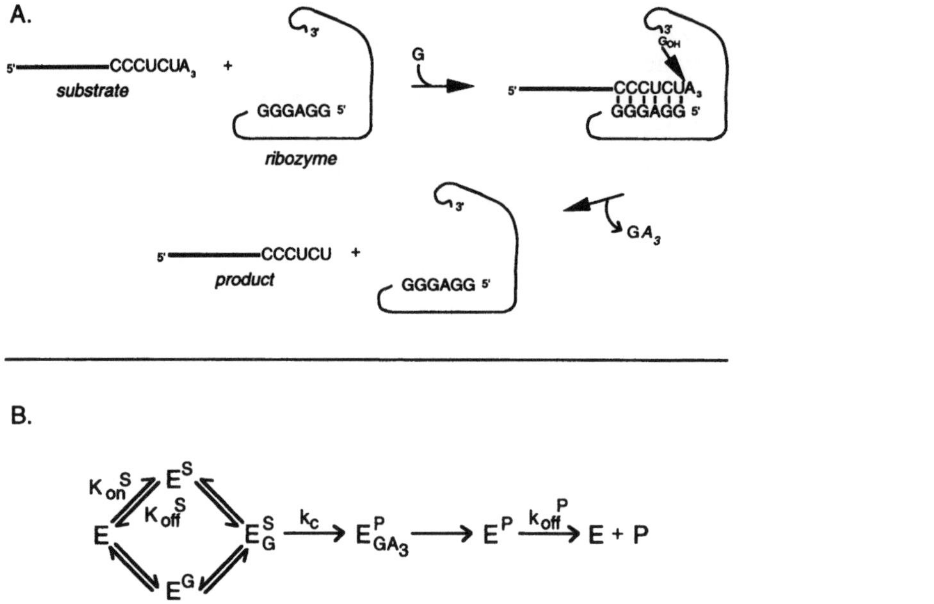

Fig. 11.3: Kinetic scheme for the endonuclease reaction catalyzed by the *Tetrahymena* ribozyme.

(A) Diagram illustrating the steps involved in ribozyme-mediated trans-cleavage of a matched RNA substrate.

(B) Kinetic pathway for the trans-cleavage reaction shown in 11.3A. The *Tetrahymena* ribozyme (E) binds both the substrate (S) and guanosine (G) to form a ternary complex (E_GS). Once bound cleavage occurs rapidly and is represented by the rate constant, k_c. Release of the downstream cleavage product, GA_3, is also rapid. Under multiple turnover conditions where S is saturating, release of the 5' or upstream cleavage product is rate limiting (k_{off}^P). Under $(k_{cat}/K_m)^S$ conditions where S is subsaturating, the rate limiting step is the binding of S (K_{on}^S).

Shown in Figure 11.3 is the kinetic pathway for the endonuclease reaction catalyzed by the L-21 ScaI form of the *Tetrahymena* ribozyme (63, 178). Under $(K_{cat}/K_m)^S$ conditions where substrate (S) is subsaturating and guanosine (G) is saturating, the rate limiting step is binding of S (K_{on}^S). Under these conditions once S and G have bound to the ribozyme the rate of chemical cleavage is fast (k_c is approximately 350 min^{-1}) while the rate of dissociation of S occurs much more slowly (K_{off}^S= 0.2 min^{-1}) (22, 63). Consequently, every RNA molecule that binds is cleaved before it can dissociate.

As noted earlier, the enhanced binding results from additional tertiary interactions that form between the ribozyme and substrate and these interactions have important implications for substrate specificity. Although substrates containing single base pair mismatches have 10^3-fold weaker binding energy than substrates that maintain base pair complementarity, the mismatched substrates still bind 10^4-fold stronger than would be predicted from base pairing alone. This results in the cleavage of perfectly matched substrates as well as the substrates containing mismatched base pairs (64). With the wild type L-21 ScaI ribozyme a decrease in the concentration of G to subsaturating levels (1uM) results in a decreased rate of cleavage of mismatched substrates, k_c^{app}=0.9 min^{-1}, and a substrate dissociation rate constant, K_{off}^S=50 min^{-1} (64). Thus under conditions of subsaturating G, the release of the mismatched substrate occurs 100-fold faster than cleavage. This is in contrast to substrates containing no mismatches in which the cleavage reaction, k_c^{app}=1.4 min^{-1} is faster than dissociation, K_{off}^S=0.2 min^{-1}. This refinement of the reaction conditions results in a ribozyme with improved substrate specificity. Mutations in the L-21 Sca I ribozyme have also been shown to yield similar results (178). Mutations that weaken tertiary interactions allow the ribozyme to detect differences in the binding energies of substrates containing mismatches compared with those that maintain base pair complementarity. However, it should be noted that while these mutants can recognize the proper substrate (high specificity), substantial cleavage occurs at incorrect cleavage sites (low fidelity).

The calculated chemical rate constant, k_c=350 min^{-1} suggests that the *Tetrahymena* ribozyme is a very good catalyst. In the absence of G, cleavage of RNA can occur by hydrolysis. The rate constant of 350 min^{-1} represents a 10^{11} rate enhancement over the estimated rate constant for the hydrolysis reaction, a rate comparable to that achieved by protein enzymes. The ribozyme cleaves RNA as quickly as it binds and is limited only by the rate of helix formation between the IGS and the substrate.

A second limitation of the L21 ScaI ribozyme is its lack of ability to function as a multiple turnover enzyme. Under k_{cat} conditions where S and G are saturating release of the 5' cleavage product from the active site is rate limiting and can be described by the rate constant, k_{cat}=0.1 min^{-1} (k_{off}^P) (63). Although the tight binding between the ribozyme and 5' exon evolved to insure the ligation of flanking exons in the self-splicing reaction of pre-rRNA, single turnover catalysis is not conducive to gene inhibition strategies. However, small mutations in specific regions of the L21 Sca I ribozyme that weaken RNA binding have been shown to accelerate the product release step (178). Similarly, substitutions which form single base pair mismatches with the IGS of the ribozyme show up to a 100-fold increase in kcat resulting from faster dissociation of the mismatched product (64) or from destabilization of the ground state enzyme-substrate complex (187).

Like the group I intron from *Tetrahymena*, the RNA subunit from RNase P and the hammerhead ribozyme both rapidly bind and cleave their respective RNA substrates. For RNase P the rate constant for the cleavage step, k_c=360 min^{-1} (single turnover) is approximately 10-fold greater than the substrate dissociation constant, K_{off}^S=39 min^{-1} (4). Similarly, for the hammerhead ribozyme the chemical rate constant, k_c=1 min^{-1} (single turnover) can be compared with a substrate dissociation constant, K_{off}^S=1.5 x 10^{-10} min^{-1} (66). Since these rate constants suggest that any substrate RNA that binds will go to product, problems with substrate specificity similar to those associated with the *Tetrahymena* ribozyme can be expected. However, unlike the intron from *Tetrahymena* binding of both substrate and product to the hammerhead ribozyme may not involve additional tertiary interactions, and their association with the ribozyme is within the range of values determined for the formation of short helices ($k_{cat}/K_M = 2.0 \times 10^7\ M^{-1}\ min^{-1}$) (single turnover) (66) .

11.3 Therapeutic Applications: Trans-Cleaving and Trans-Splicing Ribozymes

Ribozymes have the potential to become effective therapeutic agents and currently they are being tested in a wide variety of biological applications. Most applications have employed trans-cleaving hammerhead and hairpin ribozymes as inhibitors of viral gene expression. In particular, these ribozymes have been targeted at HIV-1 RNAs in infected cells to inhibit different stages of the viral life cycle (discussed below). The effectiveness of ribozymes as inhibitors of cellular gene expression in both prokaryotes (28) and eukaryotes (15, 30, 31, 139) has been examined, and in a number of studies hammerhead ribozymes have been used to reveal particular gene functions (41, 89, 140). Ribozymes are also being used to target cellular oncogenes such as the *bcr/abl* transcript (85, 109, 143, 145, 173) and c-Ha-*ras* RNAs (71, 79). As mentioned earlier, new ribozyme applications are being developed for the group I trans-splicing ribozyme from *T. thermophilia* which can be used to repair defective cellular RNAs (150) and to generate potent antivirals from viral transcripts.

The following section will focus on the therapeutic applications of trans-cleaving and trans-splicing ribozymes. Because so much effort has been devoted to developing trans-cleaving ribozymes that target HIV-1 RNA, a summary of the work in that field will be presented. Similarly, inhibition of dominant oncogenes by ribozymes will be discussed. A section will also be devoted to general principles that should be considered when engineering efficient ribozymes. In many cases these principles are based on observations from the systematic, *in vitro* kinetic analysis of RNA catalysis that have been noted over the last 15 years. A discussion of colocalization of ribozymes and target substrates within cell compartments and its potential impact on ribozyme function will also be discussed in this section. Finally, potential therapeutic applications for trans-splicing ribozymes will be presented.

11.3.1 Trans-Cleaving Ribozymes

11.3.1.1 Ribozyme-Mediated Cleavage of HIV-1 RNA

Many approaches using RNA to render cells resistant to HIV-1 infection have been employed including expression of antisense RNA (25, 142) and RNA decoys (7, 96, 97, 152). Although these approaches have been effective in suppressing HIV-1 replication in infected cells, ribozymes have two theoretical advantages as compared with other viral inhibition strategies: i) cleavage results in the direct, irreversible inactivation of the target; and ii) fewer ribozymes are expected to be required to inhibit a given target gene since a single ribozyme can catalyze multiple cleavage reactions. However, like many other antivirals, ribozymes may also be sensitive to HIV-1 sequence heterogeneity and the most effective inhibition may be attained by a combination of these strategies.

There are many stages in the viral life cycle where viral RNA intermediates may be susceptible to cleavage by ribozymes. These intermediates include incoming genomic RNAs, early mRNAs, and late transcripts (for a detailed discussion see 16). Although cleavage of incoming RNAs would prevent viral integration and therefore be highly effective in protecting cells from the virus, encapsidation of genomic RNAs within a viral core may make these transcripts inaccessible to ribozymes. It is also possible that the viral polymerase will initiate reverse transcription before the ribozyme can base pair with and cleave the target sequence. In most HIV-1 inhibition studies cleavage of incoming genomic RNAs has been assessed only semiquantitatively. Differences in the amount of proviral DNA (177) or gag mRNA (138) in ribozyme-expressing cells and controls cells has been only analyzed by PCR amplification reactions which are not internally controlled. Unfortunately, no system has been developed to date that allows for the direct detection of cleavage products in mammalian cells. Such a system would be useful in discriminating between ribozyme-mediated cleavage of incoming genomic RNA and other viral transcripts.

Early viral transcripts may prove to be the most attractive ribozyme targets for conferring resistance to HIV-1. Viral RNAs expressed at this stage of the HIV-1 life cycle, such as those encoding Tat, Rev, and Nef proteins, are less abundant and fewer catalytic RNAs may be required to be protective. Cleavage of early transcripts that inhibit the expression of regulatory proteins such as Rev should also result in the inhibition of late gene expression.

Ribozyme-mediated cleavage of late viral transcripts may not be effective in inhibiting HIV-1 replication. Although these RNAs may be accessible for ribozyme cleavage, their shear abundance would probably require that extremely high levels of ribozyme be expressed to reduce their levels in infected cells. In addition, the detrimental affects mediated by early viral regulatory proteins would not be inhibited even if late viral gene function was eliminated. An alternative strategy may be to target highly conserved sequences in the long terminal repeats (LTRs) present in all viral RNAs. This approach could result in the inhibition of both early and late transcripts.

As mentioned in the introduction to section 11.3, much work has been done to develop ribozymes that can be used as anti-HIV-1 therapeutic agents. Hammerhead and hairpin ribozymes appear to be well suited for this purpose because of their small size, simple secondary structure, and the ease with which they can be manipulated to target a specific substrate. Initially, it was shown that an anti-*gag* hammerhead ribozyme which specifically cleaved its target *in vitro* (131) could reduce HIV-1 replication in a human T cell line (138). Subsequently ribozymes have been designed which target a variety of highly conserved sequences throughout the HIV-1 genome and have been shown to inhibit HIV-1 in tissue culture (26, 40, 67, 98, 108, 154,

167, 176, 177, 182); (reviewed in 16, 129-131, 133, 181, 183) including diverse viral strains (180), as well as clinical isolates (177). Comparisons between catalytically active and inactive forms of these ribozymes have shown that it is the catalytic and not the antisense property of these RNAs that is responsible for the observed viral inhibition (67, 176, 188), although there have been reports that antisense RNAs targeted at both viral (98) and cellular (31, 139) transcripts are more effective inhibitors than ribozymes.

To more closely mimic *in vivo* infection, human peripheral blood lymphocytes have been stably transduced with a hairpin ribozyme targeting the U5 region of the HIV-1 genome. These cells were shown to resist challenge by both HIV-1 molecular clones and clinical isolates (94). More recently, macrophage-like cells which differentiated from hematopoietic stem/progenitor cells from fetal cord blood and which were stably transduced with a hairpin ribozyme targeted at the 5' leader sequence resisted infection by a macrophage-tropic virus (179). Transduction of pluripotent stem cells with HIV-1-resistance genes may represent an avenue to continually generate cells that are resistant to HIV-1 infection. Such stem cells differentiate into T-cells, monocytes and macrophages, the major targets of HIV-1 infection. Stable transduction of stem cells has been limited however because the commonly used retroviral vector gene delivery systems require that target cells be actively dividing for gene transfer to occur (see chapter on retroviral vectors).

The generation of sequence variants during HIV-1 replication (51, 55, 110, 137) has posed a major problem for immunization strategies and anti-HIV-1 drug therapies designed to suppress viral replication in infected patients. Frequent substitution of amino acids within the variable domains of the HIV-1 *env* gene has resulted in the emergence of neutralization escape mutants both in cell culture (102, 124, 126) and *in vivo* (106). Similarly, the rapid emergence of resistant viral strains has limited the effectiveness of both nucleoside (87, 88) and non-nucleoside analog reverse transcriptase inhibitors (100, 107, 125, 136). Selection for variants resistant to anti-HIV-1 ribozymes is also likely to occur since a single point mutation at the cleavage site could inhibit ribozyme-mediated cleavage. Unlike small molecule drugs which require substitutions at the protein level, even a silent point mutation can generate a ribozyme-resistant strain.

The affect of single point mutations at the cleavage site of hairpin (5'N^GHY3', where H = U, C, or A; and Y = C or U) and hammerhead (5'NUX^3', where X=C, U, or A) ribozymes has not been studied, but the more stringent sequence requirement at the cleavage site of the hairpin ribozyme could enhance its sensitivity to mutations. However, escape mutants were not detected in long-term cultures of cells expressing a hairpin ribozyme targeted at the 5' transcribed leader sequence of HIV-1 (177). One approach that has been suggested to overcome sequence heterogeneity among HIV-1 isolates is to develop multitargeted-ribozymes that cleave a given RNA target at multiple sites. These ribozymes appear to be more efficient in cleaving target substrates then a single ribozyme *in vitro* (26, 154, 169). However, *in vivo* no difference was observed in the extent of suppression of Moloney murine leukemia virus

(MoMLV) replication by single- and multiple-target-ribozymes (154). One of the most effective strategies for enhancing protection against escape mutants would be to target either single or multitarget-ribozymes at highly conserved sequences within the HIV-1 genome which do not tolerate mutations.

In summary, several studies have suggested that ribozymes can inhibit HIV-1 replication in cell culture experiments when cells are challenged with very low inoculums of HIV-1. It remains to be tested if this first generation of ribozymes can also inhibit virus replication in HIV-1-infected patients under conditions of active viremia and where a multitude of quasispecies of the virus pre-exist. Ultimately, this question will be answered with ribozymes entering into their first clinical trials in HIV-1-infected patients in the near future (172).

11.3.1.2 Ribozymes and Cancer

Genetic alteration of cellular oncogenes and the subsequent expression of defective gene products often form the basis of many types of cancers. Because ribozymes can be designed to inhibit the expression of specific genes their potential as antineoplastic agents is currently being tested. For example, hammerhead ribozymes have been reported to be able to suppress the tumorigenicity of cells harboring an activated human *ras* gene (9, 71, 80, 156) and significantly reduce tumor development and invasion of such cells when injected into nude mice (72).

More recently, the *bcr/abl* fusion transcript has been the target of many ribozyme studies (85, 86, 109, 143, 145, 173). This abnormal mRNA is transcribed from the Philadelphia chromosome which is present in 95% of patients with chronic myelogenous leukemia (CML) and in many patients with acute lymphocytic leukemia (ALL). The *bcr/abl* fusion transcript makes an attractive ribozyme target because the $p210^{bcr/abl}$ kinase it encodes is directly involved in the development of CML and ALL (36, 43, 60, 74). *In vitro* experiments have shown that the 8.5 Kb native *bcr/abl* transcript was efficiently cleaved by an anti-*bcr/abl* ribozyme (173). In CML blast crisis cell lines, expression of ribozymes targeted at *bcr/abl* mRNA was reported to be able to reduce the production of $p210^{bcr/abl}$ (86, 143, 145), and *bcr/abl* transcripts (85). In other experiments anti-*bcr/abl* ribozyme expression occurred concomitantly with a reduction in cell proliferation (85, 145).

When designing ribozymes to cleave mutated mRNAs or fusion transcripts such as those from the *bcr/abl* gene, one major concern is that the ribozyme may also cleave the homologous wild type cellular transcripts that are normally required for cell growth and differentiation. For example, cleavage of the wild type bcr mRNA by ribozymes targeted at the *bcr/abl* fusion transcript has been reported (173). Similarly, a ribozyme targeted at a mutated DHFR mRNA which is involved in the development of tumor cell resistance to methotrexate also resulted in cleavage of the wild type RNA substrate (77). Thus, ribozymes must be designed with caution when mutant cellular RNAs are the substrate targets of choice.

Once a ribozyme is in hand that can cut and inactivate an mRNA encoding a dominant mutant oncogene, the next task will be to develop an experimental system that can transfer the ribozyme containing gene to virtually all of the transformed cells in a cancer patient and once there express the ribozymes at high enough levels to cleave the vast majority of the oncogene's transcripts. This may be very difficult to achieve in practice.

11.3.1.3 Engineering Ribozymes

Unlike the *in vitro* setting, where a kinetic framework has been established from which one can understand and logically improve the reactions mediated by trans-cleaving ribozymes (discussed in section 11.2), the factors that influence ribozyme activity in the cellular millieu remain largely unknown. In order to optimize RNA catalysis *in vivo*, the individual steps of the ribozyme reactions will need to be evaluated and the parameters that limit catalysis in the cellular setting determined. Discussed below are some general principles for engineering ribozymes that have been reported in the literature and which may be important for engineering better ribozymes for gene therapy applications.

a. Substrate Specificity
The degree of target sequence specificity that can be achieved by hammerhead and hairpin ribozymes is another feature which makes catalytic RNAs attractive choices for gene inhibition studies. As described in section 11.2, the helices or stems flanking the catalytic core can be designed to form complementary base pairs with virtually any target sequence. However, the correct length and base composition of the flanking arms are critical for efficient ribozyme function. Flanking arms must be long enough to dictate that a target sequence is rare and not likely to be present in non-targeted cellular RNAs. Yet base pairing that is to extensive may result in a loss of substrate specificity (62) and a reduction in ribozyme turnover (44, 59, 109). Conversely, insufficient base pairing may result in an unstable ribozyme-substrate duplex and a reduction in cleavage (56, 78). It has been suggested that in humans a 16 nucleotide long sequence should be unique. Thus a hammerhead ribozyme with eight nucleotides in each flanking arm should recognize a unique target (130). (It is unclear how codon usage and the redundancy of protein functional domains influences this calculation, however.) In contrast, the *Tetrahymena* group I intron is expected to react with a number of transcripts in a cell because it contains only a six nucleotide internal guide sequence. Nevertheless, ribozyme-induced cellular toxicity was not observed in bacterial cells expressing the *Tetrahymena* ribozyme (150).

It should be noted that the optimal length of flanking complementary regions may differ significantly for *in vitro* and *in vivo* applications. An anti-*tat* hammerhead ribozyme containing 33 nucleotides in each flanking arm was reported to be able to efficiently inhibit HIV-1 replication in cells even though the ribozyme was not very catalytically active *in vitro* (32). The authors suggest the apparent discrepancy between intracellular and cell-free activity may be due to the radically different environment within the cell.

b. Ribozyme Expression

In several of the reports sited in this review as much as a 1000-fold molar excess of ribozyme over substrate RNA has been required to demonstrate cleavage activity both *in vitro* and *in vivo* (tissue culture) (15, 31, 83, 173). Furthermore, the extent of inhibition of HIV-1 replication in cells conferred by ribozymes (182) , RNA decoys (96, 152) , and antisense RNA (153) appears to be directly correlated with levels of expression. Similarly, elimination of p210 protein-kinase activity in a cell line expressing the *bcr/abl* fusion transcript was dependent on the level of anti-*bcr/abl* ribozyme expression (143). Based on these results it appears that one of the most critical parameters affecting the successful application of ribozymes in cellular and viral gene inhibition studies is the choice of transcription systems. Promoters that mediate stable, long term, and high levels of RNA expression will be required both to evaluate the inhibitory activity of various catalytic RNA motifs, and to sustain gene inhibition in therapeutic settings.

Depending on the particular application, use of a tissue-specific promoter, an inducible promoter, or a constitutive promoter may be most appropriate for ribozyme expression. For example, to study the role of glucokinase in the induction of the human disease, maturity-onset diabetes of the young (MODY) in an animal model, an anti-glucokinase ribozyme was expressed from the rat insulin promoter to limit the ribozyme activity to the pancreatic cells of transgenic mice (41). Inducible promoters have also been effectively used to express ribozymes targeted at various transcripts. For example, resistance to HIV-1 infection was conferred on MT-4 cells by a ribozyme that was expressed from a tat inducible promoter (a fusion thymidine kinase-trans activation -responsive (*tk*-TAR) promoter) (167). In the BL21 (DE3) strain of *E. Coli* which contains a chromosomal copy of the T7 RNA polymerase gene driven by the *lacUV5* promoter, expression of an anti-HIV-1 integrase hammerhead ribozyme could be induced by adding IPTG (isopropylthio-β-galactoside) to the growth medium. Expression of the ribozyme was shown to completely block synthesis of integrase protein (144) .

Constitutive promoters have been used most frequently to express ribozymes in mammalian cells. In recent years, several groups have reported that the highest levels of ribozyme expression (and consequently, gene inhibition) can be achieved using RNA polymerase III-based transcription units (30, 129, 143, 159, 179, 180). In addition, successful ribozyme applications with RNA polymerase II-based transcription units have also been reported (15, 89, 146).

RNA polymerase III transcription units most often consist of ribozyme sequences fused to either a tRNA or the Adenovirus VA-1 gene promoter. Because these expression cassettes are small and self-contained, they can be inserted at various positions in mammalian expression vectors used for gene transfer. RNA polymerase III promoters also offer other advantages for the expression of ribozymes. Unlike RNA polymerase II promoters, the transcripts generated by RNA polymerase III are not capped or polyadenylated allowing for the production of small ribozyme transcripts with a minimum of cis-appended sequences. This facet of RNA polymerase III expression may be advantageous because extra sequences that flank the catalytic core of a ribozyme may reduce the catalytic activity of the enzyme (163, 183). The extra sequences may cause the ribozyme to fold into an inactive structure. RNA polymerase III promoters are also active in all cell types and their normal tRNA products are synthesized in large amounts (29) which are additional features that make this an attractive expression system.

RNA polymerase III transcription units were first used in a gene therapy strategy to inhibit MoMLV replication in NIH3T3 cells (153). In this case, DNA constructs containing chimeric tRNA-driven antisense sequences were introduced into NIH3T3 cells via retroviral vectors (153). Efficient expression of the chimeric tRNA-antisense transcripts and inhibition of MoMLV replication was shown to be dependent on a novel retroviral vector (called a double copy vector) in which the chimeric tRNA gene was inserted in the vector's 3' (LTR). Similar RNA polymerase III expression systems were used to overexpress TAR-containing sequences (TAR decoys) (152), and sequences corresponding to the major Rev-binding site in the Rev response element (RRE decoys) to inhibit HIV-1 replication in a human T cell line (95, 96). This transcription system has also used to express a hammerhead ribozyme to inhibit the production of the *bcr/abl* gene product, p210, when expressed constitutively in the CML blast cell line K562 (143). Finally, hairpin ribozymes which mediate cleavage of highly conserved sequences in the HIV-1 genome have been expressed from tRNA promoters that were cloned into murine retroviral vectors. These ribozymes were shown to be able to inhibit HIV-1 replication in transiently transfected HeLa cells (180), in primary T lymphocytes (94) and in the progeny of CD^{34+} human fetal cord blood cells stably transduced with the anti-HIV-1 hairpin ribozyme gene (179).

c. Colocalization of Ribozymes and Target RNAs

One of the factors known to influence catalysis mediated by the *Tetrahymena* ribozyme is the rate at which a ribozyme encounters and binds to its substrate (discussed in section 11.2.6). It is likely that *in vivo* the frequency with which a ribozyme encounters its substrate will also be rate limiting. In fact, it is probable that substrate binding will be far more limiting *in vivo* than has been observed *in vitro*. Why? In solution in a test tube both ribozyme and substrate RNAs freely diffuse. In contrast, RNA molecules in eukaryotic cells do not appear to be randomly distributed but are localized to very specific regions both within the cell nucleus and cytoplasm (17, 75, 91-93, 123, 175). Theoretically, trafficking of ribozymes and substrates into different compartments could severely limit the efficiency of ribozyme-mediated catalysis.

To address this possibility an experimental system was devised based on two MoMLV retroviral vectors (151). The hypothesis that was tested was whether delivering a ribozyme and its substrate to the same compartment, or "colocalizing" the ribozyme and substrate, would improve the efficiency of ribozyme-mediated trans-cleavage of an RNA substrate. One MoMLV vector was used to introduce a hammerhead ribozyme designed to cleave *lacZ* RNAs into an amphotropic packaging cell line. Viral supernatants containing these vectors were then used to infect an ecotropic packaging cell line containing a retroviral vector expressing the *lacZ* target sequence. Once coexpressed inside these cells the *lacZ* target and ribozyme transcripts functioned either as mRNAs which were sorted to the cytoplasm for translation, or as genomic RNAs which were sorted to the cell surface and encapsidated into progeny virions. Ribozyme and substrate containing genomic RNAs were expected to be colocalized because they each contained the viral packaging sequence, Ψ. By contrast, ribozyme and substrate containing mRNAs were not expected to be colocalized in the cell. Therefore, it was hypothesized that if colocalization of the ribozyme and substrate was important for substrate cleavage inside of a cell then a greater reduction in the level of packaged *lacZ* RNAs would be observed as compared to any reduction in the translation of *lacZ* mRNAs. As predicted, no substantial difference in β-gal activity inside cells containing a functional hammerhead ribozyme was observed. However, the titers of β-gal virus released from the cells was reduced by 90% as compared with cells containing control vectors. These results suggest that the rate at which a ribozyme encounters and binds its substrate *in vivo* may limit its ability to mediate trans-cleavage of target RNAs and inhibit gene expression.

There are colocalization signals that could be appended to anti-HIV ribozymes that may enhance their antiviral activity. The most obvious one is the packaging signal present on HIV-1 genomic RNAs. While this approach may significantly reduce virus spread by targeting ribozymes to sites of viral assembly, the abundance of HIV-1 RNAs at such a late stage in the viral life cycle would probably require very high levels of ribozyme expression to be able to cleave all of the viral transcripts. A more suitable HIV-1 sorting signal may be the rev response element (RRE) because it should colocalize ribozymes with HIV-1 target transcripts earlier in the HIV-1 life cycle. When tethered to a ribozyme, RRE would be expected to serve a dual function as an RNA decoy and as a localization signal. As an RNA decoy, RRE has been used to short circuit the temporal switch from early to late viral gene expression in HIV-1 infected cells (95, 96). This switch is mediated by the Rev protein which binds to RRE and initiates the export of unspliced and singly spliced HIV-1 transcripts from the nucleus to the cytoplasm (33). In HIV-1 infected cells that contain ribozyme-RRE transcripts, transport of the late viral RNAs and the subsequent production of HIV-1 structural proteins would be reduced by the competitive binding of Rev to the RRE decoy. If the RRE decoy became subsaturating, then the ribozyme would be expected to be colocalized with late viral transcripts in the Rev-dependent sorting pathway since both ribozyme and HIV-1 transcripts contain the RRE localization signal. This would effectively position the ribozyme to mediate cleavage of its viral target RNA which should result in further inhibition of late viral gene expression. In addition, colocalizing ribozymes with their specific substrate RNAs may reduce cell toxicity by limiting the number of transcripts the ribozyme encounters in the cell. Concentrating a ribozyme within a specific intracellular compartment with its substrate would also be expected to enhance the ribozyme's effectiveness and reduce the overall level of ribozyme needed to inhibit the expression of a targeted RNA in a cell.

11.3.2 The Trans-splicing Ribozyme from *Tetrahymena thermophilia*

To this point, the discussion in this section has focused on the therapeutic potential of trans-cleaving ribozymes to eliminate pathogenic RNAs. Recently, it has been argued that ribozymes may also be able to alter the sequence of targeted RNAs not just destroy them (149). As mentioned earlier, the *Tetrahymena* ribozyme can catalyze a cleavage-ligation reaction in trans that is similar to its natural self-splicing reaction (section 11.2.1) (5, 68, 69). Such targeted trans-splicing reactions can potentially be used to repair mutant transcripts, and to alter viral RNAs to give them antiviral activity.

As described in the previous section, in the targeted trans-splicing reaction the *Tetrahymena* ribozyme recognizes and binds to its substrate RNA (the 5'exon) by base pairing between the IGS and sequences within the substrate (Fig. 11.2C). Following cleavage, the ribozyme splices its 3' exon onto the cleaved substrate RNA. Since the ribozyme cleaves after the sequence N_5U the only sequence requirement for the substrate is to have a uridine residue preceding the cleavage site. Thus any uridine nucleotide in any RNA molecule can serve as a target if the target site sequence is not obscured by secondary structure. The composition and the length of the RNA following the N_5U is not critical for group I ribozyme activity (184, 185). Similarly only the five nucleotides in the IGS of the ribozyme-3' exon need to be changed such that they are complementary to the five nucleotides preceding the splice site on the target RNA. Because there are no sequence requirements for the 3' exon (19) , almost any sequence can be spliced onto the 5' target transcript.

The lack of sequence requirements for the 3' exon suggests that it could be manipulated to replace a defective portion of an RNA transcript with a functional sequence (Fig. 11.4). Trans-splicing ribozymes could be designed that would cleave defective transcripts upstream of point mutations or small insertions or deletions. A 3' exon consisting of the wild type sequence could then be spliced onto the cleaved target resulting in a corrected mRNA. The first example of using a ribozyme to perform RNA repair was recently reported. The *Tetrahymena* trans-splicing ribozyme was used to correct truncated lacZ transcripts *in vitro* and in *E. Coli* (150). In order to restore the proper open reading frame for the expression of β-galactosidase in *E. Coli*, the ribozyme had to recognize and base pair with a truncated 5'*lacZ* RNA, cleave off two additional adenosine residues, and ligate a *lacZ* 3'exon sequence onto the cleaved 5' substrate RNA in the correct translational reading frame. The *Tetrahymena* ribozyme was shown to efficiently mediate trans-splicing *in vitro* and to generate translatable *lacZ* messages in *E. Coli* (150).

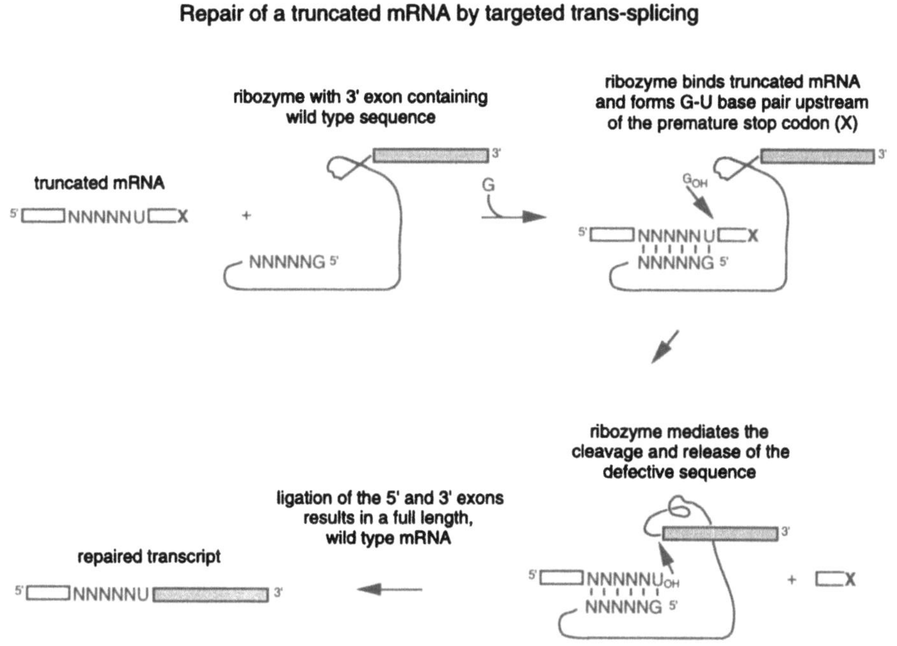

Fig. 11.4: RNA repair mediated by the *Tetrahymena* trans-splicing ribozyme.
In this illustration the *Tetrahymena* ribozyme binds to a truncated cell transcript via the IGS. Binding occurs upstream of the premature stop codon (X). Following cleavage sequences containing the termination codon are released and addition of the 3' exon to the 5' cleavage product results in a repaired transcript containing the wild type sequence.

It is not difficult to imagine how targeted trans-splicing could become useful in the treatment of inherited diseases that are caused by specific genetic mutations. For example, the genetic basis of many globin diseases is well understood and trans-splicing ribozymes could be used to correct globin transcripts that are either truncated or contain point mutations. Furthermore, because trans-splicing ribozymes act at the level of RNA as opposed to DNA, RNA repair should restore the proper expression pattern of the mutant gene. This aspect of RNA repair may make it more useful than conventional gene therapy approaches because traditional DNA add back approaches do not faithfully recapitulate the normal expression pattern of genes after gene transfer.

RNA repair may be especially useful in treating genetic diseases that are caused by the presence of deleterious proteins expressed from mutant forms of a gene. For example, a trans-splicing ribozyme that is able to repair sickle β-globin transcripts should simultaneously eliminate the production of sickle β-globin protein and restore regulated expression of the wild type β-globin gene.

Trans-splicing ribozymes may become very useful antiviral agents. Such ribozymes can be employed not to just cleave and inactivate HIV-1 transcripts but to alter their sequence to make them encode antiviral proteins. For example, the 3'exon sequence attached to an anti-HIV-1 ribozyme could be designed such that after splicing a revised viral RNA is generated that encodes a *trans*-dominant protein. It has been shown that certain mutant HIV-1 Rev and Gag proteins are not only defective, but act as *trans*-dominant inhibitors of the wild type counterparts (99, 160). In addition, the effectiveness of trans-splicing ribozymes would not solely depend on cleaving the vast majority of the viral transcripts in an infected cell because revised viral RNAs would be translated to produce anti-HIV-1 proteins which could enhance viral inhibition.

11.4 Conclusions

Trans-cleaving and trans-splicing ribozymes either alone or in combination with other therapeutic agents have the potential to restore genetic information or to eliminate it. The information presented in this chapter demonstrates how efficiently ribozymes can mediate repair of defective transcripts or cleave pathogenic RNAs *in vitro*. It is important to recognize that much of the success of the *in vitro* experiments stems not from a "try and see" approach, but from a detailed kinetic understanding of the individual steps in the ribozyme reactions. This is particularly true of the trans-cleavage reaction catalyzed by the *Tetrahymena* group I ribozyme. For example, by understanding that product release was rate limiting in a multiple turnover reaction, improved ribozymes could be logically created by weakening the binding of the ribozyme to its cleavage product. Such changes resulted in the generation of an endoribonuclease with increased turnover and greater substrate specificity (122, 178).

When transferred into a mammalian cell, a ribozyme will be required to operate in a much more complex environment. Although the results from the first generation of therapeutic ribozyme experiments are somewhat encouraging, the long term utility of ribozymes is still unclear. Many factors may limit the efficacy of therapeutic ribozymes. The sorting of RNAs into specific compartments within the cell nucleus and cytoplasm will surely influence the rate at which a ribozyme binds its substrate. Similarly, RNA binding proteins present within the cell may bind the ribozyme or substrate RNAs and influence ribozyme reaction rates in any number of ways. Thus in the long run, it will be essential to establish a kinetic framework to help us understand how the cellular environment influences ribozyme catalysis. Only with this knowledge, can we begin to logically improve ribozyme-based therapeutic agents for gene therapy applications.

Acknowledgements

Lynn Milich is supported by an Interdisciplinary Research Training Program in AIDS (2T32-AI-07392-06).

References

1. Altman, S. (1989). Ribonuclease P: an enzyme with a catalytic RNA subunit. Advances in Enzymology & Related Areas of Molecular Biology *62*, 1-36.
2. Barfod, E. T. and Cech, T. R. (1989). The conserved U.G pair in the 5' splice site duplex of a group I intron is required in the first but not the second step of self-splicing. Mol. Cell. Biol. *9*, 3657-3666.
3. Baserga, S. J. and Steitz, J. A. (1993). The diverse world of small ribonucleoproteins. In: The RNA World, R. F. Gesteland and J. F. Atkins (eds.). Cold Spring Harbor, New York: Cold Spring Harbor Laboratory Press, 359-381.
4. Beebe, J. A. and Fierke, C. A. (1994). A kinetic mechanism for cleavage of precursor tRNA catalyzed by the RNA component of *Bacillus subtilis* Ribonuclease P. Biochemistry *33*, 10294-10304.
5. Been, M. D. and Cech, T. R. (1986). One binding site determines sequence specificity of *Tetrahymena* pre-rRNA self-splicing, trans-splicing, and RNA enzyme activity. Cell *47*, 207-216.
6. Bertrand, E. L. and Rossi, J. J. (1994). Facilitation of hammerhead ribozyme catalysis by the nucleocapsid protein of HIV-1 and the heterogeneous nuclear ribonucleoprotein A1. EMBO J. *13*, 2904-2912.
7. Bevec, D., Volc-Platzer, B., Zimmermann, K., Dobrovnik, M., Hauber, J., Veres, G., and Bohnlein, E. (1994). Constitutive expression of chimeric neo-rev response element transcripts suppresses HIV-1 replication in human CD4+ T-lymphocytes. Hum. Gene Therapy *5*, 193-201.
8. Bevilacqua, P. C., Kierzek, R., Johnson, K. A., and Turner, D. H. (1992). Dynamics of ribozyme binding of substrate revealed by fluorescence-detected stopped-flow methods. Science *258*, 1355-1358.
9. Bos, J. L. (1989). *ras* oncogenes in human cancer: a review. Cancer Res. *49*, 4682-4689.
10. Branch, A. D. and Robertson, H. D. (1991). Efficient trans cleavage and a common structural motif for the ribozymes of the human hepatitis δ agent. Proc. Natl. Acad. Sci. USA *88*, 10163-10167.
11. Bratty, J., Chartrand, P., Ferbeyre, G., and Cedergren, R. (1993). The hammerhead RNA domain, a model ribozyme. Biochim. Biophys. Acta *1216*, 345-359.
12. Burke, J. M., Irvine, K. D., Kaneko, K. J., Kerker, B. J., and Oettgen, A. B. (1986). Role of conserved sequence elements 9L and 2 in self-splicing of the *Tetrahymena* ribosomal RNA precursor. Cell *45*, 167-176.
13. Buzayan, J. M., Gerlach, W. L., and Bruening, G. (1986). Non- enzymatic cleavage and ligation of RNAs complementary to a plant virus satellite RNA. Nature *323*, 349-353.
14. Buzayan, J. M., Hampel, A., and Bruening, G. (1986). Nucleotide sequence and newly formed phosphodiester bond of spontaneously ligated satellite tobacco ringspot virus RNA. Nucleic Acids Res. *14*, 9729-9743.
15. Cameron, F. H. and Jennings, P. A. (1989). Specific gene suppression by engineered ribozymes in monkey cells. Proc. Natl. Acad. Sci. USA *86*, 9139-9143.
16. Campbell, T. B. and Sullenger, B. A. (1995). Alternative approaches for the application of ribozymes as gene therapies for retroviral infections. In: Advances in Pharmacology, Vol. 33, J. T. August , M. W. Anders , F. Murad and J. T. Coyle (eds.). San Diego, California: Academic Press, Inc., 143-178.
17. Carter, K. C., Bowman, D., Carrington, W., Fogarty, K., McNeil, J. A., Fay, F. S., and Lawrence, J. B. (1993). A three dimensional view of precursor messenger RNA metabolism within the mammalian nucleus. Science *259*, 1330-1335.
18. Cech, T. R. (1992). Ribozyme engineering. Curr. Opin. Struct. Biol. *2*, 605-609.
19. Cech, T. R. (1990). Self-splicing of group I introns. Annu. Rev. Biochem. *59*, 543-568.
20. Cech, T. R. (1993). Structure and mechanism of the large catalytic RNAs: group I and group II introns and ribonuclease P. In: The RNA World, R. F. Gesteland and J. F. Atkins (eds.). Cold Spring Harbor, New York: Cold Spring Harbor Laboratory Press, 239-269.
21. Cech, T. R., Damberger, S. H., and Gutell, R. R. (1994). Representation of the secondary and tertiary structure of group I introns. Nature Struct. Biol. *1*, 273-280.
22. Cech, T. R., Herschlag, D., Piccirilli, J. A., and Pyle, A. M. (1992). RNA catalysis by a group I ribozyme: developing a model for transition state stabilization. J. Biol. Chem. *267*, 17479-17482.
23. Cech, T. R., Zaug, A. J., and Grabowski, P. J. (1981). *In vitro* splicing of the ribosomal RNA precursor of *Tetrahymena*: involvement of a guanosine necleotide in the excision of the intervening sequence. Cell *27*, 487-496.

24. Celander, D. W. and Cech, T. R. (1991). Visualizing the higher order folding of a catalytic RNA molecule. Science *251*, 401-407.
25. Chatterjee, S., Johnson, P. R., and Wong Jr., K. K. (1992). Dual-target inhibition of HIV-1 *in vitro* by means of an adeno-associated virus antisense vector. Science *258*, 1485-1488.
26. Chen, C.-J., Banerjea, A. C., Harmison, G. G., Haglund, K., and Schubert, M. (1992). Multitarget-ribozyme directed to cleave at up to nine highly conserved HIV-1 env RNA regions inhibits HIV-1 replication-potential effectiveness against most presently sequenced HIV-1 isolates. Nucleic Acids Res. *20*, 4581-4589.
27. Chowrira, B. M., Berzal-Herranz, A., and Burke, J. M. (1991). Novel guanosine requirement for catalysis by the hairpin ribozyme. Nature *354*, 320-322.
28. Churat, J. C. and Galibert, F. (1989). Can ribozymes be used to regulate procaryote gene expression? Biochem. and Biophys. Res. Commun. *162*, 1025-1029.
29. Ciliberto, G., Castagnoli, L., and Cortese, R. (1983). Transcription by RNA ploymerase III (review). Curr. Topics in Dev. Biol. *18*, 59-88.
30. Cotten, M. and Birnstiel, M. L. (1989). Ribozyme-mediated destruction of RNA *in vivo*. EMBO J. *8*, 3861-3866.
31. Cotten, M., Schaffner, G., and Birnstiel, M. L. (1989). Ribozyme, antisense RNA, and antisense DNA inhibition of U7 small nuclear ribonucleoprotein-mediated histone pre-mRNA processing *in vitro*. Mol. Cell Biol. *9*, 4479-4487.
32. Crisell, P., Thompson, S., and James, W. (1993). Inhibition of HIV-1 replication by ribozymes that show poor activity *in vitro*. Nucleic Acids Res. *21*, 5251-5255.
33. Cullen, B. and Greene, W. (1989). Regulatory pathways governing HIV-1 replication. Cell *58*, 423-426.
34. Dahm, S. C. and Uhlenbeck, O. C. (1990). Characterization of deoxy-and ribo-containing oligonucleotide substrates in the hammerhead self-cleavage reaction. Biochimie *72*, 819-823.
35. Dahm, S. C. and Uhlenbeck, O. C. (1991). Role of divalent metal ions in the hammerhead RNA cleavage reaction. Biochemistry *30*, 9464-9469.
36. Daley, G. Q., Van Etten, R. A., and Baltimore, D. (1990). Induction of chronic myelogenous leukemia in mice by the $P210^{bcr/abl}$ gene of the Philadelphia chromosome. Science *247*, 824-830.
37. Davies, R. W., Waring, R. B., Ray, J. A., Brown, T. A., and Scazzocchio, C. (1982). Making ends meet: a model for RNA splicing in fungal mitochondria. Nature *300*, 719-724.
38. Doudna, J. A., Cormack, B. P., and Szostak, J. W. (1989). RNA structure, not sequence, determines the 5' splice-site specificity of a group I intron. Proc. Natl. Acad. Sci. USA *86*, 7402-7406.
39. Downs, W. D. and Cech, T. R. (1994). A tertiary interaction in the *Tetrahymena* intron contributes to selection of the 5' splice site. Genes Dev. *8*, 1198-1211.
40. Dropulic, B., Lin, N. H., Martin, M. A., and Jeang, K.-T. (1992). Functional characterization of a U5 ribozyme: intracellular suppression of human immunodeficiency virus type 1 expression. J Virol. *66*, 1432-1441.
41. Efrat, S., Leiser, M., Wu, Y.-J., Fusco-DeMane, D., Emran, O. A., Surana, M., Jetton, T. L., Magnuson, M. A., Weir, G., and Fleischer, N. (1994). Ribozyme-mediated attenuation of pancreatic β-cell glucokinase expression in transgenic mice results in impaired glucose-induced insulin secretion. Proc. Natl. Acad. Sci. USA *91*, 2051-2055.
42. Ehrenman, K., Schroeder, R., Chandry, P. S., Hall, D. H., and Belfort, M. (1989). Sequence specificity of the P6 pairing for splicing of the group I td intron of phage T4. Nucleic Acids Res. *17*, 9147-9163.
43. Elfantry, A. G., Hariharan, I. K., and Cory, S. (1990). bcr-abl, the hallmark of chronic myeloid leukaemia in man, induces multiple haemopoietic neoplasms in mice. EMBO J. *9*, 1069-1078.
44. Fedor, M. J. and Uhlenbeck, O. C. (1990). Substrate sequence effects on hammerhead RNA catalytic efficiency. Proc. Natl. Acad. Sci. USA *87*, 1668-1672.
45. Feldstein, P. A., Buzayan, J. M., and Bruening, G. (1989). Two sequences participating in the autolytic processing of satellite tobacco ringspot virus complementary RNA. Gene *82*, 53-61.
46. Flor, P. J., Flanegan, J. B., and Cech, T. R. (1989). A conserved base pair within helix P4 of the *Tetrahymena* ribozyme helps to form the tertiary structure required for self-splicing. EMBO J. *8*, 3391-3399.
47. Forster, A. C. and Altman, S. (1990). External guide sequences for an RNA enzyme. Science *249*, 783-786.

48. Forster, A. C. and Symons, R. H. (1987). Self-cleavage of virusoid RNA is performed by the proposed 55-nucleotide active site. Cell *50*, 9-16.
49. Gerlach, W. L., Buzayan, J. M., Schneider, I. R., and Bruening, G. (1986). Satellite tobacco ringspot virus RNA: biological activity of DNA clones and their *in vitro* transcripts. Virology *151*, 172-185.
50. Goodchild, J. and Kohli, V. (1991). Ribozymes that cleave an RNA sequence from human immunodeficiency virus: the effect of flanking sequence on rate. Arch. Biochem. Biophy. *284*, 386-391.
51. Goodenow, M., Huet, T., Saurin, W., Kwok, S., Sninsky, J., and Wain-Hobson, S. (1989). HIV-1 isolates are rapidly evolving quasispecies: evidence for viral mixtures and preferred nucleotide substitutions. J. Acquir. Immune Defic. Syndr 2, 344-352.
52. Griffin Jr., E. A., Qin, Z., Michels Jr, W. J., and Pyle, A. M. (1995). Group II intron ribozymes that cleave DNA and RNA linkages with similar efficiency, and lack contacts with substrate 2'-hydroxyl groups. Chemistry & Biology 2, 761-770.
53. Guerrier-Takada, C. and Altman, S. (1984). Catalytic activity of an RNA molecule prepared by transcription *in vitro*. Science *223*, 285-286.
54. Guerrier-Takada, C., Gardiner, K., Marsh, T., Pace, N., and Altman, S. (1983). The RNA moiety of ribonuclease P is the catalytic subunit of the enzyme. Cell *35*, 849-857.
55. Hahn, B. H., Shaw, G. M., Taylor, M. E., Redfield, R. R., Markham, P. D., Salahuddin, S. Z., Wong-Staal, F., Gallo, R. C., Parks, E. S., and Parks, W. P. (1986). Genetic variation in HTLV-III/LAV over time in patients with AIDS or at risk for AIDS. Science *232*, 1548-1553.
56. Hampel, A. and Tritz, R. (1989). RNA catalytic properties of the minimum (-)sTRSV sequence. Biochemistry *28*, 4929-4933.
57. Hampel, A., Tritz, R., Hicks, M., and Cruz, P. (1990). 'Hairpin' catalytic RNA model: evidence for helices and sequence requirement for substrate RNA. Nucleic Acids Res. *18*, 299-304.
58. Haseloff, J. and Gerlach, W. L. (1989). Sequences required for self-catalysed cleavage of the satellite RNA of tobacco ringspot virus. Gene *82*, 43-52.
59. Haseloff, J. and Gerlach, W. L. (1988). Simple RNA enzymes with new and highly specific endoribonuclease activities. Nature *334*, 585-591.
60. Heisterkamp, N., Jenster, G., Ten Hoeve, J., Zovich, D., Pattengale, P. K., and Groffen, J. (1990). Acute leukaemia in *bcr/abl* transgenic mice. Nature *344*, 251-253.
61. Hendry, P., McCall, M. J., Santiago, F. S., and Jennings, P. A. (1992). A ribozyme with DNA in the hybridizing arms displays enhanced cleavage ability. Nucleic Acids Res. *20*, 5737-5741.
62. Herschlag, D. (1991). Implications of ribozyme kinetics for targeting the cleavage of specific RNA molecules *in vivo*: more isn't always better. Proc. Natl. Acad. Sci. USA *88*, 6921-6925.
63. Herschlag, D. and Cech, T. R. (1990). Catalysis of RNA cleavage by the *Tetrahymena* thermophila ribozyme. 1. kinetic description of the reaction of an RNA substrate complementary to the active site. Biochemistry *29*, 10159-10171.
64. Herschlag, D. and Cech, T. R. (1990). Catalysis of RNA cleavage by the *Tetrahymena* thermophila ribozyme. 2. kinetic description of the reaction of an RNA substrate that forms a mismatch at the active site. Biochemistry *29*, 10172-10180.
65. Herschlag, D., Khosla, M., Tsuchihashi, Z., and Karpel, R. L. (1994). An RNA chaperone activity of non-specific RNA binding proteins in hammerhead ribozyme catalysis. EMBO J. *13*, 2913-2924.
66. Hertel, K. J., Herschlag, D., and Uhlenbeck, O. C. (1994). A kinetic and thermodynamic framework for the hammerhead ribozyme reaction. Biochemistry *33*, 3374-3385.
67. Homman, M., Tzortzakaki, S., Rittner, K., Sczakiel, G., and Tabler, M. (1993). Incorporation of the catalytic domain of a hammerhead ribozyme into antisense RNA enhances its inhibitory effect on the replication of human immunodeficiency virus type 1. Nucleic Acids Res. *21*, 2809-2814.
68. Inoue, T., Sullivan, F. X., and Cech, T. R. (1985). Intermolecular exon ligation of the rRNA precursor of *Tetrahymena*: oligonucleotides can function as 5' exons. Cell *43*, 431-437.
69. Inoue, T., Sullivan, F. X., and Cech, T. R. (1986). New Reactions of the ribosomal RNA precursor of *Tetrahymena* and the mechanism of self-splicing. J. Mol. Biol. *189*, 143-165.
70. Joseph, S., Berzal-Herranz, A., Chowrira, B. M., and Butcher, S. E. (1993). Substrate selection rules for the hairpin ribozyme determined by *in vitro* selection, maturation, and analysis of mismatched substrates. Genes Dev. *7*, 130-138.
71. Kashani-Sabet, M., Funato, T., Florenes, V. A., Fodstad, O., and Scanlon, K. J. (1994). Suppression of the neoplastic phenotype *in vivo* by an anti-*ras* ribozyme. Cancer Res. *54*, 900-902.

72. Kashani-Sabet, M., Funato, T., Tone, T., Jiao, L., Wang, W., Yoshida, E., Kashfinn, B. I., Shitara, T., Wu, A. M., Moreno, J. G., Traweek, S. T., Ahlering, T. E., and Scanlon, K. J. (1992). Reversal of the malignant phenotype by an anti-*ras* ribozyme. Antisense Research and Development 2, 3-15.
73. Keese, P. and Symons, R. H. (1987). The structure of viroids and virusoids. In : Viroids and Viroid-like Pathogens, J. S. Semancik (ed.). Boca Raton, Florida: CRC Press, 1-47.
74. Kelliher, M. A., McLaughlin, J., Witte, O. N., and Rosenberg, N. (1990). Induction of a chronic myelogenous leukemia-like syndrome in mice with v-*abl* and *BCR/ABL*. Proc. Natl. Acad. Sci. USA *87*, 6649-6653.
75. Kislauskis, E. H., Li, Z., Singer, R. H., and Taneja, K. L. (1993). Isoform-specific 3'-untranslated sequences sort a-cardiac and b-cytoplasmic actin messenger RNAs to different cytoplasmic compartments. The J. Cell Biol. *123*, 165-172.
76. Knitt, D. S., Narlikar, G. J., and Herschlag, D. (1994). Dissection of the role of the conserved G-U pair in group I RNA self-splicing. Biochemistry *33*, 13864-13879.
77. Kobayashi, H., Kim, N., Halatsch, M. E., and Ohnuma, R. (1994). Specificity of ribozyme designed for mutated DHFR mRNA. Biochemical Pharmacology *47*, 1607-1613.
78. Koizumi, M., Iwai, S., and Ohtsuka, E. (1988). Construction of a series of several self-cleaving RNA duplexes using synthetic 21-mers. FEBS Lett. *228*, 228-230.
79. Koizumi, M., Kamiya, H., and Ohtsuka, E. (1993). Inhibition of c-Ha-*ras* gene expression by hammerhead ribozymes containing a stable C(UUCG)G hairpin loop. Biological and Pharmaceutical Bulletin *16*, 879-883.
80. Koizumi, M., Kamiya, H., and Ohtsuka, E. (1992). Ribozymes designed to inhibit transformation of NIH3T3 cells by the activated c-Ha-*ras* gene. Gene *117*, 179-184.
81. Kruger, K., Grabowski, P. J., Zaug, A. J., Sands, J., Gottschling, D. E., and Cech, T. R. (1982). Self-splicing RNA: autoexcision and autocyclization of the ribosomal RNA intervening sequence of *Tetrahymena*. Cell *31*, 147-157.
82. Kuo, M. Y., Sharmeen, L., Dinter-Gottlieb, G., and Taylor, J. (1988). Characterization of self-cleaving RNA sequences on the genome and antigenome of human hepatitis delta virus. J. Virol. *62*, 4439-4444.
83. L'Huillier, P. J., Davis, S. R., and Bellamy, A. R. (1992). Cytoplasmic delivery of ribozymes leads to efficient reduction in a-lactalbumin mRNA levels in C1271 mouse cells. EMBO J. *11*, 4411-4418.
84. Laggerbauer, B., Murphy, F. L., and Cech, T. R. (1994). Two major tertiary folding transitions of the *Tetrahymena* catalytic RNA. EMBO J. *13*, 2669-2676.
85. Lange, W., Cantin, E. M., Finke, J., and Dolken, G. (1993). *In vitro* and *in vivo* effects of synthetic ribozymes targeted against *BCR/ABL* mRNA. Leukemia 7, 1786-1794.
86. Lange, W., Daskalakis, M., Finke, J., and Dolken, G. (1994). Comparison of different ribozymes for efficient and specific cleavage of *BCR/ABL* related mRNAs. FEBS Lett. *338*, 175-178.
87. Larder, B. A., Darby, G., and Richman, D. D. (1989). HIV with reduced sensitivity to zidovudine isolated during prolonged therapy. Science *243*, 1731-1734.
88. Larder, B. A. and Kemp, S. D. (1989). Multiple mutations in HIV-1 reverse transcriptase confer high-level resistance to zidovudine (AZT). Science *246*, 1155-1158.
89. Larsson, S., Hotchkiss, G., Andang, M., Nyholm, T., Inzunza, J., Jansson, I., and Ahrlund-Richter, L. (1994). Reduced β2-microglobulin mRNA levels in transgenic mice expressing a designed hammerhead ribozyme. Nucleic Acids Res. *22*, 2242-2248.
90. Latham, J. A., and Clathech, T. R. (1989). Defining the inside and outside of a catalytic RNA molecule. Science *245*, 276-282.
91. Lawrence, J. B., Marselle, L. M., Byron, K. S., Johnson, C. V., Sullivan, J. L., and Singer, R. H. (1990). Subcellular localization of low-abundance human immunodeficiency virus nucleic acid sequences visualized by fluorescence in situ hybridization. Proc. Natl. Acad. Sci. USA *87*, 5420-5424.
92. Lawrence, J. B. and Singer, R. H. (1986). Intracellular localization of messenger RNAs for cytoskeletal proteins. Cell *45*, 407-415.
93. Lawrence, J. B., Singer, R. H., and Marselle, L. M. (1989). Highly localized tracks of specific transcripts within interphase nuclei visualized by in situ hybridization. Cell *57*, 493-502.
94. Leavitt, M. C., Yu, M., Yamada, O., Kraus, G., Looney, D., Poeschla, E., and Wong-Staal, F. (1994). Transfer of an anti-HIV-1 ribozyme gene into primary human lymphocytes. Hum. Gene Therapy *5*, 1115-1120.

95. Lee, S.-W., Gallardo, H. F., Gilboa, E., and Smith, C. (1994). Inhibition of human immunodeficiency virus type 1 in human T cells by a potent rev response element decoy consisting of the 13-nucleotide minimal rev-binding domain. J. Virol. *68*, 8254-8264.
96. Lee, T. C., Sullenger, B. A., Gallardo, H. F., Ungers, G. E., and Gilboa, E. (1992). Overexpression of RRE-derived sequences inhibits HIV-1 replication in CEM cells. New Biologist *4*, 66-74.
97. Lisziewicz, J., Rappaport, J., and Dhar, R. (1991). Tat-regulated production of multimerized TAR RNA inhibits HIV-1 gene expression. New Biologist *3*, 82-89.
98. Lo, K. M. S., Biasolo, M. A., Dehni, G., Palu, G., and Haseltine, W. A. (1992). Inhibition of replication of HIV-1 by retroviral vectors expressing tat-antisense and anti-tat ribozyme RNA. Virology *190*, 176-183.
99. Malim, M. H., Bohnlein, S., Hauber, J., and Cullen, B. R. (1989). Functional dissection of the HIV-1 rev trans-activator: derivation of a trans-dominant repressor of rev function. Cell *58*, 205-214.
100. Marx, J. L. (1989). Drug Resistant strains of AIDS virus found. Science *243*, 1551-1552.
101. McClain, W. H. and Seidman, J. G. (1975). Genetic perturbations that reveal tertiary conformation of tRNA precursor molecules. Nature *257*, 106-110.
102. McKeating, J. A., Gow, J., Goudsmit, J., Pearl, L. H., Mulder, C., and Weiss, R. A. (1989). Characterization of HIV-1 neutralization escape mutants. AIDS *3*, 777-784.
103. Michel, F., Jacquier, A., and Dujon, B. (1982). Comparison of fungal mitochondrial introns reveals extensive homologies in RNA secondary structure. Biochimie *64*, 867-881.
104. Murphy, F. L. and Cech, T. R. (1989). Alteration of substrate specificity for the endoribonucleolytic cleavage of RNA by the *Tetrahymena* ribozyme. Proc. Natl. Acad. Sci. USA *86*, 9218-9222.
105. Murphy, F. L. and Cech, T. R. (1993). An independently folding domain of RNA tertiary structure within the *Tetrahymena* ribozyme. Biochemistry *32*, 5291-5300.
106. Nara, P. L., Smit, L., Dunlop, N., Hatch, W., Merges, M., Waters, D., Kelliher, J., Gallo, R. C., Fischinger, P. J., and Goudsmit, J. (1990). Emergence of viruses resistant to neutralization by V3-specific antibodies in experimental human immunodeficiency virus type 1 IIIB infection of chimpanzees. J. Virol. *64*, 3779-3791.
107. Nunberg, J. H., Schleif, W. A., Boots, E. J., Obrein, J. A., Quintero, J. C., Hoffman, J. M., Emini, E. A., and Goldman, M. E. (1991). Viral resistance to human immunodeficiency virus type 1-specific pyridone reverse transcriptase inhibitors. J. Virol. *65*, 4887-4892.
108. Ojwang, J. O., Hampel, A., Looney, D. J., Wong-Staal, F., and Rappaport, J. (1992). Inhibition of human immunodeficiency virus type 1 expression by a hairpin ribozyme. Proc. Natl. Acad. Sci. USA *89*, 10802-10806.
109. Pachuk, C. J., Yoon, K., Moelling, K., and Coney, L. R. (1994). Selective cleavage of bcr-abl chimeric RNAs by a ribozyme targeted to non-contiguous sequences. Nucleic Acids Res. *22*, 301-307.
110. Pang, S., Shlesinger, Y., Daar, E. S., Moudgil, T., Ho, D. D., and Chen, I. S. (1992). Rapid generation of sequence variation during primary HIV-1 infection [published erratum appears in AIDS 1992 Jun6]. AIDS *6*, 453-60.
111. Perea, J. and Jacq, C. (1986). Role of the 5' hairpin structure in the splicing accuracy of the fourth intron of the yeast cob-box gene. EMBO J. *4*, 3281-3288.
112. Perreault, J.-P., Wu, T., Cousineau, B., Ogilvie, K. K., and Cedergren, R. (1990). Mixed deoxyribo- and ribo-oligonucleotides with catalytic activity. Nature *344*, 565-567.
113. Perrotta, A. and Been, M. D. (1991). A pseudoknot-like structure required for efficient self-cleavage of hepatitis delta virus RNA. Nature *350*, 434-436.
114. Perrotta, A. T. and Been, M. D. (1992). Cleavage of oligoribonucleotides by a ribozyme derived from the hepatitis δ virus RNA sequence. Biochemistry *31*, 16-21.
115. Perrotta, A. T. and Been, M. D. (1990). The self-cleaving domain from the genomic RNA of hepatitis delta virus: sequence requirements and the effects of denaturant. Nucleic Acids Res. *18*, 6821-6827.
116. Pley, H. W., Flaherty, K. M., and McKay, D. B. (1994). Three-dimensional structure of a hammerhead ribozyme. Nature *372*, 68-74.
117. Poeschla, E. and Wong-Staal, F. (1994). Antiviral and anticancer ribozymes. Curr. Opin. Oncology *6*, 601-606.
118. Prody, G. A., Bakos, J. T., Buzayan, J. M., Schneider, I. R., and Bruening, G. (1986). Autolytic processing of dimeric plant virus satellite RNA. Science *231*, 1577-1580.
119. Pyle, A. M. (1993). Ribozymes: a distinct class of metalloenzymes. Science *261*, 709-714.

120. Pyle, A. M., McSwiggen, J. A., and Cech, T. R. (1990). Direct measurement of oligonucleotide substrate binding to wild-type and mutant ribozymes from *Tetrahymena*. Proc. Natl. Acad. Sci. USA *87*, 8187-8191.
121. Pyle, A. M., Moran, S., Strobel, S. A., Chapman, T., Turner, D. H., and Cech, T. R. (1994). Replacement of the conserved G-U with a G-C pair at the cleavage site of the *Tetrahymena* ribozyme decreases binding, reactivity, and fidelity. Biochemistry *33*, 13856-13863.
122. Pyle, A. M., Murphy, F. L., and Cech, T. R. (1992). RNA substrate binding site in the catalytic core of the *Tetrahymena* ribozyme. Nature *358*, 123-128.
123. Raap, A. K., van de Rijke, F. M., Dirks, R. W., Sol, C. J., Boom, R., and van de Ploeg, M. (1991). Bicolor fluorescence in situ hybridization to intron and exon mRNA sequences. Exp. Cell Res. *197*, 319-322.
124. Reitz, M. S. J., Wilson, C., Naugle, C., Gallo, R. C., and Robert, G. M. (1988). Generation of a neutralization-resistant variant of HIV-1 is due to selection for a point mutation in the envelope gene. Cell *54*, 57-63.
125. Richman, D., Shih, C. K., Lowy, I., Rose, J., Pradanovich, P., Goff, S., and Griffin, J. (1991). Human immunodeficiency virus type 1 mutants resistant to nonnucleoside inhibitors of reverse transcriptase arise in tissue culture. Proc. Natl. Acad. Sci. USA *88*, 11241-11245.
126. Robert, G. M., Reitz, M. S. J., Robey, W. G., and Gallo, R. C. (1986). *In vitro* generation of an HTLV-III variant by neutralizing antibody. J. Immunol. *137*, 3306-3309.
127. Rosenstein, S. P. and Been, M. D. (1991). Evidence that genomic and antigenomic RNA self-cleaving elements from hepatitis delta virus have similar secondary structures. Nucleic Acids Res. *19*, 5409-5416.
128. Rosenstein, S. P. and Been, M. D. (1990). Self-cleavage of hepatitis delta virus genomic strand RNA is enhanced under partially denaturing conditions. Biochemistry *29*, 8011-8016.
129. Rossi, J. J. (1995). Controlled, targeted, intracellular expression of ribozymes: progress and problems. TIBTECH *13*, 301-306.
130. Rossi, J. J., Cantin, E. M., Saver, N., and Chang, P. F. (1991). The potential use of catalytic RNAs in therapy of HIV infection and other diseases. Pharmacology and Therapeutics *50*, 245-254.
131. Rossi, J. J., Cantin, E. M., Zaia, J. A., Ladne, P. A., Chen, J., Stephens, D. A., Sarver, N., and Chang, P. S. (1990). Ribozymes as therapies for AIDS. Ann. N. Y. Acad. Sci. *616*, 184-200.
132. Rossi, J. J., Elkins, D., Taylor, N., Zaia, J., and Sullivan, S. (1991). Exploring the use of antisense, enzymatic RNA molecules (ribozymes) as therapeutic agents. Antisense Research and Development *1*, 285-288.
133. Rossi, J. J., Elkins, D., Zaia, J. A., and Sullivan, S. (1992). Ribozymes as anti-HIV-1 therapeutic agents: principles, applications, and problems. AIDS Res. Hum. Retroviruses *8*, 183-189.
134. Rossi, J. J. and Sarver, N. (1992). Catalytic antisense RNA (ribozymes): their potential and use as anti-HIV-1 therapeutic agents. Advances in Experimental Medicine and Biology *312*, 95-109.
135. Ruffner, D. E., Stormo, G. D., and Uhlenbeck, O. C. (1990). Sequence requirements of the hammerhead RNA self-cleavage reaction. Biochemistry *29*, 10695-10702.
136. Saag, M. S., Emini, E. A., Laskin, O. L., Douglas, J., Lapidus, W. I., Schleif, W. A., Whitley, R. J., Hilderbrand, C., Byrnes, V. W., Kappes, J. C., Anderson, K. W., Massari, F. E., and Shaw, G. M. (1993). A short term clinical evaluation of L-697, 661, a non-nucleoside inhibitor of HIV-1 reverse transcriptase. N. Engl. J. Med. *329*, 1065-1072.
137. Saag, M. S., Hahn, B. H., Gibbons, J., Li, Y., Parks, E. S., Parks, W. P., and Shaw, G. M. (1988). Extensive variation of human immunodeficiency virus type-1 *in vivo*. Nature *334*, 440-444.
138. Sarver, N., Cantin, E. M., Chang, P. S., Zaia, J. A., Ladne, P. A., Stephens, D. A., and Rossi, J. J. (1990). Ribozymes as potential anti-HIV-1 therapeutic agents. Science *247*, 1222-1225.
139. Saxena, S. and Ackerman, E. J. (1990). Ribozymes correctly cleave a model substrate and endogenous RNA *in vivo*. J. Biol. Chem. *265*, 17106-17109.
140. Scanlon, K. J., Jiao, L., Funato, T., Wang, W., Tone, T., Rossi, J. J., and Kashani-Sabet, M. (1991). Ribozyme-mediated cleavage of c-fos mRNA reduces gene expression of DNA synthesis enzymes and metallothionein. Proc. Natl. Acad. Sci. USA *88*, 10591-10595.
141. Scott, W. G., Finch, J. T., and Klug, A. (1995). The crystal structure of an all-RNA hammerhead ribozyme: a proposed mechanism for RNA catalytic cleavage. Cell *81*, 991-1002.
142. Sczakiel, G. and Pawlita, M. (1991). Inhibition of human immunodeficiency virus type 1 replication in human T cells stably expressing antisense RNA. J. Virol. *65*, 468-472.

143. Shore, S. K., Nabissa, P. M., and Reddy, E. P. (1993). Ribozyme-mediated cleavage of the BCRABL oncogene transcript: *in vitro* cleavage of RNA and *in vivo* loss of P210 protein-kinase activity. Oncogene *8*, 3183-3188.
144. Sioud, M. and Drlica, K. (1991). Prevention of human immunodeficiency virus type 1 integrase expression in Escherichia Coli by a ribozyme. Proc. Natl. Acad. Sci. USA *88*, 7303-7307.
145. Snyder, D. S., Wu, Y., Wang, J. L., Rossi, J. J., Swiderski, P., Kaplan, B. E., and Forman, S. J. (1993). Ribozyme-mediated inhibition of bcr-abl gene expression in a Philadelphia chromosome-positive cell line. Blood *82*, 600-605.
146. Steinecke, P., Herget, T., and Schreier, P. H. (1992). Expression of a chimeric ribozyme gene results in endonucleolytic cleavage of target mRNA and a concomitant reduction of gene expression *in vivo*. EMBO J. *11*, 1525-1530.
147. Strobel, S. A. and Cech, T. R. (1993). Tertiary interactions with the internal guide sequence mediated docking of the P1 helix into the catalytic core of the *Tetrahymena* ribozyme. Biochemistry *32*, 13593-13604.
148. Strobel, S. A., Cech, T. R., Usman, N., and Beigelman, L. (1994). The 2,6-diaminopurine riboside-5-methylisocytidine wobble base pair: an isoenergetic substitution for the study of G-U pairs in RNA. Biochemistry *33*, 13824-13835.
149. Sullenger, B. A. (1995). Revising messages traveling along the cellular information superhighway. Chemistry & Biology *2*, 249-253.
150. Sullenger, B. A. and Cech, T. R. (1994). Ribozyme-mediated repair of defective mRNA by targeted *trans*-splicing. Nature *371*, 619-622.
151. Sullenger, B. A. and Cech, T. R. (1993). Tethering ribozymes to a retroviral packaging signal for destruction of viral RNA. Science *262*, 1566-1569.
152. Sullenger, B. A., Gallardo, H. F., Ungers, G. E., and Gilboa, E. (1990). Overexpression of TAR sequences renders cells resistant to human immunodeficiency virus replication. Cell *63*, 601-608.
153. Sullenger, B. A., Lee, T. C., Smith, C. A., Ungers, G. E., and Gilboa, E. (1990). Expression of chimeric tRNA-driven antisense transcripts renders NIH 3T3 cells highly resistant to moloney murine leukemia virus replication. Mol. Cell. Biol. *10*, 6512-6523.
154. Sun, L.-Q., Warrilow, D., Wang, L., Witherington, C., Macpherson, J., and Symonds, G. (1994). Ribozyme-mediated suppression of moloney murine leukemia virus and human immunodeficiency virus type 1 replication in permissive cell lines. Proc. Natl. Acad. Sci. USA *91*, 9715-9719.
155. Symons, R. H. (1992). Small catalytic RNAs. Annu. Rev. Biochem. *61*, 641-671.
156. Tabin, C. J., Bradley, S. M., Bargmann, C. I., Weinberg, R. A., Papageorge, A. G., Scolnick, E. M., Dhar, R., Lowy, D. R., and Chang, E. H. (1982). Mechanisms of activation of a human oncogene. Nature *300*, 143-149.
157. Taylor, N. R., Kaplan, B. E., Swiderski, P., Li, H., and Rossi, J. J. (1992). Chimeric DNA-RNA hammerhead ribozymes have enhanced *in vitro* catalytic efficiency and increased stability *in vivo*. Nucleic Acids Res. *20*, 4559-4565.
158. Thill, G., Vasseur, M., and Tanner, N. K. (1993). Structural and sequence elements required for the self-cleaving activity of the hepatitis delta virus ribozyme. Biochemistry *32*, 4254-4262.
159. Thompson, J. D., Ayers, D. F., Malmstrom, T. A., McKenzie, T. L., Ganousis, L., Chowrira, B. M., Couture, L., and Stinchcomb, D. T. (1995). Improved accumulation and activity of ribozymes expressed from a tRNA-based RNA polymerase III promoter. Nucleic Acids Res. *23*, 2259-2268.
160. Trono, D., Feinberg, M. B., and Baltimore, D. (1989). HIV-1 gag mutants can dominantly interfere with the replication of the wild type virus. Cell *59*, 113-120.
161. Tsuchihashi, Z., Khosla, M., and Herschlag, D. (1993). Protein enhancement of hammerhead ribozyme catalysis. Science *262*, 99-102.
162. Uhlenbeck, O. C. (1987). A small catalytic oligoribonucleotide. Nature *328*, 596-600.
163. Ventura, M., Wang, P., Ragot, T., Perricaudet, M., and Saragosti, S. (1993). Activation of HIV-specific ribozyme activity by self-cleavage. Nucleic Acids Res. *21*, 3249-3255.
164. Wang, J.-F., Downs, W. D., and Cech, T. R. (1993). Movement of the guide sequence during RNA catalysis by a group I ribozyme. Science *260*, 504-508.
165. Wang, Y.-H., Murphy, F. L., Cech, T. R., and Griffith, J. D. (1994). Visualization of a tertiary structural domain of the *Tetrahymena* group I intron by electron microscopy. J. Mol. Biol. *236*, 64-71.
166. Waring, R. B., Towner, P., Minter, S. J., and Davies, R. W. (1986). Splice-site selection by a self-splicing RNA of *Tetrahymena*. Nature *321*, 133-139.

167. Weerasinghe, M., Liem, S. E., Asad, S., Read, S. E., and Joshi, S. (1991). Resistance to human immunodeficiency virus type 1 (HIV-1) infection in human CD4+ lymphocyte-derived cell lines conferred by using retroviral vectors expressing an HIV-1 RNA-specific ribozyme. J. Virol. *65*, 5531-5534.
168. Weiss-Brummer, B., Holl, J., Schweyen, R. J., Rodel, G., and Kaudewitz, F. (1983). Processing of yeast mitochondrial RNA: involvement of intramolecular hybrids in splicing of cob intron 4 RNA by mutation and reversion. Cell *33*, 195-202.
169. von Weizsacker, F., Blum, H. E., and Wands, J. R. (1992). Cleavage of hepatitis B virus RNA by three ribozymes transcribed from a single DNA template. Biochem. and Biophys. Res. Commun. *189*, 743-748.
170. Williamson, C. L., Desai, N. M., and Burke, J. M. (1989). Compensatory mutations demonstrate that P8 and P6 are RNA secondary structure elements important for processing of a group I intron. Nucleic Acids Res. *17*, 675-689.
171. Williamson, C. L., Tierney, W. M., Kerker, B. J., and Burke, J. M. (1987). Site-directed mutagenesis of core sequence elements 9R', 9L, 9R, and 2 in self-splicing *Tetrahymena* pre-rRNA. J. Biol. Chem. *262*, 14672-14682.
172. Wong-Staal, F., Yu, M., Yamada, O., Leavitt, M., and Maruyama, A. (1994). Gene therapy for HIV-1 infection. Tenth International Conference on AIDS.
173. Wright, L., Wilson, S. B., Milliken, S., Biggs, J., and Kearney, P. (1993). Ribozyme-mediated cleavage of the *bcr/abl* transcript expressed in chronic myeloid leukemia. Exp. Hematol. *21*, 1714-1718.
174. Wu, H. N., Lin, Y. J., Lin, F. P., Makino, S., Chang, M. F., and Lai, M. M. (1989). Human hepatitis delta virus RNA subfragments contain an autocleavage activity. Proc. Natl. Acad. Sci. USA *86*, 1831-1835.
175. Xing, Y., Johnson, C. V., Dobner, P. R., and Lawrence, J. B. (1993). Higher level organization of individual gene transcription and RNA splicing. Science *259*, 1326-1330.
176. Yamada, O., Kraus, G., Leavitt, M. C., Yu, M., and Wong-Staal, F. (1994). Activity and cleavage site specificity of an anti-HIV-1 hairpin ribozyme in human T cells. Virology *205*, 121-126.
177. Yamada, O., Yu, M., Yee, J.-K., Kraus, G., Looney, D., and Wong-Staal, F. (1994). Intracellular immunization of human T cells with a hairpin ribozyme against human immunodeficiency virus type 1. Gene Therapy *1*, 38-44.
178. Young, B., Herschlag, D., and Cech, T. R. (1991). Mutations in a nonconserved sequence of the *Tetrahymena* ribozyme increase activity and specificity. Cell *67*, 1007-1019.
179. Yu, M., Leavitt, M. C., Maruyama, M., Yamada, O., Young, D., Ho, A. D., and Wong-Staal, F. (1995). Intracellular immunization of human fetal cord blood stem/progenitor cells with a ribozyme against human immunodeficiency virus type 1. Proc. Natl. Acad. of Sci. USA *92*, 669-703.
180. Yu, M., Ojwang, J., Yamada, O., Hampel, A., Rapapport, J., Looney, D., and Wong-Staal, F. (1993). A hairpin ribozyme inhibits expression of diverse strains of human immunodeficiency virus type 1. Proc. Natl. Acad. Sci. USA *90*, 6340-6344.
181. Yu, M., Poeschla, E., and Wang-Staal, F. (1994). Progress towards gene therapy for HIV infection. Gene Therapy *1*, 13-26.
182. Yu, M., Poeschla, E., Yamada, O., Degrandis, P., Leavitt, M. C., Heusch, M., Yees, J-K., Wong-Staal, F., and Hampel, A. (1995). *In vitro* and *in vivo* characterization of a second functional hairpin ribozyme against HIV-1. Virology *206*, 381-386.
183. Zaia, J. A., Chatterjee, S., Wong, K. K., Elkins, D., Taylor, N. R., and Rossi, J. J. (1992). Status of ribozyme and antisense-based developmental approaches for anti-HIV-1 therapy. Ann. N. Y. Accad. Sci. *660*, 95-106.
184. Zaug, A. J., Been, M. D., and Cech, T. R. (1986). The *Tetrahymena* ribozyme acts like an RNA restriction endonuclease. Nature *324*, 429-433.
185. Zaug, A. J. and Cech, T. R. (1986). The intervening sequence RNA of *Tetrahymena* is an enzyme. Science *231*, 470-475.
186. Zaug, A. J., Grabowski, P. J., and Cech, T. R. (1983). Autocatalytic cyclization of an excised intervening sequence RNA is a cleavage-ligation reaction. Nature *301*, 578-583.
187. Zaug, A. J., Grosshans, C. A., and Cech, T. R. (1988). Sequence-specific endoribonuclease activity of the *Tetrahymena* ribozyme: enhanced cleavage of certain oligonucleotide substrates that form mismatched ribozyme-substrate complexes. Biochemistry *27*, 8924-8931.

188. Zhou, C., Bahner, I. C., Larson, G. P., Zaia, J. A., Rossi, J. J., and Kohn, D. B. (1994). Inhibition of HIV-1 in human T-lymphocytes by retrovirally transduced anti-tat and rev hammerhead ribozymes. Gene *149*, 33-39.

III Cellular Target Systems

12. Gene Transfer and Transgene Expression in Hematopoietic Cells

Christopher Baum, Carol Stocking, Thomas Wagener, Hans-Georg Eckert and Wolfram Ostertag

12.1 Introduction

The hematopoietic system plays a key role in somatic gene therapy research. The first descriptions of retroviral gene transfer into murine hematopoietic cells date from 1983/1984 (1, 2). Only a few years later, the first gene marking experiment targeting tumor-infiltrating lymphocytes was performed in humans (3). To date, close to half of the approved gene therapy trials involve gene transfer and transgene expression in hematopoietic cells (4). Unfortunately, gene therapy concepts are more advanced than the methodology needed to fulfill their goals. This chapter aims to give an overview of current concepts and limitations in genetic manipulation of hematopoietic cells, thus hopefully providing a basis for constructive criticism of the ongoing research in the field. For reasons of space, many aspects cannot be considered in exhaustive detail. For this, the reader should refer to the literature cited.

A number of inherited and acquired disorders of hematopoietic cells cannot be sufficiently managed with conventional therapy, thus gene-based strategies are needed to compensate for this failure. Examples of inherited disorders potentially treatable by corrective gene transfer are hemoglobinopathies, immunodeficiency syndroms, lysosomal storage disorders and mucopolysaccharidoses (Tab. 12.1 and ref. therein). Acquired disorders amenable to "supplementary" gene transfer include myelosuppression induced by anticancer-chemotherapy (27, 28), and acquired immundeficiency syndrome (AIDS) due to human immunodeficiency virus (HIV) – infection (29). In addition, the ubiquity of mature blood cells render these ideal shuttles for the delivery of transgene-encoded proteins to needy organs. In the simplest situation, gene transfer into hematopoietic cells might be an effective means to shed proteins into the circulation, e.g. for correction of α-1-antitrypsin deficiency (affecting lung and liver) or bleeding disorders caused by deficiency of factor IX (30). The ability of macrophages, dendritic cells, or lymphocytes to penetrate peripheral tissues might be exploitable to transmit proteins of therapeutic capacity for the treat-

Table 12.1 A list of inherited diseases potentially curable by gene transfer in hematopoietic cells.

Group	Disease	Therapeutic Gene	Ref.
Hemoglobinopathies	β-Thalassemia	β-Globin	5
	Sickle cell anemia	γ-Globin	6
Mucopolysaccharidoses	Hurler/Scheie	α-L-iduronidase	7
	Hunter	Iduronate sulfatase	8
	Sanfillipo	N-Acetylgucosamine-6-sulphatase	9
	Morquio	Galactose-6-sulfatase	10
	Sly	β-Glucuronidase	11
	Maroteaux-Lamy	Arylsulfatase B	12
Lysosomal Storage Disorders	Metachromatic leukodystrophy	Arylsulfatase A	13
	Niemann-Pick	Acid sphingomyelinase	14
	Aspartylgucosaminuria	Apartylglucosaminidase	15
	Fabry	α-Galatosidase A	16
	Gaucher	Glucocererebrosidase	17
Inherited immuno-deficiency sydromes	SCID (autosomal)	Adenosine deaminase	18
	SCID (autosomal)	Purine nucleoside phosphorylase	19
	SCID (X-linked)	IL-2 receptor γ-chain	20
	Leukocyte adhesion deficiency	CD18	21
	Chronic granulomatous disease	gp91-*phox*	22
		p22-*phox*	23
		p47-*phox*	24
Others	Erythropoietic porphyria	Uroporphyrinogen-III-synthetase	25
	Fanconi anemia C	Fanconi anemia complementation C	26

ment of diseases affecting non-hematopoietic tissues, e.g. rheumatoid arthritis (31, 32), Duchenne muscular dystrophy (33), or hereditary storage disorders inducing multiorgan-damage (see Tab. 12.1). Finally, gene marking studies attempt to analyze the fate of cells delivered during transplantation of hematopoietic progenitors or T-lymphocytes. In some protocols, the marker gene encodes a protein enabling negative selection of transduced cells (suicide gene transfer) to allow elimination of these cells in case of pathogenicity (34).

As a consequence of the large diversity of these approaches, expression of the gene of interest is needed in different lineages and stages of hematopoiesis; furthermore, each protocol has its own specific requirements with regard to the the level and the stability of gene expression to be achieved. In contrast, owing to the hierarchical organization of the hematopoietic system, the target cell population for the actual gene transfer is much more uniform, i.e. hematopoietic progenitor cells (HPC) containing the rare population of long-term repopulating cells (LTRC). Only for the diagnosis and treatment of inherited and acquired disorders involving T-lymphocytes does a restricted lineage represent the primary target for the gene transfer.

In theory, the hierarchical organization of hematopoietic cells enables reconstitution of the entire organ system with a limited number of gene-modified LTRC. These primitive progenitors represent a rare, scattered and somewhat hidden population *in vivo*, thus necessitating enrichment, expansion and transduction to be performed *ex vivo* for reasons of both safety and efficiency. The relative low transduction frequencies and insufficient gene expression levels achieved so far in hematopoietic cells of humans and large animals have been limiting for therapeutic approaches. Therefore, the three most challenging tasks are to improve the conditions for the transduction of LTRC, to establish protocols that allow selective expansion of genetically and phenotypically modified cells, and to develop vector systems that mediate sufficient and stable gene expression *in vivo*.

We will start this review with a description of the biological target system (section 12.2) and the different sources of HPC currently under investigation (section 12.3). This will be followed by discussing more technological aspects of transduction systems and protocols (section 12.4). We will then address the pros and cons of selection systems needed to expand transduced cells (section 12.5). Finally, we will point to the efforts that need to be undertaken to achieve appropriate levels (section 12.6) and stability (section 12.7) of transgene expression *in vivo*.

12.2 The Hematopoietic System

The hematopoietic system is a highly complex multi-lineage organ with extensive turnover rates (more than 10^{11} cells daily). It is derived from a limited number (probably less than a million in human adults) of primitive stem and progenitor cells harboring enormous self-renewal and proliferation capacity (Fig. 12.1). With the exception of T-cell progenitors maturating in a thymic microenvironment, all HPC reside in the bone marrow. Close contact with stromal cells, in concert with soluble cytokines acting in a paracrine or endocrine manner, regulates proliferation and differentiation to mature cells (35-37). Differentiation commitment occurs presumably stochastically (35); cytokines, mostly acting as survival factors, trigger proliferation and lineage-maturation in accord with the needs of the organism (hybrid model of differentiation) (38).

Pluripotent hematopoietic stem cells (HSC) undergo multiple divisions for self-renewal throughout life. Intermittent periods of complete quiescence might, however, be a prerequisite for preserving pluripotency and self-renewal capacity. Consequently, less than 5% of HSC are in cycle under steady-state conditions (39). Importantly, the "age" of HSC might be defined by the number of cyclings completed, a phenomenon known as proliferative senescence or the "Hayflick limit" (40). Exhaustion of stem cell function might be due to cycle-dependent loss of telomeres (41). Activation from quiescence can result in irreversible commitment to differentiation. The agent provocateur for this unidirectional process is presumably primarily the activity of soluble cytokines. During differentiation, loss of self-renewal, restric-

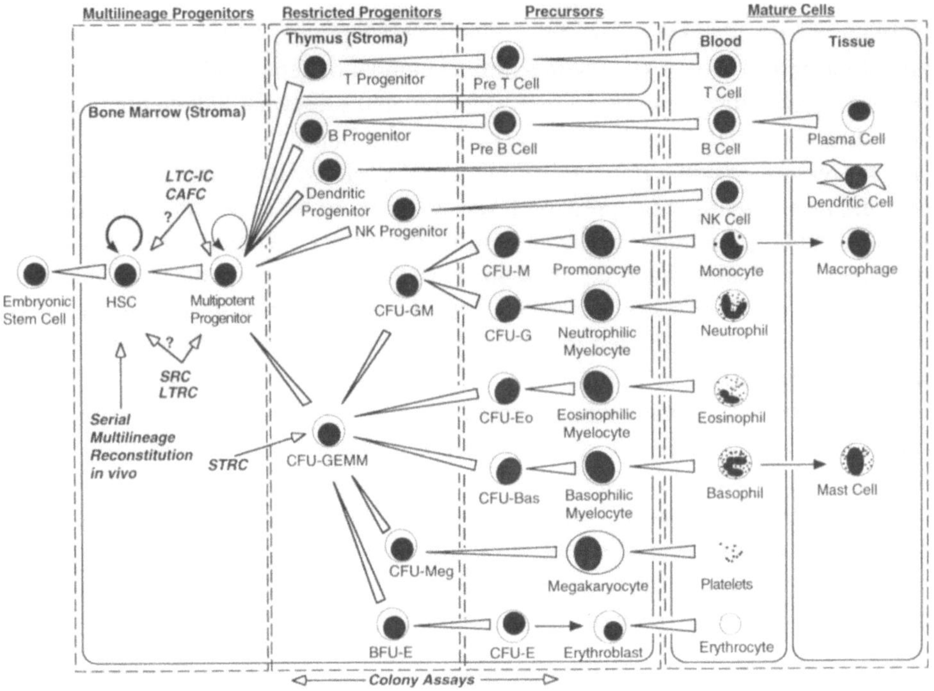

Fig. 12.1: Schematic representation of the hematopoietic system.
The developmental hierarchy and continuous multiplication of cells is symbolized by wedges between the different stages and lineages shown. Circular arrows indicate self-renewal potential. For reasons of simplicity, the number of developmental stages is reduced to those relevant for discussion. Furthermore, the four major developmental stages are delineated by dashed squares. Rounded squares circumscribe different anatomical compartments of hematopoiesis. Assay systems and resulting operational definitions of progenitor classes are given in italics, with white-headed arrows pointing to the cell populations detected in these systems (question marks indicate remaining uncertainties). HSC; hematopoietic stem cell; SRC, SCID-mouse repopulating cell; LTRC, long-term repopulating cell; STRC, short-term repopulating cell; LTC-IC, long term culture-initiating cell; CAFC, cobblestone-area forming cell. CFU, colony forming unit; BFU, burst forming unit.

tion of differentiation potential, and transient increase in proliferation capacity is observed. Maturation is characterized by entrance into the peripheral blood and distribution throughout the organism. Dendritic cells, macrophages and mast cells penetrate multiple peripheral tissues across the vessel barrier. Plasma cells preferentially reenter the bone marrow. Mature myeloid and erythroid cells have a very limited life span (days to months), while subsets of mature lymphocytes might persist for years. A number of assay systems were developed to analyze hematopoietic cells *in vivo* and *in vitro*, leading to several operationally defined classes of HPC.

HSC are defined by the dual capacity for functional self-renewal (expansion), and differentiation into erythroid, myeloid, and lymphoid lineages (42). Transfer of a single HSC is postulated to enable multi-lineage reconstitution of the complete blood system upon serial transplantations. Due to this definition, HSC can only be analyzed *in vivo*. HSC give rise to still multipotent HPC that might have more limited self-renewal capacity. In mice, and probably also in man, HSC and multipotent HPC represent roughly 1% of bone marrow progenitors or up to 0.01% of adult marrow cells (43). This population of LTRC is functionally and morphologically distinguishable from short-term repopulating cells (STRC) (44-47). Because subsets of STRC have still preserved combined lymphohematopoietic potential *in vitro* and *in vivo*, multi-lineage differentiation is not entirely predictive for long-term repopulating capacity (44, 45).

According to flow-cytometric criteria, LTRC have a distinct phenotype, i.e. Thy-1^{lo}Sca-$1^{+}$$Lin^{-}$ in mice, or $CD34^{+}$Thy-$1^{lo}$$Lin^{-}$ or $CD34^{+}$$CD38^{-}$ in humans (36). The capacity to extrude the fluorochrome rhodamine 123 further confines an immature stem cell pool, five cells of which are sufficient to reconstitute hematopoiesis in mice (46). Additional criteria for LTRC include plastic adhesiveness, reduced sedimentation velocity and lower expression of wheat-germ agglutinin binding cell surface carbohydrate (47).

Significant progress has been made to characterize human HPC *in vivo* using xenotransplantation in mice with severe combined immunodeficiency (SCID) (48, 49) or fetal sheep (50). The sheep model seemingly allows long-term observations and serial transplantations of human HPC, an important feature when addressing true stem cells. However, only few centers will be able to afford working with a large animal model. The SCID-model can be adopted in any pathogen-free facility for small animals; however, it was so far limitated by an incomplete maturation of human HPC and the hitherto exisiting failure to demonstrate serial transplantable stem cell function. Still, human SCID-mouse repopulating cells (SRC) seem to represent a very primitive progenitor cell population, and hence, can be addressed a early LTRC.

HSC are propably included in the population of long-term bone marrow culture initiating cells (LTC-IC), i.e., primitive multi-lineage progenitor cells that can be kept on stroma monolayers (51) or even, as recently reported, in defined stroma-free conditions (52), propagated and, hence, manipulated for prolonged time *in vitro*. LTC-IC correlate with "cobble-stone-area-forming cells" (CAFC), referring to a typical phenotype formed by multi-lineage progenitors in newly established stromal coculture-systems (53, 54). Like HSC, LTRC, and SRC *in vivo*, LTC-IC give rise to oligo- or monopotent, lineage-restricted progenitor and precursor cells that can be monitored in colony assays *in vitro*. This more actively proliferating population contains up to 99% of the $CD34^{+}$ cells. As lineage-restricted progenitor and precursor cells they are characterized by additional expression of lineage-specific surface markers in flow cytometry (Lin^{+}), the failure to extrude rhodamine 123, the capacity to form colonies in methylcellulose, and, most importantly, by a loss of self-renewal capacity (36). Thus, their biological properties are clearly distinct from their more primitive ancestors.

The many similarities observed *in vitro* and *in vivo* should not prompt equation of LTC-IC, CAFC, SRC, and HSC. These populations are operationally defined, and the predicitive value of the definition is strongly dependent on the complexity and the limitations of the assay system used. Flow-cytometric criteria, for example, are not predictive of stem cell function *in vivo* (46). Likewise, long-term culture initiating capacity *in vitro* and long-term repopulation capacity *in vivo* can correlate intimately, but are not necessarily interconnected.

Towards the aim of elaborating defined culture conditions for LTRC, significant progress has recently been made to develop *in vitro* systems that allow the functional dissection of mechanisms involved in stroma-progenitor interactions, cytokine responsiveness, self-renewal and differentiation of HPC, thereby providing an experimental basis to define optimal culture conditions for their maintenance and genetic engineering.

(I) Semi-immortalized embryonic stem (ES) cells can be differentiated *in vitro* to multi-lineage progenitor cells, among these maybe even true HSC (55, 56). It is to be hoped that this system can be adopted to disclose the molecular basis of commitment of multipotent progenitors. (II) Proliferation assays for LTC-IC have been developed that should facilitate the investigation of the mechanisms responsible for expansion of these cells (52, 57, 58). (III) Immortalization of hematopoiesis-supporting stroma cells and establishment of stroma-dependent hematopoietic cell lines has created defined assay systems for the analysis of stroma-dependence (59, 60). (IV) Finally, developmentally regulated gene products that control early HPC self-renewal are being defined. Increased expansion of early HPC was achieved by ectopic expression of the HOX-B4 gene, a phenomenon possibly exploitable for amplification of gene-modifed cells (61).

12.3 Sources and Enrichment of Target Cells

With bone marrow, peripheral blood (after mobilization) and umbilical cord blood, three alternative sources for HPC are at present available. Bone marrow is the classical source of HSC and HPC. The vast majority of marrow-derived LTRC is found in G_0 upon collection, requiring refined culture conditions to evoke proliferation. Furthermore, the collection procedure is cumbersome and not free of complications. In the peripheral blood, HPC are almost entirely absent under steady-state conditions. However, treatment with cytotoxic agents, like 5-fluorouracil in mice (62) or, more promising for human applications, cyclophosphamide or cytosine arabinoside (63), and/or mobilization with cytokines forces HPC to invade the peripheral blood, from where the cells can be easily, and repeatedly, harvested (64-67). Because of the necessary pretreatments, collecting HPC from the peripheral blood is primarily suited for cancer-patients receiving antineoplastic cytotoxic therapy. Mobilized peripheral blood HPC are more actively cycling, yielding perfect starting conditions for *ex vivo* expansion and transduction with retroviral vectors (section 12.4). The frequency and proliferation status of HSC in this source is, however, at present unclear and might depend on the type of regimen administered for mobilization.

Umbilical cord blood is a rich and natural source of HPC which have an elevated cloning efficiency, enhanced cytokine responsiveness and increased proliferation potential, perfect starting conditions for expansion and transduction using retroviral vectors (18, 48, 68-70) Since they can only be harvested immediately after birth, their utility for gene therapy is heavily dependent on the establishment of blood banking systems. Due to the lack of primed lymphocytes, cord blood HPC cause reduced graft-related complications in allogeneic transplantations. The only inherent limitation to the use of cord blood HPC is that the low numbers collected currently preclude engraftment of children exceeding 35 kg body weight. With the elaboration of *ex vivo* expansion systems, this hurdle may be transient in nature, thus supporting the great enthusiasm evoked by the availability of these cells (71).

All sources for HPC contain 96-99% non-clonogenic mononuclear cells. Enrichment techniques for HPC are primarily based on CD34 positivity (72). LTRC represent roughly 1% of $CD34^+$ cells, and are highly enriched in the $CD34^+CD38^-$ subpopulation, as are LTC-IC (70). In the absence of a stem cell assay for human cells, there is still some uncertainty as to what extent true HSC are included in the population of $CD34^+$ cells (73-76). Preliminary results from gene marking studies (77) are in agreement with clinical observations indicating that CD34-affinity purified HPC indeed contribute to long-term reconstitution (74-76, 78). Cloning and characterization of the murine CD34 will hopefully further aid in defining its potential as a stem cell marker (79).

12.4 Transduction Systems

Retroviral vectors are currently the best-characterized gene transfer system for HPC, with respect to both safety and efficency. They mediate stable insertion of transferred sequences in a more or less randomly elected genetic environment, with some preference for open chromatin (euchromatin) (80) and actively transcribed DNA regions (81). The plasticity of the retroviral genome allows accomodation of close to 9 kb genetic information (82), and even complete replacement of viral enhancer sequences with endogenous cis-active elements (83). Thus, retroviral vector design can take into account the individual requirements for the transcriptional control of transgenes in the therapeutic approach of interest. The principles of retroviral gene transfer technology are described elsewhere in this book (chapter 1). Primitive hematopoietic progenitor cells, however, impose some special hurdles to successful transduction with and gene expression from retroviral vectors that are to be discussed in this chapter.

One basic limitation to retroviral transduction of HPC using conventional systems based on murine leukemia viruses (MuLV) is the necessity for cell-cycle dependent breakdown of the nuclear envelope to allow nuclear entry of the viral preintegration complex (84, 85). The structural components of the preintegration complex are derived from the packaging cell. MuLV-based packaging systems are currently the only well-characterized and approved technology; thus HPC must be induced to leave their preferred quiescent status. Current protocols mostly rely on the use of cytokine cocktails to stimulate HPC prior to and during the transduction period (86-92). Pretreatment with cytotoxic agents is exploitable in gene transfer trials involving patients receiving anti-cancer chemotherapy (87, 93-95). Using these systems, gene transfer frequencies into STRC-equivalent progenitors and precursors can be above 67% (89, 95). When cells are stimulated with cytokines only, transduction efficiencies in standard protocols are usually in the range of 10-30% (94, 96). Since retroviral vectors integrate into postreplication DNA (97), one of the two daughter cells will always remain untransduced, unless multiple copies are integrated as a result of a high multiplicity of infection (MOI).

The simplest way to achieve efficient transduction of target cells is co-culture on virus producing packaging lines, but this procedure is not approved for use in human gene therapy for safety concerns. Co-culture leads to an unpredictable high MOI, thus increasing the risk for insertional mutagenesis. In addition, separation of packaging cells from HPC before reinfusion would be required to avoid delivery of immortalized cells. Furthermore, eventual occurence of and infection with contaminating replication competent retrovirus (RCR) cannot be completely prevented in a co-culture protocol, even when working with defined master cell banks. Release of RCR is indeed the most important safety-concern, although primate-models have shown that MuLV-induced lymphomas only occur in severely immunocompromised hosts and at high titer infection with RCR (98, 99).

Supernatant transductions have the great advantage of allowing GMP (good manufacturing practice) -production of defined and quality-controlled vector-stocks that can be applied at a predictable MOI. Furthermore, concentration and purification of vector-stocks can be employed to reduce a potential contamination with xenogeneic serum-components such as cytokines (affecting *ex vivo* maintenance of HPC) or, even, yet unknown pathogens (100, 101). Conventional static supernatant transduction conditions, however, suffer from comparatively low transduction efficiencies. Using centrifugation protocols (101) or flow-through conditions that elevate the likelihood of productive virus-target cell contacts (102), the efficacy of supernatant transductions might indeed be significantly elevated.

Unfortunately, most protocols designed so far to improve retroviral transduction efficiency in HPC were based on findings in methylcellulose colony assays, thus measuring predominantly STRC. A study performed in primates has shown that *in vitro* clonogenic assays have little, if any value to assess transduction efficiencies in LTRC: cytokine-combinations conventionally used for transduction of STRC like interleukin (IL)-3, IL-6 plus stem cell factor, might be detrimental for the maintenance or expansion of LTRC (103). Thus, it is not surprising to note that the highest gene transfer frequency in HPC achieved so far with retroviral vectors was observed in a gene marking trial performed in juvenile patients prestimulated with cytostatic treatment only; about 5% of *in vitro* clonogenic progenitor cells were marked after reinfusion *in vivo*, with persistence of gene-marked progenitors in some patients for at least 18 months (93). In long-term follow-ups of other human trials and large animal models, the proportion of gene-modified HPC detected usually did not exceed 0.01 to 0.1% (77, 91, 103-106). This low frequency greatly impedes the therapeutic efficacy of gene transfer in hematopoietic cells.

However, it can be predicted that improvement of *ex vivo* culture conditions for HPC, especially LTRC, will soon drastically improve transduction efficiencies. Presence of stroma, and even of purified extracellular matrix molecules, preferentially increases retroviral transduction frequency in LTC-IC (17, 107-109). In the presence of stroma 10% to 40% of LTC-IC can be transduced, as opposed to less than 1% without stroma (17, 107, 110). Accordingly, the highest gene transfer efficiency in LTRC observed so far in the canine model was achieved after repeated supernatant transduction in a stroma-based long-term culture system (111). Mechanisms responsible for that phenomenon include more appropriate production or presentation of cytokines by stroma cells, and local enrichment and presentation of retroviral particles. With the help of the recently defined proliferation assays for LTC-IC, transduction protocols will presumably become even better adapted for LTRC (52, 58).

To escape the necessity of prestimulation for retroviral insertion, vector systems based on lentiviruses like HIV or foamy viruses, the latter belonging to the family of spuma retroviridae, are emerging. Both viruses are able to infect quiescent cells (112, 113), although the transfer efficiency might still be reduced under these conditions. In HIV, the nuclear transport of the preintegration complex is guided by a nuclear localisation signal present in the viral matrix (Gag MA) protein; in addition, it requires an intact viral Vpr gene, active nuclear pores, and ATP. Importantly, the ability to infect quiescent cells might depend on the metabolic activity of the target cell, since resting primary T cells residing in a metabolically inactive G_0 state are inefficently infected by HIV (114). Parallels to early HPC are obvious, requiring formal prove that vectors based on HIV or foamy virus can transduce these cells even under unstimulated conditions.

Besides the cell-cycle limitation, primitive HPC hinder retroviral entry, reverse transcription or nuclear transport of vector sequences in a currently poorly defined way. We called this phenomenon "preintegration block" as opposed to a "postintegration block" manifesting as inefficient promoter utilization after genomic integration of the vector (115). Inefficient receptor-envelope interactions might explain the preintegration block to some extent. Thus, researchers have developed novel packaging lines utilizing alternative env-sequences from the gibbon ape leukemia virus (116), or the cat endogenous virus RD114 (117). Others have proposed pseudotyping with the G-glycoprotein of the vesicular stomatitis virus to expand the host range and to increase stability and titer of retroviral supernatants (118). Engineering of the env-gene to create fusions with ligands or variable-fragments of antibodies might offer a novel way to achieve specific and efficient targeting of defined cell populations (119). Systematic analyses of the transduction efficiences achieved with these novel packaging systems in LTRC are to be awaited.

Adeno-associated virus (AAV)-vectors might represent an alternative to retroviral vector systems. As a DNA-virus, AAV-vectors support introduction of intronic sequences that might be needed when intending incorporation of lineage-restricted control elements, as in globin gene expression for correction of β-thalassemia or sickle cell anemia (5, 120). Furthermore, AAV-vectors can persist episomally in quiescent cells, but it is not yet clear if the vector DNA also integrates. AAV-vectors were reported to mediate efficient transduction of HPC (121), but the efficiency of stable gene transfer into LTRC or HSC remains as yet unclear. Furthermore, the initial expectance of site-specific integration at chromosome 19, as observed for the parental virus, has not been fulfilled with AAV-vectors. Another limitation of the AAV-system is the reduced packaging size: only 4.7 kb are left for the gene of interest. Moreover, AAV-stocks are frequently contaminated with the replication-competent helper, adenovirus, necessitating further improvement of packaging systems to provide a safe transfer system which enables production of quality-controlled vector stocks of a predictable MOI (122).

12.5 Selection and Expansion of Transduced Cells

After *ex vivo* transduction and reinfusion of gene-modified cells, a genetic and phenotypic chimerism results in the host. The rate of chimerism, i.e., the proportion of genetically and phenotypically modified cells contributing to steady-state hematopoiesis, is the main determinant for therapeutic success. According to experiences from bone marrow transplantation studies, a chimerism with at least 10% phenotypically modified cells is to be achieved for the treatment of most inherited and acquired disorders. Only one large animal study has so far reported reconstitution with gene-modified cells at this high rate without prior myeloablation or selection of transduced cells (111). Here, maximal levels between 10% and 30% have been achieved within the first 6 months after transplantation, with a gradual decline to levels around 5% in a follow-up period of two years. The transiently elevated levels can be explained by an increased transfer efficiency in STRC, and by clonal fluctuations of stem cell functions occuring in the post-transplantation period, a pattern consistently observed in retroviral gene marking studies performed in mice (123, 124). During the post-transplantation dysequilibrium, a preferential recruitment of engrafted cells might occur due to their prestimulation during the *ex vivo* culture period. The rate of long-term expression seen after re-establishment of a hematopoietic equilibrium (more than 12 months after transplantation) is more likely to reflect the true number of gene-modified primitive progenitors in this study (111).

In man, a standard graft of one million $CD34^+$ HPC/kg body weight (125) corresponds to roughly 1% of the endogenous progenitor cell pool (126). If we assume that the gene transfer frequency in LTRC under hitherto applied transduction protocols might be in the range 0.1-10%, between 0.001% and 0.1% of the physiological pool of LTRC will be gene-modified after reinfusion (Tab. 12.2). This is in accord with the observations of most large animal studies and human trials performed so far (see section 12.4). Without selection, gene-modified cells will thus have only

Table 12.2 Frequency of gene-modified LTRC in steady state hematopoiesis is dependent on gene transfer efficiency and expansion rate.

Frequency of LTRC in un-selected and un-expanded graft	Gene transfer efficiency in LTRC	Frequency of gene-modified LTRC *without* expansion	*Ex vivo* expansion of LTRC	Frequency of gene-modified LTRC after expansion	Enrichment of gene-modified LTRC by *in vivo* selection	Number of doublings required for reconstitution
1%[a]	0.1%	0.001%	3x	0.003%	33000x	16
			10x	0.01%	10000x	14–15
			30x	0.03%	3300x	12–15
	1%	0.01%	3x	0.03%	3300x	12–13
			10x	0.1%	1000x	11
			30x	0.3%	330x	9–10
	10%	0.1%	3x	0.3%	330x	9–10
			10x	1%	100x	7
			30x	3%	33x	5
	100%	1%	3x	3%	33x	5
			10x	10%	10x	3–4
			30x	30%	3.3x	1–2

LTRC, long term repopulating cell
[a] Graft size was assumed as $1-2 \times 10^6$ CD 34^+ HPC per kg body weight

marginal contribution to long-term hematopoiesis, and thus presumably little, if any, therapeutic value.

With the exception of Fanconi-anemia or SCID-syndromes, no therapeutic gene has been described so far that might confer a selective advantage to transduced HPC under steady-state conditions. Thus, to allow selective enrichment of transduced cells to therapeutic levels, selectable marker genes have to be co-transduced. Selection and expansion can be performed *in vitro* or *in vivo* (Fig. 12.2).

Systems proposed for *in vitro* selection are primarily based on cell surface marker expression (127, 128). At low gene transfer frequencies, up to 1000x enrichment would be required *in vitro*, followed by expansion over multiple doublings to provide a graft size that enables reconstitution (Tab. 12.2). Evidence exists that *in vitro* expanded HPC maintain engraftment potential both in mice (129, 130) and in man (75, 131). However, exposure to inappropriate types and doses of cytokines *in vitro* can also favor differentiation over self-renewal, or compromise the homing efficiency of expanded "stem" cells, finally leading to engraftment failures (132).

When cells are reinfused without prior ablation of endogenous HPC, competitive repopulation occurs (Fig. 12.2). A highly expanded and selected graft consisting of 100% gene-modified HPC at best represents 10% to 30% of the endogenous progenitor fraction. With currently established protocols, the frequency might rather be

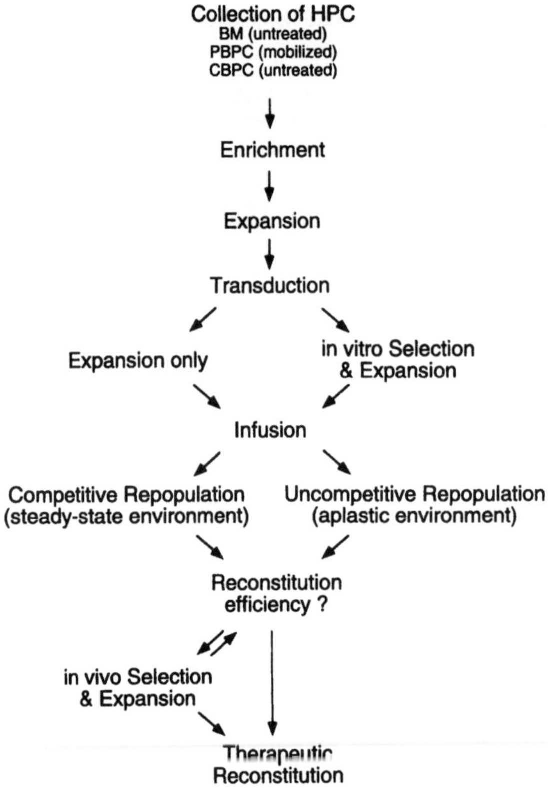

Fig. 12.2: Different ways to therapeutic reconstitution. Further discussed in section 12.5.

in the range of 0.3% to 3%, assuming 10% transduction frequency and 3 to 30 fold expansion before reinfusion (Tab. 12.2). Thus, to obtain therapeutically effective reconstitution with gene-modified cells, *in vitro*-selected cells must be reinfused in an aplastic environment (Fig. 12.2). This might represent a risky procedure in the absence of predictive indicators for a long-term reconstitution capacity of the selected graft. Moreover, extinction of transgene expression related to negative epigenetic control (silencing) occurring *in vivo* cannot be controlled using *in vitro* selection systems. In contrast, *in vivo* selection might enable step-wise expansion of genetically and phenotypically modified HPC in accord with the needs of the individual treatment. However, when only low gene transfer frequencies are obtained, or only a minority of the transduced cells expresses the selectable marker at sufficient levels, a strong and sustained proliferative pressure on transduced LTRC results *in vivo* which might cause oligoclonal repopulation, and finally, exhaustion of stem cell function (46). Thus, a good gene transfer frequency and expansion rate (see chapter 12.4) and improved vector backbones (see chapter 12.6) are equal important determinants for successful *in vivo* selection.

Table 12.3 In vivo selectable markers and drugs applicable for selection.

Gene	Immuno-genicity	Mode of Action	Drug Resistance	Cycle Dependence	Muta-genicity
Multidrug Resistance 1	no	Membrane-located efflex-plump	Taxoids, Colchicine, Vinca Alkaloids[a]	yes	no[1]
Dehydrofolate Reductase	low?	Drug insensitive mutant	Methotrexate	yes	no
O^6-alkylguanine-DNA-alkyl-transferase	no	Repairs DNA	Nitrosoureas	no	yes
Aldehyde Dehydrogenase 1	no	Metabolizes drug	Cyclophosphamide	no	yes

[a] Other drugs covered by multidrug-resistance like anthracyclines or epipodophyllotoxins may be mutagenic.

Current *in vivo* selection systems rely on so-called cancer drug resistance genes (CDRG), i.e. genes encoding proteins that protect against myeloablative side effects of drugs usually administered in anticancer-chemotherapy. Consequently, transfer and expression of CDRG in HPC has another important application in patients undergoing high-dose chemotherapy for the treatment of advanced malignancies (stem cell protection) (27, 28). Here, myelotoxic side effects most often represent *the* dose-limiting event precluding administration of curative anticancer-regimens. Furthermore, in long-term survivors of chemotherapy, mutagenic lesions induced by cytostatic agents can result in fatal myelodysplasia or secondary hematopoietic malignancies. Transfer of CDRG with retroviral vectors is anticipated to result in a protection level that progresses with each cycle of chemotherapy due to *in vivo* selection and expansion of resistant HPC. In consequence, from an estimated rate of 10% of protected HPC onwards, myelosuppression might no longer occur at a clinically significant level. Furthermore, transfer of CDRG has the unique potential of preventing even long-term side effects of chemotherapy on hematopoietic cells (27, 133). Therefore, stem cell protection might result in a markedly elevated therapeutic index of anticancer-chemotherapy. Stem cell protection studies will also provide important insight into the feasibility of selective enrichment of transduced stem cells *in vivo*.

Candidate CDRG useful for *in vivo* selection and stem cell protection are listed in Table 12.3. These include endogenous sequences like multidrug resistance 1 (mdr1), aldehyde dehydrogenase 1, or O^6-alkylguanine-DNA-alkyltransferase, retroviral transfer of which mediates resistance against drugs like taxoids (134, 135), cyclophosphamide (136), and nitrosoureas (137), respectively. Furthermore, mutant dehydrofolate reductase was successfully used for amplification of gene-modified marrow in mice (138), but it is at present unknown whether it might be immunogenic

Table 12.4 Determinants of Posttransduction/Posttransplantation Chimerism.

Level	Ex vivo Culture	Transplantation Regimen	Vector
Frequency of transduced cells	Transduction rate Expansion factor	Competitive vs. uncompetitive repopulation	Effective titer
Recruitment of transduced cells	Preservation of stem cell potential	Posttransplantation dysequilibrium	Selective advantage/ dysadvantage
Phenotype of transduced cells	No effect	No effect	Cis-active elements (level and stability of expression)

in man when expressed in hematopoietic cells. Indeed, an absolute prerequisite for *in vivo* selection of gene-modified HPC is that the expression level of the CDRG needed to mediate the selectable phenotype does not impair the biological integrity of transduced cells (immunogenicity, repopulating capacity, differentiation potential, effector functions of mature cells). Another important aspect, or even an absolute prerequisite when thinking of corrective gene transfer in patients for which conservative treatments exist, is that the side effects of the agent used for selection must be tolerable, with special concern for a potential mutagenicity. In the light of these considerations, mdr1 currently represents the most promising CDRG. In murine test systems, elevated expression of mdr1 mediates a dominant-selectable phenotype in otherwise normal HPC (139-142). Furthermore, the taxoid paclitaxel, a drug recognized by the mdr1-encoded efflux pump P-glycoprotein, is far less mutagenic then nitrosoureas or cyclophosphamide. The advantages of the latter drugs, however, would be that they act in a cell-cycle independent manner, and are thus presumably more effective at the stem cell level. Since all CDRG need fairly high expression levels to fully protect transduced cells against myeloablative drug-doses, their use is dependent on the development of improved vectors for transduction of HPC (see section 12.6).

In summary, the rate of posttransduction / posttransplantation chimerism in the host, and thus the therapeutic efficacy of the gene transfer, is determined by a combination of interdependent factors (Tab. 12.4). The general frequency of transduced cells depends on the transduction rate and the expansion factor achieved during *ex vivo* culture. The quality of the *ex vivo* culture system also influences to what extent the manipulated cells will be recruited to participate in hematopoiesis after reinfusion (preservation of stem cell potential). The transplantation regimen is also a crucial determinant for the frequency of transduced cells present in the host after reinfusion. Competitive repopulation avoids the cumbersome and highly toxic conditioning regimen, but results in a further "dilution" of gene-modified cells. Furthermore, *ex vivo* culture conditions and the type of transplantation regimen most likely influence the extent of the posttransplantation dysequilibrium, characterized by clonal fluctuations in stem cell recruitment with unpredictable chimerism during this phase. Finally, the vector determines the rate of chimerism at all three levels: the general frequency of transduced cells is dependent on the effective titer and transduction efficieny of the vector; the recruitment of transduced cells is reduced when the transgene confers a selective dysadvantage and enhanced when the transgene confers a selective advantage to stem cells (either per se or upon administration of cytotoxic agents); and the phenotype of transduced cells is actually modulated by the choice of the vector's cis-active elements which affect the level and the stability of transgene expression.

12.6 Transgene Expression

Once stable integration of a gene of interest is achieved, the cis-acting elements of the vector come into play. To enable selection, the marker gene must be under control of enhancers that permit high gene expression levels in early and late HPC. Retroviral control elements of MuLV were selected *in vivo* for high and wide-spread activity in HPC and thus seem to be perfect candidates when attempting to achieve high and unregulated gene expression levels in the hematopoietic system. Furthermore, the classical retroviral vector architecture (LTR-leader-gene-LTR, Fig. 12.3) seemingly represents the best tool for transduction of HPC (143, 144). However, conventional retroviral vectors utilizing cis-acting sequences of the lymphotropic Moloney MuLV (MoMuLV) are inefficiently expressed in immature myeloid cells (115, 145).

The retroviral cis-regulatory control elements defining the level as well as the duration of gene expression in the target cell are mainly located in the U3 region of the long terminal repeat (LTR) and in the leader-region, corresponding to the 5' untranslated sequence of the vector (Fig. 12.3). Using retroviral transfer and transient transfection of reporter gene constructs, we attempted to systematically analyze the impact of retroviral cis-regulatory sequences for gene expression in various stages and lineages of hematopoietic development (146-148). Embryonic stem cell lines were taken as a surrogate for HSC (149-151). We concentrated on cis-regulatory sequences derived from MoMuLV, the myeloproliferative sarcoma virus (MPSV), the PCC4-cell passaged MPSV (PCMV), the murine embryonic stem cell virus (MESV), the polycythaemic strain of the spleen focus-forming virus (SFFVp) and the malignant histiocytosis virus (MHSV), all known to be involved in the determination of viral host range *in vivo* and *in vitro* (115, 150, 152-160).

Our experiments corroborate earlier assumptions that the U3 region of conventional vectors based on MoMuLV is rather lymphotropic, leading to inefficient recognition in early, multi-lineage HPC and only moderate activity in more mature non-lymphoid progenitors and precursors. The deficiency of the MoMuLV U3 can be overcome by use of corresponding sequences derived from the MoMuLV-related MPSV (115). An important novel finding was that U3-regions from Friend mink cell focus-forming viruses (FMCF) such as SFFVp further exceed the activity of MPSV in all myelo-erythroid stages and lineages (146-148, 161) and, strikingly, even in

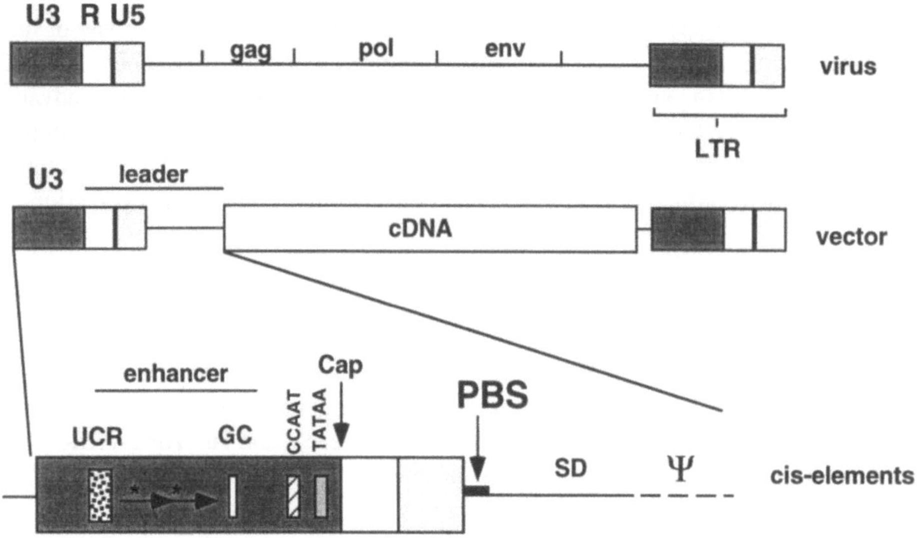

Fig. 12.3: Cis-elements of MuLV-based retroviral vectors.
On top a schematic scheme of a replication competent MuLV; shown is the proviral form. Below a typical architecture of a classical retroviral vector; the viral coding sequences are replaced with the gene of interest. Expression of this gene is still governed by viral cis-elements, especially the enhancer/promoter located in the U3-region of the long terminal repeat (LTR) and the primer binding site (PBS). The retroviral enhancer determines the strength of transcription from the retroviral CCAAT/TATAA-promoter. UCR, upstream control region, containing an Sp1-site in case of SFFVp; GC, GC-rich region, containing an Sp1-site in case of MPSV, PCMV, and MESV. The retroviral enhancer core is indicated by an asterisk; it is located in the center of a 75 base-pair stretch which is often found duplicated as a direct repeat, schematically represented as horizontal arrows. The enhancer-region from the UCR to the GC-rich part might form a hairpin or cloverleaf-like secondary structure (see section 7). Transcription is initiated at the boundary to the R region (cap-site). The PBS in the 5' untranslated region (leader) can act as a transcriptional repressor. In MESV, the PBS contains 5 point mutations that disturb access of the yet unidentified, stage-specific transcriptional repressor. SD, splice donor; Ψ, packaging signal.

embryonic stem cells, possibly related to the pluripotent HSC (149). Consequently, sustained and high-level multi-lineage gene expression in HPC from a retroviral vector controlled by the SFFVp-enhancer has been demonstrated in the murine system *in vivo* (162).

No evidence has been obtained for lineage *specificity* of retroviral enhancers; instead, they seem to be rather promiscuous and only confer gross stage and tissue *tropism* within the hematopoietic system. This might be explained by the presence of recognition sites for a variety of transcription factors involved in transcriptional control of multiple hematopoietic lineages, such as ets-family members, basic helix-loop-helix (bHLH) proteins, and AML1/core binding factor/ polyoma enhancer binding protein (PEBP) (146, 163-166). Finally, the overall activity of retroviral enhancers might be generally reduced in early HPC in comparison to more mature precursors, possibly a consequence of uncomplete enhancer utilization or active, stage-specific repression (147).

According to studies in embryonic stem cells, the MPSV U3 gains activity in primitive cells by mutations disturbing the consensus for a repressor known as embryonic LTR-binding protein targeting the upstream control region located at the 5' basis of the retroviral enhancer (154, 167) and, even more importantly, by a single point mutation creating a binding site for the ubiquitous transcriptional activator, Sp1, in the GC-rich downstream part of the retroviral enhancer (Fig. 12.3) (151, 152). FMCF-type U3 regions carry mutations in all sites known to influence the host range of murine retroviruses (146, 147, 168). Of considerable importance is the presence of two point mutations creating an Sp1 binding site in the upstream control region, and another two point mutations forming a specific and high-affinity target for PEBP in the retroviral enhancer core (146), known to be the major determinant for lineage tropism of retroviral pathogenicity in mice (158). Sp1 and PEBP might cooperate with ets-factors, bHLH-proteins and other, yet uncloned activities to achieve complete transactivation of the retroviral enhancer (169).

Stage-specific transcriptional repression of retroviral gene expression became overt when studying the influence of the sequences coincident with the retroviral primer binding site (PBS), a crucial cis-regulatory element needed to initiate the first strand synthesis during reverse transcription of the retroviral genome (Fig. 12.3). The PBS corresponding to $tRNA^{Pro}$ of conventional vectors acts as a target for transcriptional repression in early HPC (148), similarly as previously shown for embryonic cells (150). Lymphatic cells and fibroblasts are less affected by this restriction. In contrast, the PBS for $tRNA^{Gln}$ of MESV restores the retroviral enhancer activity in HPC (148), as previously shown for embryonic stem cells (150). Accordingly, vectors containing an altered PBS show better gene expression profiles in the murine system *in vivo* (144) and reduced incidence of transcriptional shut-down in human HPC-lines *in vitro* (170). However, while the PBS for $tRNA^{Pro}$ acts in a dominant negative manner in embryonic stem cells, its repressory activity in hematopoietic cells is rather relative than absolute, possibly a consequence of stronger enhancer utilization.

Following these observations, we cloned novel hybrid vectors containing the leader region including the permissive PBS of MESV and U3 regions from MPSV or the FMCF-type SFFVp. These hybrid vectors were named MPEV (for MPSV/MESV vector) and FMEV (for FMCF/MESV vector), respectively (148). Transfer of mdr1 revealed the elevated gene expression levels achieved with these vectors in HPC.

Retroviral transfer and expression of mdr1 in primary murine and human HPC by vectors based on cis-regulatory elements of MoMuLV (139) or the related Harvey murine sarcoma virus (HaMuSV) (134) results in an enhanced expression of the drug efflux pump P-glycoprotein, with no significant differences observed between these two backbones. Both vector-types confer a dominant selectable phenotype in the presence of cytotoxic agents like colchicine or paclitaxel *in vitro* (134, 135), and, in mouse models, also *in vivo* (139-142). The efficiency of protection against high doses of chemotherapeutic agents, however, was previously not systematically addressed. We observed that expression of mdr1 by FMEV and, somewhat less effectively, MPEV already significantly increases the survival of transduced primary human HPC under moderate drug doses when compared with a conventional MoMuLV-related vector. Furthermore, only FMEV and MPEV mediate high-dose resistance in primary human HPC, and thus background-free selectability (Fig. 12.4). Moreover, the *in vitro* data obtained with primary human HPC thus far showed no evidence for negative effects of the elevated mdr1-expression levels achieved with MPEV or FMEV on the biological integrity of transduced cells (161).

A first series of studies initiated in mice revealed that SF-MDR, a vector expressing mdr1 from an FMEV-backbone, also greatly improves myeloprotection *in vivo* (Fig. 12.4). Reinfusion of SF-MDR transduced HPC allowed dose-escalation and intensification of paclitaxel without limiting myelotoxicity. In contrast, the expression levels of mdr1 achieved with a conventional MoMuLV-based vector were not sufficient to achieve myeloprotection under high-dose conditions (171).

FMEV and MPEV backbones significantly improve retroviral gene expression in HPC by incorporation of appropriate enhancers and by elimination of transcriptional repressors. Further improvement of retroviral vector backbones will result from the analysis of vector-host interactions controlling post-transcriptional events. Suitable strategies include stabilization of vector-transcripts by utilization of splice signals and improvement of translation efficiency by inclusion of appropriate leader sequences (172). Thus, promising tools are arising for gene therapy protocols not requiring regulated gene expression, e.g., correction of lysosomal storage disorders or mucopolysaccharidoses, or for suppression of HIV-infection. In these cases, the therapeutic gene of interest can be linked to the selectable marker gene to provide sustained coexpression. Most work in this field was performed with the mdr1 gene coexpressed with the gene of interest via fusion gene strategies (173) or an internal ribosomal entry site (174). *Ex vivo* and *in vivo* selection might thus be applicable to select for those HPC that provide optimal gene expression levels of the therapeutic gene.

However, problems are encountered when attempting to transfer genes that need regulated expression in restricted lineages of mature hematopoietic cells. Best studied is transfer of genes for β-globin or γ-globin for correction of β-thalassemia or sickle cell anemia, respectively. Here, tissue-specific transcriptional control is achieved by inclusion of locus control regions (LCR) and enhancers of the β-globin gene. Unfortunately, LCR-sequences negatively affect the stability of retroviral vectors. Great efforts were necessary to develop high-titer vectors that preserved regulated expression of globin genes (6, 175). However, there is still a high clonal variablity of gene expression levels with these vectors, which can easily be explained by modulating effects of the random integration sites. *In vivo* selection for highly expressing clones will be difficult to achieve, because the target cell populations for

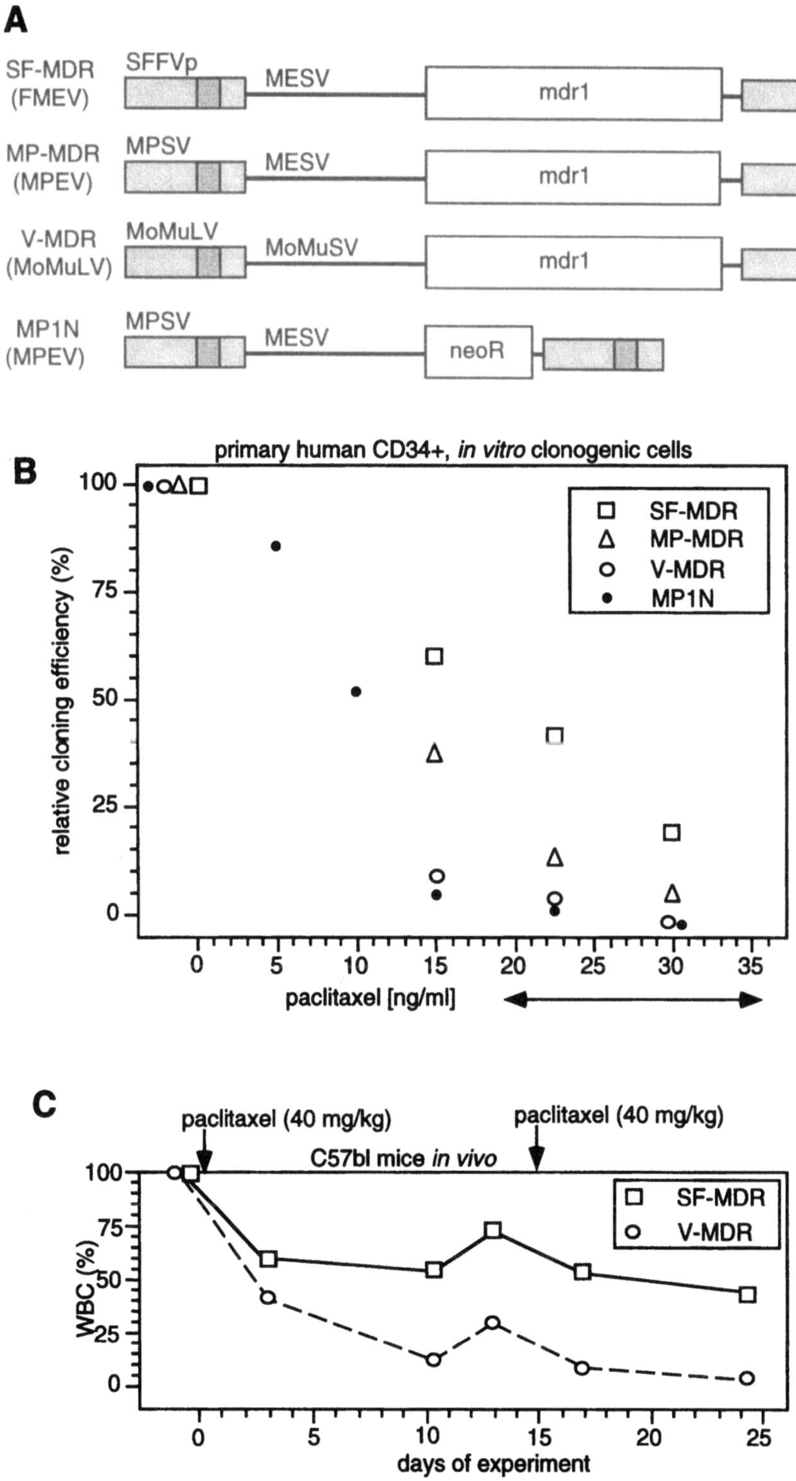
A
SF-MDR
(FMEV)
SFFVp
MESV
mdr1
MP-MDR
(MPEV)
MPSV
MESV
mdr1
V-MDR
(MoMuLV)
MoMuLV
MoMuSV
mdr1
MP1N
(MPEV)
MPSV
MESV
neoR
B
primary human CD34+, in vitro clonogenic cells
SF-MDR
MP-MDR
V-MDR
MP1N
relative cloning efficiency (%)
100
75
50
25
0
0
5
10
15
20
25
30
35
paclitaxel [ng/ml]
C
paclitaxel (40 mg/kg)
paclitaxel (40 mg/kg)
C57bl mice in vivo
SF-MDR
V-MDR
WBC (%)
100
75
50
25
0
0
5
10
15
20
25
days of experiment

◁ Fig. 12.4: The efficacy of progenitor cell selection using cancer drug resitance genes is drastically dependent on the vector-backbone.
A, Retroviral vectors for expression of the mdr1 gene. Shown are proviral forms after genomic integration. SF-MDR, an FMEV-type vector, contains the LTR of SFFVp and the leader of MESV. MP-MDR (MPEV-type) contains the LTR of MPSV and the leader of MESV. V-MDR (MoMuLV-type) contains the LTR of MoMuLV and the leader of the Moloney murine sarcoma virus (MoMuSV). MP1N contains the neomycin-phosphotransferase II gene (neoR) which does not mediate multidrug-resistance (control vector).
B, Protection against high-doses of paclitaxel in primary $CD34^+$ hHPC transduced with SF-MDR and MP-MDR and assayed in methylcellulose colony assays. The double-headed arrow delineates the peak plasma level of paclitaxel achieved during a three-hours infusion of 135-175 mg drug/m^2 body surface. The clonogenic capacity of HPC transduced with SF-MDR, MP-MDR or V-MDR under increasing drug doses depends on the gene expression levels achieved with the different vectors (modified from ref. 161).
C, SF-MDR (FMEV-backbone) greatly improves protection of HPC against an intensified regimen of paclitaxel *in vivo*. C57bl mice were engrafted with transduced syngeneic bone marrow and subjected to intraperitoneal injections of paclitaxel 6 weeks after transplantation as described in the text. The drop in the white blood cell count determined from peripheral blood reflects myelotoxicity of paclitaxel. Mean values of 4-5 sex/age matched animals. The doses of paclitaxel applied clearly are at least 4 fold higher than those formerly used for *in vivo* selection with V-MDR (139). 40 mg/kg paclitaxel in mice corresponds to roughly 360 mg/m^2 in man (139). Thus, myeloprotection with the improved vector SF-MDR, but not with V-MDR allows dose escalation in this model (modified from ref. 171).

expression of the selectable marker gene (LTRC) and for the corrective gene (late erythroblasts) are not the same. Thus, corrective gene transfer for inherited anemias can only be envisaged when vectors will be developed that escape clonal variabilities in gene expression (see section 12.7) or when *ex vivo* expansion of selected clones that confer sustained and high expression of therapeutic genes is feasible.

One final obstacle remains for patients treated for inherited deficiency syndromes: expression of the corrective protein can generate novel epitopes that might become target for immunological sensitization and rejection (176). This is an urgent problem to be addressed in appropriate animal models in order to define strategies how to avoid immunological elimination of engineered, therapeutic cells.

12.7 Transgene Silencing

Besides the difficulties encountered in expansion, transduction, and selection of transduced cells, silencing represents the fourth major hurdle to successful gene therapy in hematopoietic cells. Silencing can be defined as transcriptional shut-off of gene expression mediated by epigenetic modifications of an otherwise preserved gene. Silencing is used to control transcription of endogenous genes during hematopoietic differentiation (177). When the vector is silenced, the therapeutic effect of the transgene will be lost. This is especially problematic when considering suicide gene transfer experiments: after administration of the prodrug to be converted by the suicide-enzyme, cells in which the vector has been silenced will survive.

As outlined above, the gene transfer is targeted at an immature cell population that goes through multiple divisions and differentiation events until final maturation. Thus, silencing might be of special relevance in the hematopoietic system, especially when attempting to express genes in mature cells. Factors determining silencing events include the position of the gene within the genome and the cis-regulatory elements of the gene (178). First, silencing occurs when genes are inserted to unaccustomed positions within the genome. This process is also known as position-effect variegation (PEV) and is best studied in *Drosophila*, where it was originally described, but is likewise observed in transgenic plants and transgenic mammals (179, 180). Second, silencing is also known as the yet poorly defined mechanism which mediates shut-off of genes whose activity is superfluous for or contradicts the proper function of a cell. Thus, it is controlled by yet unidentified events acting in response to the cellular differentiation status and/or environmental signals (181). Moreover, silencing of a newly incorporated genetic sequence might in turn negatively affect neighbouring endogenous sequences, possibly representing a yet poorly defined mechanism of insertional mutagenesis (182).

With the exception of homologous recombination, transgenes introduced into a host genome, irrespective of the gene transfer method used, integrate into a novel chromosomal position. However, two lessons learned from plants and *Drosophila* indicate that the gene transfer method might well influence the incidence and dynamics of transgene silencing (179, 180). First, multiple copy integrates seem to be preferentially subjected to silencing. Second, PEV addresses incompatibilities that can occur when the transgene happens to be introduced close to or in a heterochromatin environment. Under these circumstances, the heterochromatin structure can spread across the transgene, causing its inactivation. Retroviral and AAV based gene tansfer technologies, in contrast to all non-viral transduction methods, avoid the multiple copy effect. Furthermore, retroviral systems presumably escape PEV to a certain extent, because integration preferentially occurs in transcriptionally open regions of euchromatin. However, when the cell differentiates or responds to environmental changes, previously open genomic sequences neighbouring the retroviral integration site might be subject to silencing. Then, silencing-initiating activities might also govern the integration site. Position-dependent inactivation of retroviral sequences has been clearly demonstrated in the fibroblast model (183). In the hematopoietic system, extinction of MoMuLV-vector mediated gene expression *in vivo* is additionally associated with the developmental stage of the target cell: gene transfer in LTRC is associated with a high rate of silencing, whereas transduction of more mature progenitors and pecursors is not (145). However, no correlation was observed between the position- dependent basal expression level of retroviral vectors and the long-term stability of gene expression in a non-differentiating *in vitro* system (immortalized lymphatic cells) (184).

Assuming that there is a conflict between silencing-initiating events and transcriptional activators targeting transgenic enhancers, silencing may dominate if the transgene is controlled by cis-acting elements not optimized for the transcriptional status of the host cell. Such a mechanism would not be determined by the gene transfer method but rather by the sequence of the transgene itself. Since the site of integration can not be controlled in current viral gene transfer technologies, defining the mechanisms responsible for sequence-specificity of silencing is one of the ways to escape transgene silencing. Evidence for sequence-specificity of silencing was obtained in several experimental systems. In fibroblasts, long-term expression from a retroviral vector was achieved by usage of an internal "house-keeping" promoter, whereas the MoMuLV-enhancer/promoter was silenced (185). Furthermore, two house-keeping promoters have been described where long-term expression is dependent on the presence of a binding site for the ubiquitous transcriptional activator, Sp1 (186, 187). Likewise, in embryonic stem cells otherwise silent retroviral enhancers are activated when containing an Sp1-binding site (151). Thus, subtle changes in the enhancer configuration of transgenes might drastically influence the rate of silencing. Importantly, it is not simply the transcriptional strength (188) but rather the presence of distinct cis-active elements that seemingly determines the potency for long-term expression.

Another strategy to escape silencing is to protect the transgene from environmental suppression. Interestingly, retroviruses might already have developed strategies to counteract negative position-effects. Retroviral enhancers of MuLV, for instance, create a large inverted repeat that might form a huge hairpin structure *in vivo* (189). The hairpin might function in reducing the position-dependence of enhancer recognition, and in allowing formation of complex transcription factor interactions on the highly condensed retroviral enhancer. Crucial responsive elements of retroviral enhancers indeed map to intriguing positions within the hairpin structure (146). An important role for stabilization of the hairpin structure might be ascribed to the upstream control region at the 5' base of the inverted repeat. In MoMuLV, this position is target for a putative transcriptional repressor, the embryonic LTR-binding protein, which belongs to the steroid receptor superfamily (167). In the strong enhancer of SFFVp, this position is mutated to create a binding-site for Sp1 (146, 147), which is also involved in chromatin-dependent regulation of the HIV-enhancer (190). Accordingly, preliminary data indicate that deletion of the retroviral enhancer in so-called suicide vectors might be associated with an elevated incidence of silencing (191).

Further autonomy of the retroviral enhancer might be achieved by inclusion of matrix-attachment regions (MARs), i.e. regions that are necessary to organize seperate transcriptional domains. MARs reduce the position-dependent variations of transgene activities (192). However, their potency for protecting retrovirally transduced genes in differentiating hematopoietic cells still remains to be addressed. In initial attempts to clone retroviral vectors containing MAR-sequences, we were confronted with pronounced vector-instability, most propably due to the presence of intron-like, AT-rich sequences in MARs.

A final solution to escape silencing is to develop transfer systems that allow site-specific insertion of transgenes. A first step towards this goal was the cloning of retroviral vectors which permit excision of transduced sequences using the bacterial cre/lox system (193). Using these vectors, sustained amplification of selected HPC clones might be possible after transduction with proliferation-promoting genes like HOX-B4 (61). Characterization of the integration site might allow a prediction of its transcriptional regulation during hematopoietic development. The single lox-site present after excision might be exploitable for insertion of novel sequences using the cre- recombinase. This system would allow a regulated integration event and has the additional advantage of non-viral transfection systems in that there is no DNA-size limitation. Tissue-specific control elements and MARs could thus be accomodated. In contrast to conventional non-viral transfection methods, integration might occur more efficiently, and at a predictable multiplicity (usually one gene per cell). However, considerable efforts are still needed to elaborate systems that prefer insertion over excision and to analyze potential safety hazards associated with the use of a xenogeneic recombinase system.

12.8 Conclusions

In conclusion, improving the methodology for transduction of hematopoietic cells must address the inherent complexity, potential and limitations of this fascinating system. The hierarchical organization of the hematopoietic system in principle allows for reconstitution of the complete organ with genetically modified cells, given that stable gene expression and selection for transduced cells can be achieved. On the other hand, the lack of lineage-restricted stem cells with self-renewal capacity requires the incorporation of lineage-specific promoters to ensure correction of inherited metabolic deficiences affecting defined subpopulations of mature cells such as hemoglobinopathies. Progress in gene therapy in hematopoeitic cells requires focusing on the four major hurdles currently encountered, i.e., defining the conditions for *ex vivo* expansion of LTRC, elevating the gene transfer frequency in LTRC, optimizing the methods for selection of transduced cells, and designing vectors that enable stable, persistent and sufficient expression levels of the gene of interest.

Gene marking studies will continue to provide information on safety-aspects of the gene transfer, as well as on the fate of genetically marked cells *in vivo*. The stem cell protection approach may not only occupy a central role in cancer therapy in the near future, but also provide important insight into the feasibility of selection of transduced HPC. With the improvements in transduction, expansion, and selection protocols, corrective or supplementary gene transfer for disorders not requiring regulatable gene expression will soon become feasible, rendering this approach a promising alternative to allogeneic bone marrow transplantation. Therefore, with continuous elaboration of the methodology, gene transfer and transgene expression in hematopoietic cells will maintain a central and paradigmatic role in gene therapy research.

Acknowledgements

The Heinrich-Pette-Institut is financially supported by the Freie und Hansestadt Hamburg and by the Bundesministerium für Gesundheit, Germany. The work in the lab of the authors is further supported by grants of the Deutsche Krebshilfe and the Bundesministerium für Bildung und Forschung, Germany.

References

1. Joyner, A., Keller, G., Phillips, R. A., and Bernstein, A. (1983). Retrovirus transfer of a bacterial gene into mouse hematopoietic progenitor cells. Nature *305*, 556-558.
2. Williams, D. A., Lemischka, I. R., Nathan, D. G., and Mulligan, R. C. (1984). Introduction of new genetic material into pluripotent haematopoietic stem cells of the mouse. Nature *310*, 476-480.
3. Rosenberg, S. A., Aebersold, P., Cornetta, K., Kasid, A., Morgan, R. A., Moen, R., Karson, E. M., Lotze, M. T., Yang, J. C., Topalian, S. L., et al. (1990). Gene transfer into humans – immunotherapy of patients with advanced melanoma, using tumor-infiltrating lymphocytes modified by retroviral gene transduction. N. Engl. J. Med. *323*, 570-578.
4. Brenner, M. K., McGarrity, G. J., Kohn, D. B., and Heslop, H. E. (1995). The practice of clinical gene therapy. In: Hematology 1995. Education Program of the American Society of Hematology, 93-100.
5. Einerhand, M. P. W., Antoniou, M., Zolotukhin, S., Muzyczka, N., Berns, K. I., Grosveld, F., and Valerio, D. (1995). Regulated high-level human β-globin gene expression in erythroid cells following recombinant adeno-associated virus-mediated gene transfer. Gene Ther. *2*, 336-343.
6. Leboulch, P., Huang, G. M., Humphries, R. K., Oh, Y. H., Eaves, C. J., Tuan, D. Y., and London, I. M. (1994). Mutagenesis of retroviral vectors transducing human β-globin gene and β-globin locus control region derivatives results in stable transmission of an active transcriptional structure. EMBO J. *13*, 3065-76.
7. Scott, H. S., Anson, D. S., Osborn, A. M., Nelson, P. V., Clements, P. R., Morris, C. P., and Hopwood, J. J. (1991). Human alpha-L-iduronidase: cDNA isolation and expression. Proc. Natl. Acad. Sci. USA *88*, 9695-9699.
8. Wilson, P. J., Morris, C. P., Anson, D. S., Occhiodoro, T., Bielicki, J., Clements, P. R., and Hopwood, J. J. (1990). Hunter syndrome: isolation of an iduronate-2-sulfatase cDNA clone and analysis of patient DNA. Proc. Natl. Acad. Sci. USA *87*, 8531-8535.
9. Robertson, D. A., Freeman, C., Morris, C. P., and Hopwood, J. J. (1992). A cDNA clone for human glucosamine-6-sulphatase reveals differences between arylsulphatases and non-arylsulphatases. Biochem. J. *288*, 539-544.
10. Tomatsu, S., Fukuda, S., and Masue, M. (1991). Morquio disease: isolation, characterization and expression of a full-length cDNA for human N-acetyl-galactosamine-6-sulfate sulfatase. Biochem. Biophys. Res. Commun. *2*, 677-683.
11. Marechal, V., Naffakh, N., Danos, O., and Heard, J. M. (1993). Disappearance of lysosomal storage in spleen and liver of mucopolysaccharidosis VII mice after transplantation of genetically modified bone marrow cells. Blood *82*, 1358-1365.
12. Peters, C., Rommerskirch, W., Modaressi, S., and von Figura, K. (1991). Restoration of arylsulfatase B activity in human mucopolysaccharidosis-type-VI fibroblasts by retroviral vector-mediated gene transfer. Biochem. J. *276*, 599-504.
13. Gieselmann, V., Zlotogara, J., Harris, A., Wenger, D. A., and Morris, C. P. (1994). Molecular genetics of metachromatic leukodystrophy. Hum. Mutation *4*, 233-242.
14. Levi Yeyati, P., Agmon, V., Fillat, C., Dinur, T., Dagan, A., Desnieck, R. J., Gatt, S., and Schuchmann, E. H. (1995). Fluorescence-based selection of retrovirally transduced cells in the absence of a marker gene: Direct selection of transduced type B Niemann-Pick disease cells and evidence for bystander correction. Hum. Gene Ther. *6*, 975-983.

15. Enomaa, N., Danos, O., Peltonen, L., and Jalanko, A. (1995). Correction of deficient enzyme activity in a lysosomal storage disease, aspartylglucosaminuria, by enzyme replacement and retroviral gene transfer. Hum. Gene Ther. *6*, 723-731.
16. Sugimoto, Y., Aksentijevich, I., Murray, G. J., Brady, R. O., Pastan, I., and Gottesman, M. M. (1995). Retroviral coexpression of a multidrug resistance gene (MDR1) and human alpha-galactosidase A for gene therapy of Fabry disease. Hum. Gene Ther. *6*, 905-915.
17. Wells, S., Malik, P., Pensiero, M., Kohn, D. B., and Nolta, J. A. (1995). The presence of an autologous marrow stromal cell layer increases glucocerebrosidase gene transduction of long-term culture initiating cells (LTCICs) from the bone marrow of a patient with Gaucher disease. Gene Ther. *2*, 512-520.
18. Kohn, D. B., Weinberg, K. I., Nolta, J. A., Lenarsky, C., Crooks, G. M., Hanley, M. E., Annett, G., Brooks, J. S., El-Khoureiy, A., Lawrence, K., Wells, S., Moen, R. C., Bastian, J., Williams-Herman, D. E., Elder, M., Wara, D., Bowen, T., Hershfield, M. S., Mullen, C. A., Blaese, R. M., and Parkman, R. (1995). Engraftment of gene-modified umbilical cord blood cells in neonates with adenosine deaminase deficiency. Nature Med. *1*, 1017-1023.
19. Market, M. L. (1991). Purine nucleoside phosphorylase deficiency. Immunodeficiency Rev. *3*, 45-81.
20. Puck, J. M., Deschênes, S. M., Porter, J. C., Dutra, A. S., Brown, C. J., Willard, H. F., and Henthorn, P. S. (1993). The interleukin-2 receptor gamma-chain maps to Xq13.1 and is mutated in X-linked severe combined immunodeficiency, SCIDX1. Hum. Mol. Genet. *2*, 1099-1104.
21. Wilson, R. W., Yorifuji, T., Lorenzo, I., Smith, W., Anderson, D. C., Belmont, J. W., and Beaudet, A. L. (1993). Expression of human CD18 in murine granulocytes and improved efficiency for infection of deficient human lymphoblasts. Hum. Gene Ther. *4*, 25-34.
22. Porter, C. D., Parkar, M. H., Levinsky, R. J., et al. (1993). X-linked chronic granulomatous disease: correction of NADPH oxidase defect by retrovirus-mediated expression of gp91-*phox*. Blood *82*, 2196-2202.
23. Porter, C. D., Parkar, M. H., Verhoeven, A. J., et al. (1994). p22-*phox*-deficient chronic granulomatous disease: identification of a biosynthetic intermediate of gp91-*phox* and reconstitution by retrovirus-mediated expression. Blood *84*, 2767-2775.
24. Thrasher, A., Chetty, M., Casimir, C., and Segal, A. W. (1992). Restoration of superoxide generation to a chronic granulomatous disease-derived B-cell line by retrovirus-mediated gene transfer. Blood *80*, 1125-1129.
25. Moreau-Gaudry, F., Ged, C., Barbot, C., Mazurier, F., Boiron, J. M., Bensidhoum, M., Reiffers, J., and De Verneuil, H. (1995). Correction of the enzyme defect in cultured congenital erythropoietic porphyria disease cells by retrovirus-mediated gene transfer. Hum. Gene Ther. *6*, 13-20.
26. Walsh, C. E., Mann, M. M., Emmons, R. V., Wang, S., and Liu, J. M. (1995). Transduction of CD34-enriched human peripheral and umbilical cord blood progenitors using a retroviral vector with the Fanconi anemia group C gene. J. Investig. Med. *43*, 379-385.
27. Baum, C., Margison, G. P., Eckert, H.-G., Fairbairn, L., Ostertag, W., and Rafferty, J. A. (1996). Gene transfer to augment the therapeutic index of anticancer chemotherapy. Gene Ther. *3*, 1-3.
28. Licht, T., Gottesman, M. M., and Pastan, I. (1995). Transfer of the MDR1 (multidrug resistance) gene: protection of hematopoietic cells from cytotoxic chemotherapy, and selection of transduced cells in vivo. Cytokines Mol. Ther. *1*, 11-20.
29. Yu, N., Poeschla, E., and Wong-Staal, F. (1994). Progress towards gene therapy for HIV infection. Gene Ther. *1*, 13-26.
30. Hao, Q.-L., Malik, P., Salazar, R., Tang, H., Gordon, E. M., and Kohn, D. B. (1995). Expression of biologically active human factor IX in human hematopoietic cells after retroviral vector-mediated gene transduction. Hum. Gene Ther. *6*, 873-880.
31. Chernajovski, Y., Adams, G., Podhajcer, O. L., Mueller, G. M., Robbins, P. D., and Feldman, M. (1995). Inhibition of transfer of collagen-induced arthritis into SCID mice by ex vivo infection of spleen cells with retroviruses expressing soluble tumor necrosis factor receptor. Gene Ther. *2*, 731-735.
32. Boggs, S. S., Patrene, K. D., Mueller, G. M., Evans, C. H., Doughty, L. A., and Robbins, P. D. (1995). Prolonged systemic expression of human IL-1 receptor anatagonist (hIL-1ra) in mice reconstituted with hematopoietic cells transduced with a retrovirus carrying the hIL-1ra cDNA. Gene Ther. *2*, 632-638.

33. Parrish, E. P., Cifuentes-Diaz, C., Li, Z.-L., Vicart, P., Paulin, D., Dreyfus, P. A., Peschanski, M., Harris, A. J., and Garcia, L. (1996). Targeting widespread sites of damage in dystrophic muscle: engrafted macrophages as potential shuttles. Gene Ther. *3*, 13-20.
34. Brenner, M. K. (1994). Genetic marking and manipulation of hematopoietic progenitor cells using retroviral vectors. Immunomethods *5*, 204-210.
35. Ogawa, M. (1993). Differentiation and proliferation of hematopoietic stem cells. Blood *81*, 2844-2853.
36. Morrison, S. J., Uchida, N., and Weissman, I. L. (1995). The biology of hematopoietic stem cells. Annu. Rev. Cell Dev. Biol. *11*, 35-71.
37. Metcalf, D. (1989). The molecular control of cell division, differentiation commitment and maturation in haemopoietic cells. Nature *339*, 27-30.
38. Just, U., Stocking, C., Spooncer, E., Dexter, T. M., and Ostertag, W. (1991). Expression of the GM-CSF gene after retroviral transfer in hematopoietic stem cell lines induces synchronous granulocyte-macrophage differentiation. Cell *64*, 1163-1173.
39. Morrison, S. J. and Weissman, I. L. (1994). The long-term repopulating subset of hematopoietic stem cells is deterministic and isolatable by phenotype. Immunity *1*, 661-673.
40. Hayflick, L. (1965). The limited in vitro lifetime of human diploid cell strains. Exp. Cell Res. *37*, 614-636.
41. Vaziri, H., Dragowska, W., Allsopp, R. C., Thomas, T. E., Harley, C. B., and Landsdorp, P. M. (1994). Evidence for a mitotic clock in human hematopoietic stem cells: loss of telomeric DNA with age. Proc. Natl. Acad. Sci. USA *91*, 9857-9860.
42. Siminovitch, L., McCulloch, E. A., and Till, J. E. (1963). The distribution of colony-forming cells among spleen colonies. J. Cell. Comp. Physiol. *62*, 327-336.
43. Harrison, D. E. and Zong, R.-K. (1992). The same exhaustible multilineage precursor produces both myeloid and lymphoid cells as early as 3-4 weeks after marrow transplantation. Proc. Natl. Acad. Sci. USA *89*, 10134-10138.
44. Lepault, F., Ezine, S., and Gagneralt, M. C. (1992). T- and B- lymphocyte potentials of spleen colony-forming cells. Blood *81*, 950.
45. Ball, T. C., Hirayama, F., and Ogawa, M. (1995). Lymphohematopoietic progenitors of normal mice. Blood *85*, 3086-3092.
46. Spangrude, G. J., Brooks, D. M., and Tumas, D. B. (1995). Long-term repopulation of irradiated mice with limiting numbers of purified hematopoietic stem cells: In vivo expansion of stem cell phenotype but not function. Blood *85*, 1006-1016.
47. Trevisan, M. and Iscove, N. (1995). Phenotypic analysis of murine long-term hematopoietic reconstituting cells quantitated competitively in vivo and comparison with more advanced colony-forming progeny. J. Exp. Med. *181*, 93-103.
48. Vormoor, J., Lapidot, T., Pflumio, F., Risdon, G., Patterson, B., Broxmeyer, H. E., and Dick, J. E. (1994). Immature human cord blood progenitors engraft and proliferate to high levels in severe combined immunodeficient mice. Blood *83*, 2489-2497.
49. Lapidot, T., Sirard, C., Vormoor, J., Murdoch, B., Hoang, T., Caceres, C. J., Minden, M., Paterson, B., Caligiuri, M. A., and Dick, J. E. (1994). A cell initiating human acute myeloid leukaemia after transplantation into SCID mice. Nature *367*, 645-648.
50. Zanjani, E. D., Flake, A. W., Rice, H., Hedrick, M., and Tavassoli, M. (1994). Long-term repopulating ability of xenogenieic transplanted human fetal liver hematopoietic stem cells in sheep. J. Cin. Invest. *93*, 1051-1055.
51. Dexter, T. M., Moore, M. A. S., and Sheridan, A. P. C. (1977). Maintenance of hemopoietic stem cells and production of differentiated progeny in allogeneic and semiallogeneic bone marrow chimeras in vitro. J. Exp. Med. *177*, 1612-1616.
52. Petzer, A. L., Hogge, D. E., Landdsdorp, P. M., Reid, D. S., and Eaves, C. J. (1996). Self-renewal of primitive human hematopoietic cells (long-term-culture-initiating-cells) in vitro and their expansion in defined medium. Proc. Natl. Acad. Sci. USA *93*, 1470-1474.
53. Ploemacher, R. E., van der Sluijs, J. P., Voerman, J. S. A., and Brons, N. H. C. (1989). An in vitro limiting-dilution assay of long-term repopulating hematopoietic stem cells in the mouse. Blood *84*, 3653-3659.
54. Ploemacher, R. E. (1994). Cobblestone area forming cell (CAFC) assay. In: Culture of hematopoietic cells, Freshney RI, Pragnell IB, and Freshney MG, eds. (New York: Wiley-Liss), 1-21.

55. Palacios, R., Golunski, E., and Samaridis, J. (1995). In vitro generation of hematopoietic stem cells from an embryonic stem cell line. Proc. Natl. Acad. Sci. USA *92*, 7530-7534.
56. Nakano, T., Kodama, H., and Honjo, T. (1994). Generation of lymphohematopoietic cells from embryonic stem cells in culture. Science *265*, 1098-1101.
57. Ponchio, L., Conneally, E., and Eaves, C. (1995). Quantitation of the quiescent fraction of long-term culture-initiating cells in normal human blood and marrow and the kinetics of their growth factor-stimulated entry into S-phase in vitro. Blood *86*, 3314-3321.
58. Verfaillie, C. M. and Miller, J. S. (1995). A novel single-cell proliferation assay shows that long-term culture-initiating cell (LTC-IC) maintenance over time results from the extensive proliferation of a small fraction of LTC-IC. Blood *86*, 2137-2145.
59. Nibbs, R. J. B., Itoh, K., Ostertag, W., and Harrison, P. R. (1994). Differentiation arrest and stromal cell-independent growth of murine erythroleukemia cells are associated with elevated expression of ets-related genes but not with mutation of p53. Mol. Cell. Biol. *13*, 5582-5592.
60. Itoh, K., Friel, J., Kluge, N., Kina, T., Kondo-Takaori, A., Kawamata, S., Uchiyama, T., and Ostertag, W. (1996). A novel hematopoietic mutlilineage clone, Myl-D-7, is stromal cell-dependent and supported by an alternative mechanism(s) independent of stem cell factor/c-kit interaction. Blood *87*, in press.
61. Sauvageau, G., Thorsteinsdottir, U., Eaves, C. J., Lawrence, H. J., Largman, C., Lansdorp, P. M., and Humphries, R. K. (1995). Overexpression of HOXB4 in hematopoietic cells causes the selective expansion of more primitive populations in vitro and in vivo. Genes Dev. *9*, 1753-1765.
62. Dick, J. E., Magli, M. C., Huszar, D., Phillips, R. A., and Bernstein, A. (1985). Introduction of a selectable gene in primitive stem cells capable of long-term reconstitution of the hematopoietic system of w/wv mice. Cell *42*, 71-79.
63. Wieder, R., Barak, V., and Ben-Ishay, Z. (1995). High-efficiency retroviral gene transfer into murine high-proliferative-potential cells cycle-activated by cytosine arabinosode. Hum. Gene Ther. *6*, 865-871.
64. Sutherland, D., Keating, A., Nayar, R., Anania, S., and Stewart, A. (1994). Sensitive detection and enumeration of CD34+ cells in peripheral blood by flow cytometry. Exp. Hematol. *22*, 1003.
65. Pettengell, R., Testa, N. G., Swindell, N. G., Crowther, D., and Dexter, T. M. (1993). Transplantation potential of hematopoietic cells released into the circulation during routine chemotherapy for non-Hodgkin's lymphoma. Blood *82*, 2239.
66. Socinski, M. A., Cannistra, S. A., Elias, A., Antman, K. H., Schnipper, L., and Griffin, J. D. (1988). Granulocyte-macrophage colony stimulating factor expands the circulating haemopoietic progenitor cell compartment in man. Lancet *1*, 1194-1198.
67. Gianni, A. M., Siena, S., Bregni, M., Tarella, C., Stern, A. C., Pileri, A., and Bonadonna, G. (1989). Granulocyte-macrophage colony-stimulating factor to harvest circulating haemopoietic stem cells for autotransplantation. Lancet *2*, 580-585.
68. Williams, D. A., and Moritz, T. (1994). Umbilical cord blood stem cells as targets for genetic modification: new therapeutic approaches to somatic gene therapy. Blood Cells *20*, 504-515.
69. DiGiusto, D. L., Lee, R., Moon, J., Moss, K., O'Toole, T., Voytovich, A., Webster, D., and Mule, J. J. (1996). Hematopoietic potential of cryopreserved and ex vivo manipulated umbilical cord blood progenitor cells evaluated in vitro and in vivo. Blood *87*, 1261-1271.
70. Hao, Q. L., Shah, A. J., Thiemann, F. T., Smogorzewska, E. M., and Crooks, G. M. (1995). A functional comparison of CD34 + CD38- cells in cord blood and bone marrow. Blood *86*, 3745-3753.
71. Broxmeyer, H. E. (1995). Cord blood as an alternative source for stem and progenitor cell transplantation. Curr. Opin. Pediatr. *7*, 47-55.
72. Krause, D. S., Fackler, M. J., Civin, C. I., and May, W. S. (1996). CD34: structure, biology, and clincal utility. Blood *87*, 1-13.
73. Siena, S., Bregni, M., Brando, B., Ravagnani, F., Bonadonna, G., and Gianni, A. M. (1989). Circulation of CD34+ hematopoietic stem cells in the peripheral blood of high-dose cyclophosphamide-treated patients: Enhancement by intravenous recombinant human granulocyte-macrophage colony-stimulating factor. Blood *74*, 1905-1914.
74. Shpall, E. J., Jones, R. B., Bearman, S. I., Franklin, W. A., Archer, P. G., Curiel, T., Bitter, M., Claman, H. N., Stemmer, S. M., Purdy, M., Myers, S. E., hami, L., Taffs, S., Heimfeld, S., Hallagan, J., and Berenson, R. J. (1994). Transplantation of enriched CD34-positive autologous marrow into breast cancer patients following high-dose chemotherapy: Influence of CD34-positive peripheral-blood progenitors and growth factors on engraftment. J. Clin. Oncol. *12*, 28-36.

75. Brugger, W., Heimbeld, S., Berenson, R., Mertelsmann, R., and Kanz, L. (1995). Reconstitution of hematopoiesis after high-dose chemotherapy by autologous progenitor cells generated ex vivo. N. Engl. J. Med. *333*, 283-287.
76. Civin, C. I., Trischmann, T., Davis, J., Noga, S., Cohen, K., Duffy, B., Wiley, J., Law, P., Hardwick, A., Oldham, F., and Gee, A. (1995). Highly purified CD34+ cells reconstitute hematopoiesis. Proc. Am. Soc. Clin. Oncol. *14*, 437-443.
77. Dunbar, C. E., Cottler, F. M., O'Shaughnessy, J. A., Doren, S., Carter, C., Berenson, R., Brown, S., Moen, R. C., Greenblatt, J., Stewart, F. M., and et al. (1995). Retrovirally marked CD34-enriched peripheral blood and bone marrow cells contribute to long-term engraftment after autologous transplantation. Blood *85*, 3048-3057.
78. Berenson, R. J., Bensinger, W. I., Hill, R. S., Andrews, R. G., Garcia-Lopez, J., Kalamasz, D. F., Still, B. J., Spitzer, G., Bruckner, C. D., Bernstein, I. D., and Thomas, E. D. (1991). Engraftment after infusion of CD34+ marrow cells in patients with breast cancer or neuroblastoma. Blood *77*, 1717-1722.
79. Krause, D. S., Ito, T., Fackler, M. J., Smith, O. M., Collector, M. I., Sharkis, S. J., and May, W. S. (1994). Characterization of murine CD34, a marker for hematopoietic progenitor and stem cells. Blood *84*, 691-701.
80. Rohdewohld, H., Weiher, H., Reik, W., Jaenisch, R., and Breindl, M. (1987). Retrovirus integration and chromatin structure: Moloney murine leukemia proviral integration sites map near DNase I-hypersensitive sites. J. Virol. *61*, 336-343.
81. Mooslehner, K., Karls, U., and Harbers, K. (1990). Retroviral integration sites in transgenic mov mice frequently map in the vicinity of transcribed DNA regions. J. Virol. *64*, 3056-3058.
82. Miller, A. D., Miller, D. G., Garcia, J. V., and Lynch, C. M. (1993). Use of retroviral vectors for gene transfer and expression. Methods Enzymol. *217*, 581-599.
83. Ferrari, G., Salvatori, G., Rossi, C., Cossu, G., and Mavilio, F. (1995). A retroviral vector containing a muscle-specific enhancer drives gene expressopn only in differentiated muscle fibers. Hum. Gene Ther. *6*, 733-742.
84. Roe, T., Reynolds, T., Yu, G., and Brown, P. O. (1993). Integration of murine leukemia virus DNA depends on mitosis. EMBO J. *12*, 2099-2108.
85. Miller, D. G., Adam, M. A., and Miller, A. D. (1990). Gene transfer by retrovirus vectors occurs only in cells that are actively replicating at the time of infection. Mol. Cell. Biol. *10*, 4239-4242.
86. Williams, D. A., Orkin, S. H., and Mulligan, R. C. (1986). Retrovirus-mediated transfer of human adenosine deaminase gene sequences into cells in culture and into murine hematopoietic cells in vivo. Proc. Natl. Acad. Sci. USA *83*, 2566-2570.
87. Bodine, D. M., McDonagh, K. T., Seidel, N. E., and Nienhuis, A. W. (1991). Survival and retrovirus infection of murine hematopoietic stem cells in vitro: Effects of 5-FU and method of infection. Exp. Hematol. *19*, 206-212.
88. Dick, J. E., Kamel, R. S., Murdoch, B., and Doedens, M. (1991). Gene transfer into normal human hematopoietic cells using in vitro and in vivo assays. Blood *78*, 624-634.
89. Bregni, M., Magni, M., Siena, S., Di, N. M., Bonadonna, G., and Gianni, A. M. (1992). Human peripheral blood hematopoietic progenitors are optimal targets of retroviral-mediated gene transfer. Blood *80*, 1418-1422.
90. Luskey, B. D., Rosenblatt, M., Zsebo, K., and Williams, D. A. (1992). Stem cell factor, interleukin-3, and interleukin-6 promote retroviral-mediated gene transfer into murine hematopoietic stem cells. Blood *80*, 396-402.
91. Bodine, D. M., Moritz, T., Donahue, R. E., Luskey, B. D., Kessler, S. W., Martin, D. I., Orkin, S. H., Nienhuis, A. W., and Williams, D. A. (1993). Long-term in vivo expression of a murine adenosine deaminase gene in rhesus monkey hematopoietic cells of multiple lineages after retroviral mediated gene transfer into CD34+ bone marrow cells. Blood *82*, 1975-1980.
92. Szilvassy, S. J. and Cory, S. (1994). Efficient retroviral gene transfer to purified long-term repopulating hematopoietic stem cells. Blood *84*, 74-83.
93. Brenner, M. K., Rill, D. R., Holladay, M. S., Heslop, H. E., Moen, R. C., Buschle, M., Krance, R. A., Santana, V. M., Anderson, W. F., and Ihle, J. N. (1993). Gene marking to determine whether autologous marrow infusion restores long-term haemopoiesis in cancer patients. Lancet *342*, 1134-1137.
94. Cassel, A., Cottler, F. M., Doren, S., and Dunbar, C. E. (1993). Retroviral-mediated gene transfer into CD34-enriched human peripheral blood stem cells. Exp. Hematol. *21*, 585-591.

95. Lu, M., Maruyama, M., Zhang, N., Levine, F., Friedmann, T., and Ho, A. D. (1994). High efficiency retroviral-mediated gene transduction into CD34+ cells purified from peripheral blood of breast cancer patients primed with chemotherapy and granulocyte-macrophage colony-stimulating factor. Hum. Gene Ther. *5*, 203-208.
96. Friel, J. and Ostertag, W. unpublished data.
97. Hajihosseini, M., Iavachev, L., and Price, J. (1993). Evidence that retroviruses integrate into post-replication host DNA. EMBO J. *12*, 4969-4974.
98. Anderson, W. F., McGarrity, G. J., and Moen, R. C. (1993). Report to the NIH recombinant DNA advisory committee on murine replication-competent retrovirus (RCR) assays. Hum. Gene Ther. *4*, 311-321.
99. Donahue, R. E., Kessler, S. W., Bodine, D., McDonagh, K., Dunbar, C., Goodman, S., Agricola, B., Byrne, E., Raffeld, M., Moen, R., and et al. (1992). Helper virus induced T cell lymphoma in nonhuman primates after retroviral mediated gene transfer. J. Exp. Med. *176*, 1125-35.
100. Sekhar, M., Kotani, M., Doren, S., Agarwal, R., McGarrity, G., and Dunbar, C. E. (1996). Retroviral transduction od CD34-enriched hematopoietic progenitor cells under serum-free conditions. Hum. Gene Ther. *7*, 33-38.
101. Kotani, H., Newton III, P. B., Zhang, S., Chiang, Y. L., Otto, E., Weaver, L., Blaese, R. M., Anderson, W. F., and McGarrity, G. J. (1994). Improved methods of retroviral vector transduction and production for gene therapy. Hum. Gene Ther. *5*, 19-28.
102. Chuck, A. S. and Palsson, B. O. (1996). Consistent and high rates of gene transfer can be obtained using flow-through transduction over a wide range of retroviral titers. Hum. Gene Ther. *7*, 743-750.
103. van Beusechem, V. W., Bart-Baumeister, J. A. K., Hoogerbrugge, P. M., and Valerio, D. (1995). Influence of interleukin-3, interleukin-6, and stem cell factor on retroviral transduction of Rhesus monkey CD34+ hematopoietic progenitor cells measured in vitro and in vivo. Gene Ther. *2*, 245-255.
104. Kiem, H.-P., Darovsky, B., von Kalle, C., Goehle, S., Graham, T., Miller, A. D., Storb, R., and Schuening, F. G. (1995). Long-term persistence of canine hematopoietic cells genetically marked by retrovirus vectors. Hum. Gene Ther. *7*, 89-96.
105. van Beusechem, V. W., Kukler, A., Heidt, P. J., and Valerio, D. (1992). Long-term expression of human adenosine deaminase in rhesus monkeys transplanted with retrovirus-infected bone-marrow cells. Proc. Natl. Acad. Sci. USA *89*, 7640-7644.
106. Hoogerbrugge, M., van Besusechem, V. W., Fischer, A., Debree, M., le Deist, F., Perignon, J. L., Morgan, R., Gaspar, B., Fairbanks, L. D., Skeoch, C. H., Moseley, A., Harvey, M., Levinsky, R. J., and Valerio, D. (1996). Bone marrow gene transfer in three patients with adenosine deaminase deficiency. Gene Ther. *3*, 179-183.
107. Hughes, P. F., Thacker, J. D., Hogge, D., Sutherland, H. J., Thomas, T. E., Lansdorp, P. M., Eaves, C. J., and Humphries, R. K. (1992). Retroviral gene transfer to primitive normal and leukemic hematopoietic cells using clinically applicable procedures. J. Clin. Invest. *89*, 1817-1824.
108. Moore, K. A., Deisseroth, A. B., Reading, C. L., Williams, D. E., and Belmont, J. W. (1992). Stromal support enhances cell-free retroviral vector transduction of human bone marrow long-term culture-initiating cells. Blood *79*, 1393-1399.
109. Moritz, T., Patel, V. P., and Williams, D. A. (1994). Bone marrow extracellular matrix molecules improve gene transfer into human hematopoietic cells via retroviral vectors. J. Clin. Invest. *93*, 1451-1457.
110. Eipers, P. G., Krauss, J. C., Palsson, B. O., Emerson, S. G., Todd, R. 3., and Clarke, M. F. (1995). Retroviral-mediated gene transfer in human bone marrow cells growth in continuous perfusion culture vessels. Blood *86*, 3754-3762.
111. Bienzle, D., Abrams-Ogg, A. C. G., Kruth, S. A., Ackland- Snow, J., Carter, R. F., Dick, J. E., Jacobs, R. M., Kamel-Reid, S., and Dubé, I. D. (1994). Gene transfer into hematopoietic stem cells: long-term maintenance of in vitro activated progenitors without marrow ablation. Proc. Natl. Acad. Sci. USA *91*, 350-354.
112. Naldini, L., Blömer, U., Gallay, P., Ory, D., Mulligan, R., Gage, F. H., Verma, I. M., and Trono, D. (1996). In vivo gene delivery and stable transduction of nondividing cells by a lentiviral vector. Science *272*, 263-267.
113. Russell, D. W. and Miller, A. D. (1996). Foamy virus vectors. J. Virol. *70*, 217-222.
114. Stevenson, M. (1996). Portals of entry: uncovering HIV nuclear transport pathways. Trends Cell Biol. *6*, 9-15.

115. Beck-Engeser, G., Stocking, C., Just, U., Albritton, L., Dexter, M., Spooncer, E., and Ostertag, W. (1991). Retroviral vectors related to the myeloproliferative sarcoma virus allow efficient expression in hematopoietic stem and precursor cell lines, but retroviral infection is reduced in more primitive cells. Hum. Gene Ther. *2*, 61-70.
116. Bayle, J.-Y., Johnson, L. G., St. George, J. A., Boucher, R. C., and Olsen, J. C. (1993). High-efficiency gene transfer to primary monkey airway epithelial cells with retrovirus vectors using the gibbon ape leukemia virus receptor. Hum. Gene Ther. *4*, 161-170.
117. Cosset, F.-L., Takeuchi, Y., Battini, J.-L., Weiss, R. A., and Collins, M. K. L. (1995). High-titer packaging cells producing recombinant retrovirus resistant to human serum. J. Virol. *69*, 7430-7436.
118. Burns, J. C., Friedmann, T., Driever, W., Burrascano, M., and Yee, J.-K. (1993). Vesicular stomatitis virus G glycoprotein pseudotyped retroviral vectors: Concentration to very high titer and efficient gene transfer into mammalian and nonmammalian cells. Proc. Natl. Acad. Sci. USA *90*, 8033-8037.
119. Russell, S. J. (1996). Peptide-dysplaying phages for targeted gene delivery? Nature Med. *2*, 276-277.
120. Miller, J. L. and et al. (1994). Recombinant adeno-associated virus (rAAV)-mediated expression of human gamma-globin gene in human progenitor-derived erythroid cells. Proc. Natl. Acad. Sci. USA *91*, 10183-10187.
121. Zhou, S. Z., Cooper, S., Kang, L. Y., Ruggieri, L., Heimfeld, S., Srivastava, A., and Broxmeyer, H. E. (1994). Adeno-associated virus 2-mediated high efficiency gene transfer into immature and mature subsets of hematopoietic progenitor cells in human umbilical cord blood. J. Exp. Med. *179*, 1867-1875.
122. Flotte, T. R. and Carter, B. J. (1995). Adeno-associated virus vectors for gene therapy. Gene Ther. *2*, 357-262.
123. Lemischka, I. R., Raulet, R.H., and Mulligan, R.C. (1986). Developmental potential and dynamic behaviour of hematopoietic stem cells. Cell *45*, 917-927.
124. Lemischka, I. R. (1992). What we have learned from retroviral marking of hematopoietic stem cells. Curr. Top. Microbiol. Immunol. *177*, 59-71.
125. Bender, J., To, L. B., Williams, S., and Schqartzberg, L. (1992). Defining a therapeutic dose of peripheral blood stem cells. J. Hematother. *1*, 329-334.
126. van Bekkum, D. W., and de Vries, M. J. (1967). Clinical applications of bone marrow transplantation and related experiments. In: Radiation Chimeras, van Bekum DW and de Vries MJ, eds. London: Logos Press, 193-232.
127. Pawliuk, R., Kay, R., Lansdorp, P., and Humphries, R. K. (1994). Selection of retrovirally transduced hematopoietic cells using CD24 as a marker of gene transfer. Blood *84*, 2868-2877.
128. Coneally, E., Bardy, P., Eaves, C. J., Thomas, T., Chappel, S., Shpall, E. J., and Humphries, R. K. (1996). Rapid and efficient selection of human hematopoietic cells expressing murine heat-stable antigen as an indicator of retroviral-mediated gene transfer. Blood *87*, 456-464.
129. Knobel, K. M., McNally, M. A., Berson, A. E., et al. (1994). Long-term reconstitution of mice after ex vivo expansion of bone marrow cells: differential activity of cultured bone marrow and enriched stem cell populations. Exp. Hematol. 22, 1227-1231.
130. Muench, M. O., Firpo, M. T., and Moore, M. A. (1993). Bone marrow transplantation of interleukin-1 plus kit ligand ex vivo expanded bone marrow accelerates hematopoietic reconstitution in mice without the loss of stem cell lineage and proliferation potential. Blood *81*, 3463-3469.
131. Henschler, R., Brugger, W., Luft, T., Frey, T., Mertelsmann, R., and Kanz, L. (1994). Maintenance of transplantation potential in ex vivo expanded CD34+ selected human peripheral blood progenitor cells. Blood *84*, 2898-2903.
132. Peters, S. O., Kittler, E. L. W., Ramshaw, H. S., and Quesenberry, P. J. (1996). Ex vivo expansion of murine marrow cells with interlerukin-3 (IL-3), IL-6, IL-11 and stem cell factor leads to impaired engraftment in irradiated hosts. Blood *87*, 30-37.
133. Dumenco, L. L., Allay, E., Norton, K., and Gerson, S. L. (1994). The prevention of thymic lymphomas in transgenic mice by human O6-alkylguanine-DNA-alkyltransferase. Science *259*, 219-222.
134. Ward, M., Richardson, C., Pioli, P., Smith, L., Podda, S., Goff, S., Hesdorffer, C., and Bank, A. (1994). Transfer and expression of the human multiple drug resistance gene in human CD34+ cells. Blood *84*, 1408-1414.
135. Hanania, E. G., Fu, S., Zu, Z., Hegewisch, B. S., Korbling, M., Hester, J., Durett, A., Andreeff, M., Mechetner, E., Holzmayer, T., and et al. (1995). Chemotherapy resistance to taxol in clonogenic progenitor cells following transduction of CD34 selected marrow and peripheral blood cells with a retrovirus that contains the MDR-1 chemotherapy resistance gene. Gene Ther. *2*, 285-294.

136. Magni, M., Shammah, S., Schiro, R., Mellado, W., Dalla-Favera, R., and Gianni, A. M. (1996). Induction of cyclophosphamide-resistance by aldehyde-dehydrogenase gene transfer. Blood *87*, 1097-1103.
137. Jelinék, J., Fairbairn, L. J., Dexter, M., Rafferty, J. A., Stocking, C., Ostertag, W., and Margison, G. P. (1996). Long-term protection of hematopoiesis against the cytotoxic effects of multiple doses of nitrosoureas by retrovirus-mediated expression of human O6-alkylguanine-DNA-alkyltransferase. Blood *87*, 1957-1961.
138. Corey, C. A., DeSilva, A. D., Holland, C. A., and Williams, D. A. (1990). Serial transplantation of methotrexate-resistant bone marrow: protection of murine recipients from drug toxicity by progeny of transduced stem cells. Blood *75*, 337-343.
139. Hanania, E. G., Fu, S., Roninson, I., Zu, Z., Gottesman, M. M., and Deisseroth, A. B. (1995). Resistance to taxol chemotherapy produced in mouse marrow cells by safety-modified retroviruses containing a human MDR-1 transcription unit. Gene Ther. *2*, 279-284.
140. Hanania, E. G. and Deisseroth, A. B. (1994). Serial transplantation shows that early hematopoietic precursor cells are transduced by MDR-1 retroviral vector in a mouse gene therapy model. Cancer Gene Ther. *1*, 21-25.
141. Podda, S., Ward, M., Himelstein, A., Richardson, C., dela Flor-Weiss, E., Smith, L., Gottesman, M., Pastan, I., and Bank, A. (1992). Transfer and expression of the human multiple drug resistance gene into live mice. Proc. Natl. Acad. Sci. USA *89*, 9676-9680.
142. Sorrentino, B. P., Brandt, S. J., Bodine, D., Gottesman, M., Pastan, I., Cline, A., and Nienhuis, A. W. (1992). Selection of drug-resistant bone marrow cells in vivo after retroviral transfer of human MDR1. Science *257*, 99-103.
143. Correll, P. H., Colilla, S., and Karlsson, S. (1994). Retroviral vector design for long-term expression in murine hematopoietic cells in vivo. Blood *84*, 1812-1822.
144. Riviére, I., Brose, K., and Mulligan, R. C. (1995). Effects of retroviral vector design on expression of human adenosine deaminase in murine bone marrow transplant recipients engrafted with genetically modified cells. Proc. Natl. Acad. Sci. USA *92*, 6733-6737.
145. Challita, P.-M. and Kohn, D. B. (1994). Lack of expression from a retroviral vector after transduction of murine hematopoietic stem cells is associated with methylation in vivo. Proc. Natl. Acad. Sci. USA *91*, 2567-2571.
146. Baum, C. and Ostertag, W. (1996). An Sp1-binding site in the upstream control region of the retroviral enhancer potentiates transcriptional activity of the spleen focus-forming virus (SFFVp) in a stage- and lineage independent manner. submitted
147. Baum, C., Eckert, H.-G., Stocking, C., and Ostertag, W. (1996). Activity of Friend mink cell focus-forming viruses during myelo-erythroid hematopoiesis. Exp. Hematol. *24*, 364-370.
148. Baum, C., Hegewisch-Becker, S., Eckert, H.-G., Stocking, C., and Ostertag, W. (1995). Novel retroviral vectors for efficient expression of the multidrug resistance (*mdr*-1) gene in early hematopoietic cells. J. Virol. *69*, 7541-547.
149. Baum, C., Eckert, H.-G., Laker, C., John, J., and Ostertag, W. (1996). Friend-MCF/MESV hybrid vectors (FMEV) allow efficient retroviral transduction of embryonic stem cells. submitted
150. Grez, M., Akgün, E., Hilberg, F., and Ostertag, W. (1990). Embryonic stem cell virus, a recombinant murine retrovirus with expression in embryonic stem cells. Proc. Natl. Acad. Sci. USA *87*, 9202-9206.
151. Grez, M., Zörnig, M., Nowock, J., and Ziegler, M. (1991). A single point mutation activates the Moloney murine leukemia virus long terminal repeat in embryonal stem cells. J. Virol. *65*, 4691-4698.
152. Stocking, C., Grez, M., and Ostertag, W. (1993). Regulation of retrovirus infection and expression in embryonic and hematopoietic stem cells. In: Virus Strategies. Molecular Biology and Pathogenesis, Doerfler W and Böhm P, eds. Weinheim, Germany: VCH Verlagsgesellschaft mbH, 433-455.
153. Ahlers, N., Hunt, N., Just, U., Laker, C., Ostertag, W., and Nowock, J. (1994). Selectable retrovirus vectors encoding Friend virus gp55 or erythropoietin induce polycythemia with different phenotypic expression and disease progression. J. Virol. *68*, 7235-7243.
154. Akgün, E., Ziegler, M., and Grez, M. (1991). Determinants of retrovirus gene expression in embryonal carcinoma cells. J. Virol. *65*, 382-388.
155. Friel, J., Hughes, D., Pragnell, I., Stocking, C., Laker, C., Nowock, J., Ostertag, W., and Padua, R. A. (1990). The malignant histiocytosis sarcoma virus., a recombinant of Harvey murine sarcoma

virus and Friend mink cell focus-forming virus, has acquired myeloid transformation specificity by alterations in the long terminal repeat. J. Virol. *64*, 369-378.
156. Hilberg, F., Stocking, C., Ostertag, W., and Grez, M. (1987). Functional analysis of a retroviral host range mutant: altered long terminal repeat sequences allow expression in embryonal carcinoma cells. Proc. Natl. Acad. Sci. USA *84*, 5232-5236.
157. Li, Y., Golemis, E., Hartley, J. W., and Hopkins, N. (1987). Disease specificity of nondefective Friend and Moloney murine leukemia virus is controlled by a small number of nucleotides. J. Virol. *61*, 693-700.
158. Speck, N. A., Renjifo, B. V., Golemis, E., Fredrickson, T. N., Hartley, J. W., and Hopkins, N. (1990). Mutations of the core or adjacent LVb elements of the Moloney leukemia virus enhancer alters disease specificity. Genes Dev. *4*, 223-242.
159. Stocking, C., Kollek, R., Bergholz, U., and Ostertag, W. (1985). Long terminal repeat sequences impart haematopoietic transformation properties to the myeloproliferative sarcoma virus. Proc. Natl. Acad. Sci. USA *82*, 5746-5750.
160. Stocking, C., Kollek, R., Bergholz, U., and Ostertag, W. (1986). Point mutations in the U3-region of the long terminal repeat of Moloney murine leukemia virus determine disease specificity of the myeloproliferative sarcoma virus. Virology *153*, 145-149.
161. Eckert, H.-G., Stockschläder, M., Just, U., Hegewisch-Becker, S., Grez, M., Zander, A., Ostertag, W., and Baum, C. (1996). High-dose multidrug resistance in primary human hematopoietic progenitor cells transduced with optimized retroviral vectors. Blood *88*, 3407-3415.
162. Tumas, D. B., Spangrude, G. J., Brooks, D. M., Williams, C. D., and Chesebro, B. (1996). High-frequency cell-surface expression of a foreign protein in murine hematopoietic stem cells using a new retroviral vector. Blood *87*, 509-517.
163. Lawrenz-Smith, S. C. and Thomas, C. Y. (1995). The E47 transcription factor binds to the enhancer sequences of recombinant murine leukemia viruses and influences enhancer function. J. Virol. *69*, 4142-4148.
164. Manley, N. R., O'Connell, M., Sun, W., Speck, N. A., and Hopkins, N. (1993). Two factors that bind to highly conserved sequences in mammalian type C retroviral enhancers. J. Virol. *67*, 1967-1975.
165. Zaiman, A. L., Lewis, A. F., Crute, B. E., Speck, N. A., and Lenz, J. (1995). Transcriptional activity of core binding factor α (AML1) and β subunits on murine leukemia virus enhancer cores. J. Virol. *69*, 2898-2906.
166. Sun, W., Graves, B. J., and Speck, N. A. (1995). Transactivation of the Moloney murine leukemia virus and T-cell receptor β-chain enhancers by *cbf* and *ets* requires intact binding sites for both proteins. J. Virol. *69*, 4941-4949.
167. Tsukiyama, T., Ueda, H., Hirose, S., and Niwa, O. (1992). Embryonal long terminal repeat-binding protein is a murine homolog of FTZ-F1, a member of the steroid receptor superfamily. Mol. Cell. Biol. *12*, 1286-1291.
168. Ostertag, W., Stocking, C., Johnson, G. R., Kluge, N., Kollek, R., Franz, T., and Hess, N. (1987). Transforming genes and target cells of murine spleen focus-forming viruses. Adv. Cancer Res. *48*, 193-355.
169. Baum, C. unpublished data.
170. Eckert, H.-G. and Baum, C. unpublished data.
171. Baum, C., Eckert, H.-G., Stockschläder, M., Just, U., Hegewisch-Becker, S., Hildinger, M., Uhde, A., John, J., and Ostertag, W. (1996). Improved retroviral vectors for hematopoietic stem cell protection and in vivo selection. J. Hematother. 323-329.
172. Krall, W. J., Skelton, D. C., Yu, X.-J., Riviere, I., Lehn, P., Mulligan, R. C., and Kohn, D. B. (1996). Increased levels of spliced RNA account for augmented expression from the MFG retroviral vector in hematopoietic cells. Gene Ther. *3*, 37-48.
173. Germann, U. A., Chin, K.-V., Pastan, I., and Gottesman, M. M. (1990). Retroviral transfer of a chimeric multidrug resistance-adenosine deaminase gene. FASEB J. *4*, 1501-1506.
174. Aran, J. M., Gottesman, M. M., and Pastan, I. (1994). Drug-selected coexpression of human glucocerebrosidase and P-glycoprotein using a bicistronic vector. Proc. Natl. Acad. Sci. USA *91*, 3176-3180.
175. Sadelain, M., Wang, C. H. J., Antoniou, M., Grosveld, F., and Mulligan, R. (1995). Generation of high-titer retroviral vector capable of expressing high levels of the human β-globin gene. Proc. Natl. Acad. Sci. USA *92*, 6728-6732.

176. Riddell, S. R., Elliott, M., Lewinsohn, D. A., Gilbert, M. J., Wilson, L., Manley, S. A., Lupton, S. D., Overell, R. W., Reynolds, T. C., Corey, L., and Greenburg, P. D. (1996). T-cell mediated rejection of gene-modifed HIV-specific cytotoxic T lymphocytes in HIV-infected patients. Nature Med. *2*, 216-223.
177. Lübbert, M., Brugger, W., Mertelsmann, R., and Kanz, L. (1996). Developmental regulation of myeloid gene expression and demethylation during ex vivo culture of peripheral blood progenitor cells. Blood *87*, 447-455.
178. Walters, M. C., Magis, W., Fiering, S., Eidemiller, J., Scalzo, D., Groudine, M., and Martin, D. I. K. (1996). Transcriptional enhancers act in cis to suppress position-effect variegation. Genes Dev. *19*, 185-195.
179. Meins, F. and Kunz, C. (1995). Gene silencing in transgenic plants: a heuristic autoregulation model. Curr. Top. Microbiol. Immunol. *197*, 105-120.
180. Henikoff, S. (1995). Gene silencing in *Drosophila*. Curr. Top. Microbiol. Immunol. *197*, 193-208.
181. Loo, S. and Rine, J. (1995). Silencing and heritable domains of gene expression. Annu. Rev. Cell Dev. Biol. *11*, 519-548.
182. Doerfler, W. (1995). The insertion of foreign DNA into mammalian genomes and its consequences: a concept in oncogenesis. Adv. Cancer Res. *66*, 313-344.
183. Hoeben, R. C., Migchielsen, A. A. J., van der Jagt, R. C. M., van Ormondt, H., and van der Eb, A. J. (1991). Inactivation of the Moloney murine leukemia virus long terminal repeat in murine fibroblast cell lines is associated with methylation and dependent on its chromosomal position. J. Virol. *65*, 904-912.
184. Duch, M., Paludan, K., Jorgenson, P., and Pedersen, F. S. (1994). Lack of correlation between basal expression levels and susceptibility to transcriptional shutdown among single-gene murine leukemia virus vector proviruses. J. Virol. *68*, 5596-5601.
185. Scharfmann, R., Axelrod, J. H., and Verma, I. (1991). Long-term in vivo expression of retrovirus-mediated gene transfer in mouse fibroblast implants. Proc. Natl. Acad. Sci. USA *88*, 4626-2630.
186. Macleod, D., Charlton, J., Mullins, J., and Bird, A. P. (1994). Spl sites in the mouse aprt gene promoter are required to prevent methylation of the CpG island. Genes Dev. *8*, 2282-92.
187. Brandeis, M., Frank, D., Keshet, I., Siegfried, Z., Mendelsohn, M., Nemes, A., Temper, V., Razin, A., and Cedar, H. (1994). Spl elements protect a CpG island from de novo methylation. Nature *371*, 435-437.
188. Duch, M., Paludan, K., Lovmand, J., Sorensen, M. S., Jorgensen, P., and Pedersen, F. S. (1995). The effect of selection for high-level vector expression on the genetic and functional stability of a single transcript vector derived from a low-leukemogenic murine retrovirus. Hum. Gene Ther. *6*, 289-296.
189. Clark, S. P. and Mak, T. W. (1982). Nucleotide sequences of the murine retrovirus Friend SFFVp long terminal repeats: identification of a structure with extensive dyad symmetry 5' to the TATA box. Nucl. Acid Res. *10*, 3315-3330.
190. Sheridan, P. L., Sheline, C. T., Cannon, K., Voz, M. L., Pazin, M. J., Kadonaga, J. T., and Jones, K. A. (1995). Activation of the HIV-1 enhancer by the LEF-1 HMG protein on nucleosome-assembled DNA in vitro. Genes Dev. *9*, 2090-104.
191. Bergemann, J., Lother, H., and Ostertag, W. unpublished data.
192. Phi-Van, L., von Kries, J. P., Ostertag, W., and Strätling, W. H. (1990). The chicken lysozyme matrix attachment region increases transcription from a heterologous promoter in heterologous cells and dampens position effects on the expression of transfected cells. Mol. Cell. Biol. *10*, 2302-2307.
193. Bergemann, J., Kühlcke, K., Fehse, B., Ratz, I., Ostertag, W., and Lother, H. (1995). Excision of specific DNA-sequences from integrated retroviral vectors via site-specific recombination. Nucl. Acids Res. *23*, 4451-4456.

13. The Liver as a Target for Gene Therapy

Günter Cichon and Michael Strauss

13.1 Introduction

The liver is a preferred target for gene therapy since it is the major site for a large number of synthetic, catabolic and detoxyfying pathways. Routine newborn screening for the most common inborn errors and specific investigation of risk patients allow for early diagnosis of a large number of genetic defects, but there is still a gap between advanced diagnostical methods and therapeutic options. Gene therapy has the potential to fill this gap. This article will discuss the potential use of different gene transfer methods for the development of save and efficient liver gene therapies.

13.2 Target Diseases

Poor prognosis and lack of sufficient conservative therapies makes the homozygous form of familial hypercholesterolemia an ideal target for gene therapy. Patients suffer from high serum cholesterol levels and develop marked atherosclerosis with myocardial infarction allready at childhood. The liver is not able to clear cholesterol from circulation due to a defect of the low density lipoprotein receptor (LDL receptor). 1 out of 500 individuals are suffering from heterozygous familial hypercholesterolemia. Combined dietary and pharmacological treatment could lower serum cholesterol by 40% but patients are still at increased risk for myocardial and cerebral infarction. A successful gene therapy should reduce serum cholesterol levels back to normal (see chapter 18).

Hereditary defects of enzymes involved in the uric cycle lead to hyperammonaemia and, in case of a low residual ativity, to death within days after birth. Tyrosinaemia type I, an inborn defect of fumarylacetoacetate hydrolase, leads to accumulation of succinylacetate which is toxic to hepatocytes. Patients develop early liver cirrhosis, renal insufficiency and hepatocellular carcinoma and die at childhood. The only therapeutic option for these patients would be early liver transplantation. In con-

Table 13.1 Inherited genetic diseases affecting normal liver function

Lipid metabolism	*Storage diseases*
Fam. Hypercholesterolemia (FH)	Wilson's disease
Abetalipoproteinaemia	Hurler's syndrome
Cholesterolester storage diseases	Glycogen storage disease type III and IV
	Gaucher's disease
Amino acid metabolism	Niemann-Pick disease type II
Tyrosinaemia type I	
Phenylketonuria	*others*
Cystinosis	alpha 1-antitrypsin deficiency
	Cystic fibrosis
Carbohydrate metabolism	Hemophilia A and B
Galactosaemia	
Fructosaemia	
Uric cycle	
Ornithine transcarbamylase deficiency (OTCD)	
Carbamylphosphat synthetase deficiency (CPSD)	
Argininosuccinase deficiency (ALD)	
Argininosuccinic acid synthetase deficiency (ASD)	

trast to these untreatable defects, there is a large number of inborn errors which is, to a varying extent, accessible to dietary or pharmacological intervention (Tab. 13.1).

Clotting factors in patients suffering from hemophilia A and B could be supplemented by injections of purified natural or recombinant factor VIII and IX in regular intervals, but supplementation is accompanied with a certain risk for viral infections like HIV, HBV or CMV. In many storage diseases, the liver could benefit from a local gene supplementation. The classical inborn enzyme defects of the liver like phenylketonuria, fructosemia, galactosemia and many others can be controlled quite well by dietary regimes, but diet needs to be maintained for many years and in some cases for the whole life. A reliable and efficient liver gene therapy could offer these patients a strong alternative to the current way of treatment.

13.3 The 'Ideal' Liver Vector

The prefered way of liver gene therapy would be one single application of a therapeutic vector followed by a life-long expression of the gene in the liver. Peripheral application of a vector would be convenient for the patient but requires a high liver tropism of the vector system. Expression of the therapeutic gene should be restricted to the liver and any immunological response to the vector or to the transgenic cell should be avoided. Physiological expression levels are not generally required. The number of hepatocytes in an adult human liver can be estimated to be between $5x10^{11}$ and $1x10^{12}$. The reconstitution of physiological factor VIII serum levels of about 100ng/ml in hemophiliacs will require a smaller number of transduced hepatocytes than the reduction of high serum cholesterol levels in patients suffering from familial hypercholesterolemia. Nevertheless, if a viral vector will be used, viral doses of at least $1x10^{12}$ will be required for a sufficient therapeutic effect. None of the existing gene transfer systems could fullfil all demands since in each system attractive properties are closely linked to weak points. However, there are several possibilities for improvement of most of them. Thus, the development of a number of functional liver gene transfer systems seems feasible.

13.4 Gene Transfer Systems

Recombinant viruses are efficient *in vitro* gene transfer vehicles but their *in vivo* application is restricted by immunological host defence mechanisms. Artifical transfer systems like liposomes and polylysin-DNA complexes have high transport capacities and low antigenicity. They could be directed to the liver by a receptor targeting mechanism, but these systems have only limited *in vivo* transfer rates so far. Since there is no ideal system, different viral and non-viral strategies are currently under investigation regarding application to liver-directed gene transfer.

13.4.1 Retroviruses

Retroviruses are RNA tumour viruses with a genome size of about 8 kb. Construction of recombinant viruses requires only the presence of small terminal elements in the linear genome, which leaves a transport capacity for therapeutic genes of almost 8 kb. Recombinant retroviruses are efficient gene transfer vehicles *in vitro* reaching transfer rates of close to 100% (see chapter 1). They can infect most cell types including hepatocyte lines (1, 2). The most important feature of retroviruses is their mechanism of integrating into the host genome (3). Most commonly used viral vectors are derived from murine leukemia virus. The *in vivo* application of murine leukemia viruses has two important drawbacks. Viruses are rapidly inactivated by the human complement system (4) and integration is restricted to proliferating cells (1, 2). The liver as a non-proliferating tissue is, therefore, not a very suitable target for retroviral vectors. Application of these vectors requires an *ex vivo* detour. After partial hepatectomy, hepatocytes are isolated from the resected tissue and are cultivated in the presence of retrovirus producer cells. Cultured hepatocytes normaly undergo one or two rounds of replication which allows the retroviral prereplication complex to enter the nucleus and to integrate into the host genome. After 2-3 days in culture, hepatocytes are collected and reinfused over the splenic vein. The general feasibility of this procedure has been prooven in the Watanbe rabbit (6) and in a factor IX deficient dog (7). Gene expression was detectable but too low for achieving a therapeutic effect. There are natural limits for the efficiency of this procedure and not more than 1% of the whole liver mass can be actively transduced with this approach. Five patients with homozygous familial hypercholesterolemia have been treated in a first clinical trial (8, 9, 10). In three of the five patients serum cholesterol levels dropped by 20%,

19% and 6%, the other two patients did not respond to the treatment. The limits of this procedure are obvious.

Despite of the known problems, several attemps have been made towards *in vivo* application of retroviral vectors. One approach uses delivery of the urokinase gene by a recombinant adenovirus. Urokinase expression in liver cells induces disseminated single cell necrosis which, in turn, induces proliferation in the whole liver (11). During the regenerative phase, retroviruses could be applied. In order to circumvent inactivation by the complement system, retrograde application via the cystic duct should be attempted. Since it was shown recently that HIV infects resting cells owing to its unique ability of targeting its preintegration complex to the nucleus (5), the modification of existing retroviral vector systems seems feasible. Attemps have also been made to identify the viral envelope structures which are responsible for complement activation. Elimination of these structures could be an important step towards an efficient retroviral *in vivo* approach.

13.4.2 Adenoviral Vectors

Adenoviruses are double-stranded DNA viruses with a genome size of about 35 kb. Recombinant adenoviruses could harbour therapeutic genes up to a size of 10 kb (12,13). Recombinant adenoviruses are the most efficient gene transfer vehicles at present (see chapter 2). Like retroviruses, adenoviruses can infect a broad spectrum of mammalian cells with an efficiency of almost 100%, but an important difference is the lack of integration of adenoviral DNA into the host genome. Viral genes are stabilized in an episomal state and their genome is lost rapidly in dividing cells. Recombinant adenoviruses can be produced at titers of up to $1x10^{13}$/ml (14) which is at least 3-4 orders of magnitude higher than those obtained with retroviral vectors.

Application *in vivo* is not hampert by complement problems and, after systemic application, adenoviruses of subtype 5 surprisingly exhibit a high liver tropism. After injection into the tail vein in mice, over 90% of viral DNA is found in the liver (15) and are routinely between 30 to 60% (Fig. 13.1). After portal vein infusion, transduction rates can be increased to more than 90% (16). The receptor on the surface of hepatocytes is still unknown, but integrins seem to be involved somehow (17). Integration of viral DNA into the host genome is not a prerequisite for long term expression in the liver. Gene expression after adenoviral gene transfer is stable for at leasts three months in immunodeficient mice as well as in mouse strains whose viral antigen recognition is not reactive against adenoviral antigens for different reasons (18, 19).

Adenoviral vectors have been employed for delivery of therapeutic genes in rabbits (20), dogs (21) and mice (22). The time course of gene expression is comparable between the different animals. After a remarble peak of gene expression around day 6, the expression levels fall down over the next two weeks, mostly back to the level before treatment. Peripheral application of $1x10^{13}$ virus particles in the Watanabe

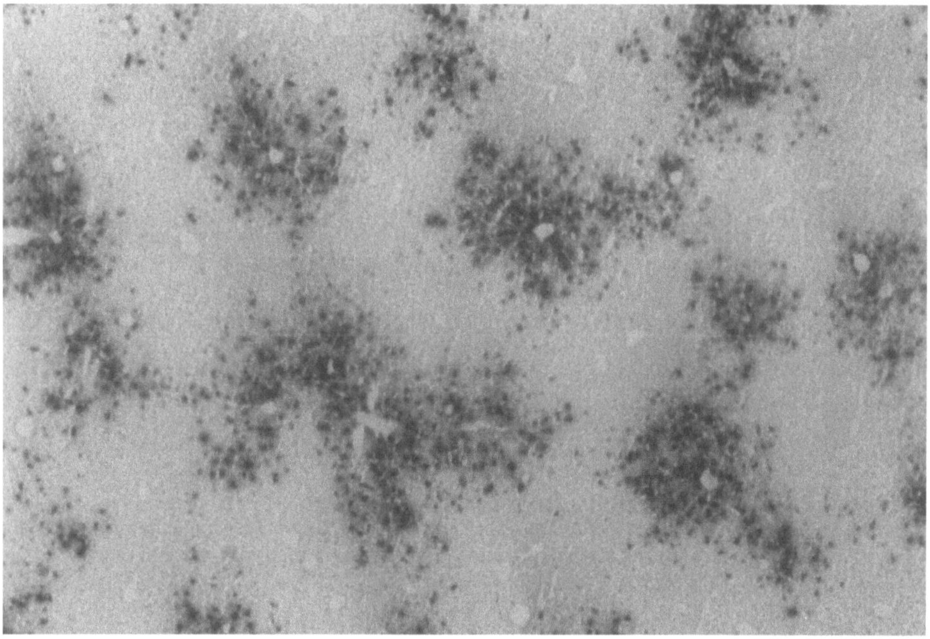

Fig. 13.1: X-gal staining of a section of mouse liver after systemic application of recombinant adenovirus ($1x10^9$) carrying the lacZ gene.

rabbit (animal model for FH) leads to an initial reduction of serum cholesterol by 60% (Fig. 13.2). However, the therapeutic effect is over after 14 days. Repeated application remains without any therapeutic effect because production of neutralising antibodies abolishes viral infection almost completely. The origin of vector antigenicity are viral capsid antigens, expression products of viral genes which remain in the vector and in some cases antigenicity of the therapeutic gene product. Another reason for the limited duration of foreign gene expression in immune competent mice or rabbits is a strong attack against transgene-expressing hepatocytes mediated by cytotoxic T-cells (23).

Several attemps have been made to reduce viral gene expression by excising or inactivating certain viral gene functions (24). Since capsid antigenicity is a general problem in the application of non-enveloped but also of enveloped viruses, attemps have also been made to block immunological defense mechanisms. These approaches will be discussed later. With regard to the high *in vivo* transduction rates, the liver tropism and the high transport capacity, recombinant adenoviruses are the strongest candidates for a reliable and effective liver gene transfer system at present.

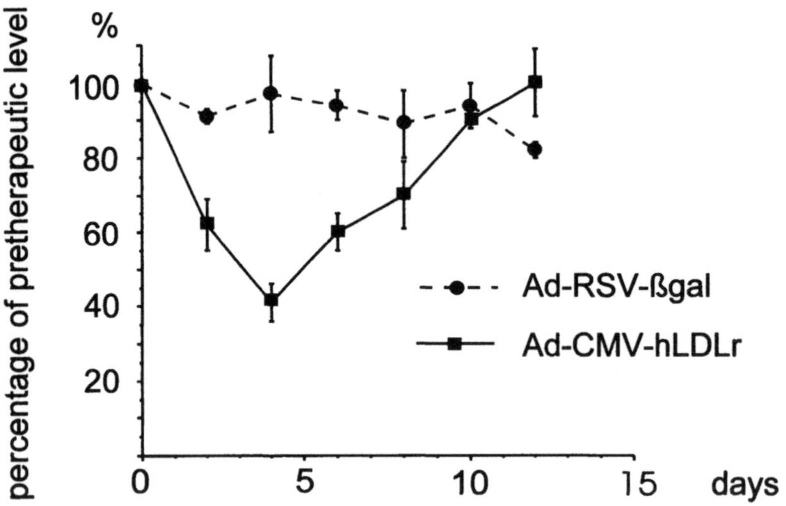

Fig. 13.2: Serum cholesterol levels in Watanabe rabbits at different time points after systemic application of $1x10^{13}$ recombinant viruses carrying the human LDL receptor gene.

13.4.3 Adeno-Associated and Herpes Virus Based Vectors

During the last three years, attempts have been made towards the development of other viral gene transfer systems. Adeno-associated viruses (AAV) are small DNA viruses with a genome size of 4.5 kb. AAV genes are frequently found in human tissues but do not seem to be associated with any pathological effect. Adeno-associated viruses share with retroviruses the mechanism of chromosomal integration (see chapter 4). Wildtyp AAV integrates into a specific locus on chromosom 19, but this property seems to be linked to the presence of the viral *rep* genes. With the removal of these genes in recombinant viruses, this useful property is lost and integration events occur by chance (25-27).

Recombinat AAV have been engineered for delivery of the CFTR gene for treatment of cystic fibrosis. Gene transfer rates into lung fibroblast lines can reach nearly 70% but about 1000 viral particles per cell are required (28). To evaluate *in vivo* efficiency, AAV was instillated into the lung of New Zealand white rabbits. Local transfer rates reached up to 50% and the transgene was still detectable after six months and did not seem to be associated with any inflammation (29). No investigations have been done to evaluate the feasibility for liver gene transfer.

Limitations for the employment of AAV vectors are their limited transport capacity and the problem of generating sufficiently high titers for liver gene transfer in humans. Since production of recombinant AAV requires helper adenoviruses, it is technically difficult to inactivate and separate helper virus without loosing much of recombinant AAV. Therefore, the prospects for AAV in liver gene therapy seems to be limited, at least in the near future.

An interesting alternative is offered by recombinant herpes simplex viruses (HSV). Since about 50% of viral genes are not essential for replication or can be supplemented in *trans*, the transport capacity of HSV-derived vectors can amount up to 70.000 base pairs. Using a so-called amplicon technique, the entire herpes genome of about 150.000 bases can be replaced by therapeutic genes. Only small sequences carrying an origin of replication and a packaging signal are required in the vector. Like adeno- and retroviruses, recombinant herpesviruses can infect a broad spectrum of mammalian cells in culture.

Apart from a certain neurotropism and the property of retrograde transneuronal transport, nothing is known about other tissue specificities so far. Application of HSV vectors *in vivo* via the portal vein has been performed in mice. It leads to reporter gene expression for about four weeks (30). Gene transfer into human primary hepatocytes seems to be very efficient (31) and the perfusion of isolated liver in Fisher rats leads to gene transfer without signs of inflamation after three days (31). However, there are reports about some toxicity of viral capsid proteins which could depress transgene expression in the target cell directly or by immunological mechansims (32). Further investigations are required to explore the prospects of herpes based vectors for liver gene therapy.

13.4.4 Receptor Targeting by Polylysin-DNA Complexes and Liposomes

As an alternative to viral vectors, artificial polylysin-DNA complexes and liposomes have a number of interesting advantages. Polylysine-DNA complexes could carry up to 48000 base pairs of DNA, and there are no known size constraints in liposomes. Both systems have a low antigenicity and can be easily prepared in large amounts. Coupling of substrate molecules like transferrin (33, 34) or asialoorosomucoid (35) to the polylysine core or of anchor proteins in the liposome membrane could direct these complexes to specific receptors on the surface of hepatocytes. This theoretically attractive approach is hampered by several problems leading to low *in vivo* transfer efficiencies. Unfortunately, a large amount of circulating complexes is cleared by the reticulo-endothelial system of the liver and the rest is destroyed after receptor-mediated uptake into the endosomes of hepatocytes.

Pharmacological inhibition of endosomal acidification by agents like chloroquin leads to increased transfer rates (36). One reason for the high transduction rates of viral vectors is their ability to escape from endosomal breakdown through an endosomolytic mechanism. In the case of adenoviruses, the drop of pH in the endosomes after receptor mediated uptake frees pentonbase structures in the viral capsid which seems to be responsible for the disruption of endosomal membranes (37). The fusogenic membrane glycoprotein of influenza virus, haemagglutinin, is another component which has been shown to enhance gene transfer rates in vitro (38). Despite the current drawbacks in *in vivo* protocols, there are many possibilities for improvement and the the non-viral systems remain strong canidates for the development of save and reliable liver gene transfer vectors (see chapter 6).

13.4.5 Other Hybrid Systems

A third non-viral approach combines viral invasion mechanisms with advantages of liver-specific DNA-packaging complexes. Adenovirus can be inactivated by psoralene and UV and still maintains its endosomolytic activity. Coupling of polylsine-DNA complexes to inactivated viruses leads to an efficient gene transfer system (39) which provides transfer rates in mammalian cell lines of up to 100%. Unfortunately, these complexes share the fate after *in vivo* application with polylysine-DNA complexes. Most of them are removed from the circulation by the reticulo-endothelial system allowing for only marginal transfection rates. It appears likely that the extended size of these particles is responsible for the limited transfer efficiency *in vivo*. There is not much space for improvement of this vector system which limits the prospect of this approach.

Another hybrid system are Sendai virus-liposome complexes. The fusion of neutral liposomes loaded with HMG1-DNA complexes to the enveloped, non-pathogenic haemagglutinating virus of Japan (Sen1dai virus) leads to a gene transfer system which is applicable for *in vivo* gene transfer protocols (40). There is a first clinical protocol in preparation which will employ this vector for delivery of an anti-proliferative gene, for blocking post dillatational proliferation of coronary endothe lial cells. Unfortunately, there are no reports about the feasibility of this strategy for liver gene transfer so far.

13.4.6 Liver-Specific Gene Expression and Extrachromosomal Stabilization of Therapeutic Genes

Expression of therapeutic genes should be restricted to the liver because expression in other organs could be disadvantageous. For example, constitutive expression of LDL receptor in myocard cells would likely overload the enzymatic pathways involved in cholesterol metabolism leading to malefunction and cell death through pathological cholesterol storage. Local application via the portal vein, liver artery or cystic duct is expected to be leaky to some extent and receptor targeting strategies are not expected to work with 100% specificity because most of the so-called liver-specific receptors are not exclusicely expressed in hepatocytes. Combination of local application of the vector and organ-specific expression seems to be the method of choice to avoid any unspecific gene expression. Several liver-specific promoters have been employed for liver-restricted expression like those of the murine albumin and human alpha 1-antitrypsin genes (41), elements of the human growth hormon promoter (42), the L promoter of pyruvate kinase (43), the FXP1 binding site of clotting factor X (44) and elements of the hepatitis B virus core promoter (45).

Another problem on the way to permanent gene expression is the loss of non-integrated, episomal genes in dividing cells. As it was shown in immundeficient mice after adenoviral gene transfer, foreign gene expression was stable for several months (18). There are medical problems imagible which require only transient expression of a gene. However, in view of a life-long supplementation, stabilisation of episomal genes is required. The ideal form of stabilisation would be the construction of a minichromosome with a mammalian origin of replication and a kinetochor which mediates symmetric distribution to daughter cells during mitosis. These elements, which characterize mammalian artificial chromosomes (MACs), span hundreds of Kb and cannot be built into a small viral delivery system. An alternative way is the employment of a viral replication system. It has been reported that presence of the replication machinery of Epstein Barr virus comprised of the oriP sequence and the EBNA1 gene, could stabilize episomal DNA in dividing hepatocytes (46). The important aspect of the Epstein-Barr virus replication mechanism is linkage of replication to the host cell division cycle. There has been some concern about the tumorogenicity of EBNA1. However, expression of EBNA1 in transgenic mice has only led to B-cell lymphomas so far (47). Another replication system which has been successfully employed in mice contains replication elements of the human papovavirus BKV (48). As with the older SV40-based replication systems, the presence of the essential element of this system, the large T antigen of BKV, may raise some concern as to the tumour-inducing potential. Further investigations are required to assess the usefulness and the safety of these replication machineries.

13.5 Immunological Aspects of Gene Transfer

There are three possible sources of antigenicity in gene transfer systems. Structural components of the transfer vehicle like viral capsid or liposomal anchor proteins, viral genes which need to remain in the vector for constructional reasons and which are coexpressed after gene transfer and the therapeutic gene product itself. Even in advanced gene transfer systems, antigenic components will be present which have the potential to induce antibody production or raise cellular defense mechanisms after MHC-I-mediated presentation on the transgenic cell. The therapeutic effect of repeated vector administration could be diminished by neutralising antibodies and need to be avoided. There are no systematic investigations as to the time course of antigen clearance after gene transfer but it is quite likely that a period of initial immunsupression for the duration of 5-10 days will be required. The removal and externalisation of viral genes from the vectors is the most important goal in current vector development for three reasons: (i) reduction of antigenicity, (ii) augmentation of transport capacity and (iii) reduction of risks of recombination. However, even in well developed vector systems, some additional genes are not dispensible like the replication machinery for episomal stabilisation. These gene products have the same fate of MHC I presentation as all other intracellular proteins and could be recognized and treated as antigens. There are several strategies for prevention of immunological defense like permanent immune supression, reduction of antigen presentation or induction of tolerance towards expected antigens. Regarding side effects and increased susceptibility for infectious diseases, a permanent immune supression with agents like cyclosporin seems to be only a second choice option. An interesting alternative would be the exploitation of viral mechanisms for escape from immune recognition. During evolution, many viruses have developed mechanisms for escaping from host immune surveillance. Adenoviruses, herpes viruses or cytomegalo viruses for example carry an arsenal of genes which interfere with immune recognition by reduction of MHC I expression, by either blocking antigen processing or supression of lymphokine receptor expression. Coexpression of these genes can lead to masking of foreign gene products and to escape from recognition. A third strategy is the induction of immune tolerance, particularly in case of an expected sensibilisation against the therapeutic gene product. There are many possibilities for interfering and disturbing the immune response of the host. The most promising strategies towards tolerance induction is temporary blocking of costimulatory signals on leukocytes during first antigen exposition and blocking of helper cell function with monoclonal antibodies or soluble ligands. Blocking the CTLA4-B7 signal pathway between leukocytes with a soluble CTLA4 has led to tolerance induction in mice (49). Many investigations have proven the effect of anti-CD4 treatment for prolonged allograft acceptance in mice and rats (50-52). An alternative option to application of pharmacological or immunological agents could be gene transfer shortly after birth because there is a pronounced tolerance for foreign antigens during the maturation phase of the immune system. Only limited attention has been drawn to tolerance induction mechanisms in gene therapy so far and careful investigations are required to evaluate their potential risks.

13.6 Single Cell Transplanation and Recolonisation Strategies

In contrast to other parenchymal cells, hepatocytes maintain a limited regeneration potential during their whole life span. Regarding the enormous metabolic workload and the susceptibility for ischemic, toxic or mechanic traumas, liver regeneration seems ingenious. It has been estimated that every liver cell could go through at least 10-20 cell cycles. This property allows for a new form of liver transplantation. There are certain hereditary metabolic defects which cause early liver cirrhosis as a consequence of pathologic accumulation of toxic metabolites. Hepatocytes whose metabolic function is restored by mutation are strongly positive selected in livers which are afflicted by such a metabolic defect. For example, the histological picture in liver sections of patients suffering from tyrosinaemia type I shows small islands of functional hepatocytes embedded in a malfunctioning surrounding. This observation led to a new therapeutic strategy.

Transgenic mice with a deficiency of fumarylacetoacetate hydrolase (FAH) show the clinical picture of tyrosinaemia type I. Pharmacological intervention aims at the block of tyrosine catabolism upstream of FAH. Treatment with 2-(2-nitro-4-trifluoromethylbenzoyl)-1,3-cyclohexandione (NTBC) reduces pathological accumulation of fumarylacetoacetat, prolongs live expectence in affected patients but cannot prevent the developement of early hepatocellular carcinoma. If these transgenic tyrosinaemia mice are treated with infusions of isolated functional hepatocytes from congenic mice after pharmacological treatment has stopped, increased degradation of malfunctioning hepatocytes occurs but leeds, at the same time, to complete repopulation of the liver by functional (transplanted) hepatocytes (53). A similar effect could be established by treatment of deteriorated liver with a retroviral vector carrying a functional FAH gene. The transduced hepatocytes are positively selected and repopulation of the liver occurs to its normal size and mass. These mice were indeed cured from their metabolic defect. However, metabolic liver malfunction does not always lead to increased degradation of hepatocytes. Therefore, positive selection of functional cells would not occur in most of the diseases. The search for inducible selection mechanisms is ongoing and immunological problems which are related to transplantation of allotype donor hepatocytes are not solved yet. In addition, it needs to be clarified to what extent regeneration can be utilized without reducing long term survival of repopulated hepatocytes. There are many open questions which need to be answered but the latter investigations are a first and rewarding step towards a new and promising direction of liver gene therapy.

13.7 Summary and Outlook

Liver gene therapy has the potential to replace liver transplantation for the treatment of metabolic diseases but the strategies are still suffering from several technical problems. The quality of the transfer vehicle is of considerable importance for the effiency of gene transfer and permanent gene expression. The most promising vectors so far are viral vectors. Among the viral vectors adenoviruses are the most efficient ones. They exhibit a high liver tropism after systemic application, reach transfer rates of more than 90% after local portal vein infusion and have a packaging capacity which is sufficient for most of the common therapeutic genes. They can even be rendered more liver specific by using a liver-specific promoter to drive foreign gene expression (54). Unfortunately, long term gene expression is prevented by a strong immune attack against transgenic liver cells but reduction of antigenicity seems feasible and the inclusion of a replication or maintainance system could allow for long term expression even without integration.

Retroviral vectors may become useful for *in vivo* gene transfer if the problems of complement-mediated inactivation and lack of infection of resting cells can finally be solved. Polylysine-DNA complexes and liposomes are the most interesting non-viral candidates for liver gene therapy because they have little antigenicity, can be easily produced in sufficient amounts and can be combined with a receptor targeting mechanism which allows for targeting of these artifical particles specifically to liver cells. The low *in vivo* transfer rates may be overcome some day. However, until this problem is solved, viral delivery systems are the prefered vectors for *in vivo* gene transfer in the liver. The search for alternative vectors is still on. We have recently shown that baculovirus can efficiently infect hepatocytes *in vitro* (55) but *in vivo* infection is also hampered be components of the complement system (56). Thus, further improvements of existing vectors as well as devlopment of new vectors is required for future *in vivo* gene therapy approaches (57).

An interesting alternative to liver-directed gene transfer is *in vivo* repopulation of transplanted or actively transduced hepatocytes. The enrichment of functional hepatocytes in a malfunctioning liver requires a positive selection for functional cells. There is only very little experience with *in vivo* selection systems so far but the recent findings will strongly stimulate research in this field of gene therapy and repopulation strategies could become an important contribution for liver gene therapy.

The great expectations which have been raised by some preclinical gene therapy studies some years ago have not been fulfilled yet, particularly regarding the time required for development of new strategies. Physiological, immunological and technical problems were clearly underestimated. However, there is no doubt that the development and availability of a safe and efficient liver gene therapy for some diseases is only a matter of time (58).

References

1. Miller, A. D. (1992). Retroviral vectors. Curr. Top. Microbiol. Immunol. *158*, 1-24.
2. Naviaux, R. K. and Verma, I. M. (1992). Retroviral vectors for persistent expression in vivo. Curr. Opin. Biotechnol. *3*, 540-547.
3. Gilboa, E. (1990). Retroviral gene transfer: applications to human therapy. Prog. Clin. Biol. Res. *352*, 301-311.
4. Russell, D. W., Berger, M. S., and Miller, A. D. (1995). The effects of human serum and cerebrospinal fluid on retroviral vectors and packaging cell lines. . Hum. Gene Ther. *6*, 635-641.
5. Bukrinsky, M. I., Haggerty, S., Dempsey, M. P., Sharova, N., Adzhubel, A., Spitz, L., Lewis, P., Goldfarb, D., Emerman, M., and Stevenson, M. (1993). A nuclear localization signal within HIV-1 matrix protein that governs infection of non-dividing cells [see comments]. Nature *365*, 666-669.
6. Chowdhury, J. R., Grossman, M., Gupta, S., Chowdhury, N. R., Baker, J. R., Jr., and Wilson, J. M. (1991). Long-term improvement of hypercholesterolemia after ex vivo gene therapy in LDLR-deficient rabbits. Science *254*, 1802-1805.
7. Kay, M. A., Rothenberg, S., Landen, C. N., Bellinger, D. A., Leland, F., Toman, C., Finegold, M., Thompson, A. R., Read, M. S., Brinkhous, K. M., and et al. (1993). In vivo gene therapy of hemophilia B: sustained partial correction in factor IX-deficient dogs [see comments]. Science *262*, 117-119.
8. Grossman, M., Raper, S. E., Kozarsky, K., Stein, E. A., Engelhardt, J. F., Muller, D., Lupien, P. J., and Wilson, J. M. (1994). Successful ex vivo gene therapy directed to liver in a patient with familial hypercholesterolaemia [see comments]. Nat. Genet. *6*, 335-341.

9. Wilson, J. M., Grossman, M., Raper, S. E., Baker, J. R., Jr., Newton, R. S., and Thoene, J. G. (1992). Ex vivo gene therapy of familial hypercholesterolemia. Hum. Gene Ther. *3*, 179-222.
10. Grossman, M., Rader, D. J., Muller, D. W., Kolansky, D. M., Kozarsky, K., Clark, B. J., 3rd, Stein, E. A., Lupien, P. J., Brewer, H. B., Jr., Raper, S. E., and et al. (1995). A pilot study of ex vivo gene therapy for homozygous familial hypercholesterolaemia [see comments]. Nat. Med. *1*, 1148-1154.
11. Lieber, A., Peeters, M. J., Gown, A., Perkins, J., and Kay, M. A. (1995). A modified urokinase plasminogen activator induces liver regeneration without bleeding. Hum. Gene Ther. *6*, 1029-1037.
12. Krougliak, V. and Graham, F. L. (1995). Development of cell lines capable of completing E1, E4, and Protein IX defective adenovirus type 5 mutants. Human Gene Therapy *6*, 1575-1586.
13. Wang, Q., Jia, X.-C., and Finer, M. H. (1995). A packaging cell line for propagation of recombinant adenovirus vectors containing two lethal gene-region deletions. Gene Therapy *2*, 775-783.
14. Kremer, E. J. and Perricaudet, M. (1995). Adenovirus and adeno-associated virus mediated gene transfer. Brit. Med. Bul. *51(1)*, 31-44.
15. Smith, T. A., Mehaffey, M. G., Kayda, D. B., Saunders, J. M., Yei, S., Trapnell, B. C., McClelland, A., and Kaleko, M. (1993). Adenovirus mediated expression of therapeutic plasma levels of human factor IX in mice. Nat. Genet. *5*, 397-402.
16. Herz, J. and Gerard, R. D. (1993). Adenovirus-mediated transfer of low density lipoprotein receptor gene acutely accelerates cholesterol clearance in normal mice. Proc. Natl. Acad. Sci. U.S.A. *90*, 2812-2816.
17. Mathias, P., Wickham, T., Moore, M., and Nemerow, G. (1994). Multiple adenovirus serotypes use alpha v integrins for infection. J. Virol. *68*, 6811-6814.
18. Yang, Y., Nunes, F. A., Berencsi, K., Gonczol, E., Engelhardt, J. F., and Wilson, J. M. (1994b). Inactivation of E2a in recombinant adenoviruses improves the prospect for gene therapy in cystic fibrosis. Nat. Genet. *7*, 362-369.
19. Barr, D., Tubb, J., Ferguson, D., Scaria, A., Lieber, A., Wilson, C., Perkins, J., and Kay, M. A. (1995). Strain related variations in adenovirally mediated transgene expression from mouse hepatocytes in vivo: comparisons between immunocompetent and immunodeficient inbred strains. Gene Ther. *2*, 151-155.
20. Li, J., Fang, B., Eisensmith, R. C., Li, X. H., Nasonkin, I., Lin Lee, Y. C., Mims, M. P., Hughes, A., Montgomery, C. D., Roberts, J. D., and et al. (1995). In vivo gene therapy for hyperlipidemia: phenotypic correction in Watanabe rabbits by hepatic delivery of the rabbit LDL receptor gene. J. Clin. Invest. *95*, 768-773.
21. Fang, B., Eisensmith, R. C., Wang, H., Kay, M. A., Cross, R. E., Landen, C. N., Gordon, G., Bellinger, D. A., Read, M. S., Hu, P. C., and et al. (1995). Gene therapy for hemophilia B: host immunosuppression prolongs the therapeutic effect of adenovirus-mediated factor IX expression. Hum. Gene Ther. *6*, 1039-1044.
22. Ishibashi, S., Brown, M. S., Goldstein, J. L., Gerard, R. D., Hammer, R. E., and Herz, J. (1993). Hypercholesterolemia in low density lipoprotein receptor knockout mice and its reversal by adenovirus-mediated gene delivery [see comments]. J. Clin. Invest. *92*, 883-893.
23. Yang, Y., Ertl, H. C., and Wilson, J. M. (1994a). MHC class I-restricted cytotoxic T lymphocytes to viral antigens destroy hepatocytes in mice infected with E1-deleted recombinant adenoviruses. Immunity. *1*, 433-442.
24. Yeh, P., Dedieu, J. F., Orsini, C., Vigne, E., Denefle, P., and Perricaudet, M. (1996) Efficient dual transcomplementation of adenovirus E1 and E4 regions from a 293-derived cell line expressing a minimal E4 functional unit. J. Virol. *70(1)*, 559-65.
25. Weitzman, M. D., Kyostio, S. R., Kotin, R. M., and Owens, R. A. (1994). Adeno-associated virus (AAV) Rep proteins mediate complex formation between AAV DNA and its integration site in human DNA. Proc. Natl. Acad. Sci. U.S.A. *91*, 5808-5812.
26. McCarty, D. M., Ryan, J. H., Zolotukhin, S., Zhou, X., and Muzyczka, N. (1994b). Interaction of the adeno-associated virus Rep protein with a sequence within the A palindrome of the viral terminal repeat. J. Virol. *68*, 4998-5006.
27. McCarty, D. M., Pereira, D. J., Zolotukhin, I., Zhou, X., Ryan, J. H., and Muzyczka, N. (1994a). Identification of linear DNA sequences that specifically bind the adeno-associated virus Rep protein. J. Virol. *68*, 4988-4997.
28. Flotte, T. R. (1992). Gene expression from adeno-associated virus vectors in airway epithelial cells. Am. J. Respir. Cell. Mol. Biol. *7*, 349-356.

29. Flotte, T. R., Afione, S. A., Conrad, C., McGrath, S. A., Solow, R., Oka, H., Zeitlin, P. L., Guggino, W. B., and Carter, B. J. (1993). Stable in vivo expression of the cystic fibrosis transmembrane conductance regulator with an adeno-associated virus vector. Proc. Natl. Acad. Sci. U.S.A. *90*, 10613-10617.
30. Miyanohara, A., Johnson, P. A., Elam, R. L., Dai, Y., Witztum, J. L., Verma, I. M., and Friedmann, T. (1992). Direct gene transfer to the liver with herpes simplex virus type 1 vectors: transient production of physiologically relevant levels of circulating factor IX. New Biol. *4*, 238-246.
31. Fong, Y. M., Federoff, H. J., Brownlee, M., Blumberg, D., Blumgart, L. H., and Brennan, M. F. (1995). Rapid and efficient gene transfer in human hepatocytes by herpes viral vectors. Hepatology *22*, 723-729.
32. Roemer, K., Johnson, P. A., and Friedmann, T. (1992). Recombination between a herpes simplex virus type 1 vector deleted for immediate early gene 3 and the infected cell genome. J. Gen. Virol. *73*, 1553-1558.
33. Wagner, E., Cotten, M., Foisner, R., and Birnstiel, M. L. (1991). Transferrin-polycation-DNA complexes: the effect of polycations on the structure of the complex and DNA delivery to cells. Proc. Natl. Acad. Sci. U.S.A. *88*, 4255-4259.
34. Lozier, J. N., Thompson, A. R., Hu, P. C., Read, M., Brinkhous, K. M., High, K. A., and Curiel, D. T. (1994). Efficient transfection of primary cells in a canine hemophilia B model using adenovirus-polylysine-DNA complexes. Hum. Gene Ther. *5*, 313-322.
35. Wu, G. Y. and Wu, C. H. (1988). Receptor-mediated gene delivery and expression in vivo. J. Biol. Chem. *263*, 14621-14624.
36. Tietz, P. S., Yamazaki, K., and LaRusso, N. F. (1990). Time-dependent effects of chloroquine on pH of hepatocyte lysosomes. Biochem. Pharmacol. *40*, 1419-1421.
37. Seth, P. (1994). Adenovirus-dependent release of choline from plasma membrane vesicles at an acidic pH is mediated by the penton base protein. J. Virol. *68*, 1204-1206.
38. Wagner, E., Plank, C., Zatloukal, K., Cotten, M., and Birnstiel, M. L. (1992a). Influenza virus hemagglutinin HA-2 N-terminal fusogenic peptides augment gene transfer by transferrin-polylysine-DNA complexes: toward a synthetic virus-like gene-transfer vehicle. Proc. Natl. Acad. Sci. U.S.A. *89*, 7934-7938.
39. Wagner, E., Zatloukal, K., Cotten, M., Kirlappos, H., Mechtler, K., Curiel, D. T., and Birnstiel, M. L. (1992b). Coupling of adenovirus to transferrin-polylysine/DNA complexes greatly enhances receptor-mediated gene delivery and expression of transfected genes. Proc. Natl. Acad. Sci. U.S.A. *89*, 6099-6103.
40. Kaneda, Y., Iwai, K., and Uchida, T. (1989). Increased expression of DNA cointroduced with nuclear protein in adult rat liver. Science *243*, 375-378.
41. Hafenrichter, D. G., Ponder, K. P., Rettinger, S. D., Kennedy, S. C., Wu, X., Saylors, R. S., and Flye, M. W. (1994). Liver-directed gene therapy: evaluation of liver specific promoter elements. J. Surg. Res. *56*, 510-517.
42. Adams, T. E. (1995). Differential expression of growth hormone receptor messenger RNA from a second promoter. Mol. Cell Endocrinol. *108*, 23-33.
43. Boquet, D., Vaulont, S., Tremp, G., Ripoche, M. A., Daegelen, D., Jami, J., Kahn, A., and Raymondjean, M. (1992). DNase-I hypersensitivity analysis of the L-type pyruvate kinase gene in rats and transgenic mice. Eur. J. Biochem. *207*, 13-21.
44. Miao, C. H., Leytus, S. P., Chung, D. W., and Davie, E. W. (1992). Liver-specific expression of the gene coding for human factor X, a blood coagulation factor. J. Biol. Chem. *267*, 7395-7401.
45. Löser, L., Sandig, V., Kirillova, I., Strauss, M. (1996). Evaluation of HBV promoters for use in hepatic gene therapy. Biol. Chem. Hoppe-Seyler, *377*, 187-193.
46. Lutfalla, G., Armbruster, L., Dequin, S., and Bertolotti, R. (1989). Construction of an EBNA-producing line of well-differentiated human hepatoma cells and of appropriate Epstein-Barr virus-based shuttle vectors. Gene *76*, 27-39.
47. Wilson, J. and Levine, A. (1993). Curr. Top. Immunol. Microbiol.
48. Thierry, A. R., Lunardi Iskandar, Y., Bryant, J. L., Rabinovich, P., Gallo, R. C., and Mahan, L. C. (1995). Systemic gene therapy: biodistribution and long-term expression of a transgene in mice. Proc. Natl. Acad. Sci. U.S.A. *92*, 9742-9746.
49. Kay, M. A., Holterman, A. X., Meuse, L., Gown, A., Ochs, H. D., Linsley, P. S., and Wilson, C. B. (1995). Long-term hepatic adenovirus-mediated gene expression in mice following CTLA4Ig administration. Nat. Genet. *11*, 191-197.

50. Volk, H. D., Siegling, A., Platzer, C., Brock, J., Emmrich, F., and Lehmann, M. (1994). Tolerance induction by anti-CD4 MAb treatment in a rat kidney transplantation model. Transplant. Proc. *26*, 3155-3156.
51. Scully, R., Qin, S., Cobbold, S., and Waldmann, H. (1994). Mechanisms in CD4 antibody-mediated transplantation tolerance: kinetics of induction, antigen dependency and role of regulatory T cells. Eur. J. Immunol. *24*, 2383-2392.
52. Kolls, J. K., Lei, D., Odom, G., Nelson, S., Summer, W. R., Gerber, M. A., Shellito, J. E. (1996). Use of transient CD4 lymphocyte depletion to prolong transgene expression of E1-deleted adenoviral vectors. Hum. Gene. Ther. *7*, 489-497.
53. Overturf, K., Aldhalimy, M., Tanguay, R., Brantly, M., Ou, C. N., Finegold, M., and Grompe, M. (1996). Hepatocytes corrected by gene therapy are selected in vivo in a murine model of hereditary tyrosinaemia type I. Nat. Genet. *12*, 266-273.
54. Sandig, V., Löser, P., Lieber, A., Kay, M. A., and Strauss, M. (1996). HBV-derived promoter directed liver-specific expression of an adenovirally transduced LDL receptor gene. Gene Therapy *3*, 1002-1009.
55. Hofmann, C., Sandig, V., Jennings, G., Rudolph, M., Schlag, P., and Strauss, M. (1995). Efficient gene transfer into human hepatocytes by baculovirus vectors. Proc. Natl. Acad. Sci. U.S.A *92*, 10099-10103.
56. Sandig, V., Hofmann, C., Steinert, S., Jennings, G., Schlag, P., and Strauss, M. (1996). Gene transfer into human liver cells and tissue by baculoviral vectors. Human Gene Therapy, 1937-1945.
57. Sandig, V. and Strauss, M. (1996). Liver-directed gene transfer: prospects for therapy. J. Molec. Med. *74*, 205-212.
58. Strauss, M. (1996). Gene transfer on its long path to clinical application. J. Molec. Med. *74*, 167-169.

14. Gene Therapy for Gaucher Disease via Genetically Engineered Primary Myoblasts

Chunming Liu, Simon Watkins, Alfred Bahnson and John A. Barranger

In recent years, a wide variety of transplantable somatic cells including fibroblasts, hepatocytes, endothelial cells, karatinocytes, lymphocytes, hematopoietic stem cells and myoblasts, have been employed as target cells for gene transfer in different kinds of models of gene transfer (1-14). Several studies in animals have indicated that myoblasts are cells with many advantages from the point of view of expression, secretion, and systemic delivery of transferred recombinant gene products that could be useful in gene therapy strategies (11, 12, 15, 16).

Myoblasts are proliferating precursor cells which exist around mature skeletal muscle fibers as satellite cells, and they have the unusual ability to proliferate and differentiate into myotubes by fusion into adjacent highly vascularized and long-term surviving mature myofibers (13, 16, 17). Myoblasts become a integral part of muscle after postnatal growth and during regeneration after tissue injury or damage (18, 19). An integrated muscle fiber that fuses with genetically modified myoblasts can stably express the gene transferred to myoblasts *ex vivo* (20). Furthermore myoblasts can be isolated, purified and expanded to large numbers in the culture, genetically manipulated *in vitro* and transplanted. Moreover, muscle has a rich vascular system that allows genetically modified myoblasts to secrete gene products into the systemic circulation. Since skeletal muscles compose 40-50% of body weight (21) in man, it is feasible and safe to prepare myoblasts from a biopsy of muscle and to genetically engineer and transplant the genetically modified myoblasts back to the study subject (14, 17, 20, 22).

In general, the application of somatic cell gene therapy should consider the following factors which include the feasibility of manipulating tissue or cells *in vitro* and their transplantation to animals, and later, to man who will finally be treated in the clinical practice of gene therapy to correct an inherited or acquired disease. These considerations would be: (1) the availability and accessibility of the target tissue; (2) the susceptibility of somatic cells to highly efficient gene transfer; (3) the stability and endurance of sufficient expression *in vitro* and *in vivo*; (4) the existence of expressed gene products in a physiological form and appropriate bioactivity; (5) feasibility of reimplantation of genetically engineered somatic cells into an appropriate site anatomically in the subjects; (6) long time survival of transplanted gene-modified somatic cell and persistence of expression *in vivo* without any effects biopharmacologically or in the risk of tumorigenesis; (7) non-immune reaction in host against grafted genetically-modified somatic cells; (8) a site with abundance of blood vascular supply in order that the recombinant gene products from the grafted tissue can be secreted into the systemic circulation. The myoblasts and muscle tissue have the greatest potential to fulfill the above requirements especially were the therapeutic product must be secreted to be effective. Myoblasts provide a convenient and feasible somatic target cell for the application of gene therapy to most inherited diseases, especially those that have shown a response to enzyme therapy (1, 16, 23-26).

Studies from our laboratory and other investigators have shown that myoblasts including the C2C12 cell line as well as primary myoblasts derived from satellite cells of muscle tissue can be efficiently transduced *in vitro* with recombinant retroviral vectors and stably express the transgene at a high level for long period of time (11-14, 27). The expressed gene products are processed and secreted extracellularly. Furthermore, the myotubes derived *in vitro* myoblasts-fused to muscle fibers express the transgene products. Studies of transplantation of genetically engineered myoblasts into muscle in mice have shown that grafted genetically-modified myoblasts can rapidly fuse into endogenous muscle fibers of the host (28-29) and persist to express the transgene and deliver the recombinant products into the system circulation for long periods of time (11-14). Some studies indicate that genetically modified myoblasts can survive *in vivo* as muscle precursor cells and as highly vascularized muscle fibers (22).

Because the C2C12 myoblasts cell line can form tumors *in vivo*, it has less practical significance for human application (17). Very recently, studies from several groups were carried out in cultures of primary myoblasts transplanted to syngeneic animals (13, 16, 17). The results are exciting and demonstrate that modified myoblasts can express and deliver factor IX and lysosomal β-glucuronidase as well as a reporter gene product, β-galactosidase, *in vitro* and *in vivo*. Most approaches isolate and purify myoblasts from newborn animals. In considering the requirements for human application, it will be almost impossible to secure HLA-matched human primary myoblasts from newborn or embryo donors. The most likely method will be to obtain human primary myoblasts from muscle biopsies from the patients themselves. From this perspective, the development of this methodology and its use in model studies of gene therapy should involve primary myoblasts derived from adult animals rather than newborn or embryo animals.

We have conducted a series of studies of gene transfer using retroviral vector to transduce mouse bone marrow cells, myoblasts, and human CD34 cells to supply glucocerebrosidase (GC) (10, 27, 30, 31) as a strategy to correct cells deficient in the enzyme. Gaucher disease is caused by inherited deficiency of GC and is characterized by abnormal lysosomal storage of glucosylceramide predominantly in macrophages (32, 33). Currently, available treatment includes enzyme replacement and allogeneic bone marrow transplantation (24, 25). Both of these approaches are efficient, but limited in application. Somatic cell-mediated gene therapy, an alternative therapeutic strategy, is potentially therapeutic for Gaucher disease. Among the somatic cells, targets that might be useful in approaching the disorder, myoblasts deserve consideration because of their unusual properties and advantages especially in terms of their ability to secrete enzyme. The success of enzyme replacement lends further support to the rationale of using transduced myoblasts in approaches to the disease.

We succeeded in constructing a new retroviral vector, originating from a simplified N2 vector called MFG. Carrying human GC, it is called MFG-GC (10) (Fig. 14.1). The human GC cDNA was inserted into a site exactly at the ATG of the deleted *env* gene. The GC gene is driven by the promoter/enhancer of the retroviral LTR (10). The MFG-GC vector was produced in both ecotropic and amphotropic package cell lines. In studies of transduction and expression in the mouse model of bone marrow transplantation, our results revealed that MFG-GC results in very high transduction of mouse bone marrow cells (10). Expression of the human transgene is life-long in mice at level of GC approximately five times that of controls. Mice transplanted with bone marrow from primary recipients at 4-8 months postprimary transplantation with MFG-GC infected bone marrow cells showed sustained expression of the GC gene at high level for greater than one year (Fig. 14.2). The peripheral blood leukocytes of those long-term secondary transplants show that the human GC gene continues to be expressed at the levels of 5-8 times above normal control levels without abatement (34). Macrophages derived form bone marrow of the long-term secondary transplants can express and secrete human GC at high levels intracellularly and extracellularly *ex vivo*. In the studies in human cells, the MFG-GC vector was applied to explore transduction and expression of human GC gene in CD34 cells from Gaucher patients and human controls (30, 31). The studies revealed that the transduction efficiency of those human CD34 cell capable of bone marrow reconstitution was greater than 50% (35).

Using myoblasts as targets in somatic cell gene transfer studies, our previous work (27) has demonstrated that C2C12 myoblasts were transduced at a high efficiency *ex vivo* with the retrovirus vector, MFG-GC. The human glucocerebrosidase gene maintained a high level of expression intracellularly and extracellularly from C2C12 myoblasts transduced with retrovirus vector MFG-GC (Fig. 14.3). Transplantation of transduced C2C12 myoblasts revealed that the gene for GC was stably expressed at the site of the implant in C3H/HeJ mouse muscle.

In our current research, primary myoblasts derived from human muscle and C3H/HeJ mouse skeletal muscle were used to conduct studies of how well transduced myoblasts could secrete GC. In these experiments, skeletal muscles from adult 4-6 week old C3H/HeJ mice or from human autopsies or biopsies were isolated, dissected free of connective tissue, and dissociated by treatment with trypsin-EDTA and collage-

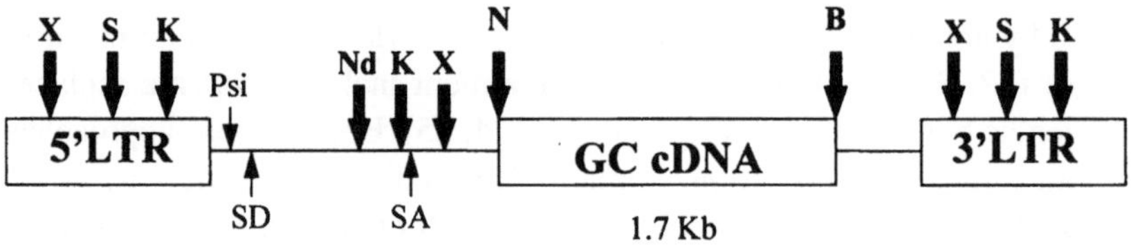

X: Xba I *S: Sst I*
K: Kpn I *Nd: Nde I*
N: Nco I *B: Bam HI*
SD: splice donor *SA: splice acceptor*

Fig. 14.1: Proviral structure of MFG-GC.
Human glucocerebrosidase cDNA was cloned into the MFG plasmid by creating a NcoI-HindIII fragment ligated to a HindIII-BclI fragment from the human cDNA and then cloned into NcoI(N)-BamHI(B) site in the MFG plasmid.

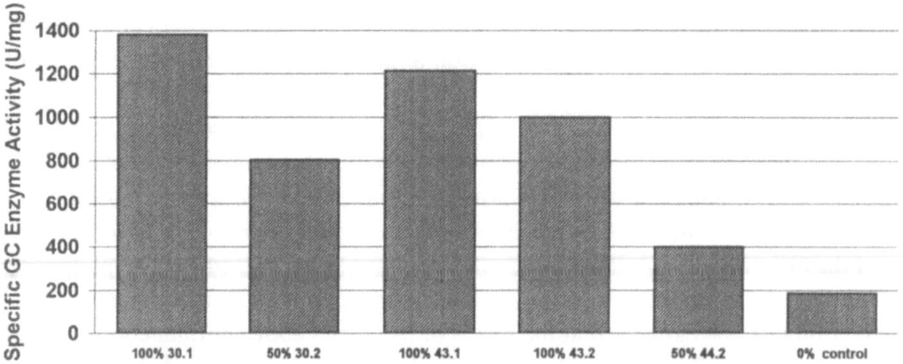

Fig. 14.2: Glucocerebrosidase enzyme activity in peripheral blood leukocytes of secondary BM transplant recipients.
These mice received bone marrow from primary recipients of transduced marrow from congenic donor mice containing a genetic marker, GPI, which enables electrophoretic determination of the proportion of donor marrow (% Donor GPI) in reconstituted myeloablated recipients.

nase. The myoblasts were harvested and seeded onto a gelatin-coated tissue culture dish with complete media [DMEM supplement with 20% fetal calf serum and basic fibroblast growth factor (bFGF)]. The human primary myoblasts were harvested, resuspended in complete media and plated. Murine primary myoblasts were cloned, purified and pooled. Human primary myoblasts were cloned by the limiting dilution assay. Murine and human primary myoblasts were identified morphologically by allowing them to differentiate into multinucleated myotubes.

A portion of primary murine and human myoblasts thus obtained were transduced *ex vivo* with the amphotropic retroviral vector MFG-GC in the presence of polybrene. For controls, murine and human primary myoblasts were infected with amphotropic MFG-lacZ carrying the bacterial B-galactosidase gene. Transduced myoblasts were harvested 48 hrs after transduction. Most of transduced myoblasts

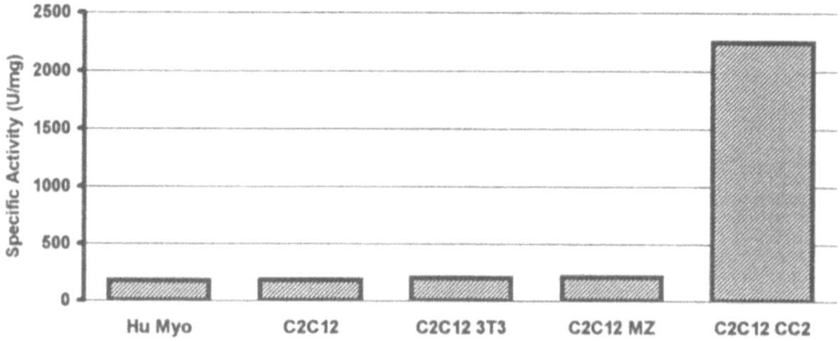

Fig. 14.3a: Intracellular glucocerebrosidase enzyme activity of myoblasts.
Hu Myo: Activity in uninfected human myoblasts.
C2C12: Activity in uninfected C2C12 murine myoblasts.
C2C12 3T3: Activity in C2C12 murine myoblasts mock-infected with supernatants from NIH3T3 cells.
C2C12 MZ: Activity in C2C12 murine myoblasts infected with supernatants from MFG-LacZ. C2C12 CC2: Activity in C2C12 murine myoblasts infected with supernatants from MFG-GC producer cells, psi-crip clone #2.

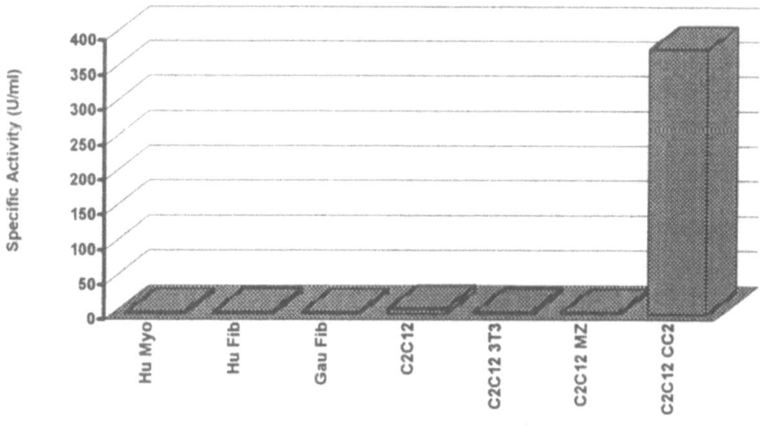

Fig. 14.3b: Extracellular secretion of glucocerebrosidase by MFG-GC transduced myoblasts. Levels of glucocerebrosidase activity in supernatants from uninfected human myoblasts (Hu Myo), uninfected human fibroblasts (Hu Fib), uninfected Gaucher fibroblasts (Gau Fib), uninfected C2C12 myoblasts (C2C12), C2C12 myoblasts mock-infected with NIH 3T3 supernatant (C2C12 3T3), C2C12 myoblasts transduced with MFG-LacZ (C2C12 MZ) and C2C12 myoblasts transduced with MFG-GC (C2C12 CC2).

were frozen at -135°C for further transplantation. A small portion of transduced myoblasts were expanded to analyze transduction efficiency by Southern blot measuring copy number of the transgene or by immunocytochemistry. Expression of human glucocerebrosidase gene in cultured cell and its secretion extracellularly was measured by enzyme assay. The results of these studies showed that primary human and murine myoblasts were successfully transduced using retrovirus vector MFG-

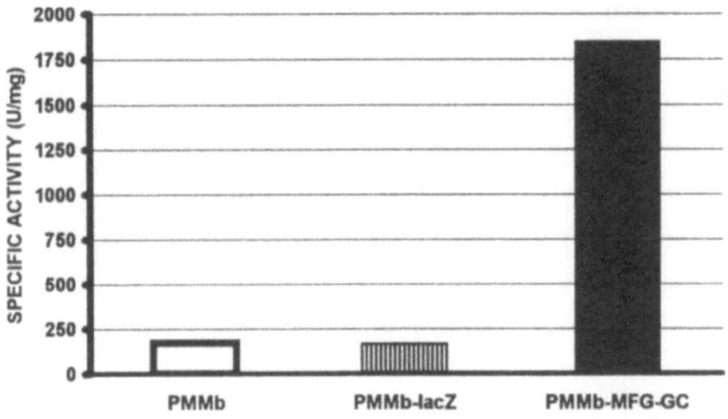

Fig. 14.4a: Intracellular glucocerebrosidase enzyme activity of MFG-GC transduced primary murine myoblasts (PMMb) *ex vivo*. Levels of intracellular glucocerebrosidase activity of non-transduced primary murine myoblasts (PMMb), of MFG-LacZ transduced primary murine myoblasts (PMMb-LacZ), and MFG-GC transduced primary murine myoblasts (PMMb-MFG-GC).

GC, estimated 2-10 copies of the transgene per myoblast. Transduction efficiency was approximately 95-100% as indicated by immunocytochemical staining using Mab 8E4 against human glucocerebrosidase. Glucocerebrosidase activity of transduced murine and human primary myoblasts were about 10.7 and 5.2 times above non-transduced and MFG-lacZ transduced controls, respectively (Fig. 14.4). Both transduced human and murine primary myoblasts secreted a high level of glucocerebrosidase activity extracellularly. The transduced primary murine myoblasts could sustain high intracellular expression and extracellular secretion of human glucocerebrosidase for up to 35 weeks (approximately 50-60 doublings) in culture, and retained the ability to differentiate into multinucleated myotubes. In order to explore the endurance of expression of the human GC gene under the same circumstance as a primary myoblast-fused into multinucleated myotubes, transduced primary murine myoblasts were allowed to differentiate into multinucleated myotubes under low concentration of fetal calf serum and without bFGF support. The results showed that multinucleated myotubes differentiated from MFG-GC transduced primary murine myoblasts continued robust expression of the GC and maintained a high level of extracellular secretion of the enzyme (Fig. 14.5).

Glucocerebrosidase is a membrane-associated lysosomal β-glucosidase (36). We have shown that under usual circumstances, the enzyme is not secreted as part of the steps in its normal biosynthesis (37). It is well known that GC can be taken up by macrophages *via* a mannose receptor (38-40). To confirm whether or not human glucocerebrosidase secreted by transduced primary murine myoblasts can be taken up by macrophages, bone marrow-derived syngenic murine macrophages were co-cultivated with transduced primary myoblasts or incubated with the supernatant collected from transduced primary murine myoblasts in culture. After co-culture for 12 hours or incubation for 2 hours, human glucocerebrosidase could be detected in

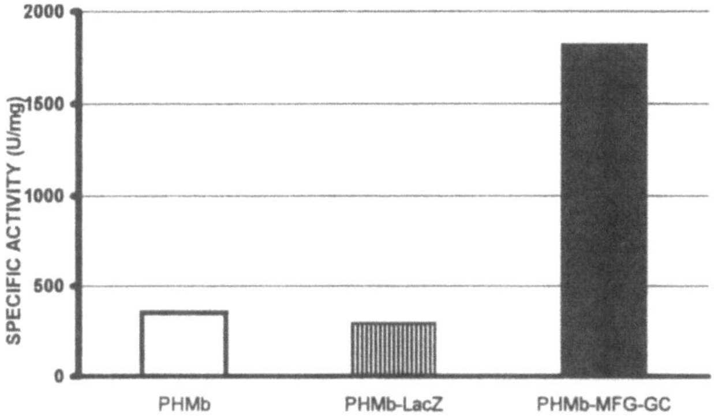

Fig. 14.4b: Intracellular glucocerebrosidase enzyme activity of MFG-GC transduced primary human myoblasts (PHMb) *ex vivo.*
Levels of intracellular glucocerebrosidase activity of non-transduced primary human myoblasts (PHMb), of MFG-lacZ transduced primary human myoblasts (PHMb-LacZ) and of MFG-GC transduced primary human myoblasts (PHMb-MFG-GC).

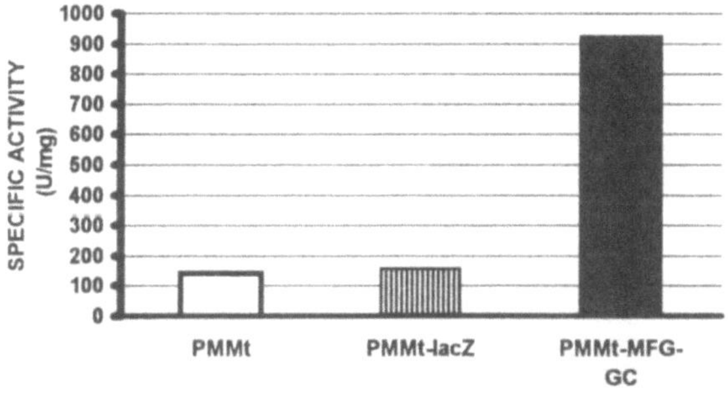

Fig. 14.5a: Intracellular glucocerebrosidase activity of myotubes differentiated from MFG-GC transduced primary murine myoblasts (PMMt) *in vitro.*
Levels of intracellular glucocerebrosidase activity of myotubes differentiated from non-transduced primary murine myoblasts (PMMt), from MFG-LacZ transduced primary murine myoblasts (PMMt-LacZ) and from MFG-GC transduced primary murine myoblasts (PMMt-MFG-GC).

macrophages by immunocytochemical assay or glucocerebrosidase activity measurements (Fig. 14.6). Furthermore, uptake by macrophages could be partly blocked by mannan.

To investigate expression of the human glucocerebrosidase gene *in vivo* and secretion of human glucocerebrosidase into circulation, MFG-GC transduced C3H/HeJ mouse primary myoblasts were implanted into the thigh muscle of syngeneic mice. Every two weeks, blood was collected *via* the retro orbital vein. Plasma was separated at 4^0 C and glucocerebrosidase activity was assayed immediately. The results revealed a high level of glucocerebrosidase activity (range from approximately $1x10^2$-$1.5x10^4$ U/ml) in plasma, beginning two weeks post-transplantation with MFG-GC transduced myoblasts and continuing over 15 weeks. Conversely,

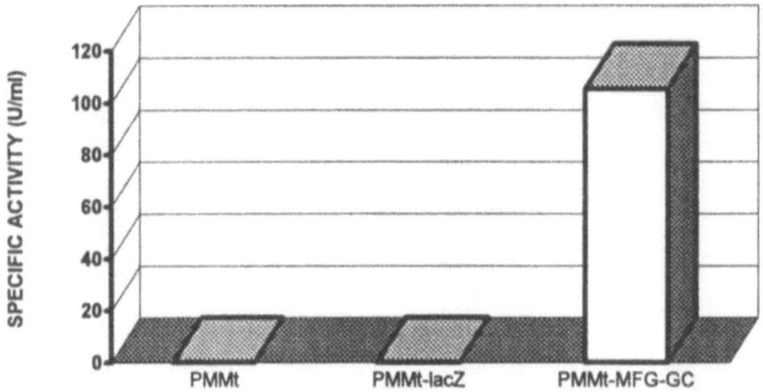

Fig. 14.5b: Extracellular glucocerebrosidase activity of myotubes differentiated from MFG-GC transduced primary murine myoblasts (PMMt) *in vitro*. Levels of extracellular glucocerebrosidase activity of myotubes differentiated from non-transduced primary murine myoblasts (PMMt), from MFG-lacZ transduced primary murine myoblasts (PMMt-LacZ) and from MFG-GC transduced primary murine myoblasts (PMMt-MFG-GC).

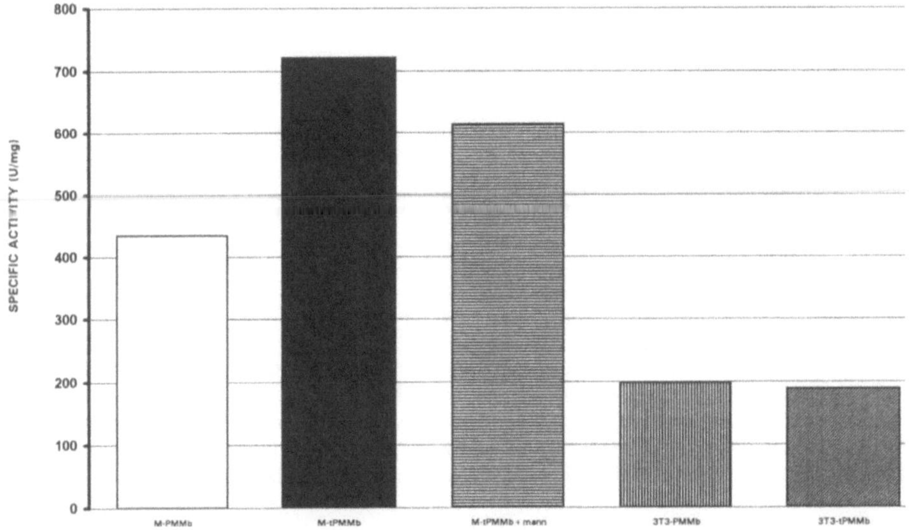

Fig. 14.6a: Intracellular glucocerebrosidase activity of BM-derived macrophages co-cultured with MFG-GC transduced primary murine myoblasts *ex vivo* for 12 hours. Intracellular glucocerebrosidase activity of BM-derived macrophages co-cultured with non-transduced primary murine myoblasts (M-PMMb), with MFG-GC transduced primary murine myoblasts (M-tPMMb), with MFG-GC transduced primary murine myoblasts plus 3mg/ml of mannan (M-tPMMb+mann), with. intracellular glucocerebrosidase activity of NIH 3T3 co-cultured with non-transduced primary murine myoblasts (3T3-PMMb) and with MFG-GC transduced primary murine myoblasts (3T3-tPMMb).

non-transplanted control mice and mice transplanted with MFG-lacZ transduced primary myoblasts showed no enzyme activity in their plasma. Glucocerebrosidase activity of livers of the implanted mice at 45 days post-transplantation was about 3 fold

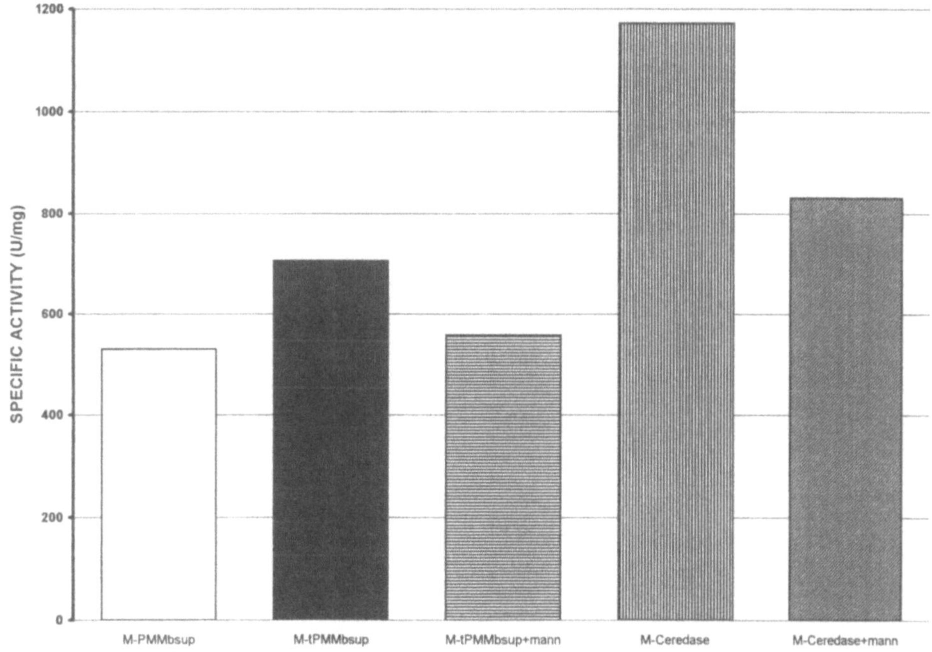

Fig. 14.6b: Intracellular glucocerebrosidase activity of BM-derived macrophages incubated with supernatants from the cultures of MFG-GC transduced primary murine myoblasts *in vitro* for 2 hours. Intracellular glucocerebrosidase activity of BM-derived macrophages incubated with supernatants from non-transduced primary murine myoblasts (M-PMMbsup), with MFG-GC transduced primary murine myoblasts (M-tPMMbsup), with MFG-GC transduced primary murine myoblasts plus 3mg/ml of mannan (M-tPMMbsup+mann), with a concentration 0.1 IU/ml of Ceredase® (M-Ceredase) and with a concentration 0.1 IU/ml of Ceredase® plus 3mg/ml of Mannan (M-Ceredase+mannan).

higher than the non-transplanted control mice or mice transplanted with MFG-lacZ transduced primary myoblasts (Fig. 14.7).

Immunohistochemistry revealed that expression of human glucocerebrosidase in muscle at the site of implantation persisted for at least 90 days post-transplantation. More importantly, the immunohistochemical analysis demonstrated that the human glucocerebrosidase secreted by MFG-GC transduced primary myoblasts implanted in muscle of C3H/HeJ mice was captured and taken up by macrophages in distant organs including liver, spleen and bone marrow.

Myoblasts are skeletal muscle precursor cells which originate from mesodermal stem cells known as satellite cell which exist around mature multinucleated muscle fibers. Myoblasts do not proliferate and divide but can be activated by injury or damage into a phase of proliferation and differentiation and fuse with adjacent myofibers to result in regeneration. To explore the fate of genetically modified primary myoblasts transplanted intramuscularly in syngenic mice, a experiment of re-isolation

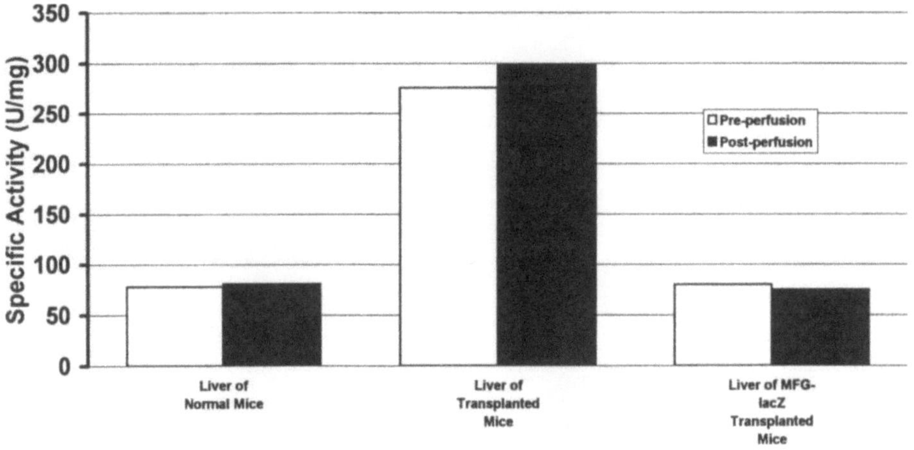

Fig. 14.7: Glucocerebrosidase enzyme activity of livers of C3H/HeJ mice transplanted with MFG-GC transduced syngenetic primary myoblasts at 45 day post-transplantation. Enzyme activity was measured at pre-perfusion and post-perfusion with PBS.

1. Liver of normal mice: Glucocerebrosidase activity of livers of C3H/HeJ mice of non-transplantation.
2. Liver of transplanted mice: Glucocerebrosidase activity of livers of C3H/HeJ mice of transplanted with MFG-GC transduced syngenetic primary myoblasts.
3. Liver of transplanted mice: Glucocerebrosidase activity of livers of C3H/HeJ mice of transplanted with MFG-LacZ transduced syngenetic primary myoblasts.

and culture of MFG-GC transduced primary myoblasts transplanted into mice was conducted. The transplanted muscles with MFG-GC transduced primary myoblasts, from mice at day 10, 15, 30, 60, 90 of post-transplantation, were dissected free of fat and connective tissue, minced, and dissociated as described previously. The re-isolated myoblasts prepared from transplanted mice were plated into a gelatin-coated tissue culture. Signal cell clones (50-100 cells) formed at about one week and cells from the clones were further cultured for another week to test the potential to differentiate into myotubes. The dishes were fixed with 100% methanol, and immunocytochemistry was performed with MAb 8E4 against human glucocerebrosidase. The experiments showed that transplanted transduced primary myoblasts could not only fuse into existing myofibers and expressed human glucocerebrosidase, but they could survive as precursor cells in the implanted muscle. The presence of transduced myoblasts that have not fused suggests that they remain in the muscle as myoblasts and continue to express human glucocerebrosidase gene in the transplanted mice. The reisolated myoblasts continued to express human glucocerebrosidase and secreted it extracellularly.

These studies using primary human and murine myoblasts suggest that primary myoblasts are valuable in studying gene therapy for patients with Gaucher disease as a method to produce glucocerebrosidase to complement their deficient enzyme. Further work will focus upon transplantation of transduced human primary myoblasts to SCID mice to study expression and delivery of the recombinant human glucocerebrosidase to target organs and cells, and the effect of transplanting these cells in the transgenic Gaucher mouse model of Gaucher disease.

Acknowledgement

The work reviewed in this manuscript was supported by grants from the National Institutes of Health RO1-DK43709, PO1-DK44935, RO1-DK45110, MO1-RR00084, MO1-RR00056-28, the National Gaucher Foundation and Genzyme Corporation.

References

1. Moullier, P., Bohl, D., Heard, J. M., and Danos, O. (1993). Correction of lysosomal storage in liver and spleen of MPS VII mice by implantation of genetically-modified skin fibroblasts. Nat. Genet. *4*, 154-159.
2. Moullier, P., Marechal, V., Danos, O., and Heard, J. M. (1993). Continuous systemic secretion of a lysosomal enzyme by genetically-modified mouse skin fibroblasts. Transplantation. *56*(2), 427-432.
3. Cristiano, R. J., Smith, L. C., Kay, M. A., Brinkley, B. R., and Woo, S. L. (1993). Hepatic gene therapy: efficient gene delivery and expression in primary hepatocytes utilizing a conjugated adenovirus-DNA complex. Proc. Natl. Acad. Sci. USA. *90*(24), 11548-11552.
4. Lal, B., Induti, R. R., Couraud, P. O., Goldstein, G. W., and Laterra, J. (1994). Endothelial cell implantation and survival within experimental gliomas. Proc. Natl. Acad. Sci. USA. *91*(21), 9695-9699.
5. Vogel, J. C. (1993). Keratinocyte gene therapy. Arch. Dermatol. *129*(11), 1478-1483.
6. Gerrard, A. J., Hudson, D. L., Brownlee, G. G., and Watt, F. M. (1993). Towards gene therapy for haemophilia B using primary human keratinocytes. Nat. Genet. *3*(2), 180-183.
7. Fauser, A. A. (1991). Long-term expression of gene introduction into normal human T-lymphocytes by retroviral-mediated gene transfer. J. Cell. Biochem. *45*(4), 353-358.
8. Sutkowski, N., Kou, M. L., Varela-Echavarria, A., and Doughherty, R. Y. (1994). A murine model for B-lymphocyte somatic cell gene therapy. Proc. Natl. Acad. Sci. USA. *91*(19), 8875-8879.
9. Lehn, P. M. (1987). Gene therapy using bone marrow transplantation. Bone Marrow Transplantation. *1*(3), 243-258.
10. Ohashi, T., Boggs, S., Robbins, P., Bahnson, A., Patrene, K., Wei, F. S., Wei, J. F., Lucht, L., Fei, Y., Clark, S., Kimak, M., He, H., Mowery-Rushton, P., and Barranger, J. A. (1992). Efficient transfer and sustained high expression of the human glucocerebrosidase gene in mice and the functional macrophages following transplantation of bone marrow transduced by retroviral vector. Proc. Natl. Acad. Sci. USA. *89*, 11332-11336.
11. Barr, E. and Leiden, J. M. (1991). Systemic delivery of recombinant proteins by genetically modified myoblasts. Science. *254*, 1507-1509.
12. Dhawan, J., Pan, L.C., Pavlath, G. K., Travis, M. A., Lanctot, A. M., and Blau, H. M. (1991). Systemic delivery of human growth hormone by injection of genetically engineering myoblasts. Science. *254*, 1509-1512
13. Yao, S. N. and Kurachi, K. (1992). Expression of human factor IX in mice after injection of genetically modified myoblasts. Proc. Natl. Acad. Sci. USA. *89*(8), 3357-3361.
14. Dai, Y., Roman, M., Naviaux, R. K., and Verma, I. M. (1992). Gene therapy via primary myoblasts: long-term expression of factors IX protein following transplantation *in vivo*. Proc. Natl. Acad. Sci. USA. *89*(22), 10892-10895.
15. Roman, M., Axelrod, J. H., Dai, Y., Naviaux, R. K., Friedmann, T., and Verma, I. M. (1992). Circulating human or canine factor IX from retrovirally transduced primary myoblasts and established myoblast cell lines grafted into murine skeletal muscle. Somat. Cell Mol. Genet. *18*(3), 247-258.

16. Naffakh, N., Pinset, C., Montarras, D., Pastoret, C., Danos, O., and Heard, J. M. (1993). Transplantation of adult-derived myoblasts in mice following gene transfer. Neuromuscl. Disord. *3*(5/6), 413-417.
17. Rando, T. A. and Blau, H. M. (1994). Primary mouse myoblast purification, characterization, and transplantation for cell-mediated gene therapy. J. Cell Biol. *125*(6), 1275-1287.
18. Watt, D. J. (1990). A comparison of long-term survival of muscle precursor cell suspensions and minced muscle allografts in the non-tolerant mouse. Adv. Exp. Med. Biol. *280*, 35-38.
19. Watt, D. J., Morgan, J. E., and Partridge, T. A. (1991). Allografts of muscle precursor cells persist in the non-tolerized host. Neuromuscl. Disord. *1*, 345-355.
20. Blau, H. M., Dhawan, J., and Pavlath, G. K. (1993). Myoblasts in pattern formation and gene therapy. Trends. Genet. *9*(8), 269-274.
21. Tortora, G. J. and Anagnostakos, N. P. (1984). Principles of anatomy and physiology. Chapter 11, 4th ed. New York: Harper & Row, 223-264.
22. Yao, S.-N. and Kurachi, K. (1993). Implanted myoblasts not only fuse with myofibers but also survive as muscle precursor cells. J. Cell Sci. *105*(Pt4), 957-963.
23. Marechal, V., Naffakh, N. V., Danos, O., and Heard, J. M. (1993). Disappearance of lysosomal storage in spleen and liver of mucopolysaccharidosis VII mice after transplantation of genetically modified bone marrow cells. Blood. *82*(4), 1358-1365.
24. Barranger, J. A., Ohashi, T., Hong, C. M., Tomich, J., Aerts, J. F. G. M., Tager, J. M., Nolta, J. A., Sender, L. S., Weiler, S., and Kohn, D. B. (1989). Molecular pathology and therapy of Gaucher disease. Japanese J. Inherit. Metab. Dis. *51*, 45-71.
25. Rappeport, J. M., Barranger, J. A., and Ginns, E. I. (1986). Bone marrow transplantation in Gaucher disease. In: *Birth Defects*: *Original Articles Series*. March of Dimes Birth Defects Foundation, Vol. 22, No. 1. New York: Alan R. Liss, Inc., 101-109.
26. Hershfield, M. S., Chaffee, S., and Sorensen, R. V. (1993). Enzyme replacement therapy with PEC-ADA in adenosine deaminase deficiency: Overview and case report of three patients including two now receiving gene therapy. Pediatr. Res. *33*, 42.
27. Bansal, V., Mowery-Rushton, P., Lucht, L., Li, J., Bahnson, A., Watkins, S. C., and Barranger, J. A. (1994). Transduction, expression, and secretion of human glucocerebrosidase by murine myoblasts. Ann. N. Y. Acad. Sci. *176*, 307-319.
28. Morgan, J. E., Watt., D. J., Sloper, J. C., and Partridge, T. A. (1988). Partial correction of an inherited biochemical defect of skeletal muscle by grafts of normal muscle precursor cells. J. Neurol. Sci. *86*, 137-147.
29. Partridge, T. A. (1991). Myoblast transfer: A possible therapy for inherited myopathies. Muscle. Nerve *14*, 197-212.
30. Bahnson, A., Nimgaonkar, M., Fei, Y., Boggs, S., Robbins, P., Ohashi, P., Dunigan, J., Li, J., Ball, E. D., and Barranger, J. A. (1994). Transduction of CD34+ enriched cord blood and Gaucher bone marrow cells by a retroviral vector carrying the glucocerebrosidase gene. Gene Therapy *1*, 176-184.
31. Nimgaonkar, M., Bahnson, A., Boggs, S., Dunigan, J., Ball, E. D. and Barranger, J. A. (1994). Transduction of mobilized peripheral blood CD34+ cells with the glucocerebrosidase gene. Gene Therapy *1*(3), 201-207.
32. Barranger, J. A. and Ginns, E. I. (1989). Glucosylceramide Lipidoses: Gaucher Disease. In: *The Metabolic Basis of Inherited Disease*. Chapter 67, 6th ed. McGraw-Hill Information Services Company: Health professions Division, 1699-1720.
33. Marchand F. (1907). Uber soq. idiopathische splenomegalie-typus Gaucher. Muench Med. Wochenschr. *54*, 1102.
34. Nimgaonkar, M., Bahnson, A., Kemp, A., Lancia, J., Mannion-Henderson, J., Boggs, S., Mohney, T., Dunigan, J., Barranger, J. A., and Ball, E. D. (1995). Long term expression of the glucocerebrosidase gene in mouse and human hematopoietic progenitors. Leukemia. *9*(1), 38-42.
35. Bahnson, A., Dunigan, J., Baysal, B., Mohney, T., Atchison, R. W., Nimgaonkar, M., Ball, E. D., and Barranger, J. A. (1995). Centrifugal enhancement of retroviral-mediated gene transfer. J. Virol. Meth. *54*, 131-143.
36. van Dongen, J. M., Willemsen, R., Ginns, E. I., Sips, H. J., Tager, J. M., Barranger, J. A., and Resuer, A. J. (1986). The subcellular localization of soluble and membrane-bound lysosomal enzymes in I-cell fibroblasts: A comparative immunocytochemical study. Eur. J. Cell Biol. *39*, 179-186.
37. Erickson, A. H., Ginns, E. I., and Barranger, J. A. (1985). Biosynthesis of the lysosomal enzyme, glucocerebrosidase. J. Biol. Chem. *260*, 14319-143324.

38. Furbish, F. S., Steer, C. J., Krett, N. L, and Barranger, J. A. (1981). Uptake and distribution of placental glucocerebrosidase in rat hepatic cells and effects of sequential deglycosylation. Biochim Biophys Acta. *673*(4), 425-434.
39. Murray, G. J. (1987). Lectin-specific targeting of lysosomal enzymes to reticuloendothial cells. Methods Enzymol. *149*, 25-42.
40. Murray, G. J. and Jin, F.-S. (1995). Immunoelectron microscopic localization of mannose-terminal glucocerebrosidase in lysosomes of rat liver Kupffer cells. J. Histochem. Cytochem. *43*(2), 149-158.

15. In vivo Secretion of Therapeutic Proteins from Neo-Organs

Delphine Bohl and Jean Michel Heard

The delivery of therapeutic proteins in human patients is currently performed by injecting pharmacological preparations of material extracted from animal tissues or from recombinant bacteria, yeasts or animal cells. The manufacturing of these drugs requires highly optimized production and purification processes, as well as sophisticated quality controls. Long term treatments imply repeated injections, which in some cases lead to almost unaffordable costs. The development of efficient gene transfer vehicles, which allow the introduction and stable expression of foreign genes into mammalian tissues, might circumvent the limitations of classical biotechnology products. In that respect, the permanent systemic delivery of therapeutic proteins from engineered tissues appears as one of the most attractive applications of gene therapy.

The in vivo engraftment of engineered drug-secreting cells in animals can be performed in various ways. Non or weakly tumorigenic cells can be simply injected intraperitoneally, subcutaneaously or under the kidney capsule of syngeneic recipients, where they persist for some period of time, delivering a therapeutic protein in the serum transiently (1). The long term secretion of human growth hormone (hGH) has been obtained by grafting genetically-modified syngeneic primary fibroblasts under the kidney capsule of rabbits (2). Primary or immortalized cells from non-autologous origin can also be grafted into immunocompetent animals, provided that they are enclosed into microcapsules which protect them from immune rejection. The immune protection is not effective for xenogeneic cells. Long term systemic secretion of therapeutic proteins like adenosine deaminase (ADA) (3), hGH (4-6) or factor IX (7) has been reported using immortalized cell lines encapsulated in hollow fibers or capsules. The technique appears very versatile. It should allow future large scale manufacturing at relatively low costs. However, currently available polymers may induce strong inflammatory reaction in certain tissues, while the survival of allogeneic cells and the maintenance of their secretory capacities have been documented in a few situations only and for limited length of time. The biocompatibility and long term tolerance of a variety of polymers is currently under intensive investigations in order to bring this technology to the clinic.

A very different strategy consists in incorporating engineered autologous cells in neo-formed tissues, from which they can deliver the therapeutic protein to the systemic circulation. Gene transfer is performed in immortalized or primary myoblast cultures. Cells are further transplanted back in regenerating muscles of syngeneic or immuno-incompetent recipient animals, where they form genetically-modified myofibers. Thus, skeletal muscles have been used as a source of hGH (8, 9), factor IX (10, 11), erythropoietin (12-16) or lysosomal enzyme (16). Alternatively, procedures have been described for creating novel organs. The goal is to reproduce, as faithfully as possible, the structure and function of a vascularized tissue in which genetically-modified cells can survive for very long periods of time, thus providing a permanent source of the therapeutic protein. We will review the various approaches that have been proposed for the preparation of such neo-organs, and the applications of the methods for the treatment of genetic or acquired diseases.

15.1 Conception and Realisation of Neo-Organs

Neo-organs are engrafted structures that are expected to provide a suitable environment for the long term survival of cells and the efficient delivery of a therapeutic protein in the blood stream. Such a requirement implies that the graft is well vascularized and persists for long periods of time without necrosis or fibrosis. It is also desirable that the implant is precisely localized, and eventually removable and replaceable.

15.1.1 From Skin Equivalents to Neo-Organs

Historically, the first attempts to reconstitute tissue structures from cultured cells were performed by Bell et al., using a derm equivalent covered with keratinocytes (17). A derm equivalent consists in dermal fibroblasts mixed with collagen, serum and medium. A collagen lattice is formed, which is contracted by fibroblasts. Keratinocytes can be overlayed, forming a skin equivalent. After grafting in syngeneic recipients, skin equivalents heal within seven days. The junctions with surrounding tissues become difficult to detect after a few months. A network of capillary vessels develops rapidly, and the graft persists for the lifetime of the animal. Allogeneic fibroblasts can be accepted in skin equivalent transplants, provided that class II antigen bearing cells have been eliminated previously (18).

The obvious clinical application of skin equivalent is the treatment of severe burns. The technique is now largely used in human patients (19). Since both keratinocytes and fibroblasts can be manipulated in vitro easily and transduced by physical methods or by retrovirus vectors efficiently, skin equivalent also appeared as a possible technique for grafting genetically-modified cells.

Keratinocytes were first considered as a target for gene transfer. Using a retrovirus vector, Morgan et al. have expressed the hGH cDNA and a selective marker in human keratinocytes (20). They documented the secretion of the hormone in engrafted nude mice and concluded that the transfer of the protein from the epidermis to the serum was rather inefficient. In contrast to these data, Fenjves et al. reported that the engraftment of unmodified human epidermis (21) or of human keratinocytes engineered with a retroviral vector in order to secrete a recombinant form of ApoE (22), allow the secretion of significant amounts of ApoE in the serum of recipient nude mice. A short term delivery of human factor IX from retrovirus vector-transduced human keratinocytes has also been reported (23). Other attempts to engraft genetically-modified keratinocytes in skin equivalent involved non-secreted reporter proteins (24, 25).

The other component of skin equivalent, dermal fibroblasts, has been intensively investigated as a potential tissue from which recombinant proteins could be secreted into the circulation. Derm fibroblasts are obtained from a skin biopsy, grown as a primary culture during which gene transfer is performed using a retrovirus vector. After cell amplification, cultures are trypsinized and the fibroblasts are embedded into a collagen gel, with the aim to form a tissue-like structure. In the absence of epidermal covering, derm equivalents are implanted subcutaneously. The first demonstration of the feasibility of using derm equivalents for gene therapy purpose was provided by St. Louis and Verma (26). In normal mice implanted subcutaneously with primary embryo fibroblasts engineered to secrete human factor IX, they showed that the clotting factor was delivered to the blood during a couple of weeks. Similar observations were reported by Palmer et al. (27). Further investigations revealed that the progressive disappearance of the protein from the blood resulted from a shut off of the expression, secondary to promoter extinction (28, 29). Long term persistence of lacZ expressing cells was obtained when viral promoters were exchanged for internal housekeeping gene promoters (28). Using a similar approach in which factor IX was expressed from a retrovirus LTR, a chinese group reported the long term secretion of therapeutic levels of the protein in rabbits and in two hemophiliac patients (30-33).

Implants of genetically-modified fibroblasts were also prepared by using solid artificial supports instead of a collagen lattice. Proposed artificial materials included either permanent structures made of nylon matrice (34), microcarrier beads (35), or biodegradable fibers (36). The secretion of human transferin during 12 weeks was documented after the engraftment of immortalized NIH 3T3 fibroblasts seeded on nylon matrices implanted subcutaneously in mice (34).

With the aim to stimulate neo-vessel formation, angiogenic factors have been included in the implanted material. St. Louis and Verma used gelfoam sponges coated with basic fibroblast growth factor (bFGF). The material was inserted in the vicinity of collagen lattices containing genetically-modified fibroblasts (26). Also using bFGF-coated gelfoam sponges, Thompson et al. observed the constitution of neo-vessels in implants cemented in the peritoneal cavity, the kidney, the spleen or the abdominal wall of rats (37). Rat hepatoma cells marked with a retrovirus vector remained viable when implanted into bFGF-coated gelfoam sponges, whereas they disappeared from uncoated sponges. In other experiments, Thompson et al. also obtained the formation of organoid neovascular structures by implanting polytetrafluoroethylene (PTFE) fibers previously coated with collagen, heparin and bFGF, in the peritoneal cavity of rats (38). PTFE fibers induced a local chronic inflammatory response and an intense stimulation of neo-vessel formation, leading to the constitution of highly vascularized granulomas. Normal rat hepatocytes injected into these preformed neo-organs persisted for several weeks (38). Two groups reported the correction of hyperbilirubinemia after the transplantation of normal isogeneic hepatocytes in bilirubin glucuronosyl transferase deficient rats (Gunn rats), but this result could not be reproduced by Thompson et al. (38-40). Using the same technique in the mouse, Moullier et al. reported the long term delivery of a lysosomal enzyme (human β-glucuronidase) and its capture by distant organs (41).

15.1.2 Realisation and Characteristics of Neo-Organs

In order to improve the reproducibility of neo-organ formation, bFGF-coated PTFE fibers were introduced in derm-equivalents (42). The combination of these two techniques allows the ex-vivo preparation of implants which are further introduced in the peritoneal cavity. Derm fibroblasts are grown as a primary culture during which retrovirus-mediated gene transfer can be easily performed. After amplification, cells are embedded into a collagen lattice in the same time as bFGF-coated PTFE fibers. Contraction of the collagen lattice occurs normally, leading to a compact structure which can be easily manipulated for surgical implantation (Fig. 15.1a). Neo-vessels develops rapidly in the recipient animal (Fig. 15.1b), and after a few weeks, the histological examination of the implant reveals a mesenchymal tissue very similar to a normal derm (Fig. 15.1c). The inflammatory reaction is limited to the presence of a few giant cells lining PTFE fibers, but lymphocytes and granulocytes are rarely observed. Neo-organs conserve the same macroscopical and histological aspect over very long periods of time (Fig. 15.1d). In mice, they have been maintained for 14 months in the absence of fibrotic or necrotic reaction. Neo-organs implanted in dogs were still functional after one year (45).

As reported for derm equivalent implanted subcutaneously (17), fibroblasts stop dividing rapidly after inclusion in a collagen matrix (D. Bohl, unpublished results). This was documented by using a fluorescent dye incorporated into cell membranes. In cells grown in culture flasks, division led to a progressive lost of fluorescence. In contrast, fluorescence persisted at the same level for ten days in fibroblasts embedded in collagen lattices kept in vitro, indicating that no more mitosis occurred. Few cell divisions took place within the few days following in vivo implantation, as previously observed for derm equivalent (17), afterwhat cells turned back to a non-proliferative state.

The phenomenons which take place in vivo during the constitution of neo-organs remain mostly undocumented. The development of a local inflammatory response to PTFE fibers is likely to be required for the stimulation of neo-vessel formation.

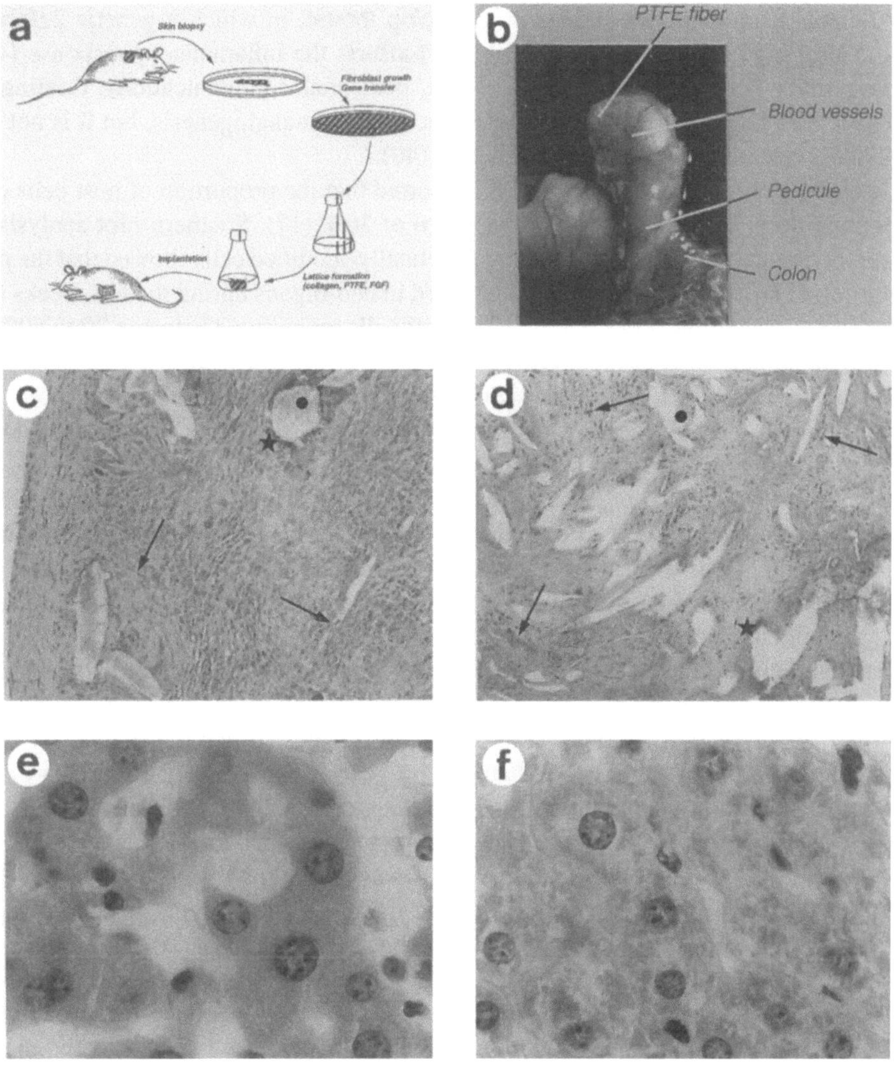

Fig. 15.1a: Schematic protocol for the constitution of neo-organs and the implantation in recipient mice.

Fig. 15.1b: Macroscopic view of a neo-organ 24 days after implantation in the peritoneal cavity of a mouse. PTFE fibers and neo-vascularization are indicated.

Fig. 15.1c, d: Histological sections of neo-organs resected one month (1c) or fourteen months (1d) after implantation in the peritoneal cavity of mice. Neo-vessels (arrows), giant cells (stars) and synthetic material (circles) are indicated. (Goldner's trichrome, magnification 10x).

Fig. 15.1e: Histological section of the liver of a control MPSVII mouse showing distension of lysosomes in Kupffer cells (Haemtoxylin eosin, magnification 40x).

Fig. 15.1f: Histological section of the liver of a MPSVII mouse bearing β-glucuronidase secreting neo-organs for two months (haematoxylin eosin, magnification 40x).

Attempts to implant neo-organs in the op/op mouse, in which a genetic defect of a macrophage stimulating factor (M-CSF) affects the inflammatory response (44), have never been successful (N. Naffakh, personnal communication). Coating of PTFE with heparin and bFGF probably facilitate neo-angiogenesis, but it is not absolutely required for neo-organ formation (40).

Using a chromosomal marker, Bell reported that the proportion of host cells colonizing derm equivalents was in the range of 30% (17). Southern blot analysis of intraperitoneal neo-organs containing genetically-modified cells showed that the proportion of host cells progressively increased in neo-organs during the few weeks following implantation. After two months, host cells represented between 20 to 50% of the total number of cells in neo-organs (D. Bohl, unpublished data). Colonizing cells are mostly fibroblasts, smooth muscle cells and endothelial cells.

Although the long term persistence of neo-organ structures is now well documented, the long term survival of implanted cells is more questionable. This can be examined by using genetically-marked cells. As discussed below, the secretion of erythropoietin (Epo) is a convenient marker of cell survival in the living animal. In mice implanted with neo-organs secreting murine Epo, Naffakh et al. observed that the elevation of the hematocrit was stable for six months, suggesting that most of the implanted cells survived (43). Beyond that time, hematocrit values decreased progressively. After ten months, hematocrits were still elevated, but half less than at six months. Since no immunological response developed against the murine Epo in implanted mice, the reduction of Epo levels suggested that the number of Epo-secreting cells had decreased. Whether Epo gene expression had been turn off or whether a fraction of implanted cells had been eliminated is still not known. It is however likely that implanted cells are heterogeneous, including short life elements with limited self-renewal potential, which may be progressively lost over time. If it would be feasible, the selection of long living cells with high proliferative potential would probably improve the long term survival of engrafted fibroblasts in neo-organs.

An interesting feature of neo-organs is the possibility of removing the implant when necessary. In their experiments in MPS VII mice treated with β-glucuronidase secreting cells, Moullier et al. have shown that storage lesions rapidly reappeared after excision of neo-organs (42). This indicates that no or very few genetically-modified fibroblasts had migrated outside from the implant to colonize distant tissues. Similar indications were obtained in implanted dogs by PCR analysis looking for vector sequences in distant tissues (45).

15.1.3 Gene Transfer Procedures and Vehicles

The realisation of neo-organ implies the in vitro amplification of primary cells obtained from a tissue biopsy. Gene transfer can be performed immediately after the initiation of the culture, after which genetically-modified cells are amplified until the required number is available. Retrovirus vectors are especially well suited for such a gene transfer procedure. They efficiently infect rapidly dividing cells, and because they stably integrate their genome in the chromosome of target cells, foreign DNA sequences are transmitted to the entire progeny. Non integrative vectors, like adenovirus vectors and non-viral vehicles associating plasmidic DNA with protein and/or lipids imply that gene transfer takes place after cell amplification. High transduction efficiency is therefore required in order to transduce a significant proportion of the cells. This can be achieved by adenovirus vectors easily (46), whereas non-viral vectors require the selection of transduced cells (47).

As discussed above, gene expression may be turned off after transduced cells are implanted in vivo, at least in the mouse. The use of constitutive promoters, the expression of which is required whatever the differentiation status of the cells, is therefore highly recommended for mouse experiments. Housekeeping gene promoters, like the dihydrofolate reductase gene promoter or the phosphoglycerate kinase gene promoter allowed very long term in vivo expression (28, 42). Whether promoter shut off takes place in other animal species is not documented. Fibroblasts engrafted in rabbits or humans seem to have been able to express foreign genes from the metallothionein gene promoter or from the retroviral LTR promoter for prolonged period of time (2, 31, 32).

15.2 Secretion of Therapeutic Proteins from Neo-Organs

15.2.1 Secretion of Lysosomal Enzymes for the Treatment of Mucopolysaccharidosis (MPS)

MPS form a group of 6 lysosomal storage diseases characterized by an abnormal development of the skeleton. These inborn errors of the metabolism are caused by a genetic deficiency of a lysosomal acid hydrolase, which leads to the accumulation of insoluble material in lysosomes (48). In MPS the deficiency affects the degradation of proteoglycans. Insoluble products accumulate in lysosomes, thus impairing the normal development of the most affected tissues, which in MPS are bones and joints. Later on, a progressive alteration of lung, heart and brain functions occurs. In severe forms, life expectancy is less than ten years. Most of lysosomal enzyme molecules bind the mannose-6-phosphate receptors which direct them to the lysosome. However, a fraction of the enzyme is secreted in the extracellular medium and recaptured by cell surface mannose-6-phosphate receptors, providing a possibility of cross-correcting deficient cells from circulating enzyme. This is the rational for enzyme replacement therapies of MPS, which can be performed by bone marrow transplantation, by infusing purified enzyme preparations, or by gene therapy. The possibility of using autologous fibroblasts engineered to secrete the missing lysosomal enzyme has been explored for the treatment of MPS. The gusmps mouse is deficient in β-glucuronidase and presents a disease similar to human MPS VII. It is a useful model to evaluate MPS potential treatments. Correction of the disease phenotype has been reported after allogeneic bone marrow transplantation, gene transfer into hematopoietic stem cells and infusion of purified β-glucuronidase (see reviews in (49, 50)).

The human β-glucuronidase cDNA under the control of the mouse PGK-1 promoter, has been transferred with a retroviral vector into primary fibroblasts prepared from MPS VII mouse skin. The modified cells secreting 100-200 units of β-glucuronidase/10^6 cells/24 hours were included into collagen/PTFE neo-organs and implanted into the peritoneal cavity of 14 syngeneic MPS VII animals (42). Each animal received two lattices containing 10 to 15×10^6 fibroblasts and diminished its urinary GAG levels within a few days. Beta-glucuronidase activity was found in the liver, spleen, lung, brain, kidney, heart and bone marrow of the implanted animals which were sacrificed after 10 to 155 days. The highest activity levels (between 4 and 6% of normal) were found in the liver and spleen and a complete correction of the lysosomal storage lesions was observed in these organs (Fig. 15.1e, 1f). Although the enzyme activity in the liver appeared concentrated into tissue macrophages, the storage lesions were corrected in hepatocytes as well. This suggested that corrected macrophages could act in locally clearing out substrates which were not degraded by the hepatocytes. Intraperitonal neo-organs containing autologous cells secreting human β-glucuronidase have also been implanted in normal dogs (45). The procedure developed in mice has been scaled up in order to prepare implants made of 10^9 cells and a minimally invasive surgical procedure was used for installing them into the peritoneal cavity. One to six neo-organs have been introduced in small pocket prepared in the peritoneal sheet forming the omentum. The grafts were tolerated perfectly well by the animals. Chronic inflammatory reaction was not observed in the peritoneal cavity. During the one year follow-up after implantation, the uptake of the secreted human enzyme was demonstrated in liver biopsies after heat inactivation of the endogenous activity. These studies indicate that the procedure could be transferred to a clinical setting.

MPS I (Hurler disease) is much more frequent than MPS VII with estimates varying between 1:50 000 and 1:100 000 live birth. Genotype/phenotype correlation has begun to be established and pre or perinatal diagnosis is possible. The mechanisms of synthesis, processing, secretion and uptake of β-glucuronidase and α-L-Iduronidase are closely related, and the observations made in gene transfer experiments involving MPS VII models can be extrapolated to MPS I. The secretion of human α-L-Iduronidase from vascularised neo-organs and its secretion by distant organs could be demonstrated in nude mice (51). Furthermore, gene therapy can also be tested in MPS I dogs (52). Phase I/II trials in MPS I patients will have to assess the feasibility of the procedure, its tolerance by the patient, the efficiency and duration of enzyme secretion and the effect on the course of the disease.

15.2.2 Secretion of Erythropoietin (Epo) for the Treatment of Epo-Responsive Anemias

In recent years, recombinant Human Erythropoietin (rHuEpo) has became widely used as a replacement therapy for patients in whom a normal erythropoiesis can be stimulated (53). Epo is the glycoprotein hormone regulating erythropoiesis in mammals (for review, see Koury and Bondurant (54)). It is secreted mainly by kidney peritubular cells in response to hypoxia (55), binds a specific cell surface receptor on erythroid progenitor cells, thus promoting their proliferation and terminal differentiation into erythrocytes (56).

Epo is used to treat chronic anemias resulting from renal failure, where kidney cells naturally producing the hormone are destroyed. It is also administered to patients with inflammatory syndroms created by rheumatoid arthritis, cancers, or infectious diseases such as AIDS. The hormone has a half-life in the circulation of about eight hours, and long-term treatments require one to three weekly injections of 50-500 units/kg. The synthesis of γ-globin in erythropoietic cells is stimulated in the presence of Epo. This provided a rationale for the treatment of hemoglobinopathies in which β-globin chain synthesis is affected (β-thalassemia or sickle-cell anemia). A therapeutic effect of repeated injections of rHuEpo in association with hydroxyurea has been demonstrated in patients with sickle cell disease, with increased HbF synthesis, decreased polymerization of abnormal hemoglobin (HbS), and improvement of rheologic properties of erythrocytes (57). rHuEpo was also administered to patients with thalassemia intermedia, in whom anemia was improved (58, 59). The treatment of these hemoglobinopathies required doses about 10 fold higher than for acquired anemias (for a review on the use of rHuEpo in humans, see (53)).

Alternative approaches for prolonged in vivo delivery of Epo through gene transfer have been actively investigated during recent years. Following gene transfer, the secreted hormone present in the serum can be quantified using sensitive biological or immunological assays. Importantly, the biological effect of Epo secretion can be easily monitored over time in experimental animals by measuring the hematocrit, which directly reflects the red cell mass. Thus, beyond its clinical relevance, Epo gene transfer has become a valuable paradigm in which to study the in vivo delivery of therapeutic proteins, because of its convenience as an experimental tool.

In rodents, implants of primary fibroblasts (43), myoblasts (14, 16) and vascular smooth muscle cells (60) have also been investigated as a source of Epo secretion. Following infection with recombinant retroviruses, these cell types are induced to secrete high levels of biologically active Epo in culture. Transduced fibroblasts were implanted into syngeneic mice as neo-organs (43). The hematocrit of animals implanted with 10^6 to 10^7 engineered cells increased from an initial value around 45% up to 60%-80%, and persisted at this level over a ten month observation period. However, the progressive decline of the hematocrit observed between six and ten months suggested that the neo-organ environment may not be optimal for the self-renewal of transduced fibroblasts. When fibroblasts were transduced with recombinant adenoviral vectors instead of a retroviral vector, an elevation of the hematocrit was also observed, although over shorter periods of time (up to six weeks) (46).

The levels of Epo secretion achieved in these studies could possibly provide useful therapy. By extrapolation from their results in mice, Naffakh et al. estimated that a therapeutic effect could be obtained in patients with hemoglobinopathies by implanting 10^{10} genetically-modified autologous fibroblasts, while 10^9 would be enough for correcting the anemia associated with chronic renal failure (43). That seems feasible, since neo-organs containing 6×10^9 engineered skin fibroblasts were successfully implanted in the omentum of dogs and kept secreting a recombinant protein for one year (45). The therapeutic efficiency of Epo delivery from engineered cells is currently being evaluated in the murine model for β-thalassemia.

15.2.3 Perspectives for the Treatment of HIV-Infection

The inhibitory effects of soluble CD4 (sCD4) have been largely documented in vitro. In contrast, attempts to obtain in vivo anti-viral effects by injecting sCD4 in infected rhesus monkeys (61, 62) or in AIDS patients (63, 64) were unsuccessful. Pharmacokinetics studies performed in AIDS patients treated with sCD4 revealed a short half-life of the product and its diffusion in a rectricted body compartment (63). Taken together with studies documenting the conditions required for an anti-viral effect to be observed in vitro (65), these data suggested that in human patients treated with sCD4 inhibitory concentrations of the drug have probably never been reached in organ tissues. Although several others explanations for the failure of sCD4 trials have been proposed, the biological interest of sCD4 needs an appropriate delivery system to be assessed. A permanent systemic secretion has been considered as a possible mean to improve sCD4 delivery to body compartments where HIV replicates. This was obtained by using retrovirus-mediated gene transfer into fibroblasts and the implantation of neo-organs in human CD4 transgenic mice (66). However, serum sCD4 levels were far below what it is required for an anti-viral effect, and they progressively decreased over a two months period, probably due to an immune response against the truncated CD4 protein.

Neutralizing anti-gp120 antibodies or chemokine analogues might also be of interest for the treatment of HIV infection. Investigations aimed at obtaining the permanent delivery of these substances from engineered cells implanted into neo-organs are currently performed.

15.2.4 Perspectives for the Treatment of Hemophilias

Hemophilias are obvious targets for gene therapies aimed at providing a continuous delivery of factor VIII or factor IX. The benefit of regular infusions of low doses of factor VIII in hemophiliac children has proved to be efficient in preventing the bleeding episodes and improving the well-being of patients (67).

As discussed above, the delivery of factor IX has been obtained in animals receiving genetically-modified fibroblasts implanted in microcapsules (7), derm equivalents (26, 27, 30-33), collagen-coated microcarriers (35) or genetically-modified myoblasts engrafted in skeletal muscles (10, 11, 15). Liver cells have also been used as a source of clotting factor in factor IX deficient dogs (68). No experiment involving the use of neo-organs has been reported until now.

One report describes the transient secretion of therapeutic levels of human factor VIII from primary fibroblasts transduced with a retroviral vector and implanted in SCID mice in neo-organs (69). Transient expression resulted from the use of the viral LTR promoter, which was turned off after in vivo engraftment of the cells. Due to the large size of the factor VIII cDNA (9.2 kb), which impairs its insertion in retrovirus vectors, a modified version of the clotting factor has been used, in which the B domain is deleted. Since this molecule may be immunogenic in human patients, previous investigations in the dog model of factor VIII deficiency is required before human trials can be proposed with the retroviral approach. This supposse the isolation of the dog factor VIII cDNA and the construction of the corresponding retrovirus vector. Non-viral vectors, which can accomodate large fragments of DNA, still do not allow efficient gene transfer and sustained expression of foreign genes (70).

15.3 Immunological Response against Neo-Organ Components and Secreted Proteins

As it is often the case in gene therapy approaches, the therapeutic protein may be recognized as antigenic by the host. Antigenicity may result from the absence of the normal protein in the deficient recipient or from modifications introduced in the natural molecule to improve its function or to facilitate its use as a therapeutic agent.

There is no report of cytotoxic response directed against neo-organ components or genetically-modified cells. The absence of infiltration of neo-organ tissue by lymphocytes or phagocytic cells suggests that both genetically-modified cells, artificial fibers and collagen are relatively well tolerated by the recipient organism. However, the existence of a cell-mediated immune response has not been investigated directly, and only few molecules have been secreted from neo-organs until now.

A strong immunization against the secreted molecule has been documented for several proteins of xenogeneic origin. High titers of circulating antibodies were detected in mice implanted with fibroblasts secreting the human β-glucuronidase or the human sCD4 (41, 42, 45, 66). Treatment of recipient animals with cyclosporin prevents the appearance of antibodies. Surprisingly, in animals which received human β-glucuronidase secreting cells, the efficiency of protein delivery to distant tissues, as well as the secretion and uptake of the lysosomal enzyme were equivalent in both cyclosporin treated and untreated recipients. This observation suggests that immunization against the secreted protein may not be systematically deleterious for the functionality and the persistence of the implant.

15.4 Conclusions

Among the various techniques which have been proposed for the in vivo delivery of therapeutic proteins, neo-organs will probably be one of the first to be evaluated in a human setting. A clinical trial involving neo-organs is justified by the expectation that this technique will be at low risk for patients, and by the fact that a long term delivery of a therapeutic molecule has been documented with neo-organs implanted in dogs. Moreover, there are a number of question marks regarding the actual therapeutic efficacy of the method and the persistence of implants over very long periods of time in humans, which can be answered by a clinical trial only.

In regard to the other approaches that have been proposed for the in vivo implantation of genetically-modifed cells, neo-organs present the advantage of forming a tissue-like structure (in contrast with microencapsulation), the constitution of which does not require previous destruction of the recipient tissue (in contrast with muscle cell engraftment). Such characteristics are compatible with a long term maintenance of the implant, while the procedure may be repeated easily. The possibility of removing the implant may also be a desirable feature for introducing genetically-modified cells in humans.

Among the potential applications of protein delivery from engineered tissues, there are a number of diseases which will require a regulated expression of the therapeutic gene. Diabetis is a paradygm of such a disease. Treating these diseases by implanting engineered cells will suppose that complex regulatory pathways can be reproduced by gene transfer in target cells which are not the natural hosts for those regulations. Tremendous difficulties must be overcome before this goal can be achieved.

References

1. Selden, R. F., Skoskiewicz, M. J., Burke Howie, K., Russell, P. S., and Goodman, H. M. (1987). Implantation of genetically engineered fibroblasts into mice: implications for gene therapy. Science. *236*, 714-718.
2. Heartlein, M. W., Roman, V. A., Jiang, J. L., Sellers, J. W., Zuliani, A. M., Treco, D. A., and Selden, R. F. (1994). Long-term production and delivery of human growth hormone in vivo. Proc. Natl. Acad. Sci. USA. *91*, 10967-10971.
3. Hughes, M., Vassilakos, A., Andrew, D. W., Hortelano, G., Belmont, J. W., and Chang, P. L. (1994). Delivery of a secretable adenosine deaminase through microcapsules: a novel approach to somatic gene therapy. Hum. Gene Ther. *5*, 1445-1455.

4. al-Hendy, A., Hortelano, G., Tannenbaum, G. S., and Chang, P. L. (1995). Correction of the growth defect in dwarf mice with non autologous microencapsulated myoblasts an alternate approach to somatic gene therapy. Hum. Gene Ther. *6*, 165-175.
5. Tai, I. T. and Sun, A. M. (1993). Microencapsulation of recombinant cells: a new delivery system for gene therapy. FASEB J. *7*, 1061-1069.
6. Chang, P. L., Shen, N., and Westcott, A. J. (1993). Delivery of recombinant gene products with microencapsulated cells in vivo. Human Gene Ther. *4*, 433-440.
7. Liu, H. W., Ofosu, F. A., and Chang, P. L. (1993). Expression of human factor IX by microencapsulated recombinant fibroblasts. Hum. Gene Ther. *4*, 291-301.
8. Dhawan, J., Pan, L. C., Pavlath, G. K., Travis, M. A., Lanctot, A. M., and Blau, H. M. (1991). Systemic delivery of human growth hormone by injection of genetically-modified myoblasts. Science. *254*, 1509-1512.
9. Barr, E. and Leiden, J. F. (1991). Systemic delivery of recombinant proteins by genetically-modified myoblasts. Science. *254*, 1507-1509.
10. Dai, Y., Roman, M., Naviaux, R. K., and Verma, I. M. (1992). Gene therapy via primary myoblasts: Long-term expression of factor IX protein following transplantation in vivo. Proc. Natl. Acad. Sci. USA. *89*, 10892-10895.
11. Roman, M., Axelrod, J. H., Dai, Y., Naviaux, R. K., Friedmann, T., and Verma, I. M. (1992). Circulating human or canine factor IX from retrovirally transduced primary myoblasts and established myoblast cell lines grafted into murine skeletal muscle. Somat. Cell Mol. Genet. *18*, 247-258.
12. Tripathy, S. K., Goldwasser, E., Lu, M. M., Barr, E., and Leiden, J. M. (1994). Stable delivery of physiologic levels of recombinant erythropoietin to the systemic circulation by intramuscular injection of replication-defective adenovirus. Proc . Natl. Acad. Sci. USA. *91*, 11557-11561.
13. Hamamori, Y., Samal, B., Tian, J., and Kedes, L. (1995). Myoblast transfer of human erythropoietin gene in a mouse model of renal failure. J. Clin. Invest. *95*, 1808-1813.
14. Hamamori, Y., Samal, B., Tian, J., and Kedes, L. (1994). Persistent erythropoiesis by myoblast transfer of erythropoietin cDNA. Human Gene Ther. *5*, 1349-1356.
15. Yao, S. N., Smith, K. J., and Kurachi, K. (1994). Primary myoblast-mediated gene transfer: persistent expression of human factor IX in mice. Gene Ther. *1*, 99-107.
16. Naffakh, N., Pinset, C., Montarras, D., Li, Z., Paulin, D., Danos, O., and Heard, J. M. (1996). Long-term secretion of therapeutic proteins from genetically-modified skeletal muscles. Hum. Gene Therapy. *7*, 11-21.
17. Bell, E., Sher, S., Hull, B., Merrill, C., Rosen, S., Chamson, A., Asselineau, D., Dubertret, L., Coulomb, B., Lapierre, C., Nusgens, B., and Neveux, Y. (1983). The reconstitution of living skin. J. Invest. Dermatol. *81*, 2s-10s.
18. Sher, S. E., Hull, B. A., Rosen, S., Church, D., Friedman, L., and Bell, E. (1983). Acceptance of allogeneic fibroblasts in skin equivalent transplants. Transplantation. *36*, 552-555.
19. Still, J. M., Orlet, H. K., and Law, E. J. (1994). Use of cultured epidermal autografts in the treatment of large burns. Burns. *20*, 539-541.
20. Morgan, J. R., Barrandon, Y., Green, H., and Mulligan, R. C. (1987). Expression of an exogenous growth factor hormone gene by transplantable human epidermal cells. Science. *244*, 1476-1479.
21. Fenjves, E. S., Gordon, D. A., Pershing, L. K., Williams, D. L., and Taichman, L. B. (1989). Systemic delivery of apolipoprotein E secreted from grafts of epidermal keratinocytes: implications for epidermal function and gene therapy. Proc. Natl. Acad. Sci. USA. *86*, 8803-8807.
22. Fenjves, E. S., Smith, J., Zaradic, S., and Taichman, L. B. (1994). Systemic delivery of secreted protein by grafts of epidermal keratinocytes: prospects for keratinocytes gene therapy. Human Gene Ther. *5*, 1241-1248.
23. Gerrard, A. J., Hudson, D. L., Brownlee, G. G., and Watt, F. M. (1993). Towards gene therapy for haemophilia B using primary human keratinocytes. Nature Genetics. *3*, 180-183.
24. Flowers, M. E. D., Stockschlader, M. A. R., Schuening, F. G., Niederwieser, D., Hackman, R., Miller, A. D., and Storb, R. (1990). Long-term transplantation of canine keratinocytes made resistant to G418 through retrovirus-mediated gene transfer. Proc. Nat. Acad. Sci. USA. *87*, 2349-2353.
25. Stockschläder, M. A. R., Schuening, F. G., Graham, T. C., and Storb, R. (1994). Transplantation of retrovirus-transduced canine keratinocytes expressing the β-galactosidase gene. Gene Therapy. *1*, 317-322.
26. St Louis, D. and Verma, I. M. (1988). An alternative approach to somatic cell gene therapy. Proc. Natl. Acad. Sci. USA. *85*, 3150-3154.

27. Palmer, T. D., Tompson, A. R., and Miller, A. D. (1989). Production of human factor IX in animals by genetically modified skin fibroblasts: potential therapy for hemophilia B. Blood. *73*, 438-445.
28. Scharfmann, R., Axelrod, J. H., and Verma, I. M. (1991). Long-term in vivo expression of retrovirus-mediated gene transfer in mouse fibroblast implants. Proc. Natl. Acad. Sci. USA. *88*, 4626-4630.
29. Palmer, T. D., Rosman, G. J., Osborne, W. R. A., and Miller, A. D. (1991). Genetically modified skin fibroblasts persist long after transplantation but gradually inactivate introduced genes. Proc. Natl. Acad. Sci. USA. *88*, 1330-1334.
30. Dai, Y. F., Qiu, X. F., Xue, J. L., Hsueh, J. L., and Liu, Z. D. (1992). High efficient transfer and expression of human clotting factor IX cDNA in cultured human primary skin fibroblasts from hemophilia B patients by retroviral vectors. Sci. China B. *35*, 183-193.
31. Lu, D. R., Zhou, J. M., Zheng, B., Qiu, X. F., Wang, J. M., Meng, P. L., Han, F. L., Ming, B. H., and Wang, X. P. (1993). Stage I clinical trial of gene therapy for hemophilia. Sci. China B. *36*, 1342-1351.
32. Zhou, J. M., Qiu, X. F., Lu, D. R., Lu, J. Y., and Xue, J. L. (1993). Long term expression of human factor IX cDNA in rabbits. Sci. China B. *36*, 1333-1341.
33. Zhou, J. M., Dai, Y. F., Qiu, X. F., Hou, G. Y., Akira, Y., and Xue, J. L. (1993). Expression of human factor IX cDNA in mice by implants of genetically-modified skin fibroblasts from a hemophila B patient. Sci. China B. *36*, 1082-1092.
34. Petersen, M. J., Kaplan, J., Jorgensen, C. M., Schmidt, L. A., Li, L., Morgan, J. R., Kwan, M. K., and Krueger, G. G. (1995). Sustained production of human transferrin by transduced fibroblasts implanted into athymic mice: a model for somatic gene therapy. J. Invest. Dermatol. *104*, 171-176.
35. Neuzil, D., Holzman, M., Rozga, J., Levine, J., Moscioni, A. D., and Demetriou, A. A. (1993). Fibroblast transplantation in rats: transduction and function of foreign genes. J. Surg. Res. *54*, 631-637.
36. Gilbert, J. C., Takada, T., Stein, J. E., Langer, R., and Vacanti, J. P. (1993). Cell transplantation of genetically-modifed cells on biodegradable polymer scaffolds in syngeneic rats. Transplantation. *56*, 423-427.
37. Thompson, J. A., Anderson, K. D., DiPietro, J. M., Zwiebel, J. A., Zametta, M., Anderson, W. F., and Maciag, T. (1988). Site-directed neovessel formation in vivo. Science. *241*, 1349-1352.
38. Thompson, J. A., Haudenschild, C. C., Anderson, K. D., Di Pietro, J. M., Anderson, W. F., and Maciag, T. (1989). Heparin-binding growth factor 1 induces the formation of organoid neovascular structures in vivo. Proc. Natl. Acad. Sci. USA. *86*, 7928-7932.
39. Thompson, J. A., Haudenschild, C. C., Anderson, K. D., Di Pietro, J. M., Anderson, W. F., and Maciag, T. (1992). Heparin-binding growth factor 1 induces the formation of organoid neovascular structures in vivo. Correction. Proc. Natl. Acad. Sci. USA. *89*, 7849.
40. Borel-Rinkes, I. H., Bijma, A. M., Kappers, W. A., Sinaasappel, M., Hoek, F. J., Jansen, P. L., Valerio, D., and Terpstra, O. T. (1992). Evidence of metabolic activity of adult and fetal rat hepatocytes transplanted into solid supports. Transplantation. *54*, 210-214.
41. Moullier, P., Maréchal, V., Danos, O., and Heard, J. M. (1993). Continuous systemic secretion of a lysosomal enzyme by genetically-modified mouse skin fibroblasts. Transplantation. *56*, 427-432.
42. Moullier, P., Bohl, D., Heard, J. M., and Danos, O. (1993). Correction of lysosomal storage in the liver and spleen of MPS VII mice by implantation of genetically-modified skin fibroblasts. Nature Genet. *4*, 154-159.
43. Naffakh, N., Villeval, J. L., Moullier, P., Henri, A., Rouyer-Fessard, P., Danos, O., Heard, J. M., and Beuzard, Y. (1995). Sustained delivery of erythtropoietin in mice by autologous implants of genetically-modified skin fibroblasts. Proc. Natl. Acad. Sci. USA. *92*, 3194-3198.
44. Wiktor-Jedrzejzak, W. A., Bartocci, A., Ferrante, A. W., Ahmed-Ansari, A., Sell, K., Pollard, J. W., and Stanley, E. R. (1990). Total absence of colony-stimulating factor 1 in the macrophage-deficient osteopetrosic (op/op) mouse. Proc. Natl. Acad. Sci. USA. *87*, 4828-4832.
45. Moullier, P., Bohl, D., Cardoso, J., Heard, J. M., and Danos, O. (1995). Long term delivery of lysosomal enzyme by genetically-modified fibroblasts in dogs. Nature Medicine *1*, 353-358.
46. Descamps, V., Blumenfeld, N., Perricaudet, M., Beuzard, Y., and E.J., Kremer (1995). Organoids direct systemic expression of erythropoietin in mice. Gene Therapy. *2*, 411-417.
47. Cotten, M., Wagner, E., and Birnstiel, M. L. (1993). Receptor-mediated transport of DNA into eukaryotic cells. Methods in Enzymology. *217*, 618-645.
48. Neufeld, E. F. (1991). Lysosomal storage diseases. Annu. Rev. Biochem. *60*, 257-280.
49. Danos, O. and Heard, J. M. (1995). Gene therapy of mucopolysaccharidosis in Molecular and Cell Biology of Human Gene Therapeutics. Dickson, G. (eds). Chapman and Hall: London, 350-367.

50. Salvetti, A., Heard, J. M., and Danos, O. (1995). Gene therapy of lysosomal storage diseases. British medical Bulletin. *51*, 106-122.
51. Salvetti, A., Moullier, P., Cornet, V., Brooks, D., Danos, O., and Heard, J. M. (1995). In vivo delivery of human α-L-iduronidase in mice implanted with neo-organs. Human Gene Ther. *6*, 1153-1159.
52. Shull, R. M., Munger, R. J., Spellacy, E., Hall, C. W., Constantopoulos, G., and Neufeld, E. (1982). Canine α-L-Iduronidase deficiency: a model of Mucopolysaccharidosis I. Amer. J. Pathol. *109*, 244-248.
53. Adamson, J. W. and Ersbach, J. W. (1990). The use of recombinant human erythropoietin in humans. Molecular control of hemopoiesis. Wiley: Chichester (Ciba Foundation Symposium 148), 186-200.
54. Koury, M. J. and Bondurant, M. C. (1992). The molecular mechanism of erythropoietin action. Eur. J. Biochem. *210*, 649-663.
55. Koury, S. T., Bondurant, M. J., and Koury, M. J. (1988). Localization of erythropoietin synthetizing cells in murine kidney by in situ hybridization. Blood. *71*, 524-527.
56. Erslev, A. J. (1991). Erythropoietin. N. Engl. J. Med. *324*, 1339-1344.
57. Rodgers, G. P., Dover, G. J., Uyesaka, N., Noguchi, C. T., Schechter, A. N., and Nienhuis, A. W. (1993). Augmentation by erythropoietin of the fetal-hemoglobin response to hydroxyurea in sickle cell disease. N. Engl. J. Med. *328*, 73-76.
58. Rachmilewitz, E. A., Goldfarb, A., and Dover, G. (1988). Administration of erythropoietin to patients with β-thalassemia intermedia: a preliminary trial. Blood. *78*, 1145-1147.
59. Olivieri, N. F., Freedman, M. H., Perrine, S. P., Dover, G. J., and Sheridan, B. (1992). Trial of recombinant erythropoietin: three patients with thalassemia intermedia. Blood. *80*, 3258-3260.
60. Osborne, W. A., Ramesh, N., Lau, S., Clowes, M. M., Dale, D. C., Clowes, A. W., and XXX (1995). Gene therapy for long-term expression of erythropoietin in rats. Proc. Natl. Acad. Sci. USA. *92*, 8055-8058.
61. Watanabe, M., Reimann, K. A., De Long, P. A., Liu, T., Fisher, R. A., and Letvin, N. L. (1989). Effect of recombinant soluble CD4 in rhesus monkeys infected with simian immunodeficienct virus of macaques. Nature. *337*, 267-269.
62. Watanabe, M., Chen, Z. W., Tsubota, H., Lord, C. I., Levine, C. G., and Letvin, N. L. (1991). Soluble human CD4 elicits an antibody response in rhesus monkeys that inhibits simian immunodeficiency virus replication. Proc. Natl. Acad. Sci. USA. *88*, 120-124.
63. Kahn, J. O., Allan, D., Hodges, T. L., Kaplan, L., Arri, C. J., Fitch, H. F., Izu, A. E., Mordenti, J., Sherwin, S. A., Groopman, J. E., and Volberding, P. A. (1990). The safety and pharmacokinetics of recombinant soluble CD4 (rCD4) in subjects with the acquired immunodeficiency syndrome (AIDS) and AIDS-related complex. Ann. Int. Med. *112*, 254-261.
64. Schooley, R. T., Merigan, T. C., Gaut, P., Hirsch, M. S., Holodniy, M., Flynn, T., Liu, S., Byington, R. E., Henochowicz, S., Gubish, E., Spriggs, D., Kufe, D., Schindler, J., Dawson, A., Thomas, D., Hanson, D. G., Letwin, B., Liu, T., Gulinello, J., Kennedy, S., Fisher, R., and Ho, D. D. (1990). Recombinant soluble CD4 therapy in patients with the acquired immunodeficiency syndrome (AIDS) and AIDS-related complex. Ann. Int. Med. *112*, 247-253.
65. Dimitrov, D. S., Willey, R. L., Martin, M. A., and Blumenthal, R. (1992). Kinetics of HIV-1 interactions with sCD4 and CD4+ cells: implications for inhibition of virus infection and initial steps of virus entry into cells. Virology. *187*, 398-406.
66. Valere, T., Bohl, D., Klatzmann, D., Danos, O., Sonigo, P., and Heard, J. M. (1995). Continuous secretion of soluble CD4 in mice. Gene Therapy. *2*, 197-202.
67. Nilsson, I. M., Berntrop, E., Lofqvist, T. and Pettersson, H. (1992). Twenty-five years' experience of prophylactic treatment in severe hemophilia A and B. J. Int. Med. *232*, 25-32.
68. Kay, M. A., Rothenberg, S., Landen, C. N., Bellinger, D. A., Leland, F., Toman, C., Finegold, M., Thompson, A. R., Read, M. S., Brinkhous, K. M., and Woo, S. L. C. (1993). In vivo gene therapy of hemophilia B: sustained partial correction in factor IX-deficient dogs. Science. *262*, 117-119.
69. Dwarki, V. J., Belloni, P., Nijjar, T., Smith, J., Couto, L., Rabier, M., Clift, S., Berns, A., and Cohen, L. K. (1994). Gene therapy for hemophilia A: production of therapeutic levels of human factor VIII in vivo in mice. Proc. Natl. Acad. Sci. USA. *92*, 1023-1027.
70. Zatloukal, K., Cotten, M., Berger, M., Schmidt, W., Wagner, E., and Birnstiel, M. L. (1994). In vivo production of human factor VIII in mice after implantation of primary fibroblasts transfected by receptor-mediated, adenovirus-augmentented gene delivery. Proc. Natl. Acad. Sci. USA. *91*, 5148-5152.

IV Genetic Diseases

16. Gene Therapy for Cystic Fibrosis – Strategies, Problems and Perspectives

Charles Coutelle

16.1 Reasons for Gene Therapy of Cystic Fibrosis

Among the relatively few monogenetic diseases presently considered approachable by gene therapy Cystic Fibrosis (CF) is probably the most intensively investigated condition. More than 10 clinical trials using different vector systems to target the airway epithelia are presently under way or have been concluded in the USA, the UK and France (1-17). Several factors contribute to this prominence and the paradigmatic role cystic fibrosis has assumed: with a frequency of about 1 in 2,000 newborns CF is one of the most common autosomal recessive monogenetic disease in Caucasians. Despite the constantly improving treatment which has dramatically increased the life expectancy to an average of currently about 30 years, it still remains a serious life shortening condition for which no curative therapy is available to date. The continuous improvement in the management of CF has, however, created an atmosphere of hope and willingness among patients and their families to participate in new experimental treatments and charities such as the American CF Foundation, the British CF Trust and the French AFLM are playing a major role in supporting research into new therapeutic strategies including gene therapy. Although CF is a multi-organ disease affecting virtually all exocrine epithelia due to a defect in the ion and water transport of these cells, its main pathological manifestation occurs in the pancreas, the gut and the airway system. One of the great therapeutic advances of the last two decades has been the treatment of the intestinal symptoms by pancreatic enzyme replacement. In contrast comparatively little progress has been made in the management of pulmonary CF-disease, which due to repeated infections often with drug resistant pathogens leads to progressive emphysema, brochiectasis and cardio-pulmonary deficiency as the major cause of death (18). On the other hand, the airway system is relatively easily accessible for therapeutic approaches by inhalation or brochioscopy and would even allow repeated administration of a therapeutic gene construct.

CFTR (Cystic Fibrosis Transmembrane Regulator), the protein which is defective in CF, is expressed only in very minute amounts in normal airway epithelia (19, 20). Furthermore cell culture studies, in which normal and CF-cells were mixed, suggest that only about 10% of cells expressing normal CFTR are needed for correction of the ion transport defect in all cells of a monolayer (21), probably due to chloride fluxes through gap junctions (22). It was, therefore, suggested that only relatively low levels of expression of a therapeutic gene sequence might be needed to correct the defect. In addition to these favourable circumstances a noninvasive *in vivo* test has been developed on the basis of the increased negative electrical potential across the nasal epithelia of CF patients caused by the underlying ion transport defect (23). This test allows discrimination between normal subjects and CF patients (24-26) and has become widely used as a surrogate endpoint in the current clinical gene therapy protocols for CF.

16.2 The CFTR Gene, its Protein Product and Function

One of the main prerequisites for gene therapy of an inherited disease is the availability and functional analysis of the defective gene. The gene, which if mutated, causes CF encodes the cystic fibrosis transmembrane conductance regulator (CFTR) protein. It is situated on the long arm of chromosome 7 (7q22), has a size of 250 kb and was cloned in 1989. The CFTR gene consists of 27 exons giving rise to an mRNA of 6.2 kb which is translated into the 170 KD (1480AA) CFTR-protein (27-29).

CFTR is an integral membrane component of the apical portion of many epithelial cells (19). One of its main functions is that of a cAMP activated channel, allowing chloride conductance across the epithelial cell layer (30, 31). CFTR seems also to be involved in the regulation of the pH of intracellular vesicles (32) and to function as a negative regulator of the amiloride sensitive Na^+-transport, which is responsible for sodium absorption (33, 34). The movement of chloride from the interstitial compartment to the luminal side of the epithelial cell layer is carefully balanced by that of other ions, in particular sodium and bicarbonate, which in turn determine the movement of water (33, 35). Mutations in the CFTR gene which are responsible for disturbances in the cAMP mediated chloride conductance lead, therefore, not only to alterations in the ion balance reflected by changes in the electrical potential across the epithelia but also to a severe impairment of water fluxes through the epithelial sheath (33, 35). This results in an increased concentration of salt in the sweat, and of enzymes in the pancreatic secretions as well as to dehydration of mucous in the secretions of the gut and the airways. Concentrated pancreatic secretions and mucous obstruction are major factors causing pancreas fibrosis. The molecular basis for the repeated infection of the pulmonary epithelia is still not clear although mucous obstruction is a contributing factor. Recent animal experiments have suggested, that the pulmonary pathology may be primarily due to the bacterial infections rather than to the impaired ion and water transport (36). Imundo et al. (37) have made the increased intracellular pH in CF epithelia (32) responsible for the under-asialysation of cellular gangliosides on the epithelial surface which increases binding of pseudomonas and may, thereby, support the chronic pulmonary infection. In contrast, Pier et al. (38) regard the lack of bacterial internalization into epithelial cells and subsequent clearance from the airways by desquamation as the main contributor to the chronic bacterial infection. This hypothesis is supported by the observation of a CFTR dependent internalization of *Pseudomonas aeruginosa* via a bacterial lipopolysaccharide and by animal experiments showing that suppression of bacterial ingestion by the airway epithelium inhibited bacterial clearance from the lungs (38).

16.3 Preclinical Approaches to CF-Gene Therapy

Ideally, somatic gene therapy for CF should permanently replace the mutated CFTR gene by its normal counterpart in the affected cells. However, for technical reasons, this strategy, which would require highly effective homologous recombination *in vivo* is presently not feasible. Complementation of the two mutated CFTR alleles by addition of the normal gene sequence to the cells of affected individuals is, therefore, the aim of all present CF gene therapy strategies. The principle feasibility of this approach has been shown by extensive *in vitro* and *in vivo* studies including investigations on mouse models for CF using different vector systems. In particular adenoviral vectors, cationic lipid/DNA-complexes and adeno-associated virus have been used in animal experiments to show that human CFTRcDNA can be introduced *in vivo* into airway epithelia and is expressed at detectable (39-47) and even functional levels (48, 49).

16.3.1 Nonintegrative Vector Systems

Adenovirus and DNA/lipid vectors do not normally lead to integration of the transgene sequence into the host genome. They can, therefore, only confer transient gene expression because the vector will eventually be lost by cell division and death or by degradation of the vector. However, given the severity of CF and the time and effort spent daily for its symptomatic treatment, a causal therapeutic effect even lasting only for a few days but which could be sustained by repeated application over a lifetime would be a dramatic achievement.

Adenovirus vectors were the first gene transfer systems to be used in clinical gene therapy trials for CF. Adenovirus has a genome size of about 36kb. It can package about 38 kb DNA containing a foreign DNA insert of up to about 7.5kb. It infects lung epithelia naturally, can be produced at high titres and shows a high level of infectivity in cell culture and *in vivo*. The adenoviral vectors which have been constructed for gene therapy are derivatives of the wild type adenovirus types 2 and/or

5 and belong to the subgroup C of this virus for which no oncogenic potential has been observed in any species, including man (50, 51). All first generation adenoviral vectors are deleted in the E1 gene region and in most cases also in the E3 gene region, to gain additional space for gene insertion. The E1 genes are responsible for virus replication and for expression of the E2 functions which in turn play an important role in the regulation of viral protein expression. It was, thus, assumed, that the E1 deletion might be a sufficient safety feature to make adenovirus constructs applicable for clinical use in gene therapy. Such replication deficient virus has to be grown in the human embryonic kidney helper cell-line 293 which expresses the E1 function constitutively (52), to allow packaging of the virus. The deficient virus is able to infect the target cells and to express the transgene sequence in them but it cannot spread by further replication and infection of other cells. To maintain this safety feature all clinical grade preparations are rigorously checked to exclude the presence of replication competent revertants (2).

Different first generation adenovirus vectors containing human CFTR cDNA or marker genes were applied by instillation or inhalation to the airways of rodents (40, 45, 53-55) and primates (39-41, 47, 56-58) to assess safety and transgene expression in preparation for human trials. Adenovirus mediated human CFTR mRNA was observed for up to 6 weeks in the airway epithelium of cotton rats (45), but since adenovirus usually does not integrate into the host genome, it will eventually be lost (41) by degradation, the regular replacement of epithelial cells or by host reactions against the transfected cells (59). Permanent transgene expression would, therefore, require repeated reinfection which causes immune reactions against the adenovirus and the transgene product (55). Transient and usually mild to moderate virus-dose dependent inflammatory reactions have been observed by several investigators in rodents and primates (39, 47, 54-56, 58) and explain some of the problems seen in the human trials.

Cationic lipid DNA complexes (60-63) have been used successfully in nonviral approaches for the introduction of DNA into cells *in vitro* and *in vivo*. In contrast to anionic or neutral liposomes they do not encapsulate the DNA into the aqueous phase inside a lipid bilayer but form complexes through interaction of the positive charges of the lipid headgroups and the negative phosphate backbone of the DNA. The rather ill defined morphology and size of these complexes (64, 65) may account for some of the still unsolved problems in their transfection properties. These complexes have, however, no theoretical packaging related size constraints with respect to the DNA and have indeed mediated transfection of various expressing plasmids and even DNA of 650kb yeast artificial chromosomes (YACs) into cells (66). Since these DNA-lipid complexes have no replicative capability most of the health and safety concerns associated with virus vectors do not apply. Furthermore, lipid formulations containing hydrolizable bonds have been synthesised to reduce general cytotoxicity. Such lipids are naturally degraded after fusion with the cell membrane and delivery of the complexed DNA. One of these preparations composed of 3β[N-(N',N'-dimethylaminoethane-carbamoyl] cholesterol/dioleoyl phosphatidylethanolamin/(DC-Chol/DOPE) (67) shows efficient *in vitro* transfection (67, 68). No histological, biochemical or functional abnormalities were observed after intravenous administration of DC-Chol/DOPE-plasmid DNA to mice (69) and repeated intravenous, intraperitoneal or subcutaneous injection of these complexes caused no autoimmune reaction or transfer of DNA into gonadal cells (70). DC-Chol/DOPE-plasmid DNA has also been well tolerated during application in gene therapy trials for melanoma (71-73).

Application of different liposome plasmid cDNA complexes *in vivo* into the airways of mice was successfully performed by instillation (46, 48) and nebulisation (49, 74). Genes expressing different marker proteins such as luciferase, β-galactosidase (46) or chloramphenicol-acetyl-transferase (CAT) (74), as well as human CFTR cDNA (46), were transferred *in vivo* in complex with DOTMA/DOPE (Lipofectin) (N-[1-(2,3-dioleyloxy)propyl]-N,N,N-trimethylammonium chloride/dioleoyl phosphatidyl-ethanolamine) (48) and DC-Chol/DOPE (49) to the lungs of two different transgenic CF mouse strains by intratracheal instillation and nebulization respectively. Successful correction of the cAMP-dependent chloride transport defect in the airways of these animals was detected *in vitro* and *in vivo* by electrophysiological measurements (48, 49). Application of DC-Chol/DOPE to the nasal epithelium of normal subjects *in vivo* showed no toxicity (75) and was recently also used in our first human trial to transfer CFTR cDNA into nasal epithelial cells of CF patients *in vivo* (11).

16.3.2 Integrative Vectors Systems

Retroviruses and adeno-associated viruses are able to integrate into the host genome and could, therefore, in principle, serve as vector systems for permanent gene delivery.

Retroviral vectors are based on replication defective Malony murine leukaemia virus (for review see 76, 77). They can carry about 10 kb of a foreign gene and were used in one of the first *in vitro* experiments to demonstrate functional correction of the cellular CF defect by introduction of normal CFTR cDNA into CF cells (78). The main advantage of this vector system is the possibility of permanent correction by integration of the retroviral vector containing the gene of interest into the host genome, provided stem cells can be infected and expression shutdown avoided. The random integration carries, however, the very remote risk of mutagenesis and oncogenesis (76, 79). Present retroviral vectors need dividing cells for infection and have thus not been considered for *in vivo* CF-gene therapy. In addition the complex anatomy of the organs affected by CF does not allow an *ex vivo* gene therapy approach in which proliferating primary cells from the affected organ are transfected with the corrective gene construct in culture and are then transplanted back into the patient. However, the airway epithelia of CF patients have an increased proliferation rate and may thus be amenable to transfection with retroviruses (80). Retroviral vectors may also be used for fetal gene therapy strategies (see below 81, 82).

Adeno-associated virus vectors (AAV) (for review see 83, 84) has a pulmonary tropism, like adenovirus, and has recently been considered for *in vivo* gene therapy of cystic fibrosis (10). The wild type virus is able to integrate preferentially into a region of chromosome 19 which appears to be functionally silent thereby reducing the risk of virus induced mutagenesis and oncogenesis by random integration. However, this preferred integration does not seem to be retained in the modified virus-vectors. Furthermore, AAV and AAV vectors need adenovirus as a helper for propagation in culture and the subsequent generation of high titre adenovirus free AAV-preparations for clinical use is still difficult to achieve. Indeed it has recently been shown, that AAV even requires some degree of adenovirus contamination for effective formation of its double stranded form and subsequent transgene expression which seriously limits the use of present AAV vectors for gene therapy (85). AAV has a small genome of 5kb and can, therefore, only accommodate an insert of foreign DNA up to about 4.5kb after deletion of its rep and VP-genes. Flotte et al. (86) have been able to introduce an N-terminal truncated human CFTR cDNA into an AAV-vector and have used it to express functional CFTR in a human bronchial CF cell line. *In vivo* instillation of the AAV-CFTR construct into the lungs of rabbits directed CFTR-protein expression for up to six months (44). Clinical trials to test this vector system are presently under way (10).

16.4 The First Clinical Trials for CF

Based on the extensive investigations in cell culture, tissue samples, and on normal and CF experimental animals several groups in the USA and the UK started about three years ago with phase 1 clinical trials for somatic gene therapy of the pulmonary manifestations of cystic fibrosis. In the USA all initial trials were conducted using adenoviral vectors while our team in Britain consisting of the CF gene therapy groups at the Royal Brompton Hospital/National Heart and Lung Institute, London, the Department of Biochemistry and Molecular Genetics, St. Mary's Hospital Medical School/Imperial College, London and the MRC Human Genetics Unit in Edinburgh was the first to apply cationic lipids for transfer of the CFTRcDNA to the nasal epithelium of CF-patients. The results of four of the American adenoviral studies (15-17, 40, 87) and of our lipid-based trial (11) have been published. In the meantime at least nine further studies using adenoviral, cationic lipid or adeno-associated vector systems have been completed or are under way in the USA, the UK and France. All present trials use the respiratory route as certainly the most convenient means of delivery; especially since repeated application will be necessary using the present vector systems. All patients included in these trials are adults with moderate manifestation of CF.

16.4.1 Safety, Ethical and Legal Considerations

None of the present trials intends to provide definitive treatment, even to the very small number of adult volunteer patients involved who are informed (written consent) about the purely investigative nature of these studies. The main objective of the trials is to obtain information on the safety and toxicity of the constructs. However, most protocols include tests which allow to assess efficacy and dose response by measuring the correction of the ion transport defect *in vivo* , as a surrogate end point, on a very small area of the airway epithelia of the patients and by molecular studies to detect vector, CFTR mRNA and CFTR-protein on epithelial cells from biopsies and brushings of the airways.

Although somatic gene therapy does not create new ethical problems (88) ethical approval for individual protocols has to be obtained because it is presently still research towards a novel form of treatment or, at best, an experimental treatment. In the USA approval for these trials has to be given by the NIH Recombinant Advisory Committee (RAC) on the basis of extensive written documentation and a public hearing and by the Food and Drug Administration (FDA). The documents submitted to the NIH RAC by the different U.S. groups for the Cystic Fibrosis trials describe the disease, its molecular and cellular basis, its current therapy and prognosis and justify the relevance of gene therapy for this disease. They describe the structure, biology or chemistry of the vector systems. Experimental data of *in vitro* tests and *in vivo* experiments in rodents and primates supporting the efficacy and safety of these constructs and a detailed description of the production and quality control of the recombinant vectors are provided. Various safety aspects regarding the potential of the vector systems or the transgene sequence to provoke adverse reactions in the host, to transfect germline cells or to contaminate the environment are considered. The virus vectors are specifically analysed with respect to possible replicative, recombinatory and oncogenic potentials and their ability to integrate into the host genome. The protocols for the proposed phase I trials are described in detail, giving an exact account of the procedures to be followed and defining the criteria for patient selection and for assessment of the results with respect to safety and efficacy. Special consideration is given to confidentiality and informed consent, follow up of the patients and the reporting of adverse effects (1-3, 5-9).

Approval for our lipid based phase I clinical trial was given by the Committee on the Ethics of Gene Therapy, (now replaced by the Gene Therapy Advisory Committee) after consideration by the local ethics committees of the participating institutions and cross reference to the Advisory Committee on Gene Manipulation and the Department of the Environment [89]. DNA for the trial was prepared at the MRC Human Genome Mapping Project Resource Centre, at Northwick Park Hospital, according to a detailed protocol agreed with the Medicines Control Agency. The cationic lipid DC-Chol/DOPE was prepared using FDA approved procedures by Drs. Huang and Gao (University of Pittsburgh) (67).

16.4.2 Testing for Efficacy

All phase I clinical studies for gene transfer into airway epithelial cells of CF patients which have been published to date have used both molecular markers and electrophysiological measurements to determine the efficiency of transgene expression. Four of them are adenovirus based and one describes gene transfer using lipofection. The assessment of CFTR gene transfer and expression has often given inconsistant results when using plasmid DNA-PCR, CFTR mRNA specific reverse transcription PCR (RT-PCR) and CFTR immuno-histochemistry on epithelial brushings and biopsies. This may be due to inhomogeneous transfection *in situ* in combination with low expression rates and the very small amount of biopsy material available for analysis.

Changes in the ion transport behaviour can be monitored by measuring the electrical potential difference on the surface of nasal epithelial cells. This non-invasive procedure is used to obtain baseline electrical potential differences and to monitor the reaction to pharmacological stimulation for each CF patient before treatment. Moreover, it also allows to determine the time course of gene expression after gene application. These measurements monitor the basal electrical potential difference between the luminal and interstitial side of the epithelial layer. This is largely determined by the concentration-ratio of the luminal to the interstitial Na^+, which is strongly influenced by the efflux of chloride through the CFTR-channel and probably also by a regulatory effect of CFTR on the sodium-pump (34, 90). The normal basal nasal potential difference (PD) of about -20mV is, therefore, predominantly due to Na^+-transport from the luminal surface across the epithelial barrier into the interstitium and to a lesser extent by the chloride efflux through the CFTR-channel. This negative PD is raised to about -40 to -50 mV by an increased inwards Na^+-transport in CF-epithelia and the inhibition of CFTR mediated chloride efflux (26). Inhibition of Na^+-influx by amiloride, a sodium pump inhibitor, reduces this negative PD strongly in CF and less markedly in normal epithelia (25, 26, 91).

In normal subjects lowering of the chloride concentration on the luminal epithelial surface by instillation of a low chloride solution increases the outwards Cl^--conductance through an intact CFTR-channel following the concentration gradient and, thereby, increases the negative PD (25, 26). This difference is further increased by activation of the CFTR channel function by cAMP-generating drugs such as forskolin, isoprenaline or terbutaline. In contrast, in CF, the chloride conductance does not respond to either cAMP or low chloride stimulation. Two recent studies analyse the different protocols for measurements of nasal PD and discuss the pitfalls of this technique. If performed properly, PD-measurements can clearly discriminate between normal and CF patients and detect the corrective effect of CFTR cDNA transfer and expression on the electrophysiological defect in CF-patients (25, 26).

16.4.3 Adenovirus Trials

Several different first generation adenovirus recombinants lacking the E1 adenoviral functions have been used in clinical phase 1 gene therapy trials for CF. The Ad2/CFTR-1 vector (92) used by the teams of Welsh and Smith (93) expresses the CFTR-cDNA under control of adenovirus 2 E1a-enhancer and promotor. The E1 region coding sequence has been deleted and replaced by the CFTR cDNA. It also contains the E1b 3'-intron, -untranslated region and -polyadenylation signal. The adenoviral E3 gene has been retained in this vector. Since the gp19 protein expressed from the E3 region prevents MHC class I antigen presentation on the cell surface (94) T-cell response against cells infected with such adenovirus should be inhibited and virus induced inflammatory reactions reduced. In the study conducted by Crystal's group (15, 16, 95) an E1/E3 deleted adenoviral vector, AdCFTR, was used which expresses the human CFTRcDNA under regulatory control of the adenovirus C type 2 major late promotor with its three tripartite leader sequences at the 5' end, and the SV40 polyadenylation site at the 3' end (45). Knowles et al. (17) used an E1 deleted adenovirus serotype 5 construct, H5.020CBCFTR (synonymous to Ad5-CB-CFTR), expressing the CFTRcDNA under control of a CMV enhancer and chicken β-actin promotor (1). In a more recent repeated application trial Zabner et al. (87) used a E1/E4 deleted vector (Ad2/CFTR-2) (7).

In their first trial Zabner et al. (93) applied increasing doses from $2x10^6$ to $5x10^7$ infectious units of Ad2/CFTR-1 to the nasal epithelium of three CF patients. All three patients showed a decrease of the baseline electrical potential difference and an increase in the response to cAMP-agonists as expected for a corrective effect. The changes in the baseline potential differences lasted for about three weeks, while cAMP stimulation was only detectable for about ten days. No replication competent adenovirus was found in the epithelium of the nose, pharynx, blood, urine or stool. No adenovirus-related pathology was observed and no increase in antibodies against adenovirus was detected up to 35 days after treatment (93). In a similar investigation performed by Crystal's group (15) nine patients were given a single nasal application of increasing AdCFTR doses ($2x10^5$ to $2x10^{8.5}$ pfu in 0.2ml Ringer solution). All patients involved in this trial were negative for endogenous adenovirus 2/5 E1a sequences. They harboured no viral pathogens, had low baseline levels of anti-adenovirus antibodies and their respiratory epithelia did not promote AdCFTR replication when infected *in vitro* (16). All individuals in this trial served as their own controls with measurements made in the nostril to be treated before AdCFTR administration and in the contra-lateral untreated nostril. Measurements of the baseline PD as well as PD changes in response to amiloride and after administration of isoproterenol in low chloride solution were performed. All electrophysiological parameters investigated on the treated nostrils showed on average over two weeks significant improvements towards normal values in comparison to pretreatment while no change was seen in the nontreated nostrils. No differences in the treated and nontreated nostrils were observed in response to ATP suggesting that the epithelium remained functionally intact (15).

Unfortunately both these trials were performed unblinded and their positive electrophysiological results could not be confirmed in a recent double blind, vector controlled study by Knowles et al. (17) who administered the adenovirus CFTR vector in increasing doses up to 2×10^{10} pfu to the nasal epithelium of their patients. CFTR gene expression was detected by RT-PCR only at doses between 2×10^{8} to 2×10^{10} pfu in five of nine patients but in none of the three patients receiving 2×10^{7}pfu. An *in situ* signal indicating the presence of transgene CFTRmRNA was observed only in one of 3 patients receiving 2×10^{9}pfu. The authors attribute the failure to detect any electrophysiological correction in their patients to the very low efficiency of adenovirus transfection of the airway epithelia (96). This is most likely due to the lack of the adenovirus penton-binding αv- and α5- integrins on the surface of the columnar cells of the differentiated airway epithelium (97). Using 8×10^{8} pfu adenovirus delivered *in vivo* to the nasal epithelium of CF mice, Grubb et al. (96) observed only a relatively inefficient adenovirus mediated correction of the chloride conductance and no effect on the amiloride sensitive Na^+-transport. A similar observation was made by Goldman et al. (98) using the xenograft model system of fully differentiated bronchial epithelia from a CF patient transplanted into nude mice. Infection with an adenovirus CFTR cDNA construct at an moi of 100 was sufficient to reconstitute the cAMP stimulated chloride channel to near normal levels, despite transfection of only about 5% of cells whereas the correction of Na^+ hyperabsorption was only partial and variable (98). Johnson et al. (99) found that the Na^+-hyperabsorption in primary human CF epithelial cells could only be corrected by application of a virus dose which transfected these cells to almost 100% (as monitored by an adenovirus lacz control). Chloride fluxes through gap-junctions between CFTR-expressing and nonexpressing cells have been made responsible for the correction of the CFTR chloride mediated electrical potential even at a low level of CFTR gene expression (21). In contrast CFTR probably regulates the Na^+-transport only in CFTR expressing cells and, therefore, much higher levels of CFTR gene transfection which lead to CFTR expression in virtually all cells may be needed to correct the overall Na^+-flux which is mainly responsible for the basal potential difference (100). It could, hence, be argued that only the short-term chloride mediated changes in the electrical potentials observed by Zabner et al. (93) are CFTR-mediated while the longer lasting changes in the basal potential may be due to tissue damage, which is known to reduce the basal potential difference (26). This seems to be substantiated by the second trial from this group using repeated administration of an E1/E4 deleted vector (87) in which much higher virus doses had to be applied to observe relatively small changes in the electrical potentials only in the low chloride/terbutaline experiments. These responses seem to by less pronounced after the second administration.

Since no signs of inflammation were observed in the initial nasal studies applying adenovirus doses up to 10^{7}pfu (16) Crystal's group extended their trial in three patients by a subsequent single bronchoscopic delivery using increasing AdCFTR doses (two patients 20ml, one patient 5ml) into a large bronchus one day after the nasal application (16). They observed no recombination or complementation, no shedding of the vector, no generation of replication competent virus, and no rise of neutralising antibody titres in any of their patients after this treatment. Transgene mRNA and protein expression was found in the nasal epithelia of one of the patients and protein was also detected in the bronchial epithelium of another one. At doses up to 2x 10^7pfu no vector related adverse effects were seen in three patients. However, about 12 hours after bronchial application of 10^9 pfu (20ml) to a fourth patient a transient systemic and pulmonary syndrome was observed characterised by headache, fatigue, fever, tachycardia, hypotension, dyspnoe and deterioration of several functional lung tests. Bronchial breath sounds and roentgenographic infiltrates were detectable in the region of the right lower and middle lobes. All subjective and objective clinical symptoms were resolved within two weeks after antipyretic and antibiotic treatment combined with support of the circulatory and respiratory system. No adenovirus could be detected in nasal, pharyngeal, rectal, blood or urine samples and no contaminant was found after rechecking the vector preparation. The follow up of this patient for 6-12 months demonstrated no long term adverse effects (16). Transient and usually mild dose dependent inflammatory reactions had been observed in animal experiments after adenovirus application using doses between 10^7-10^{11} pfu (39, 47, 54, 56). However, a severe reaction very similar in its symptoms to the incidence observed by Crystal has lead to the death of a monkey after administration of 10^{12} pfu (Welsh, Williamsburg CF-Conference 1994). Because of these adenovirus induced inflammatory reactions Knowles et al. (17) concluded, that simply increasing the adenovirus dose could not be a strategy to overcome the above described efficiency problem in the infection of differentiated airway epithelium with adenovirus.

Significantly increased levels of IL-6 in the affected patient and moderate increases in the other three patients were found in the initial bronchial administration study (16). It was, therefore, suggested that the observed severe clinical syndrome may have been mediated by this cytokine. Further investigations on five more patients receiving bronchial virus doses of 2x10^5 to 2x10^8pfu showed a dose dependent increase of IL-6 in the serum within the first hours of virus administration. Since the release of this cytokine was not observed from cultured airway epithelial cells of normal control probants after adenovirus infection the authors suggest that the severe reaction was most likely due to a virus induced release of the cytokine from inflammatory cells present in the CF airways (95). This would, however, not quite explain the severe reaction in the normal monkey described above. Adenovirus has been shown to induce the production of several cytokines among them IL-6 in the lung of mice after pulmonary infection (101). IL-6, which is known as a pleiotropic modulator of many immune functions including the activation of a T cell response to alloantigens appears as soon as 6 hours after virus administration, before the early phase of inflammatory cellular infiltration (101). Another cytokine contributing to the inflammatory response to adenovirus infection seems to be IL-8. It is induced by replication deficient adenovirus in airway epithelial cells in *vitro* (102) and *in vivo* (58, 103) and stimulates neutrophil infiltration which has been described as initial reaction to wild-type and E1 deleted adenovirus infection (54).

16.4.4 Cationic Lipid/DNA Trials

The first cationic lipid mediated clinical phase I gene therapy trial for CF (11) used the lipid mixture of DC-Chol/DOPE (67) and the CFTRcDNA plasmid, pSVCFTR containing the full length CFTRcDNA under SV40 promotor control (49). This trial was conducted double blind and placebo controlled on 15 adult CF patients by spray administration of the test substance in 200µl portions to each nostril in 10-15 min. intervals over a period up to seven hours. The placebo group received the lipid only (three patients 500µg, three patients 1500µg). The DNA group was given increasing doses of the CFTR cDNA/lipid complex in a DNA to liposome ratio of 1:5 (DNA 10, 100, and 300µg per nostril). Nasal electrical potential differences were read from both nostrils on days -14, -7, 0, 1, 3, 5, 7, 11, 18 and 28. Nasal brushings and biopsies were taken on day 4 from one nostril and were analysed for signs of toxicity or inflammation by histology, and for plasmid DNA and CFTR-mRNA by PCR and RT-PCR. In addition a broad range of laboratory investigations and two lung function tests were performed. The results of this study showed no treatment related local or general adverse reactions and no induction of antibodies against the administered DNA. The data on DNA and RNA detection were ambiguous. Although individual reductions of the basal electrical potential difference were seen in patients before and after lipid/DNA treatment, they were, overall, not statistically significant. The amiloride induced reduction in the potential difference before and after treatment was statistically significant ($p=0.02$) in the group of lipid/DNA treated patients in contrast to that in the placebo group. After perfusion of the nasal epithelium with low chloride solution a PD increase towards normalization was observed in the lipid/DNA treated group ($p=0.08$) but not in patients given placebo. The maximum PD difference before and after treatment in this group was seen on day 3 ($p=0.05$). These effects were not dose dependent. No effect was observed when isoprenaline was used as cAMP stimulating agent. All described changes became detectable at day 1, showed a maximum at days 3-4 and disappeared latest by day 7, with strong variations of their extent in individual patients. As already discussed in the case of the adenovirus trials the Na^+ mediated changes in PD (baseline PD and amiloride induced changes) have to be judged with caution because of the effects of cell damage on this parameter, but the measurements in low chloride conditions can be regarded as indication of a slight and transient correction of the chloride channel function after CFTRcDNA administration.

Three other phase 1 gene therapy trials for CF using the cationic lipids DC-Chol/DOPE (12) DIMRIE/DOPE (5) and DOTMA (14) have been performed or in evaluation of the results.

16.5 Assessment of Present Gene Therapy Vector Systems for CF

Both, the adenovirus and the liposome based trials have illustrated the advantages and problems associated with each of these gene transfer systems. They have shown the principle ability of both vector systems to transfer genes into the affected airway cells *in vivo* and to obtain, at best, variable and transient correction of the cellular defects caused by CF. In both cases the human nasal epithelium has proven a good model system for the testing of gene transfer constructs and to overcome present problems. The main and very important lesson of these first clinical trials is, however, that much more basic research on the cell and molecular biology of the gene transfer systems and on transgene expression is needed to achieve really therapeutic levels of sustained CFTR gene expression.

16.5.1 Adenoviral Vectors

The following problems have been identified in relation to the use of the current first generation adenoviral vectors (104):

- The potential of the replication-defective recombinant virus to recombine with endogenous viral sequences (105, 106) generating replication competent wild type virus; or to be transcomplemented by wild type virus (107) or by expressed endogenous E1a DNA leading to replication of the recombinant construct.
- The induction of vector mediated immune responses (59, 108), which may be the underlying cause for vector mediated inflammation (16, 47, 54, 56) and toxicity causing apoptotic cell death and reduced recruitment of cells into S phase (104) as well as inactivation of the adenovirus-construct by neutralizing antibodies (59).
- The poor transfection of differentiated pseudostratified epithelia in the proximal airways by adenovirus because of lack of the adenovirus penton-RGD binding $\alpha v\beta 5$ integrins in the differentiated columnar cells (97).

Most of these problems are being presently intensively investigated in several groups. Thorough analysis of any adenovirus preparation intended for clinical application is performed to detect and avoid contamination with wild-type revertants resulting from recombination with E1 sequences from the helper cell line (106). In addition patients to be included in adenoviral trials are tested for endogenous E1 sequences in their airway epithelia (2). A further attenuation of the AdVCFTR vector has been achieved by introduction of a mutation in the cis acting regulatory sequence controlling encapsidation (107), which will reduce its ability to package in the event of wild type virus E1-gene complementation in trans.

The inflammatory and immune reactions caused by adenovirus vector administration into the airways have been studied using several animal models including different mice strains with selective defects of the components of their immune system. These investigations have shown, that both humoral and cellular immune reactions are involved (59, 109, 110). Administration of high doses of wild-type, recombinant and even inactivated adenovirus leads to pulmonary infiltrates consisting of macrophages and lymphocytes with neutrophils appearing transiently (47, 56, 101) and to elevated levels of IL-1, IL-6, and TNF in the lung (101) and IL-6 in the blood (95, 101). The humoral response to viral proteins generates neutralizing antibodies (40, 53, 111), which prohibit successful gene transfection by repeated vector administration. These antibodies are produced by B cells in reaction to uptake of the administered virus followed by MHC class II viral protein processing and antigen presentation. The antigen presentation leads to a stimulation of $CD4^+$ TH-cells which in turn induce an expansion of the specific B cell pool. These B-cells are then responsible for an excessive production of anti-adenoviral antibodies, which rapidly eliminate the adenovirus vector thereby preventing successful expression by repeated vector administration (59, 110, 112).

The cellular reaction against adenovirus vector administration to the airways appears to be predominantly a TH1 helper cell mediated accumulation of Class I restricted $CD8^+$ cytotoxic T lymphocytes (CTL) in the lungs (59). These CTLs have been made responsible for destruction of the adenovirus transfected epithelia followed by repopulation of the bronchus with non-transfected cells, thereby severely limiting persistent expression of the transfected transgene (59). Incomplete shutdown of adenoviral gene expression, even after E1 removal, is thought to cause this cellular reaction (59, 109) since E2a expression and low level of fibre and even hexon has been detected in different cell systems after infection with first generation adenovirus constructs (109, 113, 114). Based on this hypothesis several strategies have been devised to prevent leaky viral gene expression. An obvious solution would be the construction of new adenovirus vectors in which not only E1 but also other viral genes, in particular E2a, are eliminated or inactivated (108, 115, 116). The main obstacle to this approach is the need of these gene functions (like the E1 functions) for growth and generation of infectious virus. Second generation vectors have, thus, been constructed in which the E2a DNA binding protein (DBP) is functional and allows virus production at 32°C in the packaging 293 cell-line, but is rendered inactive at 39°C by introduction of a temperature sensitive mutation (108). Instillation of such second generation CFTRcDNA vector into the airways of mice was associated with high transgene expression up to 21 days, a time at which expression from first generation vectors is no longer detectable. The expression was associated with substantially less inflammation, a dramatic reduction in the infiltration of $CD8^+$ cells into the epithelium and a lower proliferation rate of airway epithelial cells as compared to the effects of first generation vectors (108). A preclinical study comparing the effect of instillation of first and second generation adenovirus vectors containing either marker genes or cCFTRcDNA into the lung of nonhuman primates showed prolonged recombinant gene persistence and expression as well as reduced perivascular inflammatory reactions to the second generation vectors (116).

However, in non of these systems are the results as impressive as observed with first generation adenovirus vectors in fully immunodeficient animals where expression is stable for over 2 years without inflammation. A CTL response to E2a caused by partial activity of the mutant DBP at the intermediate temperature of 37°C and/or direct toxicity of the input viral proteins have been suggested as explanation for this observation (108).

As a next step to improving adenoviral vectors, third generation constructs are being developed in which the E2 (117) or E4 genes (7) are partially deleted in addition to E1. E4 plays a major regulatory role in the viral life cycle including virus replication and late gene expression (118, 119). Other approches aim at the deletion of all viral genes (120, 121). Again, such deletion viruses require either helper virus or the parallel construction of a packaging cell-line providing the deleted functions in trans (117, 122, 123). It appears that only the open reading frames 6 and 7 of the adenovirus E4 gene are necessary to complement E4 (123).

Unfortunately, it is very likely that none of these constructs will give complete absence of immunological reactions since even incomplete and UV-inactivated adenovirus from which no gene expression could be detected induces pulmonary inflammation at high dose application (124). Therefore immuno-suppression to prevent both humoral and cytotoxic responses is investigated as an alternative or additional strategy to deal with the inflammatory reactions provoked by adenoviral vectors. Steroids, cyclosporin, the immunosuppressant FK506 or antilymphocyte antibodies have shown a dramatic reduction of immune reactions (115, 125-128).

Injection of adenovirus infected cells into the thymus has been tried as an alternative approach to eliminate the CTL response in mice. This procedure has lead to a marked reduction in adenovirus-specific CTLs and to prolonged hepatic transgene expression after intravenous challenge with adenovirus recombinants. It is assumed, that T-lymphocyte tolerance is induced when developing thymocytes encounter the injected antigen. However, this does not eliminate the production of neutralizing antibodies by the host after the intravenous virus injection (129).

Selective suppression of the production of neutralizing antibodies allowing efficient readministration of recombinant virus has been achieved by co-administration of IFNγ (or IL-12, which stimulates IFNγ production) together with the adenovirus into the airways of mice (112). INFγ reduces $CD4^+TH2$ cell activation which is needed for B cell expansion and stimulation of antibody-production. Kay et al. (130) have used an antibody to specifically block the B cell CD28 site of interaction with $CD4^+TH2$ cells, thereby inhibiting B cell stimulation. They achieved a marked attenuation of anti-vector antibody production and prolonged transgene expression in liver cells over 5 months. Therefore, an approach to adenovirus mediated gene therapy of CF based on the administration of a short-acting immune modulator at the time of vector administration may be sufficient to overcome humoral immunreactions. In addition, vector constructs conferring high transgene expression should allow a substantial reduction of the actual virus load and may in combination with repeated immune suppression/modulation help to overcome the discussed problems even with multiple virus applications. An other, however, fairly limited approach to avoid neutralising antibodies is to change the serotype of the adenovirus vector in consecutive administrations (131).

So far no immune response against the CFTR transgenic protein has been reported in the current investigations and this may not become a major problem in CF gene therapy because the defect is almost exclusively due to point mutations or very small deletions. However recent studies using adenovirus mediated gene transgene expression of erythropoietin (132) and LDL-receptor (133) in transgenic knockout mice for these proteins have shown significant immune reactions to the respective transgene proteins which significantly limit the corrective capability of this approach.

A major and initially unappreciated problem influencing the efficiency of adenovirus for transfection of airway epithelia seems to be the lack of the αvβ3 and αvβ5 integrins and a predominance of α9β1 integrins on differentiated columnar epithelia of the proximal airways (97, 134, 135). These integrins bind to the adenoviral penton-RGD motif and play a major role in the internalization of the common serotypes of this virus (136). Integrin αvβ5 seems also to play a major, albeit ill understood, role in the release of the adenovirus from endosomes (137). Although alternative routes of entry of adenovirus seem to exist, lack of the integrins αvβ3 and αvβ5 is probably the main reason for the rather poor *in vivo* transfection rates of differentiated columnar cells in the proximal airways and especially the nose (55, 96, 97, 135). In contrast, undifferentiated cells investigated on a xenograft model *in vivo* express the αvβ5 integrin and are readily infected by adenovirus (97). The natural upregulation of these integrins is probably the reason for the selective adenovirus transfection of the outgrowing cells of regenerating airway epithelial cultures (138) and may also occur *in vivo* during regeneration of the damaged CF-epithelia. Another approach suggested to overcome these targeting problems would be to investigate the αvβ5 integrin independent pathways used for instance by the type 40 adenovirus which does not have a penton RGD-sequence (97). Alternatively adenovirus can be modified by replacement of the RGD motif to target specific cells (139). This approach could be adapted to increase the transfection of differentiated airway epithelial cells. Drug induced upregulation of the αvβ3 and αvβ5 integrins (140) has also been suggested as a means to increase adenovirus infection in the upper airways (97) but infecion of the lower airways which would be the main target for gene therapy of CF may after all not be such a problem (135).

16.5.2 Cationic Lipid Mediated Gene Transfer

Cationic lipid mediated gene transfer *in vivo* has so far shown no toxic or immunogenic reactions and first investigations also showed no reduction of transgene expression following repeated administration (141, 142). This is particularly important with respect to the possibility of prolonging and enhancing transgene expression by repeated administration.

This completely artificial transfection system is, however, much less effective than adenoviruses. Attempts to increase cationic lipid mediated transgene expression require intensive research to determine the rate limiting step(s) of the gene transfer and expression cascade (entry, lysosome escape, nuclear targeting and transgene transcription and translation) and to devise constructs which are able to overcome potential barriers.

Several recent *in vitro* studies have tried to determine optimal conditions for cationic lipid mediated transfection (68, 141, 143). Not unexpectedly, great differences in transgene expression were observed between different cell lines and lipid preparations. As a general rule, a lipid to DNA ratio resulting in a moderate positive net charge seems to be most advantageous for *in vitro* cell transfection; most likely because of charge interaction with the negative cell surface. Transgene expression can be enhanced by increasing the lipid and DNA concentration to a level just above which toxicity (143) and/or precipitation counteract this effect.

Recent investigations indicate that cationic lipids are internalised after cell binding by an endocytosis process, rather than by external fusion with the cell membrane (144-146). Endocytosis is enhanced by inclusion of cholesterol in the lipid mixtures (147). This entry mechanism is probably strongly influenced by the structural features of the lipid /DNA complexes and could presumably be greatly improved by avoiding the structural heterogeneity of present formulations. Since endocytic cell entry of the cationic liposome/DNA complex is a relatively slow process prolonging contact time between the lipid DNA-complexes and cells also increases transgene expression (68, 143, 146, 148). This appears to differ somewhat from cell-line to cell-line and from lipid to lipid in relation to the applied concentration. Raja-Walia et al. (148) found maximal transfection of endothelial cells with LipofecAMINE at about 3-4 hours (0.6mgDNA: 8mgLipofectAMINE/per cell culture well). Caplen et al. (68) observed a constant increase of transgene expression up to 24 hours after transfection in two airway epithelial cell lines at a low DNA-DC-Chol/DOPE concentration (2.8mgDNA/ml:14mg DC-Chol/DOPE). About the same expression level was reached as a maximum as early as 1hour after transfection with the 10-fold DNA/lipid concentration and was reduced in one of the two cell lines by further increasing the concentration 100-fold, probably due to cell toxicity. Zabner et al. (146) found that about 6 hours of incubation *in vitro* where needed to achieve uptake of fluorescently-labelled DNA complexed to DIMRIE/DOPE into the majority of cells of various lines. Using gold labelled DNA to follow the fate of the DNA/lipid complexes by electron-microscopy these authors detected the label after 24 hours in large perinuclear vesicles but not free in the cytoplasm or the nucleus. The DNA-complex containing endosomes did not seem to fuse with the lysosomal compartment and it appears, therefore, that release of the complex from the endocytic compartment and transfer to the nucleus are very critical steps determining the efficiency of transgene expression. The presence of DOPE in most cationic lipid mixtures is essential for fusion of the cationic lipids with the mammalian cell membrane after endocytosis and thereby for release from the endosome (144, 149). The mechanism of DOPE mediated endosomal release is not entirely clear but appears to be pH independent in a range of pH 4.5 to 7.0 (145). Endosomolysis by the protonable cationic compounds lipopolyspermine (150) and polyethylenimine (151) is attributed to their capacity to continuously buffer the flow of protons into the ageing endosomes. This is followed by an influx of chloride ions and water which leads to swelling and finally rupture of these structures (151). Amphiphilic peptides (152) and adenovirus (148, 153) have also been used as endosomolytic agents in combination with cationic lipids. A dramatic increase in lipofection efficiency observed in by addition of transferrin to the transfection complex may also be due to escape from the endosome (154).

Lipopolyspermine also has a high affinity for chromatin (150) and may support nuclear transfer. Nuclear proteins such as HMG (155, 156) and histones (157) are other candidate for such function and the enhancement of lipid mediated gene expression by adenovirus (148, 153) may be due to endosomolysis as well as nuclear transfer. An other important event in this cascade is the release of plasmid DNA from the DNA/lipid complex for transcription in the nucleus. Complexing of a marker gene with a cationic lipid at the optimal lipid/DNA ratio for *in vitro* transfection of 5:1 resulted in marked inhibition of transgene expression after injection into the nucleus of Xenopus oocytes in comparison to naked DNA or a 1:1 complex (146). Finally, the rate and kinetics of transcription are very much dependent on the regulatory sequences used in the expression constructs such as promoters, enhancers and introns.

Most of these investigations have been performed in cell culture, and it remains to be analysed how valid these conclusions are for *in vivo* transfection and gene expression. For instance optimal lipid/DNA ratios and net charge are influenced by the surrounding cellular and humoral conditions and will have to be adjusted according to the route of *in vivo* administration.

Specific cell targeting of the CFTR expressing epithelial cells represents an unsolved problem for all gene transfer systems. It is even not entirely clear which airway epithelial cells to target. In the longer run and in combination with methods to achieve permanent transgene delivery targeting of stem cells would be required and should be combined with specific expression targeting. Surfactant-associated protein B (158), lectin (157) and antibodies against the polymeric immunoglobulin receptor, which is expressed at the basolateral surface of airway epithelial cells (159) linked to polylysine have been used for delivery targeting of the airway epithelium. This principle may also be adaptable for the targeting of cationic lipids.

16.6 Other Aspects of Gene Therapy for CF

Presently, the above described problems and approaches dominate the research efforts towards more efficient and less toxic delivery systems for gene therapy of CF. At this stage of development it seems prudent to continue to use the excellent model of the nasal epithelium as a test field for development of novel vector systems, for the analysis of the effects of repeated application and to introduce more sensitive and reliable *in vivo* detection methods for transgene expression. The great advantage of the nasal epithelial model is, that it allows repeated and continuous monitoring of the effect of gene delivery without the strain of general anaesthesia and bronchoscopy, which is certainly justified once a gene-delivery system with real therapeutic potential has been developed. An interesting intermediate model which may allow to evaluate therapeutic effects by monitoring clinical parameters is the application of gene transfer vectors to the maxillary sinus, as recently suggested by Welsh et al. (7).

In all trials using the nasal epithelium as model system for CF gene therapy, repeated measurements of the electrical potential differences across the epithelial sheath have served as the best surrogate endpoint to measure efficacy and duration of the CFTR expression. Detection of the CFTR transgene and its expression products has so far not been very reliable in most cases. Improvements in this field (160) and the introduction of new methods such as SPQ (161) and other single cell analysis techniques (162) are, therefore, of great importance. Finally it should not be forgotten, that CF is a systemic disease and that, although the pulmonary condition remains its most serious manifestation, affections of other organ systems such as the biliary tract, the pancreas and the gut should be kept in mind with respect to treatment by gene therapy.

16.6.1 Pulmonary Delivery

So far all attempts towards gene therapy for CF have been oriented towards the pulmonary manifestations of CF through airway delivery. However, even when an effective and nontoxic procedure for gene transfer of the airway epithelia has been developed it may not be easy to reach the target cells in the airways because of the mucous obstruction characteristic for CF. A combination of gene therapy with mucolytic treatment may, thus, be necessary. Alternatively, application of the gene transfer vectors via the circulation may be needed. This would require the gene-delivery vector to penetrate the different cell layers of the blood vessels and the interstitial tissue to reach the cells of the multi-straited airway epithelia from the basolateral side. Topical catheter application of the constructs to the arterial vessels serving the airways would be necessary to avoid loss of the construct by transfection of irrelevant cells. This route is, however, rather complicated since these vessels branch off at several sites of the *Aorta thoracica* and from the *Aa.intercostales*. Delivery targeting specifically to the airway epithelia would make vascular delivery more effective, and first results using antibodies against the polymeric immunoglobulin receptor as targeting ligands seem indeed very promising (159).

16.6.2 Other Organ Systems

Specific cell targeting may also apply to delivery via the blood stream to the sites of intestinal manifestation of CF, in particular the pancreas and biliary system, provided suitable targeting ligands can be identified. Enteral delivery may be possible by development of encapsulated vector systems which are set free in the bowel.

Hepatobiliary cirrhosis is a major untreatable, and sometimes lethal, clinical manifestation of cystic fibrosis (CF) caused by mucus obstruction of the bile ducts in the biliary epithelium which are the primary hepatic cells in which CFTRmRNA and protein has been detected (163, 164). Liver cirrhosis has been found in up to 5% (165) of all CF patients but about 72% of them show already signs of focal biliary cirrhosis (166). Many patients who now live into adulthood, due to improved treatment of the lung and pancreas manifestations, may develop this severe extrapulmonary disease particularly if the initial approaches to pulmonary gene therapy eventually lead to therapeutic success. Application to the biliary system is possible by biliary canulation. Successful transfection of the biliary epithelium with adenovirus vectors *in vivo* has been achieved by retrograde cannulation of the bile duct of rats (164), therefore, third generation adenovirus-vector systems will also be of interest for therapeutic application to the biliary system.

16.6.3 Fetal Gene Therapy for CF

Despite the early onset of CF and its systemic manifestation affecting several organs most current efforts towards somatic gene therapy are concentrating on the treatment of adult patients for CF lung disease. CFTR protein is already detected very early during normal human embryogenesis and may, therefore, play an important physiological role in human fetal development. It is thus possible that the earlier treatment is initiated the more likely it will be to prevent irreversible organ damage in CF patients. The main aim of fetal somatic gene therapy for CF would, therefore, be primary prevention of the disease manifestation by introduction of a normal CFTR gene at a time of fetal development at which the target cells can be easily and perhaps stably transfected. By applying gene therapy vector systems at mid to late pregnancy into the amniotic cavity or into the fetal circulation it may be possible to deliver treatment sufficiently early to prevent the development of prenatal tissue damage. In addition it may allow otherwise inaccessible organ or cell systems to be reached and to target still expanding stem cell populations, thereby enabling permanent correction with an integrating vector. Furthermore gene delivery to the fetus could also avoid immune sensitisation, which would give the possibility of repeated treatments with non-integrating vector systems after birth (for review see (82, 167). Such immunotolerance has indeed been found after a second vector administration to neonatal mice (168).

The simplest approach to fetal gene therapy would be the administration of the normal human CFTR gene into the amniotic-fluid surrounding the fetus. The amniotic fluid is swallowed by the foetus, adsorbed through the gut and excreted through the kidneys. The injected gene construct should, therefore, reach the gut, and perhaps also the pancreatic and bile duct cells. The foetus also performs "breathing" movements which may allow the gene construct to reach the lung cells.

First investigations *ex vivo* indicate that undifferentiated human fetal lung cells are susceptible to AdV-mediated gene transfer (169). Our own studies have shown that several viral and lipid based gene therapy systems are able to transfect fetal culture cells in the presence of amniotic fluid (170). Recent attempts towards fetal gene therapy in sheep have used quite elaborate surgical interventions to instil recombinant retroviruses (81) or adenovirus vectors (171, 172) into the fetal airways *in utero.* Expression of marker-transgenes including the human CFTR gene was detected in the airway or alveolar epithelia. The requirement for surgical instillation and repeated application, the relatively low number of transfected cells and toxic reactions which prevented in some cases a follow-up to term, indicate the need for further improvement of the technology and vector systems. Transfection of the airways and intestinal tract in sheep and mice (173) and of the airways in rats (174) has recently been achieved by intraamniotic injection of adenovirus-recombinants. A report on transplacental fetal transfection of multiple organ systems including DNA transfer to the germline by lipopolyamine/DNA complexes injected into the tail vein of mice (175) could so far not be reproduced in other laboratories including our own.

Given the many unknown aspects of fetal gene therapy it is essential to extensively investigate this new approach to gene therapy in animal models for specific diseases as well as to improve on the technology of delivery and to assess efficacy and possible side effects including the inadventitious transfection of germ line cells. Cystic fibrosis may also in this respect play a pathfinder role towards a broader application of this approach to other diseases.

16.7 Outlook

In conclusion, many problems have still to be solved before gene therapy for CF will become a real therapeutic option. However, the great advancements made since the discovery of the CFTR gene only eight years ago in understanding the pathophysiology of CF, establishing its molecular diagnosis, and investigating new forms of treatment based on knowledge about its molecular pathology indicate the great potential of basic research to solve the problems of this disease and encourage us to continue our work on this new approach towards a cure for cystic fibrosis.

Acknowledgements

The work of the Cystic Fibrosis Gene Therapy Research Group, Imperial College Medical School at St Mary's is supported by the Muller Bequest, the Cystic Fibrosis Trust, the Medical Research Council, the AFLM/AFM and the Sir Julius Thorn Charitable Trust.

References

1. Boucher, R. C., Knowles, M. R., Johnson, L. G., Olson, J. C., Pickles, R., Wilson, J. M., Engelhardt, J. F., Yang, Y., and Grossman, M. (1994). Clinical protocol: Gene Therapy for cystic fibrosis using E1-deleted adenovirus: A phase I trial in the nasal cavity. Hum. Gene Ther. *5,* 615-639.
2. Crystal, R. G., Jaffe, A. H., Brody, S., Mastrangeli, A., McElvaney, N. G., Rosenfeld, M., Chu, C.-S., Danel, C., Hay, J., and Eissa, T. (1995). Clinical protocol: A phase 1 study, in cystic fibrosis patients, of the safty, toxicity, and biological efficacy of a single administration of a replication deficient, recombinant adenovirus carrying the cDNA of the normal cystic fibrosis transmembrane conductance regulator gene in the lung. Hum. Gene Ther. *6,* 643-666.
3. Crystal, R. G., Mastrangeli, A., Sanders, A., Cooke, J., King, T., Gilbert, F., Henschke, C., Pascal, W., Herena, J., Harvey, B.-G., Hirschowitz, E., Diaz, D., Russi, T., Pacheco, F., Sikand, V., and Brion, P. (1995). Clinical protocol: Evaluation of repeat administration of a replication deficient recombinant adenovirus containing the human cystic fibrosis transmembrane conductance regulator cDNA to the airways of individuals with cystic fibrosis. Hum. Gene Ther. *6,* 667-703.
4. Dorkin, H. I. (1995). Adenovirus mediated gene transfer for cystic fibrosis. Safety of single administration in the lung (lobar instillation) Table: Human GeneMarker/Therapy Clinical Protocols. Hum. Gene Ther. *6,* 548.
5. Sorscher, E. J., Logan, J. J., Frizzell, R. A., Lyrene, R. K., Bebok, Z., Dong, J. Y., Duvall, M. D., Felgner, P. L., Matalon, S., Walker, L., and Wiatrak, B. J. (1994). Clinical Protocol: Gene therapy for cystic fibrosis using cationic liposome mediated gene transfer : a phase 1 trial of safty and efficacy in the nasal airway. Hum. Gene Ther. *5,* 1259-1277.
6. Welsh, M. J., Smith, A. E., Zabner, J., Rich, D. P., Scott, M. G., Gregory, R. J., Pratt, B. M., and Moscicki, R. A. (1994). Clinical protocol: Cystic fibrosis gene therapy using an adenovirus vector: in vivo safety and efficacy in nasal epithelium. Hum. Gene Ther. *5,* 209-219.
7. Welsh, M. J., Zabner, J., Scott, M. G., Smith, A. E., Moscicki, R., and Wadsworth, S. (1995). Clinical protocol: Adenovirus-mediated gene transfer for cystic fibrosis. Part A. Safty of dose and repeat administration in the nasal epithelium. Part B. Clinical efficacy in the maxillary sinus. Hum. Gene Ther. *6,* 205-218.
8. Wilmott, R. W., Whitsett, J. A., Trapnell, B., Wert, S., Baughman, R., Cuppoletti, J., and Tolstoshev, P. (1994). Clinical protocol; Gene therapy for cystic fibrosis utilizing a replication deficient recombinant adenovirus vector to deliver the human cystic fibrosis transmembrane conductance regulator cDNA to the airways. A phase I study. Hum. Gene Ther. *5,* 1019-1057.
9. Wilson, J. M., Engelhardt, J. F., Grossman, M., Simon, R. H., and Yang, Y. (1994). Clinical protocol: Gene therapy of cystic fibrosis lung disease using E1 deleted adenoviruses: A phase 1 trial. Hum. Gene Ther. *5,* 501-519.
10. Flotte, T. R., Carter B., Conrad, C., Guggino, W. B., Reynolds, T., Rosenstein, B., Taylor, G., Walden, S., and Wetzel, R. (1995). A phase 1 study of an adeno-asscoieted virus-CFTR gene vector in adult CF patients with mild lung disease. Hum. Gene Ther *7,* 1145-1159.
11. Caplen, N. J., Alton, E. W. F. W., Middelton, P. G., Dorin, J. R., Stevenson, B. J., Gao, X., Durham, S., Jeffrey, P. K., Hodson, M. E., Coutelle, C., Huang, L., Porteous, D. J., Williamson, R., and Geddes, D. M. (1995). Liposome-mediated CFTR gene transfer to the nasal epithelium of patients with cystic fibrosis. Nature Medicine *1,* 39-46.
12. Gill, D. R., Southern, K. W., Mofford, K. A., Seddon, T., Huang, L., Sorgi, F., Thompson, A., MacVanish, L. J., Ratcliff, R., Bilton, D., Lane, D. L., Littlewood, J. M., Webb, A. K., Middleton, P. G.,

Colledge, W. H., Cuthbert, A. W., Evans, M. J., Higgins, C. J., Hyde, S. C. (1996). A placebo controlled study of liposome-mediated gene transfer to the nasal epithelium of patients with cystic fibrosis. Gene Therapy *4*, 199-209.

13. Bellon, G., Michel-Calemard, L., Thouvenot, D., Jagneaux, V., Poitevin, F., Malcus, C., Accart, N., Layani, M. P., Aymard, M., Bernon, H., Bienvenu, J., Courtney, M., Doring, G., Gilly, B., Gilly, R., Lamy, D., Levrey, H., Morel, Y., Paulin, C., Perraud, F., Rodillon, L., Sene, C., So, S., Touraine-Moulin, F., Schatz, C., Pavirani, A. (1997) Aerosol administration of a recombinant adenovirus expressing CFTR to cystic fibrosis patients: a phase I clinical trial. Hum Gene Ther *8* 15-25.
14. Porteous, D. J., Dorin, J. R., McLachlan, G., Davidson-Smith, H., Davidson, H., Stevenson, B., Carothers, A. D., Wallace, W. A. H., Morales, S., Hones, C., Kallmeyer, G., Michelis, U., Naujoks, K., Ho, L.-P., Samways, J., Imrie, M., Greening, A., Innes, J. A. (1997). Evidence for safty and efficacy of DOTAP-cationic liposome mediated CFTR gene transfer to the nasal epithelium of patients with cystic fibrosis. Gene Ther. *4*, 210-218.
15. Hay, J. G., McElvaney, N. G., Herena, J., and Crystal, R. G. (1995). Modification of nasal epithelial potential differences of individuals with cystic fibrois consequent to local administration of a normal CFTR cDNA adenovirus gene transfer vector. Hum. Gene Ther. *6*, 1487-1496.
16. Crystal, R. G., McElvaney, N. G., Rosenfeld, M. A., Chu, C.-S., Mastrangeli, A., Hay, J. G., Brody, S. L., Jaffe, H. A., Eissa, N. T., and Danel, C. (1994). Administration of an adenovirus containing the human CFTR cDNA to the respiratory tract of individuals with cystic fibrosis. Nature Genet. *8*, 42-51.
17. Knowles, M. R., Hohneker, K. W., Zhou, Z., Olson, J. C., Noah, T. L., Hu, P.-C., Leigh, M. W., Engelhardt, J. F., Edwards, L. J., Jones, K. R., Grossman, M., Wilson, J. M., Johnson, L. G., and Boucher, R. C. (1995). A controlled study of adenoviral-vector-mediated gene transfer in the nasal epithelium of patients with cystic fibrosis. N. Engl. J. Med. *333*, 823-831.
18. Bye, M. R., Ewig, J. M., and Quittell, L. M. (1994). Cystic Fibrosis. Lung *172*, 251-270.
19. Crawford, I., Maloney, P. C., Zeitlin, P. L., Guggino, W. B., Hyde, S. C., Turley, H., Gatter, K. C., Harris, A., and Higgins, C. F. (1991). Immunocytochemical localization of the cystic fibrosis gene product CFTR. Proc. Natl. Acad. Sci. USA *88*, 9262-9266.
20. Trapnell, B. C., Chu, C.-S., Paakko, P. K., Banks, T. C., Yoshimura, K., Ferrans, V. J., Chernick, M. S., and Crystal, R. G. (1991). Expression of the cystic fibrosis transmembrane regulator gene in the respiratory tract of normal individuals and individuals with cystic fibrosis. Proc. Natl. Acad. Sci. USA *88*, 6565-6569.
21. Johnson, L. G., Olsen, J. C., Sarkadi, B., Moore, K. L., Swanstrom, R., and Boucher, R. C. (1992). Efficiency of gene transfer for restoration of normal airway epithelial function in cystic fibrosis. Nature Genetics *2*, 21-25.
22. Gilula, N. B., Reeves, O. R., and Steinbach, A. (1972). Metabolic coupling, ionic coupling and cell contacts. Nature *235*, 262-265.
23. Knowles, M. R., Carson, J. L., Collier, A. M., Gatzy, J. T., and Boucher, R. C. (1981). Measurements of nasal transepithelial electric potential differences in normal human subjects *in vivo*. Am Rev Respir Dis *124*, 484-490.
24. Alton, E. W. F. W., Hay, J. G., Munro, C., and Geddes, D. M. (1987). Measurement of nasal potential difference in adult cystic fibrosis, Young's syndrome, and bronchiectasis. Thorax *42*, 815-817.
25. Middleton, P. G., Geddes, D. M., and Alton, E. W. F. W. (1994). Protocols for the *in vivo* measurement of the ion transport defects in cystic fibrosis nasal epithelium. Eur.Resp.J. *7*, 2050-2056.
26. Knowles, M. R., Paradiso, A. M., and Boucher, R. C. (1995). *In vivo* nasal potential difference: Techniques and protocols for assessing efficacy of gene transfer in cystic fibrosis. Hum. Gene Ther. *6*, 445-455.
27. Riordan, J. R., Rommens, J. M., Kerem, B.-S., Alon, N., Rozmahel, R., Grzelczak, Z., Zielenski, J., Lok, S., Plavsic, N., Chou, J.-L., Drumm, M. L., Iannuzzi, M. C., Collins, F. S., and Tsui, L.-C. (1989). Identification of the cystic fibrosis gene: cloning and characterization of complementary DNA. Science *245*, 1066-1073.
28. Rommens, J. M., Iannuzzi, M. C., Kerem, B.-S., Drumm, M. L., Melmer, G., Dean, M., Rozmahel, R., Cole, J. L., Kennedy, D., Hidaka, N., Zsiga, M., Buchwald, M., Riordan, J. R., Tsui, L.-C., and Collins, F. S. (1989). Identification of the cystic fibrosis gene: chromosome walking and jumping. Science *245*, 1059-1065.

29. Kerem, B.-S., Rommens, J. M., Buchanan, J. A., Markiewicz, D., Cox, T. K., Chakravarti, A., Buchwald, M., and Tsui, L.-C. (1989). Identification of the cystic fibrosis gene: Genetic analysis. Science *245,* 1073-1080.
30. Bear, C. E., Li, C., Kartner, N., Bridges, R. J., Jensen, T. J., Ramjeesingh, M., and Riordan, J. R. (1992). Purification and functional reconstitution of the cystic fibrosis transmembrane conductance regulator (CFTR). Cell *68,* 809-818.
31. Welsh, M. J., Anderson, M. P., Rich, D. P., Berger, H. A., Denning , G. M., Ostedgaard, L. S., Sheppard, D. N., Cheng, S. H., Gregory, R. J., and Smith, A. E. (1992). Cystic fibrosis transmembrane conductance regulator: A chloride channel with novel regulation. Neuron *8,* 821-829.
32. Barasch, J. and Al-Awqati, Q. (1993). Defective acidification of the biosynthetic pathway in cystic fibrosis. J. Cell Sci *Supplm.17,* 229-233.
33. Jiang, C., Finkbeiner, W. E., Widdicombe, H. J., McCray, P. B., and Miller, S. S. (1993). Altered fluid transport accross airway epithelium in cystic fibrosis. Science *262,* 424-427.
34. Stutts, M. J., Cannessa, C. M., Olsen, J. C., Hamrick, M., Cohn, J. A., Rossier, J. B., and Boucher, R. C. (1995). CFTR as a cAMP-dependent regulator of sodium channels. Science *269,* 847-850.
35. Kopelman, H., Corey, M., Gaskin, K., Durie, P., Weizman, Z., and Forestner, G. (1988). Impaired chloride secretion, as well as bicarbonate secretion, underlies the fluid secretory defect in the cystic fibrosis pancreas. Gatroenterol. *95,* 349-355.
36. Davidson, D. J., Dorin, J. R., McLachlan, G., Ranaldi, V., Lamb, D., Doherty, C., Govan, J., and Porteous, D. J. (1995). Lung disease in the cystic fibrosis mouse exposed to bacterial pathogens. Nature Genet. *9,* 351-357.
37. Imundo, L., Barasch, J., Prince, A., and Al-Awqati, Q. (1995). Cystic fibrosis epithelial cells have a receptor for pathogenic bacteria on their apical surface. Proc. Natl. Acad. Sci. *92,* 3019-3023.
38. Pier, G. R., Grout, M., Zaidi, T. S., Olsen, J. C., Johnson, L. G., Yankaskas, J. R., and Goldberg, J. B. (1996). Role of mutant CFTR in hypersusceptibility of cystic fibrosis patients to lung infections. Science *271,* 64-67.
39. Bout, A., Imler, J.-L., Schultz, H., Perricaudet, M., Zurcher, C., Herbrink, P., Valerio, D., and Pavirani, A. (1994). *In vivo* adenovirus-mediated transfer of human CFTR cDNA to rhesus monkey airway epithelium: efficacy, toxicity and safety. Gene Ther *1,* 385-394.
40. Zabner, J., Petersen, D. M., Puga, A. P., Graham, S. M., Couture, L. A., Keyes, L. D., Lukason, M. J., St.George, J. A., Gregory, R. J., Smith, A. E., and Welsh, M. J. (1994). Safety and efficacy of repetitive adenovirus-mediated transfer of CFTR cDNA to airway epithelia of primates and cotton rats. Nature Genet. *6,* 75-83.
41. Engelhardt, J. F., Simon, R. H., Yang, Y., Zepeda, M., Weber-Pendelton, S., Doranz, B., Grossman, M., and Wilson, J. M. (1993). Adenovirus mediated transfer of the CFTR gene to lung of nonhuman primates: biological efficacy study. Hum. Gene Ther. *4,* 759-769.
42. Logan, J. J., Bebok, Z., Walker, L. C., Peng, S., Felgner, P. L., Siegal, G. P., Frizzell, R., Dong, J., Howard, M., Matalon, S., Lindsey, J. R., DuVall, M., and Sorscher, E. J. (1995). Cationic lipids for reporter gene and CFTR transfer to rat pulmonary epithelium. Gene Ther. *2,* 38-49.
43. Sené, C., Bout, A., Imler, J.-L., Schultz, H., Willemot, J.-M., Henneble, V., Zurcher, C., Valerio, D., Lamy, D., and Pavirani, A. (1995). Aerosol-mediated delivery of recombinant adenovirus to the airways of nonhuman primates. Hum. Gene Therapy *6,* 1587-1593.
44. Flotte, T. R., Afione, S. A., Conrad, C., McGrath, S. A., Solow, R., Oka, H., Zeitlin, P. L., Guggino, W. B., and Carter, B. J. (1993). Stable *in vivo* expression of the cystic fibrosis transmembrane conductance regulator with an adeno-associated virus vector. Proc. Natl. Acad. Sci. USA *90,* 10613-10617.
45. Rosenfeld, M. A., Yoshimura, K., Trapnell, B. C., Yoneyama, K., Rosenthal, E. R., Dalemans, W., Fukayama, M., Bargon, J., Stier, L. E., Stratford-Perricaudet, L., Perricaudet, M., Guggino, W. B., Pavirani, A., Lecocq, J.-P., and Crystal, R. G. (1992). *In vivo* transfer of the human cystic fibrosis transmembrane conductance regulator gene to the airway epithelium. Cell *68,* 143-155.
46. Yoshimura, K., Rosenfeld, M. A., Nakamura, H., Scherer, E. M., Pavirani, A., Lecocq, J.-P., and Crystal, R. G. (1992). Expression of the human cystic fibrosis transmembrane conductance regulator gene in the mouse lung after *in vivo* intratracheal plasmid-mediated gene transfer. Nucleic Acids Res. *20,* 3233-3240.
47. Brody, S. L., Metzger, M., Danel, C., Rosenfeld, M. A., and Crystal, R. G. (1994). Acute responses of non-human primates to airway delivery of an adenovirus vector containing the human cystic fibrosis transmembrane conductance regulator cDNA. Hum. Gene Ther. *5,* 821-836.

48. Hyde, S. C., Gill, D. R., Higgins, C. F., Trezise, A. E. O., MacVinish, L. J., Cuthbert, A. W., Ratcliff, R., Evans, M. J., and Colledge, W. H. (1993). Correction of the ion transport defect in cystic fibrosis transgenic mice by gene therapy. Nature *362,* 250-255.
49. Alton, E. W. F. W., Middleton , P. G., Caplen, N. J., Smith, S. N., Steel, D. M., Munkonge, F. M., Jeffery, P. K., Geddes, D. M., Hart, S. L., Williamson, R., Fasold, K. I., Miller, A. D., Dickinson, P., Stevenson, B. J., McLachlan, G., Dorin, J. R., and Porteous, D. J. (1993). Non-invasive liposome-mediated gene delivery can correct the ion transport defect in cystic fibrosis mutant mice. Nature Genet. *5,* 135-142.
50. Strauss, S. E. (1984). Adenovirus infections in humans. In: Adenoviruses, Ed. Ginsberg, H. S. New York, London: Plenum Press, 451-496.
51. Horowitz, M. S. (1990). Adenoviruses. In: Virology, Eds. Fields, B. N. & Knipe, D. M. New York: Raven Press, Ltd, 1723-1740.
52. Graham, F. L., Smiley, J., Russel, W. C., and Nairn, R. (1977). Characteristics of a human cell line transformed by DNA from human adenovirus type 5. J.Gen.Virol. *36,* 59-72.
53. Yei, S., Mittereder, N., Tang, K., O'Sullivan, C., and Trapnell, B. C. (1994). Adenovirus-mediated gene transfer for cystic fibrosis: Quantitative evaluation of repeated *in vivo* vector administration to the lung. Gene Ther. *1,* 192-200.
54. Yei, S., Mittereder, N., Wert, S., Whitsett, J. A., Wilmott, R. W., and Trapnell, B. C. (1994). *In vivo* evaluation of the safety of adenovirus-mediated transfer of the human cystic fibrosis transmembrane conductance regulator cDNA to the lung. Hum. Gene Ther. *5,* 731-744.
55. Dong, J.-Y., Wang, D., van Ginkel, F. W., Pascual, D. W., and Frizzell, R. A. (1996). Systematic analysis of repeated gene delivery into animal lungs with a recombinant adenovirus vector. Hum Gene Ther *7,* 319-331.
56. Simon, R. H., Engelhardt, J. F., Yang, Y., Zepeda, M., Weber-Pendelton, S., Grossman, M., and Wilson, J. M. (1993). Adenovirus mediated transfer of the CFTR to lung of nonhuman primates: toxicity study. Hum.Gene Ther. *4,* 771-780.
57. Bout, A., Perricaudet, M., Baskin, G., Imler, J.-L., Scholte, B. J., Pavirani, A., and Valerio, D. (1994). Lung gene therapy: *in vivo* adenovirus-mediated gene transfer to rhesus monkey airway epithelium. Hum. Gene Ther. *5,* 3-10.
58. Wilmott, R. W., Amin, R. S., Perez, C. R., Wert, S., Keller, G., Boivin, G. P., Hirsch, R., De Inocencio, J., Lu, P., Reising, S. F., Yei, S., Whitsett, J. A., and Trapnell, B. C. (1996). Safety of adenovirus-mediated transfer of the human cystic fibrosis transmembrane conductance regulator cDNA to the lungs of nonhuman primates. Hum. Gene Ther. *7,* 301-318.
59. Yang, Y., Li, Q., Ertl, H. C. J., and Wilson, J. M. (1995). Cellular and humoral immune responses to viral antigens create barriers to lung-directed gene therapy with recombinant adenoviruses. J.Virol. *69,* 2004-2015.
60. Felgner, P. L., Gadek, T. R., Holm, M., Roman, R., Chan, H. W., Wenz, M., Northrop, J. P., G. M. Ringold, G. M., and Danielsen, M. (1987). Lipofection: A highly efficient, lipid-mediated DNA-transfection procedure. Proc. Natl. Acad. Sci. USA *84,* 7413-7417.
61. Felgner, J. H., Kumar, R., Sridhar, C. N., Wheeler, C. J., Tsai, Y. J., Border, R., Ramsey, P., Martin, M., and Felgner, P. L. (1994). Enhanced gene delivery and mechanism studies with a novel series of cationic lipid formulations. J.Biol. Chem. *269,* 2550-2561.
62. Behr, J.-P. (1994). Gene transfer with synthetic cationic amphiphiles: Prospects for gene therapy. Bioconj. Chem. *5,* 382-389.
63. Gao, X.,Huang, L. (1995). Cationic liposome-mediated gene transfer. Gene Ther. *2,* 710-722.
64. Gershon, H., Ghirlando, R., Guttman, S. B., and Minsky, A. (1993). Mode of formation and structural features of DNA-cationic liposome complexes used for transfection. Biochemistry *32,* 7143-7151.
65. Sternberg, B., Sorgi, F. L., and Huang, L. (1994). New structures in complex formation between DNA and cationic liposomes visualized by freeze-fracture electron microscopy. FEBS-L. *356,* 361-366.
66. Lamb, B. T., Sisoda, S. S., Lawler, A. M., Slunt, H. H., Kitt, C. A., Kearnes, W. G., Pearson, P. L., Price, D. L., and Gearhart, J. D. (1993). Introduction and expression of the 400 kilobase precursor amyloid protein gene in transgenic mice. Nature Genet. *5,* 22-29.
67. Gao, X. and Huang, L. (1991). A novel cationic liposome reagent for efficient transfection of mammalian cells. Biochem. Biophys. Res. Commun. *179,* 280-285.
68. Caplen, N. J., Kinrade , E., Sorgi, F., Gao, X., Gruenert, D., Geddes, D., Coutelle, C., Huang, L., Alton, E. W. F. W., and Williamson, R. (1995). *In vitro* liposome-mediated DNA transfection of epithelial cell lines using the cationic liposome DC-Chol/DOPE. Gene Ther. *2,* 603-613.

69. Stewart, M. J., Plautz, G. E., Del Buono, L., Yang, Z. Y., Xu, L., Gao, X., Huang, L., Nabel, E. G., and Nabel, G. J. (1992). Gene transfer *in vivo* with DNA-liposome complexes: Safety and acute toxicity in mice. Human. Gene Ther. *3,* 267-275.
70. Nabel, E. G., Gordon, D., Yang, Z.-Y., Ling, X., San, H., Plautz, G. E., Wu, B.-Y., Gao, X., Huang, L., and Nabel, G. J. (1992). Gene transfer *in vivo* with DNA-liposome complexes: lack of autoimmunity and gonadal localization. Hum. Gene Ther. *3,* 649-656.
71. Nabel, G. J., Nabel, E. G., Yang, Z.-Y., Fox, B. A., Plautz, G. E., Gao, X., Huang, L., Shu, S., Gordon, D., and Chang, A. E. (1993). Direct gene transfer with DNA-liposome complexes in melanoma: Expression, biological activity and lack of toxicity in humans. Proc. Natl. Acad.Sci. USA *90,* 11307-11311.
72. Nabel, G. J., Chang, A., Nabel, E. G., Plautz, G., Fox, B. A., Huang, L., and Shu, S. (1992). Clinical protocol: Immunotherapy of malignancy by *in vivo* gene transfer into tumors. Hum. Gene Ther. *3,* 399-410.
73. Nabel, E. G., Yang, Z., Muller, D., Chang, A. E., Gao, X., Huang, L., Cho, K. J., and Nabel, G. J. (1994). Safety and toxicity of catheter gene delivery to the pulmonary vasculature in a patient with metastatic melanoma. Hum. Gene Ther. *5,* 1089-1094.
74. Stribling, R., Brunette, E., Liggitt, D., Gaensler, K., and Debs, R. (1992). Aerosol gene delivery *in vivo*. Proc.Natl.Acad.Sci.USA *89,* 11277-11281.
75. Middleton, P. G., Caplen, N. J., Gao, X., Huang, L., Gaya, H., Geddes, D. M., and J.Alton, E. W. F. W. (1994). Nasal application of the cationic liposome DC-Chol:DOPE does not alter ion transport, lung function or bacterial growth. Eur. Resp.J. *7,* 442-445.
76. McLachlin, J. R., Cornetta, K., Eglitis, M. A., and Anderson, W. F. (1990). Retroviral-mediated gene transfer. Prog. Nucleic Acid Res. Mol. Biol. *38,* 91-135.
77. Roemer, K. and Friedmann, T. (1992). Concepts and Strategies for human gene therapy. Eur.J.Biochem. *208,* 211-225.
78. Drumm, M. L., Pope, H. A., Cliff, W. H., Rommens, J. M., Marvin, S. A., Tsui, L.-C., Collins, F. S., Frizzell, R. A., and Wilson, J. M. (1990). Correction of the cystic fibrosis defect *in vitro* by retrovirus-mediated gene transfer. Cell *62,* 1227-1233.
79. Anderson, W. F. (1993). What about those monkeys that got T-cell lymphoma? Hum. Gene Ther. *4,* 1-2.
80. Leigh, M. W., Kylander, J. E., Yankaskas, J. R., and Boucher, R. C. (1995). Cell proliferation in bronchial epithelium and submucosal glands of cystic fibrosis patients. Am.J.Resp.Cell Mol.Biol. *12,* 605-612.
81. Pitt, B. R., Schwarz, M. A., Pilewski, J. M., Nakayama, D., Mueller, G. M., Robbins, P. D., Watkins, S. A., Albertine, K. H., and Bland, R. D. (1995). Retrovirus-mediated gene transfer in lungs of living fetal sheep. Gene Ther. *2,* 344-350.
82. Coutelle, C., Douar, A.-M., College, W. H., and Froster, U. (1995). The challenge of fetal gene therapy. Nature Med. *1,* 864-866.
83. Carter, B. J. (1992). Adeno-associated virus vectors. Current Opinion in Biotechnology *3,* 533-539.
84. Kotin, R. M. (1994). Prospects for the use of adeno-associated virus as a vector for human gene therapy. Hum. Gene Ther. *5,* 793-801.
85. Fischer, K. J., Gao, G.-P., Weitzman, M. D., DeMatteo, R., Burda, J. F., and Wilson, J. F. (1996). Transduction with recombinant adeno-associated virus for gene therapy is limited by leading strand synthesis. J.Virol. *70,* 520-532.
86. Flotte, T. R., Afione, S. A., Solow, R., Drumm, M. L., Markakis, D., Guggino, W. B., Zeitlin, P. L., and Carter, B. J. (1993). Expression of the cystic fibrosis transmembrane conductance regulator from a novel adeno-associated virus promoter. J.Biol.Chem. *268,* 3781-3790.
87. Zabner, J., Ramsay, B. W., Meeker, D. P., Aitken, M. L., Balfour, R. P., Gibson, R. L., Lauspach, J., Moscicki, R. A., Richards, S. M., Standaert, T. A., Williams-Warren, J., Wadsworth, S. C., Smith, A. E., and Welsh, M. J. (1996). Repeat administration of an adenovirus vector encoding cystic fibrosis transmembrane conductance regulator to the nasal epithelium of patients. J.Clin.Invest. *97,* 1504-1511.
88. Clothier, C. M. Report of the Committee on the Ethics of Gene Therapy. 1992, HMSO: .
89. Caplen, N. J., Gao, X., Hayes, P., Elaswarapu, R., Fisher, G., Kinrade, E., Chakera, A., Schorr, J., Hughes, B., Dorin, J. R., Porteous, D. J., Alton, E. W. F. W., Geddes, D. M., Coutelle, C., Williamson, R., Huang, L., and Gilchrist, C. (1994). Gene therapy for cystic fibrosis in humans by liposome mediated DNA transfer: regulatory process and production of resources. Gene Therapy *1,* 139-147.

90. Chinet, T. C., Fullton, J. M., Yankaskas, J. R., Boucher, R. C., and Stutts, M. J. (1994). Mechanism of sodium hyperabsorption in cultured cystic fibrosis nasal epithelium: a patch clamp study. Am.J.Physiol. *35,* C1061-C1068.
91. Grubb, B. R., Vick, R. N., and Boucher, R. C. (1994). Hyperabsorption of Na^+ and raised Ca^{2+}-mediated Cl^- secretion in nasal epithelia of CF mice. Am. J.Physiol. *266,* C1478-C1483.
92. Rich, D. P., Couture, L. A., Cardoza, L. M., Guiggio, V. M., Armentano, D., Espino, P. C., Hehir, K., Welsh, M. J., Smith, A. E., and Gregory, R. G. (1993). Development and analysis of recombinant adenovirus for gene therapy of cystic fibrosis. Hum. Gene Ther. *4,* 461-476.
93. Zabner, J., Couture, L. A., Gregory, R. J., Graham, S. M., Smith, A. E., and Welsh, M. J. (1993). Adenovirus-mediated gene transfer transiently corrects the transport defect in nasal epithelia of patients with cystic fibrosis. Cell *75,* 207-216.
94. Wold, W. S. M., and Gooding, L. R. (1991). Region E3 of adenovirus: a cassette of genes involved in host immunosurveillance and virus-cell interactions. Virology *184,* 1-8.
95. McElvaney, N. G., and Crystal, R. G. (1995). IL-6 release and airway administration of human CFTR cDNA adenovirus vector. Nature Medicine *1,* 182-184.
96. Grubb, B. R., Pickles, R. J., Ye, H., Yankaskas, J. R., Vick, R. N., Engelhardt, J. F., Wilson, J. M., Johnson, L. G., and Boucher, R. C. (1994). Inefficient gene transfer by adenovirus vector to cystic fibrosis airway epithelia of mice and humans. Nature *371,* 802-806.
97. Goldman, M. J. and Wilson, J. M. (1995). Expression of $\alpha_v\beta_5$ integrin is necessary for efficient adenovirus-mediated gene transfer in the human airway. J.Virol. *69,* 5951-5958.
98. Goldman, M. J., Yang, Y., and Wilson, J. M. (1995). Gene therapy in a xenograft model of cystic fibrosis lung corrects chloride transport more effectively than the sodium defect. Nature Genet. *1995,* 126-131.
99. Johnson, L. G., Boyles, S. E., Wilson, J., and Boucher, R. C. (1995). Normalization of raised sodium absorption and raised calcium-mediated chloride secretion by adenovirus-mediated expression of cystic fibrosis transmembrane conductance regulator in primary human cystic fibrosis airway epithelial cells. J.Clin. Invest. *95,* 1377-1382.
100. Boucher, R. C. (1996). Current status of CF gene therapy. TIG *12,* 81-84.
101. Ginsberg, H. S., Moldawer, L. L., Sehgal, P. B., Redington, M., Kilian, P. L., Chanock, R. M., and Prince, G. A. (1991). A mouse model for investigating the molecular pathogenesis of adenovirus pneumonia. Proc.Natl.Acad.Sci.USA *88,* 1651-1655.
102. Amin, R., Wilmott, R., Schwarz, Y., Trapnell, B., and Stark, J. (1995). Replication-defficient adenovirus induces expression of interleukin-8 by airway epithelial cells *in vitro*. Hum. Gene Ther. *6,* 145-153.
103. Wilmott, R. W., Amin, R. S., Perez, C. R., Wert, S. E., Whitsett, J. A., and Trapnell, B. C. (1994). Transfer of the CFTRcDNA to the lung of non-human primates by a recombinant adenovirus: A study of toxicity. Pediatr. Res. *35,* 400A (Abstract 2383).
104. Teramoto, S., Johnson.L.G., Huang, W., Leigh, M. W., Boucher, R. C. (1995). The effect of adenoviral vector infection on cell proliferation in cultured primary human airway epithelial cells. Hum. Gene Ther. *6,* 1045-1053.
105. Eissa, N. T., Chu, C.-S., Danel, C., and Crystal, R. G. (1994). Evaluation of the respiratory epithlium of normals and individuals with cystic fibrosis for the presence of adenovirus E1a sequences relevant to the use of E1a⁻ adenovirus vectors for gene therapy of the respiratory manifestations of cystic fibrosis. Hum. Gene Ther. *5,* 1105-1114.
106. Lochmüller, H., Jani, A., Huard, J., Prescott, S., Simenau, M., Massie, B., Karpati, G., and Acsadi, G. (1994). Emergence of early region 1-containg replication -competent adenovirus in stocks of replication-defective adenovirus recombinants (DE1+DE3) during multiple passages in 293 cells. Hum. Gene Ther. *5,* 1485-1491.
107. Imler, J.-L., Bout, A., Dreyer, D., Dieterlé, A., Schultz, H., Valerio, D., Mehtali, M., and Pavirani, A. (1995). *Trans*-complementation of E1-deleted adenovirus: A new vector to reduce the possibility of codissimination of wild-type and recombinant adenoviruses. Hum. Gene Ther. *6,* 711-721.
108. Yang, Y., Nunes, F. A., Berencsi, K., Gönczöl, E., Engelhardt, J. F., and Wilson, J. M. (1994). Inactivation of E2a in recombinant adenoviruses improves the prospect for gene therapy in cystic fibrosis. Nature Genet. *7,* 362-369.
109. Yang, Y., Nunes, F. A., Berencsi, K., Furth, E. E., Gönczöl, E., and Wilson, J. M. (1994). Cellular immunity to viral antigens limits E1-deleted adenoviruses for gene therapy. Proc.Natl.Acad.Sci. USA *91,* 4407-4411.

110. Wilson, C. and Kay, M. A. (1995). Immunomodulation to enhance gene therapy. Nature Med. *1*, 887-889.
111. van Ginkel, F. W., Lui, C., Simecka, J. W., Dong, J.-Y., Greenway, T., Frizzell, R. A., Kiyono, H., McGhee, J. R., and Pascual, D. W. (1995). Intratracheal gene delivery with adenoviral vector induces elevated systemic IgG and mucosal IgA antibodies to adenovirus and β-galactosidase. Hum. Gene Therap. *6*, 895-903.
112. Yang, Y., Trinchieri, G., and Wilson, J. M. (1995). Recombinant IL-12 prevents formation of blocking IgA antibodies to recombinant adenovirus and allows repeated gene therapy to mouse lung. Nature Med. *1*, 890-893.
113. Engelhardt, J. F., Yang, Y., Stratford-Perricaudet, L. D., Allen, E. D., Kozarasky, K., Perricaudet, M., Yankaskas, J. R., and Wilson, J. M. (1993). Direct gene transfer of human CFTR into human bronchial epithelia of xenografts with E1-deleted adenoviruses. Nature Genet. *4*, 27-34.
114. Dai, Y., Schwarz, E. M., Gu, D., Zang, W.-W., Sarvetnick, N., and Verma, I. M. (1995). Cellular and humoral immune responses to adenoviral vectors containing factor IX: Tolerization of factor IX and vector antigens allows for long-term expression. Proc. Natl. Acad. Sci USA *92*, 1401-1405.
115. Engelhardt, J. F., Litzky, L., and Wilson, J. M. (1994). Prolonged transgene expression in cotton rat lung with recombinant adenovirus defective in E2a. Hum. Gene Ther. *5*, 1217-1229.
116. Goldman, M. J., Litzky, L. A., Engelhardt, J. F., and Wilson, J. M. (1995). Transfer of the CFTR gene to the lung of nonhuman primates with E1-deleted, E2a-defective recombinant adenovirus : A preclinical toxicology study. Hum.Gene Ther. *6*, 839-851.
117. Gorziglia, M. I., Kadan, M. J., Yei, S., Lim, J., Lee, G. M., Luthra, R., and Trapnell, B. C. (1996). Elimination of both E1 and E2a from adenovirus vectors further improves prospects for in vivo human gene therapy. J.Virol *70*, 417-4178.
118. Halbert, D. N., Cutt, J. R., and Shenks, T. (1985). Adenovirus early region 4 encodes functions required for efficient DNA replication, late gene expression and host cell shutoff. J.Virol. *56*, 250-257.
119. Weinberg, D. H. and Ketner, G. (1986). Adenoviral early region 4 is required for efficient viral DNA replication and for late gene expression. J.Virol. *57*, 833-838.
120. Fischer, K. J., Choi, H., Burda, J., Chen, S.-J., and Wilson, J. F. (1996). Recombinant adenovirus deleted of all genes for gene therapy of cystic fibrosis. Virology *217*, 11-22.
121. Kochanek, S., Clemens, P. R., Mitani, K., Chen, H. H., Chan, S., and Caskey, C. T. (1996). A new adenoviral vector : Replacement of all viral coding sequences with 28kb of DNA independently expressing both full-length dystrophin and β-galactosidase. Proc.Natl.Acad.Sci. *93*, 5731-5752.
122. Krougliak, V. and Graham, F. L. (1995). Development of cell lines capable of complementing E1, E4, and protein IX defective adenovirus type 5 mutants. Hum. Gene Ther. *6*, 1575-1586.
123. Yeh, P., Dedieu, J.-F., Orsini, C., Vigne, E., Denefle, P., and Perricaudet, M. (1995). Construction of E1E4 deleted adenoviruses for human gene therapy. Gene Ther. *2 Supplement 1*, S3 abstract 7.
124. McCoy, R. D., Davidson, B. L., Roessler, B. J., Huffnagel, G. B., Janich, S. L., Laing, T. J., and Simon, R. H. (1995). Pulmonary inflammation induced by incomplete or inactivated adenoviral particles. Hum. Gene Ther. *6*, 1553-1560.
125. Zsengeller, Z. K., Wert, S. E., Hull, W. M., Hu, X., Yei, S., Trapnell, B. C., and Whitsett, J. A. (1995). Persistance of replication-deficient adenovirus-mediated gene transfer in lungs of immune-deficient (nu/nu) mice. Hum.Gene Ther. *6*, 457-467.
126. Vilquin, J.-T., Guerette, B., Kinoshita, I., Roy, B., Goulet, M., Gravel, C., Roy, R., and Tremblay, J. P. (1995). FK506 immunosuppression to control the immune reactions triggerd by first-generation adenovirus-mediated gene transfer. Hum. Gene Ther. *6*, 1391-1401.
127. Fang, B., Eisensmith, R. C., Wang, H., Kay, M. A., Cross, R. E., Landen, C. N., Gordon, G., Bellinger, D. A., Read, M. S., Hu, P. C., Brinkhous, K. M., and Woo, S. L. C. (1995). Gene therapy for Hemophilia B: Host immunosuppression prolongs the therapeutic effect of adenovirus-mediated Factor IX expression. Hum. Gene Therap. *6*, 1039-1044.
128. DeMatteo, R. P., Markmann, J. F., Kozarsky, K. F., Barker, C. F., and Raper, S. E. (1996). Prolongation of adenoviral transgene exression in mouse liver by T lymphocyte subset depletion. Gene Therap. *3*, 4-12.
129. DeMatteo, R. P., Raper, S. E., Ahn, M., Fisher, K. J., Burke, C., Radu, A., Widera, G., Claytor, B. R., Barker, C. F., and Markmann, J. F. (1995). Gene transfer to the thymus; a means of abrogating the immune response to recombinant adenovirus. Annals of Surgery *222*, 229-242.

130. Kay, M. A., Holterman, A.-X., Meuse, L., Gwon, A., Ochs, H. D., Linsley, P. S., and Wilson, C. B. (1995). Long term hepatic adenovirus-mediated gene expression in mice following CTLA4Ig administration. Nature Genet. *11,* 191-196.
131. Kass-Eisler, A., Leinwand, L., Gall J., Bloom, B., and Falk-Pedersen, E. (1996). Circumventing the immune response to adenovirus-mediated gene therapy. Gene Ther. *3,* 154-162.
132. Tripathy, S. K., Black, H. B., Goldwasser, E., and Leiden, J. M. (1996). Immune response to transgene-encoded proteins limit the stablility of gene expression after injection of replication-defective adenovirus vectors. Nature Med. *2,* 545-550.
133. Kozarsky, K. F., Jooss, K., Donahee, M., Strauss III, J. F., and Wilson, J. M. (1996). Effective treatment of familial hyperchlestrolaemia in the mouse model using adenovirus-mediated transfer of the VLDL receptor gene. Nature Genet. *13,* 54-62.
134. Weinacker, A., Ferrando, R., Elliott, M., Hogg, J., Balmes, J., and Sheppard, D. (1995). Distribution of integrins αvβ6 and α9β1 and their known ligands fibronectin and tenascin in human airways. Am.J.Respir.Cell Mol.Biol. *12,* 547-557.
135. Goldman, M. J. and Wilson, J. M. (1996). Gradient of RGD dependent entry of adenoviral vector in nasal and intrapulmonary epithelia: implications of cystic fibrosis. *3,* 811-818.
136. Wickham, T. J., Mathias, P., Cheresh, D. A., and Nemerow, G. R. (1993). Integrins αvβ3 and αvβ5 promote adenovirus internalization but not virus attachment. Cell *73,* 309-319.
137. Wickham, T. J., Filardo, E. J., Cheresh, D. A., and Nemerow, G. R. (1994). Integrin αvβ5 selectively promotes adenovirus mediated cell membrane permeabilization. J.Cell Biol. *127,* 257-264.
138. Dupuit, F., Zahm, J.-M., Pierrot, D., Brezillon, S., Bonnet, N., Imler, J.-L., Pavirani, A., and Puchelle, E. (1995). Regenerating cells in human airway surface epithlium represent preferential targets for recombinant adenovirus. Hum. Gene Ther. *6,* 1185-1193.
139. Wickham, T. J., Carrion, M. E., and Kovesdi, I. (1995). Targeting of adenovirus penton base to new receptor through replacement of its RGD motif with other receptor-specific peptide motifs. *2,* 750-756.
140. Huang, S., Endo, R. I., and Nemerow, G. R. (1995). Upregulation of integrins αvβ3 and αvβ5 on human monocytes and T lymphocytes facilitates adenovirus-mediated gene delivery. J.Virol. *69,* 2257-2263.
141. McLachlan, G., Davidson, D. J., Stevenson, B. J., Dickinson, P., Davidson-Smith, H., Dorin, J. R., and Porteous, D. J. (1995). Evaluation *in vitro* and *in vivo* of cationic liposome -expression construct complexes for cystic fibrosis gene therapy. Gene Ther. *2,* 614-622.
142. Liu, Y., Liggitt, D., Zhong, W., Tu, G., Gaensler, K., and Debs, R. (1995). Cationic liposome-mediated gene delivery. J.Biol.Chem. *270,* 24864-24870.
143. Fasbender, A. J., Zabner, J., and Welsh, M. J. (1995). Optimization of cationic lipid-mediated gene transfer to airway epithelia. Am.J.Physiol. *269,* L45-L51.
144. Zhou, X. and Huang, L. (1994). DNA transfection mediated by cationic liposomes containing lipopolylysine: characterization and mechanisms of action. Bioch. Biophys.Acta *1189,* 195-203.
145. Wrobel, I. and Collins, D. (1995). Fusion of cationic liposomes with mammalian cells occurs after endocytosis. Biochem. Biophys.Acta *1235,* 296-304.
146. Zabner, J., Fasbender, A. J., Moninger, T., Poellinger, K. A., and Welsh, M. J. (1995). Cellular and molecular barriers to gene transfer by a cationic lipid. J.Bol.Chem. *270,* 18997-19007.
147. Bennett, M. J., Nantz, M. H., Balasubramaniam, R. P., Gruenert, D. C., and Malone, R. W. (1995). Cholesterol enhances cationic liposome-mediated DNA transfection of human respiratory epithelial cells. Biosci. Reports *15,* 47-53.
148. Raja-Walia, R., Webber, J., Naftilan, J., Chapman, G. D., and Naftilan, A. J. (1995). Enhancement of liposome-mediated gene transfer into vascular tissue by replication deficient adenovirus. Gene Ther. *2,* 521-530.
149. Farhood, H., Serbina, N., and Huang, L. (1995). The role of dioleoyl phosphatidylethanolamine in cationic liposome mediated gene transfer. Biochem. Biophys Acta *1235,* 289-295.
150. Remy, J.-S., Sirlin, C., Vierling, P., and Behr, J.-P. (1994). Gene transfer with a series of lipophilic DNA-binding molecules. Bioconjugate Chem. *5,* 647-654.
151. Boussif, O., Lezoualc'h, F., Zanta, M. A., Mergny, M. D., Scherman, D., Demeneix, B., and Behr, J. P. (1995). A versatile vector for gene and oligonucleotide transfer into cells in culture and in vivo: polyethyleneimine. Proc.Natl.Acad.Sci.USA *92,* 7297-7301.
152. Kamata, H., Yagisawa, H., Takahashi, S., and Hirata, H. (1994). Amphipathic peptides enhance the efficiency of liposome-mediated DNA-transfection. Nucl.Acids Res. *22,* 536-537.

153. Yoshimura, K., Rosenfeld, M. A., Seth, P., and Crystal, R. G. (1993). Adenovirus-mediated augmentation of cell transfection with unmodified plasmid vectors. J.Biol Chem. *268,* 2300-2303.
154. Cheng, P.-W. (1996). Receptor ligand-facilitated gene transfer: enhancement of lipososme mediated gene transfer and expression by transferrin. Human Gene Ther. *7,* 275-281.
155. Böttger, M., Vogel, F., Platzer, M., Kiessling, U., Grade, K., and Strauss, M. (1988). Condensation of vector DNA by the chromosomal protein HMG1 results in efficient transfection. Biochem. Biophys.Acta. *950,* 221-228.
156. Kaneda, Y., Iwai, K., and Uchida, T. (1989). Increased expression of DNA cointroduced with nuclear protein in adult rat liver. Science *243,* 375-378.
157. Yin, W. and Cheng, P.-W. (1994). Lectin conjugate-directed gene transfer to airway epithelial cells. Bioch. Biophys. Res.Commun. *205,* 826-833.
158. Baatz, J. E., Bruno, M. D., Ciraolo, P. J., Glasser, S. W., Stripp, B. R., Smyth, K. L., and Korfhagen, T. R. (1994). Utilization of modified surfactant-associated protein B for delivery of DNA to airway cells in culture. Proc. Natl.Acad. Sci. USA *91,* 2547-2551.
159. Ferkol, T., Perales, J. C., Eckman, E., Kaetzel, C. S., Hanson, R. W., and Davis, P. B. (1995). Gene transfer into the airway epithelium of animals by targeting the polymeric immunoglobulin receptor. J.Clin.Invest. *95,* 493-502.
160. Hart, S. L., Mayall, E., Munkonge, F. M., Frost, A., Huang, L., Vassiliou, M., Williamson R., Alton, E. W. F. W., and Coutelle, C. (1995). The introduction of two silent mutations into a CFTR cDNA construct allows improved detection of exogenous mRNA in gene transfer experiments. Hum. Molec.Genet. *4,* 1597-1602.
161. Stern, M., Munkonge, F. M., Caplen, N. J., Sorgi, F., Huang, L., Geddes, D. M., and Alton, E. W. F. W. (1995). Quantitative fluorescence measurements of choride secretion in native airway epithelium from CF and non-CF subjects. Gene Therap. *2,* 766-774.
162. Renier, M., Tamanini, A., Nicolis, E., Rolfini, R., Imler, J.-L., Pavirani, A., and Cabrini, G. (1995). Use of a membrane potential-sensitive probe to assess biological expression of the cystic fibrosis transmembrane regulator. Hum. Gene Therap. *6,* 1275-1283.
163. Cohn, J. A., Strong, T. V., Picciotto, M. R., Nairn, A. C., Collins F.S., and Fitz, J. G. (1993). Localization of the cystic fibrosis transmembrane regulator in human bile duct epithelial cells. Gastroenterology *105,* 1857-1864.
164. Yang, Y., Raper, S. E., Cohn, J. A., Engelhardt, J. F., and Wilson, J. M. (1993). An approach for treating the hepatobiliar disease of cystic fibrosis by somatic gene transfer. Proc.Natl.Acad.Sci. USA *90,* 4601-4605.
165. Ramakrishna, J. and Grand, R. J. (1994). Gene therapy for exocrine pancreatic insufficiency. Gastroenterology *106,* 1711-1713.
166. Vawter, G. F. and Shwachman, H. (1979). Cystic fibrosis in adults: an autopsy study. Pathol Ann. *14,* 357-382.
167. Douar, A.-M., Themis, M., and Coutelle, C. (1996). Fetal gene therapy. Human Molecul. Reprod. *2,* 633-641.
168. Walter, J., You, Q., Hagstrom, J. N., Sands, M., and High, K. A. (1996). Successful expression of factor IX following repreat administration of an adenoviral vector in mice. Proc.Natl.Acad.Sci. *93,* 3056-3061.
169. Ballard, P. L., Zepeda, M. L., Schwartz, M., Lopez, N., and Wilson, J. M. (1995). Adenovirus-mediated gene transfer to human fetal lung *ex vivo.* Am J. Physiol. *268,* L839-L845.
170. Douar, A.-M., Themis, M., Friedman, T., and Coutelle, C. (1996). Effect of amniotic fluid on cationic lipid mediated transfection and viral infection. Gene Therap. *3,* 789-796.
171. McCray, P. B., Armstrong, K., Zabner, J., Miller, D. W., Koretzky, G. A., Couture, L., Robillard, J. E., Smith, A. E., and Welsh, M. J. (1995). Adenoviral-mediated gene transfer to fetal pulmonary epithelia *in vitro* and *in vivo.* J.Clin.Invest. *95,* 2620-2632.
172. Vincent, M. C., Trapnell, B., Baughman, R. P., Wert, S. E., Whitsett, J. A., and Iwamoto, H. S. (1995). Adenovirus-mediated gene transfer to the respiratory tract of fetal sheep *in utero.* Hum. Gene Ther. *6,* 1019-1028.
173. Holzinger, A., Trapnell, B. C., Weaver, T. E., Whitsett, J. A., and Iwamoto, H. S. (1995). Intraamniotic administration of an adenovirus vector for gene transfer to fetal sheep and mouse tissue. Pediatr.Res. *38,* 844-850.
174. Sekhon, H. S. and Larson, J. E. (1995). *In utero* gene transfer into the pulmonary epithelium. Nature Med. *1,* 1201-1203.

175. Tsukamoto, M., Ochíya, T., Yoshida, S., Sugimura, T., and Terada, M. (1995). Gene transfer and expression in progeny after intravenous DNA injection into pregnant mice. Nature Genet. *9*, 243-248.

17. Gene Therapy for Dystrophin Deficiency

George Dickson

17.1 Background and Concepts

Duchenne muscular dystrophy (DMD) is an X-linked disease involving the progressive degeneration of skeletal muscle and was first described in 1868 by the Englishman Meryon (1). In recent years it has become apparent that DMD is in fact a complex disorder involving not only skeletal muscle but also heart, central and peripheral nervous system and smooth muscle tissues (2). The predominant pathology is however associated with necrosis of skeletal muscle fibres and subsequent phagocyte activity, leucocyte infiltration, interstitial fibrosis, fat deposition and disruption of normal skeletal muscle structure. The huge 3 Mb DMD gene contains at least 79 exons and eight tissue-specific promoter elements spanning ~2% of the X chromosome and encoding in skeletal muscle a 427kD membrane-associated protein called dystrophin (3, 4). Mutations which disrupt the gene reading frame result in the absence of dystrophin and yield the severe DMD phenotype, whereas deletions maintaining the gene reading frame result in milder allelic disease termed Becker muscular dystrophy (BMD). One interesting feature of DMD is that more than two-thirds of patients actually exhibit a small proportion of so-called dystrophin-positive revertant myofibers in which exon skipping or other mechanisms have restored the reading frame. In vitro studies have recently demonstrated splicing modulation of mutated dystrophin transcripts by chemically-modified synthetic antisense oligoribonucleotides resulting in restoration of dystrophin mRNA open reading frames (5). In addition to classic gene replacement therapies this type of oligonucleotide pharmaceutical might offer strategies to specifically restore correct open reading frames in mutated dystrophin gene transcripts in selected cases of DMD patients.

In common with a number of other tissues, the DMD gene product in muscle is now known to associate with the myofibre sarcolemma via interactions with cortical actin microfilaments and with a transmembrane protein/glycoprotein complex linked to the extracellular matrix (6, 7). The dystrophin-associated glycoprotein complex (DAGC) includes the extracellular proteoglycan, α-dystroglycan, which binds

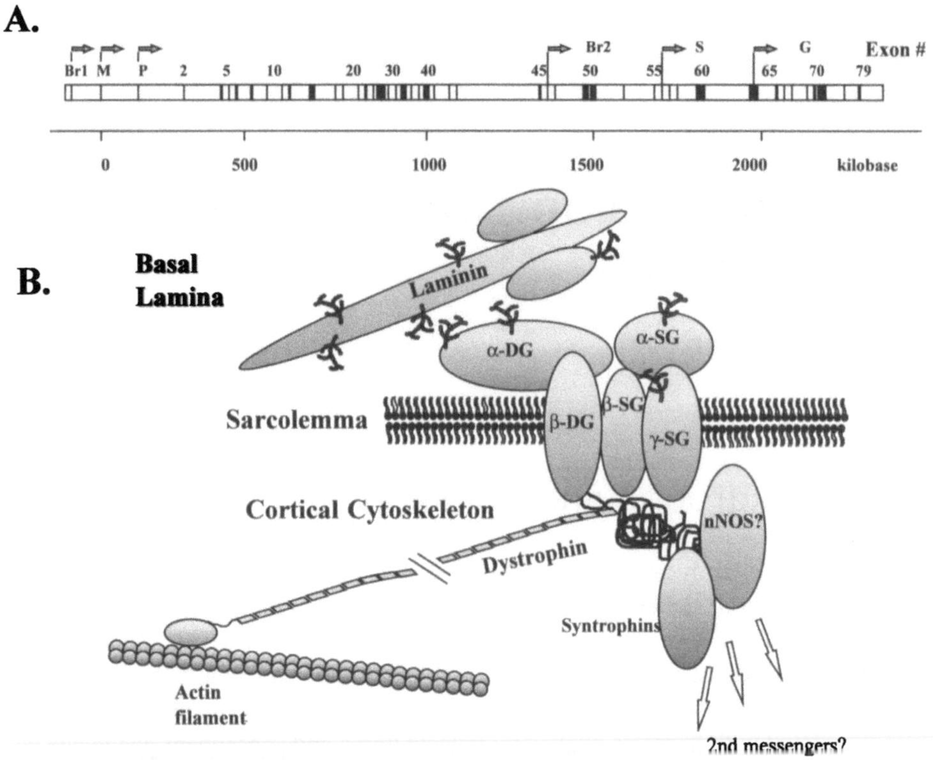

Fig. 17.1: Schematic representations of the DMD gene and the dystrophin-associated glycoprotein complex.

A. A linear map of the human DMD gene indicating exon positions and promoter elements active in muscle (M), brain (Br1 and 2), Purkinje cells (P), Schwann cells (S) and CNS glial cells (G).

B. Model of the dystrophin-associated glycoprotein complex showing the relationships between the dystroglycan (DG) and sarcoglycan (SG) components as well as syntrophins and the neuronal-type nitric oxide synthetase (nNOS).

to merosin in myofibre basal laminae, as well as a number of other integral and cytoplasmic membrane proteins: β-dystroglycan; α-, β- and γ-sarcoglycans; syntrophin and nitric oxide synthetase isoforms. The DAGC is thought to provide a physical link and may mediate a signalling pathway between the extracellular matrix and the sarcolemmal cytoskeleton whose disruption is now known to be associated with a variety of human diseases in addition to DMD and BMD: the absence of merosin can cause congenital muscular dystrophy, and various autosomal recessive and limb-girdle muscular dystrophies have now been shown to involve sarcoglycan mutations (7).

Table 17.1 Comparison of Dystrophin cDNA expression in transgenic *mdx* mice

	Full-length dystrophin	Becker-type dystrophin
Promoter elements	skeletal actin/ MCK	skeletal actin/ MCK
Levels effective	> 20% of wild-type	> 30% of wild-type
Generalised necrosis	eliminated	eliminated
Diaphragm necrosis	eliminated	highly reduced
Serum creatine kinase	normalised	normalised
Force generation	normalised	normalised
Cardiovascular effects	unclear	unclear

17.2 Transgenic Animal Studies

The recent clinical and molecular observations on the dystrophinopathies and the DAGC have underlined the potential importance of restoring biochemical linkage between the myofibre cytoskeleton and the extracellular matrix if gene therapy of DMD is to be effective. Transgenic animal studies using dystrophin-deficient *mdx* mice have confirmed this hypothesis and the therapeutic efficacy of a variety of different dystrophin minigene constructs (Tab. 17.1). Despite early reports to the contrary the *mdx* mouse has proved to be an excellent model for DMD. Young *mdx* mice initially show histopathological changes similar to those in early DMD, variation in fiber size, central nucleation, cellular infiltration and phagocytosis. At later stages in much older animals gross interstitial fibrosis, myofibre loss and muscle weakness becomes clearly apparent. Human dystrophin expression in transgenic *mdx* skeletal muscle results in a complete correction of the biochemical and pathological phenotype of these animals (8, 9), and interestingly, massive overexpression of dystrophin shows no obvious toxic effects (10). While these data strongly suggest that dystrophin gene replacement may be an effective strategy to treat DMD, the size of the wild-type dystrophin cDNA has precluded its insertion into common recombinant viral vectors (11). Alternative smaller recombinant dystrophin minigenes have therefore been isolated based on BMD phenotypes. In particular England et al. (12) reported a case of an extremely mild BMD patient expressing an almost half-sized 229kD dystrophin deleted for exons 17 through 48. To evaluate the therapeutic potential of this minidystrophin, the corresponding 6.3kb cDNA was expressed in transgenic *mdx* mice and yielded a high degree of protection with restoration of the DAGC at the sarcolemma (9,13). One interesting observation here has been that ~20-30% of normal levels of recombinant dystrophin expression can virtually normalise muscle pathology, a value in line with studies on mosaic DMD carriers. Given that some patients already express 2-5% of revertant dystrophin it is apparent that screening of chemical libraries for activators of the dystrophin promoter could provide the basis of another alternative gene therapeutic strategy for DMD.

Table 17.2 Human mini-dystrophin expression after intramuscular transfer in *mdx* mice

	PLASMID DNA	RETROVIRUS	ADENOVIRUS
Dose/muscle	400mg	10^4 cfu	10^9 pfu
Optimal age	4-6 weeks	4-6 weeks	neonate
Target cells	myofibres	myoblasts	myofibres
DNA form	episomal	integrated	episomal
Dys+ *mdx* fibres	1–5%	3–12%	5–50%
Longevity	> 6 months	> 9 months	> 6 months
Functional	Yes	Yes	Yes
1° Antibody	No	Yes (mice)	Yes
CTL Response	No	No	Yes

17.3 DNA-Mediated Gene Transfer

In reference to somatic gene therapy the skeletal musculature in man can constitute over 50% of the body mass and a range of anatomical, physiological and cellular features has made this tissue an attractive candidate for gene transfer procedures. Skeletal muscle is a stable postmitotic tissue with high regenerative capacity, good vascularity and easy access. In addition to DMD the transfer of genetic material to skeletal muscle has been proposed for other inherited neuromuscular diseases, for inborn errors of metabolism, for common acquired disorders such as diabetes mellitus, hyperlipidaemia and cardiovascular disease, and for vaccination against a number of infectious agents (14). The simplest, most inexpensive and possibly the safest muscle gene transfer procedure is the direct administration of pure DNA to skeletal muscle in vivo by direct intramuscular injection of supercoiled plasmid (14, 15) (Tab. 17.2). Although the efficiency is low, injected DNA remains extrachromosomal, the post-mitotic nature of muscle fibres allows long-term expression and functional dystrophin minigenes have been transferred to *mdx* skeletal muscle by this means. If the efficiency of DNA-mediated muscle gene transfer could be enhanced via complexing with polycations, receptor-mediated targetting proteins or chemically-inactivated viruses (16, 17, 18) this approach could provide an important therapeutic strategy. Intense interest has also focused on the use of cationic liposomes as vehicles for the transfer of recombinant genes into a variety of tissues both in-vitro and in-vivo (19, 20, 21). Zhu and colleagues have recently reported (20) that a single systemic injection of an expression plasmid/cationic liposome complex into adult mice efficiently transfected many different tissues. In a recent report, Trivedi and Dickson (21) showed efficient transfer of dystrophin cDNA into primary muscle cultures using liposome vehicles.

17.4 Adenoviral Vector Systems

Genetically-engineered human adenoviruses have been proposed as particularly attractive vectors for DMD gene therapy since they can infect quiescent myofibres as well as dividing myoblast cells and can be purified and concentrated to high titre making systemic clinical administration feasible (22, 23, 24) (Fig. 17.2A). The adenoviral genome is relatively easy to manipulate, does not normally undergo rearrangements and rarely integrates into the host chromosome. The present generation of adenoviral vectors have however a tight size limit for DNA packaging of approximately 105% of wild type. Sharp reductions in viral yield and stability are observed above this level (25, 26). With replication-defective adenoviral vectors deleted for E1, E3 and parts of the E4 regions the capacity for foreign DNA is restricted to about 7.8kb and cannot accommodate full-length dystrophin minigenes. However the 6.3kb human minidystrophin cDNA based on a BMD phenotype (see above) has been used to construct E1/E3-deleted recombinant adenovirus vectors (22, 23) (Dickson et al., unpublished data). In vivo studies involving intramuscular injections into neonatal *mdx* mice have shown that these adenovirus vectors can efficiently direct the synthesis of significant amounts of the 229kD minidystrophin with 5-50% of fibres showing correct sarcolemmal localisation (Tab. 17.2). In one study dystrophin-transduced muscle showed only ~12% regeneration compared with over 60% in age-matched control *mdx* muscle clearly demonstrating functional efficacy. Under these conditions dystrophin expression was reported up to six months after virus injection with no adverse histopathological alterations or cytotoxic immune responses (24). The route of adenovirus administration is a major factor in terms of transduction efficiencies to different tissues. Following systemic administration of a recombinant adenovirus vector expression of transgenes in a variety of different organs and tissues is generally found. Kass-Eisler and colleagues (27) have reported that the most effective and widespread method of gene transfer was via intracardiac injection with equivalent efficiencies as via the intramuscular route. Infectivity of muscle and other organs is also very dependant on the maturity of the tissue and/ or age of the treated animal. Immature or regenerating muscle is efficiently transduced but mature muscle fibres are not, even at high titres (28, 29) (Dickson et al., unpublished data). This may correlate with surface density of available adenovirus internalisation receptors but immunological factors also play a dominant role. Significantly increased efficiencies of minidystrophin adenoviral infection is observed in adult *mdx* mice if abundant muscle regeneration is present and in particular if immunodeficient animals are used (Dickson et al., unpublished data). Second generation adenovirus vectors are being designed to ablate expression of immunodominant viral proteins focusing on the E2a gene (30) and evidence suggests that gp19 encoded by the E3 region is important in avoidance of the host immune surveillance (31). Furthermore, in terms of clinical safety, the deletion of more than one essential regulatory element in viral vectors would be advantageous to minimise recombination events leading

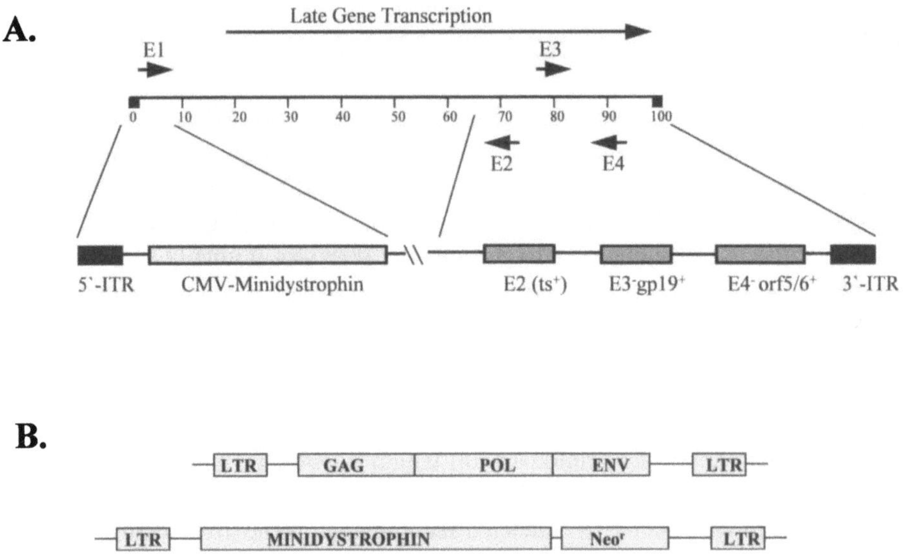

Fig. 17.2: Adenoviral and retroviral vectors carrying minidystrophin gene constructs.

A. Adenovirus-based gene transfer vectors. A simplified transcriptional map of the 36kb adenovirus genome is shown conventionally divided into 100 map units. Adenovirus exhibits an exceptionally complex and regulated transcriptional cascade involving multiple alternative splicing events. Transcription of E1 products activates all the other early gene regions and late transcription then occurs following replication of the viral genome. Recombinant adenoviral vectors often incorporate foreign DNA into the deleted E1 region to yield replication-defective particles. Alterations in E2, E3 and E4 regions to increase vector capacity or further inhibit relication potential have also been utilised.

B. Retrovirus-based gene transfer vectors. The genome of murine leukaemia virus (MLV) is represented showing the viral long terminal repeats (LTRs) which direct replication, integration and packaging, and the GAG, POL and ENV structural genes which encode core and envelope proteins and reverse transcriptase/integrase activities. In MLV-based vectors GAGPOL and ENV regions are replaced by foreign DNA and retroviral plasmid constructs are used to transfect packaging cells which supply the viral proteins in trans and release replication-defective retroviral vector particles.

to replication-competant adenovirus during production scale-up. Presently, a number of laboratories are working towards production of vectors which are gp19+ and E1, E3 and E4 defective (32, 33, 34). Furthermore attempts are underway to develop vectors in which much larger portions or indeed the entire adenoviral structural gene complement has been deleted leaving only minimal sequence elements necessary for replication and packaging (35) (Dickson et al., unpublished data). Such vectors have been proposed to potentially provide up to 35kb insert capacity which would allow the use of complex promoter elements and linked reporter genes, and for example the full-length dystrophin cDNA. In addition minimal vector systems would eliminate secondary cell-mediated immune responses to infected cells. At present such systems require so-called helper virus for capsid assembly and packaging but complementation strategies using packaging-defective viral genomes are being tested.

17.5 Retroviral Vector Systems

Recombinant viruses based on mouse C-type retroviruses are another vector system used for gene transfer in animal and human muscle cells (36) (Fig. 17.2B). Here the retroviral genome is stably integrated into host cell DNA and has the potential for indefinite expression of the recombinant protein. These vectors are genetically engineered so that all proteins necessary for viral replication can be provided in trans, generating infectious particles that cannot produce immunogenic viral products once infection has occurred. The main limitations of retroviral vectors are again the insert capacity (9-10kb), low titre, instability and the fact that only mitotic cells can be transduced. Retroviral vectors have been used to infect muscle cells in vitro and expression of the transgene has been detected after myoblast fusion into myotubes, indicating that profound changes in cell transcription pattern do not necessarily affect retroviral gene expression (37, 38). Following this initial in-vitro study performed with a reporter gene, a recombinant retrovirus carrying the 229kD dystrophin cDNA has been constructed and used to infect *mdx* myoblasts, resulting in the expression and correct localization of the protein at the sarcolemma of the skeletal muscle myotubes (39). Studies in various mouse models have shown that myoblasts transduced in vitro with retroviral vectors are able to fuse with pre-existing myofibers when transplanted in vivo and have been successfully applied in a mouse model for ex-vivo gene therapy for DMD (40, 41). However the expression of the recombinant gene after implantation in vivo is somehow transient in primary myoblasts and sustained expression has been obtained only using heterologous enhancer-promoter sequences (42).

Direct in-vivo administration of recombinant retrovirus has also been attempted (43). Although post-mitotic myofibres are refractory to retroviral infection, satellite cells activated during the process of muscle regeneration can be targeted (Tab. 17.2). Direct injection of viral particles into regenerating skeletal muscle results however in relatively low efficiency of transduction, possibly due to the limited temporal availability and interference by immune factors which reduce further the already short half-life of retroviruses in vivo (44). Some of these problems have been overcome by direct implantation of mitotically-inactivated retroviral producer cells into regenerating skeletal muscle, resulting in a much higher efficiency of transduction and in the infection of satellite cells which contribute to new fibre formation (44, 45). Compared to the ex-vivo approach, direct in-vivo infection of satellite cells by retroviral vectors results in long-term expression of the recombinant gene and insertion of muscle specific enhancers within the wild type enhancer may result in a muscle specific transcription of the vector (46). Other major issues still remain to be solved before retroviral-mediated gene transfer to skeletal muscle can be attempted in humans. With regard to the ex-vivo approach, long-term expression of the recombinant protein after myoblast transplantation is necessary as well as the capability of growing clinically useful quantities of fusion-competent myoblasts. With regard to the in-vivo approach, the efficiency of transduction needs to be further improved and animal models such as xmd dogs in which the rate of spontaneous muscle degeneration-regeneration is probably more close to DMD need to be carefully examined. Until recently one major obstacle to direct in vivo retroviral gene therapy has been the sensitivity of virions to human complement-mediated lysis (47). However retroviral packaging cells of human origin and expressing RD114 feline leukaemia virus envelope glycoproteins instead of the usual amphotropic or ecotropic glycoproteins yield retroviral pseudotypes resistant to inactivation by human complement (48, 49), and drugs or monoclonal antibodies able to bind and inactivate specific complement factors are available (50, 51). Finally a recent study has suggested that blood-borne macrophage/monocyte cells target rapidly to sites of sporadic regeneration in dystrophic muscle (52). Such cells engineered in vitro to produce minidystrophin retroviral vectors might provide a systemic infusion vehicle for widespread targetting of gene transfer to pathological skeletal muscles in vivo.

17.6 Future Perspectives

Although there are many hurdles which have as yet to be overcome on the route to developing acceptable gene therapy protocols targetted to the skeletal musculature, it is encouraging to note the speed at which progress is being made. It is unlikely that a single delivery system will provide the perfect panacea for gene therapy however advances in viral vector development provide hope for future treatment protocols. A better understanding of basic muscle cell biology is also clearly necessary for the development and improvement of therapeutic reagents and treatment strategies, and issues of stability of expression, immune responses and viral mobilisation need to be carefully examined. Finally, it must be emphasized that much of the *in vivo* work published to date has pertained to mouse model systems and it will be essential to reproduce and extend these studies in large animal models before plans for human clinical trials can be sensibly put forward and considered.

Acknowledgements

This work is supported by the Muscular Dystrophy Group of Great Britain and Northern Ireland, the Leopold Muller Trust, the Medical Research Council, the Genzyme Corporation and Boehringer Mannheim.

Abbreviations:

BMD, Becker Muscular dystrophy
DMD, Duchenne muscular dystrophy
pfu, plaque-forming units.

References

1. Meryon, E. (1852). On granular and fatty degeneration of voluntary muscle. Med.-Chir. Trans. (London) *35*, 73-84.
2. Emery, A. E. H. (1993). Duchenne Muscular Dystrophy (2nd edn.), Oxford Monographs on Medical Genetics 24. Oxford: Oxford University Press.
3. Ahn, A. H. and Kunkel, L. M. (1993). The structural and functional diversity of dystrophin. Nature Genet. *3*, 283-291.
4. Brown, S. C. and Dickson, G. (1995). Duchenne Muscular Dystrophy. In: Molecular and Cell Biology of Human Gene Therapeutics, G. Dickson (ed.). London: Chapman and Hall, 261-280.
5. Takeshima, Y., Nihio, H., Sakamoto, H., Nakamura, H., and Matsuo, M. (1995). Modulation of in vitro splicing of the upstream intron by modifying an intraexon sequence which is deleted from the dystrophin gene in dystrophin Kobe. J. Clin. Invest. *95*, 515-520.
6. Campbell, K. P. (1995). Three muscular dystrophies: loss of cytoskeleton-extracellular matrix linkage. Cell *80*, 675-679.
7. Worton, R. (1995). Muscular dystrophies: diseases of the dystrophin-glycoprotein complex. Science *270*, 755-756.
8. Wells, D.J., Wells, K.E., Walsh, F.S., Davies, K.E., Goldspink, G., Love, D.R., Chan-Thomas, P., Dunckley, M.G., Piper, T., and Dickson, G. (1992). Human dystrophin expression corrects the myopathic phenotype in transgenic *mdx* mice. Hum. Mol. Genet. *1*, 35-40.
9. Wells, D. J., Wells, K. E., Asante, E. A., Turner, G., Sunada, Y., Campbell, K. P., Walsh, F. S., and Dickson, G. (1995). Expression of human full-length and minidystrophin in transgenic *mdx* mice: implications for gene therapy of Duchenne Muscular Dystrophy. Hum. Mol. Genet. *4*, 1245-1250.
10. Cox, G.A., Cole, N.M., Matsumura, K., Phelps, S.F., Hauschka, S.D., Campbell, K.P., Faulkner, J.A., and Chamberlain, J.S. (1993). Overexpression of dystrophin in transgenic *mdx* mice eliminates dystrophic symptoms without toxicity. Nature *364*, 725-729.
11. Dickson, G., Love, D.R., Davies, K.E., Wells, K.E., Piper, T.A., and Walsh, F.S. (1991). Human dystrophin gene transfer: production and expression of a functional recombinant DNA-based gene. Hum. Genet. *88*, 53-58.
12. England, S.B., Nicholson, L. V. B., Johnson, M. A., Forrest, S. M., Love, D. R., Zubrzycka-Gaarn, E. E., Bulman, D. E., Harris, J. B., and Davies, K. E. (1990). Very mild muscular dystrophy associated with deletion of 46% of dystrophin. Nature *343*, 180-182.
13. Phelps, S. E., Hauser, M. A., Cole, N. M., Rafael, J. A., Hinkle, R. T., Faulkner, J. A., and Chamberlain, J. S. (1995). Expression of full-length and truncated dystrophin mini-genes in transgenic *mdx* mice. Hum. Mol. Genet. *4*, 1251-1258.
14. Dickson, G. (1995). Molecular and Cell Biology of Human Gene Therapeutics. London: Chapman and Hall.
15. Acsadi, G., Dickson, G., Love, D. R., Jani, A., Walsh, F. S., Gurusinghe, A., Wolff, J. A., and Davies, K. E. (1991). Human dystrophin expression in *mdx* mice after intramuscular injection of DNA constructs. Nature *352*, 815-818.
16. Kawai, S. and Nishizawa, N. (1984). A new procedure for DNA transfection with polycations and DMSO. Mol. Cell Biol. *4*, 1172-1174.
17. Wu, G. Y., Wilson, J. M., and Wu, G. Y. (1989). Targetting genes: delivery and persistent expression of a foreign gene driven by mammalian regulatory elements. J. Biol. Chem. *269*, 16985-16987.
18. Wagner, E., Zatloukal, K., Cotten, M., Kirlappos, H., Mechtler, K., Curiel, D. T., and Birnstiel, M.L. (1992). Coupling of adenovirus to transferrin-polylysine/DNA complexex greatly enhances

receptor-mediated gene delivery and expression of transfected genes. Proc. Natl. Acad. Sci. USA. *89*, 6099-6103.

19. Alton, E. W. F. W., Middleton, P. J., Caplen, N. J., Smith, S. N., Steel, S., Munkonge, F. M., Jeffrey, P. K., Geddes, D. M., Hart, S. L., Williamson, R., Fasold, K. I., Miller, A. D., Dickinson, P., Stevenson, B. J., McLachlan, G., Dorin, J. D., and Porteous, D. J. (1993). Non-invasive liposome-mediated gene delivery can correct the ion transport defect in cystic fibrosis mutant mice. Nat. Genet. *5*, 135-142.
20. Zhu, N., Liggitt, D., Liu, Y., and Debs, R. (1993). Systemic gene expression after intravenous DNA delivery into adult mice. Science. *261*, 209-211.
21. Trivedi, R.A. and Dickson, G. (1995). Liposome-mediated gene transfer into normal and dystrophin-deficient mouse myoblasts. J. Neurochem. *64*, 2230-2238.
22. Ragot, T., Vincent, N., Chafey, P., Vigne, E., Gilgenkrantz, H., Couton, D., Cartaud, J., Briand, P., Kaplan, J-C., Perricaudet, M., and Kahn, A. (1993). Efficient adenovirus-mediated transfer of a human minidystrophin gene to skeletal muscle of *mdx* mice. Nature *361*, 647-650.
23. Acsadi, G., Massie, B., and Jani, A. (1995). Adenovirus-mediated gene transfer into striated muscles. J. Mol. Med. *73*, 165-180.
24. Vincent, N., Ragot, T., Gilgenkrantz, H., Couton, D., Chafey, P., Gregoire, A., Briand, P., Kaplan, J-C., Kahn, A., and Perricaudet, M. (1993). Long-term correction of mouse dystrophic degeneration by adenovirus-mediated transfer of a minidystrophin gene. Nature Genet. *5*, 130-134.
25. Bett, A. J., Prevec, L., and Graham, F.L. (1993). Packaging capacity and stability of adenovirus type 5 vectors. J. Virol. *67*, 5911-5921.
26. Bett, A. J., Haddara, W., Prevec, L., and Graham, F. L. (1994). An efficient and flexible system for construction of adenovirus vectors with inserts or deletions in early regions 1 and 3. Proc. Natl. Acad. Sci. USA. *91*, 8802-8806.
27. Kass-Eisler, A., Falck-Pedersen, E., Elfenbein, D. H., Alvira, M., Buttrick, P. M., and Leiwand, L. A. (1994). The impact of developmental stage, route of administration and the immune system on adenovirus-mediated gene transfer. Gene ther. *1*, 395-402.
28. Acsadi, G., Jani, A., Massie, B., Simoneau, M., Holland, P., Blaschuk, K., and Karpati, G. (1994). A differential efficiency of adenovirus-mediated *in vivo* gene transfer into skeletal muscle cells of different maturity. Hum. Mol. Genet. *3*, 579-584.
29. Acsadi, G., Jani, A. Huard, A., Blaschuk, K., Massie, B., Holland, P., Lochmüller, H., and Karpati, G. (1994). Cultured human myoblasts and myotubes show markedly different transducibility by replication-defective adenovirus recombinants. Gene Therapy. *1*, 338-340.
30. Yang, Y., Nunes, F. A., Berensci, K., Gönczol, E., Engelhardt, J. F., and Wilson, J. M. (1994). Inactivation of E2a in recombinant adenoviruses improves the prospect for gene therapy for cystic fibrosis. Nat. Genet. *7*, 362-369.
31. Gooding, L. R. (1992). Virus proteins that counteract host immune systems. Cell, *71*, 5-7.
32. Kremer, E. J. and Perricaudet, M. (1995). Adenovirus and adeno-associated virus mediated gene transfer. Brit. Med. Bull. *51*, 31-44.
33. Armentano, D., Sookdeo, C. C., Hehir, K. M., Gregory, R. J., St.George, J. A., Prince, G. A., Wadsworth, S. C., and Smith, A. E. (1995). Characterisation of an adenovirus gene transfer vector containing an E4 deletion. Hum. Gene Ther. *6*, 1343-1353.
34. Krougliak, V. and Graham, F.L. (1995). Development of cell lines capable of complementing E1, E4 and protein IX defective type 5 mutants. Hum. Gene Ther. *6*, 1575-1568.
35. Mitani, K., Graham, F. L., Caskey, C. T., and Kochanek, S. (1995). Rescue, propagation, and partial purification of a helper virus-dependent adenovirus vector. Proc. Natl. Acad. Sci. USA *92*, 3854-3858.
36. Fassati, A., Dunckley, M. G., and Dickson, G. (1995). Retroviral Vectors. In: Molecular and cell biology of human gene therapeutics, G. Dickson (ed.). London: Chapman and Hall, 1-19.
37. Dhawan J., Pan L. C., Pavlath G. K., Travis M. A., Lanctot A. M., and Blau H. M (1991). Systemic delivery of human growth hormone by injection of genetically engineered myoblasts. Science *254*, 1509-1512.
38. Smith, B. F., Hoffman, R. K., Giger, U., and Wolfe, J. H. (1990). Genes transferred by retrovirus vectors into normal and mutant myoblasts in primary cultures are expressed in myotubes. Mol. Cell. Biol. *10*, 3268-3271.

39. Dunckley, M. G., Love, D. R., Davies, K. E., Walsh, F. S., Morris, G. E., and Dickson, G. (1992). Retroviral-mediated transfer of a dystrophin minigene into *mdx* mouse myoblasts *in vitro*. FEBS Lett. *296*, 128-134.
40. Naffakh, N., Pinset, C., Montarras, D., Pastoret, C., Danos, O., and Heard, J. M. (1994). Transplantation of adult-derived myoblasts in mice following gene transfer. Neuromusc Disorders *3*, 413-417.
41. Salvatori, G., Ferrari G., Mezzogiorno, A., Servidei, S., Coletta, M., Tonali, P., Giavazzi, R., Cossu, G., and Mavilio, F. (1993). Retroviral vector-mediated gene transfer into human primary myogenic cells leads to expression in muscle fibers *in vivo*. Hum. Gene Ther. *4*, 713-722.
42. Dai, Y., Roman M., Naviaux, R. K., and Verma, I. M (1992). Gene therapy via primary myoblasts: Long-term expression of factor IX protein following transplantation *in vivo*. Proc. Natl. Acad. Sci. USA. *89*, 10892-10895.
43. Dunckley, M.G., Wells, D.J., Walsh, F.S., and Dickson, G. (1993). Direct retroviral-mediated transfer of a dystrophin minigene into *mdx* mouse muscle *in vivo*. Hum. Mol. Genet. *2*, 717-723.
44. Fassati, A., Wells, D. J., Walsh, F. S., and Dickson, G. (1995). Efficiency of *in vivo* gene transfer using murine retroviral vectors is strain-dependent in mice. Hum. Gene Ther. *6*, 1177-1183.
45. Fassati, A., Wells, D. J., Walsh, F. S., and Dickson, G. (1996). Recombinant retroviruses are safe and efficient vectors for long-term *in vivo* gene transfer into mouse skeletal muscle. Hum. Gene Ther.
46. Ferrari, G., Salvatori, G., Rossi, C., Cossu, G., and Mavilio, F. (1995). A retroviral vector containing a muscle-specific enhancer drives gene expression only in differentiated muscle fibers. Hum. Gene Ther. *6*, 733-742.
47. Welsh, R. M., Cooper, N. R., Jensen, F. C., and Oldstone, M. B. A. (1975). Human serum lyses RNA tumour viruses. Nature *257*, 612-614.
48. Takeuchi, Y., Cosset, F. L., Lachmann, P. J., Okada, H., Weiss, R. A., and Collins, M. K. L. (1994). Type-C retrovirus inactivation by human complement is determined by both the viral genome and the producer cell. J. Virol., *68*, 8001-8007.
49. Cosset, F. L., Takeuchi, Y., Battini, J. L., Weiss, R. A., and Collins, M. K. L. (1995). High-titer packaging cells producing recombinant retroviruses resistant to human complement. J. Virol. *69*, 7430-7436.
50. Rother, R. P., Squinto, S. P., Mason, J. M., and Rollins, S. A. (1995). Protection of retroviral vector particles in human blood through complement inhibition. Hum. Gene Ther. *6*, 429-435.
51. Matis, L. A. and Rollins, S. A. (1995). Complement-specific antibodies: designing novel anti-inflammatories. Nat. Med. *8*, 839-842.
52. Parrish, E. R., Cifuentes-Diaz, C., Li, Z.-L., Vicart, P. Paulin, D., Dreyfus, P. A., Peschanski, M., Harris, J. A., and Garcia, L. (1996). Targetting widespread sites of damage in dystrophic muscle: erngrafted macrophages as potential shuttles. Gene Ther. *3*, 13-21.

18. Gene Therapy For Familial Hypercholesterolemia

Michael Gotthardt and Herbert Schuster

18.1 Introduction

Familial Hypercholesterolemia (FH) is an inherited disorder of lipid metabolism characterized by an elevated concentration of LDL due to reduced LDL receptor activity, and deposition of cholesterol in tendons, skin and arteries. With a prevalence of 1:500 it is one of the most common inherited diseases. The wide variety in clinical expression ranges from myocardial infarction at as early as 6 months of age to lack of clinical manifestations. FH necessitates therapeutic interventions which range from dietary fat restriction to surgery and plasma apheresis. For severely affected patients there is a great need for a generally available treatment that does not require weekly invasive sessions as for example apheresis, or major surgery and continued immunosupression, as in liver transplantation.

Gene therapy has the potential to meet these requirements in the near future. Since FH is a monogenetic disease, the cDNA coding for the LDL receptor gene is available, considerable experience in gene transfer to the liver has been gathered, and different animal models are available and characterized to optimize gene transfer and expression. Data from animal experiments, as well as results of the clinical protocol for *ex vivo* gene therapy of FH, clearly demonstrate that we are still at the beginning of the process of improving vector systems and understanding and interfering with the reactions of the organism against gene delivery and expression. The aim of this chapter is to introduce clinical features and molecular mechanisms of the disease and possible diagnostic and therapeutic procedures with emphasis on gene therapy including feasibility, evaluation of different gene delivery systems available, necessary supportive measures, and future directions.

18.2 Familial Hypercholesterolemia (FH)

18.2.1 Definition

FH is an autosomal dominant disorder characterized by an elevation in serum cholesterol bound to low-density lipoprotein (LDL) and is therefore one of the conditions producing the hyperlipoproteinemia type II phenotype. FH is caused by mutations and/or deletions in the LDL receptor (LDLR) gene on chromosome 19.

18.2.2 Phenotype

Heterozygotes develop tendon xanthomas, corneal arcus, and coronary artery disease; the last usually becomes evident in the fourth or fifth decade of life. Homozygotes develop these features at an accelerated rate, with coronary artery disease before 10 years of age resulting in death before the age of 30 years. Homozygotes are sometimes recognized by the presence of plantar xanthomas located between the great and second toes.

18.2.2.1 Biochemistry

Normally, LDL is bound at the cell membrane and taken into the cell, ending up in lysosomes where the protein is degraded and the cholesterol is made available by cholesterol ester hydrolysation. The process results in the suppression of microsomal enzyme, 3-hydroxy-3-methylglutaryl coenzyme A (HMG CoA) reductase, which is the rate-limiting step in cholesterol de novo synthesis. At the same time, a reciprocal stimulation of cholesterol ester synthesis, via acyl-coenzyme A cholesterol acyltransferase (ACAT), takes place. When cholesterol is not internalized, HMG CoA reductase is not suppressed and hypercholesterolemia results.

The LDL receptor binds two ligands, apoB-100 the sole protein of LDL and apoE, found in VLDL, IDL, chylomicron remnants and subclasses of HDL. Receptor-independent removal of LDL is possible and accounts for 30% of deposition in normal subjects. Receptor-independent removal is less efficient than the receptor-dependent pathway, and probably involves the liver as well as extrahepatic tissues in equal parts. Macrophages are involved and seem to help eliminate LDL via the macrophage scavenger receptor, binding acetylated and oxidized LDL. Inside the cell, cholesterol of lysosomal or nonlysosomal origin is stored after esterification by ACAT in lipid droplets, from which the cholesterol can be released by neutral cholesterol ester hydrolase (NCEH) for synthesis of steroid hormones or vitamin D, transport to membranes, incorporation into lipoprotein particles, or degradation. Hepatic conversion of cholesterol into bile acids involves more than 15 different enzymes localized in the endoplasmic reticulum and the cytosol that are responsible for converting the steroid nucleus into a more polar compound, and mitochondrial and peroxisomal enzymes, to degrade the steroid side chain.

18.2.2.2 Genomic- and Protein Structure

The LDL receptor is synthesized as a 120 kd glycoprotein precursor that undergoes change to a 160 kd mature glycoprotein through the addition of oligosaccharides. The LDL receptor consists of a cysteine-rich ligand binding domain, an EGF precursor homology domain with N-linked sugars, the O-linked sugar domain, a transmembrane domain containing 22 hydrophobic amino acids, and a cytoplasmic domain with a signal sequence for localizing the receptor in coated pits, as well as one for targeting the receptor to sinusoidal surface of polarized hepatocytes. The gene for the LDL receptor was mapped to 19p13.1-p13.3 by in situ hybridization (Lindgren et al., 1986), is more than 45 kb long, and contains 18 exons, most of which correlate with functional domains previously defined at the protein level. Of the 18 exons, 13 encode protein sequences that are homologous to sequences in other proteins such as the C9 component of complement, the precursor for EGF, and three proteins involved in the blood clotting system, namely factor IX, factor X, and protein C.

18.2.2.3 Regulation of Gene Expression

Regulation of LDL-Receptor expression occurs at transcriptional level. A 10 bp sequence, the sterol response element (SRE), contained in the promoter of the LDL-receptor as well as HMG-CoA reductase, mediates cholesterol responsiveness of transcription by interaction with SRE binding proteins (SREBP-1 and -2). These appear to be ER-membrane bound transcription factors. The protease SCA (SREBP cleavage activity) can release the transcriptionally active domain if the cleavage site is accessible. This process is probably dependent on cholesterol composition of the ER-membrane (Wang et al., 1994; 1995). Lack of intracellular cholesterol would in that way lead to a lower cholesterol content of the ER-membrane with better accessibility of the SREBP cleavage site. After cleavage, the transcriptionally active domain can bind the sterol response elements and induce transcription of the LDL receptor and HMG-CoA reductase.

18.2.3 Genotype

Both receptor-absent and receptor-defective mutants occur. There is also an internalization mutant of the LDL receptor that binds LDL but is unable to facilitate passage of LDL to the inside of the cell (Anderson et al., 1977). Class I mutations (null alleles) lead to less than 2 percent of the normal amount of high affinity binding, uptake, and degradation of ^{125}I-LDL. In cells with double class I mutations, LDL does not suppress HMG-CoA reductase activity or cholesterol synthesis, nor does it activate cholesterol formation. With class II mutations (transport defective alleles), transport to Golgi is markedly delayed or abolished, with the molecular mass of the receptor remaining at 120 kD. Class IIA alleles produce a protein that fails to be transported out of the ER. The more common class IIB mutation leads to reduced transport of a variable portion of the receptor to the cell membrane via Golgi. Normal binding of LDL, but not beta-VLDL, requires at least one cysteine-rich repeat in the EGF precursor homology domain. Binding defective alleles result from class III mutations. Class IV Mutations lead to internalization deficiency, with mutations involving the cytoplasmic domain alone in class IVA. Class IVB alleles produce truncated receptors that lack the membrane spanning domain as well as the cytoplasmic tail, so that most of these molecules are secreted from the cell, with 10% remaining adherent, but not able to internalize bound LDL. In class V mutations (recycling defective alleles), the EGF precursor homology domain mediates acid-dependent dissociation of receptor and ligand in the endsosome, an essential event for receptor recycling. Promoter mutations lead to only moderate decrease in LDL receptor activity.

18.2.4 Population Genetics

In most populations, the frequency of homozygote FH is 1 in a million, and the frequency of heterozygotes not less than 1 in 500. Among survivors of myocardial infarction, the frequency of heterozygotes is about 1 in 20. The highest total frequency of FH is found in two different populations in South Africa, namely in the Ashkenazic Jews (1 in 67) and in the Afrikaners. Both groups together feature an incidence of about 1 in 100. In the French Canadian population, the frequency of FH has been estimated at 1 in 270 with only 5 mutations comprising 76% of the mutant receptor alleles. A total of 4 mutations are responsible for most cases of FH in Middle East. In such populations, screening for FH at the DNA level is reasonable. In North America and in Europe (with exception of Finland), a wider variety of mutations responsible for FH exists.

18.2.5 Diagnosis

FH is one of several disorders included in the designation familial type II Hyperlipoproteinemia. FH must be distinguished from other causes of elevations in LDL cholesterol by demonstration of a decrease in LDL receptor activity, the documentation of a mutation in the LDL receptor gene, or the presence of findings such as tendon xanthomas, autosomal dominant transmission, and expression in childhood.

18.2.5.1 Differential Diagnosis

Diagnosis in the homozygote usually causes no difficulty. The finding of juvenile atherosclerosis and cutaneous xanthomas, as well as a plasma cholesterol level exceeding 650 mg/ml in a nonjaundiced child, is virtually pathognomonic. Sometimes differentiation between severely affected heterozygote form and a homozygote with a mutation causing only partially impaired receptor function, can be difficult and may require molecular diagnosis. A defect in cholesterol uptake leading to the same clinical picture and elevation of plasma LDL up to 700 mg/ml, probably due to lack of discrimination of different sterols for esterification, is the cause of phytosterolemia. This disease differs from FH in that it features a response to dietary restriction of cholesterol and to the administration of oral cholestyramine.

The diagnosis of heterozygotic FH begins with the documentation of elevated LDL cholesterol levels. However, an elevated LDL cholesterol alone is not a very sensitive indicator of FH. The majority of type 2 lipoprotein patterns in patients is attributable to familial combined hyperlipidemia, caused by environmental and multiple genetic factors. However, family history and clinical features such as tendon xanthomas, that never occur in familial combined hyperlipidemia, can help with the differential diagnosis. Familial defective apolipoprotein B-100 (FDB) cannot be differentiated from FH solely on clinical grounds. FDB can be diagnosed or ruled out only by direct detection of the mutation at amino acid position 3500 in Apo B-100. Rare causes of an FH-like phenotype can be due to increased production of LDL and monoclonal gammopathy producing an immunoglobulin reacting with the LDL receptor.

18.2.5.2 Laboratory Diagnosis

Functional tests for quantifying LDL receptor activity in cultured fibroblasts are based on binding, uptake and proteolytic degradation of ^{125}I-labeled LDL, or measurement of metabolic activities suppressed or stimulated by LDL (e.g. LDL-mediated stimulation of [^{14}C] oleate incorporation into cellular cholesteryl [^{14}C] oleate. In families with FH, 90% of affected versus nonaffected individuals can be distinguished from one another on the basis of LDL values alone. The number of receptors can be quantified by immunoblotting or by immunoprecipitation. Alternative methods are lymphocyte proliferation assays and flow cytometry with fluorescence labeled LDL. In populations where particular mutations are frequent, such as Afrikaners, French Canadians, and Finns, mutations can be detected directly using PCR-based methods. Prenatal diagnosis is possible by quantitative assessment of LDL receptor activity in cultured amnion fluid cells and measurement of cholesterol in fetal blood. More reliable direct analysis of the receptor mutation can be done as early as week eight of gestation using tissue obtained from chorionic villus biopsy.

18.2.6 Therapy

Dietary restriction of fats is generally recommended for any person with hypercholesterolemia, e.g. cholesterol intake limited to 150 mg/day total fat, and saturated fat ($<7\%$ of total calories). In heterozygotes, this intervention can result in up to a

20% decrease in plasma cholesterol concentrations. Drug therapy should be initiated if LDL cholesterol values remain elevated.

18.2.6.1 Pharmacology

First line drugs in pharmacological treatment of FH are bile acid resins (cholestyramine and cholestipol), which alone lead to 15-20% LDL reduction, or when combined with nicotinic acid up to 40% decrease in LDL cholesterol levels in heterozygotes. Bile resins act by reducing reabsorption of bile acids, an effect which is partly compensated by increased production of cholesterol in the liver. Their action can also be synergistically influenced by cholesterol synthesis inhibitors such as lovastatin, simvastatin, and pravastatin. The inhibition of HMG-CoA reductase leads to increased expression of the LDL receptor, as well as HMG-CoA reductase. However, even with a combination of HMG CoA reductase inhibitor and bile resin, cholesterol levels do not generally decrease by more than 50 to 60%. Possible side effects are liver toxicity and rhabdomyolysis, sometimes severe enough to produce renal failure. In postmenopausal women with heterozygous FH, estrogen replacement therapy can help decrease LDL-cholesterol by 14%.

Macrophage colony stimulator stimulant treatment is currently being evaluated for patients with refractory FH. Preliminary results show LDL lowering with little toxicity. Moderate effects on cholesterol levels have been shown for probucol; however, the drug may provide a more important risk reduction based on its anti-oxidant effects. Probucol reduces arteriosclerosis in the WHHL hyperlipidemic rabbit and causes regression of cutaneous and tendon xanthomas, probably related to its ability to block oxidation of LDL. For a detailed review on drug therapy of FH, the reader is referred to a comprehensive pharmacology textbook (such as: Goodman and Gilman's "The Pharmacological Basis of Therapeutics").

FH homozygotes are generally resistant to treatments effective in heterozygotes, except for those that produce partially functional LDL receptors. Such patients usually require surgery, removal of LDL via plasma exchange, LDL apheresis or a combination of these treatments.

18.2.6.2 Surgery and Direct Interventions

Portocaval anastomosis will reduce cholesterol levels from 25 to 50%. Although well tolerated in the face of normal liver function, this operation is not associated with sufficient reduction in LDL cholesterol concentrations to be used as a sole therapy in homozygotes. The most successful approach for severely affected FH patients is direct removal of LDL from the plasma. Plasma exchanged for normal plasma or albumin every two weeks combined with oral nicotinic acid can reduce plasma cholesterol by about 50% on a long term basis. The procedure is well tolerated, but lifelong repetition can be emotionally difficult. In the past few years, plasma exchange therapy has been replaced by LDL apheresis, where plasma is passed over columns that absorb apo B-100-containing lipoproteins (via heparin agarose, dextran sulfate agarose or anti LDL-antibody agarose). Another approach is called double-filtration plasma exchange with the first filter separating the plasma and the second filter selectively trapping VLDL, LDL and IDL. Major problems are the requirement of weekly treatments, removal of numerous other plasma proteins, as well as removal of HDL with heparin sepharose. Furthermore, an increased precipitation of fibrinogen as well as aggregation of platelets with dextrans occur. Therefore, alternative column materials have been developed. For example, an immobilized C-reactive protein is capable of binding to cholesterol or cholesterol-containing materials.

A total of five homozygotes have been treated with liver transplantation, leading to dramatic decrease in plasma LDL concentrations. However, the operation is hazardous in patients with advanced atherosclerosis and the lifelong need for immunosuppression carries an additional substantial risk. This state-of-affairs emphasizes the potential value of hepatic-directed gene therapy.

18.2.6.3 Gene Therapy

Whether or not a genetic disease is principally eligible for gene therapy is a question of the risk/benefit ratio. The possible benefit must be compared to potential harm; different forms of treatment must be compared as well as no treatment at all. Additional consideration must be paid to ethical concerns and financial issues. It is most important to providc an individual gcnc thcrapy that fits thc special requirements of the genetic defect. Thus, several strategies might be applicable to the same disease. In designing a strategy for gene therapy, one has to take into account several factors: 1. the target tissue, expression level and regulation required, 2. the possibility of *ex vivo* versus *in vivo* gene therapy, 3. repeated administration versus long term expression, and 4. obstacles to gene delivery and expression in terms of interference with immune system, intracellular degradation, and transcription regulation or imprinting.

The primary basis of a genetic treatment is a genetic diagnosis. First, a specific diagnosis assists in deciding whether introduction of a functional gene will be sufficient or if replacement of the defective gene will be necessary. The latter entails eliminating the defective RNA/protein, together with complete restoration of normal protein function. Second, if replacement is desired, a genetic diagnosis can help to select the most appropriate targeting vector. Third, a protein that is missing renders the introduced therapeutic protein as foreign to the immune system, while replacement of a nonfunctional protein would not have this effect.

FH is a disease with a broad spectrum of underlying genetic defects providing several opportunities to interfere on a genetic basis. Furthermore, because of the advanced understanding of the pathophysiology, biochemistry, and genetics of the disease, together with the existence of animal models and an already conducted gene therapy trial, FH is a suitable model to show possibilities and limits for genetic treatment in general.

18.3 Animal Models

The natural animal model for FH is the Watanabe heritable hyperlipidemic rabbit, which has a genetic defect in the LDL receptor, leading to atherosclerosis. The mutant receptor in the Watanabe heritable hyperlipidemic rabbit is not transported to the cell surface at a normal rate (Yamamoto et al., 1986). This defect arises from an in-frame deletion of 12 nucleotides that eliminates 4 amino acids from the cysteine-rich ligand binding domain of the LDL receptor, as shown by cloning and sequencing of complementary cDNAs from normal and Watanabe rabbits. A similar mutation was found in a patient with FH, whose receptor also failed to be transported to the cell surface (Yamamoto et al., 1986). These findings suggest that mechanisms may exist that prevent surface expression of improperly folded proteins with unpaired or improperly bonded cysteine residues. Probucol, originally developed as an antioxidant, prevents the progression of atherosclerosis in the Watanabe rabbit by limiting oxidative LDL modification and foam cell transformation of macrophages (Kita et al., 1987).

In addition, two families of rhesus monkeys have been shown to feature hypercholesterolemia due to deficiency in the LDL receptor (Scanu et al., 1988; Hummel et al., 1990). The defect results from a nonsense mutation in exon 6, changing codon 284 from TGG to TAG. LDLR RNA is reduced by about 50% on quantitative analysis of RNA obtained at liver biopsy in heterozygous animals.

The influence of LDL receptor expression on lipid metabolism has been shown using transgenic mice derived from fertilized eggs injected with the LDL receptor gene under control of the mouse metallothioneine-I promoter. Overexpression of LDL receptors causes elimination of both apo E and apo B, the 2 ligands of the LDL receptor, from the plasma (Hofmann et al., 1988). Protection against diet-induced hypercholesterolemia, even with unregulated over expression of LDL receptors, could be shown with mice transgenic for the LDL receptor gene under a transferrin promoter segment (Yokode et al., 1990).

The Watanabe rabbit was used for the development of liver-directed gene therapy based on transplantation of autologous hepatocytes that had been genetically corrected *ex vivo*, with recombinant DNA (Chowdhury et al., 1991). Animals transplanted with LDLR-transduced autologous hepatocytes demonstrated a 30 to 50% decrease in total serum cholesterol that persisted for the duration of the experiment (122 days). Recombinant-derived LDLR RNA was harvested from tissues with no diminution for up to 6.5 months after transplantation. This experiment can be seen as a first hint that an *ex vivo* approach for gene therapy of FH is not suitable, first in terms of sufficient reduction in cholesterol levels; and, second in terms of long-term gene expression. Gene transfer via adenovirus (Wilson et al., 1992c) restored cholesterol levels to near normal values in the Watanabe rabbit; however, the effect lasted for only a few weeks.

A new animal model for homozygous FH was developed through targeted disruption of the LDLR gene in mice (Ishibashi et al., 1993). Homozygous LDL receptor-deficient mice showed delayed clearance of VLDL, intermediate density lipoproteins (IDL), and LDL from the plasma. As a result, total plasma cholesterol levels increased about twofold in homozygous deficient mice compared to wild type mice. Adult mice did not exhibit gross evidence of xanthomatosis and the extent of their aortic atherosclerosis was minimal. On the other hand, in mice homozygous for the targeted disruption of the LDLR gene which were fed a high cholesterol diet, total plasma cholesterol increased from 250 to more than 1,500 mg/dl. In wild type littermates fed the same diet, total plasma cholesterol values remained less than 160 mg/dl. After 7 months, the homozygous deficient mice developed massive xanthomatous infiltration of the skin and subcutaneous tissues. The aorta and coronary ostia exhibited gross atheromata, and the aortic valve leaflets were thickened by cholesterol-laden macrophages. Reversal of the phenotype was accomplished by adenovirus mediated gene transfer of the LDL receptor driven by a CMV promoter.

18.4 Clinical Protocol

Wilson et al. (Wilson et al., 1992) presented a detailed clinical protocol for the *ex vivo* gene therapy of FH. The approach they proposed was to treat homozygous FH patients with symptomatic coronary artery disease. These patients have a relatively poor prognosis but can tolerate a noncardiac surgical procedure with acceptable risks. Wilson et al. planned to recover hepatocytes from these patients and to reimplant them after genetic correction by means of a retrovirus-mediated gene transfer. The first patient they treated was a 20-year-old woman from Quebec City, Canada. In June 1992, she underwent operation at the University of Michigan Medical Center. From the 250 grams of removed liver, cells were harvested into eight hundred 10 cm^2 Petri dishes. The cells were incubated with a retrovirus containing the normal LDL receptor gene. About 25% of the cells had taken up the virus, when they were delivered back to the liver via injection through a Hickman catheter previously inserted into the mesenteric vein. The only post-procedural irregularity was transient portal hypertension. After 18 months, there was no pathology apparent in a 50 mg liver biopsy specimen. Less easy than evaluating the side effects of the treatment is the assessment of the therapeutic effect. After 18 months, there were single cells showing transcription of the LDL receptor gene (1:1,000 to 1:10,000), which is of limited help in assessing the success of the procedure for two reasons: first, transduced cells are not evenly distributed throughout the liver; second, expression of the protein was not determined. Nevertheless, the serum LDL level was lowered approximately 18% compared with pretreatment levels. This decrease was associated with a decline in the LDL:HDL cholesterol ratio. This effect cannot clearly be attributed to increased LDL receptor expression after gene transfer, since, after surgery and under close medical surveillance it is always possible that patients are more careful with their diets. Moreover, not only increased LDL receptor activity, but also diminished secretion of VLDL leads to reduction of LDL levels. Finally, the patient received lovastatin, which can account for the reported effects in a patient with a residual receptor activity. However, four more patients had undergone this treatment by March, 1994 (Grossman et al., 1995).

Patients with the French Canadian missense mutation in exon 3 (leading only to reduced activity) showed a much higher post-treatment receptor activity than loss of function mutations, at similar transfection efficiencies. This state-of-affairs makes lovastatin treatment rather than gene delivery likely responsible for reductions in cholesterol in these patients. Furthermore, lack of effect on serum LDL in patients with a loss-of-function mutation demonstrates that partial liver resection does not have a nonspecific effect on serum LDL values. Taken together, the question arises whether conducting an experimental therapy with the potential risk of thrombus formation and portal hypertension, risk of bleeding etc. is warranted on the basis of preliminary results from animal experiments. The question is underscored when considering the fact that the same procedure did not lead to satisfying results. Finally, health care planners will be interested to note that the $75,000 cost per patient is a conservative estimate of things to come.

Criteria for the evaluation of FH-directed gene therapy have been proposed by Brown, Goldstein, Havel and Steinberg (Brown et al., 1994). The criteria are as follows: 1. before patient studies are undertaken, animal experiments should provide clear-cut evidence for persistent activity of the exogenously transferred LDL receptor gene in mediating the removal of LDL from plasma; 2. any observed reduction of plasma LDL cholesterol should result from an increased activity of LDL receptors; 3. the increased receptor activity should be shown to result from expression of the exogenously derived gene, rather than from activation of the patient's own receptors in response to surgical or other manipulation of the liver. These criteria remain both an admonishment and a challenge.

18.4.1 Prospects

The first challenge is to get gene therapy for FH to work in the most appropriate animal models. The efficiency of the treatment has to be demonstrated based on an LDL turnover test, performed before and after the procedure. The step of proposing a clinical protocol can be taken can be taken only after the efficacy of the protocol has been shown. From the experience of animal experiments and the clinical trials, there is a clear indication for alternative strategies addressing two major problems: 1. gene expression (long term and appropriate level) and, 2. economic considerations (cost and availability).

18.5 Gene Therapy

18.5.1 *Ex Vivo* Gene Therapy

The *ex-vivo* approach has not been shown to work efficiently in the animal model or in the clinical protocol conducted, despite the potential of life-long persistence of the donor cells (Ponder et al., 1991). Transfected cells in vitro usually show constant levels of expression over long periods of time; however, when returned in vivo, they often lose expression over a few weeks (Richards et al., 1993; Palmer et al., 1991), which possibly reflects methylation and deletion problems. The use of cellular promoters instead of viral promoters and construction of self inactivating vectors, which lack viral enhancer and promoter elements, could help prolong *in vivo* expression (Yee et al., 1987). Further improvement can be achieved via efficiency of gene transfer and number of transplanted cells.

Transfer efficiency can be increased using different gene delivery systems, increasing the number of viruses/particles/DNA employed, or by combination of different gene delivery systems (Lieber et al., 1995b). Optimization would have to be performed with a vector system that integrates the therapeutic gene or is able to replicate, so that persistence of the therapeutic gene is guaranteed. This safeguard is necessary since the possibility of repeated application is limited unless the hepatocytes can be propagated in vitro. This feat has not been achieved for more than 3 weeks despite extensive work in different laboratories (Bader et al., 1992; Li et al., 1993; Gerlach et al., 1994). Nevertheless, the problem of increased risk of portal hypertension with delivery of a greater number of cells into portal circulation, or even after intrasplenic injection (Gupta et al., 1993), requires a different approach to reintroduce the modified cells. An example that moves in the direction of a hybrid artificial organ, is the use of fibroblasts cultured on an artificial matrix. The resultant construction is termed an organoid (Moullier et al., 1993).

Transplantation of an organoid consisting of LDL receptor-expressing hepatocytes would require acess to blood flow for clearance of lipoprotein particles, adequate supply of nutritients and hormones, and access to the biliary system to enable elimination of toxic products. With the likely requirement of different cell types (hepatocytes, fibroblasts, endothelial cells etc.), a particular structure and vessels for connection to the vascular system, hepatic organoids are far from being established. Additionally, the organoid approach would mean an enormous financial input in terms of tissue culture supply and quality control.

Basically, the use of *ex vivo* gene therapy for treatment of FH requires the possibility of growing hepatocytes in culture and the possibility of assembling a hepatic organoid or replacing an adequate portion of the liver with modified hepatocytes. Technical problems and financial issues make it highly unlikely that hepatic organoids will ever lead to an acceptable clinical protocol.

18.5.2 *In Vivo* Gene Therapy

The problems encountered with the *ex vivo* approach make an *in vivo* approach the method of choice for FH-directed gene therapy. With the LDL receptor cDNA and genomic clones available, there are a wide variety of possible strategies ranging from establishing the DNA construct, perfecting gene delivery, to supporting long term gene expression. Major goals for genetic treatment of inherited disorders are sufficient expression levels and long term expression of the therapeutic gene. In FH, neither tissue-specific expression nor regulation of the LDL receptor gene expression is required, so that overexpression of the transgene is satisfactory. Two strategies can be applied to achieve restoration of protein function. Introduction of an expression cassette can be employed. Such a cassette consists of a suitable promoter and the human LDL receptor with a poly A signal. The cassette is integrated randomly into the genome or it may persist extrachromosomally, integrated into a replication system to prevent loss of the transgene. For treatment based on homologous recombination, a targeting vector is required consisting of a genomic region overlapping the defect, possibly with a positive and/or negative selectable element (Morrow et al., 1993).

18.5.2.1 Gene Transfer

Since the liver is the target organ, gene transfer can take place via the circulation through various avenues: 1. systemic administration (Smith et al., 1993), 2. portal vein injection or perfusion (Kozarsky et al., 1994), 3. intrasplenic injection, or 4. hepatic artery injection. Alternative approaches include the biliary system or direct injection into the hepatic parenchyma (Hickman et al., 1994). The possibility of recirculating the vector with the therapeutic gene exists, as well as temporal occlusion of the efferent vessels to increase transfer efficiency. Prior to vector delivery, it is necessary to determine weather or not the vector system or vector complex tolerates the vehicle. Furthermore, strategies such as preperfusion with normal saline, heparin etc. must be considered. Since the beginning of *in vivo* gene delivery methodology, many different transfer vehicles have been devised, most of which have been shown suitable for liver-directed gene delivery.

Retroviral vectors have been used in 80% of clinical gene therapy trials, based upon *ex vivo* transduction, with no associated toxicity, despite lack of targeted integration and a high error rate of reverse transcriptase. Among other cells, hepatocytes can efficiently be infected (up to 80% of mouse – and 40% of human hepatocytes) (Pages et al., 1995).

Lack of tropism is no disadvantage for gene therapy of FH, except for loss of liver-directed gene delivery to surrounding tissue. Therefore, binding and infection of all kinds of human cells can be compensated by either selective perfusion of the liver, *ex vivo* gene delivery, application of a greater amount of virus, or by modifying the envelope proteins to target hepatocytes more specifically via ligand/receptor interaction. The latter implies altering tropism by incorporation of foreign or hybrid envelope proteins (Salmons et al., 1993; Neda et al., 1991).

Titers of 10^6-10^7 particles/ml are standard, up to 10^{10} possible by optimized vector production, storage, and concentration. Major disadvantages are packaging capacity of less than 8 kb maximum and the requirement of dividing cells for infection, that can be achieved by partial hepatectomy, in-flow occlusion technology, or induction of liver regeneration by urokinase gene transfer (Kay et al., 1992; Rettinger et al., 1993; Lieber et al., 1995c). Lentiviral vectors, based on HIV or caprine arthritis-encephalitis virus (CAEV) may be of value for the efficient transduction of quiescent cells (Lewis et al., 1994). Persistence of gene expression has been shown lasting from as little as six months (Kay et al., 1992), to over one year (Rettinger et al., 1994), and up to the life span of the recipient system (Branchereau et al., 1994).

The most highly touted advantage of adenovirus-mediated gene transfer is that wild type adenoviruses are safe, as shown by vaccination with unattenuated adenovirus and lack of tumor production of vectors Ad2 and Ad5 in rodents (Horwitz, 1990). Human wildtype adenovirus readily infects almost any cell type *in vitro* and *in vivo*, dividing cells as well as quiescent cells, with high efficiency. The titers can be as high as $5x10^{11}$ pfu/ml. The virus rarely integrates and has been shown to lead to expression for up to one year in immunocompetent mice (Stratford-Perricaudet et al., 1992a,b) with a marked variability in the persistence of gene expression in different strains (Barr et al., 1995). In humans, adenovirus infections have been reported, which may persist for up to two years (Horwitz, 1990).

Different adenoviral vectors have been used to reduce high cholesterol levels in the mouse and rabbit FH model (Ishibashi et al., 1993; Kozarsky et al., 1994; Li et al., 1995) and have been used to increase LDL receptor production in hepatocytes from FH Patients (Kozarsky et al., 1993). Sufficient hepatocytes can be transfected with the transgenes to achieve near normal cholesterol levels; however, the major drawback has been maintaining a long term effect. The opposite is likely to be the case for adenovirus-mediated homologous recombination (Wang et al., 1993). Expression from the physiologic locus should last longer than from artificial expression cassettes not integrated into the genome; however, the transfection frequency has been too low to achieve metabolically relevant expression levels. This mediocre performance can be improved as soon as adenoviral vectors with a greater capacity for foreign sequences are available (Morrow et al., 1993).

Current adenoviral vectors with deletions in the E1 and E3 regions can package 7-8 kb, that can be increased by another 2 kb with a deletion in the E4 (Armentano et al., 1995). Further improvement will be possible when nonessential viral sequences are eliminated, thereby retaining only the therapeutic sequences, the packaging signal, and the terminal repeats. This could be achieved by means of sequence-specific recombination and/or the use of helper viruses/improved packaging cell lines. In terms of safety issues, one must take into account the possibility of slow accumulation of toxic viral encoded protein, shedding of recombinant adenovirus, and anaphylactic shock, with minor acute toxicity in the animal model (Zhang et al., 1995) and the potential for replication secondary to transcomplementation due to virus-like factors. Currently, the phenomenon has been only shown in tissue culture systems with a human tumor cell line, but never in primary cultures (Jones et al., 1979). In conclusion, adenoviral vectors are a safe and efficient gene transfer vehicle with the potential of transferring large pieces of DNA into a wide variety of target cells.

In addition to adenoviral and retroviral vectors, some less extensively characterized viral vectors exist that have been used for liver directed gene transfer. Adeno-associated virus (AAV) is capable of infecting hepatocytes *in vivo*, with kinetics of gene expression yet to be determined (Flotte et al., 1995). However, repeated administration is possible (Lebkowski et al., 1988). The major drawbacks are difficult to prepare viral stocks and contamination with wildtype AAV and helper virus. The maximum packaging capacity is only 4.7 kb (Muzyczka, 1992) and safety issues have not yet fully been elucidated. The AAV approach is also likely to present problems in terms of interference with prenatal development and oncogenesis (Botquin et al., 1994; Walz et al., 1992). Other viral vector systems of potential use in liver directed gene transfer are based on herpes simplex virus and baculovirus (Lu et al., 1995; Hofmann et al., 1995a).

Its central position in metabolism with clearing all kinds of particles from the plasma makes the liver an ideal target for receptor mediated gene delivery. Most research has targeted the hepatocyte-specific asialoglycoprotein and the transferrin receptor. Using asialo-orosomucoid polylysine (ASOR-PL), large DNA fragments (up to 48 kb) can be transferred. The complexes are inexpensive and simple to produce and are not highly immunogenic. After intravenous injection, 85% of the complexes are taken up by the liver within 10 minutes. However, application of ASOR-PL in the Watanabe rabbit has led to only transient reductions in cholesterol levels (Wilson et al., 1992c). After endocytosis, intracellular trafficking leads to degradation in the acidified endosome. Investigators have attempted to circumvent this problem by adding the antimalarial drug chloroquine, an inhibitor of endosomal acidification (Tietz et al., 1990), using empty adenovirus capsids (Cotten et al., 1992) or synthetic peptides derived from the fusogenic influenza virus membrane glycoprotein and hemagglutinin (HA-2) (Wagner et al., 1992). No acute or chronic toxicity was observed; however, formation of antibodies was associated with repetitive administration of the ASO-PL DNA complexes in mice (Stankovics et al. 1994). Possible alternatives with ligand function are surface proteins derived from the hepatitis virus and from malaria plasmodia (Hofmann et al., 1995b; Ding et al., 1995). To complex DNA and to avoid degradation, different cationic polymers could also be employed. Polyethyleneimine (Boussif et al., 1995) is an example; however *in vivo* data are presently only available for administration into the central nervous system.

Several studies have shown the effectiveness of polycationic liposome delivery of DNA *in vivo*, with the liver being one of the target organs (Zhu et al., 1993; Takehara et al., 1995). Similar to polymer complexes, polycationic liposomes are an attractive alternative to viral transfer methods. The liposomes avoid problems with size constraints, they offer lower immunogenicity, and are cheaper to prepare in bulk.

The lipid combinations DC-chol/DOPE and DMRIE/DOPE have received approval for human trials. DMRIE/DOPE-mediated transfection appears to be two to seven times more efficient than DC-chol/DOPE (Middleton et al., 1994). Together with novel lipid formulations, further improvements are possible by enhanced packaging of the DNA with polymers (Remy et al., 1995) or DNA binding proteins (Kaneda et al., 1989b), and by the addition of fusogenic peptides (e.g. hemagglutinating virus of Japan) to liposome complexes (Kaneda et al., 1989a).

Similar to retrovirus-mediated gene delivery, the proliferative state of the cell population targeted has a profound effect on transfection efficiency (Takeshita et al., 1994) probably due to disruption of the nuclear membrane in mitosis. Presently, major limitations to gene expression after liposomal delivery is movement of DNA from the cytoplasm to the nucleus, and dissociation of lipid and DNA before transcription can occur (Zabner et al., 1995). Direct delivery of DNA without viral or chemical packaging has been shown by direct injection into the liver using an ordinary syringe (Hickman et al., 1995; Malone et al., 1994), or by using a jet injector, which administers the DNA as a pressurized droplet stream (Furth et al., 1992). Both approaches lead to only a small number of transfected cells and expression for up to more than 60 days. Thus, particle bombardment does not lead to a significant improvement in hepatic transfer and expression levels decline to 1-5% of peak levels after one week (Cheng et al., 1993). Nevertheless, stable gene transfer to different cell lines has been demonstrated (Yang et al., 1990), as well as a fivefold greater efficiency than lipofection, calcium phosphate co-precipitation, or electroporation in a tissue culture system (Thompson et al., 1993).

18.5.3 Supportive Measures

18.5.3.1 Induction of Cell Division

For direct injection, retrovirus infection, receptor-mediated gene transfer, and lipofection, the proliferative state of a cell population has a profound effect on transfection efficiency (Davis et al., 1993; Wilson et al., 1993a; Takeshita et al., 1994; Boussif et al., 1995). Cell proliferation probably interferes by the release of DNA-stabilizing nuclear factors into the cytoplasm, when the nuclear membrane is disrupted. Regenerating hepatic cells can be artificially induced by administering chloroform, (Larson et al., 1995), carbon tetrachloride (Maldonado et al., 1994), 1,4-dichlorobenzene (Eldridge et al., 1992), or thioacetamide (Ledda-Columbano et al., 1991). Alternative approaches include the infusion of hepatocyte growth factor (Roos et al., 1995), or partial hepatectomy (Kay et al., 1993). Hepatic inflow occlusion (Rettinger et al., 1993) has also been employed. Suicide genes have been introduced in conjunction with a drug (Caruso et al., 1993) or by themselves (Lieber et al., 1995c). Induction of hepatic regeneration by urokinase gene transfer has been shown to work 5 to 10-fold more efficiently than partial hepatectomy, has less side effects, and is less invasive.

18.5.3.2 Induction of Tolerance

Immunoreaction against the delivery system, as well as the therapeutic protein are the major obstacles to long term gene expression. Since there is no interspecies consistency in immunoreactions against delivery systems, a Phase I clinical trial would be necessary to determine what kind of immune response is elicited in humans. Preliminary data favor the neutralizing antibody pathway for adenovirus elimination (at least for Ad 1,2,5 and 6) (Sakamoto et al., 1995). For adenoviral gene transfer, delivery would have to be optimized in the most compatible animal models.

Cytotoxic T lymphocytes (CTL) eliminate adenovirus-infected mouse hepatocytes (Yang et al., 1995a,b). Therefore, nu/nu mice, lacking an immune system, and cyclosporin treated immunocompetent mice, show improved foreign gene expression (Zsengeller et al., 1995). Nevertheless, nude mice lacking a CTL response can also suppress expression after 6-8 weeks, probably due to neutralizing antibodies (Dai et al., 1995). In the rabbit model, cyclosporin does not help prolong expression (Kozarsky et al., 1993), because the major pathway of eliminating infected hepatocytes is via neutralizing antibodies. Monkeys tolerated adenoviral gene transfer to the lung without an increase in neutralizing antibodies; however, dose dependent infiltration of inflammatory cells, primarily lymphocytes, occurred without being clinically apparent (Brody et al., 1994). Other pharmacological agents besides cyclosporin have been used for induction of tolerance in adenovirus-mediated gene delivery (Lochmuller et al., 1995) and for induction of transplantation tolerance, as reviewed elsewhere (Kahan, 1992; Thomson, 1994).

Different viruses have developed means to escape the immune system, e.g. by interfering with antigen processing and MHC class I-restricted presentation, thereby helping circumvent the immune response via CTL. In Herpes simplex virus, this escape has been accomplished by ICP47 (Fruh et al., 1995), in Epstein-Barr Virus by EBNA-1 (Levitskaya et al., 1995), and in adenovirus by gp19k, a protein encoded by the early region E3 that anchors MHC I to the endoplasmatic reticulum membrane (Rawle et al., 1989). This mechanism is impeded in adenovirus type 5 vectors, that have been E3- deleted to increase packaging capacity (Bett et al., 1993). Attachment of gp19k or ICP47 to the gene delivery system or their active domains fused to the therapeutic protein may help in diminishing immune reactions, as has been shown for the gly-ala repeats of EBNA-1 (Levitskaya et al., 1995). In autoimmune disease and transplantation, alternative ways to induce tolerance are being developed ranging from soluble receptors to antibodies against cytokines (Gaston et al., 1994; Alegre et al., 1995), over thymic (Baumann et al., 1995; Knechtle et al., 1994) to oral tolerance (Hafler et al., 1995; Weiner et al. 1994).

The following have been applied to or may be useful in gene therapy related projects: 1. application of cyclosporin A improves longevity of expression (Dai et al., 1995), 2. coadministration of recombinant adenovirus with soluble CTLA4Ig, that interferes with interaction of T cell and antigen presenting cell, leads to persistent adenoviral gene expression in mice without long-term immunosuppression (Kay et al., 1995); and 3. adenoviral gene transfer to the thymus results in induction of immunologic unresponsiveness to the vector and might be extended to other proteins that the virus could be engineered to encode (DeMatteo et al., 1995).

For liver directed gene therapy, application via portal vein could have a significant advantage due to induction of peripheral tolerance. Such peripheral tolerance has been shown for delivery of hepatocytes expressing hepatitis B virus surface antigen (HBs-Ag) via the portal vein. In this example, antibody production was not elicited even after rechallenging with HBs-Ag in highly immunogenic form. In contrast, subcutaneous or intraperitoneal transplantation of HBs-Ag transgenic hepatocytes results in a much different response (Vemuru et al., 1992). Expression of HBsAg has been shown to persist life-long in hepatocytes transplanted to the liver via portal vein or spleen, whereas in all recipients transplanted into peritoneal cavity or dorsal foot pad, serum HBsAg eventually disappeared after 2-6 weeks (Gupta et al., 1994). Whether or not a specific vector or therapeutic protein can induce tolerance after application via portal vein, must be individually determined. Tolerance could be dependent on the possibility of the antigen behaving as a superantigen (Blankson et al., 1995; Ochi et al., 1993). Dose and maintenance of expression and coadministration of immunomodulators certainly play an important role (Sabapathy et al., 1994; Gorczynski et al., 1995).

In addition to humoral and cellular immune response, for some vectors the complement system must be taken into consideration (Takeuchi et al., 1994). Possible interventions could be based on heparin, (Weiler et al., 1992), monoclonal antibodies against complement factors (Rother et al., 1995), or on different proteins such as vitronectin, clusterin, homologous restriction factor etc., or soluble complement receptors (Couser et al., 1995). A more mechanical method to escape immune response is the microencapsultion of the genetically modified cells, leading to improved transgene expression *in vivo*. An example is the administration of human growth hormone in mice for over 115 days (Tai et al., 1993; Liu et al., 1993). For treatment of FH, cells able to metabolize cholesterol could be delivered packed in immune-protective membranes to permit uptake of LDL and secretion of the products of cholesterol metabolism.

The most general method to specifically suppress undesired immunological reactions is by gene delivery as early as possible, to acquaint the developing immune system with the therapeutic protein, as well as the vector. Even in newborns, there seems to be still enough plasticity remaining to allow the immune system to less efficiently counteract gene transfer vectors. For instance, long-term expression from an adenovirus-vector was observed for more than one year in muscle cells and presumably hepatocytes in the mouse model (Stratford-Perricaudet et al., 1990). This prolonged expression may have been related to administering the recombinant adenovirus to newborns, thereby interfering with the normal immune response. Prenatal gene delivery could possibly lead to more efficient induction of tolerance, thereby prolonging expression even further.

18.5.3.3 Prenatal Gene Transfer

Until now, different approaches to prenatal gene therapy have been utilized, most of which targeted lung or hematopoietic cells using retro- or adenovirus-mediated gene delivery. Major drawbacks result from the delivery procedure rather than from the vector itself. Follow-up, prior to term delivery is also a problem (Pitt et al., 1995; Clapp et al., 1991). Furthermore, current E1-deleted adenoviral vectors may engender antibody production of fetal origin and cellular infiltration may lead to gradual elimination of the virus after amniotic injection (McCray et al., 1995). Better results were obtained with the *ex vivo* delivery of retrovirus-infected hematopoietic cells with long term expression for more than two years after birth (Kantoff et al., 1989). Targeting hepatocytes makes the stage of hepatic development crucial for vector delivery. The liver after all, plays a major role in hematopoiesis, primarlily during mid-gestation. Therefore, delivery to hematopoietic tissue as well as hepatocytes can result from direct hepatic injection (Clapp et al., 1991), placental injection (Jaenisch, 1980), or injection into the peritoneal cavity (Hatzoglou et al., 1990).

Delivery via the maternal circulation could be an attractive alternative to embryonal, placental, umbilical chord or amnionic injection. Each of these approaches is associated with an invasive procedure and a potential risk to the fetus. The intravenous injection of plasmid/lipopolyamine expression complexes into pregnant mice resulted in successful gene transfer into the embryos (Tsukamoto et al., 1995). The plasmids persisted for at least one month after birth (40 days after injection), with larger amounts contained in heart, lung, liver, and skeletal muscle. Thereafter, expression gradually decreased together with amount of plasmid. Neither integration into the genome, nor germ line transmission, were detected. These results have apparently not yet been confirmed by other laboratories.

18.5.3.4 Homologous Recombination

For homologous recombination in gene therapy, two major applications exist. First, repair of a defective gene can be accomplished by exchanging the mutated genomic sequence for the wild type sequence contained in a targeting vector. A second approach is the elimination (knock-out) of proteins essential for graft rejection (Coffman et al., 1993), in order to generate universal donor cells that can be used for transplantation across histocompatibility barriers (Arbones et al., 1994). For *ex vivo* treatment of FH, these donor cells could be hepatocytes expressing the LDL receptor. Unfortunately, escape from the immune system goes hand-in-hand with lack of ability to eliminate the transplanted cells in the case of viral infection or malignant transformation. Together with problems in propagation and reintroduction *in vivo*, these obstacles preclude an *ex vivo* approach with non-immunogenic hepatocytes in FH patients.

The major advantage of repairing the defect on a genomic level is that physiological expression is obtained and shutdown of gene expression is less likely than from an artificial expression cassette. Until now, this feat has only been accomplished in embryonic stem cell tissue culture systems (Savransky et al., 1994) and in primary cells (Arbones et al., 1994). Delivery of the targeting construct has mainly been achieved by electroporation or with adenovirus vectors (Wang et al., 1993; Mitani et al., 1995). Targeting efficiency is dependent on a lot of factors, including length and degree of homology, genomic location and strand breaks, the position in cell cycle, etc. The overall frequency of recombination after electroporation is less than 10^{-4} with a ratio of homologous recombination (HR) to random integration (RI) as high as 3/1. This ratio is even less with adenovirus vectors (10^{-6} at HR/RI ratios of 1/5) (Morrow et al., 1993). Even with a significant improvement in recombination frequency and possibility of the selection of gene targeting events (Hanson et al., 1995;Wu et al., 1988), the problem of hepatocyte propagation and delivery would persist, unless homologous recombination could be performed *in vivo*.

18.6 Clinical Protocol

The therapeutic goal in FH patients can be divided into two major components; namely, elimination of LDL cholesterol from the blood and suppression of endogenous cholesterol synthesis. These aims can be accomplished by an exogenously delivered LDL receptor. An alternative strategy could be the elimination of LDL via "plasma apheresis" at the molecular level. This goal may be achieved by interfering with intracellular trafficking of LDL in gut or renal epithelium, in conjunction with inhibition of cholesterol synthesis using HMG-CoA reductase inhibitors or possibly ribozymes to reduce expression of proteins involved in cholesterol synthesis or its regulation. The elimination of LDL cholesterol does not vary much between species. Nevertheless, investigation of cholesterol metabolism in different animals with regard to alternative degradation pathways (enzymes for cholesterol breakdown, conjugation or different cholesterol transporters) could help establish new gene therapeutic strategies.

Physiologically, the pathway for elimination of cholesterol in humans involves degradation in hepatocytes and elimination of degradation products via the bile. Other pathways of elimination can be used by supplying an engineered LDL receptor to transport LDL particles across gut or renal epithelium or to mediate excretion via sweat glands. Minor modifications in the cytoplasmic domain of the LDL receptor reduced its ability to return to the basolateral surface following endocytosis, resulting in receptor transcytosis from basolateral endosomes to the apical plasma membrane in canine kidney cells (Matter et al., 1993). How this state of affairs could reduce cholesterol levels and cause possible side effects, awaits experimental animal studies.

Fusion proteins containing the ligand binding domain of the LDL receptor and the cytoplasmatic domain of proteins transcytosed in renal cells, could provide an alternative answer (Casanova et al., 1990). With only elimination of LDL particles increased, endogenous cholesterol synthesis would work against the therapeutic effect. Therefore, coadministration of HMG-CoA reductase inhibitors is necessary. Co-delivery of a ribozyme directed against a critical component in the regulation of cholesterol synthesis (HMG CoA reductase, SREBP, SCA etc.), selected from a ribozyme expression library, might provide an alternative (Lieber and Strauss, 1995). The ribozyme strategy in-and-of-itself will not be an alternative to HMG CoA reductase inhibitors and other pharmacological agents in treatment of moderately elevated cholesterol levels. However, gene therapy conducted in patients with delivery of two therapeutic sequences instead of one does not necessarily imply an additional burden in terms of side effects, safety or economy. Performed efficiently, the replacement of the mutated LDL receptor, either by repairing the genomic defect, or by supplementing a functional receptor, would be the strategy of choice. Principal problems remain the gene delivery, efficiency of homologous recombination, and persistent expression. Until now, many types of vectors used in gene therapy have been applied in the treatment of FH. The animal models and tissue culture experiments in human cells, have been the most promising to date.

The maintenance of therapeutic sequences can result from integration into the genome, as in retrovirus-mediated gene delivery, or AAV, and less efficiently for other delivery systems including endosomal replication. A prerequisite for extrachromosomal replication is a circularized DNA containing an origin of replication and a DNA binding protein to mediate replication and maintenance in human cells. Two different replication systems have been used in mammalian cells. The Epstein-Barr virus oriP and EBNA-1 gene (Yates et al., 1985; Sun et al., 1994) and BK virus sequences (Cooper et al., 1993). These constructs have mediated replication in cell lines of different species for plasmids as long as 330 kb in the EBV-derived system, and with *in vivo* replication in mice shown for the BK virus system (Thierry et al., 1995). Replication systems and homologous recombination hold great promise for future gene therapy strategies, with possible support from selection systems adopted for application *in vivo*. The receptor defect in FH patients is an ideal prerequisite for selection, since an antagonist coupled to LDL could prevent cell death from a hepatotoxin (e.g. galactosamine) in LDL receptor-expressing hepatocytes (Wu et al., 1988). Maintaining selection pressure could lead to gradual repopulation of the liver with LDL receptor-expressing hepatocytes. A wide variety of expression systems exist, with viral promoters generally shorter than cellular promoters. These systems mediate a higher level but a shorter term of gene expression. Tissue specific expression of the LDL receptor in FH patients could be advantageous if gene delivery is not targeted to the liver and if LDL receptor overexpression in cells other than hepatocytes leads to side effects (e.g. increased susceptibility to vascular lesions after receptor delivery to the endothelium). Therefore, any expression cassette must be tested in tissue culture and in the animal model prior to use in a clinical protocol.

In over 80% of clinical protocols approved to date, viral vectors have been the delivery system of choice. In evolution, viruses have been selected to exhibit short term gene expression and to exploit cells. Similarly, cells have been selected to ward off viral gene expression, infection and persistence of the virus genome. Thus, it is not surprising that major modifications will be required for long term expression and/or the possibility of repeated application of viral vectors. Vector development is one of the greatest needs in gene therapy. Viral vectors are extremely costly to produce in adequate amounts. Quality control presents problems with regard to storage and transport. Nevertheless, some viral features such as replication, breaking up endosomes, passing the nuclear membrane, binding MHC-I to reduce immune reactions against targeted cells, can be applied in some nonviral vector systems. Such systems are based on complex formation (e.g. with polymers, lipids, dendrimers, beads etc.) and physical delivery (jet injection / particle bombardment). It is difficult to predict which combination of vector systems and application methods will eventually be the most successful gene therapy approach for FH. With regard to economy and safety issues, as well as efficiency of delivery, either a replication system persisting extrachromosomally or a targeting construct for homologous recombination, hold the most promise. These approaches must be integrated in an artificial particle with DNA binding and with endosome and nuclear membrane disruption ability. They must be delivered via the portal vein in combination with *in vivo* selection. If such multiple criteria can be met, an acceptable protocol may be developed.

However, the problem of immune tolerance induction and the nature of the vector system still remain to be solved. To induce portal tolerance, the vector together with lymphocytes or fibroblasts engineered to express the therapeutic gene and antigens derived from the vector from earlier biopsies, could be co-injected into the portal vein. The lymphocytes or fibroblasts would have to have been harvested earlier. Another possibility would be the induction of tolerance by injecting the pregnant mother intravenously with a plasmid containing an expression cassette for the therapeutic gene and possibly vector antigens packed in liposomes. Until now, no practicable and reducible method for induction of tolerance has been developed.

18.7 Summary

FH is a disease with a simple underlying defect requiring complex therapeutic decisions. Particularly in severely affected individuals, the lack of a satisfactory treatment has led to high hopes for gene therapeutic strategies. Unfortunately, currently available gene transfer systems have neither the ability to effect stable gene transfer and expression, nor do they allow for repeated applications. Knowledge accrued from several years of vector development and *in vivo* gene delivery has helped to point out possible approaches for improvements in liver-directed gene transfer. Similarly, progress has been made in circumventing adverse reactions in response to foreign gene expression. Moreover, different animal models have been created and characterized to evaluate new gene therapy strategies for treatment of FH. Nevertheless, major obstacles must still be overcome before gene therapy for FH becomes a reality.

References

Anderson, R. G., Goldstein, J. L., and Brown, M. S. (1977). A mutation that impairs the ability of lipoprotein receptors to localize in coated pits on the cell surface of human fibroblasts. Nature *270*, 695-699.

Arbones, M. L., Austin, H. A., Capon, D. J., and Greenburg, G. (1994). Gene targeting in normal somatic cells: inactivation of the interferon-gamma receptor in myoblasts. Nat. Genet. *6*, 90-97.

Armentano, D., Sookdeo, C. C., Hehir, K. M., Gregory, R. J., St.George, J. A., Prince, G. A., Wadsworth, S. C., and Smith, A. E. (1995). Characterization of an adenovirus gene transfer vector containing an E4 deletion. Hum. Gene Ther. *6*, 1343-1353.

Bader, A., Rinkes, I. H., Closs, E. I., Ryan, C. M., Toner, M., Cunningham, J. M., Tompkins, R. G., and Yarmush, M. L. (1992). A stable long-term hepatocyte culture system for studies of physiologic processes: cytokine stimulation of the acute phase response in rat and human hepatocytes. Biotechnol. Prog. *8*, 219-225.

Barr, D., Tubb, J., Ferguson, D., Scaria, A., Lieber, A., Wilson, C., Perkins, J., and Kay, M. A. (1995). Strain related variations in adenovirally mediated transgene expression from mouse hepatocytes in vivo: comparisons between immunocompetent and immunodeficient inbred strains. Gene Ther. *2*, 151-155.

Baumann, E. E., Buckingham, F., and Herold, K. C. (1995). Intrathymic transplantation of islet antigen affects CD8+ diabetogenic T-cells resulting in tolerance to autoimmune IDDM. Diabetes *44*, 871-877.

Bett, A. J., Prevec, L., and Graham, F. L. (1993). Packaging capacity and stability of human adenovirus type 5 vectors. J. Virol. *67*, 5911-5921.

Blankson, J. N., Loh, D. Y., and Morse, S. S. (1995). Superantigens and conventional antigens induce different responses in alpha beta T-cell receptor transgenic mice. Immunology *85*, 57-62.

Botquin, V., Cid Arregui, A., and Schlehofer, J. R. (1994). Adeno-associated virus type 2 interferes with early development of mouse embryos. J. Gen. Virol. *75*, 2655-2662.

Boussif, O., Lezoualc'h, F., Zanta, M. A., Mergny, M. D., Scherman, D., Demeneix, B., and Behr, J. P. (1995). A versatile vector for gene and oligonucleotide transfer into cells in culture and in vivo: polyethylenimine. Proc. Natl. Acad. Sci. U. S. A. *92*, 7297-7301.

Branchereau, S., Calise, D., and Ferry, N. (1994). Factors influencing retroviral-mediated gene transfer into hepatocytes in vivo. Hum. Gene Ther. *5*, 803-808.

Brown, M. S., Goldstein, J. L., Havel, R. J., and Steinberg, D. (1994). Gene therapy for cholesterol [letter; comment]. Nat. Genet. *7*, 349-350.

Caruso, M., Panis, Y., Gagandeep, S., Houssin, D., Salzmann, J. L., and Klatzmann, D. (1993). Regression of established macroscopic liver metastases after in situ transduction of a suicide gene. Proc. Natl. Acad. Sci. U. S. A. *90*, 7024-7028.

Casanova, J. E., Breitfeld, P. P., Ross, S. A., and Mostov, K. E. (1990). Phosphorylation of the polymeric immunoglobulin receptor required for its efficient transcytosis. Science *248*, 742-745.

Cheng, L., Ziegelhoffer, P. R., and Yang, N. S. (1993). In vivo promoter activity and transgene expression in mammalian somatic tissues evaluated by using particle bombardment. Proc. Natl. Acad. Sci. U. S. A. *90*, 4455-4459.

Chowdhury, J. R., Grossman, M., Gupta, S., Chowdhury, N. R., Baker, J. R.,Jr., and Wilson, J. M. (1991). Long-term improvement of hypercholesterolemia after ex vivo gene therapy in LDLR-deficient rabbits. Science *254*, 1802-1805.

Clapp, D. W., Dumenco, L. L., Hatzoglou, M., and Gerson, S. L. (1991). Fetal liver hematopoietic stem cells as a target for in utero retroviral gene transfer. Blood *78*, 1132-1139.

Coffman, T., Geier, S., Ibrahim, S., Griffiths, R., Spurney, R., Smithies, O., Koller, B., and Sanfilippo, F. (1993). Improved renal function in mouse kidney allografts lacking MHC class I antigens. J. Immunol. *151*, 425-435.

Cooper, M. J. and Miron, S. (1993). Efficient episomal expression vector for human transitional carcinoma cells. Hum. Gene Ther. *4*, 557-566.

Cotten, M., Wagner, E., Zatloukal, K., Phillips, S., Curiel, D. T., and Birnstiel, M. L. (1992). High-efficiency receptor-mediated delivery of small and large (48 kilobase gene constructs using the endosome-disruption activity of defective or chemically inactivated adenovirus particles. Proc. Natl. Acad. Sci. U. S. A. *89*, 6094-6098.

Couser, W. G., Johnson, R. J., Young, B. A., Yeh, C. G., Toth, C. A., and Rudolph, A. R. (1995). The effects of soluble recombinant complement receptor 1 on complement-mediated experimental glomerulonephritis. J. Am. Soc. Nephrol. *5*, 1888-1894.

Dai, Y., Schwarz, E. M., Gu, D., Zhang, W. W., Sarvetnick, N., and Verma, I. M. (1995). Cellular and humoral immune responses to adenoviral vectors containing factor IX gene: tolerization of factor IX and vector antigens allows for long-term expression. Proc. Natl. Acad. Sci. U. S. A. *92*, 1401-1405.
Davis, H. L., Whalen, R. G., and Demeneix, B. A. (1993). Direct gene transfer into skeletal muscle in vivo: factors affecting efficiency of transfer and stability of expression. Hum. Gene Ther. *4*, 151-159.
DeMatteo, R. P., Raper, S. E., Ahn, M., Fisher, K. J., Burke, C., Radu, A., Widera, G., Claytor, B. R., Barker, C. F., and Markmann, J. F. (1995). Gene transfer to the thymus. A means of abrogating the immune response to recombinant adenovirus. Ann. Surg. *222*, 229-239.
Eldridge, S. R., Goldsworthy, T. L., Popp, J. A., and Butterworth, B. E. (1992). Mitogenic stimulation of hepatocellular proliferation in rodents following *1*,4-dichlorobenzene administration. Carcinogenesis *13*, 409-415.
Flotte, T. R., Barraza Ortiz, X., Solow, R., Afione, S. A., Carter, B. J., and Guggino, W. B. (1995). An improved system for packaging recombinant adeno-associated virus vectors capable of in vivo transduction. Gene Ther. *2*, 29-37.
Fruh, K., Ahn, K., Djaballah, H., Sempe, P., van Endert, P. M., Tampe, R., Peterson, P. A., and Yang, Y. (1995). A viral inhibitor of peptide transporters for antigen presentation. Nature *375*, 415-418.
Furth, P. A., Shamay, A., Wall, R. J., and Hennighausen, L. (1992). Gene transfer into somatic tissues by jet injection. Anal. Biochem. *205*, 365-368.
Gaston, R. S. (1994). Cytokines and transplantation: a clinical perspective. Transplant. Sci. *4* Suppl *1*, S9-19.
Gerlach, J. C., Encke, J., Hole, O., Muller, C., Ryan, C. J., and Neuhaus, P. (1994). Bioreactor for a larger scale hepatocyte in vitro perfusion. Transplantation *58*, 984-988.
Gorczynski, R. M. and Chen, Z. (1995). Tolerance after portal venous immunization is enhanced by anti-IFN-gamma and is maintained only with persistence of antigen. Transplant. Proc. *27*, 227-229.
Grossman, M., Rader, D. J., Muller, D. W., Kolansky, D. M., Kozarsky, K., Clark, B. J.,3rd, Stein, E. A., Lupien, P. J., and Brewer, H. B.,Jr. (1995). A pilot study of ex vivo gene therapy for homozygous familial hypercholesterolaemia [see comments]. Nat. Med. *1*, 1148-1154.
Gupta, S., Vemuru, R. P., Lee, C. D., Yerneni, P. R., Aragona, E., and Burk, R. D. (1994). Hepatocytes exhibit superior transgene expression after transplantation into liver and spleen compared with peritoneal cavity or dorsal fat pad: implications for hepatic gene therapy. Hum. Gene Ther. *5*, 959-967.
Gupta, S., Yerneni, P. R., Vemuru, R. P., Lee, C. D., Yellin, E. L., and Bhargava, K. K. (1993). Studies on the safety of intrasplenic hepatocyte transplantation: relevance to ex vivo gene therapy and liver repopulation in acute hepatic failure. Hum. Gene Ther. *4*, 249-257.
Hafenrichter, D. G., Ponder, K. P., Rettinger, S. D., Kennedy, S. C., Wu, X., Saylors, R. S., and Flye, M. W. (1994). Liver-directed gene therapy: evaluation of liver specific promoter elements. J. Surg. Res. *56*, 510-517.
Hafler, D. A. and Weiner, H. L. (1995). Antigen-specific immunosuppression: oral tolerance for the treatment of autoimmune disease. Chem. Immunol. *60*, 126-149.
Hanson, K. D. and Sedivy, J. M. (1995). Analysis of biological selections for high-efficiency gene targeting. Mol. Cell Biol. *15*, 45-51.
Hatzoglou, M., Lamers, W., Bosch, F., Wynshaw Boris, A., Clapp, D. W., and Hanson, R. W. (1990). Hepatic gene transfer in animals using retroviruses containing the promoter from the gene for phosphoenolpyruvate carboxykinase. J. Biol. Chem. *265*, 17285-17293.
Hickman, M. A., Malone, R. W., Lehmann Bruinsma, K., Sih, T. R., Knoell, D., Szoka, F. C., Walzem, R., Carlson, D. M., and Powell, J. S. (1994). Gene expression following direct injection of DNA into liver. Hum. Gene Ther. *5*, 1477-1483.
Hofmann, C., Sandig, V., Kirillova, I., Jennings, G., Rudolph, M., Schlag, P., and Strauss, M. (1995). Hepatocyte-specific binding of L/S-HBV particles expressed in insect cells. Biol. Chem. Hoppe Seyler *376*, 173-178.
Hofmann, S. L., Russell, D. W., Brown, M. S., Goldstein, J. L., and Hammer, R. E. (1988). Overexpression of low density lipoprotein (LDL) receptor eliminates LDL from plasma in transgenic mice. Science *239*, 1277-1281.
Horwitz, M. S., in Fields, B. N., Knipe, D. M., et al, eds. Virology, 2nd ed. New York, Raven Press, 1990, 1723-1741.
Hummel, M., Li, Z. G., Pfaffinger, D., Neven, L., and Scanu, A. M. (1990). Familial hypercholesterolemia in a rhesus monkey pedigree: molecular basis of low density lipoprotein receptor deficiency. Proc. Natl. Acad. Sci. U. S. A. *87*, 3122-3126.

Ishibashi, S., Brown, M. S., Goldstein, J. L., Gerard, R. D., Hammer, R. E., and Herz, J. (1993). Hypercholesterolemia in low density lipoprotein receptor knockout mice and its reversal by adenovirus-mediated gene delivery [see comments]. J. Clin. Invest. *92*, 883-893.

Jaenisch, R. (1980). Retroviruses and embryogenesis: microinjection of Moloney leukemia virus into midgestation mouse embryos. Cell *19*, 181-188.

Jones, N. and Shenk, T. (1979). Isolation of adenovirus type 5 host range deletion mutants defective for transformation of rat embryo cells. Cell *17*, 683-689.

Kahan, B. D. (1992). Immunosuppressive therapy. Curr. Opin. Immunol. *4*, 553-560.

Kaneda, Y., Iwai, K., and Uchida, T. (1989). Introduction and expression of the human insulin gene in adult rat liver. J. Biol. Chem. *264*, 12126-12129.

Kaneda, Y., Iwai, K., and Uchida, T. (1989). Increased expression of DNA cointroduced with nuclear protein in adult rat liver. Science *243*, 375-378.

Kantoff, P. W., Flake, A. W., Eglitis, M. A., Scharf, S., Bond, S., Gilboa, E., Erlich, H., Harrison, M. R., Zanjani, E. D., and Anderson, W. F. (1989). In utero gene transfer and expression: a sheep transplantation model. Blood *73*, 1066-1073.

Kay, M. A., Holterman, A. X., Meuse, L., Gown, A., Ochs, H. D., Linsley, P. S., and Wilson, C. B. (1995). Long-term hepatic adenovirus-mediated gene expression in mice following CTLA4Ig administration. Nat. Genet. *11*, 191-197.

Kay, M. A., Li, Q., Liu, T. J., Leland, F., Toman, C., Finegold, M., and Woo, S. L. (1992). Hepatic gene therapy: persistent expression of human alpha 1-antitrypsin in mice after direct gene delivery in vivo. Hum. Gene Ther. *3*, 641-647.

Kita, T., Nagano, Y., Yokode, M., Ishii, K., Kume, N., Ooshima, A., Yoshida, H., and Kawai, C. (1987). Probucol prevents the progression of atherosclerosis in Watanabe heritable hyperlipidemic rabbit, an animal model for familial hypercholesterolemia. Proc. Natl. Acad. Sci. U. S. A. *84*, 5928-5931.

Knechtle, S. J., Wang, J., Jiao, S., Geissler, E. K., Sumimoto, R., and Wolff, J. (1994). Induction of specific tolerance by intrathymic injection of recipient muscle cells transfected with donor class I major histocompatibility complex [see comments]. Transplantation *57*, 990-996.

Kozarsky, K., Grossman, M., and Wilson, J. M. (1993). Adenovirus-mediated correction of the genetic defect in hepatocytes from patients with familial hypercholesterolemia. Somat. Cell Mol. Genet. *19*, 449-458.

Kozarsky, K. F., McKinley, D. R., Austin, L. L., Raper, S. E., Stratford Perricaudet, L. D., and Wilson, J. M. (1994). In vivo correction of low density lipoprotein receptor deficiency in the Watanabe heritable hyperlipidemic rabbit with recombinant adenoviruses. J. Biol. Chem. *269*, 13695-13702.

Larson, J. L., Wolf, D. C., and Butterworth, B. E. (1995). Induced regenerative cell proliferation in livers and kidneys of male F-344 rats given chloroform in corn oil by gavage or ad libitum in drinking water. Toxicology *95*, 73-86.

Lazjuk, G., Zolotukhina, T., Kirillova, I., Lurie, I., Novikova, I., and Abasheva, G. (1987). Prenatal morphology of partial monosomy 18q. Zentralbl. Gynakol. *109*, 126-129.

Lebkowski, J. S., McNally, M. M., Okarma, T. B., and Lerch, L. B. (1988). Adeno-associated virus: a vector system for efficient introduction and integration of DNA into a variety of mammalian cell types. Mol. Cell Biol. *8*, 3988-3996.

Ledda Columbano, G. M., Coni, P., Curto, M., Giacomini, L., Faa, G., Oliverio, S., Piacentini, M., and Columbano, A. (1991). Induction of two different modes of cell death, apoptosis and necrosis, in rat liver after a single dose of thioacetamide. Am. J. Pathol. *139*, 1099-1109.

Lewis, P. F. and Emerman, M. (1994). Passage through mitosis is required for oncoretroviruses but not for the human immunodeficiency virus. J. Virol. *68*, 510-516.

Li, A. P., Barker, G., Beck, D., Colburn, S., Monsell, R., and Pellegrin, C. (1993). Culturing of primary hepatocytes as entrapped aggregates in a packed bed bioreactor: a potential bioartificial liver. In Vitro Cell Dev. Biol. *29A*, 249-254.

Li, J., Fang, B., Eisensmith, R. C., Li, X. H., Nasonkin, I., Lin Lee, Y. C., Mims, M. P., Hughes, A., Montgomery, C. D., Roberts, J. D., and et al. (1995). In vivo gene therapy for hyperlipidemia: phenotypic correction in Watanabe rabbits by hepatic delivery of the rabbit LDL receptor gene. J. Clin. Invest. *95*, 768-773.

Lieber, A. and Strauss, M. (1995). Selection of efficient cleavage sites in target RNAs by using a ribozyme expression library. Mol. Cell Biol. *15*, 540-551.

Lieber, A., Vrancken Peeters, M. J., and Kay, M. A. (1995). Adenovirus-mediated transfer of the amphotropic retrovirus receptor cDNA increases retroviral transduction in cultured cells. Hum. Gene Ther. *6*, 5-11.

Lieber, A., Vrancken Peeters, M. J., Meuse, L., Fausto, N., Perkins, J., and Kay, M. A. (1995). Adenovirus-mediated urokinase gene transfer induces liver regeneration and allows for efficient retrovirus transduction of hepatocytes in vivo. Proc. Natl. Acad. Sci. U. S. A. *92*, 6210-6214.

Lindgren, V., Luskey, K. L., Russell, D. W., and Francke, U. (1985). Human genes involved in cholesterol metabolism: chromosomal mapping of the loci for the low density lipoprotein receptor and 3-hydroxy-3-methylglutaryl-coenzyme A reductase with cDNA probes. Proc. Natl. Acad. Sci. U. S. A. *82*, 8567-8571.

Liu, H. W., Ofosu, F. A., and Chang, P. L. (1993). Expression of human factor IX by microencapsulated recombinant fibroblasts. Hum. Gene Ther. *4*, 291-301.

Lochmuller, H., Petrof, B. J., Allen, C., Prescott, S., Massie, B., and Karpati, G. (1995). Immunosuppression by FK506 markedly prolongs expression of adenovirus-delivered transgene in skeletal muscles of adult dystrophic [mdx] mice. Biochem. Biophys. Res. Commun. *213*, 569-574.

Lu, B., Gupta, S., and Federoff, H. (1995). Ex vivo hepatic gene transfer in mouse using a defective herpes simplex virus-1 vector. .Hepatology *21*, 752-759.

Maldonado, V., Chan, L., Melendez, J., Rincon, A. R., Zhu, H. J., and Panduro, A. (1994). Regulation of apo B mRNA expression in liver and intestine during liver regeneration induced by CCl_4. Biochim. Biophys. Acta *1211*, 1-6.

Malone, R. W., Hickman, M. A., Lehmann Bruinsma, K., Sih, T. R., Walzem, R., Carlson, D. M., and Powell, J. S. (1994). Dexamethasone enhancement of gene expression after direct hepatic DNA injection. J. Biol. Chem. *269*, 29903-29907.

Matter, K., Whitney, J. A., Yamamoto, E. M., and Mellman, I. (1993). Common signals control low density lipoprotein receptor sorting in endosomes and the Golgi complex of MDCK cells. Cell *74*, 1053-1064.

McCray, P. B.,Jr., Armstrong, K., Zabner, J., Miller, D. W., Koretzky, G. A., Couture, L., Robillard, J. E., Smith, A. E., and Welsh, M. J. (1995). Adenoviral-mediated gene transfer to fetal pulmonary epithelia in vitro and in vivo. J. Clin. Invest. *95*, 2620-2632.

Middleton, P. G., Caplen, N. J., Gao, X., Huang, L., Gaya, H., Geddes, D. M., and Alton, E. W. (1994). Nasal application of the cationic liposome DC-Chol:DOPE does not alter ion transport, lung function or bacterial growth [published erratum appears in Eur Respir J 1994 May;7(5):1024]. Eur. Respir. J. *7*, 442-445.

Mitani, K., Wakamiya, M., Hasty, P., Graham, F. L., Bradley, A., and Caskey, C. T. (1995). Gene targeting in mouse embryonic stem cells with an adenoviral vector. Somat. Cell Mol. Genet. *21*, 221-231.

Morrow, B. and Kucherlapati, R. (1993). Gene targeting in mammalian cells by homologous recombination. Curr. Opin. Biotechnol. *4*, 577-582.

Moullier, P., Marechal, V., Danos, O., and Heard, J. M. (1993). Continuous systemic secretion of a lysosomal enzyme by genetically modified mouse skin fibroblasts. Transplantation *56*, 427-432.

Muzyczka, N. (1992). Use of adeno-associated virus as a general transduction vector for mammalian cells. Curr. Top. Microbiol. Immunol. *158*, 97-129.

Neda, H., Wu, C. H., and Wu, G. Y. (1991). Chemical modification of an ecotropic murine leukemia virus results in redirection of its target cell specificity. J. Biol. Chem. *266*, 14143-14146.

Ochi, A., Yuh, K., and Migita, K. (1993). Not every superantigen induces tolerance in vivo. Semin. Immunol. *5*, 57-63.

Pages, J. C., Andreoletti, M., Bennoun, M., Vons, C., Elcheroth, J., Lehn, P., Houssin, D., Chapman, J., Briand, P., Benarous, R., and et al. (1995). Efficient retroviral-mediated gene transfer into primary culture of murine and human hepatocytes: expression of the LDL receptor. Hum. Gene Ther. *6*, 21-30.

Palmer, T. D., Rosman, G. J., Osborne, W. R., and Miller, A. D. (1991). Genetically modified skin fibroblasts persist long after transplantation but gradually inactivate introduced genes. Proc. Natl. Acad. Sci. U. S. A. *88*, 1330-1334.

Ponder, K. P., Gupta, S., Leland, F., Darlington, G., Finegold, M., DeMayo, J., Ledley, F. D., Chowdhury, J. R., and Woo, S. L. (1991). Mouse hepatocytes migrate to liver parenchyma and function indefinitely after intrasplenic transplantation. Proc. Natl. Acad. Sci. U. S. A. *88*, 1217-1221.

Rawle, F. C., Tollefson, A. E., Wold, W. S., and Gooding, L. R. (1989). Mouse anti-adenovirus cytotoxic T lymphocytes. Inhibition of lysis by E3 gp19K but not E3 14.7K. J. Immunol. *143*, 2031-2037.

Remy, J. S., Kichler, A., Mordvinov, V., Schuber, F., and Behr, J. P. (1995). Targeted gene transfer into hepatoma cells with lipopolyamine-condensed DNA particles presenting galactose ligands: a stage toward artificial viruses. Proc. Natl. Acad. Sci. U. S. A. *92*, 1744-1748.

Rettinger, S. D., Ponder, K. P., Saylors, R. L., Kennedy, S. C., Hafenrichter, D. G., and Flye, M. W. (1993). In vivo hepatocyte transduction with retrovirus during in-flow occlusion. J. Surg. Res. *54*, 418-425.

Richards, C. A. and Huber, B. E. (1993). Generation of a transgenic model for retrovirus-mediated gene therapy for hepatocellular carcinoma is thwarted by the lack of transgene expression. Hum. Gene Ther. *4*, 143-150.

Sabapathy, T. K., Hwang, L. A., and Hui, K. M. (1994). Differential effect of staphylococcal enterotoxin B upon the induction of tolerance on peripheral CD4+V beta 8+ and CD8+V beta 8+ T cells. Cell Immunol. *158*, 83-95.

Salmons, B. and Gunzburg, W. H. (1993). Targeting of retroviral vectors for gene therapy. Hum. Gene Ther. *4*, 129-141.

Savransky, E., Hytiroglou, P., Harpaz, N., Thung, S. N., and Johnson, E. M. (1994). Correcting the PiZ defect in the alpha 1-antitrypsin gene of human cells by targeted homologous recombination. Lab. Invest. *70*, 676-683.

Scanu, A. M., Khalil, A., Neven, L., Tidore, M., Dawson, G., Pfaffinger, D., Jackson, E., Carey, K. D., McGill, H. C., and Fless, G. M. (1988). Genetically determined hypercholesterolemia in a rhesus monkey family due to a deficiency of the LDL receptor. J. Lipid Res. *29*, 1671-1681.

Smith, T. A., Mehaffey, M. G., Kayda, D. B., Saunders, J. M., Yei, S., Trapnell, B. C., McClelland, A., and Kaleko, M. (1993). Adenovirus mediated expression of therapeutic plasma levels of human factor IX in mice. Nat. Genet. *5*, 397-402.

Stankovics, J., Crane, A. M., Andrews, E., Wu, C. H., Wu, G. Y., and Ledley, F. D. (1994). Overexpression of human methylmalonyl CoA mutase in mice after in vivo gene transfer with asialoglycoprotein/polylysine/DNA complexes. Hum. Gene Ther. *5*, 1095-1104.

Stratford Perricaudet, L. D., Briand, P., and Perricaudet, M. (1992). Feasibility of adenovirus-mediated gene transfer in vivo. Bone Marrow Transplant. 9 Suppl *1*, 151-152.

Stratford Perricaudet, L. D., Levrero, M., Chasse, J. F., Perricaudet, M., and Briand, P. (1990). Evaluation of the transfer and expression in mice of an enzyme-encoding gene using a human adenovirus vector. Hum. Gene Ther. *1*, 241-256.

Stratford Perricaudet, L. D., Makeh, I., Perricaudet, M., and Briand, P. (1992). Widespread long-term gene transfer to mouse skeletal muscles and heart. J. Clin. Invest. *90*, 626-630.

Sun, T. Q., Fernstermacher, D. A., and Vos, J. M. (1994). Human artificial episomal chromosomes for cloning large DNA fragments in human cells [published erratum appears in Nat Genet 1994 8(4):410]. Nat. Genet. *8*, 33-41.

Tai, I. T. and Sun, A. M. (1993). Microencapsulation of recombinant cells: a new delivery system for gene therapy. FASEB J. *7*, 1061-1069.

Takehara, T., Hayashi, N., Miyamoto, Y., Yamamoto, M., Mita, E., Fusamoto, H., and Kamada, T. (1995). Expression of the hepatitis C virus genome in rat liver after cationic liposome-mediated in vivo gene transfer. Hepatology *21*, 746-751.

Takeshita, S., Gal, D., Leclerc, G., Pickering, J. G., Riessen, R., Weir, L., and Isner, J. M. (1994). Increased gene expression after liposome-mediated arterial gene transfer associated with intimal smooth muscle cell proliferation. In vitro and in vivo findings in a rabbit model of vascular injury. J. Clin. Invest. *93*, 652-661.

Takeuchi, Y., Cosset, F. L., Lachmann, P. J., Okada, H., Weiss, R. A., and Collins, M. K. (1994). Type C retrovirus inactivation by human complement is determined by both the viral genome and the producer cell. J. Virol. *68*, 8001-8007.

Thierry, A. R., Lunardi Iskandar, Y., Bryant, J. L., Rabinovich, P., Gallo, R. C., and Mahan, L. C. (1995). Systemic gene therapy: biodistribution and long-term expression of a transgene in mice. Proc. Natl. Acad. Sci. U. S. A. *92*, 9742-9746.

Thompson, T. A., Gould, M. N., Burkholder, J. K., and Yang, N. S. (1993). Transient promoter activity in primary rat mammary epithelial cells evaluated using particle bombardment gene transfer. In Vitro Cell Dev. Biol. 29A, 165-170.

Thomson, A. W. (1994). Immunosuppressive drugs and the induction of transplantation tolerance. Transpl. Immunol. 2, 263-270.

Tietz, P. S., Yamazaki, K., and LaRusso, N. F. (1990). Time-dependent effects of chloroquine on pH of hepatocyte lysosomes. Biochem. Pharmacol. *40*, 1419-1421.

Tsukamoto, M., Ochiya, T., Yoshida, S., Sugimura, T., and Terada, M. (1995). Gene transfer and expression in progeny after intravenous DNA injection into pregnant mice. Nat. Genet. *9*, 243-248.

Vemuru, R. P., Davidson, A., Aragona, E., Chowdhury, J. R., Burk, R. D., and Gupta, S. (1992). Immune tolerance to a defined heterologous antigen after intrasplenic hepatocyte transplantation: implications for gene therapy. FASEB J. *6*, 2836-2842.

Wagner, E., Plank, C., Zatloukal, K., Cotten, M., and Birnstiel, M. L. (1992). Influenza virus hemagglutinin HA-2 N-terminal fusogenic peptides augment gene transfer by transferrin-polylysine-DNA complexes: toward a synthetic virus-like gene-transfer vehicle. Proc. Natl. Acad. Sci. U. S. A. *89*, 7934-7938.

Walz, C., Schlehofer, J. R., Flentje, M., Rudat, V., and zur Hausen, H. (1992). Adeno-associated virus sensitizes HeLa cell tumors to gamma rays. J. Virol. *66*, 5651-5657.

Wang, Q. and Taylor, M. W. (1993). Correction of a deletion mutant by gene targeting with an adenovirus vector. Mol. Cell Biol. *13*, 918-927.

Wang, X., Pai, J. T., Wiedenfeld, E. A., Medina, J. C., Slaughter, C. A., Goldstein, J. L., and Brown, M. S. (1995). Purification of an interleukin-1 beta converting enzyme-related cysteine protease that cleaves sterol regulatory element-binding proteins between the leucine zipper and transmembrane domains. J. Biol. Chem. *270*, 18044-18050.

Wang, X., Sato, R., Brown, M. S., Hua, X., and Goldstein, J. L. (1994). SREBP-*1*, a membrane-bound transcription factor released by sterol-regulated proteolysis [see comments]. Cell *77*, 53-62.

Weiler, J. M., Edens, R. E., Linhardt, R. J., and Kapelanski, D. P. (1992). Heparin and modified heparin inhibit complement activation in vivo. J. Immunol. *148*, 3210-3215.

Weiner, H. L., Friedman, A., Miller, A., Khoury, S. J., al Sabbagh, A., Santos, L., Sayegh, M., Nussenblatt, R. B., Trentham, D. E., and Hafler, D. A. (1994). Oral tolerance: immunologic mechanisms and treatment of animal and human organ-specific autoimmune diseases by oral administration of autoantigens. Annu. Rev. Immunol. *12*, 809-837.

Wilson, J. M., Grossman, M., Cabrera, J. A., Wu, C. H., and Wu, G. Y. (1992). A novel mechanism for achieving transgene persistence in vivo after somatic gene transfer into hepatocytes. J. Biol. Chem. *267*, 11483-11489.

Wilson, J. M., Grossman, M., Raper, S. E., Baker, J. R.,Jr., Newton, R. S., and Thoene, J. G. (1992). Ex vivo gene therapy of familial hypercholesterolemia. Hum. Gene Ther. *3*, 179-222.

Wilson, J. M., Grossman, M., Wu, C. H., Chowdhury, N. R., Wu, G. Y., and Chowdhury, J. R. (1992). Hepatocyte-directed gene transfer in vivo leads to transient improvement of hypercholesterolemia in low density lipoprotein receptor-deficient rabbits. J. Biol. Chem. *267*, 963-967.

Wissing, M., Alegre, M. L., Abramowicz, D., Willems, F., Donckier, V., Vanderhaeghen, M. L., Leo, O., and Goldman, M. (1995). Emergence of Th2-like cells after anti-CD3 monoclonal antibody administration in mice. Transplant. Proc. *27*, 378-379.

Wu, G. Y., Keegan Rogers, V., Franklin, S., Midford, S., and Wu, C. H. (1988). Targeted antagonism of galactosamine toxicity in normal rat hepatocytes in vitro. J. Biol. Chem. *263*, 4719-4723.

Yamamoto, T., Bishop, R. W., Brown, M. S., Goldstein, J. L., and Russell, D. W. (1986). Deletion in cysteine-rich region of LDL receptor impedes transport to cell surface in WHHL rabbit. Science *232*, 1230-1237.

Yang, N. S., Burkholder, J., Roberts, B., Martinell, B., and McCabe, D. (1990). In vivo and in vitro gene transfer to mammalian somatic cells by particle bombardment. Proc. Natl. Acad. Sci. U. S. A. *87*, 9568-9572.

Yang, Y. and Wilson, J. M. (1995). Clearance of adenovirus-infected hepatocytes by MHC class I-restricted CD4+ CTLs in vivo. J. Immunol. *155*, 2564-2570.

Yates, J. L., Warren, N., and Sugden, B. (1985). Stable replication of plasmids derived from Epstein-Barr virus in various mammalian cells. Nature *313*, 812-815.

Yee, J. K., Moores, J. C., Jolly, D. J., Wolff, J. A., Respess, J. G., and Friedmann, T. (1987). Gene expression from transcriptionally disabled retroviral vectors. Proc. Natl. Acad. Sci. U. S. A. *84*, 5197-5201.

Yokode, M., Hammer, R. E., Ishibashi, S., Brown, M. S., and Goldstein, J. L. (1990). Diet-induced hypercholesterolemia in mice: prevention by overexpression of LDL receptors. Science *250*, 1273-1275.

Zabner, J., Fasbender, A. J., Moninger, T., Poellinger, K. A., and Welsh, M. J. (1995). Cellular and molecular barriers to gene transfer by a cationic lipid. J. Biol. Chem. *270*, 18997-19007.

Zhang, W. W., Alemany, R., Wang, J., Koch, P. E., Ordonez, N. G., and Roth, J. A. (1995). Safety evaluation of Ad5CMV-p53 in vitro and in vivo. Hum. Gene Ther. *6*, 155-164.

Zhu, N., Liggitt, D., Liu, Y., and Debs, R. (1993). Systemic gene expression after intravenous DNA delivery into adult mice. Science *261*, 209-211.

19. Therapeutic Strategies for the Lysosomal Storage Disorders

John A. Barranger, Michael J. Vallor, Randall Learish, Alfred Bahnson, Maya Nimgaonkar, Edward Ball, Jane Mannion-Henderson, Trina Mohney, James Dunigan, Margeret Beeler, Joseph Mierski, Jason Lancia, Amy Kemp, and Erin Rice

19.1 Introduction

The lysosomal storage disorders are a group of inherited disorders for which gene therapy offers a potential cure. With few exceptions, the genes coding the lysosomal enzymes have been cloned and characterized. An important part of the rationale for the gene therapy approach is based on the demonstration that the gene products are therapeutic. Studies in patients with Gaucher disease and in several animal models of lysosomal storage disorders demonstrate that enzyme replacement reverses the visceral complications of lysosomal storage (1-7). Furthermore, bone marrow transplantation (BMT) studies suggest that hematopoietic cells may be appropriate targets for gene transfer and may alleviate the signs and symptoms in some of the lysosomal storage disorders (Tab. 19.1). However, based on available information, the response to genetic correction of bone marrow is likely to be different in each of the disorders. For example, BMT in the mannosidosis cat model results in reversal of both visceral and central nervous system manifestations of the disease(8), but in the globoid cell leudodystrophy mouse BMT has limited effects on the neurological complications of the disorder (9). It is now clear that the results will be highly dependent on the pathobiology specific to each disorder. In developing gene therapy strategies for the lysosomal storage disorders, a primary concern is gene delivery to a target cell in which genetic correction is likely to be efficacious and is feasible. Different strategies will need to be taken to evaluate the potential of gene transfer as a therapeutic tool. The strategies are summarized in Table 19.2. Moreover, studies in animal models will be helpful in predicting the expected results in man.

Table 19.1 Lyosomal Storage Disorders: Results of Bone Marrow Transplantation

Enzyme Deficiency	Disorder	Results
Iduronidase	Human MPS 1	Improved intellectual function; decreased storage in liver, spleen; improved skeletal deformity and mobility
Iduronidase	Canine MPS 1	Improved longevity, mobility; decreased storage and excretion of GAG
Arylsulfatase A	Human MLD	Infantile, late infantile forms – not improved; Juvenile form – improved longevity and intellectual function
Fucosidase	Canine fucosidosis	Decreased storage and fucosyl compounds; improvement in CNS and PNS
α-Mannosidase	Human Mannosidosis	Improved life span; stable neurologic signs for about 7 months
α-Mannosidase	Feline Mannosidosis	Improved storage in viscera and brain; improved neurologic function
Arylsulfatase B	Human MPS VI	Decreased visceral storage; improvement in cardiovascular system and skeleton
Galactocerebrosidase	Human globoid cell leukodystrophy	Some improvement in longevity, decreased storage in brain
Galactocerebrosidase	Murine globoid cell leukodystrophy	Increased longevity; decreased storage in brain
Cerebroside β-galactosidase	GM1 gangliosidosis	Decreased storage; increased longevity; improved CNS
Glucocerebrosidase	Glucosylceramide lipidosis	Decreased visceromegaly; improved longevity in Type 1 disease
Sphingomyelase	Sphingomyelin lipidosis	Decreased storage in viscera; no change in CNS
Glucuronidase	Canine MPS VII	Improved longevity; increased mobility
Glucuronidase	Murine MPS VII	Increased longevity;decreased visceral storage, no change in CNS, PNS or eye

19.2 Animal Models of Lysosomal Storage Disorders

Lysosomal storage disorders are caused by the deficiency of a catabolic enzyme. Localized within the organelle by a highly specialized transport system, the enzymes reside within the lysosome, and are responsible for the degradation of macromolec-

Table 19.2 Gene Transfer Strategies

1. *Ex vivo* gene transfer to transplantable cells **involved** in the **pathogenesis**
   Distribution *in vivo*: Widely or locally
   Examples: Bone marrow, hepatocytes, oligodendrocyte
2. *Ex vivo* gene transfer to transplantable cells capable of secreting a **therapeutic protein**
   Distribution *in vivo*: Usually locally (mini-organ)
   Examples: Fibroblasts, myoblasts, bone marrow
3. *In vivo* **targeted** gene transfer to a single organ in which genetic correction has **broad** therapeutic effects on **multiple** organs
   Distribution *in vivo*: Single organ
   Example: Liver
4. *In vivo* **direct** gene transfer locally to correct a major manifestation of the disease
   Distribution *in vivo*: Local
   Examples: Brain, vascular endothelium

ular polymers or complex molecules. Failure to degrade these molecules results in storage of the compounds and disease. The principle compounds involved in the lysosomal storage disorders are glycosaminoglycans (mucopolysaccharide), glycolipids, and glycoproteins (10, 11). Substrate accumulation is followed by cell dysfunction and death, and the development of severe and often ultimately lethal symptoms in the patient, particularly when the central nervous system is involved. Many of these enzymatic deficiencies are also found in animals (Tab. 19.3) (12-29), providing defined and controlled models for study. Bone marrow transplantation has been studied in some of these models, and correlations have been drawn to the human diseases (Tab. 19.1). These results are useful in considering strategies designed to transfer genes to bone marrow stem cells (20-31). The majority of these animal models are found naturally in domestic species, where selective inbreeding promotes the production of homozygotes for recessive mutant alleles. Many examples were discovered in pet and farm animals (32). For diseases not yet discovered in animals, the advent of gene targeting in pluripotent mouse embryonic stem (ES) cells (33) has allowed the experimental development of at least three murine examples of human lysosomal storage conditions; glucocerebroside storage (Gaucher disease), cerebroside sulfatide storage (metachromatic leukodystrophy), and hexosaminidase A deficiency (Tay Sachs disease) (15-17). Additionally, gene targeted murine models of sphingomyelinase deficiency (Niemann-Pick disease), mucopolysaccharidosis VI, and ceramide trihexosidase deficiency (Fabry disease) have been reported (18-20), but are not yet fully described in the published literature

19.2.1 The Gaucher Mouse – Interpretation of Pathology

One of the most well studied lysosomal storage diseases , Gaucher disease is caused by the deficiency of the glucocerebrosidase (GC) (34). Presentation is usually not

Table 19.3 Some Animal Models of Lysosomal Storage Diseases

Species	Enzyme Deficiency	Human Disease Counterpart[ref]
Rodent-Mouse	β-glucuronidase	Mucopolysaccharidosis VII[4] (Sly Disease)
	Galactosylceramidase	Globoid Cell Leukodystrophy[5]
	Unknown, Cholesterol transport	Niemann Pick Disease Type C[6]
	Arylsulfatase A *	Metachromatic Leukodystrophy[7]
	Glucocerebrosidase*	Gaucher Disease[8]
	Hexosaminidase A*	Tay Sachs Disease[9]
	Sphingomyelinase*	Niemann-Pick Disease[10]
	Arylsulfatase B*	Mucopolysaccharoidosis VI[11] (Marotoux Lamy Disease)
	Ceramide Trihexosidase*	Fabry Disease[12]
Rodent-Rat	Lysosomal Acid Lipase	Wolman's Disease[13]
Cat	α-Mannosidase	α-Mannosidosis[14]
	β-Hexosaminidase A	Type II GM2 Gangliosidosis[15]
	Arylsulfatase B	Mucopolysaccharoidosis VI[16]
Dog	β-Galactosidase	GM1 Gangliosidosis[17]
	Fucosidase,	Fucosidosis[18]
	Iduronidase	Hurler Disease[19]
Sheep	β-Galactosidase and α-Neuraminidase	GM1 Gangliosidosis[20]
Goat	β-Mannosidase	β-Mannosidosis[21]
Bovine	α-Mannosidase	α-Mannosidosis[22]

associated with neurological symptoms in most patients (type 1 disease), but there is typically liver, spleen, and skeletal involvement. Storage of the GC substrate, glucocerebroside, is largely confined to the cells of the reticuloendothelial system, particularly tissue macrophages. The disease is variable in the severity of symptoms, and age of onset, with patients sometimes being diagnosed in late adulthood. The correlation of genotype and phenotype is inexact. Despite these limitations, effective molecular treatments are available which can partly or wholly ameliorate the disease symptoms. These include both enzyme replacement therapy (2, 3), and bone marrow transplantation (BMT) (35). However, in spite of our understanding of the causes of Gaucher disease, many uncertainties remain concerning the fundamental pathobiology of the disease process, and the cellular mechanisms of its treatment. For these reasons, an animal model is desirable. However, since no naturally occurring example of glucocerebroside storage in animals has been described, a glucocerebrosidase deficient mouse was produced using gene targeting techniques. It was the first animal model of a lysosomal storage disorder produced using these techniques (16). A naturally occurring canine example of Gaucher disease was described in the early 1970s, but was not able to be propagated (36). This dog exhibited neurovisceral storage of glucocerebroside, but no organomegaly or skeletal involvement. The gene-targeted glucocerebrosidase deficient mouse was described in 1992. In this gene "knockout" approach, the neomycin phosphotransferase (neo^r) gene was introduced into exons 9 and 10 of the mouse GC gene. Homozygotes with this disruption exhibited cyanosis, poor feeding and movement, thin skin with rugations, and death in the first hours of life (16). Glucocerebroside storage was seen in macrophages of the liver, bone marrow, spleen and brain, but there was not appreciable organomegaly. The short life span of these mice probably did not allow time for the development of more profound organomegaly, and bone lesions. However, adult mice which are treated with the glucocerebrosidase inhibitor, conduritol B epoxide, develop visceral glucocerebroside storage cells, but never develop organomegaly (37). It is therefore possible that glucocerebroside metabolism and the physiologic effects of its storage may be different in mice and humans. Indeed, the first animal model developed with gene targeting was for Lesch Nyhan disease (hypoxanthine phosphoribosyl transferase deficiency), in which it was found that homozygotes for the gene disruption exhibited no phenotype at all (38). Inhibition of a related metabolic pathway was required for the Lesch Nyhan mouse model to display the overt pathology associated with the human disease (39).

The transgenic glucocerebrosidase deficient mouse model is representative of a rare subset of very severe Gaucher disease (40). The skin rugations in particular are interesting as they have been observed only in these cases, which are not typical of Type 2 (neurologic) disease. It has been hypothesized that GC activity may be important in maintaining skin permeability barrier integrity (38). Thus, while this model may not exhibit all of the commonly observed symptoms of Gaucher disease, it may provide some insight into glucocerebroside metabolism that was previously unknown.

The lethal phenotype of the GC gene disruption precludes the use of this model in most studies of gene transfer and cell transplantation. In an attempt to abrogate this lethal phenotype, PCR mutagenesis was used to install the most common human Gaucher disease mutation (N370S) directly onto the mouse gene (41). The selectable neo^r gene was located to the region just outside the 3' end of the mouse GC gene so that the gene function would not be destroyed. Homozygotes for this mutant GC locus die and are resorbed *in utero,* due to the disruption of a previously unknown gene, metaxin, which lies less than a kilobase downstream of the mouse GC gene (42). The metaxin gene was not affected in the first gene targeted mouse, and heterozygotes of both targeted loci were mated. These matings produced mice with one functional metaxin gene and one mutant, but otherwise intact GC gene. An important and unanticipated finding was that the N370S mutant GC was not found to rescue the lethal phenotype of the GC gene disruption, and these mice exhibited a pathology identical to mice with the homozygous gene disruption (42). This outcome may indicate that other important genes may have been affected during homologous recombination, either in the original knockout (8), or in the disruption involving both the GC and metaxin genes (42). The GC and metaxin genes lie in a tight group along with the thrombospondin 3 gene (42) – other genes may also lie closely upstream of GC. Furthermore, it is also possible that the N370S mutation affects the murine enzyme more severely than the human protein (the genes are only 80% homologous between species (43)). Finally, there may be differences between the human and murine physiologies, such that the mutant enzyme is rendered nonfunctional in mice. Rescue of the lethal phenotype may be better with a human mutant transgene (44) rather than trying to impose this mutation directly on the murine protein.

So far, none of the animal models of Gaucher disease have exhibited the organomegaly and skeletal involvement prominent in the human disease. It may be that organomegaly and skeletal phenotypes will not be seen in Gaucher disease in non-primate systems. If these models are to be used to examine Gaucher disease therapies, simple clearance of the stored lipid may be the only experimental parameter available to judge clinical effectiveness. It can be seen, however, that interspecies differences in physiology may complicate the use of animal models of lysosomal storage diseases, and that definitions of specific diseases as they present in the animal, may have to be modified. Indeed, the gene targeting models of arylsulfatase A and hexosaminidase A deficiencies also exhibit only lipid storage, with no other early physiological changes. It can be seen that the effects of enzyme deficiency in humans and the corresponding animal models may be very different, particularly in the case of gene targeting where the complete absence of enzyme is not biochemically reflective of most disease states. Even in these models, dissimilar pathologies in human and animal systems may offer insights into aspects of lysosomal storage diseases and their metabolic and physiological effects that are not apparent in most human cases.

Animal models of lysosomal storage disease provide a powerful investigative tool, particularly when the animal phenotype closely matches the disease symptoms in humans as in the MPS VII mouse (45). A variety of therapeutic strategies may be assessed without resorting to human clinical experimentation. Caution must be exercised in the use of animal models, however, as animal physiologies may differ from human, leading to unexpected observations. This is especially evident in gene targeted mouse models of lysosomal storage disease, considering the lethal results of glucocerebrosidase deficiency in mice, and the lack of overt morphological effects in the hexosaminidase A and arylsulfatase A-deficient mouse models. However, as is seen in the skin abnormalities of the GC-deficient mouse, animal models which differ from the human phenotype, can provide additional insight into the disease pathology that is not possible in humans.

19.3 Therapeutic Strategies

19.3.1 Enzyme Therapy for Gaucher Disease

Prior to our characterization of the oligosaccharide side chains of glucocerebrosidase as 'complex type' with the sequence of R-asp-GlcNac-GlcNac-Man-GlcNac-GalNeuNac (46), we had deduced that this sequence was present in the oligosaccharide by sequential exoglycosidase treatments and in vivo lectin binding studies (1). These studies were carried out to explore the possibility that a mannose-specific lectin existed on non-parenchymal cells (Kupffer cells and endothelial cells) in liver that could facilitate the uptake of glucocerebrosidase by receptor-mediated endocytosis. The conclusions from these studies were that a mannose-specific receptor did bind glucocerebrosidase on which mannose residues were revealed by partial degradation of the oligosaccharide by exoglycosidase treatment. These results have recently been confirmed (47). This system could be utilized to import glucocerebrosidase into the non-parenchymal cell population of liver, resulting in a 10-40 fold increase in glucocerebrosidase enzyme activity in these cells (1) These results led to a clinical trial of mannose-terminated glucocerebrosidase in a single patient with Gaucher disease. In this study, profound anemia was reversed by infusions of 10mg/week of mannose-terminated glucocerebrosidase (2, 34). This amount of enzyme is a calculated full replacement of the enzyme present in liver. Subsequently, these clinical results were confirmed in a study of 12 patients who received 60 I.U/kg of enzyme every two weeks (3). Recently, it has been suggested that a dose of 30 U/kg per month or less would be as efficacious as higher doses if given in small infusions three times each week (48). This conclusion has been questioned because the hematologic responses at this dose and regimen were inferior (49, 50). Furthermore, only a few patients were studied. However, a few patients did have a reduction in liver size in this study. A larger study has been reported (51) in which the smaller dose, more frequent regimen appeared to be effective. The ensuing controversies regarding dosing have not shed much light on the issues. Recently, the Office of Medical Applications of Research conducted a Technology Assessment Conference on Gaucher Disease at the National Institutes of Health in Bethesda, Maryland in March, 1995. The panel of 15 reviewers concluded that it was not possible to determine what dose of glucocerebrosidase should be recommended. The panel report states: 'Clearly, the choice of dosage and frequency of enzyme administration will have to be adjusted individually while monitoring each patient's progress.' A recent report from the manufacturer presents data that the majority of patients can tolerate a reduction of the initial dose from 120 U/kg to a maintenance dose of 60 U/kg monthly without regression of symptoms (52). Whether further reductions of dose are possible must be answered by additional study.

Until recently there has been little data on how enzyme therapy might effect patients with neuronopathic Gaucher disease (Types 2 and 3 patients). Theoretically, enzyme replacement might be beneficial in these cases. There are two reasons for this conclusion. First, animal models and neurologically affected patients with lysosomal storage diseases can improve following bone marrow transplantation. These results indicated that activity of the enzyme in or from transplanted cells is causing the storage material to decrease in the brain. Second, the glucosylceramide that accumulates in the brain of neurologically affected patients is probably largely derived from sources outside the central nervous system. Therefore, a decrease in substrate load over a long time may result in a decrease in brain glucosylceramide. Enzyme replacement with glucocerebrosidase has been shown to result in a decrease in total glucosylceramide content in extraneural tissues. Given these observations, there is a small chance that enzyme replacement will benefit patients with neurological involvement.

For these reasons, we initiated a clinical study of patients with Types 2 and 3 Gaucher disease. We have entered three patients with the rare Type 3 phenotype on this protocol. Two of the three patients have genotypes that are correlated with the neuronopathic forms of Gaucher disease, i.e. 444/444 or 444/?. In two patients treated for longer than 30 months there has been no progression of neurological symptoms. The third patient shows no progression after a year of treatment. Other groups studying enzyme treatment of neuronopathic Gaucher disease indicate that progression of neurological symptoms appears to be slower in patients receiving glucocerebrosidase infusions (53, 54). Unpublished data suggests that there is little or no effect of enzyme infusions, even if given intrathecally, in type 2 cases.

19.3.2 Transgenic Production of Large Amounts of Glucocerebrosidase

In an effort to render enzyme replacement therapy more available and cost-effective, dairy animals may be useful as 'bio-reactors' for enzyme production. Introduction of the human GC gene into transgenic dairy cattle under the control of a mammary-specific promoter could allow the excretion of the human enzyme into cow's milk. Extraction of pharmaceutically useful amounts of GC from milk should be greatly simplified because of both the liquid nature of milk (free soluble protein) and the great production capacity of dairy cattle (300 kg milk protein/animal per year (55)). This approach would abrogate the problems both of low abundance and extraction from solid tissue. Recombinant mammalian cell culture production systems are typically hampered by low expression, and recombinant bacterial production – though capable of large yields – lack the ability to add oligosaccharide side chains that many mammalian proteins (including GC) require for biological activity. Furthermore, recombinant fermentation production systems often require complex and labor/material intensive fermentation and purification processes.

The cost of large-scale GC production in dairy animals is speculative at this time. It has been estimated for other proteins that transgenic production will be 5-12 fold more efficient in production costs than cell culture (55). Expense will be incurred in the initial production of founder animals for any transgenic line. Also, the downstream purification can be complex, depending on the nature of the recombinant protein (56). However, transgenic mice are now producing foreign proteins in their milk at up to 20-30% of the total milk protein content. If GC transgene expression in cattle can be attained at only 10% of total milk protein then up to 30kg of recombinant GC may be collected from each animal annually (55). This represents about 8650 liters of milk per animal per year (cow's milk protein content is 3.4% of the total mass of milk (56)). Several groups have achieved success in producing human proteins in the milk of rodents and large farm animals. α-1-Antitryrsin was secreted at levels of up to 7 mg/ml in mice (57), and at up to 35 g/l in sheep (58). Sheep have also been used to produce human clotting factor IX at 25 ng/ml (59). Human longer-acting tissue phaminogen activator (LAtPA) has been produced in goat milk at 3Ýg/ml (60), and human IL-2 has been produced at up to 430 ng/ml in rabbits (61). The human lactoferrin gene has been successfully introduced into dairy cattle, although no information as to the levels of expression obtained in milk (62). Transgenic pigs have exhibited 1 g/l expression of whey acidic protein (WAP – a rodent milk protein) in their milk (63). Thus transgenic dairy animals may provide considerable savings both in the high concentration of recombinant GC in their milk, and in the relatively low process volume.

19.4 Gene Transfer Studies

19.4.1 Gaucher Disease

Glucocerebrosidase (GC) (E.C. 3.2.1.45) is a monomeric, membrane – associated, hydrophobic glycoprotein with a molecular mass of 65000 daltons. Human GC protein contains 497 amino acids. The enzyme is translated as a precursor protein with a 19 amino acid hydrophobic signal peptide which directs its co-translational insertion into the lumen of the endoplasmic reticulum-golgi-lysosome complex (64). GC acts at the acidic pH of the lysosomal environment to hydrolyze beta-glucosidic linkages in complex lipids. The catalytic activity is increased by detergents, lipids and a naturally occurring activator protein known as sphingolipid activator protein-2 (SAP-2 or saposin C) (65, 66). Human GC cDNA was first cloned from a hepatoma cDNA library expressed in lambda gt11 by screening with a polyclonal antibody developed to purified human placental GC (67). Subsequent characterizations of isolated GC CDNA clones have described the complete nucleotide sequence (68, 69). The glucocerebrosidase gene has been localized to human chromosome 1 q21 by *in situ* hybridization (70). Analysis of genomic clones has shown the GC gene to consist of 11 exons and 10 introns spanning approximately 7KB. A pseudogene exists in tandem with the active gene (70-72).

Deficiency of glucocerebrosidase is the cause of Gaucher disease (73, 74). More than thirty mutations in the gene are known, but only two are common (75). These two account for approximately 80% of the mutant alleles (75). A third mutation has recently been reported and is believed to occur in 20% of type 1 Gaucher patients (76). The several diseases grouped under the eponym of GD are inherited in an autosomal recessive pattern and each disease is caused by a deficiency in the lysosomal acid hydrolase responsible for the catabolism of glucosylceramide. Although glucocerebrosidase is deficient in all cells, for unknown reasons, the accumulation of glucosylceramide occurs virtually only in macrophages. Storage within only one cell type is unique among the lysosomal storage diseases. This characteristic of the pathobiology of Gaucher disease has allowed the development of two successful treatment strategies based on the correction of enzyme deficiency in macrophages. The first approach to be of clinical benefit was allogeneic bone marrow transplantation (BMT) which results in the repopulation of affected tissues with enzyme-competent macrophages. Transplanted patients improve after engraftment of donor marrow and gradually resolve their organomegaly (77). The second approach to result in clinical improvement in Gaucher disease patients was macrophage-targeted enzyme replacement (ERT). This method took advantage of naturally occurring mannose receptors on macrophages and the exposition of accessible mannose residues in the oligosaccharides of glucocerebrosidase to effect an efficient, targeted delivery of the enzyme to macrophages. Patients have responded well to this treatment and many of the related clinical problems have resolved. Both of these approaches are novel and are important because they provide some means of therapy where none previously existed. However, both approaches have significant limitations. Allogeneic bone marrow transplantation carries morbidity and mortality risks that are unacceptable for many patients. Furthermore, matched donors do not exist for the majority of patients. Macrophage-targeted enzyme replacement is a life-long treatment and extremely therapy. Controversy exists over the amount of enzyme and frequency of administration necessary to produce a clinical response. Clearly, additional clinical studies are needed to clarify this issue.

Despite their limitations, the success of BMT and ERT has demonstrated that enzymatic correction of only one cell type – the macrophage – results in effective therapy. This is important from the point of view of developing gene therapy for Gaucher disease because these approaches demonstrate the potential efficacy of gene transfer. The fact that marrow transplantation is effective allows one to consider bone marrow stem cells as an appropriate target for gene transfer. Furthermore, because expression of the GC gene in the macrophage lineage alone may be sufficient for therapy, one could consider, as less permanent alternatives, gene transfer to committed macrophage precursors, peripheral blood monocytes, or to cultures of bone marrow capable of producing enzyme competent macrophage precursors. High efficiency gene transfer vectors that result in sustained expression of the gene in transduced cells would be valuable in the development of these approaches to gene therapy.

To be an effective permanent treatment, gene therapy requires the transfer and sustained expression of genes in cells important to the pathogenesis of the disease. Sufficient expression of a transduced glucocerebrosidase gene in the progeny of pluripotent bone marrow stem cells would correct the deficiency of the enzyme in all cell lineages including monocytes/macrophages. Experience from allogeneic marrow transplantation and macrophage-targeted enzyme replacement supports the idea that gene transfer would be curative provided adequate expression of the GC gene were accomplished in a sufficient number of macrophages. Much experience has been gained with systems to evaluate the efficiency of gene transfer and expression in bone marrow stem cells using retroviral vectors (78, 79). Most of these studies have been conducted in mice where conditions to accomplish efficient gene transfer to hematopoietic stem cells have been defined (78-86). Recent data suggest that 20% or more of the stem cells can be transduced and survive to repopulate marrow (80). For the mouse, critical parameters for efficient retroviral gene transfer and repopulation of bone marrow stem cells including high titer virus producer cell lines (VPL), pretreatment of mice with 5-fluorouracil (5-FU) to initiate stem cell cycling (87-88), pre-culture of bone marrow with growth factors including IL-3 and IL-6 (33), and extended exposure to infection by coculture of the cells with the VPL for 2-6 days with growth factors (80). Several studies have shown that the GC gene can be transferred to mouse bone marrow cells, but these have not demonstrated sustained expression in animals (90-94). A vector that results in sustained expression of the human GC gene *in vivo* would be highly desirable. We have recently reported the development of a retroviral vector with this capability (30). Many fewer studies of gene transfer have been conducted in larger animals and it is apparent that experimental conditions have not yet been fully optimized (79, 95, 96). Conditions that improve the efficiency of gene transfer to human hematopoietic stem cells and at the same time permit their survival as marrow repopulating cells would be useful to the development of gene therapy and would be applicable to the treatment of Gaucher disease as well as other hematopoietic disorders (97, 98). Since retroviral vectors require dividing cells to infect and integrate the potentially therapeutic gene (99, 100), culture systems and a defined set of cytokines that initiate and support human hematopoietic stem cell cycling would be important in increasing the percentage of stem cells transduced. Experience to date in non-human primates and man indicates that stem cell transduction will be less efficient than in murine systems. If this proves to be true, the utility of transplanting shorter-lived transduced progenitors will need to be studied and evaluated.

Models may be useful for the assessment of transduction efficiency and expression of the human GC gene in human cells *in vivo*. The severe combined immunodeficient mouse (SCID) has recently been used to study human hematopoiesis (98). In this model 50% or more of the bone marrow of the mouse is of human origin and persists for several months. Retroviral infection of human bone marrow or enriched stem cells *in vitro* followed by their transplantation into the SCID mouse has permitted the assessment of the efficiency of transduction and persistence of expression of transferred genes in human hematopoietic cells *in vivo*. This model system may be useful in evaluating these parameters in studies of the transfer of the GC gene to human hematopoietic cells. Further, transgenic models of human genetic disease are being developed to study the potential for gene therapy. Several groups including our own are pursuing the mouse model of Gaucher disease by attempting to "knock out" and/or target mutations to the endogenous mouse GC gene using homologous recombination in mouse embryonic stem cells.

As the first step toward the goal of gene therapy for Gaucher disease, we cloned the gene for glucocerebrosidase and demonstrated that the cDNA in a retroviral vector, pWE-GC, could correct the deficiency of GC in fibroblasts from a patient with Gaucher disease (67, 69, 102, 103). Earlier we had demonstrated that allogeneic bone marrow transplantation resulted in the reversal of the disease process. These results supported the hypothesis that gene therapy might be used to treat the disease by transfer of the GC gene to bone marrow cells. In addition, we demonstrated that targeting GC enzyme to macrophages is an effective treatment. This result supported the hypothesis that replacement of even a portion of the activity of the deficient enzyme is adequate to treat the disease, i.e. the gene product is therapeutic. It also provides some encouragement that reconstitution of only a part of the normal enzyme capacity of a tissue by its repopulation with transduced bone marrow derived cells will be sufficient to treat the disease. We concluded from these studies that gene transfer was possible as a treatment for Gaucher disease, provided that some fraction of tissue macrophages could be transduced with and would express the GC gene.

19.4.1.1 Synopsis of Data from Mouse Bone Marrow Transplants

Using a simplified retroviral vector, MFG-GC, reconstitution studies were performed in mice with transduced whole bone marrow (30). These data provide evidence that the retroviral vector can efficiently transduce hematopoietic stem cells in the murine system. The presence of the human GC gene at copy numbers of 1-2 per mouse genome in primary CFU-S, in hematopoietic tissues of mice surviving up to 8 months after reconstitution, and in every spleen colony in secondary recipients demonstrates that primitive cells in the bone marrow were efficiently transduced and repopulated the tissues of the mice. It is unlikely that long lived committed progenitor cells were responsible for this effect considering the length of time after reconstitution. Comparisons to an earlier vector (N2-SV-GC) demonstrated a similar transduction efficiency in primary CFU-S and the tissues of long term reconstituted mice. This result indicates that high transduction efficiency is not a unique property of the MFG-GC vector per se, nor is it necessary in the murine system to develop a titer of greater than 10^6 particles/ml to achieve transduction efficiencies of hematopoietic stem cells that approach 100%. More likely, this high efficiency transduction resulted from the bone marrow culture procedures. The result was probably due to the combination of 5-FU pretreatment, extended pre-culture of the marrow, the use of cytokines (IL-3, IL-6, and stem cell factor) in culture, as well as the coculture of the target marrow with a viral producer line of related animal origin which may elaborate stem cell supporting cytokines.

Sustained, high expression *in vivo* of the transferred human GC gene is a unique feature of the MFG-GC vector. Secondary transplantations resulted in sustained expression of the transgene in white blood cells for more than one year of 2-4 times the background GC activity. The N2-SV-GC vector by comparison resulted in poor expression in spleen colonies and essentially no expression of the gene product in the tissues of long term reconstituted animals. This could have been the result of promoter suppression or down regulation of the heterologous SV40 promoter. We have not examined the reasons for this failure in detail because of the success of the MFG-GC vector. However, the difference is not only simplification of the vector. Several laboratories have observed that simplified vectors carrying the human GC gene failed to result in sustained expression of the gene *in vivo* (90-94). While expression *in vitro* or short term *in vivo* could be achieved, expression of the GC gene was rapidly lost *in vivo*. A theoretical explanation of our results is that the insertion of the passenger gene in the MFG vector exactly at the Nco I site of the deleted *env* gene is critical to the generation of a stable transcript that is more efficiently translated. Experimental evidence for this conclusion has not yet been provided.

Demonstration of efficient transduction and high expression of the transferred human GC gene in the monocyte/macrophage lineage demonstrated in these studies is important to the consideration of gene therapy for GD. As discussed earlier, the macrophage is a primary target for the therapy of GD. The results provided here demonstrate that macrophage progeny derived from transduced HSC repopulate bone marrow, liver, spleen, and lung. These are the organs primarily involved in the pathology of the most common phenotype of GD. This is best appreciated by the results in liver. The copy number of the human GC gene in liver of 0.1/genome is congruous with the proportion of bone marrow derived macrophages (Kupffer cells) comprising the liver (~15%) (104, 105). The enzymatic activity in the livers of reconstituted animals was nearly twice the background activity indicating that macrophages carrying and expressing the human gene repopulated the organ. Direct studies of macrophages cultured from the animals demonstrated that cells of this lineage carried and efficiently expressed the human gene.

Although it is satisfying to have overcome the problem of low expression, possible problems related to overexpression of the gene product in cells might be a cause for concern when considering the potential application of gene therapy for GD. However, several lines of evidence suggest that the increase in GC expression measured in this study should not be problematic. First, the animals in these experiments appeared to be normal for up to 8 months. Furthermore, GC is a constitutively expressed house keeping enzyme that is not highly regulated. Some variation in its activity occurs normally and should be tolerated. In addition, in animal studies of enzyme replacement using macrophage targeted enzyme, the glucocerebrosidase activity of Kupffer cells was increased by more than 10 fold without harmful side effects (1). Moreover, chronic administration of 10-100mg of enzyme twice a month to more than 1500 patients for periods of up to 13 years has not resulted in any significant clinical problems.

19.4.1.2 Transfer and Expression of the GC Gene to Human Peripheral Blood Stem Cells (PBSC)

Results of initial experiments with CD34+ enriched PB cells from G-CSF primed lymphoma patients show that transduction occurs and leads to expression of the glucocerebrosidase gene in these cells and their progeny using the MFG-GC retroviral vector. Enzyme activities were measured in cells harvested shortly after infection in three different mixtures of cytokines: 1) IL-3, IL-6, SCF, and GM-CSF; 2) IL-3, IL-6, and SCF; and 3) SCF and PIXY. In each case the enzyme activity in the MFG-GC infected group was more than 80% higher than the level of endogenous activity in noninfected cells grown under the same conditions. Control samples infected with a retroviral vector containing the bacterial *lacZ* gene did not differ significantly from noninfected cells, eliminating the possibility that the increased expression from MFG-GC was a result of nonspecific viral exposure or infection conditions. Infected cells were expanded *in vitro* for two weeks following infection, and a sustained elevation was observed in enzyme activity in the experimental group *versus* control cells. Southern blot hybridization of DNA from these cells suggested that a transduction efficiency of 5 to 15% had occurred.

19.4.1.3 Transduction of Gaucher and Non Gaucher Bone Marrow CD34+ Cells

Bone marrow cells were obtained from a Gaucher patient at the time of knee replacement surgery. A cell suspension prepared from intermedulary aspirates and discarded wafers of bone and cartilage was separated by Ficol density gradient. A total of 5×10^9 nucleated cells applied to a CellPro column yielded 2×10^7 CD34+ enriched cells. Among the enriched cells, 30-40% were large mononuclear cells characteristic of Gaucher disease; perhaps the unusual physical properties of these cells contributed to their retention on the column. The enriched cells were shown to be 44% CD34+ by FACS analysis.

Nonfrozen cells were prestimulated in IL-3, IL-6, PIXY and SCF, and cells were infected 6 times over 5 days. Following transduction and one week of expansion in PIXY and SCF, the nonadherent hematopoietic cells expressed marginally elevated levels of enzyme activity above noninfected controls (transduced nonadherent = 22-25 U/mg *versus* nontransduced nonadherent = 13-15 U/mg). Three weeks after infection, the adherent cells established monolayers in the original culture dishes, and interestingly, these adherent cells expressed much elevated levels of enzyme activity (transduced adherent > 400 U/mg *versus* nontransduced adherent < 50 U/mg). Adherent cells became detached to some extent during harvests of nonadherent cells, confounding results of enzyme assays in this experiment. In subsequent experiments, nonadherent cells were analyzed prior to outgrowth of an adherent monolayer, and/or following transfer and expansion in secondary dishes, which contained few adherent cells and did not establish significant monolayers. It is not known why adherent cells were transduced and nonadherent cells were not transduced in these experiments.

Subsequent transduction of cryopreserved Gaucher bone marrow CD34+ cells was performed following a 48 hour prestimulation period in IL-3, IL-6, and SCF, using four or five viral supernatant exposures over four days. The resulting GC enzyme activity of nonadherent transduced cells was increase 10 to 40 fold above levels in nontransduced Gaucher cells in two experiments. These enzyme levels were about 3 fold higher than normal enzyme activity levels in nonGaucher bone marrow.

Parallel transduction experiments were conducted using CD34+ enriched bone marrow cells from nonGaucher patients, resulting in similar enhancement of GC enzyme activity above normal levels.

In order to examine the contribution of each individual supernatant exposure in a series of infections, experiments were conducted in which portions of the cells were removed prior to subsequent infections. These cells (nonGaucher bone marrow and Gaucher bone marrow) were expanded without further exposure to vector-containing supernatant, and were harvested and compared for enzyme expression. Additionally, Gaucher bone marrow CD34+ cells were exposed to vector-containing supernatants twice a day for two days in comparison to once a day for four days to determine the effectiveness of a shortened infection time period. The resulting enzyme activities indicated that increased enzyme expression follows each exposure, but they suggest that either the number of infections and/or the time period for exposure could be reduced to yield adequate transduction and expression.

The presence of the normal GC enzyme in cells from nonGaucher individuals precludes the use of immunocytochemistry for visualization of vector expression because the endogenous expression is already high. However, the mutant genes present in the donor Gaucher patient (genotype 444/X produce enzyme which is not detectable by this particular immunocytochemical staining procedure. Therefore we were able to use immunocytochemical staining of transduced Gaucher bone marrow cells to estimate transduction efficiency based upon the strong expression from the MFG-GC vector in transduced Gaucher bone marrow cells. Similarly cultured BM cells (nontransduced) from a nonGaucher patient show a slightly less intense level of staining.

Immunocytochemical Staining for GC Expression in Transduced Gaucher Bone Marrow Cells

	Expansion in IL-3, IL-6, SCF	Expansion in GCT-CM, SCF
MFG-GC Transduced	21 +/- 1.5%	13 +/- 1.6%
Non transduced	0	0

GC+/total cells × 100%; mead +/- std error (n = 10 fields)

Infection of cells was to be conducted in medium containing IL-3 + IL-6 + SCF, but expansion of cells was compared between two different cytokine mixtures: 1) continuation of IL-3 + IL-6 + SCF or 2) giant cells tumor conditioned medium (GCT-CM) + SCF. The latter mixture contains G-CSF, GM-CSF, IL-6, and additional factors, and is considerably less costly than the recombinant cytokines. It appeared that GCT-CM favored expansion of polymorphonuclear cells (PMN), which were clearly more abundant in the cytospin preparations. These PMN cells were uniformly negative for GC expression in immunocytochemically stained cytospins of nonGaucher and transduced Gaucher bone marrow cells.

It is possible that the more abundant nonexpressing PMN in samples containing GCT-CM diluted the specific GC activity to lower average values than those measured for cells expanded in IL-3, IL-6, and SCF. A parallel reduction in Southern hybridization intensity implies that granulocyte precursers were less frequently transduced than monocytoid precursers. The latter appeared to have been more proliferative in the medium containing IL-3 + IL-6 + SCF based upon cytospin preparations of the resulting expanded progeny. Additionally, it appeared that the cells which express sufficient glucocerebrosidase for positive immunocytochemical detection were of similar morphology (monocytoid) in both nonGaucher and transduced Gaucher bone marrow.

Transduced cells from several experiments were expanded in sufficient numbers to yield DNA for Southern blot hybridization. Comparison of vector-sequence bands from transduced cells to similarly digested DNA spiked with clonal DNA of known vector-copy number indicated that approximately 2 copies of the vector were integrated per 10 expanded cells. These data imply that 20 of the progenitors were transduced with at least a single copy of the vector, a result which is consistent with the immunocytochemical staining presented above. The transduction efficiency of CD34 cells in the experiments described above was in the range of 20%. PCR analyses of cord blood CD34 cells for the transgene transduced by these conventional protocols demonstrated a transduction efficiency of CFU-GM in this same range. Using centrifugation as a method to enhance transduction, we have been able to increase the average transduction efficiency to 50% (106, 108).

19.4.1.4 Clinical Protocol of Gene Transfer in Patients with Gaucher Disease

The data obtained in these pre-clinical studies provided substantial information that the MFG-GC vector at a titer of 10^6 particles/ml using the centrifugation promoted infection of CD34 cells resulted in high transduction efficiency and expression of the GC transgene. In an IRB-approved protocol, we studied the ability of this vector to transduce CD34 cells mobilized from the bone marrow by G-CSF. The use of mobilized cells was chosen because the bone marrow of patients with Gaucher disease is often packed with storage cells and frequently displays significant fibrosis. We predicted that sufficient bone marrow for an autologous transplantation of genetically corrected CD34 cells may be difficult to obtain. The first question to be answered in this approach was the responsiveness of the bone marrow to G-SCF in patients with a storage process that involves the marrow. In a study of three patients two of which had biopsy proven bone marrow involvement, it was demonstrated that G-CSF at a dose of 5-10 ug/kg provided sufficient mobilization to result in the collection of 1-2 $\times 10^6$ genetically corrected CD34 cells. The amount of cells is sufficient to reconsistute a myeloablated patient. An important aim of the first clinical study in this disease is to assess the ability of this dose of cells to competitively engraft patients who have not been myeloablated. Should this competitive engraftment be successful, it would be a major step in gene transfer to hematopoietic stem cells for many inherited diseases in terms of reduced risks and safety for the patient.

In these patients and in studies of three candidates in the clinical trial (see Fig. 19.1), transduction efficiency of the mobilized CD34 cells was 8-87%. The enzymatic activity of the transduced CD34 cell enriched product was 9-50 fold greater than the baseline deficient activity in these patients' cells. These results are satisfying in that the logistics of transducing a reconstituting dose of genetically corrected cells require a technical compromise. The maximum transduction efficiency obtained in CD34 cells is at a ratio of 3×10^5 cells/1.5 ml of supernatant in a 15ml round bottom tube (106). A dose of approximately 10^8 CD34 cells per transplant given an average enrichment of 50% would require several thousand tubes to be spun for 4 hours. This is not practical. Therefore, we investigated the use of blood transfer bags in a centrifuge with a flat carrier. Since surface area seems a critical feature for optimum transduction, this approach was studied. While less transduction efficiency and consequently less expression of the transgene is obtained, 10^7 cells can be infected satisfactorily in a single 150 ml bag. This method is being employed in the clinical trial described. Analyses of peripheral blood leukocytes for GC activity and carriage of the transgene in the MFG-GC vector sequences are in progress. In addition, the transduction of bone marrow reconstituting, genetically corrected CD34 cells is being estimated in the SCID – hu bone model. Furthermore, estimates of stem cell transduction using the high proliferation potential (HPP-CFU) assay are being made using the transduced CD34 cells prepared for the transplantation of the study candidates.

19.5 Metachromatic Leukodystrophy

Metachromatic leukodystrophy (MLD) is an autosomal recessive disorder of sulfatide metabolism adversely affecting the formation and maintenance of myelin. Individuals with MLD are deficient in the activity of arylsulfatase A (ASA), a lysosomal enzyme which hydrolyzes galactosyl sulfatide (cerebroside sulfate). This single gene defect gives rise to lysosomal storage and accumulation of sulfatide in the white matter of the central nervous system (CNS) and peripheral nerves, and to a lesser extent the visceral organs. It is believed that sulfatide-induced changes are responsible

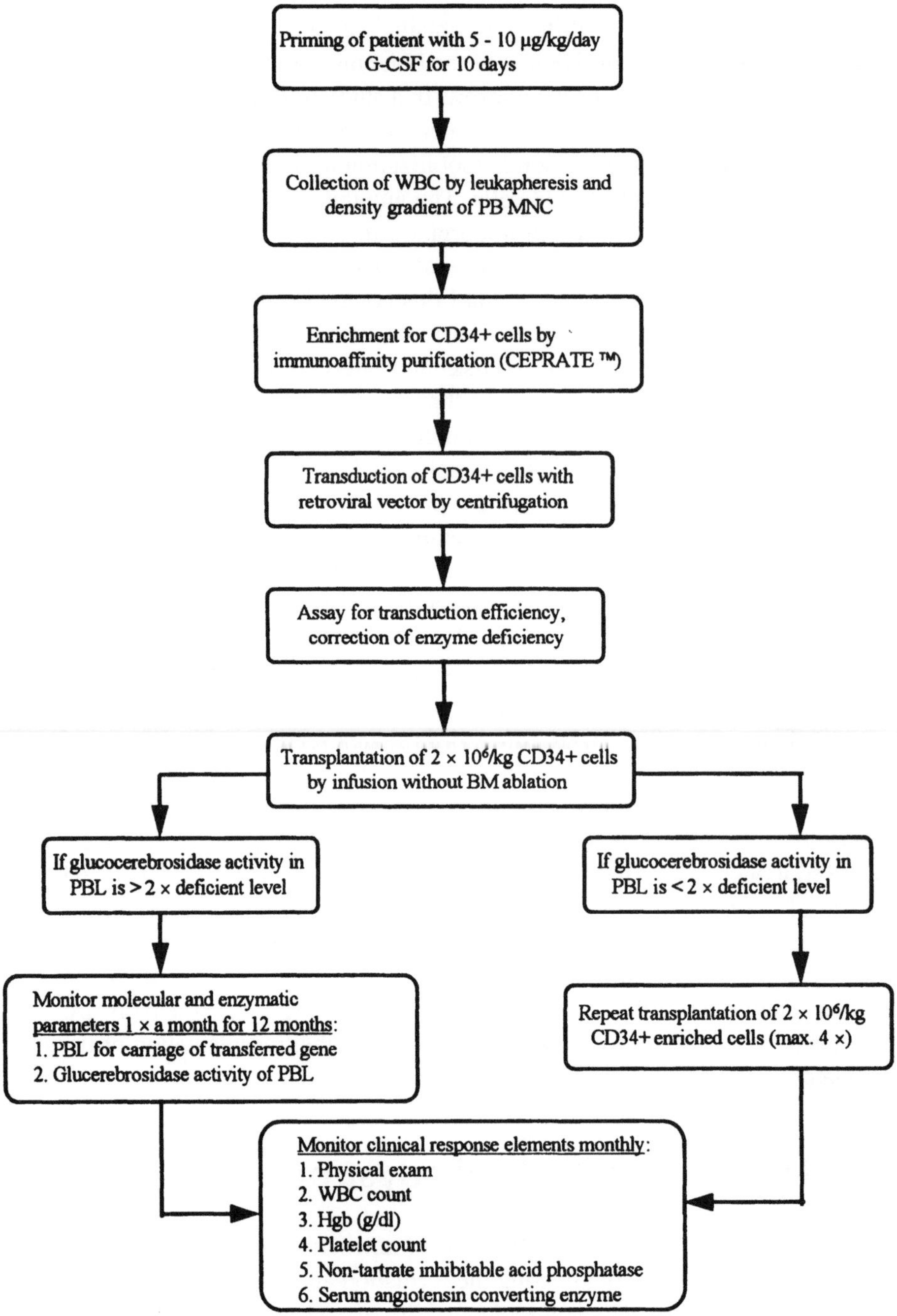

Fig. 19.1: Flow diagram for the Clinical Trial of Gene Therapy for Gaucher Disease

for the loss of myelination in the CNS, and the range of ensuing neurologic deficits which often result in death (109).

Molecular cloning of the human cDNA and determination of the gene structure (110-112) has led to the identification and characterization of many mutant alleles and correlations of genotype with MLD phenotype (113-117). A pseudodeficiency allele exists which results in low enzyme activity *in vitro*, but with no clinical phenotype (118, 119). These reduced activities, which are also sometimes found in heterozygote carriers, suggest that a low enzyme activity is sufficient to prevent lysosomal storage of sulfatides and that a small amount of enzymatic correction of the defect may be therapeutic in MLD.

Bone marrow transplantation (BMT) has been carried out in several human lysosomal storage diseases, and while normalized levels of enzymatic activity are typically observed in peripheral blood leukocytes after successful BMT, reports have varied concerning improvements in the neuropsychological status of patients with CNS involvement (120-122). Others have reported that BMT has resulted in clinical improvement in a limited number of individuals with juvenile MLD, however no improvement has been noted with BMT in infantile or late infantile type MLD (123, 124).

Anatomic and clinical improvement have been reported following BMT in animal models of lysosomal storage disease with CNS involvement (125-127). These reports support previous histochemical studies in suggesting that BM-derived cells are able to migrate and gain entry to the developing and adult brain (128-130). However, the number of these cells, and their ability to provide a "cross-correction" of enzyme remains unclear and requires further study. The stage of development of the CNS, and therefore the age of the recipient animal, is an important factor in these studies (131). Thus, the value of BMT as a single modality of treatment for MLD remains controversial.

Because of its accessibility and the clinical experience in transplantation, the bone marrow (BM) has become a desired target for gene transfer. Recent data suggest that a relatively small number of hematopoetic stem cells (HSC) are sufficient to reconstitute the BM, and these cells can be efficiently and stably transduced by retroviruses and made to express a transgene for the life of the recipient animal (132-134). Furthermore, Krall, et al., have reported that in the brains of mice transplanted with retrovirally transduced BM, 20% of microglia were positive for the presence of a transgene 3 to 4 months post-BMT (135).

Retrovirus-mediated gene transfer and transplantation has provided evidence that genetically-corrected cells reverse the visceral storage in the MPS VII mouse model of β-glucuronidase deficiency (136, 137). The brain pathology, however, remained unchanged in these mice, detection of enzyme in brain was equivocal, and no improvement in the behavioral abnormalities was observed (138, 139).

These studies suggest a possible role for gene transfer and transplantation of bone marrow in the treatment of storage diseases with CNS involvement. Several possible mechanisms may exist which could make such a strategy efficacious: 1) Degradation of storage material may occur by genetically-corrected enzyme-competent cells derived from the transplanted bone marrow ("sink" effect), thereby reducing the possibility of accumulation of the storage material in brain. 2) The transgene product (enzyme) may be secreted from BM-derived brain macrophages to neighboring cells (glia). 3) Circulating enzyme may be endocytosed by macrophages destined for the CNS.

Previous work in our lab has shown that following transplantation of bone marrow transduced by a retroviral vector that incorporates the glucocerebrosidase cDNA (MFG-GC), the transgene expressed at high levels throughout the lifetime of reconstituted mice (30). More recently, we have shown that transduction of bone marrow with a vector that carries the arylsulfatase A cDNA (MFG-ASA) resulted in highly efficient gene transfer, as demonstrated by the presence of vector DNA in 90% of spleen colonies 12 days after BMT. The vector sequence was detected in macrophage cultures obtained from transduced donor BM cells, and subsequently in the BM, spleen, lung, and liver, and brain of long term reconstituted mice (4 months after BMT), indicating that BM-derived cells had migrated to and repopulated the viscera and the brain. Furthermore, the DNA signal was present in 100% of the spleen colonies of mice transplanted secondarily with bone marrow from long term reconstituted mice, indicating not only successful reconstitution, but evidence for transduction of the pluripotent HSC in the primary recipients. Enzyme activity in the brains of mice 4 months after BMT was unchanged as compared to non-transduced control samples, probably as a result of the small number of BM-derived cells that enter the brain. However, approximately 2 out of 3 spleen colonies obtained from secondarily transplanted mice expressed arysulfatase A above control levels, indicating that the MFG vector was continuing to express the transgene (140).

The *in vivo* results obtained in these studies of MFG-ASA are nearly identical to those obtained with MFG-GC. Thus, transduction and expression of a transgene in this system are not specific to the transferred gene, but rather are features of the vector. The simplified design of the MFG construct, use of the wild type 5' LTR, and splice donor/acceptor sites are believed to be important for expression of the transgene from this vector.

The oligodendrocyte is the ultimate target for a therapy to treat MLD. Numerous studies have been made in attempt to culture and transplant oligodendrocytes. These cells have been transplanted into mouse models of demyelinating disease and shown to be able to survive, migrate and myelinate in the brain (141, 142). Mouse and rat neural progenitor cells have been identified which have the potential to differentiate into glial or neuronal cell types (143, 144). These cells also retain functional properties of neuronal and glial cells *in vivo*. When the CG-4 cell line (expressing a LacZ marker gene) was transplanted into the spinal cords of myelin-deficient rats, expression of LacZ was obtained in cells that had migrated along the spinal cord and showed the ability to myelinate neighboring neurons (145). Snyder injected retrovirally-transduced C.17 cells expressing the β-glucuronidase gene into the brains of newborn MPS VII mice (146). The donor cells expressed the LacZ transgene in both neuronal and glial cell types, and there was evidence of a reduction of lysosomal storage in host cells, presumably by a cross-corrective mechanism.

These data point to the intriguing possibility of establishing a human cell line which is both transplantable to the central nervous system and one that maintains the functional properties of oligodendrocytes. Such a cell line would offer the potential of treating MLD by gene transfer to, and transplantation of, the pathologically important cell type. It is also possible that cell types, such as myoblasts or fibroblasts, engineered to secrete the deficient enzyme, could be transplanted into the CNS and provide a cross-corrective supply of ASA.

The availability of a retroviral vector that efficiently transduces HSC or other transplantable cells, e.g. oligodendroglia or oligospheres (147), will be very useful in studying a variety of therapeutic approaches. The MFG-ASA vector provides this resource. It could be utilized to study many questions in the ASA deficient mouse, recently reported by Gieselmann (148), that are important in developing treatments for MLD.

References

1. Furbish, F. S., Steer, C. J., Krett, N. L., and Barranger, J. A. (1981). Uptake and distribution of placental glucocerebrosidase in rat hepatic cells and effects of sequential deglycosylation. Biochim Biophys Acta *673*, 425-434.
2. Barranger, J. A., Ohashi, T., Hong, C. M., Tomich, J., Aerts, J. F. G. M., Tager, J. M., Nolta, J.A., Sender, L. S., Weiler, S., and Kohn, D. B. (1989). Molecular pathology and therapy of Gaucher disease. Jap J Inher Met Dis *51*, 45.
3. Barton, N. W., Brady, R. O., Dambrosia, J. M., Bisceglie, A. M., Doppelt, S. H., Hill, S. C., Mankin, H. J., Murray, G. J., Parker, R. I., Argoff, C. E., Grewal, R. P., and Yu, K. T. (1991). Replacement therapy for inherited enzyme deficiency: macrophage targeted glucocerebrosidase for Gaucher's disease. N Eng J Med *324*, 1464.
4. Sands, M. S., Vogler, C., Kyle, J. W., Grubb, J. H., Levy, B., Galvin, N., Sly, W. S., and Birkenmeier, E. H. (1994). Enzyme replacement therapy for murine mucopolysaccharidosis type VII. J Clin Invest *93*, 2324-2331.
5. Moullier, P., Bohl, D., Heard, J. M., and Danos, O. (1993). Correction of lysosomal storage in the liver and spleen of MPS VII mice by implantation of genetically modified skin fibroblasts. Nature Gent *4*, 154-159.
6. Shull, R. M., Kakkis, E. D., McEntee, M. F., Kania, S. A., Jonas, A. J., and Neufeld, E. F. (1994). Enzyme replacement in a canine model of Hurler syndrome. Proc Natl Acad Sci USA *91*, 12937-12941.
7. Peters, C., Evers, M., Saftig, P., Isbrandt, D., Schmahl, W., Schmidt, P., and von Figura, K. (1995). Molecular pathogenesis of the Maroteaux-Lamy syndrome and generation of a mouse model for studies towards somatic gene therapy. Third Symposium on Gene Therapy, Berlin, Germany, April.
8. Walkley, S. U., Thrall, M. A., Dobrenis, K., Huang, M., March, P. A., Siegel, D. A., and Wurzelmann, S. (1994). Bone marrow transplantation corrects the enzyme defect in neurons of the central nervous system in a lysosomal storage disease. Proc Natl Acad Sci USA *91*, 2970.
9. Yeager, A. M., Shinn, C., Shinohara, M., and Pardol, D. M. (1993). Hematopoietic cell transplantation in the twitcher mouse. Transplantation *56*, 185-190.
10. In Barranger, J. A. and Brady, R. O. (eds.). (1984). Molecular Basis of Lysosomal Storage Disorders. Orlando: Academic Press Ltd, xvii-xix.
11. In Scriver, C. R., Beaudet, A. L., Sly, W. S., and Valle, D. (eds.). (1995). The Metabolic and Molecular Basis of Inherited Disease. New York: McGraw Hill Inc., see specific chapters.

12. Birkenmeier, E. H., Davisson, M. T., Beamer, W. G., Ganschow, R. E., Vogler, C. A., Gwynn, B., Lyford, K. A., Maltais, L. M., and Wawrzyniak, C. J. (1989). Murine mucopolysaccharidosis type VII. Characterization of a mouse with β glucuronidase deficiency. J Clin Invest *83*, 1258.
13. Duchen, L. W., Eicher, E. M., Jacobs, J. M., Scaravilli, F., and Teixeira, F. (1980). A globoid type of leukodystrophy in the mouse: The mutant twitcher. In: Neurological Mutations Affecting Myelination, Bauman, N. (ed.). Amsterdam: Elsevier, 107.
14. Pentchev, P. G., Boothe, A. D., Kruth, H. S., Weintroub, H., Stivers, J., and Brady, R. O. (1984). A genetic storage disorder in BALB/C mice with a metabolic block in esterification of exogenous cholesterol. J Biol Chem *259(9)*, 5784.
15. Gieselman, V., personal comm.
16. Tybulewicz, V. L. J., Tremblay, M. L., LaMarca, M. E., Willemsen, R., Stubblefield, B. K., Winfield, S., Zablocka, B., Sidransky, E., Martin, B. M., Huang, S. P., Mintzer, K. A., Westphal, H., Mulligan, R. C., and Ginns, E. I. (1992). Animal model of Gaucher's disease from targeted disruption of the mouse glucocerebrosidase gene. Nature *357*, 407.
17. Yamanaka, S., Johnson, M. D., Grinberg, A., Westphal, H., Crawley, J. N., Taniike, M., Suzuki, K., and Proia R. L. (1994). Targeted disruption of the hex A gene results in mice with biochemical and pathologic features of Tay-Sachs disease. Proc Nat Acad Sci USA *91(21)*, 9975.
18. personal comm.
19. personal comm.
20. personal comm.
21. Kuriyama, M., Yoshida, H., Suzuki, M., Fujiyama, J., and Igata, A. (1990). Lysosomal acid lipase deficiency in rats: lipid analyses and lipase activities in liver and spleen. J Lipid Res *31*, 1605.
22. Muldoon, L. L., Neuwelt, E. A., Pagel, M. A., and Weiss, D. L. (1994). Characterization of the molecular defect in a feline model for Type II GM2-Gangliosidosis (Sandhoff Disease). Am J Pathol *144(5)*, 1109.
23. Jezyk, P. F., Haskins, M. E., Patterson, D. F., Mellman, W. J., and Greenstein, M. (1977). Mucopolysaccharidosis in a cat with arylsulfatase B deficiency: A model of Maroteaux-Lemay syndrome. Science *198*, 834.
24. Alroy, J., Orgad, U., De Gasperi, R., Richard, R., Warren, C. D., Knowles K., Thalhammer, J. G., and Raghaven, S. S. (1992). Canine GM1 gangliosidosis. A clinical, morphologic, histochemical and biochemical comparison of two different models. Am J Pathol *140*, 675.
25. Healy, P. J., Farrow, B. R. H., Nicholas, F. W., Hedberg, K., and Ratcliffe, R. (1984). Canine fucosidosis: a bichemical and genetic investigation. Res Vetrin Sci *36*, 354.
26. Shull, R. M., Helman, R. G., Spellacy, E. Constantopoulous, G., Munger R. J., and Neufeld, E. F. (1984). Morphologic and biochemical studies of canine mucopolysaccharidosis I. Am J Pathol *114*, 487.
27. Prieur, D. J., Ahern Rindell, A. J., and Murnane, R.D. (1991). Ovine GM1-gangliosidosis. Am J Pathol *139*, 1511.
28. Jones, M. Z., Cunningham, J. G., Dade, A. W., Dawson, G., Laine, R. A., Williams, C. S. F., Alessi, D. M., Mostowski, U. V., and Vorrow, J. R. (1982). In: Animal models of inherited metabolic diseases. New York: Alan R. Liss Inc., 165.
29. Jolly, R. D. and Thompson, K. G. (1978). The pathology of bovine mannosidosis. Vet Pathol *15*, 141.
30. Ohashi, T., Boggs, S. S., Robbins, P. D., Bahnson, A. B., Patrene, K., Wei, F. S., Wei, J. F., Li, J., Lucht, L., Fei, Y., Clark, S., Kimak, M., He, H., Mowery-Rushton, P., and Barranger, J. A. (1992). Efficient transfer and sustained high expression of the human glucocerebrosidase gene in mice and their functional macrophages following transplantation of bone marrow transduced by a retroviral vector. Proc Natl Acad Sci USA *89*, 11332.
31. Nimgaonkar, M, Mierski, J, Beeler, M, Kemp, A, Lancia, J, Mannion-Henderson, J, Mohney, T, Bahnson, A., Rice, E., Ball, E. D., and Barranger, J. A. (in press). Cytokine mobilization of peripheral blood stem cells in patients with Gaucher disease with a view to gene therapy. Exp. Hematol.
32. Desnick, R. J., McGovern, M. M., Schuchman, E. H., and Haskins, M.E. (1982). In: Animal models of human metabolic diseases. New York: Alan R. Liss Inc., 27.
33. Mansour, S. L., Thomas, K. R., and Capecchi, M. R. (1988). Disruption of the proto-oncogene *int-2* in mouse embryo derived stem cells: a general strategy for targeting mutations to non-selectable genes. Nature *336*, 348.
34. Barranger, J. A. and Ginns, E. H (1989). In: The Metabolic Basis of Inherited Disease, Scriver, C. R., Beaudet, A. L., Sly, W. S., and Valle, D. (eds.). New York: McGraw Hill Inc., 2641.

35. Rappeport, J. M., Barranger, J. A., and Ginns, E.H. (1986). Bone marrow transplantation in Gaucher disease. In: Birth Defects: Original Article Series. March of Dimes Birth Defects Foundation. New York: Alan R. Liss Inc., *22*, 101.
36. Hartley, W. J. and Blakemore, W. F. (1973). Neurovisceral glucocerebroside storage (Gaucher's disease) in a dog. Vet Pathol *10*, 191.
37. Adachi, M. and Volk, B. W. (1977). Gaucher disease in mice induced by conduritol β epoxide: morphologic features. Arch. Pathol. Lab. Med. *101*, 255.
38. Finger, S., Heavans, R. P., Sirinathsinghji, D. J. S., Kuehn, M. R., and Dunnett, S. B. (1988). Behavioral and neurochemical evaluation of a transgenic mouse model of Lesch-Nyhan syndrome. J Neurol Sci *86*, 203.
39. Wu, C. L. and Melton, D. W. (1993). Production of a mouse model for Lesch-Nyhan syndrome in hypoxanthine phosphoribosyltransferase-resistant mice. Nature Genetics *3*, 235.
40. Sidransky, E., Sherer, D. M., and Ginns, E.I. (1992). Gaucher disease in the neonate: a distinct Gaucher phenotype ia analogous to a mouse model created by targeted disruption of the glucocerebrosidase gene. Pediatric Res *32(4)*, 494.
41. LaMarca, M. E., Yoshikawa, H., McKinney, C. E., Stubblefield, B. K., Winfield, S., Carmon, L., Martin, B. M., Sidransky, E., and Ginns, E. I. (1993). Targeting the common N370S Gaucher point mutation to the glucocerebrosidase gene in murine embryonic stem cells. Am J. Hum. Genet. *53(3)*, Abs. 917.
42. Bornstein, P., McKinney, C. E., LaMarca, M. E., Winfield, S., Shingu, T., Devaralu, S., Vos, H. L., and Ginns, E. I. (1995). Metaxin, a gene contiguous to both thrombospondin 3 and glucocerebrosidase, is required for embryonic development in the mouse: Implications for Gaucher disease. Proc Nat Acad Sci *92*, 4547.
43. O'Neill, R., Toshihou, T., Kozak, C. A., and Brady, R. O. (1989). Comparison of the chromosomal localization of murrine and human glucocerebrosidase genes and of deduced amino acid sequences. Proc Nat Acad Sci *86*, 5049.
44. Wei, F. S., Vallor, M. J., Wei, J. F., and Barranger, J. A. (1994). Mouse model of Gaucher disease created by targeted disruption of murin glucocerebrosidase (GC) gene and transgenic expression of of a mutant human GC gene. Abs, Cold Spring Harbor, NY, September.
45. Birkenmeier, E. H., Barker, J. E., Vogler, C. A., Kyle, J. W., Sly, W. S., Gwynn, B., Levy, B., and Pegors, C. (1991). Increased life span and correction of metabolic defects in in murine mucopolysaccharidosis type VII after syngeneic bone marrow transplantation. Blood *78*, 3081.
46. Takasaki, S., Murray, G. J., Furbish, F. S. et al. (1984). Structure of the *N*-asparagine-linked oligosaccharide units of human placental glucocerebrosidase. J Biol Chem *259*, 10112-10117.
47. Murray, G. J. and Jin, F.-S. (1995). Immunoelectron microscopic localization of mannose-terminal glucocerebrosidase in lysosomes of rat liver Kupffer cells. J Histol Cytol *43*, 149-158.
48. Figueroa, M. D., Rosenbloom, B. E., Kay, A. C. et al. (1992). A less costly regimen of alglucerase to treat Gaucher's disease. N Engl J Med *327*, 1632-1636.
49. Grabowski, G. S. (1992). Treatment of Gaucher's disease. N Engl J Med *328*, 1565.
50. Barton, N. W., Brady, R. O., and Dambrosia, J. M. (1993). Treatment of Gaucher's disease. N Engl J Med *328*, 1564-1565.
51. Zimran, A., Kay, A., Gelbart, T. et al. (1991). Gaucher disease: clinical, laboratory, radiologic, and genetic features of 53 patients. Am J Med *49*, 855-859.
52. Fitzpatrick, M., Moscicki, R., and Jarvis, M. (1993). International Collaborative Gaucher Group Registry Program. Clinical Summary Support, December.
53. Erickson, A. H., Johansson, K., Mansson, J. E., and Svennerholm, L. (1993). Enzyme replacement therapy of infantile Gaucher disease. Neuropediatrics *24*, 237-238.
54. Bembi, B., Zanatta, M., Carrozzi, M. et al. (1994). Enzyme replacement treatment in type 1 and type 3 Gaucher's disease. Lancet *344*, 1679-1682.
55. Hodgson, J. (1992). Whole animals for wholesale protein production. Bio Technology *10*, 863-865.
56. Wilkins, T. D. and Velander, O. (1992). Isolation of recombinant proteins from milk. J Cell Biochem *49*, 333-338.
57. Archibald, A. L., McClenaghan, M., Hornsey, V. et al. (1990). High level expression of biologically active human α_1 antitrypsin in the milk of transgenic mice. Proc Natl Acad Sci USA *87*, 5178-5182.
58. Wright, G., Carver, A., Cotton, D. et al. (1991). High level expression of active human α_1 antitrypsin in the milk of transgenic sheep. Bio Technology *9*, 830-834.

59. Clark, A. J., Bessos, H., Bishop, J. O. et al. (1989). Expression of human anti-hemophilic factor IX in the milk of transgenic sheep. Bio Technology *7*, 487-492.
60. Ebert, K. M., Selgrath, J. P., DiTullio, P. et al. (1991). Transgenic production of a variant of human tissue-type plasminogen activator in goat milk: generation of transgenic goats and analysis of expression. Bio Technology *9*, 935-938.
61. Buhler, T. H., Bruye(ac)re, T. H., Went, D. F. et al. (1990). Rabbit β-casein promoter directs secretion of human interleukin-2 into the milk of transgenic rabbits. Bio Technology *8*, 140-143.
62. Krimpenfor, P., Rademakers, A., Eyestone, W. et al. (1991). Generation of transgenic dairy cattle using in vitro embryo production. Bio Technology *9*, 844.
63. Wall, R. J., Pursel, V. G., Sharnay, A. et al. (1991). High level synthesis of a heterologous milk protein in the mammary gland of transgenic swine. Proc Natl Acad Sci USA *88*, 1696-1700.
64. Erickson, A. H., Ginns, E. I., Barranger, J. A. (1985). Biosynthesis of the lysosomal enzyme glucocerebrosidase. J Biol Chem *260*, 14319-14324.
65. Ho, M. W. and O'Brien, J. S. (1971). Gaucher's disease and deficiency of 'acid'-glucosidase and reconstitution of enzyme activity in vitro. Proc Natl Acad Sci USA *68(11)*, 2810-2813.
66. O'Brien, J. S., Kretz, K. A., Dewji, N., Wenger, D. A., Esch, F., and Fluharty, A. L. (1988). Coding of two sphingolipid activator proteins (SAP-1 and SAP-2) by the same genetic locus. Science *241(4869)*, 1098-1101.
67. Ginns, E. I., Choudary, P. V., Martin, B. M., Winfield, S., Stubblefield, B., Mayor, J., Markle-Lehman, D., Murray, G. J., Bowers, L. A., and Barranger, J. A. (1984). Isolation of cDNA clones for human beta glucocerebrosidase using the lambda gt11 expression system. Biochem Biophys Atca Commun *123(2)*, 574-580.
68. Sorge, J., West, C., Westwood, B., and Beutler, E. (1985). Molecular cloning and nucleotide sequence of human glucocerebrosidase cDNA. Proc Natl Acad Sci USA *82*, 7289-7293.
69. Tsuji, S., Choudary, P. V., Martin, B. M., Winfield, S., Barranger, J. A., and Ginns, E. I. (1986). Nucleotide sequence of cDNA containing the complete coding sequence for human lysosomal glucocerebrosidase. J Biol Chem *261*, 50-53.
70. Ginns, E. I., Choudary, P. V., Tsuji, S., Martin, B. M., Stubblefield, B. K., Sawyer, J., Hozier, J., and Barranger, J. A. (1985). Gene mapping and leader polypeptide sequence of human glucocerebrosidase: implications for Gaucher disease. Proc Natl Acad Sci USA *82*, 7107-7115.
71. Tsuji, S., Choudary, P. V., Martin, B. M., Stubblefield, B. K., Mayor, J. A., Barranger, J. A., and Ginns, G. I. (1987). A mutation in the human glucocerebrosidase gene in neuropathic Gaucher's disease. New Engl J Med *316*, 570-575.
72. Horowitz, M., Wilder, S., Horowitz, Z., Reiner, O., Gelbart T., and Beutler, E. (1989). The human glucocerebrosidase gene and pseudogene: structure and evolution. Genomics *4(1)*, 87-96.
73. Patrick, A. D. (1965). A deficiency of glucocerebrosidase in Gaucher's disease. Biochem J. *97*, 17C.
74. Brady, R. O., Kanfer, J. N., and Shapiro, D. (1965). Metabolism of glucocerebrosides II: evidence of an enzymatic deficiency in Gaucher's disease. Biochem Biophys Res Commun. *18*, 221-225.
75. Tsuji, S., Martin, B. M., Barranger, J. A., Stubblefield, B., Lamarca, M., and Ginns, E. I. (1988). Genetic heterogeneity in type 1 Gaucher disease: multiple genotypes in Ashkenazic and non-Ashkenazic individuals [published erratum appears in Proc Natl Acad Sci USA (1988) Aug; 85(15), 5708]. Proc Natl Acad Sci USA *85(7)*, 2349-2352.
76. Beutler, E., Gelbard, T., Kuhl, W., Sorge, J., and West, C. (1991). Identification of the second common Jewish Gaucher disease mutation makes possible population based screening for the heterozygous state. Proc Natl Acad Sci USA *88*, 10544-19547.
77. Hobbs, J. R., Jones, K. H., Shaw, P. J., Lindsay, I., and Hancock, M. (1987). Beneficial effect of pre-transplant splenectomy of displacement bone marrow transplantation for Gaucher's syndrome. Lancet *1*, 1111-1115.
78. Miller, A. D. (1990). Progress towards human gene therapy. Blood *76*, 271-278.
79. Miller, A. D. (1992). Human gene therapy comes of age. Nature *357*, 455-460.
80. Bodine, D. M., McDonagh, K. T., Seidel, N. E., and Nienhuis, A. W. (1991). Survival and retrovirus infection of murine hematopoeitic stem cells in vitro and effects of 5-FU and method of infection. Exp. Hematol. *19*, 206-212.
81. Weatherall, D. J. (1991). Gene therapy in perspective. Nature *349*, 275-276.
82. Lim, B., Apperley, J. F., Orkin, S.,H., and Williams D.A. (1989). Long-term expression of human adenosine deaminase in mice transplanted with retrovirus-infected hematopoietic stem cells. Proc Natl Acad Sci USA *86*, 8892-8896.

83. Szilvassy, S. J., Fraser, C. C., Eaves, C. J., Landsdorp, P. M. Eaves, A. C., and Humphries, R. K. (1989). Retrovirus-mediated gene transfer to purified hematopoietic stem cells with long-term lympho-myelopoietic repopulating ability. Proc Natl Acad Sci USA *86*, 8798-8802.
84. Fraser, C. C., Eaves, C. J. Svilvassy, S. J., and Humphries, K. (1990). Expansion in vitro of retrovirally marked totipotent hematopoietic stem cells. Blood *76(6)*, 1071-1076.
85. Capel, B., Hawley, R. G., and Mintz, B. (1990). Long- and short-lived murine hematopoietic stem cell clones individually identified with retroviral integration markers. Blood *75*, 2267-2270.
86. Van Zant, G., Chen, J. J., and Scott-Micus, K. (1991). Developmental potential of hematopoietic stem cells determined using retrovirally marked allogeneic marrow. Blood *77*, 756-763.
87. Harrison, D. E. and Lerner, C. P. (1991). Most primitive hematopoietic stem cells are stimulated to cycle rapidly after treatment with 5-fluorouracil. Blood *78*, 1237-1240.
88. Hodgson, G. S. and Bradley, T. R. (1979). Properties of hematopoietic stem cells surviving 5-fluorouracil treatment: evidence for a pre-CFU-S cell? Nature *281*, 381-382.
89. Bodine, D. M., Karlson, S., and Neinhuis, A.W. (1989). Combination of interleukin-3 and 6 reserves stem cell function in culture and enhances retrovirus-mediated gene transfer into hematopoietic stem cells. Proc Natl Acad Sci USA *86(22)*, 8897-8901.
90. Nolta, J. A., Sender, L. S., Barranger, J. A., and Kohn, D. (1990). Expression of human glucocerebrosidase in murine long-term bone marrow cultures after retroviral vector mediated transfer. Blood *75*, 787-791.
91. Correll, P. H., Fink, J. K., Brady, R. O., Perry, L. K., and Karlsson, S. (1989). Production of human glucocerebrosidase in mice after retroviral gene transfer into multipotential hematopoietic progenitor cells. Proc Natl Acad Sci USA *86*, 8912-8916.
92. Kohn, D. B., Nolta, J., Weinthal, J., Bahner, I., Yu, X. J., Lilley, J., and Crooks, G. M. (1991). Toward gene therapy for Gaucher disease. Hum Gene Ther *2(2)*, 101-105.
93. Weinthal, J., Nolta, J., Yu, X. J., Lilley, J., Ubribe L., and Kohn, D. B. (1991). Expression of human glucocerebrosidase following retroviral vector-mediated transduction of murine hematopoietic stem cells. Bone Marrow Transplantation *8(5)*, 403-412.
94. Correll, P. H., Kew, Y., Perry, L. K., Brady, R. O., Fink, J. K., and Karlsson, S. (1990). Expression of human glucocerebrosidase in long-term mice following retroviral-mediated gene transfer into hematopoietic stem cells. Hum Gene Ther *1(3)*, 227-287.
95. Carter, R. F., Abrams-Ogg, A. G. C., Dick, J. E., Kruth, S. A., Vali, V. E., Kamel-Reid, S., and Dube, I. D. (1992). Autologous transplantation of canine long-term marrow culture cells genetically marked by retroviral vectors. Blood *79(2)*, 356-364.
96. Barranger, J. A. (in press). Hematopoietic cell gene transfer. Gene Therapy.
97. Hughes, P. F. D., Eaves, C. J., Hogge, D. E., and Humphries, R. K. (1989). High efficiency gene transfer to human hematopoietic cells maintained in long-term marrow culture. Blood *74*, 1915-1925.
98. Dick, J. E., Kamel-Reid, S., Murdoch, B., and Doedens, M. (1991). Gene transfer into normal human hematopoietic cells using in vitro and in vivo assays. Blood *78*, 624-634.
99. Miller, D. G., Adam, M. A., and Miller, A. D. (1990). Gene transfer by retrovirus vectors occurs only in cells that are actively replicating at time of infection. Mol Cell Biol *10*, 4293-4342.
100. Scherdin, U., Rhodes, K., and Breindl, M. (1990). Transcriptionally active genome regions are preferred targets for retrovirus integration. Journal of Virology *64(2)*, 907-912.
101. Lapidot, T., Pflumio, F., Doedenos, M., Murdoch, B., Williams, D. E., and Dick, J. E. (1992). Cytokine stimulation of multilineage hematopoiesis from immature human cells engrafted in SCID mice. Science *225*, 1137-1141.
102. Choudary, P. V., Barranger, J. A., Tsuji, S., Mayor, J., LaMarca, M. E., Cepko, C. L., and Mulligan, R. C. (1986). Retrovirus-mediated transfer of the human glucocerebrosidase gene to Gaucher fibroblasts. Mol Biol Med *3(3)*, 293-299.
103. Choudary, P. V., Tsuji, S., Martin, B. M., Guild, B. C., Mulligan, R. C., Murrary, G. J., Barranger, J. A., and Ginns, E. I. (1986). The molecular biology of Gaucher disease and the potential for gene therapy. Cold Spring Harbor Symposium on Quantitative Biology 2, 1047-1052.
104. Barranger, J. A. (1984). Marrow transplantation in genetic disease. N Engl J Med. *311(28)*, 1629-1631.
105. Barranger, J. A., Pentchev, P. G., Furbish, F. S., Steer, C. J., Jones, E. A., and Brady, R. O. (1978). Biochem Biophys Res Commun. *83*, 1055-1060.

106. Bahnson, A. B., Dunigan, J. T., Baysal, B. E., Mohney, T., Atchison, R. W., Nimgaonkar, M. T., Ball, E. D., and Barranger, J. A. (1995). Centrifugal enhancement of retroviral mediated gene transfer. Journal of Virological Methods *54*, 131-143.
107. Mannion-Henderson, J., Kemp, A., Mohney, T., Nimgaonkar, M., Lancia, J., Beeler, M. T., Mierski, J., Bahnson, A. B., Ball, E. D., and Barranger, J. A. (1995). Efficient Retroviral Mediated transfer of the glucocerebrosidase gene in $CD34^+$ enriched umbilical cord blood human hemapoietic progenitors. Experimental Hematology *23*, 1628-1632.
108. Nimgaonkar, M., Mierski, J., Beeler, M., Kemp, A., Lancia, J., Mannion-Henderson, J., Mohney, T., Bahnson, A., Rice, E., Ball, E. D., and Barranger, J. A. (1996). Experimental Hematology *24*, 1633-1641.
109. Kolodny, E. H. and Fluharty, A.(1995). In: Metabolic Basis of Inherited Disease, Vol. 2, edition 7. New York, NY: McGraw-Hill, 2693-2739.
110. Stein, C., Gieselmann, V., Kreysing, J., Schmidt, B., Pohlmann, R., Waheed, A., Meyer, H. E., O'Brien, J. S., and von Figura, K.(1989). J. Biol. Chem. *264*, 1252-1259.
111. Kreysing. J., von Figura, K., and Gieselmann, V. (1990). Structure of the arylsulfatase A gene. Eur. J. Biochem. *191*, 627.
112. Kreysing, J., Polten, A., Hess, B., von Figura, K., Menz, K., Steiner, F., and Gieselmann, V. (1994). Structure of the mouse arylsulfatase A gene and cDNA. Genomics *19*, 249-256.
113. Polten, A., Fluharty, A. L., Fluharty, C. B., Kappler, J., von Figura, K., and Gieselmann, V. (1991). Molecular basis of different forms of metachromatic leukodystrophy [see comments]. N. Eng. J. Med. *324*, 18-22.
114. Gieselmann, V., Polten, A., Kreysing, J., and von Figura, K. (1994). Molecular genetics of metachromatic leukodystrophy. J. Inher. Metab. Dis. *17*, 500-509.
115. Fluharty, A. L., Fluharty, C. B., Bohne, W., von Figura, K., and Gieselmann, V. (1991). Two new arylsulfatase A (ARSA) mutations in a juvenile metachromatic leukodystrophy (MLD) patient. Am J. of Hum. Genet. *49(6)*, 1340-1350.
116. Kappler, J., von Figura, K., and Gieselmann, V. (1992). Late-onset metachromatic leukodystrophy: molecular pathology in two siblings. Ann. Neurol. *31(3)*, 256-261.
117. Bohne, W., von Figura, K., and Gieselmann, V. (1991). An 11-bp deletion in the arylsulfatase A gene of a patient with late infantile metachromatic leukodystrophy. Hum. Genet. *87*, 155-158.
118. Kappler, J., Leinekugel, P., Conzelmann, E., Kleijer, W. J., Kohlschutter, A., Tonnesen, T., Rochel, M., Freycon, F., and Propping, P. (1991). Genotype-phenotype relationship in various degrees of arylsulfatase A deficiency. Hum. Genet. *86*, 463-470.
119. Gieselmann, V., Fluharty, A. L., Tonnesen, T., and von Figura, K. (1991). Mutations in the arylsulfatase A pseudodeficiency allele causing metachromatic leukodystrophy. Am. J. Hum. Genet. *49*, 407-413.
120. Whitley, C. B., Belani, K. G., Chang, P. N., Summers, C. G., Blazar, B. R., Tsai, M. Y., Latchaw, R. E, Ramsay, N. K, Kersey, J. H. (1993). Long-term outcome of Hurler syndrome following bone marrow transplantation. Am. J. Med. Gen. *46(2)*, 209-218.
121. Bergstrom, S. K., Quinn, J. J., Greenstein, R., and Ascensao, J. (1994). Long-term follow-up of a patient transplanted for Hunter's disease type IIB: a case report and literature review. [Review] Bone Marrow Transpl. *14(4)*, 653-658.
122. Vellodi, A., Young, E., New, M., Pot-Mees, C., and Hugh-Jones, K. (1992). Bone marrow transplantation for Sanfilippo disease type B. J. Inher. Met. Dis. *15(6)*, 911-918.
123. Krivit, W., Shapiro, E., Kennedy, W., et al. (1990). Treatment of late infantile metachromatic leukodystrophy by bone marrow transplantation New Eng. J. Med. *322(1)*, 28-32.
124. Pridjian, G., Humbert, J., Willis, J., and Shapira, E. (1994). Presymptomatic late-infantile metachromatic leukodystrophy treated with bone marrow transplantation. J Pediatrics *125*, 755-758.
125. Poorthuis, B. J., Romme, A. E., Willemsen, R., and Wagemaker, G. (1994). Bone marrow transplantation has a significant effect on enzyme levels and storage of glycosaminoglycans in tissues and in isolated hepatocytes of mucopolysaccharidosis type VII mice. Pediatric Res. *36*, 187-193.
126. Walkley, S. U., Thrall, M. A., Dobrenis, K. et al. (1994). Bone marrow transplantation corrects the enzyme defect in neurons of the central nervous system in a lysosomal storage disease. Proc Natl Acad Sci USA *91(8)*, 2970-2974.
127. Taylor, R. M., Farrow, B. R. H., and Stewart, G. J. (1992). Amelioration of clinical disease following bone marrow transplantation in fucosidase deficient dogs. Am. J. Hum. Gen. *42*, 628-632.

128. Lassmann, H. and Hickey, W. F. (1993). Radiation bone marrow chimeras as a tool to study microglia turnover in normal brain and inflammation. Clin. Neuropath. *12*, 284-285.
129. de Groot, C. J., Huppes, W., Sminia, T., Kraal, G., and Dijkstra, C. D. (1992). Determination of the origin and nature of brain macrophages and microglial cells in mouse central nervous system, using non-radioactive in situ hybridization and immunoperoxidase techniques. GLIA *6*, 301-309.
130. Unger, E. R., Sung, J. H., Manivel, J. C., Chenggis, M. L., Blazar, B. R., Krivit, W. (1993). Male donor-derived cells in the brains of female sex-mismatched bone marrow transplant recipients: a Y chromosome specific in situ hybridization study. J. Neuropath. & Exp. Neurol. *52*, 460-470.
131. Sands, M. S., Barker, J. E., Vogler, C., Levy, B., Gwynn, B., Galvin, N., Sly, W. S., and Birkenmeier, E. (1993). Treatment of murine mucopolysaccharidosis type VII by syngeneic bone marrow transplantation in neonates. Lab Investigation. *68*, 676-686.
132. Spain, L. M., Mulligan, R. C. Purification and characterization of retrovirally transduced hematopoietic stem cells. Proc Natl Acad Sci USA *89*, 3790-3794.
133. Ohashi, T., Boggs, S., Robbins, P. et al. (1992). Efficient transfer and sustained expression of the human glucocerebrosidase gene in mice and their functional macrophages following transplantation of bone marrow transduced by a retroviral vector. Proc Natl Acad Sci USA *89*, 11332-11336.
134. Szilvassy, S. J. and Cory, S. (1994). Efficient retroviral gene transfer to purified long-term repopulating hematopoietic stem cells. Blood *84*, 74-83.
135. Krall, W. J., Challita, P. M., Perlmutter, L. S., Skelton, D. C., and Kohn, D. B. (1994). Cells expressing human glucocerebrosidase from a retroviral vector repopulate macrophages and central nervous system microglia after murine bone marrow transplantation. Blood *83*, 2737-2748.
136. Birkenmeier, E. H. (1991). Correction of murine mucopolysaccharidosis type VII (MPS VII) by bone marrow transplantation and gene transfer therapy. Hum. Gene Ther. *2*, 113-116.
137. Marechal, V., Naffakh, N., Danos, O., and Heard, J. M. (1993). Disappearance of lysosomal storage in spleen and liver of mucopolysaccharidosis VII mice after transplantation of genetically modified bone marrow cells. Blood *82*, 1358-1365.
138. Moullier, P., Bohl, D., Heard, J. M., and Danos, O. (1993). Correction of lysosomal storage in the liver and spleen of MPS VII mice by implantation of genetically modified skin fibroblasts [see comments]. Nature Genet *4*, 154-159.
139. Bastedo, L., Sands, M. S., Lambert, D. T., Pisa, M. A., Birkenmeier, E., and Chang, P. L. (1994). Behavioral consequences of bone marrow transplantation in the treatment of murine mucopolysaccharidosis type VII. J. Clin. Investigation *94*, 1180-1186.
140. Learish, R., Ohashi, T., Robbins, P., Bahnson, A., Boggs, S., Patrene, K., Schwartz, B., Gieselmann, V., and Barranger, J. A. (in press). Retroviral gene transfer and sustained expression of human arylsulfatase A. Gene Therapy.
141. Groves, A. K., Barnett, S. C., Franklin, R. J., Crang, A. J., Mayer, M., Blakemore, W. F., Noble, M. (1993). Repair of demyelinated lesions by transplantation of purified O-2A progenitor cells [see comments]. Nature *362*, 453-455.
142. Lachapelle, F., Duhamel-Clerin, E., Gansmuller, A., Baron-Von Evercooren, A., Villarroya, H., and Gumpel, M. (1994). Transplanted transgenically marked oligodendrocytes survive, migrate and myelinate in the normal mouse brain as they do in the shiverer mouse brain. Eur. J. Neuroscience *6*, 814-824.
143. Snyder, E. Y., Deitcher, D. L., Walsh, C., Arnold-Aldea, S., Hartwieg, E. A., and Cepko, C. L. (1992). Multipotent neural cell lines can engraft and participate in development of mouse cerebellum. Cell *68*, 33-51.
144. Louis, J. C., Magal, E., Muir, D., Manthorpe, M., and Varon, S. (1992). CG-4, a new bipotential glial cell line from rat brain, is capable of differentiating in vitro into either mature oligodendrocytes or type-2 astrocytes. J. Neurosci. Res. *31*, 193-204.
145. Tontsch, U., Archer, D. R., Dubois-Dalcq, M., and Duncan, I. D. (1994). Transplantation of an oligodendrocyte cell line leading to extensive myelination. Proc Natl Acad Sci USA *91*, 11616-11620.
146. Snyder, E. Y., Taylor, R. M., and Wolfe, J. H. (1995). Neural progenitor cell engraftment corrects lysosomal storage throughout the MPS VII mouse brain. Nature *374*, 367-370.
147. Avellana-Adalid, V., Lachapelle, F., Rougon, G., and Baron-Von Evercooren, A. (1994). Annual Meeting of the Society for Neouroscience. Miami, FLA, November.
148. Gieselmann, V. (1994). International Symposium on Demyelinating Diseases, Bonn, West Germany, September.

V Non-Genetic and Infectious Diseases

20. Gene Therapy for Arthritis

Steven C. Ghivizzani, Richard Kang, Christopher H. Evans and Paul D. Robbins

20.1 Introduction

Over 30 million Americans suffer from at least one of the various types of arthritis, making it the most prevalent pathologic condition among those over 65 years of age. Of the arthritides, rheumatoid arthritis (RA) currently accounts for approximately five million individuals, with nearly 150,000 new cases diagnosed each year in the USA (1). RA is a chronic, progressive inflammatory disease which primarily affects the joints. Although the precise etiology of RA is unknown, it is possibly initiated through contact with an infectious agent which activates an immune response against auto-antigens in the joint tissues. This self-recognition stimulates excessive local production of various inflammatory cytokines such as interleukin-1 (IL-1) and tumor necrosis factor α (TNFα), as well as other cellular factors. The elevated levels of these proteins drive the disease process. Large numbers of leukocytes infiltrate the thin synovial lining of the joint capsule, as well as the synovial fluid which normally serves as a lubricant for the articulating surfaces. The matrix of the cartilages is degraded, while synthesis of new matrix for repair is inhibited. The synovial lining, normally 2-3 cell layers deep, becomes massively thickened and hypercellular. The hypertrophied synovium becomes activated; acquiring a transformed phenotype, it forms an invasive pannus which erodes the cartilage, tendons and subchondral bone. Over time, the progressive erosion can result in total joint failure. All forms of arthritis, including RA are presently incurable, with the final recourse for many patients being joint replacement.

While numerous compounds have been used in the treatment of RA, effective and targeted drug delivery is a primary drawback of existing anti-arthritic therapeutics. Therapeutic agents introduced into the bloodstream can only enter the joint space from the vascular system of the synovium. The absence of a mechanism for selectively transporting solutes from the synovial capillaries (2) requires that drugs passively diffuse from the synovium through the joint cavity to the articular cartilage, which is avascular. These deficiencies may be worsened in cases of chronic RA where synovial vascularity is markedly reduced (3). Therefore, repeated systemic administration of large concentrations of anti-arthritic drugs is required to achieve stable therapeutic levels within the joint. Additionally, drug delivery through the blood limits the range of useful agents to small diffusible molecules, thereby excluding most proteins (4, 5). This is an unfortunate circumstance, in light of the increasing number of proteins being identified with anti-arthritic potential (6-11). Although direct intraarticular injection obviates these limitations to drug delivery, its usefulness is severely reduced by the short intraarticular half-life of most materials. Indeed many agents injected into the joint space are cleared in less than 30 minutes (12).

20.2 Strategies for Treatment of RA with Gene Therapy

20.2.1 Local vs. Systemic Gene Delivery

Although certain HLA alleles have been associated with a predisposition for RA, it is not generally considered to be a genetic disease. RA is, however, a good candidate for treatment by gene therapy based upon the transfer of genes encoding therapeutic proteins, either locally to the intraarticular tissues such as the synovial lining, or systemically to extraarticular tissues (13-19). Therefore, rather than replacing a defective gene product, gene transfer technology is utilized as a targeted, biological drug delivery system. If stable insertion and persistent, controlled expression of these genes could be achieved, a long-lasting therapeutic effect might be possible.

With the local approach to gene therapy many of the problems associated with current therapeutics for RA would be alleviated. The joint itself becomes the site of drug production, allowing for the intraarticular accumulation of gene products at the site of disease. Nontarget organs would have reduced exposure to various immunosuppressive agents or gene delivery vehicles. Treating numerous individual sites may, at the outset, appear rather tedious. In practice, however, it would be no more difficult than current practices of intraarticular injections of steroids, and is certainly less costly, invasive and time-consuming than the present recourse of multiple joint replacements.

RA is, however, a systemic polyarticular disease, and local delivery of therapeutic genes to synovial lining cells could at best only halt progression of the disease in genetically modified joints. Successful and complete treatment of RA may also require a more general approach involving not only the delivery of genes locally but to extraarticular sites as well, where secreted gene products could gain access to the systemic circulation. Recently, it has been shown that expression of a soluble receptor for TNFα from genetically modified T-cells was able to block the transfer of collagen induced arthritis from a DBA to a SCID mouse (20). However, a systemic approach, while having the potential advantage of treating multiple joints as well as extraarticular sites simultaneously, would greatly increase the possibility of adverse

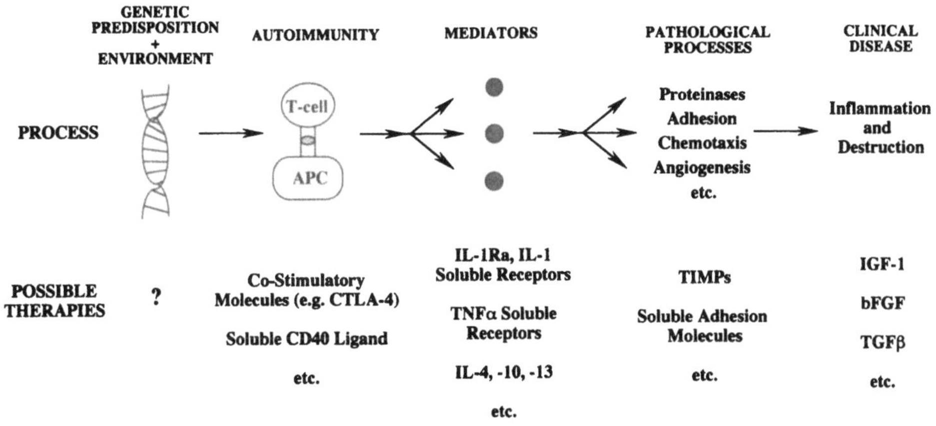

Fig. 20.1: Pathogenic progression of rheumatoid arthritis and possible therapeutic targets. Current research suggests that RA is an autoimmune disease whose etiology is complex. However, the delivery of genes encoding therapeutic proteins that block principal mediators or critical points along the progression of the disease may have a therapeutic effect.

side effects from exposure to circulating gene products and/or gene transfer vectors. Since local delivery has the advantages of increased safety and the ability to concentrate therapeutic gene products at the sites of major pathology, we have initially focused on this approach.

20.2.2 Proteins with Anti-arthritic Potential

The precise mechanism which initiates the progression of RA is poorly understood. Current research suggests that there may in fact be several, each unique to the individual and/or their HLA phenotype (21). Therefore, it is at this time impossible to design a therapeutic gene delivery scheme which would combat the disease at its source. However, even without knowing the exact etiology of RA it may be possible to interrupt the inflammatory or proliferative cascade at common key points, resulting in the down-regulation of the entire process. To this end there are several different strategies that might be used (Fig. 20.1).

First would be to block the inflammatory response at the cellular level by inhibiting the stimulatory pathways of immune cells. For example, the overexpression of molecules such as soluble CTLA4 or CD28 could serve to interrupt the activation of helper-T-cells by antigen presenting cells by interfering with the costimulatory pathway (22). Tissue transplantation studies suggest that this type of strategy may produce anergy in antigen-specific T-cell populations (23). Additionally, high secreted levels of the soluble form of CD40, the B-cell receptor for CD40 ligand, could also serve to inhibit B-cell differentiation and maturation by competing for CD40 ligand on the surface of T-helper cells and blocking interaction between the two cell types (24).

A second approach would be to inhibit the intraarticular activity of the principal inflammatory cytokines. Certain proinflammatory cytokines like IL-1 and TNFα have been shown to induce severe articular inflammation as well as stimulate cartilage breakdown and inhibit cartilage repair. Agents which are able to effectively neutralize the activity of these molecules should have therapeutic value if sufficient levels are introduced into the joint space. Inhibitors of IL-1 include interleukin-1 receptor antagonist (IL-1Ra or IRAP) (25) and soluble forms of IL-1 receptors types I and II (26, 27). IL-1Ra binds to cell-surface receptors for IL-1 and blocks their ability to transmit intracellular signals. The soluble receptors bind directly to IL-1 rendering it inactive. Likewise, soluble secreted forms of the cellular receptors for TNFα, p55 and p75, have been shown to abrogate TNFα activity (10, 28). Each of these proteins has been shown to have therapeutic efficacy in animal models when delivered by conventional methods and are currently in preclinical trials. Additionally, a number of lymphokines such as IL-4, IL-10 and IL-13 have promising anti-inflammatory properties, and exploitation of these offers an alternative strategy to directly blocking IL-1 or TNFα directly.

A third option would be to inhibit the activity of proteinases. Once activated, synovial fibroblasts, chondrocytes and macrophages secrete large quantities of collagenase and stromelysin which are capable of destroying almost all matrix proteins in articular cartilage and bone. Antagonists of these mediators include proteinase inhibitors such as the tissue inhibitors of metalloproteinases, TIMPs -1 and -2 (29).

An additional therapeutic strategy, of importance to osteoarthritis as well as RA, is the introduction of proteins which would stimulate the growth and repair of the damaged joint tissue, in particular the articular cartilage. Certain growth factors such as insulin like growth factor (IGF-1) (30), transforming growth factor β (TGFβ) (31), basic fibroblastic growth factor (bFGF) (32) and bone morphogenic protein-2 (BMP-2) (33) may be of use in stimulating new cartilage synthesis and repair.

20.3 Gene Transfer to the Synovium

An attractive tissue to target for the local delivery of therapeutic genes is the synovial lining of the joint. It has a very large surface area which is further increased due to the presence of folds or villi. The synovium also captures particles introduced into the joint. Synovial fibroblasts are easily isolated and propagated in culture, and following transplantation by intraarticular injection, readily engraft into the lining. Synovium contains at least two distinct cell types, type A synovial cells which closely resemble macrophages, and type B synovial cells which are fibroblastic. Unlike other tissues that line body cavities, synovium lacks a basement membrane, allowing synoviocytes and their secreted products direct contact with the joint space (34).

There are two basic routes for delivering genes to synovial tissue. The first method involves transfer in situ, whereby the gene is introduced by direct injection into the joint space. When a suitable vector system is used, the genetic material is taken up and expressed by either or both synovial cell types. The second method involves *ex vivo* transfer, where synovial tissue is surgically removed from the joint and the synoviocytes grown in culture. The foreign genetic material is introduced *in vitro* and the genetically modified synoviocytes transplanted to the recipient joint by intraarticular injection, where they engraft into the synovium. Since passaged synovial cultures predominantly consist of type B synovial cells, the *ex vivo* approach tends to utilize selectively this cell type. Distinct advantages of the *ex vivo* method are: 1) safety; since all genetic manipulations are performed *in vitro*, gene transfer vectors are not introduced into the patient and 2) selection; use of marker genes, such as the neomycin phosphotransferase gene *(neo)*, enable selection in culture, ensuring that a high percentage of transplanted cells express the gene of interest.

20.4 Gene Delivery Systems

Various types of viral and non-viral delivery systems have been evaluated for use in gene transfer to the synovium (35) (Fig. 20.2). Successful gene transfer and expression has been achieved with several viral systems; each, however, has certain advantages and limitations in its application. Amphotropic retroviruses (36, 37) readily infect synovial cells in culture and enable long-term, stable expression of biologically active proteins *in vitro*. Following transplantation by intraarticular injection into the rabbit knee, synoviocytes transduced *in vitro*, however, are able to maintain transgene expression *in vivo* for 5-6 weeks (15). The requirement for active cell division for retroviral infection prevents infection of mitotically quiescent synovial cells by direct intraarticular injection of retroviral particles in most cases, although *in vivo* intraarticular transduction of synoviocytes *in vivo* has been observed with high titer retroviral preparations ($>10^8$ pfu/ml) (Ghivizzani, S. C., Evans, C. H., and Robbins P. D., personal observations).

Adenoviral (38) vectors are able to infect synovial cells efficiently both *in vivo* and in culture. Since adenoviral DNA remains episomal after infection, transgene expression is lost upon subsequent passage of infected cultures. *In vivo*, β-galactosidase (*lacZ*) transgene expression in the synovium has been observed up to 8 weeks post-infection (39). A secretable transgene product, human IL-1Ra (hIL-1Ra), has been observed up to 4 weeks post-infection of the rabbit knee, though expression was noted to decline sharply after 7 days (40). Adenoviral infection *in vivo*, though, has been observed to provoke leukocytic infiltration into the joint space as well as thickening of the synovial lining. Since this response can be detected as early as 3 days and as late as 14 days post-infection, the inflammation is likely due to a combination of irritation to large numbers of viral particles and/or an immune recognition of the transduced cells due to production of endogenous adenoviral proteins (35, 41). The ability of the current generation of adenoviral vectors to provoke an immune response is a definite hindrance to the treatment of an inflammatory disease such as

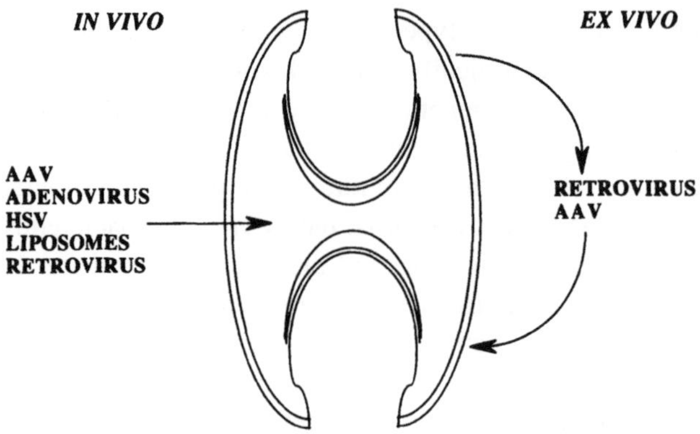

Fig. 20.2: Vectors used in gene transfer to the synovium. Retroviruses are well suited to gene delivery to the synovium by the *ex vivo* method. However, the best vector for direct use *in vivo* is less clear. Adenovirus and herpes simplex virus readily infect synoviocytes *in vivo*, but the current generations of these vectors are not yet suited for use inhumans. Future progress requires further development of these vectors as well as the evaluation of other vectors, such as adeno-associated virus (AAV)(63).

RA. Adenoviral vectors may be useful in the laboratory setting, though, allowing for rapid short term screening of therapeutic gene products in animal models.

Like adenovirus, herpes simplex virus (HSV) (42) also readily infects synovial cells *in vitro* and *in vivo*. However, HSV infection appears to be cytotoxic to synovial cells, killing all infected cells in culture within 7 days. *In vivo*, HSV-*lacZ* transgene expression is readily observable up to three days post-infection, but at seven days detectable gene expression is lost. PCR analyses associate the loss of gene expression with the loss of the HSV genome from synovial tissue, suggesting that HSV infection is also toxic *in vivo* (35).

Of the nonviral systems, gene delivery via cationic liposomes has been the most promising. Although somewhat variable for each experiment, certain cationic liposomes (43) have been shown to transfect up to 30% of the synoviocytes in primary cell cultures (35). Since lipofected DNA remains episomal, transgene expression is lost upon passage. Liposomes will also transfect synovial cells *in vivo*, but expression is also transient, persisting from 2-3 days. Naked DNA has also been shown to be taken up *in vivo* at a low frequency, probably by type A synovial cells, and expressed for up to three days.

20.5 Gene Therapy for Arthritis in the Rabbit Knee Joint

We have been using the rabbit knee as a model to test the efficacy of intraarticular gene transfer and expression. The rabbit knee joint is approximately the same size as the human metacarpophalangeal joint, a common site of RA. To date, the most success has been achieved with the *ex vivo* method of gene transfer using a retroviral vector (MFG) (44, 45). Early experiments demonstrated that both primary synovial cell cultures as well as a rabbit synovial fibroblast cell line, HIG-82 (46), could be readily transduced *in vitro* by amphotropic retroviral vectors to express stably marker genes *lacZ* and *neo.* Later work showed that the gene encoding hIL-1Ra also could be stably expressed and secreted at high levels by these cell cultures (15). IL-1Ra is a 17 kDa protein which preferentially binds to cell surface IL-1 type I receptors and competitively inhibits their interaction with interleukin-1 (IL-1) (47). IL-1 has been implicated as a key mediator of inflammatory arthritis in various animal models, including arthritis of the rabbit knee. Intraarticular injection of recombinant IL-1 induces a leukocytic infiltrate, synovial cell hyperplasia, cartilage breakdown and inhibition of cartilage matrix synthesis (48-50). Using a retroviral vector expressing IL-1Ra (MFG-IRAP) (15), synovial cells in culture could be genetically modified to express up to 500 ng IL-1Ra per 10^6 cells per 72 hours. Allogeneic transplantation of transduced HIG-82 and primary synovial cell cultures into rabbit knees resulted in expression of detectable levels of IL-1Ra protein for about two weeks. In contrast, transduced primary synovial cultures autografted into their respective rabbit donors continued to express Il-1Ra at diminishing levels for at least five weeks after transplantation (15).

To determine if IL-1Ra was expressed in the knee joint at therapeutic levels, transduced and naive synoviocytes were transplanted into opposite knees of rabbits three days prior to intraarticular injections of recombinant human IL-1 protein. In rabbit knees transplanted with IL-1Ra+ cells, the leukocytosis normally produced by injection of IL-1 was strongly suppressed (14). Inhibition was greatest at the lowest doses of injected IL-1, consistent with the competitive mechanism through which IL-1Ra antagonizes IL-1. This result demonstrated that the IL-1Ra produced intraarticularly as a result of gene transfer was biologically active and could protect the joint from the inflammatory response provoked by IL-1 (14, 15).

The therapeutic effect of intraarticularly expressed IL-1Ra was further examined in an antigen-induced model of arthritis. In this model, rabbits were first sensitized to ovalbumin by intradermal injections; arthritis was then initiated in one knee of the sensitized rabbits by the intraarticular injection of 5 mg of ovalbumin (51). The following day, rabbits received an intraarticular injection of approximately 10^7 autologous synoviocytes previously infected with a retrovirus carrying either the IL-1Ra gene or as a control, the *lacZ* marker gene. Four days later, the rabbits were sacrificed, knee joints lavaged, and the lavage fluid examined for leukocytic infiltrate as well as levels of glycosaminoglycans (GAG) which are released into the synovial fluid following cartilage degradation. Articular cartilage was also shaved from the femoral condyles and used to measure the rates of matrix synthesis. In the presence of the IL-1Ra gene, all pathophysiological parameters associated with antigen-induced arthritis (51, 52) were significantly reduced. Virtually complete inhibition of leukocyte infiltration was observed, accompanied by significant protection from both cartilage breakdown and inhibition of cartilage matrix synthesis (53).

Similarly, Makarov et al. (1996) (54) have demonstrated therapeutic efficacy following the *ex vivo* delivery of the hIL-1Ra gene to the joints of rats with recurrent bacterial cell wall-induced arthritis. Furthermore, Roessler et al. (1995) (40) have shown that the *in vivo* adenoviral mediated transfer of hIL-1Ra to rabbit knees can block the intraarticular pathogenic effects of subsequent injections of recombinant hIL-1.

These results provide encouraging evidence that the transfer of genes expressing therapeutic proteins to the joint could form the basis for a viable method for the treatment of human arthritis. Since RA is a complex disease involving the interaction of multiple cytokines and cellular factors, experiments are currently underway to evaluate the therapeutic efficacy of other types of anti-inflammatory or immunomodulatory gene products. Those currently being considered include various cytokines (IL-4, IL-10 and viral IL-10), cytokine inhibitors (soluble TNFα and IL-1 receptors), as well as metalloproteinase inhibitors (TIMP-1 and TIMP-2). Other agents such as soluble CD28, which blocks T-cell activation, and soluble anti-adhesion molecules, which inhibit cell-cell contact and cell-matrix interactions, should also be of therapeutic value. Due to the complexity of RA, it is likely that specific combinations of these potentially therapeutic proteins will need to be used to halt progression of the disease and to promote repair of the joint.

20.6 Clinical Protocol for Gene Therapy of Arthritis

The promising results obtained using the rabbit knee model have led to the submission, and recent FDA approval, of a clinical protocol to test the safety, feasibility and efficacy of treating RA by gene therapy. This protocol involves the infection of human synovial cells with a retroviral vector carrying the IL-1Ra gene (MFG-IRAP) and autologous transplantation of the modified cells into the metacarpophalangeal (MCP) joints of patients scheduled to undergo MCP joint replacement. The synovial cells to be cultured for subsequent infection and transplantation will be isolated from a different joint, such as the wrist, which has been scheduled for arthroplasty on a date prior to replacement of the MCP joints. One week following autologous transplantation of the genetically modified synovial cells, the MCP arthroplasty will be performed, and the excised joints will be analyzed for IL-1Ra expression as well as any biological changes that have occurred in response to the transgene product. Each patient will also be monitored in follow-up examinations for adverse side effects of the gene transfer procedure. By using this approach, the safety and feasibility of treating arthritis by local gene transfer will undergo preliminary evaluation in human subjects without patients having to undergo additional surgical procedures. It should be noted that this protocol is not designed to treat arthritis; the joints into which genetically modified cells are to be injected will be at the terminal stage of disease. Rather, the trial is designed to demonstrate successful gene transfer and expression as well as the safety of the procedure. Therapeutic efficacy will be determined in later trials, using gene transfer to joints at an earlier stage of arthritis.

20.7 Gene Transfer to Articular Chondrocytes

Other tissues of the joint are also being evaluated for their utility as targets for gene delivery, not only for the treatment of RA but for osteoarthritis and repair of ligament and meniscal lesions. Among these, considerable progress has been achieved with gene transfer to chondrocytes (55), the cells which reside within, and synthesize the cartilage matrix.

Due to the limited ability of cartilage to heal itself, surgical methods have been developed for the treatment of full thickness defects in articular cartilage (56, 57). Some of these procedures involve the transplantation of autologous or allogeneic chondrocytes within a three-dimensional matrix into the damaged region. The success of such techniques, though, has been variable and limited in that the repaired region does not return to normal cartilage either histologically, biochemically or biomechanically. This can likely be attributed to an inappropriate cytokine environment during the repair process. However, if an exogenous gene or combination of genes could be inserted into the chondrocytes prior to transplantation, it may be possible to express therapeutic proteins or growth factors *in vivo* at the appropriate level and period of time to stimulate better repair.

In order to determine the feasibility of using gene therapy to augment repair of articular cartilage defects, articular chondrocytes were tested for their susceptibility to *in vitro* transduction by amphotropic retrovirus. Chondrocytes from rabbit knees and shoulders were isolated, cultured and infected with retrovirus encoding the *lacZ* and *neo* genes. Following selection in media containing G418, virtually all of the surviving cells were found to stain positive for *lacZ* activity. Furthermore, these G418 selected, *lacZ*+ chondrocyte cultures could be maintained for several passages. After it was determined that chondrocytes could be successfully transduced, the duration of transgene expression *in vivo* was evaluated. For these experiments, allogeneic G418 resistant, *lacZ*+ chondrocyte cultures were incorporated into type I bovine collagen gels at 2×10^6 cells/ml and implanted into full thickness defects in the medial femoral condyles of rabbits. At weekly intervals, animals were sacrificed and the recovered implants analyzed for cells expressing the *lacZ* transgene. At 1 and 2 weeks post-transplant, *lacZ*+ cells were readily cultured from recovered gels. However, in frozen sections of the gels, *lacZ*+ chondrocytes could be identified for as long as 4 weeks post-transplant. However, the number of visible *lacZ*+ cells was noted to gradually diminish with time (55).

These findings demonstrate that retrovirally transduced chondrocytes can continue to express a transgene after transplantation *in vivo* within a collagen matrix, and as such permit optimism for the use of gene transfer to chondrocytes as a means to enhance the repair of full-thickness cartilage defects. The question of which transgene would best facilitate this repair is currently under investigation. Growth factors such as IGF-1, TGFβ and BMP-2 are currently being tested in the rabbit model system.

20.8 Animal Models of Arthritis Induced by Gene Transfer

There are currently in existence several animal models which closely simulate the pathologies observed in humans with RA, such as antigen-induced arthritis, streptococcal cell wall arthritis in rats and collagen-induced arthritis in DBA mice (51, 58, 59). The pathophysiological mechanisms of action of these models, however, are complex and poorly understood. Animal models of this type are thus of limited utility in studying the responses of joints to individual mediators of human arthritis. Since injected agents can be rapidly cleared from the joint space, the sustained delivery of individual agents to the joints of normal animals or animals with models of RA, requires the use of invasive pumping devices which themselves provoke an inflammatory response (60). As a consequence, it is still unclear which individual mediators contribute to which pathologies; this is a major impediment to the logical design of anti-arthritic drugs. Gene transfer techniques, on the contrary, offer the possibility of generating products of interest *in situ* in order to study the intraarticular responses they provoke. Such approaches will not only help to define the roles of individual gene products in the arthritic process, but also permit better evaluation of anti-arthritic drugs designed to antagonize the actions of specific mediators of disease.

Initial studies to evaluate the utility of the gene delivery approach have involved the transfer of the human IL-1β (hIL-1β) gene into rabbit knee joints. As with experiments for the treatment of arthritis, these procedures have also employed an *ex vivo* method. In preliminary experiments, both the HIG-82 synovial cell line and autologous primary synoviocyte cultures were transduced *in vitro* with amphotropic retrovirus containing the cDNAs for hIL-1β and *neo*. Following selection in medium containing G418 both types of cultures secreted similar levels of hIL-1β *in vitro*; between 100 and 200 ng per 10^6 cells per 48 hours. Following transplantation of 10^7 transduced cells into rabbit knees, analysis of recovered lavage fluids showed that the autografted synoviocytes produced approximately five-fold higher levels of hIL-1β than the allografted HIG-82 cells, so much so that the elevated IL-1β in rabbits receiving autografts proved fatal to 2 of 3 animals tested.

In subsequent experiments, 75% fewer autografted synoviocytes were used to test the effects of chronic intraarticular overexpression of hIL-1β. hIL-1β+ synoviocytes were autografted into the left knees of rabbits while naive synoviocytes were injected into the right knees. Groups of rabbits were then sacrificed at weekly intervals and analyzed for levels of expression and pathological responses to hIL-1β. ELISA measurements showed that expression of the hIL-1β transgene persisted for over two weeks. During this period knees receiving hIL-1β+ cells developed a severe monoarticular arthritis with such intraarticular pathologies as cartilage breakdown and inhibition of matrix synthesis, leukocytosis of the synovium and synovial fluid (>10^8 WBC/ml of lavage fluid), synovial hyperplasia and hypertrophy, neo-vascularization and fibrosis, as well as the formation of lymphoid follicles and highly aggressive pannus which eroded into periarticular bone. Systemic pathologies included fever, elevated erythrocyte sedimentation rate, and loss of weight. Interestingly, with the exception of bone and cartilage erosions from invasive pannus, all the pathological sequelae of IL-1β overexpression were quickly reversed following the loss of transgene expression at between 2-4 weeks post-transplantation. Contralateral control knees receiving naive synoviocytes remained normal throughout the experiment (61).

The results of these experiments clearly illustrate the capacity of IL-1β to serve as an effector molecule in RA, demonstrating that sustained synovial expression of IL-1 can produce all the major pathologies associated with human RA. Furthermore, the rapidity with which improvements occurred following loss of transgene expression, suggests that IL-1 does not initiate an irreversible cascade of pathological events that are perpetuated in its absence, but instead that maintenance of chronic inflammation is reliant upon constitutive overproduction of mediators such as IL-1. This observation suggests that agents which successfully block the intraarticular activity of principal inflammatory agents will be therapeutic in the treatment of this disease.

The success of these initial studies serves to encourage further development of the gene transfer approach for generating animal models for arthritis. Similar to experiments involving hIL-1β, gene transfer is also being used to evaluate the roles of other agents involved with disease processes in RA, such as stromelysin and nitric oxide (from inducible nitric oxide synthase). In addition to understanding the role of specific proteins in the disease process, animals models in which specific arthritis-inducing proteins are constitutively expressed intraarticularly could also be used for evaluating the efficacy of specific anti-arthritic drugs *in vivo*.

Acknowledgements

This work was supported in part by NIH grants DK44935 and CA59371, GenVec, Inc. and Orthogen Gmbh.

References

1. Harris, E. (1993). Clinical features of rheumatoid arthritis. In: Textbook of Rheumatology, W. N. Kelley et al. (eds.). New York: Saunders Co, 874-911.
2. Wallis, W.J. and Simkin, P. A. (1983). Antirheumatic drug levels in human synovial fluid and synovial tissue: observations on extravascular pharmacokinetics. Clin. Pharmacokinet. *8*, 496-522.

3. Stevens, C. R., Blake, D. R., Merry, P., Revell, P. A., and Levick, J. R. (1991). A comparative study by morphometry of the microvasculature in normal and rheumatoid synovium. Arthritis Rheum. *34*, 1508-1513.
4. Levick, J. R. (1981). Permeability of rheumatoid and normal human synovium to specific plasma proteins. Arthritis Rheum. *24*, 1550-1560.
5. Wallis, W. J., Simkin, P. A., and Nelp, W. B. (1987). Protein traffic in human synovial effusions. Arthritis Rheum. *30*, 57-63.
6. Wahl, S. M., Wong, H., and McCartney-Francis, N. (1989). Role of growth factors in inflammation and repair. J. Cell. Biochem. *40*, 193-199.
7. Arend, W. P and Dayer, J. M. (1995). Inhibition of the production and effects of interleukin-1 and tumor necrosis factor a rheumatoid arthritis. Arthritis Rheum. *38*, 151-160.
8. Smith, R. J., Chin, J. E., Sam, L. M., and Justen, J. M. (1991). Biologic effects of an interleukin-1 receptor antagonist protein on interleukin-l stimulated cartilage erosion and chondrocyte responsiveness. Arthritis Rheum. *34*, 78-83.
9. Vannier, E., Miller, E. C., and Dinarello, C. A. (1992). Coordinated antiinflammatory effects of interleukin-4: interleukin-4 suppresses interleukin-1 production but up-regulates gene expression and synthesis of interleukin-1 receptor antagonist. Proc. Natl. Acad. Sci. USA *89*, 4076-4080.
10. Piguet, P. F., Grau, C. V., Vesin, C., Loetscher, H., Gentz, R., and Lesslauer, W. (1992). Evolution of collagen arthritis in mice is arrested by treatment with anti-tumour necrosis factor (TNF) antibody or a recombinant soluble TNF receptor. Immunology *77*, 510-514.
11. van de Loo, F. A. J., Arntz, O. J., Otterness, I. G., and van de Berg, W. B. (1992). Protection against cartilage proteoglycan synthesis inhibition by antiinterleukin 1 antibodies in experimental arthritis. J. Rheum. *19*, 348-356.
12. Dingle, J. T., Thomas, D. P., King, B., and Bard, D. R. (1987). *In vivo* studies of articular tissue damage mediated by catabolin/interleukin-1. Ann. Rheum. Dis. *46*, 527-533.
13. Bandara, G., Robbins, P. D., Georgescu, H. I., Mueller, G. M., Glorioso, J. C., and Evans, C. H. (1992). Gene transfer to synoviocytes: prospects for gene treatment of arthritis. DNA Cell Biol. *11*, 227-231.
14. Hung, G. L., Galea-Lauri, J., Mueller, G. M., Georgescu, H. I., Larkin, L. A., Tindal, M., Robbins, P. D., and Evans, C. H. (1993). Suppression of intraarticular responses to interleukin-1 by transfer of the interleukin-1 receptor antagonist gene to synovium. Gene Therapy *1*, 64-69.
15. Bandara, G., Mueller, G. M., Galea-Lauri, J., Tindal, M. H., Georgescu, H. I., Suchanek, M. K., Hung, G. L., Glorioso, J. C., Robbins, P. D., and Evans, C. H. (1993). Intraarticular expression of biologically active interleukin 1-receptor-antagonist protein by *ex vivo* gene transfer. Proc. Natl. Acad. Sci. USA *90*, 10764-10768.
16. Evans, C.H. and Robbins, P. D. (1994). The interleukin-1 receptor antagonist and its delivery by gene transfer. Receptor *4*, 9-15.
17. Evans, C. H. and Robbins, P. D. (1994). Gene therapy for arthritis. In: Gene Therapeutics: Methods and Applications of Direct Gene Transfer, J. A. Wolff (ed.). Boston: Birkhauser, 320-343.
18. Evans, C.H. and Robbins, P. D. (1994). Gene therapy for arthritis. J. Rheumat. *21*, 779-782.
19. Evans, C. H. and Robbins, P. D. (1995). Gene Therapy as a treatment for arthritis. Exp. Opin. Invest. Drugs *4*, 843-852.
20. Chernajovsky, Y., Adams, G., Podhajcer, O. L., Mueller, G. M., Robbins, P. D., and Feldmann, M. (1995). Inhibition of transfer of collagen-induced arthritis into SCID mice by *ex vivo* infection of spleen cells with retroviruses expressing soluble tumor necrosis factor receptor. Gene Therapy 2, 731-735.
21. Wicks, I., McColl, G., and Harrison, L. (1994). New perspectives on rheumatoid arthritis. Immunol. Today *15*, 554-557.
22. Finck, B. K., Linsley, P. S., and Wofsy, D. (1994). Treatment of murine lupus with CTLA4Ig. Science *265*, 1225-1227.
23. Mueller, D. L., Jenkins, M. K., and Schwartz, R. H. (1989). Clonal expansion versus functional clonal inactivation: a costimulatory signalling pathway determines the outcome of T cell antigen receptor occupancy. Annu. Rev. Immunol. 7, 445-480.
24. Durie, F. H., Fava, R. A., Foy, T. M., Aruffo, A., Ledbetter, J. A., and Noelle, R. J. (1993). Prevention of collagen-induced arthritis with an antibody to gp39, the ligand for CD40. Science *261*, 1328-1331.
25. Arend, W. P. (1993). Interleukin-1 receptor antagonist. Adv. Immunol. *54*, 167-227.

26. Dower, S. K., Wignall, J. M., Schooley, K., McMahan, C. J., Jackson, J. L., Prickett, K. S., Lipton, S., Cosman, D., and Sims, J. E. (1989). Retention of ligand binding activity by the extracellular domain of the IL-1 receptor. J. Immunol. *142*, 4314-4320.
27. Colotta, F., Dower, S. K., Sims. J. E., and Mantovani, A. (1994). The type II "decoy" receptor: a novel regulatory pathway for interleukin-1. Immunol. Today *15*, 580-583.
28. Wooley, P. H., Dutcher, J., Widmer, M. B., and Gillis, S. (1993). The influence of a recombinant human soluble tumor necrosis factor receptor FC fusion protein on Type II collagen induced arthritis in mice. J. Immunol. *151*, 6602-6607.
29. Banda, M. J. and Howard, E. W. (1991). Secreted inhibitors of mammalian metalloproetinases. Ann. N. Y. Acad. Sci. *624*, 109-115.
30. Hardingham, T. E., Bayliss, M. T., Rayan, V., and Noble, D. P. (1992). Effects of growth factors and cytokines on proteoglycan turnover in articular cartilage. Brit. J. Rheum. *31*, Suppl. 1, 1-6.
31. Hunziker, E. B. and Rosenberg, L. (1994). Induction of repair in partial thickness articular cartilage lesions by timed release of TGFb. Trans. Orthop. Res. Soc. *19*, 236.
32. Cuevas, R., Burgos, J., and Baird, A. (1988). Basic fibroblast growth factor (FGF) promotes cartilage repair *in vivo*. Biochem. and Biophys. Res. Commun. *156*, 611-618.
33. Reddi, A. H. and Cunningham, N. S. (1993). Initiation and promotion of bone differentiation by bone morphogenetic proteins. J. Bone Min. Res. *8*, Suppl. 2, 499-502.
34. Hung, G. L. and Evans, C. H. (1994). Synovium. In: Knee Surgery, Vol. 1, F.H. Fu, C.D. Harner, and K.G. Vince (eds.). **Place of Publication:** Willams and Wilkins, 141-154.
35. Nita, I., Ghivizzani, S. C., Galea-Lauri, J., Bandara, G., Georgescu, H. I., Robbins, P. D. and Evans, C. H. (1996). Direct gene delivery to synovium: an evaluation of potential vectors *in vitro* and *in vivo*. Arthritis Rheum. (in press).
36. Miller, A. D. (1992). Human gene therapy comes of age. Nature *357*, 455-460.
37. Mulligan, R. C. (1991). Gene transfer and gene therapy. Principles, prospects and perspective. In: Etiology of Human Disease at the DNA Level. J. Lindstern and U. Petterson, (eds.). New York: Raven Press, Ltd., 143-189.
38. Rosenfeld, M. A., Yoshimura, K. Trapnell, B. C., Yoneyama, K., Rosenthal, E. R., Dalemans, W., Fukayama, M., Bargon, J., Stier, L. E., Strattford-Perricaudet, L., Perricaudet, M., Guggino, W. B., Pavirani, A., Lecocq, J. P., and Crystal, R. G. (1992). *In vivo* transfer of the human cystic fibrosis transmembrane condutance regulator gene to the airway epithelium. Cell *68*, 143-155.
39. Roessler, B. J., Allen, E. D., Wilson, J. M., Hartman, J. W., and Davidson, B. L. (1993). Adenoviral mediated gene transfrer to rabbit synovium *in vivo*. J. Clin. Invest. *92*, 1085-1092.
40. Roessler, B. J., Hartman, J. W., Vallance, D. K., Latta, J. M., Janich, S. L., and Davidson, B. L. (1995). Inhibition of interleukin-1-induced effects in synoviocytes transduced with the human IL-1 receptor antagonist cDNA using an adenoviral vector. Hum. Gene Ther. *6*, 402-406.
41. Ghivizzani, S. C., Kang, R., Hatton, C., Emmert, A., Evans, C. H., and Robbins, P. D. (1996). Suppression of antigen-induced arthritis in rabbits using andenoviral- mediated transfer of the human soluble interleukin-1 receptor type I. (in preparation).
42. Glorioso, J. C., DeLuca, N. A., Goins, W. F., and Fink, D. J. (1994). Development of Herpes simplex virus vectors for gene transfer to the central nervous system. In: Gene Therapeutics, J. A. Wolf, (ed.). Boston: Birkhauser, Chapter 15.
43. Gao, X. and Huang, L. (1991). A novel cationic liposome reagent for efficient transfection of mammalian cells. Biochem. Biophys. Res. Comm. *179*, 280-285.
44. Dranoff, G. (1993). Vaccination with irradiated tumor cells engineered to secrete murine granulocyte-macrophage colony-stimulating factor stimulates potent, specific, and long-lasting antitumor immunity. Proc. Natl. Acad. Sci. USA *90*, 3539-3543.
45. Robbins, P. D., Tahara, H., Mueller, G. M., Hung, G., Bahnson, A., Zitvogel, L., Galea-Lauri, J., Ohashi, T., Patrene, K., Boggs, S., Evans, C. H., Barranger, J. A., and Lotze, T. M. (1993). Retroviral vectors for use in human gene therapy for cancer. Ann. N. Y. Acad. Sci. *716*, 72-89.
46. Georgescu, H. I., Mendelow, D., and Evans, C. H. (1988). HIG-82: An established cell line from rabbit periarticular soft tissue which retains the "activatable" phenotype. *In vitro 24*, 1015-1022.
47. Arend, W. P. (1991). Interleukin-1 receptor antagonist. A new member of the interleukin-1 family. J. Clin. Invest. *88*, 1445-1451.
48. Pettipher, E. R., Higgs, G. A., and Henderson, B. (1986). Interleukin-1 induces leukocyte infiltration and cartilage proteoglycan degradation in the synovial joint. Proc. Natl. Acad. Sci. USA *83*, 8749-8753.

49. Arner, E. C., DiMeo, T. M., Ruhl, D. M., and Pratta, M. A. (1989). *In vivo* studies on the effects of human recombinant interleukin 1b on articular cartilage. Agents and Actions *27*, 254-257.
50. McDonnell, J., Hoerrner, L. A., Lark, M. W., Harper, C., Dey, T., Lobner, J., Eiermann, B., Kazazis, D., Singer, I. I., and Moore, V. L. (1992). Recombinant human interleukin-1B-induced increase in levels of proteoglycans, stromelysin and leukocytes in rabbit synovial fluid. Arthritis Rheum. *35*, 799-805.
51. Consden, R., Doble, A., Glynn, L. E., and Nind, A. P. (1971). Production of a chronic arthritis with ovalbumin: its retention in the rabbit knee joint. Ann. Rheum. Dis. *30*, 307-315.
52. Pettipher, E. R., Henderson, B., Hardingham, T., and Ratcliffe, A. (1989). Cartilage proteoglycan depletion in acute and chronic antigen-induced arthritis. Arthritis Rheum. *32*, 501-507.
53. Otani, K., Nita, I., Macaulay, W., Georgescu, H. I., Mueller, G. M., Robbins, P. D., and Evans, C. H. (1996). Suppression of antigen-induced arthritis in rabbits by *ex vivo* gene therapy. J. Immunol. (in press).
54. Makarov, S. S., Olsen J. C., Jonston, W. N., Anderle, S. K., Brown, R. R., Baldwin, A. S., Haskill, S., and Schwab, J. H. (1996). Suppression of experimental arthrits by gene transfer of interleukin 1 receptor antagonist cDNA. Proc. Natl. Acad. Sci. USA *93*, 402-406.
55. Kang, R., Marui, T., Nita, I., Georgescu, H. I., Suh, J. K., Robbins, P. D., and Evans, C. H. (1996). Gene therapy for full-thickness articular cartilage defects. (submitted).
56. Bentley, G. and Greer, R. B. (1971). Homotransplantation of isolated epiphyseal and articular cartilage chondrocytes into joint surfaces. Nature *230*, 385-389.
57. Brittberg, M. Lindahl, A., Nilsson, A., Ohlsson, C., Isaksson, O., and Peterson, L. (1994). Treatment of deep cartilage defects in the knee with autologous chondrocyte transplantation. N. Engl. J. Med. *331*, 889-895.
58. Wilder, R. L., Lafyatis, R., Yocum, D. E., Case, J. P., Kumkumian, G. K., and Remmers, E. F. (1989). Mechanisms of bone and cartilage destruction in rheumatoid arthritis: lessons from the streptococcal cell wall arthritis model in LEW/N rats. Clin. Exp. Rheum. *7*, Suppl. 3, 123-127.
59. Boissier, M. C., Feng, X. Z., Carlioz, A., Roudier, R., and Fournier, C. (1987). Experimental autoimmune arthritis in mice. I. Homologous type II collagen is responsible for self-perpetuating chronic polyarthritis. Ann. Rheum. Dis. *46*, 691-700.
60. Feige, U., Karbowski, C., Rordorf-Adam, C., and Pataki, A. (1989). Arthritis induced by continuous infusion of hr-interleukin-1a into the rabbit knee joint. Int. J. Tiss. Reac. *XI(5)*, 225-238.
62. Ghivizzani, S. C., Kang, R., Georgescu, H. I., Nita, I., Jaffurs, D., Engle, J. M., Watkins, S. C., Tindal, M. H., Suchanek, M. K., McKenzie, L. R., Evans, C. H., and Robbins, P. D. (1996). Constitutive intraarticular express;ion of human IL-1β following gene transfer to rabbit synovium produces all major pathologies of human rheumatoid arthritis. (submitted).
63. Kotin, R. M. (1994). Prospectys for the use of adenoassociated virus as a vector for human gene therapy. Human Gene Ther. *5*, 793-801.

21. Gene Therapy for HIV-1 Disease

Ernst Böhnlein

21.1 HIV-1: A New Challenge for Medical Research

In the early 1980s, alarmingly increasing numbers of patients with immune deficiency symptoms were reported in metropolitan US clinics. This patient group comprised recipients of blood products (e.g. factor VIII) and homosexual men, and led to the proposal that this "new disease syndrome" might be the result of a blood borne virus. In 1983, successful isolation of a novel human retrovirus from patients with the acquired immune deficiency syndrome (AIDS) was reported by two groups (10, 85). Subsequent studies led to the notion that this new virus, renamed human immunodeficiency virus (HIV; 51), was indeed the etiologic agent causing the acquired immune deficiency syndrome (AIDS; 86, 125). HIV-1 disease is characterized by an extended viral incubation period between the time of primary infection and first clinical symptoms. HIV-1 predominantly infects cells of the hematopoietic system, in particular CD4-positive T lymphocytes, macrophages, dendritic cells, and microglial cells of the central nervous system (126). HIV-1 infection depletes these immune effector cells which gradually develops into immune deficiencies eventually leading to opportunistic infections (i.e. candida, PCP, CMV, etc.), a hallmark of AIDS, and malignancies (i.e. lymphomas, Kaposi sarcoma). In addition to the lympho-hematopoietic tissues, HIV-1 disease also involves the central nervous system (126).

The rapid spread of AIDS in American and European metropolitan areas posed a new challenge for the public health systems in the US and in Western Europe, and extensive programs were launched to support basic and clinical AIDS research. Initially, the scientific and medical communities were confident that effective treatment modalities would be developed in a short time frame. In addition to the successful vaccination campaigns against smallpox and polio, this belief was further supported by newly available powerful molecular biology techniques which led to rapid determination of the complete nucleotide sequence (179, 231) and hence, the molecular structure of HIV-1. The efforts of many laboratories worldwide contributed to this molecular unraveling of HIV-1, which became the best understood retrovirus to date (57, 91). However, this knowledge could not be readily translated into effective treatments for patients infected with HIV-1. In particular, efforts to develop vaccines have failed until now. Antiviral drugs directed against enzymes essential for retroviral replication such as reverse transcriptase (RT; 119) and Protease (113, 143) have been developed. Clinically, monotherapies rapidly lead to a selective advantage of pre-existing minor HIV-1 variants resistant to the drug (144). Combinations of two anti-retroviral drugs also lead to drug-resistant HIV-1 isolates (54, 118). At present, triple-drug clinical trials (combinations of 2 RT inhibitors and 1 Protease inhibitor) are ongoing. Inital results look promising (53) but it is too early to assess whether this treatment modality will reduce HIV-1 replication sufficiently to prevent resistance development and to reverse disease progression and restore immune function. In addition to efficacy and development of resistant strains, bio-availability, adverse effects and the need to administer these anti-retroviral drugs for extended time periods are issues which will impact on the medical need for gene therapy.

The infidelity of reverse transcriptase during HIV-1 replication results in a vast number of retroviral variants (149). This genetic plasticity accounts for the rapid selection of drug resistant mutants resulting in only transient reduction in plasma viremia and in similarly short-lived increases in peripheral CD4-positive T lymphocytes (99, 173, 240). However, the results of these initial drug studies also support the concept that inhibition of HIV-1 replication is a critical step towards the restoration of immune function in AIDS patients.

21.2 Gene Therapy: An Alternative Approach

By 1988, multiple developments led to the proposal of "intracellular immunization" (i.e. gene therapy) as a clinical alternative for AIDS (9, 197). First, it was evident that the development of effective HIV-1 vaccines and anti-retroviral drugs with durable clinical benefit would be more difficult than initially anticipated. This task became even more difficult in the light of the genetic plasticity of HIV-1 as demonstrated by the large number of "quasispecies" detectable in AIDS patients (149). Secondly, the reverse transcribed provirus integrates into the chromosome of infected cells. HIV-1 infected patients are mosaic (i.e. contain both HIV-1 infected and non-infected cells) and hence, AIDS can be considered an acquired genetic disease amenable to gene therapy. Thirdly, gene delivery technologies became available and gene therapy was at the verge of becoming a clinical reality both for late stage cancer patients (192, 193) and children suffering from ADA immune deficiency (22, 23). Lastly, advances in HIV-1 molecular biology offered opportunities to interfere with the retroviral replication cycle at different stages (57, 91).

21.2.1 HIV-1 Gene Therapy: Clinical Issues

Although HIV-1 is the most studied and, in molecular terms, the best understood retrovirus, many open questions remain, in particular with respect to virus-host interactions and viral-induced pathology. More than a decade after the successful isolation of HIV-1 (10, 85) many pathogenic mechanisms have been suggested (89, 126) but the actual contribution of any of these proposed mechanisms to disease progression is still controversial (62). HIV-1 was considered to be a "latent" virus for many years but recent data contradict this view. To the contrary, viral loads in lymph nodes (170) and spleen (46) in the asymptomatic phase are very high. Kinetic studies in peripheral blood also demonstrate vigorous viral replication and massive T cell turnover (99, 173, 240). These findings suggest an active and effective immune response during the early stages of disease. Therefore, HIV-1 has a clinically-latent phase but does not appear to be a latent virus at the cellular level. If the number of newly generated peripheral CD4-positive T cells is similar to the number of cells lost to HIV-1 infection, a progressive reduction of CD4-positive T cells is expected as was observed for the "clinical latency" phase of the disease. The viral burden is a major factor which can influence this peripheral T cell homeostasis and very recent data indicate that plasma viremia is a better and more reliable prognostic marker for disease progression (147, 195) than the traditionally-used CD4 cell count.

These clinical findings suggest that effective AIDS therapies must retain the peripheral T cell homeostasis for extended periods of time to arrest disease progression. Theoretically, this goal can be achieved in several ways. First, reduction in viral load will change the equilibrium in favour of the infected host and reverse disease progression. The transient increase in peripheral T cells in patients treated with antiretroviral drugs (99, 240) supports this hypothesis. An improved immune response should reduce the number of virus-producing cells and hence the viral burden, resulting in a similar clinical outcome. Alternatively, the number of peripheral T cell counts could be increased. However, these cells would be targets for new HIV-1 infection in the absence of a concomitant improved immune response. Most likely, this scenario would result in a "new homeostasis" at higher viral burden and hence, more rapid disease progression (147, 195). To circumvent this problem, the newly generated peripheral T cells could be made "resistant" to HIV-mediated cytotoxic effects by gene modification which will be discussed in detail in the following sections. The peripheral T cell homeostasis comprises both T cell differentiation from hematopoietic stem/progenitor cells as well as expansion of peripheral T cells. Even early in the infection, generation of new T lymphocytes from hematopoietic progenitor cells might be required to completely restore immune function [e.g. full T cell receptor (TCR) repertoire] in HIV-1 infected individuals. For these reasons, gene therapy approaches that can protect cells throughout the T cell differentiation pathway are ideal "intracellular immunization" strategies.

21.2.2 Genetic Immune Restoration

Two different approaches to restore the immune system of HIV-1 infected patients by gene intervention therapies have been pursued. In one approach, recombinant DNA molecules are used to induce an immune response ("DNA vaccination"). The other strategy is generally referred to as adoptive immunotherapy and approaches are discussed including genetic modification of HIV-1 specific cytotoxic T lymphocytes (CTLs) *in vitro*.

The serendipitous finding that intramuscular injection of "naked" DNA can result in functional gene expression (243) offered a novel and attractive alternative to deliver therapeutic gene products *in vivo*. However, the efficacy of this approach is low and transient and DNA immunization might be the major application of this new technology. DNA immunizations are more attractive than "standard" vaccination strategies predominantly because the risk of contamination with "active" virus is virtually non-existent. In addition, large scale manufacture of DNA is well established and production costs compare favourably to recombinant proteins. The proof of principle was demonstrated in mouse models for influenza (core antigen; 221) and HIV-1 (gp160 envelope protein; 235, 236). Subsequently, a phase I clinical protocol based on HIV-1 gp160 has been approved and is currently ongoing in Europe. A more customized vaccination approach was developed by an independent group which transduced skin fibroblasts from HIV-1 patients *ex vivo* with a retrovirus encoding the HIV-1 envelope (Env) protein (237). Injection of such gene-modified cells induced specific CTL responses in mice (237) subsequently leading to a phase I clinical trial to evaluate safety of this procedure (87). Subsequently, this group modified their initial protocol by direct injection of the recombinant retrovirus intramuscularly (101, 252). At present, an open label phase I/II study (98) is under way to address both safety and efficacy of this novel technology. As a treatment modality for already-infected individuals, it is presently unclear whether these genetic approaches can elicit an effective immune response beyond that presently observed during "normal" disease progression. However, DNA vaccinations could become an effective prophylactic treatment regimen for many infectious diseases including AIDS based on the low risk, low development costs and versatility (e.g. possibility to express many viral antigens) of the technology.

An alternative approach to boost the immune system is based on the selection of activated CTLs *in vitro*. These adoptive immune therapy protocols include gene modification predominantly for two reasons, *ex vivo* and *in vivo* selection. In particular, the hygromycin-thymidine kinase fusion protein (HyTK; 183, 234) can be used as a positive selection marker during *ex vivo* expansion. Secondly, the TK-portion of the fusion gene serves as a safety measure and allows elimination of the reinfused gene-modified CTLs if required, using the pro-drug ganciclovir. For HIV-1 specific approaches, incorporation of an antiviral gene, which would give these CTLs a survival advantage in the presence of infectious virus (see next chapters) could also be considered. This strategy has been successfully employed to prevent CMV infection in patients after bone marrow transplantation (183). Using the same approach to establish CTLs against HIV-1 Env proteins *in vitro*, no reduction in viral parameters was detected in AIDS patients after reinfusion (184). Furthermore, the authors failed to detect the infused cells. However, all patients had a detectable immune response to both components of the HyTK selectable marker, suggesting that the immune competence against foreign antigens in AIDS patients is more effective than in the CMV bone marrow transplantation patients (186). However, these very same patients evidently are unable to raise an effective immune response against HIV-1 specific antigens resulting in viral eradication. At present, it is unclear whether this immune response is specific for CTL-based approaches or inherent to all protein-based gene therapy strategies. In particular, gene products delivered via the hematopoietic stem cell should induce tolerance and not an immune response against the newly acquired protein. One particular form of adoptive immunotherapy attempts to circumvent the problem by introducing a novel signalling pathway. Patient derived CD8-positive T cells as well as NK cells can be genetically modified to express a chimeric receptor molecule (188, 216). An HIV-1 antigen-specific extracellular domain is linked to the intracellular signalling domain of the T cell receptor ζ-chain (188) and it is proposed that this approach can induce an effective immunotherapy in HIV-1 infected patients. After evaluating the safety of infusion of non-gene-modified T lymphocytes from the healthy into the HIV-1 infected individual (232), a phase I clinical trial has been initiated in syngeneic twins to assess the safety of this approach (233).

21.2.3 Intracellular Gene Therapy Approaches

21.2.3.1 "Intracellular Immunization"

In contrast to the genetic vaccination strategies relying on immune stimulation to achieve systemic antiviral efficacy, the common goal of all "intracellular immunization" strategies is to inhibit HIV-1 replication at the cellular level. As a consequence, viral spread (cell-free and cell-associated) will be attenuated or ideally completely arrested resulting in reduced viral burden. Results from recent antiviral drug studies suggest that in the effective phase (before drug-resistant mutants appear), this reduction in viral burden is accompanied by an increase in CD4-positive T lymphocytes (99, 240). These clinical data confirm the initial "intracellular immunization" hypothesis (9), in particular if the gene-modified cells expressing sufficient levels of the therapeutic gene have a pronounced survival and/or proliferation advantage.

Understanding of the HIV-1 replication cycle and the underlying molecular pathways is critical towards developing effective as well as specific gene intervention strategies for AIDS. The infectious cycle begins with the budding of immature virions from productively-infected cells (Fig. 1). Intra-virion proteolytic processing of the Gag-pol precursor proteins condenses the viral core and renders the viral particle infectious. These mature viral particles then initiate another round of infec-

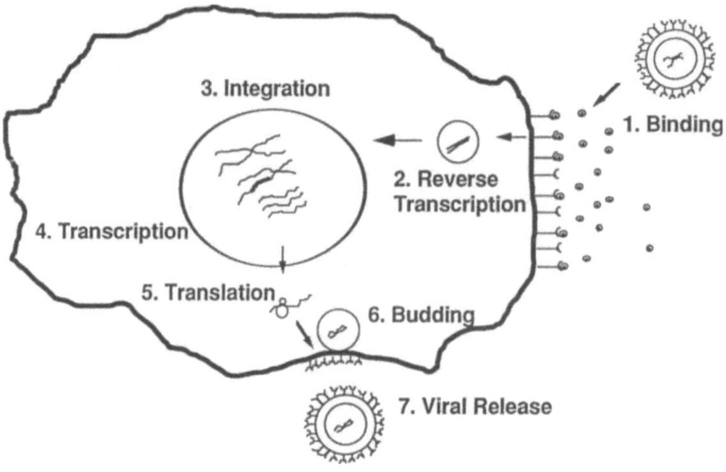

Fig. 21.1: HIV-1 Replication Cycle.

tion by binding to the CD4 surface antigen (61). Reverse transcription commences inside the virus particle and after penetration of the cell membrane, the fully reverse transcribed retroviral provirus integrates into the cellular genome mediated by a retroviral-encoded protein (Integrase).

Many studies over the last decade demonstrated that the CD4-surface receptor is required but not sufficient for HIV-infection (241). The observation that β-chemokines (RANTES, MIP-1α, MIP-1β) can suppress HIV-1 replication (50) revived the search for the HIV-1 co-receptor. A G-protein-coupled orphan receptor of the seven-transmembrane receptor superfamily was first identified as the functional co-receptor for T-cell tropic HIV-1 isolates (77). Subsequently, the CCR-5 β-chemokine receptor (3, 65, 67) and other members of this receptor family (47, 66) were shown to be required for HIV-1 infection of macrophages. Interestingly, CCR5 receptor mutants are very frequent in caucasians (ca. 20% heterozygotes) and have been linked to resistance to HIV-1 infection in long-term non-progressors (130, 196).

Transcription of the integrated proviral genome is complex and has been studied in detail (15, 58, 76, 191). The proviral genome is flanked by the long terminal repeats (LTR). The U3 region of the upstream LTR contains all cis-acting elements (enhancer, promoter) required for transcriptional initiation. The downstream LTR contains the poly-adenylation signal and cis-elements for proper RNA processing. The HIV-1 provirus comprises the Gag and Env genes which encode the viral capsid and envelope structural proteins required for particle formation. Interference with the HIV-1 structural genes prevents infectious particle formation. The pol reading frame encodes all enzymatic functions required for the retroviral life cycle including reverse transcriptase, Integrase, Protease and RNAse H. Drugs targeted against the RT and Protease function demonstrate that these retroviral genes are suitable targets for gene intervention strategies (99, 240). In addition to these universal retroviral gene functions, HIV-1 as a complex retrovirus encodes additional gene products (56, 57) required for HIV-1 replication *in vivo*. Some of these genes, referred to as accessory proteins (vif, vpu, vpr and nef), are not essential for replication *in vitro* and it is presently unclear whether these genes are useful targets for gene intervention strategies. In contrast, several approaches against the regulatory genes, Tat and Rev, have been proposed. Tat is a nuclear protein that specifically binds to a cis-acting RNA element (TAR) present on the 5' end of all HIV-1 messenger RNAs resulting in significant upregulation of HIV-1 transcription. These primary HIV-1 transcripts are readily spliced and are predominantly translated into non-structural proteins (e.g. Nef, Tat and Rev). The Rev regulatory gene product acts at the post-transcriptional level and switches the HIV-1 replication from the early (regulatory gene products) into the late phase (structural proteins and enzyme functions). Like Tat, Rev also binds to a RNA-encoded cis-element (Rev Response element, RRE) which is strategically located in the intron of the Env gene (134). Hence, Rev can only bind to incompletely- or unspliced HIV-1 transcripts. The Rev/RRE complex interacts with cellular co-factor(s) to translocate the unspliced HIV-1 transcripts into the cytoplasm (26, 84, 194) where they can be translated into the Gag-pol and Env gene products. In addition, two copies of the full-length retroviral transcripts serve as retroviral genome and are packaged into the budding virus particles. The Rev/RRE regulatory pathway is highly conserved in all lentiviruses (57) which reflects the critical role of the Rev protein for the coordinated expression of the HIV-1 genome during the replication cycle. The structural Gag-pol and Env proteins can now assemble at the cell membrane and form new budding virions to complete the infection cycle (Fig. 1).

21.2.3.2 Therapeutic Genes

Many intracellular immunization strategies for HIV-1 disease have been suggested ranging from approaches specific for the HIV-1 virus (e.g. RNA decoys for HIV-1 regulatory genes, trans-dominant negative proteins) to more generally applicable methods (e.g. ribozymes, antisense). The proof of principle for most proposed strategies has been demonstrated in transient systems, in many instances relying predominantly on surrogate experimental assays instead of directly measuring inhibition of HIV-1 replication. The vast number of gene therapy approaches in combination with the multitude of experimental systems to demonstrate efficacy make it very difficult to assess the therapeutic antiviral potential of any individual strategy. However, some approaches have been used by several independent investigators and can be used as reference points. In particular, several dominant-negative variants of the HIV-1 Rev gene have been described (135, 136, 148) and have been demonstrated to be effective by several research groups (6, 17, 75, 137, 178, 208, 225). The first approved clinical HIV gene therapy trial evaluates the safety and efficacy of this strategy in peripheral blood lymphocytes (163).

In the following chapters, therapeutic genes are discussed based on the nature of the approach (i.e. RNA-based or protein-based). At the expense of comprehensive descriptions of the various strategies, the focus of this review is to discuss the advantages and disadvantages of various intracellular immunization approaches.

Conditional Toxins

Several approaches have been proposed with the objective to specifically eliminate newly HIV-1 infected cells. In these "suicide gene" strategies, cells are modified with constructs encoding cytotoxic gene products such as diphtheria toxin or the Herpes simplex virus-derived thymidine kinase (HSV-TK) protein which can convert the prodrug ganciclovir into a cytotoxic compound. Specific cell killing of infected cells and reduction of HIV-1 replication could be demonstrated in tissue culture systems (31, 41, 59). Constitutive expression of the cytotoxic gene products even at low levels poses a risk and hence, inducible transgene expression is an important aspect of these strategies. Considering the HIV-1 LTR-basal expression in the absence of the inducing principle (15), cytotoxic effects in particular in primary hematopoietic cells must be addressed carefully. For these reasons, indirect approaches relying on prodrug activation are more attractive. Even if cytotoxcity could be controlled at the transcriptional level, it remains unclear whether killing of newly HIV-1 infected cells would be clinically more effective than blocking of viral replication, preventing viral spread and/or viral-induced cytotoxicity. Even if successful, the number of functional T cells necessary to restore a patient's immune system is unlikely to increase which is the major concern with this type of approach.

RNA-Based Intracellular Immunization Strategies

In theory, nucleic-acid based strategies are less likely to induce immune responses than protein-based approaches and are therefore attractive for the development of intracellular immunization therapies. In HIV-1 disease, predominantly intracellular antisense, ribozymes and RNA decoy expression have been studied.

Antisense

Viruses, including HIV-1, with their limited genome size offer attractive targets for antisense-based gene intervention strategies. In 1990, feasibility to interfere with HIV-1 function by intracellular antisense transcripts was demonstrated in microinjection experiments in SW480 fibroblast cells *in vitro* (199). This initial observation was subsequently confirmed in stably gene-modified CD4-positive cell lines (104, 200). Considering the genetic variability of HIV-1, antisense-mediated inhibition of several heterologous HIV-1 strains *in vitro* provided further preclinical evidence for the utility of these strategies (181). These initial studies also indicated that not all intracellularly-expressed antisense transcripts were effective (180). Most successful studies have concentrated on the 5' end of the HIV-1 genome (i.e. Gag, leader sequence) as well as the critical Tat and Rev regulatory proteins (105, 131). In addition to these target sequences, one study reported successful inhibition of HIV-1 replication by intracellular expression of antisense sequences to the HIV-1 packaging sequence (52) and to TAR, the cis-acting RNA sequence of the Tat trans-activator protein (48). Using a recombinant AAV vector as delivery vehicle, intracellular expression of sequences complementary to the 5' leader sequence including the TAR element efficiently and persistenly blocked HIV-1 replication in a human CD4-positive T cell line (43).

The mode of action in many of these antisense-based strategies has not been studied in detail but it is generally assumed that complementary transcripts will form double-stranded RNA molecules intracellularly. This can prevent translation of the target mRNA. Alternatively, these double-stranded RNA molecules might be rapidly degraded. In a recent study, anti-HIV-1 efficacy of intracellular ψ-Gag antisense expression was demonstrated. In this study, the concomitant reduction of intracellular steady-state levels of both the target (HIV-1 genomic transcript) and the recombinant antisense ψ-Gag RNA has been demonstrated (226). However, it is unclear what parameters are critical for intracellular heteroduplex formation. In particular, transcript length and base composition can influence the secondary structure and hence, antiviral efficacy. Subcellular localization is also not well studied, although most RNA polymerase II (Pol_{II})-derived transcripts are predominantly detected in the cytoplasmic compartment. RNA polymerase III (Pol_{III})-promoter derived transcripts can be directed into specific intracellular compartments (100) but such transcripts have a different set of problems including stability and functional secondary structure. The antiviral efficacy of short intracellular antisense transcripts appears to be less pronounced (181) and of a transient nature (105). Also, the genetic variety of HIV-1 quasispecies poses a problem for antisense-based strategies, in particular for short intracellular transcripts which might fail to anneal sufficiently to the mutated target RNA.

The serendipitous process of finding effective antisense sequences might explain why no clinical "intracellular immunization" protocol has yet been submitted despite the large number of preclinical reports demonstrating efficacy against HIV-1. Nevertheless, antisense approaches are still attractive and many of the above mentioned issues can be overcome by combining several antisense transcripts in one vector. In the future, improved gene delivery and gene expression technology will allow high intracellular steady-state expression levels of multiple antisense transcripts, further reducing the probability of escape mutants. The availability of highly divergent HIV-1 strains also allows better assessment of the risk associated with a particular antisense strategy with respect to appearance and selection of escape mutants in *in vitro* experiments.

Ribozymes

Antisense-based strategies require at least equimolar intracellular steady-state levels relative to the target RNA. Theoretically, ribozymes have an advantage over antisense strategies and require lower effective intracellular concentrations because of their inherent enzymatic activity. A hammerhead-ribozyme (97) directed against HIV-1 Gag provided the first experimental evidence that intracellular expression of ribozymes can block HIV-1 replication (198). Since then, studies describing ribozymes targeted at different regions of the HIV-1 genome including the leader sequence (239), Gag (213), Env (44), Integrase (202), vif (132), Tat and Rev(250) have been reported. In addition to these hammerhead-type ribozymes, a hairpin ribozyme (93) directed against the 5' leader sequence showed efficacy against several HIV-1 strains (168, 247), in T cell lines (245) in primary T cells (120), and in macrophage-like cells derived from umbilical cord blood hematopoietic progenitor cells (248). Preliminary evidence suggests that intracellular expression of this hairpin ribozyme acts on the incoming virus, reducing the proviral copy numbers 50-100 fold (245). The same research group also generated a second hairpin ribozyme directed against the HIV-1 pol reading frame and demonstrated efficacy in T cell lines (249).

Ribozymes share many of the problems inherent to antisense approaches including escape mutants and expression of incorrectly-structured and hence, non-functional RNAs. Although antiviral efficacy of ribozymes has been demonstrated in *in vitro* experiments, the reduction of HIV-1 replication has not yet been correlated with intracellular steady-state ribozyme expression. This lack of correlation is mostly due to technical limitations (i.e. quantitate low RNA expression levels at single cell resolution) and more studies to address these basic issues will be required. Of note, Thompson and colleagues described a series of PolIII-ribozyme cassettes and demonstrated high intracellular expression levels (215). In another report, three Pol_{III}-promoter cassettes have been compared with respect to subcellular localization and intracellular steady-state expression levels (100).

RNA Decoys

Critical steps in the HIV-1 replication cycle rely on interactions between retroviral proteins and virus-encoded RNA transcripts. Similar to all retroviruses, HIV-1 must package two copies of the full-length mRNA into the budding particle. The interaction between the Gag-pol precursor protein and the packaging signal (ψ) must be specific considering the intracellular abundance of RNA transcripts. Signal sequences in the 5' region of the transcript have been identified (49, 112) but do not appear to be sufficient for specific and efficient packaging (16, 182). In addition to the packaging function, the two main trans-acting factors (Tat, Rev) require HIV-1 encoded cis-acting RNA sequences (TAR, RRE: 57) and are targets for RNA decoy approaches. Overexpression of the TAR sequence, the target of the Tat transcriptional activator protein, was the first demonstration of intracellular immunization against HIV-1 (209). In extension of the initial study, the TAR decoy Pol_{III} tRNA promoter expression cassette was modified resulting in higher steady-state levels of the RNA decoy and more pronounced antiviral effficacy (123). The Rev trans-activator protein also relies on the ability to bind to a structured RNA sequence, named Rev response element (RRE). Overexpression of short PolIII transcripts encoding a subfragment of the RRE sequence can inhibit HIV-1 replication (121). More recently, the RRE-specific sequence was minimized to 13 nucleotides while retaining antiviral efficacy in clonal cell lines but HIV-1 replication was not significantly inhibited in uncloned transduced T cell populations (122). RRE decoys were also expressed as part of a recombinant neomycin-phosphotransferase RNA. Stable expression of this chimeric transcript blocked Rev function in stably transfected cells (253) and inhibited HIV-1 replication in the CD4-positive CEM T cell line (18).

In addition to these natural ligands, RNA molecules can be selected *in vitro* for high binding specificity and affinity to HIV-1 derived proteins (74, 218) which could be exploited to block specific gene functions. Such artificial ligands were derived against the HIV-1 RT (219), Integrase (4), Rev (102, 220, 88), and Tat (220) proteins. The RT-specific ligands blocked function of the HIV-1 enzyme but not of other retroviruses *in vitro* (219). One artificial ligand could functionally replace RRE in an *in vitro* assay (211). These results are promising and open new venues for HIV-1 gene therapy. To date, successful inhibition of HIV-1 replication by an intracellularly expressed artificial ligand in gene-modified T cells has not been reported. In addition, this line of approach has a theoretical disadvantage. Natural ligands have evolved and serve a biological function which is reflected in the restricted genetic variability of both the cis-acting RNA and the trans-acting protein. This selective pressure does not exist for "non-natural" ligands and viral proteins can mutate away from the artificial decoy as long as they retain their ability to interact with the natural ligand. Once efficacy has been demonstrated intracellularly, experiments to study escape mutants from artificial ligands will be feasible. It will be interesting to learn whether such escape mutants can be generated and whether they appear with accelerated kinetics relative to "natural" ligands.

As a group, RNA-based strategies rely on intracellular base-pairing of nucleotides and hence, are highly specific tools. This advantage can also become a major limitation if the target RNA is very heterogenous as is the case with HIV-1 transcripts. Longer transcripts should be less prone to escape mutants since more nucleotides are available for the intracellular annealing process. Our understanding of this intracellular process is very limited and many compounding parameters including local transcript concentration, secondary structure, and the impact of RNA binding proteins need to be studied in more detail to design more effective intracellular approaches. The delineation of the most effective structure of a particular RNA-based strategy is currently poorly understood and a time-consuming process. Combinations of multiple effective RNA-based strategies (antisense, ribozymes, decoys) in one vector offer an attractive alternative with improved antiviral efficacy and lower risk of escape mutants. However, antisense transcripts are degraded when bound to the target RNA (226) and hence, combination vectors must be developed which allow independent expression of the individual RNA-based strategies.

Protein-Based Intracellular Immunization Strategies

Trans-Dominant Negative Viral Proteins

HIV-1 requires several critical retrovirally-encoded gene functions to complete the replication cycle including the Tat and Rev trans-activating proteins (57). Mutational analysis of the Rev protein revealed mutant proteins which were not only functionally inactive but also blocked wildtype protein function in trans (135, 136). These trans-dominant (or dominant-negative) proteins could block Rev function and HIV-1 replication in transient transfection studies (135). In 1992, two research groups independently demonstrated that stable expression of a trans-dominant Rev mutation, RevM10, inhibits HIV-1 replication in retroviral-transduced CD4-positive human T cells (17, 137). Subsequently, these studies were confirmed by others (6, 178). Intracellular RevM10 expression also reduces HIV-1 replication in chronically-infected cells (75) and in primary peripheral blood lymphocytes (83, 225). These data formed the basis for the first RAC and FDA approved phase I clinical "intracellular immunization" study in AIDS (163). Preliminary results from this clinical study are exciting and demonstrate longer persistence of RevM10 gene-modified T cell in AIDS patients than control cells (242). Trans-dominant Tat mutant proteins which can block Tat-mediated gene expression *in vitro* have also been described (90, 156, 171). Recently, a chimeric protein combining both trans-dominant Tat and Rev proteins, termed Trev, was shown to block HIV-1 replication *in vitro* (1) but it is currently unclear whether this strategy is more effective than intracellular expression of trans-dominant Rev proteins.

In addition to the HIV-1 regulatory proteins, mutant structural proteins can be exploited as intracellular immunization strategies. Mutant HIV-1 Gag proteins which can not be processed properly by the retroviral Protease or can not assemble correctly thus interfering with virion maturation and rendering the secreted particles non-infectious have been described (217). Intracellular expression of mutant HIV-1 Env proteins were shown to prevent gp120-mediated cytopathic effects and to reduce virus replication *in vitro* (36, 37). An HIV-2 envelope protein deletion mutant eliminated specific CD4-binding of the produced virions and was shown to reduce viral spread (207). Another interesting experimental result was exploited by Matsuda et al. (145). They constructed a chimeric provirus, replacing the HIV-1 vpr gene with the HIV-2 specific vpx open reading frame. This chimeric virus failed to productively replicate in SupT1 cells. Subsequently, it was also demonstrated that co-expression of vpx inhibited HIV-1 Bru strain replication *in vitro* (145).

Retroviral enzyme functions can also be developed as a basis for gene intervention strategies. An HIV-1 Protease mutant unable to form heterodimers was described (5). Co-expression of this mutant with wildtype protein in transient transfection studies demonstrated impaired Protease activity and reduced virus maturation (5, 190). More recently, the antiviral efficacy of this approach was also demonstrated in HIV-1 infection experiments in CD4-positive CEM cells stably expressing the mutant Protease (107). Interestingly, this gene intervention strategy was effective in blocking an HIV-1 isolate which was resistant to various small molecule Protease inhibitors and provides the first evidence that gene therapy might still be effective when resistance to anti-retroviral drug therapy has developed.

Gene-modified cells expressing "novel" trans-dominant proteins could induce an immune response resulting in the elimination of the transduced cell population *in vivo*. This is particularly true for strategies based on the structural proteins, Gag and Env, which represent the most immunogenic HIV-1 antigens. Precedence for this has been reported in HIV-1 seropositve patients in adoptive immunotherapy studies (185) whereas the same group had not observed a similar reaction in immunosuppressed bone marrow recipients (183). At present, it is unclear whether immune responses will become a limiting factor for protein-based gene therapies in AIDS and the currently ongoing phase I trial based on RevM10 (163, 242) will provide important clinical information.

Intracellular Antibody Fragments

Successful intracellular expression of antibody fragments in mammalian cells (20) and inhibition of cellular functions (p21ras) in Xenopus oocytes (21) set the stage for application of this technology to HIV-1 gene therapy. This technology relies on chimeric molecules comprised of the heavy and light chain variable fragments generated by PCR amplification from hybridomas fused via a short linker amino acid bridge (71). The first single chain variable fragment (sFv) was directed against the HIV-1 Env protein (45, 140) and blocked spread of virus by endocytoplasmic retention of the sFv-Env complexes similar to the KDEL-CD4 mutants (38, 39). Overexpression of the sFv appears to be non-toxic in tissue culture. In the meantime, sFvs against Tat (150), Rev (68-70), RT (201) and Integrase (127) have been described. Of note, intracellulatr expression of sFvs against Integrase and RT act on the incoming virus reducing the number of integrated proviruses. This strategic advantage over late-acting intracellular immunization strategies (i.e. td Gag) makes these approaches very attractive for further development. In addition to sFvs, one report also describes inhibition of HIV-1 replication by intracellular expression of antibody fragments (133).

Similar to trans-dominant negative proteins, intracellular expression of sFvs might induce immune responses resulting in the elimination of the gene-modified cells. Generation of sFvs based on human antibodies amplified from HIV-1 patients (174 would clearly be an improvement on the currently available technology which is based on murine monoclonal antibodies. SFvs are highly specific reagents recognizing particular epitopes and this specificity might become a limitation of clinical utility. Epitope mapping of the sFvs and analysis of the existing HIV-1 isolate database could identify regions of the HIV-1 genome which are highly conserved, minimizing the risk of escape mutants. Also, bivalent molecules directed against two epitopes are in principle possible and could increase the antiviral spectrum.

Other Protein-Based Approaches

Exogenous soluble CD4 (sCD4) protein blocked infection of laboratory-adapted HIV-1 isolates in tissue culture systems (203). Intracellular expression of sCD4 was also effective in preventing HIV-1 replication *in vitro* (157). Recently, a retrovirus was constructed encoding a more stable and Tat-inducible sCD4-immunoglobulin fusion protein (159). However, the therapeutic benefit of these strategies is questionable since primary isolates are more resistant to sCD4-mediated neutralizitaion (60, 169). In another approach, a mutant CD4 HIV-1 receptor molecule containing the

KDEL (single letter amino acid code) endoplasmatic reticulum (ER) retention signal was expressed intracellularly. As a result, the mutant CD4 protein effectively trapped the gp160/120 retroviral glycoproteins in the ER and prevented gp120/CD4 mediated syncitia formation (38) and infectious virus production in tissue culture (39).

In addition to trans-dominant versions of retroviral proteins, mutant cellular co-factors also offer possibilities to interfere with the HIV-1 replication cycle. In particular, cellular co-factors for the Rev protein have been studied and cellular proteins have been identified (26, 84, 194). Mutants of eIF-5A have been described which could inhibit HIV-1 replication in clonal (19) and mixed CD4-positive T cell populations (108). If protein-based strategies elicit immune responses, intracellular expression of mutant cellular co-factors (19, 108) should be less likely subject to immune recognition and could circumvent this problem. Whether overexpression of such mutants is associated with cytotoxicity awaits further studies.

The first interferon-mediated gene therapy approach for HIV-1 was based on the intracellular expression of the α2-interferon gene (12). Subsequently. transduction of tissue culture cells with a retroviral vector encoding interferon-beta (IFN-B) was shown to reduce retroviral infection of both a MLV-based lacZ vector and HIV-1 virus (227, 228). The data suggests that this approach acts very early in the retroviral life cycle (prior to reverse transcription) and reduces integration of proviral genomes. HIV-1 replication was also reduced in primary cells from healthy normal and HIV-1 infected patients transduced with the IFN-B vector. These *in vitro* data are encouraging but cytotoxic effects of IFN-B production/secretion in/from transduced primary cells, in particular hematopoietic stem cells, need to be addressed to assess the safety of such approaches.

21.2.3.3 Target Cells

The choice of target cell as vehicle for HIV-1 gene therapy approaches is critical towards the development of clinically-relevant treatments. To date, many gene therapy phase I clinical protocols have evaluated the safety of gene transfer into peripheral T cells. In particular, the ADA deficiency protocol which has received broad attention, could demonstrate clinical benefit. Intriguingly, the number of gene-marked peripheral T cells remained stable for extended periods of time in the absence of continuously repeated infusions (24, 25). This clinical result is promising for the development of T cell based gene therapy protocols although it might reflect the pronounced selective advantage of ADA+ T cells in SCID-ADA patients. This encouraging result is further supported by independent findings that hematopoietic stem cell gene transfer can also be effectively performed in this patient population (29, 114). In an extension of the initial T cell-based phase I SCID-ADA protocol, umbilical cord blood derived stem cells from three prenatally diagnosed SCID-ADA patients were gene-modified and reinfused without conditioning. Follow-up data from this study suggest that the marking in peripheral T cells is significantly higher than in any other blood lineages or in the bone marrow compartment (114) reflecting the prominent selective advantage in the T cell lineage. Bordignon and his colleagues addressed these clinical issues in a protocol designed to directly compare gene delivery via T cells and hematopoietic stem cells (28). In this clinical protocol, two distinguishable vectors were used to transduce peripheral T cells and bone marrow derived hematopoietic stem cells. Early in this study, the majority of marked peripheral T cells carried the vector used for the *in vitro* T cell transductions. However, T cells derived from gene-modified bone marrow stem cells took over and represent the majority of the peripheral T cells three years after reinfusion (29). These data provide the first evidence that hematopoietic stem cell gene transfer can be effective even if the selective advantage exists only for the gene-modified differentiated blood cells. The data also indicates that infusion of gene modified T cells might alleviate clinical symptoms as a bridging therapy until sufficient hematopoietic stem cell-derived T lymphocytes become available.

Although the loss of CD4-positive lymphocytes has been used for many years as the clinical marker for disease progression, HIV-1 also infects other blood-derived cells including macrophages and microglial cells in the brain. Currently, several peripheral T cell-based clinical phase I AIDS protocols have been approved by the Recombinant Advisory Committee (RAC) and FDA (Tab. 21.1). The first "intracellular immunization" study for AIDS is currently conducted at the University of Michigan. In this clinical protocol, a dominant-negative mutant (RevM10) of the HIV-1 Rev wildtype protein is introduced into peripheral blood T lymphocytes of HIV-1 infected patients using a biolistics approach (i.e. gene delivery by plasmid DNA-coated gold particles) as well as amphotropic retroviral gene delivery (163). Preliminary data from this first trial indicate a selective survival advantage of the RevM10-expressing T lymphocytes over cells modified with a control vector not expressing the RevM10 gene product (242). However, the data from this study also demonstrate that gene-modified T cells in this patient population are short-lived.

The results from this pioneering study are exciting and provide clinical evidence that intracellular expression of therapeutic genes can provide a survival advantage in HIV-1 positive patients. However, several shortcomings of the technology might prevent the clinical utility of this approach. First, transduction of T cells *in vitro* requires stimulation of the cells which also activates expression of HIV-1 proviral sequences (25, 162). To overcome this problem, the current transduction protocols are performed in the presence of non-nucleoside reverse transcriptase inhibitors which selectively inhibit HIV-1 replication but do not prevent reverse transcription of the recombinant MLV vector genome required for successful retroviral transduction. The appearance of drug-resistant isolates in patients treated with such drugs might become a practical problem and could preclude access to this technology for a fraction of otherwise eligible patients. Also, the large scale expansion of HIV-1 infected T cells presents a biosafety hazard to the operator but T cell expansion technologies are developed to prevent/reduce this risk (134). The logistics of this approach are challenging (*in vitro* expansion, multiple treatments, etc.) and might pose a developmental hurdle towards clinical utility.

Table 21.1 HIV-1 Gene Therapy Clinical Protocol

Principal Investigator	Protocol Type	Clinical Phase	Gene Delivery Vehicle	Therapeutic Gene	Target Cell	Status	Clinical Protocol
Greenberg PD	AIT	Phase I	Retrovirus	NA (Hygro-TK)	CD8 (autologous)	NIH-RAC. FDA	Ref. 184
Riddell SR	AIT	Phase I	Retrovirus	NA (Neo-TK)	CD8 (autologous)	FDA review	NA
Walker RE	AIT/GM	Phase I/II	Retrovirus	NA (Neo)	CD8 (Syngeneic)	NIH-RAC, FDA	Ref. 232
Walker RE	AIT	Phase I/II	Retrovirus	NA (CD4-ζ-chain receptor)	CD8 (Syngeneic)	NIH-RAC, FDA	Ref. 233
Galpin JE	GV	Phase I/II	Retrovirus	Env (Rev)	Fibroblast (autologous)	NIH-RAC, FDA	Ref. 87
Conant M	GV	Phase I/II	Retrovirus	Env (Rev)	IM injection	FDA review	
Parenti D	GV	Phase II	Retrovirus	Env (Rev)	IM injection	NIH-RAC, FDA	NA
Nabel GJ	GT	Phase I	Retrovirus, Biolistics	RevM10	CD4 (autologous)	NIH-RAC, FDA	Ref. 163
Wong-Staal F	GT	Phase I	Retrovirus	Hairpin ribozyme	CD4 (autologous)	NIH-RAC	NA
Pomerantz RJ	GT	Phase I	Retrovirus	anti-Rev sFv	CD4 (autologous)	NIH-RAC, FDA	NA
Marasco WA	GT	Phase I	Retrovirus	anti-Env sFv	CD4 (autologous)	NA	NA
Morgan R	GT	Phase I	Retrovirus	Td Rev, TAR antisense	CD4 (Syngeneic)	NA	Ref. 160
Rosenblatt JD	GT	Phase I	Retrovirus	Ribozyme	CD34+	FDA review	NA

AIT: Adoptive immunotherapy
GV: Genetic vaccination
IM: Intra-muscular
GM: Gene marking
GT: Gene therapy ("intracellular immunization")
NA: Not applicable/not available

HIV-1 infects several blood cell lineages (176) but the above mentioned technologies have only been successfully demonstrated for peripheral T cells. Although the cellular reservoirs for HIV-1 in patients are not fully understood, it has been clearly demonstrated that macrophages as well as other hematopoietic stem cell derived cells (e.g. microglia, dendritic cells) can harbor the virus in AIDS patients (176). For this reason, hematopoietic stem cells (HSC) have been proposed as targets for clinical AIDS gene therapy protocols (9). This approach also offers the opportunity to restore immune function (i.e. a complete T cell repertoire) which might already be depleted even in early stage patients. However, this line of clinical research also faces a set of hurdles. Technically, large-scale GMP-isolation of this rare cell population is more difficult than for more abundant cell types, and reliable clinical enrichment processes need to be developed. HIV-1 infection of hematopoietic stem cells is still controversial (64, 79, 80, 206, 230) and, not unlike T cell approaches, the purity of the enriched cell population (e.g. removal of HIV-1 infected cells) before re-infusion may become a regulatory issue. Whether normal T cell differentiation from hematopoietic stem/progenitor cells is functional in HIV-1 infected individuals is an open scientific question. The majority of hematopoietic stem cells are mitotically quiescent and hence, not good targets for retroviral vectors, the most advanced gene delivery technology yielding stable transfer. The preclinical development of improved gene delivery protocols is also hampered by the lack of adequate *in vitro* assays for human hematopoietic stem cells. Most gene transfer experiments are currently evaluated based on gene transfer frequencies into clonogenic assays measuring committed progenitors [colony forming units (CFU)] or more primitive cells [long term culture initiating cells (LTC-IC)]. A few pioneering HSC gene therapy studies have been initiated both in monogeneic diseases (SCID-ADA; 28, 114) and in cancer (32-34, 63, 72). The preliminary results from these initial studies demonstrate technical feasibility albeit yielding only very low gene transfer frequencies (<1 per 10^4 peripheral cells is marked) demonstrating the need for more appropriate *in vitro* HSC assays to improve gene transfer into this rare cell population (35, 73).

21.2.3.4 Gene Delivery

The selection of target cells directly impacts the choice of gene transfer technology. All AIDS intracellular immunization strategies utilize retroviral gene delivery *in vitro* (Tab. 21.1). This technology results in integration of the transgene into the cellular genome. Consequently, gene marked progeny cells carry and potentially express the therapeutic gene. This feature is particularly attractive for stem cell based approaches but also for long-lived T lymphocytes with the potential to proliferate in response to antigenic stimuli. Retroviral gene delivery has been intensively studied and the available reagents including vectors (103, 151, 158, 167,238) and packaging cell lines (55, 139, 142, 152, 154, 186) make this technology relatively safe (30, 214). However, two areas of concern have been identified. First, retroviruses have the potential for insertional mutagenesis (13, 14, 205). Inactivation of essential housekeeping genes is the most likely event which can not be measured since such cells will not proliferate. More critical is the transcriptional activation of neighboring cellular oncogenes by the transcriptional regulatory elements located in the Moloney virus LTR. So far, no incidents have been reported in any of the ongoing or completed clinical studies. Recombination events leading to replication competent retroviruses (RCR) represent the other area of concern. The development of sensitive *in vitro* assay systems to detect RCR has helped reduce this biosafety risk (82, 92). However, the "state of the art" detection technology is time-consuming and labor intense. Mandatory rigorous testing of clinical retroviral batches and gene-modified cells is very expensive and prevents more studies, in particular by academic centers. The fact that presently only a few laboratories are equipped and certified to perform these critical assays is a logistical issue for the rapid development of clinical gene therapies.

Although several studies could demonstrate successful retroviral-mediated gene transfer into hematopoietic stem cells in murine (95, 109, 187, 222), non-human primates (110, 223, 224) and human hematopoietic stem cells (42, 175, 229), this technology is not well-suited for such mitotically quiescent cells. Alternative vectors based on lentiviral genomes (164) might be developed into more effective HSC gene delivery tools.

Adeno-associated viral (AAV) vectors (161, 189) offer an alternative to retroviral gene transfer. Site-specific integration of wild-type AAV virus is a particularly attractive feature of this technology (116). Unfortunately, the currently available recombinant vectors appear to integrate more randomly than the wildtype virus and it is unclear whether the associated risks are lower than with retroviruses. Successful AAV-mediated gene transfer into hematopoietic stem cells has been demonstrated by several groups (155, 251). Although integration has been demonstrated (78), it is controversial whether recombinant AAV vectors are/remain integrated and more studies are required to address this issue fully. In contrast to retroviral gene delivery, no stable AAV packaging cell lines have been described in the literature to date. In particular the AAV Rep function appears to be cytotoxic in many cell types. Wild-type AAV has a 5 kb genome and the recombinant vectors share this size limitation. Unlike retroviral vectors, AAV particles are very stable. Taken together, AAV vector technology has many attractive and promising features but significant developmental efforts are required to make AAV vectors a clinical alternative to retroviral vectors for HSC gene therapy.

In addition to viral vector systems, physical methods have been considered for *ex vivo* gene transfer. In particular DNA injection and the gene gun technology (biolistics) have received considerable attention after reports that gene-modified peripheral T cells could be detected in patients treated with the gene gun using DNA-covered gold particles (242). The limited potential of these technologies to result in stable integration restricts their use in approaches relying on proliferation of the gene-modified target cells after re-infusion. However, these strategies can find applications in DNA vaccination approaches.

21.2.3.5 Clinical Trials

Several clinical protocols for HIV-1 disease have been submitted for review to the NIH recombinant advisory committee (RAC) and/or the FDA (Tab. 21.1). So far, no adverse effects were reported from any of the approved and initiated AIDS gene therapy trials. Approximately half of the clinical protocols represent adoptive immunotherapy or genetic vaccination strategies. In addition to the feasibility of genetic vaccinations and adoptive immunotherapies in AIDS patients, these clinical studies will provide basic knowledge concerning mechanisms of immune reconstitution in HIV-1 infected individuals.

To date, only one intracellular immunization protocol using gene-modified peripheral blood lymphocytes has been initiated (163). Several patients have already received gene-modified cells and the initial preliminary results suggest that RevM10-expressing cells have a selective advantage (242) in this patient population. A protocol using intracellular expression of hairpin ribozymes in primary T cells has been approved and will be evaluated clinically in the near future. Another RAC-approved protocol by Morgan et al. will evaluate safety of the first vector encoding two antiviral genes (td Rev, TAR-antisense) in the syngeneic setting of identical twins (160). Additional protocols using other antiviral genes and also exploring gene transfer into hematopoietic stem cells (Rosenblatt JD; Tab. 21.1) will soon enter the clinical phase. In the meantime, results from the first clinical studies will become available and will provide important information with respect to safety, efficacy and possible toxicity, including adverse immune induction. This knowledge will be extremely valuable for the rational design and development of improved second-generation clinical protocols employing improved gene delivery methods and antiviral gene combinations.

21.3 Future Directions

The antiviral efficacy of many HIV-1 gene therapy strategies has been demonstrated in various *in vitro* assay systems. An impressive number of approaches has been developed far enough to demonstrate antiviral efficacy in CD4-positive T cell lines. A smaller number has been evaluated in primary peripheral blood lymphocytes or mononuclear cells (70, 83, 120, 169, 176, 204, 210, 225, 226, 228) and antiviral efficacy in macrophages derived from transduced hematopoietic progenitor cells has only been reported in three instances (7, 208, 249).

Future HIV-1 gene therapy activities must address the potential cytotocixity of a particular strategy and balance it against the predicted antiviral efficacy, particularly if therapeutic benefit is limited by the ability of HIV-1 to generate escape mutants. New antiviral genes will be developed; specifically strategies which act on the incoming virus (201, 245) should be pursued intensively. Intracellular expression of the antiviral genes can be improved, notably, inducible systems might offer advantages with respect to cytotoxicity and immune function. Improved gene delivery systems will be required to make gene therapy a clinical reality with "injectable vectors" being the ultimate goal towards treatment of large patient numbers.

The cytotoxicity of most AIDS gene therapy strategies in primary cells has not been studied yet. It has been demonstrated for the trans-dominant RevM10 gene that expression of the gene product has no detectable adverse effects in transduced human peripheral blood lymphocytes (83) or in hematopoietic progenitor cells indicated by their engraftment potential in the SCID-hu Thy/Liv model (175). Using this *in vivo* model (146, 165, 172), functional T cells can be derived from transduced hematopoietic stem cells with sufficient RevM10 expression to prevent replication of HIV-1 isolates (27). It is unknown whether nucleic acid-based strategies can result in cytotoxicity but it was postulated that intracellular expression of RNA decoys (e.g. TAR) might sequester cellular binding proteins. In principle, intracellularly expressed ribozymes or antisense sequences have the potential to anneal to cellular transcripts with lower specificity than to the target sequence but no cytotoxic effects have been reported. Mutant variants of cellular factors required for HIV-1 replication (e.g. Rev cellular cofactors) are attractive candidates since they might circumvent cytotoxicity altogether (108). Also, as cellular proteins, these approaches are less likely to induce immune responses. A better understanding of the molecular interplay between HIV-1 genes/gene products (RNA, proteins) might open up new opportunities to develop highly specific ways to interfere genetically with HIV-1 replication.

The rapid development of drug resistance has been a disappointing experience in the initial clinical drug monotherapy studies. So far, none of the gene therapy approaches have been evaluated to address whether and how fast HIV-1 escape mutants will develop. It is very unlikely that such escape mutants will be detected in the planned or ongoing phase I clinical studies in the near future based on the fact that gene therapies act differently than drug therapies. Pharmacological compounds act systemically, and place selective pressure on the replicating virus both extracellularly and intracellularly as long as effective drug doses are maintained. This "new" environment offers a selective advantage to pre-existing variants which then become the major quasi-species (99, 240). In contrast, gene therapies act on the cellular level and as long as only a small fraction of cells express effective concentrations of the antiviral gene product, HIV-1 can replicate uninhibited and without selective pressure in these cells (i.e. non-gene-modified, insufficient expression of the therapeutic gene). In contrast to the clinical setting, the development of HIV-1 escape mutants can be studied in the laboratory by passaging the virus in clonal cell lines expressing the antiviral gene. In some instances, a more direct approach can be taken. Ribozymes have a defined target sequence. Considering the available database on HIV-1 isolates, the efficacy of intracellularly expressed ribozymes in blocking replication of HIV-1 isolates carrying mutations in the ribozyme target sequences can easily be analyzed. Similarly, short antisense transcripts can be tested against mutant HIV-1 isolates with varying degrees of homology. Intracellularly expressed sFvs interact with a specific target peptide and again, efficacy against HIV-1 variants can be directly determined.

HIV-1 gene therapy strategies which can prevent integration of the incoming virus have the potential to prevent *de novo* infection of protected cells and eventually can eradicate the virus from the patient, the ultimate goal of antiviral therapy. Cells equipped with such strategies will not express any HIV-1 gene products. In contrast, post-integration strategies (i.e. td Rev, antisense approaches) are only effective in suppressing infectious virion production. However, early HIV-1 gene products (Tat, Rev, Nef) are continously expressed and might have cytotoxic effects. Strategies which rely on the production of non-infectious virions (e.g. td Gag) might be even less desirable and could stimulate a response of the remaining immune system. The antiviral efficacy of multiple antiviral genes combined in one vector should be cumulative and therefore less likely to generate escape mutants. However, the currently available vector technology needs to be improved to be able to express multiple antiviral genes/gene products simultaneously, independently, and at effective levels.

This is particularly relevant for antisense strategies which can lead to degradation of the antisense RNA if effective (226). Current retroviral vectors rely on the full-length transcript or internal promoters for transgene expression. However, the complete therapeutic transcript encompassing the antisense sequence will be degraded. It will be important to design retroviral vector systems which can independently express a protein-based strategy and a therapeutic antisense transcript. Double copy vectors (96) have the potential to express RNAs independently but instability of this vector design has been reported (106, 138) and future development will be needed to generate safe and clinically useful reagents. Improved vector designs including modified promoter elements to limit positional effects would be desirable and would improve the overall efficacy. Inducible systems are also of interest since these strategies might prevent unwanted negative side effects if the antiviral gene is only expressed when needed (i.e. after HIV-1 infection; 128, 129, 204). The disadvantage of this approach is that strategies acting early in the HIV-1 replication cycle preventing or reducing proviral integration are not compatible with HIV-1 inducible expression systems.

The current gene delivery technology is sufficient to prove the feasibility of HIV-1 gene therapy which focuses predominantly on antiviral efficacy. Obviously, improvement of the gene transfer component in particular into hematopoietic stem cells is required for HIV-1 gene therapy to become a useful clinical treatment. Retroviral gene transfer into this primitive cell population is feasible and has been demonstrated in animal models including mice (109, 117, 187, 222) and non-human primates (110, 223, 224). Advanced technologies including refined vector manufacturing (81, 115) and improved transduction protocols (8, 115), already incrementally augment gene delivery into human hematopoietic stem and progenitor cells. Pseudotyping with other, non-amphotropic envelope proteins (i.e. VSV-G; 246) in combination with non-MLV-based vectors (11) might further improve HSC gene transfer efficacies. In addition, incorporation of *in vivo* selectable marker genes such as MDR-1 could be used to give the transduced HSC a selective advantage as was demonstrated in mouse models (95). Lentiviral-based delivery systems can infect quiescent cells (164) and several groups are working on HIV-based vector and packaging cell systems (40, 177). HIV-based delivery systems can be used to generate fusion proteins with antiviral gene products (i.e. RNAse) to efficiently incorporate this antiviral entity into HIV-1 particles (244) similar to previously published reports (166). Exposure to DNA damaging agents might improve gene delivery into mitotically-quiescent cells for lentiviral or retroviral-based gene delivery systems similar to reports on AAV vectors (2). However, the risks associated with such treatments must be carefully assessed and significant efforts will be required to develop these new technologies to meet the safety standards currently available and required for clinical amphotropic retroviral gene delivery. Targeted gene delivery using modified retroviral particles to interact with specific cell surface markers (94, 111, 141) could improve gene delivery into the target cells of choice which might be particularly relevant for *in vivo* gene delivery ("injectable vectors"). Furthermore, such *in vivo* gene delivery systems could be combined with homologous recombination approaches to specifically alter the genomic composition of a subset of relevant cells. For example, site-specific mutations could be inserted into the β-chemokine receptors to render T lymphocytes and macrophages of HIV-1 infected patients partially resistant. To achieve such longterm goals, the efficacy of both injectable vectors and homologous recombination must be improved beyond the current state of the art technologies.

Unless the currently evaluated multiple drug combinations are highly effective and can be tolerated for extended time periods, the medical need for gene therapy in HIV-1 disease exists. In the absence of appropriate animal models to study HIV-1 gene therapy approaches, phase I clinical trials are required to evaluate new gene therapy regimens. However, antiviral efficacy and cytotoxicity of any proposed gene therapy strategies must be thoroughly studied in adequate *in vitro* systems before entering the clinical phase. At present, the pathology of HIV-1 disease is still not fully understood and thus, AIDS clinical trials, including gene therapy studies, offer the only experimental recourse to address fundamental questions concerning the disease including: 1.) Will reducing the viral burden have long term clinical benefits and allow restoration of a functional immune system? 2.) Can *ex vivo* manipulated cells repopulate HIV-1 infected immune tissues including thymus, lymph nodes and bone marrow? 3.) Can gene-modified hematopoietic stem or progenitor cells differentiate into functional T lymphocytes and macrophages in HIV-1 infected individuals? 4.) Are gene-modified cells resistant to direct and indirect HIV-1 induced cell killing *in vivo*? All ongoing and planned gene therapy studies in AIDS will address aspects of these questions and will provide the scientific foundation for future clinical gene therapy development programs in HIV-1 disease. As in most other disease areas, improvement of gene transfer and gene expression will be required for effective clinical gene therapies. More importantly, less complex gene therapy technologies which are resistant to complement inactivation (212) will allow *in vivo* delivery and ultimately provide access of gene therapies to a large HIV-1 infected patient population.

Acknowledgements

I would like to thank my colleagues R. Pomerantz, H. Kaneshima, M. Bonyhadi and U. Junker for their helpful discussions and constructive review of this chapter. I am also very grateful to M. Böhnlein for exquisite editorial input and Sandy Swanson for excellent secretarial support.

References

1. Aguilar-Cordova, E., Chinen, J., Donehower, L. A., Harper, J.W., Rice, A. P., Butel, J. S., and Belmont, J. W. (1995). Inhibition of HIV-1 by a double transdominant fusion gene. Gene Ther. *2*, 181-186.
2. Alexander, I. E., Russell, D. W., and Miller, A. D. (1994). DNA-damaging agents greatly increase the transduction of nondividing cells by adeno-associated virus vectors. J. Virol. *68*, 8282-8287.
3. Alkhatib, G., Combadiere, C., Broder, C. C., Feng, Y., Kennedy, P. E., Murphy, P. M., and Berger, E. A. (1996). CC CKR5: A RANTES, MIP-1α, MIP-1β receptor as a fusion cofactor for macrophage-tropic HIV-1. Science *272*, 1955-1958.
4. Allen, P., Worland, S., and Gold, L. (1995). Isolation of high-affinity RNA ligands to HIV-1 integrase from a random pool. Virology *209*, 327-336.
5. Babé, L. M., Rosé, J., and Craik, C. S. (1995). Trans-dominant inhibitory human immunodeficiency virus type 1 protease monomers prevent protease activation and virion maturation. Proc. Natl. Acad. Sci. USA *92*, 10069-10073.
6. Bahner, I., Zhou, C., Yu, X.-J., Hao, Q.-L., Guatelli, J. C., and Kohn, D. B. (1993). Comparison of trans-Dominant Inhibitory Mutant Human Immunodeficiency Virus Type 1 Genes Expressed by Retroviral Vectors in Human T Lymphocytes. J. Virol. *67*, 199-3207.
7. Bahner, I., Kearns, K., Hao, Q.-L., Smogorzewska, E. M., and Kohn, D. B. (1996). Transduction of human CD34+ hematopoietic progenitor cells by a retroviral vector expressing an RRE decoy inhibits human immunodeficiency virus type 1 replication in myelomonocytic cells produced in long-term culture. J Virol. *20*, 4352-4360.
8. Bahnson, A. B., Dunigan, J. T., Baysal, B. E., and Mohney, T. (1995). Atchison, R. W., Nimgaonkar, M. T., Ball, E. D., and Barranger, J. A. (eds.) Centrifugal enhancement of retroviral mediated gene transfer. J. Virol. Meth. *54*, 131-143.
9. Baltimore, D. (1988). Intracellular immunization. Nature *335*, 395-396.
10. Barre-Sinoussi, F., Chermann, J. C., Rey, F., Nugeyre, M. T., Chamaret, J., Gruest, J., Dauguet, C., Axler-Blin, C., Vezinet-Brun, F., Rouzioux, C., Rozenbaum, W., and Montagnier, L. (1983). Isolation of a T-lymphotropic retrovirus from a patient at risk for acquired immunodeficiency syndrome (AIDS). Science *220*, 868-871.
11. Baum, C., Hegewisch-Becker, S., Eckert, H. G., Stocking, C., and Ostertag, W. (1995). Novel retroviral vectors for efficient expression of the multidrug resistance (mdr-1) gene in early hematopoietic cells. J. Virol. *69*, 7541-7547.
12. Bednarik, D. P., Mosca, J. D., Rai, N. B. K., and Pitha, P. M. (1989). Inhibition of human immunodeficiency virus (HIV) replication by HIV-trans-activated α2-interferon. Proc Natl Acad Sci USA *86*, 4958-4962.
13. Ben-David, Y., Giddens, E. B., Letwin, K., and Bernstein, A. (1991). Erythroleukemia induction by Friend murine leukemia viru: insertional activation of a new member of the ets gene family, Fli-1, closely linked to c-ets-1. Genes & Dev. *5*, 908-918.
14. Ben-David, Y., Lavigueur, A., Cheong, G. Y., and Bernstein, A. (1990). Insertional inactivation of the p53 gene during friend leukemia: a new strategy for identifying tumor suppressor genes. New Biol. *2*, 1015-1023.
15. Berkhout, B. and Jeang, K. T. (1992). Functional roles for the TATA promoter and enhancers in basal and Tat-induced expression of the human immunodeficiency virus type 1 long terminal repeat. J. Virol. *66*, 139-149.

16. Berkovitz, R. D., Hammarskjöld, M. L., Helga-Maria, C., Rekosh, D., and Goff, S. P. (1995). 5" regions of HIV-1 RNAs are not sufficient for encapsidation: implications for the HIV-1 packaging signal. Virology *212*, 718-723.
17. Bevec, D., Dobrovnik, M., Hauber, J., and Böhnlein, E. (1992). Inhibition of HIV-1 Replication in Human T-cells by Retroviral-Mediated Gene Transfer of a Dominant-Negative Rev Trans-Activator. Proc. Natl. Acad. Sci. USA *89*, 9870-9874.
18. Bevec, D., Volc-Platzer, B., Zimmermann, K., Dobrovnik, M., Hauber, J., Veres, G., and Böhnlein, E. (1993). Constitutive Expression of Chimeric Neo-Rev Response Element Transcripts Suppresses HIV-1 Replication in Human CD4+ T Lymphocytes. Hum Gene Ther. *5*, 193-201.
19. Bevec, D., Jaksche, H., Oft, M., Wöhl, T., Himmelspach, M., Pacher, A., Schebesta, M., Koettnitz, K., Dobrovnik, M., Csonga, R., Lottspeich, .F, and Hauber, J. (1996). Inhibition of HIV-1 replication in lymphocytes by mutants of the Rev cofactor eIF-5A. Science *271*, 1858-1860.
20. Biocca, S., Neuberger, M. S., and Cattaneo, A. (1990). Expression and targeting of intracellular antibodies in mammalian cells. EMBO J. *9*, 101-108.
21. Biocca, S., Pierandrei-Amaldi, P., and Cattaneo, A. (1993). Intracellular expression of anti-p21ras single chain Fv fragments inhibits meiotic maturation of xenopus oocyte. Biochem. Biophys. Res. Comm. *197*, 422-427.
22. Blaese, R. M. (1990). Treatment of severe combined immune deficiency (SCID) due to adenosine deaminase (ADA) with autologous lymphocytes transduced with a human ADA gene. Hum Gene Ther. *1*, 327-362.
23. Blaese, R. M. (1993). Treatment of severe combined immune deficiency (SCID) due to adenosine deaminase (ADA) with CD34+ selected autologous peripheral blood cells transduced with a human ADA gene. Hum Gene Ther. *4*, 521-527.
24. Blaese, R. M., Culver, K. W., Miller, A. D., Carter, C. S., Fleisher, T., Clerici, M., Shearer, G., Chang, L., Chiang, Y., Tolstoshev, P. et al. (1995). T lymphocyte-directed gene therapy for ADA-SCID: initial trial results after 4 years. Science *270*, 475-480.
25. Böhnlein, E., Lowenthal, J. W., Siekevitz, M., Ballard, D. W., Franza, B. R., and Greene, W. C. (1988). The same inducible nuclear protein(s) regulates mitogen activation of both the interleukin-2 receptor-alpha gene and type 1 HIV. Cell *53*, 827-836.
26. Bogerd, H. P., Fridell, R. A., Madore, S., and Cullen, B. R. (1995). Identification of a novel cellular cofactor for the Rev/Rex class of retroviral regulatory proteins. Cell *82*, 485-494.
27. Bonyhadi, M. L., Auten, J., Moss, M., Voytovich, A., Kalfoglou, C., Plavec, I., Forestell, S., Böhnlein, E., and Kaneshima, H. (1997). Efficient Expression of RevM10 in Human Hematopoietic Stem Cell Derived T cells: A Preclinical Feasibility Study. J. Virol. *71*, 4707-4716.
28. Bordignon, C., Mavilio, F., Ferrari, G., Servida, P., Ugazio, A. G., Notarangelo, L. D., Gilboa, E., Rossini, S., O'Reilly, R. J., Smith, C. A. et al. (1993). Transfer of the ADA gene into bone marrow cells and peripheral blood lymphocytes for the treatment of patients affected by ADA-deficient SCID. Hum Gene Ther. *4*, 513-520.
29. Bordignon, C., Notarangelo, L. D., Nobili, N., Ferrari, G., Casorati, G., Panina, P., Mazzolari, E., Maggioni, D., Rossi, C., Servida, P. et al. (1995). Gene therapy in peripheral blood lymphocytes and bone marrow for ADA-immunodeficient patients. Science *270*, 470-475.
30. Boris-Lawrie, K. and Temin, H. M. (1994). The retroviral vector. Replication cycle and safety considerations for retrovirus mediated gene therapy. Ann. NY Acad. Sci. *716*, 59-70.
31. Brady, H. J. M., Miles, C. G., Pennington, D. J., and Dzierzak, E. A. (1994). Specific ablation of human immunodeficiency virus Tat-expressing cells by conditionally toxic retroviruses. Proc. Natl. Acad. Sci. USA *91*, 365-369.
32. Brenner, M. K. (1991). Autologous bone marrow transplant for children with acute myelogenous leukemia in first remission: use of marker genes to investigate the biology of marrow reconstitution and the mechanism of relapse. Hum Gene Ther. *2*, 137-159.
33. Brenner, M. K. (1991). A phase I trial of high dose carboplation and epotoside with autologous marrow support for treatment of stage d neuroblastoma in first remission: use of marker genes to investigate the biology of marrow reconstitution and the mechanism of relapse. Hum Gene Ther. *2*, 257-272.
34. Brenner, M. K. (1991). A phase I trial of high dose carboplation and epotoside with autologous marrow support for treatment of relapse/refractory neuroblastoma withouit apparent bone marrow involvement: use of marker genes to investigate the biology of marrow reconstitution and the mechanism of relapse. Hum Gene Ther. *2*, 273-286.

35. Brenner, M. K. (1995). Human somatic gene therapy: progress and problems. J. Int. Med. *237*, 229-239.
36. Buschacher, G. L., Freed, E. O., and Panganiban, A. T. (1992). Cells induced to express a human immunodeficiency virus type 1 envelope gene mutant inhibit the spread of wild-type virus. Hum Gene Ther. *3*, 391-397.
37. Buchschacher, G. L., Freed, E. O., and Panganiban, A. T. (1995). Effects of second-site mutations on dominant interference by a human immunodeficiency virus type 1 envelope glycoprotein mutant. J. Virol. *69*, 1344-1348.
38. Buonocore, L. and Rose, J. K. (1990). Prevention of HIV-1 glycoprotein transport by soluble CD4 retained in the endoplasmatic reticulum. Nature *345*, 625-628.
39. Buonocore, L. and Rose, J. K. (1993). Blockade of human immunodeficiency virus type 1 production in CD4+ T cells by an intracellular CD4 expressed under control of the long terminal repeat. Proc. Natl. Acad. Sci. USA *90*, 2695-2699.
40. Carroll, R., Lin, J. T., Dacquel, E. J., Mosca, J. D., Burke, D. S., and St. Louis, D.C. (1994). A human immunodeficiency virus type 1 (HIV-1)-based retroviral vector system utilizing stable HIV-1 packaging cell lines. J. Virol. *68*, 6047-6051.
41. Caruso, M. and Klatzmann, D. (1992). Selective killing of CD4+ cells harboring a human immunodeficiency virus-inducible suicide gene prevents viral spread in an infected cell population. Proc. Natl. Acad. Sci. USA *89*, 182-186.
42. Champseix, C., Marechal, V., Khazaal, I., Schwartz, O., Fournier, S., Schlegel, N., Dranoff, G., Danos, O., Blot, .P, Vilmer, E., Herad, J.-M., Peault, B., and Lehn, P. (1996). A cell surface maker gene transferred with a retroviral vector into CD34+ cord blood cells is expressed by their T-cell progeny in the SCID-hu thymus. Blood *88*, 107-113.
43. Chatterjee, S., Johnson, P. R., and Wong, K. K. (1992). Dual-target inhibition of HIV-1 *in vitro* by means of an Adeno-Associated virus antisense vector. Science *258*, 1485-1488.
44. Chen, C. J., Banerjea, A. C., Harmison, G. G., Haglund, K., and Schubert, M. (1992). Multitarget-ribozyme directed to cleave at up to nine highly conserved HIV-1 env RNA regions inhibits HIV-1 replication – potential effectiveness against most presently sequenced HIV-1 isolates. Nucl. Acids. Res. *20*, 4581-4589.
45. Chen, S.-Y., Bagley, J., and Marasco, W. A. (1994). Intracellular antibodies as a new class of therapeutic molecules for gene therapy. Hum Gene Ther *5*, 595-601.
46. Cheynier, R., Henrichwark, S., Hadida, F., Pelletier, E., Oksenhendler, E., Autran, B., and Wain-Hobson, S. (1994). HIV and T cell expansion in splenic white pulps is accompanied by infiltration of HIV-specific cytotoxic T lymphocytes. Cell *78*, 373-387.
47. Choe, H., Farzan, M., Sun, Y., Sullivan, N., Rollins, B., Ponath, P. D., Wu, L., Mackay, C. R., LaRosa, G., Newman, W., Gerard, N., Gerard, C., and Sodroski, J. (1996). The β-chemokine receptors CCR3 and CCR5 facilitat infection by primary HIV-1 isolates. Cell *85*, 1135-1148.
48. Chuah, M. K., Vandendriessche, T., Chang, H. K., Ensoli, B., and Morgan, R. A. (1994). Inhibition of human immunodeficiency virus type-1 by retroviral vectors expressing antisense-TAR. Hum Gene Ther. *5*, 1467-1475.
49. Clever, J., Sassetti, C., and Parslow, T. G. (1995). RNA secondary structure and binding sites for gag gene poroducts in the 5' packaging signal of human immunodeficiency virus type 1. J. Virol. *69*, 2101-2109.
50. Cocchi, F., DeVico, A. L., Garzino-Demo, A., Arya, S. K., Gallo, R. C., and Lusso, P. (1995). Identification of RANTES, MIP-1(, and MIP-1(as the major HIV-suppressive factors produced by CD8+ T cells. Science *270*, 1811-1815.
51. Coffin, J., Haase, A., Levy, J. A., Montagnier, L., Oroszlan, S., Teich, N., Temin, H., Toyoshima, K., Varmus, H., Vogt, P., and Weiss, R. (1986). Human immunodeficiency viruses. Science *232*, 697.
52. Cohli, H., Fan, B., Joshi, R. L., Ramezani, A., Li, X., and Joshi, S. (1994). Inhibition of HIV-1 multiplication in a human CD4+ lymphocytic cell line expressing antisense and sense RNA molecules containing HIV-1 packaging signal and rev response element(s). Antisense Res. and Dev. *4*, 19-29.
53. Collier, A. C., Coombs, R. W., Schoenfeld, D. A., Bassett, R. L., Timpone, J., Baruch, A., Jones, M., Facey, K., Whitacre, C., McAuliffe, V. J., Friedman, H. M., Merigan, T. C., Reichman, R. C., Hooper, C., and Corey, L. (1996). Treatment of human immunodeficiency virus infection with saquinavir, zidovudine, and zalcitabine. N Engl J Med *334*, 1011-1017.
54. Condra, J. H., Schleif, W. A., Blahy, O. M., Gadryelski, L. J., Graham, D. J., Quintero, J. C., Rhodes, A., Robbins, H. L,, Roth, E., Shivaprakash, M., Titus, D., Yang, T., Teppler, H., Squires, K. E.,

Deutsch, P. J., and Emini, E. A. (1995). *In vivo* emergence of HIV-1 variants resistant to multiple protease inhibitors. Nature *374*, 569-571.
55. Cosset, F. L., Takeuchi, Y., Battini, J. L., Weiss, R. A., and Collins, M. K. L. (1995). High-titer packaging cells producing recombinant retroviruses resistant to human serum. J. Virol. *69*, 7430-7436.
56. Cullen, B. R. and Greene, W. C. (1990). Functions of the auxilliary gene products of the human immunodeficiency virus type 1. Virology *178*, 1-5.
57. Cullen, B. R. (1991). Human Immunodeficiency Virus as a Prototypic Complex Retrovirus. J. Virol. *65*, 1053-1056.
58. Cullen, B. R. (1991). Regulation of HIV-1 gene expression. FASEB J. *5*, 2361-2368.
59. Curiel, T. J., Cook, D. R., Wang, Y., Hahn, B. H., Ghosh, S. K., and Harrison, G.S. (1993). Long-term inhibition of clinical and laboratory human immunodeficiency virus strains in human T-cell lines containing an HIV-regulated diphtheria toxin A chain gene. Hum Gene Ther. *4*, 741-747.
60. Daar, E. S., Li, X. L., Moudgil, T., and Ho, D. D. (1990). High concentartion of recombinant soluble CD4 are required to neutralize primary human immunodeficiency virus type 1 isolates. Proc Natl Acad Sci USA *87*, 6574-6578.
61. Dalgleish, A. G., Beverley, P. C., Clapham, P. R., Crawford, D. H., Greaves, M. F., and Weiss, R. A. (1984). The CD4 (T4) antigen is an essential component of the receptor for the AIDS retrovirus. Nature *312*, 763-767.
62. Dalgleish, A. G., Moyle, G. J., Easterbrook, P., and Gazzard, B. G. (1995). AIDS: clinical and scientific issues past, present and future. Q. J. Nucl. Med. *39*, 156-162.
63. Deisseroth, A. B. (1991). Autologous bone marrow transplantation for chronic myelogenous leukemia with retroviral markers are used to discriminate between relapse which arises from systemic disease remaining after preparative therapy versus relapse to residual leukemic cells in autologous marrow: a pilot trial. Hum Gene Ther. *2*, 359-376.
64. DeLuca, A., Teofili, L., Antinori, A., Iovino, M. S., Mencarini, P., Visconti, E., Tamburrini, E., Leone, G., and Ortona, L. L. (1993). Haemapoietic CD34+ progenitor cells are not infected by HIV-1 *in vivo* but show impaired clonogenesis. Brit. J. Haematol. *85*, 20-24.
65. Deng, H., Liu, R., Ellmeier, W., Choe, S., Unutmaz, D., Burkhart, M., Di Marzio, P., Marmon, S., Sutton, R. E., Hill, C. M., Davis, C. B., Peiper, S. C., Schall, T. J., Littman, D. R., and Landau, N. R. (1996). Identification of a major co-receptor for primary isolates of HIV-1. Nature *381*, 661-666.
66. Doranz, B. J., Rucker, J., Yi, Y., Smyth, R. J., Samson, M., Peiper, S. C., Parmentier, M., Collman, R. G., and Doms, R. W. (1996). A dual-tropic primary HIV-1 isolate that uses fusin and the β-chemokine recptors CKR-5, CKR-3, and CKR-2b as fusion cofactors. Cell 85, 1149-1158.
67. Dragic, T., Litwin, V., Allaway, G. P., Martin, S. R., Huang, Y., Nagashima, K. A., Cayanan, C., Maddon, P. J., Koup, R. A., Moore, J. P., and Paxton, W. A. (1996). HIV-1 entry into CD4+ cells is mediated by the chemokine receptor CC-CKR-5. Nature *381*, 667-673.
68. Duan, L., Bagasra, O., Laughlin, M. A., Oakes, J. W., and Pomerantz, R. J. (1994). Comparison of Trans-Dominant Inhibitory Mutant Human Potent inhibition of human immunodeficiency virus type 1 replication by an intracellular anti-Rev single chain antibody. Proc Natl Acad Sci USA *91*, 5075-5079.
69. Duan, L., Zhang, H., Oakes, J. W., Bagasra, O., and Pomerantz, R. J. (1994). Molecular and virological effects of intracellular anti-rev single-chain variable fragments on the expression of various human immunodeficiency virus-1 strains. Hum Gene Ther. *5*, 1315-1324.
70. Duan, L., Zhu, M., Bagasra, O., and Pomerantz, R. J. (1995). Intracellular immunization against HIV-1 infection of human T lymphocytes: utility of anti-Rev single-chain variable fragments. Hum Gene Ther. *6*, 1561-1573.
71. Duan, L. and Pomerantz, R. J. (1996). Intracellular antibodies for HIV-1 gene therapy. Science & Medicine *3*, 24-33.
72. Dunbar, C. E., Cottler-Fox, M., O'Shaughnessy, J. A., Doren, S., Carter, C., Berenson, R., Brown, S., Moen, R. C., Greenblatt, J., Stewart, F. M. et al. (1995). Retrovirally marked CD34-enriched peripheral blood and bone marrow cells contribute to long-term engraftment after autologous transplantation. Blood *85*, 3048-3057.
73. Dunbar, C.E. and Emmons RV (1994). Gene transfer into hematopoietic progenitor and stem cells: progress and problems. Stem Cells Dayt. 12: 563-576.

74. Ellington A. D. and Szostak, J. W. (1990). *In vitro* selection of RNA molecules that bind specific ligands. Nature *346*, 818-822.
75. Escaich, S., Kalfoglou, C., Plavec, I., Kaushal, S., Mosca, J. D., and Böhnlein, E. (1995). RevM10-mediated Inhibition of HIV-1 Replication in Chronically-infected T Cells. Hum Gene Ther. *6*, 625-634.
76. Felber, B. K. and Pavlakis, G. N. (1993). Molecular biology of HIV-1: Positive and negative regulatory elements important for virus expression. AIDS *7*, 51-62.
77. Feng, Y., Broder, C. C., Kennedy, P. E., and Berger, E. A. (1996). HIV-1 entry cofactor: functional cDNA cloning of a seven-transmembrane G protein-coupled receptor. Science *272*, 872-876.
78. Fischer-Adams, G., Wong, K. K., Podsakoff, G., Forman, S. J., and Chatterjee, S. (1996). Integration of Adeno-associated virus vectors in CD34+ human hematopoietic progenitor cells after transduction. Blood *88*, 492-504.
79. Folks, T.,M., Kessler, S. W., Orenstein, J. M., Justement, J. S., Jaffe, E. S., and Fauci, A. S. (1988). Infection and replication of HIV-1 in purified progenitor cells of normal human bone marrow. Science *242*, 919-922.
80. Folks, T. M. (1991). Human immunodeficiency virus in bone marrow: still more questions than answers. Blood *77*, 1625-1626.
81. Forestell, S. P., Böhnlein, E., and Rigg, R. J. (1995). Retroviral End-Point Titer is not Predictive of Gene Transfer Efficiency: Implications for Vector Production. Gene Ther. *2*, 1-8.
82. Forestell, S. P., Dando, J. S., Böhnlein, E., and Rigg, R. J. (1996). Improved Detection of Replication Competent Retrovirus. Journal of Virological Methods *60*, 171-180.
83. Fox, B. A., Woffendin, C., Yang, Z.-Y., San, H., Ranga, U., Gordon, D., Osterholzer, J., and Nabel, G. J. (1995). Genetic modification of human peripheral blood lymphocytes with a transdominant negative form of rev: safety and toxicity. Hum Gene Ther. *6*, 997-1004.
84. Fritz, C. C., Zapp, M. L., and Green, M. R. (1995). A human nucleoporin-like protein that specifically interacts with HIV Rev. Nature *376*, 530-533.
85. Gallo, R. C., Sarin, P. S., Gelmann, E. P., Robert-Guroff, M., Richardson, E., Kalyanaraman, V. S., Mann, D., Sidhu, G. D., Stahl, R. E., Zolla-Pazner, S., Leibowitch, J., and Popovic, M. (1983). Isolation of human T-cell leukemia virus in acquired immune deficiency syndrome (AIDS). Science *220*, 865-867.
86. Gallo, R. C., Salahuddin, S. Z., Popovic, M., Shearer, G. M., Kaplan, M., Haynes, B. F., Palker, T. J., Redfield, R., Oleske, J., and Safai, B. (1984). Frequent detection and isolation of cytopathic retroviruses (HTLV-III) from patients with AIDS and at risk for AIDS. Science *244*, 500-503.
87. Galpin, J. E., Casciato, D. A., and Richards, S. B. (1994). A phase I clinical trial to evaluate the safety and biological activity of HIV-IT (TAF) (HIV-1IIIBenv-transduced, autologous fibroblasts in asymptomatic HIV-1 infected subjects. Hum Gene Ther. *5*, 997-1017.
88. Giver, L., Bartel, D. P., Zapp, M. L., Green, M. R., and Ellington, A. D. (1993). Selection and design of high-affinity RNA ligands for HIV-1 Rev. Gene *137*, 19-24.
89. Gougeon, M. L., Colizzi, V., Dalgleish, A. G., and Montagnier, L. (1993). New concepts in AIDS pathogenesis. AIDS Res. Human Retroviruses *3(9)*, 287-289.
90. Green, M., Ishino, M., and Lowenstein, P. M. (1989). Mutational analysis of HIV-1 tat minimal domain peptides: identification of trans-dominant mutants that suppress HIV-LTR-driven gene expression. Cell *58*, 215-223.
91. Greene, W.C. (1990). Regulation of HIV-1 gene expression. Ann. Rev. Immunology *8*, 453-475.
92. Haapala, D. K., Robey, W. G., Oroszlan, S. D., and Tsai, W. P. (1985). Isolation from cats of an endogenous type C virus with a novel envelope glycoprotein. J. Virol. *53*, 827-833.
93. Hampel, A., Tritz, R., Hicks, M., and Cruz, P. (1990). Hairpin catalytic RNA model: evidence for for helices and sequence requirement for target RNA. Nucl. Acids Res. *18*, 299-304.
94. Han, X., Kasahara, N., and Kan, Y. W. (1995). Ligand-directed retroviral targeting of human breast cancer cells. Proc Natl Acad Sci USA *92*, 9747-9751.
95. Hanania, E. G. and Deisseroth, A. B. (1994). Serial transplantation shows that early hematopoietic precursor cells are transduced by MDR-1 retroviral vector in a mouse gene therapy model. Cancer Gene Ther. *1*, 21-25.
96. Hantzopoulos, P. A., Sullenger, B. A., Ungers, G., and Gilboa, E. (1986). Improved gene expression upon transfer of the adenosine transaminase minigene outside the transcriptional unit of a retroviral vector. Proc Natl Acad Sci USA *86*, 3519-3523.

97. Haselhoff, J. and Gerlach, W. I. (1988). Simple RNA enzymes with new and highly specific endoribonuclease activities. Nature *334*, 585-591.
98. Haubrich, R. and McClutchan, J.A. (1995). An open label, phase I/II clinical trial to evaluate the safety and biological activity of HIV-IT (V) (HIV-1IIIBenv/rev retroviral vector) in HIV-1-infected Subjects. Hum Gene Ther. *6*, 941-955.
99. Ho, D. D., Neumann, A. U., Perelson, A. S., Chen, W., Leonard, J. M., Markowitz, M. (1995). Rapid turnover of plasma virions and CD4 lymphocytes in HIV-1 infection. Nature *373*, 123-126.
100. Ilves, H., Barske, C., Junker, U., Böhnlein, E., and Veres, G. (1996). Retroviral vectors designed for targeted RNA Polymerase III-driven expression: a comparative study. Gene *171*, 203-208.
101. Irwin, M. J., Laube, L. S., Lee, V., Austin, M., Chada, S., Anderson, C. G., Townsend, K., Jolly, D. J., and Warner, J. F. (1994). Direct injection of a recombinant retroviral vector induces human immunodeficiency virus-specific immune responses in mice and nonhuman primates. J. Virol. *68*, 5036-5044.
102. Jensen, K. B., Green, L., MacDougal-Waugh, S., and Tuerk, C. (1994). Characterization of an *in vitro*-selected RNA ligand to the HIV-1 Rev protein. J. Mol. Biol. *235*, 237-247.
103. Jolly D. (1994). Viral vector system for gene therapy. Cancer Gene Ther *1*, 51-64.
104. Joshi, S., Van Brunschot, A., Asad, S., Van Der Elst, I., Read, S. E., and Bernstein, A. (1991). Inhibition of human immunodeficiency virus type 1 multiplication by antisense and sense RNA expression. J. Virol. *65*, 5524-5530.
105. Junker, U., Rittner, K., Homann, M., Bevec, D., Böhnlein, E., and Sczakiel, G. (1994). Polymerase III-driven transcription of chimeric tRNA-antisense RNA genes reduces replication of the human immunodeficiency virus type 1 in human T-cell lines. Antisense Res. & Dev. *4*, 165-172.
106. Junker, U., Böhnlein, E., and Veres, G. (1995). Genetic Instability of a MoMLV-Based Double Copy Vector Designed for HIV-1 Gene Therapy. Gene Ther. *2*, 639-646.
107. Junker, U., Escaich, S., Plavec, I., McPhee, F., Rosé, J., Craik, C. S., and Böhnlein, E. (1996). Intracellular Expression of HIV-1 Protease Variants Inhibits Replication of Wild-type And Protease Inhibitor Resistant HIV-1 Strains in Human T-Cell Lines. J. Virol. *70*, 7765-7772.
108. Junker, U., Bevec, D., Barske, C., Kalfoglou, C., Escaich, S., Dobrovnik, M., Hauber, J., and Böhnlein, E. (1996). Intracellular Expression of Cellular eIF5A Mutants Inhibits HIV-1 Replication in Human T Cells: A New Approach for AIDS Gene Therapy. Hum Gene Ther. *7*, 1861-1869.
109. Kaleko, M., Garcia, J. V., Osborne, W. R., and Miller, A. D. (1990). Expression of human adenosine deaminase in mice after transplantation of genetically-modified bone marrow. Blood *75*, 1733-1741.
110. Kantoff, P. W., Gillio, A. P., McLachlin, J. R., Bordignon, C., Eglitis, M. A., Kernan, N. A., Moen, R. C., Kohn, D. B., Yu, S. F., Karlsson, S., Zwiebel, J. A., Gilboa, E., Blaese, R. M., Nienhuis, A., O'Reilly, R. J., and Anderson, W. F. (1987). Expression of human adenosine deaminase in nonhuman primates after retrovirus-mediated gene transfer. J. Exp. Med. *166*, 219-234.
111. Kasahara, N., Dozy, A. M., and Kan, Y. W. (1994). Tissue-specific targeting of retroviral vectors through ligand-receptor interactions. Science *266*, 1373-1376.
112. Kaye, J. F., Richardson, J. H., and Lever, A. M. (1995). Cis-acting sequences involved in human immunodeficiency virus type 1 RNA packaging. J. Virol. *69*, 6588-6592.
113. Kempf, D. J., Marsh, K. C., Paul, D. E., Knigge, M. .F., Norbeck, D. W., Kohlbrenner, W. E., Codacovi, L., Vasavanonda, S., Bryant, P., Wang, X. C., Wideburg, N. E., Clement, J. J., Plattner, J. J., and Erickson, J. (1991). Antiviral and pharmacokinetic properties of C2 symmetric inhibitors of the human immunodeficiency virus type 1 protease. Antimicrob. Agents Chemother. *35*, 2209-2214.
114. Kohn, D.B., Weinberg, K. I., Nolta, J. A., Heiss, L. N., Lenarsky, C., Crooks, G. .M., Hanley, M. E., Annett, G., Brooks, J. S., el Khoureiy, A. et al. (1995). Engraftment of gene-modified umbilical cord blood cells in neonates with adenosine deaminase deficiency. Nature Med. *1*, 1017-1023.
115. Kotani, H., Newton, P. B., Zhang, S., Chiang, Y. L., Otto, E., Weaver, L., Blaese, R. M., Anderson, W. F., and McGarrity, G. J. (1994). Improved methods of retroviral vector transduction and production for gene therapy. Hum Gene Ther. *5*, 19-28.
116. Kotin, R. M., Siniscalco, M., Samulski, R. J., Zhu, X. D., Hunter, L., Laughlin, C. A., McLaughlin, S., Muzyczka, N., Rocchi, M., and Berns, K. I. (1990). Site-specific integration by adeno-associated virus. Proc. Natl. Acad. Sci. USA *87*, 2211-2215.
117. Krall, W. J., Challita, P. M., Perlmutter, L. S., Skelton, D. C., and Kohn, D. B. (1994). Cells expressing human glucocerebrosidase from a retroviral vector repopulate macrophages and central nervous system microglia after murine bone marrow transplantation. Blood *83*, 2737-2748.

118. Larder, B. A., Kellam, P., and Kemp, S. D. (1993). Convergent combination therapy can select viable multidrug resistant HIV-1 *in vitro*. Nature *365*, 451-453.
119. Larder, B. A. (1993). Inhibitors of HIV reverse transcriptase as antiviral agents and drug resistance. In: Skalka, A. M and Goff, S. P. (eds.), Reverse transcriptases. New York: Cold Spring Harbor Laboratory, 205-222.
120. Leavitt, M. C., Yu, M., Yamada, O., Kraus, G., Looney, D., Poeschla, E., Wong-Staal, F. (1994). Transfer of an anti-HIV-1 ribozyme gene into primary human lymphocytes. Hum Gene Ther. *5*, 1115-1120.
121. Lee, T., Sullenger, B. A., Gallardo, H. F., Ungers, G. F., and Gilboa, E. (1992). Overexpression of RRE-derived sequences inhibits HIV-1 replication in CEM cells. New Biol. *4*, 66-74.
122. Lee, T., Gallardo, H. F., Gilboa, E., and Smith, C. (1994). Inhibition of human immunodeficiency virus type 1 in human T cells by a potent Rev-response element decoy consisiting of the 13-nucleotide minimal Rev-binding domain. J. Virol. *68*, 8254-8264.
123. Lee, S. W., Gallardo, H. F., Gaspar, O., Smith, C., and Gilboa, E. (1995). Inhibition of HIV-1 in CEM cells by a potent TAR decoy. Gene Ther. *2*, 377-384.
124. Levine, B. L., Mosca, J. D., Riley, J. L., Carroll, R. G., Vahey, M. T., Jagodzinski, L. L., Wagner, K. F., Mayers, D. L., Burke, D. S., Weislow, O. S., St Louis, D. C., and June, C. H. (1996). Antiviral effect and *ex vivo* CD4+ T cell proliferation in HIV-positive patients as a result of CD28 costimulation. Science *272*, 1939-1943.
125. Levy, J. A., Hollander, H., Shimabukuro, J., Mills, J., and Kaminsky, L. (1994). Isolation of lymphocytopathic retroviruses from San Francisco patients with AIDS. Science *225*, 840-842.
126. Levy, J. A. (1993). Pathogenesis of human immunideficiency virus infection. Microbiological Reviews *57(1)*, 183-289.
127. Levy-Mintz, P., Duan, L., Zhang, H., Hu, B., Dornadula, G., Zhu, M., Kulkosky, J., Bizub-Bender, D., Skalka, A. M., and Pomerantz, R. J. (in press). Intracellular expression of single-chain variable fragments (SFv) to inhibit early stages of the viral life0cycle by targeting HIV-1 integrase. J. Virol.
128. Lisziewicz, J., Rappaport, J., and Dhar, R. (1991). Tat-regulated production of multimerized TAR RNA inhibits HIV-1 gene expression. New Biol. *3*, 82-89.
129. Lisziewicz, J., Sun, D., Smythe, J., Lusso, P., Lori, F., Louie, A., Markham, P., Rossi, J., Reitz, M. S., and Gallo, R. C. (1993). Inhibition of human immunodeficiency virus type 1 replication by regulated expression of a polymeric Tat activation response RNA decoy as a strategy for gene therapy in AIDS. Proc. Natl. Acad. Sci. USA *90*, 8000-8004.
130. Liu, R., Paxton, W. A., Choe, S., Ceradini, D., Martin, S. R., Horuk, R., MacDonald, M. E., Stuhlmann. H., Koup, R. A., and Landau, N. R. (1996). Homozygous defect in HIV-1 coreceptor accounts for resistance of some multiply-exposed individuals to HIV-1 infection. Cell *86*, 367-377.
131. Lo, K. M., Biasolo, M. A., Dehni, G., Palu, G., and Haseltine, W. A. (1992). Inhibition of replication of HIV-1 by retroviral vectors expressing tat-antisense and anti-tat ribozyme RNA. Virology *190*, 176-183.
132. Lorentzen, E. U., Wieland, U., Kuhn, J. E., and Braun, R. W. (1991). *In vitro* cleavage of HIV-1 vif RNA by a synthetic ribozyme. Virus Genes *5*, 17-23.
133. Maciejewski, J. P., Weichold, F. F., Young, N. S., Cara, A., Zella, D., Reitz, M. S., and Gallo, R. C. (1995). Intracellular expression of antibody fragments directed against HIV-1 reverse transcriptase prevents HIV-1 infection *in vitro*. Nature Med. *1*, 667-673.
134. Malim, M. H., Hauber, J., Le, S.-Y., Maizel, J. V., and Cullen, B. R. (1989). The HIV-1 rev transactivator acts through a structured target sequence to activate nuclear export of unspliced viral mRNA. Nature *338*, 254-257.
135. Malim, M. H., Böhnlein, S., Hauber, J., and Cullen, B.R. (1989). Functional dissection of the HIV-1 Rev trans-activator-derivation of a trans-dominant repressor of Rev function. Cell *58*, 205-214.
136. Malim, M. H., McCarn, D. F., Tiley, L. S., and Cullen, B. R. (1991). Mutational definition of the human immunodeficiency virus type 1 Rev activation domain. J. Virol. *65*, 4248-4254.
137. Malim, M. H., Freimuth, W. W., Liu, J., Boyle, T. J., Lylerly, H. K., Cullen, B. R., and Nabel, G. J. (1992). Stable expression of transdominant rev protein in human T cells inhibits human immunodeficiency virus replication. J. Exp. Med. *176*, 1197-1201.
138. Malvilio, F., Ferrari, G., Rossini, S., Nobili, N., Bonini, C., Casorati, G., Traversari, G., and Bordignon, C. (1994). Peripheral blood lymphocytes as target cells of retroviral vector-mediated gene transfer. Blood *83*, 1988-1997.

139. Mann, R., Mulligan, R. C., and Baltimore, D. (1983). Construction of a retrovirus packaging mutant and its use to produce helper-free defective retrovirus. Cell *33*, 153-159.
140. Marasco, W. A., Hasletine, W. A, and Chen, S. Y. (1993). Design, intracellular expression, and activity of a human anti-human immunodeficiency virus type 1 gp120 single-chain antibody. Proc. Natl. Acad. Sci. USA *90*, 7889-7893.
141. Marin, M., Noel, D., Valsesia-Wittman, S., Brockly, F., Etienne-Julan, M., Russell, S., Cosset, F. L., and Piechaczyk, M. (1996). Targeted infection of human cells via major histocompatibility complex class I molecules by Moloney murine leukemia virus-derived viruses displaying single-chain antibody fragment-envelope fusion proteins. J. Virol. *70*, 2957-2962.
142. Markovitz, D., Goff, S., and Bank, A. (1988). Construction and use of a safe and efficient amphotropic packaging cell line. Virology *167*, 400-406.
143. Markowitz, M., Saag, M., Powderly, W. G., Hurley, A. M., Hsu, A., Valdes, J. M., Henry, D., Sattler, F., La Marca, A., Leonard, J. M., and Ho, D. D. (1995). A preliminary study of ritonavir, an inhibitor of HIV-1 protease, to treat HIV-1 infection. New Engl. J. Med. *333*, 1534-1539.
144. Markowitz, M., Mo, H., Kempf, D. J., Norbeck, D. W., Bhat, T. N., Erickson, J. W., and Ho, D. D. (1995). Selection and analysis of human immunodeficiency virus type 1 variants with increased resistance to ABT-*538*, a novel protease inhibitor. J. Virol. *69*, 701-706.
145. Matsuda, Z., Yu, X., Yu, Q.-C., Lee, T.-H., and Essex, M. (1983). A virion-specific inhibitory molecule with therapeutic potential for human immunodeficiency virus type 1. Proc. Natl. Acad. Sci. USA *90*, 3544-3448.
146. McCune, J. M., Namikawa, R., Kaneshima, H., Shultz, L. D., Lieberman, M., and Weissman, I. L. (1988). The SCID-hu mouse: murine model for the analysis of human hematolymphoid differentiation and function. Science *241*, 1632-1639.
147. Mellors, J. W., Rinaldo, C. R., Gupta, P., White, R. M., Todd, J. A., and Kingsley, L. A. (1996). Prognosis in HIV-1 infection predicted by the quantity of virus in plasma. Science *272*, 1167-1170.
148. Mermer, B., Felber, B. K., Campbell, M., and Pavlakis, G. N. (1990). Identification of trans-dominant HIV rev protein mutants by direct transfer of bacterially produced proteins into human cells. Nucl. Acids Res. *18*, 2037-2044.
149. Meyerhans, A., Cheynier, R., Albert, J., Seth, M., Kwok, S., Sninsky, L., Morfeldt-Manson, L., Asjo, B., and Wain-Hobson, S. (1989). Temporal fluctuations in HIV quasispecies *in vivo* are not reflected by sequential HIV isolations. Cell *58*, 901-910.
150. Mhashilkar, A. M., Bagley, J., Chen, S. Y., Szilvay, A. M., Helland, D. G., and Marasco, W. A. (1995). Inhibition of HIV-1 tat-mediated LTR transactivation and HIV-1 infection by anti-Tat single chain intrabodies. EMBO J. *14*, 1542-1551.
151. Miller, A. D. and Rosman, G. J. (1989). Improved retroviral vectors for gene transfer and expression. Bio/Tech. *7*, 980-988.
152. Miller AD, and Buttimore, C. (1986). Redesign of retrovirus packaging cell lines to avoid recombination leading to helper virus production. Mol. Cell. Biol. *6*, 2895-2902.
153. Miller, A. D. (1990). Retrovirus packaging cells. Human Gene Ther. *1*, 5-14.
154. Miller, A. D., Garcia, J. V., von Suhr, N., Lynch, C. M., Wilson, C., and Eiden, M. V. (1991). Construction and properties of retrovirus packaging cells based on gibbon ape leukemia virus. J. Virol *65*, 2220-2224.
155. Miller, J. L., Donahue, R. E., Sellers, S. E., Samulski, R. J., Young, N. S., and Nienhuis, A. W. (1994). Recombinant adeno-associated virus (rAAV)-mediated expression of a human gamma-globin gene in human progenitor-derived erythroid cells. Proc. Natl. Acad. Sci. USA *91*, 10183-10187.
156. Modesti, N., Garcia, J., Debouck, C., Peterlin, M., and Gaynor, R. (1991). Trans-dominant Tat mutants with alterations in the basic domain inhibit HIV-1 gene expression. New Biol. *3*, 759-768.
157. Morgan, R. A., Looney, D. J., Muenchau, D. D., Wong-Staal, F., Gallo, R. C., and Anderson, W. F. (1990). Retroviral vectors expressing soluble CD4: a potential gene therapy for AIDS. AIDS Res. Hum. Retroviruses *6*, 183-191.
158. Morgan, R. A, Couture, L., Elroy-Stein, O., Ragheb, J., Moss, B., and Anderson, F. W. (1992). Retroviral vectors containing putative internal ribosome entry sites: development of a polycistronic gene transfer system and applications to human gene therapy. Nucl. Acids Res. *20*, 1293-1299.
159. Morgan, R. A., Baier-Bitterlich, G., Ragheb, J. A., Wong-Staal, F., Gallo, R. C., and Anderson, F. W. (1994). Further evaluation of soluble CD4 as an anti-HIV type 1 gene therapy: demonstration of protection of primary human peripheral blood lymphocytes from infection by HIV type 1. AIDS Res. Hum. Retroviruses *10*, 1507-1515.

160. Morgan, R. A. and Walker, R. (1996). Gene therapy for AIDS using retroviral mediated gene transfer to deliver HIV-1 antisense TAR and transdominant Rev protein genes to syngeneic lymphocytes in HIV-1 infected identical twins. Hum Gene Ther *7*, 1281-1306.
161. Muzyczka, N. (1992). Use of adeno-associated virus as a general transduction vector for mammalian cells. Curr. Top. Microbiol. Immunol. *158*, 97-129.
162. Nabel, G. J. and Baltimore, D. (1987). An inducible transcription factor activates expression of human immunodeficiency virus in T cells. Nature *326*, 711-713.
163. Nabel, G. J., Fox, B. A., Post, L., Thompson, C. B., and Woffendin, C. (1994). A molecular genetic intervention for AIDS – effects of a transdominant negative form of Rev. Hum. Gene Ther. *5*, 79-92.
164. Naldini, L., Blomer, U., Gallay, P., Ory, D., Mulligan, R., Gage, F. H., Verma, I., and Trono, D. (1996). *In vivo* gene delivery and stable transduction of nondividing cells by a lentiviral vector. Science *272*, 263-266.
165. Namikawa, R., Kaneshima, H., Lieberman, M., Weissman, I. L., and McCune, J. M. (1988). Infection of the SCID-hu mouse by HIV-1. Science *242*, 1684-1686.
166. Natsoulis, G. and Boeke, J. D. (1991). New antiviral strategy using capsid-nuclease fusion proteins. Nature *352*, 632-635.
167. Naviaux, R. K. and Verma, I. M. (1992). Retroviral vectors for persistent expression *in vivo*. Current Opin Biotech. *3*, 540-547.
168. Ojwang, J. O., Hampel, A., Looney, D. J., Wong-Staal, F., and Rappaport, J. (1992). Inhibition of human immunodeficiency virus type 1 expression by a hairpin ribozyme. Proc. Natl. Acad. Sci. USA *89*, 10802-10806.
169. Orloff, S. L., Kennedy, M. S., Belperron, A. A., Maddon, P. J., and McDougal. (1993). Two mechanisms of soluble CD4 (sCD4)-mediated inhibition of human immunodeficiency virus type 1 (HIV-1) infectivity and their relation to primary HIV-1 isolates with reduced sensitivity to sCD4. J. Virol. *67*, 1461-1471.
170. Pantaleo, G., Graziosi, C., Demarest, J. F., Butini, L., Montroni, M., Fox, C. H., Orenstein, J. M., Kotler, D. P., and Fauci, A. S. (1993). HIV infection is active and progressive in lymphoid tissue during the clinically latent stage of disease. Nature *362*, 355-358.
171. Pearson, L., Garcia, J., Wu, F., Modesti, N., Nelson, J., and Gaynor, R. (1990). A trans-dominant tat mutant that inhibits tat-induced gene expression from the HIV LTR. Proc. Natl. Acad. Sci. USA *87*, 5079-5083.
172. Peault, B., Weissman, I. L., Baum, C., McCune, J. M., and Tsukamoto, A. (1991). Lymphoid reconstitution of the human fetal thymus in SCID mice with CD34+ precursor cells. J. Exp. Med. *174*, 1283-1286.
173. Perelson, A. S., Neumann, A. U., Markowitz, M., Leonard, J. M., and Ho, D.D. (1996). HIV-1 dynamics *in vivo*: virion clearance rate, infected cell life span, and viral generation time. Science *271*, 1582-1586.
174. Pilkington, G. R., Duan, L., Zhu, M., Keil, W., and Pomerantz, R. J. (1996). Recombinant human Fab antibody fragments to HIV-1 Rev and Tat regulatory proteins: Direct selection from a combinatorial phage display library. Mol. Immun. *4*, 439-450.
175. Plavec, I., Voytovich, A., Moss, K., Webster, D., Hanley. M.-B., Escaich, S., Ho, K. E., Bönlein, E., and DiGiusto, D. L. (1996). ustained retroviral gene marking and expression in lymphoid and myeloid cells derived from transduced hematopoietic progenitor cells. Gene Ther. *3*, 717-724.
176. Plavec, I., Agarwal, M., Ho, K. E., Pineda, M., Auten, J., Baker, J., Matsuzaki, H., Escaich, S., and Böhnlein, E. (1997). High levels of trans-dominant RevM10 protein are required for inhibition of HIV-1 replication in cell lines and primary T cells: implications for gene therapy of AIDS. Gene Ther. *4*, 128-139.
177. Poznansky, M., Lever, A., Bergeron, L., Haseltine, W., and Sodroski, J. (1991). Gene transfer into human lymphocytes by a defective human immunodeficiency virus type 1 vector. J. Virol. *65*, 532-536.
178. Ragheb, J. A., Bressler, P., Daucher, M., Chiang, L., Chuah, M. K., Vandendriessche, T., and Morgan, R.A. (1995). Analysis of trans-dominant mutants of the HIV type 1 Rev protein for their ability to inhibit Rev function, HIV type 1 replication, and their use as anti-HIV gene therapeutics. AIDS Res Hum Retroviruses *11*, 1343-1353.
179. Ratner, L., Fisher, A., Jagodzinski, L. L., Mitsuya, H., Liou, R. S., Gallo, R. C., and Wong-Staal, F. (1987). Complete nucleotide sequences of functional clones of the AIDS virus. AIDS Res. Hum. Retroviruses *3*, 57-69.

180. Rhodes, A. and James, W. (1990). Inhibition of human immunodeficiency virus replication in cell culture by endogenously synthesized antisense RNA. J. Gen. Virol. *71*, 1965-1974.
181. Rhodes, A. and James, W. (1991). Inhibition of heterologous strains of HIV by antisense RNA. AIDS *5*, 145-151.
182. Richardson, J. H., Child, L. A., and Lever, A. M. (1993). Packaging of human immunodeficiency virus type 1 RNArequires cis-acting sequences outside the 5' leader region. J. Virol. *67*, 3997-4005.
183. Riddell, S. R., Watanabe, K. S., Goodrich, J. M., Li, C. R., Agha, M. E., and Greenberg, P. D. (1992). Restoration of viral immunity in immunodeficient humans by adoptive transfer of T cell clones. Science *257*, 238-241.
184. Riddell, S. R., Greenberg, P. D., Overell, R. W., Loughram, T. P., Gilbert, M. J., Lupton, S. D., Agosti, J., Scheeler, S., Cooms, R. W., and Cores, I. (1993). Phase I study of cellular adoptive immunotherapy using genetically modified CD8+ HIV-specific T cells for HIV seropositive patients undergoing allogeneic bone marrow transplant. Hum Gene Ther. *3*, 319-338.
185. Riddell, S. R., Elliott, M., Lewinsohn, D. A., Gilbert, M. J., Wilson, L., Manley, S. A., Lupton, S. D., Overell, R. W., Reynolds, T. C., Corey, L., and Greenberg, P. D. (1996). T-cell mediated rejection of gene-modified HIV-specific cytotoxic T lymphocytes in HIV-infected patients. Nature Med. *2*, 216-223.
186. Rigg, R., Dando, J., Chen, J., Plavec, I., and Böhnlein, E. (1996). A novel human amphotropic retroviral packaging cell line: high titer, complement-resistance and improved safety. Virology *218*, 290-295.
187. Rivière, I., Brose K., and Mulligan, R. C. (1995). Effect of retroviral vector design on expression of human adenosine deaminase in murine bone marrow transplant recipients engrafted with genetically modified cells. Proc. Natl. Acad. Sci. USA *92*, 6733-6737.
188. Roberts, M. R., Qin, L., Zhang, D., Smith, D. H., Tran, A.-C., Dull, T. J., Groopman, J. E., Capon, D. J., Byrn, R. A., and Finer, M. H. (1994). Targeting of human immunodeficiency virus-infected cells by CD8+ T lymphocytes armed with universal T-cell receptors. Blood *84*, 2878-2889.
189. Rolling, F. and Samulski, R. J. (1995). AAV as a viral vector for human gene therapy. Generation of recombinant virus. Mol. Biotechnol. *3*, 9-15.
190. Rosé, J. R., Babé, L. M., and Craik, C. S. (1995). Defining the level of human immunodeficiency virus type 1 (HIV-1) protease activity required for HIV-1 particle maturation and infectivity. J. Virol. *69*, 2751-2758.
191. Rosen, C. A. (1991). Regulation of HIV gene expression by RNA-protein interactions. Trends in Gen. *7*, 9-14.
192. Rosenberg, S. A. (1990). The treatment of patients with advanced cancer using cyclophosphamide, interleukin-2 and tumor infiltrating lymphocytes. Hum. Gene Ther. *1*, 73-92.
193. Rosenberg, S. A. (1990). Gene therapy of patients with advanced cancer using tumor infiltrating lymphocytes transduced with the gene coding for tumor necrosis factor. Hum Gene Ther. *1*, 441-480.
194. Ruhl, M., Himmelspach, M., Bahr, G. M., Hammerschmid, F., Jaksche, H., Wolff, B., Aschauer, H., Farrington, G. K., Probst, H., Bevec, D., and Hauber, J. (1993). Eukaryotic initiation factor 5A is a cellular target of the human immunodeficiency virus type 1 Rev activation domain mediating trans-activation. J. Cell Biol. *123*, 1309-1320.
195. Saag, M. S., Holodnyi, M., Kuritzkes, D. R., O'Brien, W. A., Coombs, R., Poscher, M. E., Jacobsen, D. M., Shaw, G. M., Richman, D. D., and Volberding, P. A. (1996). HIV viral load markers in clinical practice. Nature Med. 2, 625-629.
196. Samson, M., Libert, F., Doranz, B. J., Rucker, J., Liesnard, C., Farber, C.-M., Saragosti, S., Lapoumeroulie, C., Cognaux, J., Forceille, C., Muyldermans, G., Verhofstede, C., Burtonboy, G., Georges, M., Imai, T., Rana, S., Yi, Y., Smyth, R. J., Collman, R. G., Doms, R. W., Vassart, G., and Parmentier, M. (1996). Resistance to HIV-1 infection in caucasian individuals bearing mutant alleles of the CCR-5 chemokine receptor gene. Nature *382*, 722-725.
197. Sanford, J. C. (1988). Applying the PDR Principle to AIDS. J. Theor. Biol. *130*, 469-480.
198. Sarver, N., Cantin, E. M., Chang, P. S., Zaia, J. A., Ladne, P. A., Stephens, D. A., and Rossi, J. J. (1990). Ribozymes as potential anti-HIV-1 therapeutic agents. Science *247*, 1222-1225.
199. Sczakiel, G, Pawlita, M., and Kleinheinz A. (1990). Specific inhibition of human immunodeficiency virus replication by RNA transcribed in sense and antisense orientation from the 5'-leader/gag region. Biochem. Biophys. Res. Comm. *169*, 643-651.

200. Sczakiel, G. and Pawlita, M. (1991). Inhibition of human immunodeficiency virus type 1 replication in human T cells stably expressing antisense RNA. J. Virol. *65*, 468-472.
201. Shaheen, F., Duan, L., Zhu, M., Bagasra, O., and Pomerantz, R.J. (1996). Targeting human immunodeficiency virus type 1 reverse transcriptase by intracellular expression of single-chain variable fragments to inhibit early stages of the viral life cycle. J. Virol. *70*, 3392-3400.
202. Sioud, M. and Drlica, K. (1991). Prevention of human immunodeficiency virus type 1 integrase expression in E. coli by a ribozyme. Proc. Natl. Acad. Sci. USA *88*, 7303-7307.
203. Smith, D. H., Byrn, R. A., Marsters, S. A., Gregory, T., Groopman, J. E., and Capon, D. J. (1987). Blocking of HIV-1 infectivity by a soluble, secreted form of the CD4 antigen. Science *238*, 1704-1707.
204. Smythe, J. A., Sun, D., Thomson, M., Markham, P. D., Reitz, M. S., Gallo, R. C., and Lisziewicz, J. (1994). A rev-inducible mutant gag gene stably transferred into T lymphocytes: an approach to gene therapy against human immunodeficiency virus type 1 infection. Proc. Natl. Acad. Sci. USA *91*, 3657-3661.
205. Stamps, A. C., Davies, S. C., Burman, J., and O'Hare, M. J. (1994). Analysis of proviral integration in human mammary epithelial cell lines immortalized by retroviral infection with a temperature-sensitive SV40 T-antigen construct. Int. J. Cancer *57*, 865-874.
206. Stanley, S. K., Kessler, S. W., Justement, J.S., Schnittmann, S. M., Greenhouse, J. J., Brown, C. C., Musongela, L., Musey, K., Kapita, B., and Fauci, A.S. (1992). CD34+ bone marrow cells are infected with HIV in a subset of seropositive individuals. J. Immunol. *149*, 689-697.
207. Steffy, K. R. and Wong-Staal, F. (1993). Transdominant inhibition of wild-type human immunodeficiency virus type 2 replication by an envelope deletion mutant. J. Virol. *67*, 1854-1859.
208. Su, L., Lee, R., Bonyhadi, M., Matsuzaki, H., Forestell, S., Escaich, S., Böhnlein, E., and Kaneshima, H. (1997). Hematopoietic Stem Cell-Based Gene Therapy For AIDS: Efficient Transduction and Expression of RevM10 in Myeloid Cells *In vivo* And *In vitro*. Blood *89*, 2283-2290.
209. Sullenger, B. A, Gallardo, H. F., Ungers, G. E., and Gilboa, E. (1990). Overexpression of TAR sequences renders cells resistant to human immunodeficiency virus replication. Cell *63*, 601-608.
210. Sun, L. Q., Pyati, J., Smythe, J., Wang, L., MacPherson, J., Gerlach, W., and Symonds, G. (1995). Resistance to human immunodeficiency virus type 1 infection conferred by transduction of human peripheral blood lymphocytes with ribozymes, antisense, or polymeric trans-activation response element constructs. Proc. Natl. Acad. Sci. USA *92*, 7272-7276.
211. Symensma, T. L., Giver, L., Zapp, M., Takle, G. B., and Ellington, A. D. (1996). RNA aptamers selected to bind human immunodeficiency virus type 1 Rev *in vitro* are Rev responsive *in vivo*. J. Virol. *70*, 179-187.
212. Takeuchi, Y., Cosset, F. L., Lachmann, P. J., Okada, H., Weiss, R. A., and Collins, M. K. L. (1994). Type C retrovirus inactivation by human complement is determined by both the viral genome and the producer cell. J. Virol *68*, 8001-8007.
213. Taylor, N. R. and Rossi, J. J. (1991). Ribozyme-mediated cleavage of an HIV-1 gag RNA: the effects of nontargeted sequences and secondary structure on ribozyme cleavage activity. Antisense Res. Dev. *1*, 173-186.
214. Temin, H. M. (1990). Safety considerations in somatic gene therapy of human disease with retrovirus vectors. Hum Gene Ther. *1*, 111-123.
215. Thompson, J. D., Ayers, D. F., Malmstrom, T. A., McKenzie, T. L., Ganousis, L., Chowrira B. M., Couture, L., and Stinchcomb, D. T. (1995). Improved accumulation and activity of ribozymes expressed from a tRNA-based RNA polymerase III promoter. Nucl. Acids Res. *23*, 2259-2268.
216. Tran, A.-C., Zhang, D., Byrn, R., and Roberts, M. R. (1995). Chimeric z-receptors direct human natural killer (NK) effector functiopn to permit killing of NK-resistant tumor cells and HIV-infected T lymphocytes. J. Immunol. *155*, 1000-1009.
217. Trono, D., Feinberg, M. B., and Baltimore, D. (1989). HIV-1 gag mutants can dominantly interfere with the replication of the wild-type virus. Cell *59*, 113-120.
218. Tuerk, C. and Gold, L. (1990). Systematic evolution of ligands by exponential enrichment: RNA ligands to bacteriophage T4 DNA polymerase. Science *249*, 505-510.
219. Tuerk, C., MacDougal, S., and Gold, L. (1992). RNA pseudoknots that inhibit human immunodeficiency virus type 1 reverse transcriptase. Proc. Natl. Acad. Sci. USA *89*, 6988-6992.
220. Tuerk, C. and MacDougal-Waugh, S. (1993). *In vitro* evolution of functional nucleic acids: high affinity RNA ligands of HIV-1 proteins. Gene *137*, 33-39.

221. Ulmer, J. B., Donnelly, J. J., Parker, S. E., Rhodes, G. H., Felgner, P. L., Dwarki, V. J., Gromkowski, S. H., Deck, R. R., DeWitt, C. M., Friedman, A. et al. (1993). Heterologous protection against influenza by injection of DNA encoding a viral protein. Science *259*, 1745-1749.
222. Van Beusechem, V. W., Kukler, A., Einerhand, M. P., Bakx, T. A., van der Eb, A. J., van Bekkum, D. W., and Valerio, D. (1990). Expression of human adenosine deaminase in mice transplanted with hemopoietic stem cells infected with amphotropic retroviruses. J. Exp. Med. *172*, 729-736.
223. Van Beusechem, V.W., Kukler, A., Heidt, P. J., and Valerio, D. (1992). Long-term expression of human adenosine deaminase in rhesus monkeys transplanted with retrovirus-infected bone marrow cells. Proc. Natl. Acad. Sci. USA *89*, 7640-7644.
224. Van Beusechem, V. W., Bart-Baumeister, J. A., Bakx, T. A., Kapstein, L. C., Levinsky, R. J., and Valerio, D. (1994). Gene transfer into nonhuman primate CD34+CD11b- bone marrow progenitor cells capable of repopulating lymphoid and myeloid lineages. Hum Gene Ther. *5*, 295-305.
225. Vandendriessche, T., Chuah, M. K., Chiang, L., Chang, H. K., Ensoli, B., and Morgan, R. A. (1995). Inhibition of clinical human immunodeficiency virus (HIV) type 1 isolates in primary CD4+ T lymphocytes by retroviral vectors expressing anti-HIV genes. J. Virol. *69*, 4045-4052.
226. Veres, G., Escaich, S., Barske, C., Baker, J., and Böhnlein, E. (1996). Constitutive intracellular Expression of Gag Antisense RNA Specifically and Effectively Inhibits HIV-1 Replication in Human CD4-Positive T Cells. J. Virol. *70*, 8792-8800.
227. Vieillard, V., Lauret, E., Rousseau, V., and DeMayer, E. (1994). Blocking of retroviral infection at a step prior to reverse transcription in cells transformed to constitutively express interferon beta. Proc. Natl. Acad. Sci. USA *91*, 2689-2693.
228. Vieillard, V., Lauret, E., Maguer, V., Jacomet, C., Rozenbaum, W., Gazzolo, L., and DeMayer, E. (1995). Autocrine interferon-beta synthesis for gene therapy of HIV infection: increased resistance to HIV-1 in lymphocytes from healthy and HIV-infected individuals. AIDS *9*, 1221-1228.
229. VonKalle, C., Kiem, H. P., Goehle, S., Darovsky, B., Heimfeld, S., Torok-Strob, B., Strob, R., and Schuening, F. G. (1994). Increased gene transfer into human hematopoietic progenitor cells by extended *in vitro* exposure to a pseudotyped retroviral vector. Blood *84*, 2890-2897.
230. Von Laer, D., Hufert, F. T., Fenner, T. E., Schwander, S., Dietrich, M., Schmitz, H., and Kern, P. (1990). CD34+ Hematopoietic Progenitor cells are not a major reservoir of the human immunodeficiency virus. Blood *76*, 1281-1286.
231. Wain-Hobson, S., Sonigo, P., Danos, O., Cole, S., and Alizon, M. (1985). Nucleotide sequence of the AIDS virus LAV. Cell *40*, 9-17.
232. Walker, R. E., Blaese, R. M., Carter, C. S., Chang, L., Klein, H., Lane, H. C., Leitman, S. F., and Mullen, C. A. (1993). A study of the safety and survival of the adoptive transfer of genetically marked syngeneic lymphocytes in HIV infected identical twins. Hum Gene Ther. *4*, 659-680.
233. Walker, R. E., Blaese, R. M., Davey, R. T., Falloon, J., Mullen, C., Polis, M. A., Finer, M., McGuire, B., Roberts, M., and Zsebo, K. (1996). A phase I/II pilot study of the safety of the adoptive transfer of syngeneic gene-modified cytotoxic T lymphocytes in HIV-infected identical twins. Hum Gene Ther. *7*, 367-400.
234. Walter, E. A., Greenberg, P. D., Gilbert, M. J., Finch, R. J., Watanabe, K. S., Thomas, E. D., and Riddell, S. R. (1995). Reconstitution of cellular immunity against cytomegalovirus in recipients of allogeneic bone marrow by transfer of T-cell clones from the donor. N. Engl. J. Med. *333*, 1038-1044.
235. Wang, B., Ugen, K. E., Srikatan, V., Agadjanyan, M. G., Dang, K., Refaeli, Y., Sato, A. I., Boyer, J., Williams, W. V., and Weiner, D. B. (1993). Gene inoculation generates immune responses against human immunodeficiency virus type 1. Proc. Natl. Acad. Sci. USA *90(9)*, 4156-4160.
236. Wang, B., Merva, M., Dang, K., Ugen, K. E., Boyer, J., Williams, W. V., and Weiner, D. B. (1994). DNA inoculation induces protective *in vivo* immune responses against cellular challenge with HIV-1 antigen-expressing cells. AIDS Res. Hum Retroviruses *10*, 35-41.
237. Warner, J. F., Anderson, C. G., Laube, L., Jolly, D. J., Townsend, K., Chada, S., and St. Louis, D. (1991). Induction of HIV-specific CTL and antibody responses in mice using retroviral vector-transduced cells. AIDS Res. Hum. Retroviruses *7*, 645-655.
238. Warner, J. F., Jolly, D., Mento, S., Galpin, J., Haubrich, R., and Merritt, J. (1995). Retroviral vectors for HIV immunotherapy. Ann. NY Acad. Sci. *772*, 105-116.
239. Weerasinghe, M., Liem, S. E., Asad, S., Read, S. E., and Joshi, S. (1991). Resistance to human immunodeficiency virus type 1 (HIV-1) in human CD4+ lymphocyte-derived cell lines conferred by using retroviral vectors expressing an HIV-1 RNA-specific ribozyme. J. Virol. *65*, 5531-5534.

240. Wei, X., Ghosh, S. K., Taylor, M. E., Johnson, V. A., Emini, E. A., Deutsch, P., Lifson, J. D., Bonhoeffer, S., Nowak, M. A., Hahn, B. H., Saag, M. S., Shaw, G. M. (1995). Viral dynamics in human immunodeficiency virus type 1 infection. Nature *373*, 117-122.
241. Weiss, R. A. and Clapham, P. R. (1996). Hot fusion of HIV. Nature *381*, 647-648.
242. Woffendin, C., Ranga, U., Yang, Z.-Y., Xu, L., and Nabel, G. J. (1996). Expression of a protective gene prolongs survival of T cells in human immunodeficiency virus infected patients. Proc. Natl. Acad. Sci. USA *93*, 2889-2894.
243. Wolff, J. A., Malone, R. W., Williams, P., Chong, W., Acsadi, G., Jani, A., and Felgner, P. L. (1990). Direct gene transfer into mouse muscle *in vivo*. Science *247*, 1465-1468.
244. Wu, X., Liu, H., Xiao, H., Kim, J., Seshaiah, P., Natsoulis, G., Boeke, J. D., Hahn, B. H., and Kappes, J. C. (1995). Targeting foreign proteins to human immunodeficiency virus particles via fusion with vpr and vpx. J. Virol. *69*, 3389-3398.
245. Yamada, O., Yu, M., Yee, J. K., Kraus, G., Looney, D., and Wong-Staal, F. (1994). Intracellular immunization of human T-cells with a hairpin ribozyme against human immunodeficiency virus type 1. Gene Ther. *1*, 38-45.
246. Yee, J. K., Miyanohara, A., LaPorte, P., Bouic, K., Berns, K., and Friedmann, T. (1994). A general method for the generation of high-titer, pantropic retroviral vectors: highly efficient infection of primary hepatocytes. Proc. Natl. Acad. Sci. USA *91*, 9564-9568.
247. Yu, M., Ojwang, J., Yamada, O., Hampel, A., Rappaport, J., Looney, D., and Wong-Staal, F. (1993). A hairpin ribozyme inhibits expression of diverse strains of human immunodeficiency virus type 1. Proc. Natl. Acad. Sci. USA *90*, 6340-6344.
248. Yu, M., Poeschla, E., Yamada, O., Degrandis, P., Leavitt, M. C., Heusch, M., Yees, J. K., Wong-Staal, F., and Hampel, A. (1995). *In vitro* and *in vivo* characterization of a second functional hairpin ribozyme against HIV-1. Virology *206*, 381-386.
249. Yu, M., Leavitt, M. C., Maruyama, M., Yamada, O., Young, D., Ho, A. D., Wong-Staal, F. (1995). Intracellular immunization of human fetal cord blood stem/progenitor cells with a ribozyme against human immunodeficiency virus type 1. Proc. Natl. Acad. Sci. USA *92*, 699-703.
250. Zhou, C., Bahner, I., Larson, G. P., Zaia, J. A., Rossi, J. J., and Kohn, D. B. (1994). Inhibition of HIV-1 in human T lymphocytes by retrovirally transduced anti-tat and rev hammerhead ribozymes. Gene *149*, 33-39.
251. Zhou, S. Z., Broxmeyer, H. E., Cooper, S., Harrington, M. A., and Srivasta, A. (1993). Adeno-associated virus 2-mediated gene transfer in murine hematopoietic progenitor cells. Exp. Hematol. *21*, 928-933.
252. Ziegner, U. H., Peters, G., Jolly, D. J., Mento, S. J., Galpin, J., Prussak, C. E., Barber, J. R., Harnett, D. E., Bohart, C., Klump, W. et al. (1995). Cytotoxic T-lymphocyte induction in asymptomatic HIV-1-infected patients immunized with Retrovector-transduced autologous fibroblasts expressing HIV-1 IIIB Env/Rev proteins. AIDS *9*, 43-50.
253. Zimmermann, K., Weber, S., Dobrovnik, M., Hauber, J., and E. Böhnlein, E. (1992). Expression of Chimeric Rev Response Element Sequences Interferes with HIV-1 Rev Function. Hum Gene Ther. *3*, 155-161.

VI Cancer

22. Suicide Genes for the Treatment of Cancer

Kenneth W. Culver

22.1 Introduction

The effectiveness of conventional cancer therapies is limited not only by unsatisfactory efficacy, but also by toxicity. Many drugs damage the immune system, further exacerbating the underlying tumor-induced suppression of host immunity that prevents immunologic destruction of the tumor. Advances in molecular biology suggest that gene transfer has the potential to be a significant advance in cancer therapy because toxicities to the immune system are minimal allowing for the potential development of a synergy between gene transfer and a concomitant antitumor immune response. Suicide genes are a promising class of biologics that have shown substantial efficacy and limited toxicity in cancer studies in rodents (1). Moving from preclinical animal models to successful human trials is a daunting task even with the great promise that gene transfer encompasses. This chapter will discuss the preclinical to clinical transition uses "suicide" gene transfer for cancer therapy, with a particular focus on the use of the Herpes Simplex Thymidine Kinase (HS-tk) gene.

22.2 Mechanisms of Suicide Gene Action

22.2.1 General Concepts

The ultimate goal of cancer therapy is to selectively destroy malignant cells. Suicide genes are well suited for this function because they confer a selective vulnerability for destruction of the genetically-modified cells by a minimally toxic prodrug (2). The transfer of suicide genes themselves can be an uneventful process, since these genes typically produce intracellular enzymes that do not significantly affect cellular homeostasis. However, upon the administration of a prodrug that is converted to a far more toxic drug by the gene product, lethal cellular damage is inflicted to the gene-containing cells (3). For gene therapists, the goal is to use suicide genes

Table 22.1 Examples of prodrugs activated by herpes thymidine kinases

Prodrug	Enzyme activator	Toxic byproduct
Acyclovir (ACV)	HS-tk	ACV-triphosphate
1-β-D-arabinofuranosylthymine (araT)	HS-tk	araT-triphosphate
(E)-5-(2-Iodovinyl)-2'-deoxyuridine (IVDU)	HS-tk	IVDU-triphosphate
5-iodo-5'-amino-2'-5'-dideoxyuridine (AIU)	HS-tk	AIU-triphosphate
Ganciclovir (GCV)	HS-tk	GCV-triphosphate

to selectively destroy tumor cells, eliminating collateral damage to normal tissues, especially to the immune system. There are a number of different types of suicide genes in development. Only two have been approved for human clinical trials.

22.2.2 The Herpes Simplex-Thymidine Kinase (HS-tk) Gene

The HS-tk gene was first cloned in 1979 from the type I virus (4). This gene is now the most used suicide gene in preclinical and clinical studies around the world because it produces target cell destruction with limited non-specific tissue toxicity (1). HS-tk catalyzes the monophosphorylation of several drugs, including the FDA approved anti-herpes drugs acyclovir (ACV) and ganciclovir (GCV) (Tab. 22.1) (5, 6, 7). Cellular kinases then convert the monophosphate (MP) forms of the drugs to the diphosphate (DP) and triphosphate (TP) forms that incorporate into DNA as a nucleoside analog. The ACV-TP and GCV-TP inhibit DNA polymerase from continuing beyond the analog leading to DNA fragmentation (Fig. 22.1). As a result, when a Herpes Simplex virus infected cell is exposed to ACV or GCV, the phosphorylated, incorporated drug product leads to fragmentation of the DNA, there by killing the herpes virus infected cell (8, 9, 10). Host cellular thymidine kinases do not phosphorylate ACV or GCV, only the MP and DP forms, limiting cellular toxicity (11).

The same method can be used to destroy malignant cells (12). Transfer of the HS-tk gene into tumor cells in vitro results in their destruction with treatment by ACV or GCV either in vitro or after reimplantation into mice (13). GCV has a more potent antitumor effect than ACV and other analogs in animal tumor model systems (14). This is probably due to a greater uptake of GCV into Herpes simplex infected cells and the fact that GCV is a better substrate for both viral and host kinase enzymes, while being only minimally phosphorylated in HS-tk negative cells (15).

This method of cell destruction requires that the cell be actively dividing in order to incorporate the triphosphate derivatives in the cellular DNA. Since tumor cells are usually the most actively dividing cell type in most tissues, this feature provides for some level of selectivity for tumor cell destruction in vivo. Based upon these types of preclinical studies and the fact that GCV is FDA approved for the treatment of

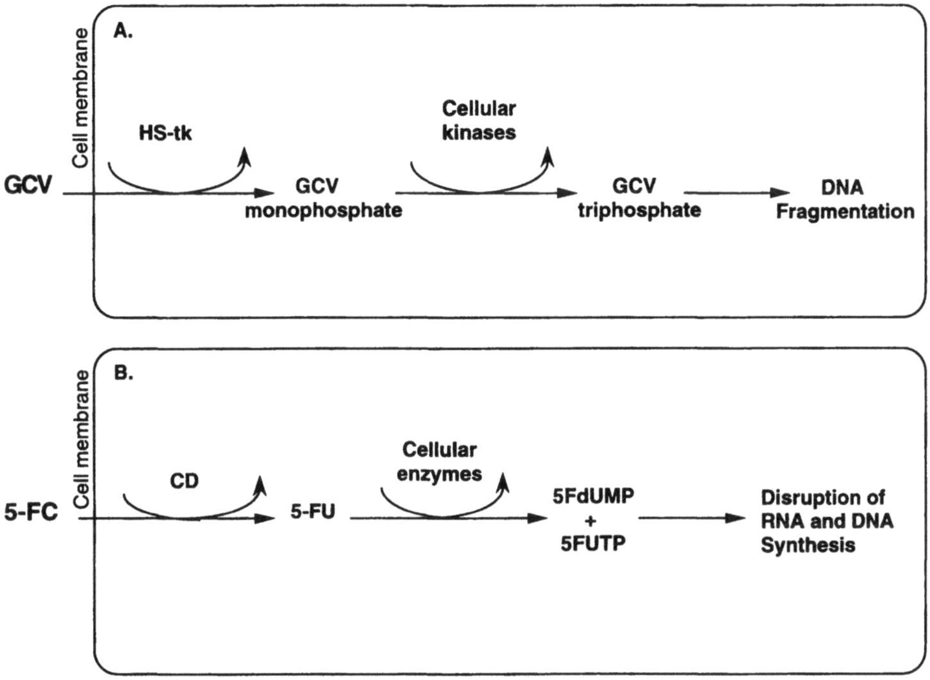

Fig. 22.1: The HS-tk/GCV and CD/5-FC suicide gene systems.

A. Following the intravenous administration of GCV, the HS-tk enzyme converts the drug to GCV-MP, an intracellular product. Cellular kinases convert the GCV-MP to GCV-TP. GCV-TP destroys tumor cells by incorporating into the elongating DNA chain and inducing chain termination. A leading hypothesis for the mechanism of the bystander effect is the passage of GCV-TP through gap junctions to result in destruction of adjacent HS-tk negative cells.

B. The systemic administration of 5-FC results in the direct conversion of the drug to 5-FU within CD gene containing cells. Several different enzymatic pathways produce 5-FU metabolites including a monophosphate form (5FdUMP) and a triphosphate form (5FUTP) that results in cell death. The bystander effect with this system is thought to result from the diffusion of 5-FU and its metabolites into the extracellular space destroying tumor cells in the local area.

herpes virus infections, GCV has been used exclusively with HS-tk gene transfer in approved human gene therapy clinical trials.

22.2.3 Cytosine Deaminase (CD)

Cytosine deaminase is the second suicide gene approved for human clinical trials. It is normally present in bacteria and fungi, but not mammalian cells. The CD enzyme deaminates the FDA approved antifungal drug, 5-fluorocytidine (5-FC), into 5-fluorouracil (5-FU), an FDA approved chemotherapeutic for cancer (Fig. 22.1) (16). 5-FU and its metabolites inhibit thymidylate synthetase and incorporate into RNA resulting in cell death. Therefore, when the CD gene is inserted into a variety of different tumor cells, the genetically-altered cells generated their own chemotherapy from 5-FC leading to their destruction in vitro and in mice (17, 18). However, since the action of 5-FU is not cell cycle specific like HS-tk, investigators have been attempting to limit gene expression to tumor cells using tissue-specific promoters. One example is the carcinoembryonic antigen (CEA) gene that is expressed almost exclusively in hepatocellular carcinoma cells. Insertion of a CD gene under control of a CEA promoter into CEA positive and negative tumor cells resulted in CEA tumor-specific gene expression and cell destruction (19). There is one approved trial in the U.S.A., which is designed to directly inject recombinant adenovirus containing the CD gene into the liver of patients with hepatic metastases. The goal of the phase I trial is to assess the toxicities associated with high efficiency CD gene transfer and production of 5-FU in the liver in patients with metastatic colorectal cancer.

22.2.4 Other Suicide Genes

There are a number of other "suicide" genes in addition to HS-tk and CD that might be useful in the treatment of solid tumors (Tab. 22.2). (3, 20 – 31). Most of these genes were initially developed for attachment to monoclonal antibodies. The goal was to use the specificity of the antibody to selectively deliver the enzyme. Systemic administration of the prodrug would then result in selective destruction of the enzymatically altered cells. However, this approach has been hampered by inefficient delivery of the antibody-enzyme conjugate. While gene therapy also has a delivery problem, the transfer of genes has several advantages including the control of gene expression, the possibility of producing substantially greater amounts of enzyme per cell and a longer duration of enzyme activity compared to previous enzyme transfer/antibody-targeting approaches.

22.3 Preclinical Studies with HS-tk and CD

Insertion of the HS-tk gene into tumor cells was shown to result in tumor cell death in response to treatment with ACV ten years ago (12). This finding has been confirmed by many investigators in a variety different tumor model systems (1, 32). The early studies used ex vivo gene transfer, followed by reimplantation of the altered cells (33). While this was a useful model system, it has limited potential for systemic clinical use since it appears that the HS-tk system is only capable of eliminating the treated tumor mass and not distant tumor deposits (34). More recently, clinically applicable in vivo gene transfer systems using retroviral and adenoviral vectors have been developed (35, 36). The development of HS-tk gene transfer for human clinical application has focused primarily on brain tumors because they are known to remain localized for extended periods of time (37). Since most patients will die of local progression of the tumor, as opposed to distant metastases, a local therapy could potentially provide significant clinical benefit.

Table 22.2 Examples of suicide genes under investigation

Prodrug	Enzyme activator	Toxic byproduct
Adriamycin-glucuronide	β-glucuronidase	Adriamycin
Amygdalin	β-glucosidase	Cyanide
Cyclophosphamide	Cytochrome P450 2B1	4-hydroxy cyclophosphamide
Cytosine arabinoside (araC)-5'-galactoside	E. coli β-galactosidase	Cytosine arabinoside (araC)
Daunomycin-glucuronide	β-glucuronidase	Daunomycin
Deoxyadenosine analogs	PNP	Toxic adenine analogs
5'-Deoxy-5-fluorouridine	Thymidine phosphorylase	5-fluorouracil (5-FU)
Doxorubicin-phosphate	Alkaline phosphatase	Doxorubicin
Doxorubicin-phenoxyacetamide	Penicillin amidase	Doxorubicin
Doxorubicin-cephalosporin	β-lactamase	Doxorubicin
Epirubicin-glucuronide	β-glucuronidase	Epirubicin
Etoposide-phosphate	Alkaline phosphatase	Etoposide
5-fluorocytosine (5-FC)	Cytosine deaminase	5-fluorouracil (5-FU)
Hypoxanthine	Xanthine oxidase	Hydrogen peroxide, OH & O2 radicals
Methotrexate-alanine	Carboxypeptidase A	Methotrexate
Mitomycin phosphate	Alkaline phosphatase	Mitomycin C
Melphalan-phenoxyacetamide	Penicillin amidase	Melphalan
Phenylenediamine mustard-cephalosporin	β-lactamase	Phenylenediamine mustard
Phenol mustard-glucuronide	β-glucuronidase	Phenol mustard
6-thioxanthine	Xanthine-guaninephosphoribosyl transferase (XGPRT)	6-thioguanosine mono-phosphate
Xanthine	Xanthine oxidase	Hydrogen peroxide, OH & O2 radicals

The first in vivo gene transfer method that advanced into the clinic used murine retroviral vectors. These vectors are produced by transfecting NIH 3T3 mouse fibroblast cells with the genes required for production of an infectious, HS-tk containing replication-deficient virus particle (38). The vector particles bud from the genetically engineered cells called vector producer cells (VPC) and are harvested from the overlying tissue culture medium. Initial studies focused on the direct injection of vector-containing supernate, but the gene transfer efficiency was very low (1-3%) (39). However, the implantation of HS-tk VPC directly into the tumor mass resulted in a gene transfer efficiency of 10-55%. Since most cells are not actively dividing at the same time, the direct injection of VPC that are actively producing vectors for 7-14 days within the tumor offers a greater opportunity for gene transfer into more slowly dividing tumor cells. Each of the approved human gene therapy trials that are injecting HS-tk VPC use safety-modified NIH 3T3 cells (PA317) to minimize the risk of recombination that could lead to the production of replication-competent virus particles.

Retroviral vectors were initially chosen for use in human brain tumors because they require a proliferating target cell population for integration and vector gene expression. In the brain, the tumor is the predominant mitotic cell type, potentially allowing selective gene transfer into tumor cells. Other clinically applicable vector systems under development at that time did not have inherent selectivity for tumor cells. Safety of the delivery system was assessed in a series of rodent and non-primate experiments that confirmed the selective nature of the in vivo gene transfer into tumor cells without detectable spread of the vector to normal brain tissue (39, 40).

The stereotactic injection of experimental rat 9L brain tumors with HS-tk VPC demonstrated complete destruction of microscopic tumor in 80% of rats with GCV administration (41). In survival experiments, the injection of HS-tk VPC into the rat C6 glioma resulted in long term survival (> 8 months) with GCV, while all control animals (No GCV) died within seven weeks (42). There were no associated systemic toxicities related to the gene transfer with this form of in vivo gene delivery (43). Interestingly, complete tumor ablation occurred in these animal despite the fact that less than 100% of the tumor cells contained the HS-tk gene (44). This finding, the destruction of the adjacent HS-tk negative tumor cells has been termed the "bystander" tumor killing effect (41). Since the possibility of 100% gene transfer in vivo is currently not possible with any delivery system, the bystander effect is critical for the successful elimination of tumors with current delivery methods.

The second in vivo gene transfer system utilizes the direct, intratumoral injection of recombinant adenoviral vectors containing the HS-tk or CD gene. These are non-integrating vectors that infect tumor cells regardless of their state of proliferation. Therefore, the direct injection of HS-tk vectors into rodent brain tumors results in transduction of both normal and malignant tissues. However, treatment with GCV only resulted in damage to malignant cells (36). These results suggest that the antitumor activity of the HS-tk/GCV system requires proliferating cells. Consequently, tissue-specific promoters may not be required as originally hypothesized. In addition to these impressive safety studies in rodents, adenoviral vectors were able to achieve antitumor efficacies comparable to the injection of VPC. Both retroviral and adenoviral vectors have also been used to transfer the CD gene into subcutaneous tumors in mice (18, 45). Formation of a subcutaneous tumor +/- CD found that 47-100% of the tumors resolved depending on the tumor cell type, while the CD negative tumors all grew despite 5-FC administration (18).

Issues of efficient gene delivery remain a significant problem for adenoviral vectors since they appear to only infect cells in the local area of injection like retroviral vectors. However, adenovirus vectors do express HS-tk at much higher level than retroviral vectors in the same cell type which may markedly increase their direct therapeutic effect and the bystander tumor cell killing (46, 47). Therefore, the use of adenoviral vectors may achieve a greater antitumor effect with the same dose of GCV or allow the use of a decreased dose of GCV if toxicities limit clinical application.

22.4 The Bystander Tumor Killing Effect

A first step in the evaluation of the HS-tk in vivo bystander effect was to determine its potency. Mixtures (1:1 and 1:10) of HS-tk transduced and non-transduced tumor cells, were injected subcutaneously into mice. Once tumor formation was measurable, GCV was administered IP. This treatment resulted in complete tumor ablation in 14 of 15 animals at a 1:1 mixture (41). If only 10% (1:10) of the tumor cells were HS-tk positive, 9 out of 15 animals were tumor free five weeks later. This was the first published observation that a potent antitumor bystander effect was associated with the HS-tk/GCV system. A number of possible mechanisms underlying the bystander effect have been proposed.

The leading hypothesis states that the phosphorylated derivatives of GCV that are formed in the transduced cells pass through gap junctions into adjacent tumor cells where they induce DNA fragmentation (Fig. 22.1) (48, 49). Other hypotheses include: 1) phagocytosis of fragments of HS-tk destroyed cells ("apoptotic vesicles") by neighboring tumor cells resulting their destruction (50, 51); 2) The destruction of intratumoral vascular endothelium due to transfer of the HS-tk gene (52). Since intratumoral endothelial cells proliferate much faster than normal in response to the tumor-derived angiogenesis factors, they are susceptible to destruction by GCV-TP; and 3) The development of antitumor immunity (53). No investigator have published evidence of a soluble factor (phosphorylated derivatives of GCV do not freely pass through cell membranes) that is released by intact cells that results in the death of adjacent cells (54). Elucidation of all of the mechanisms involved in the bystander effect may provide an opportunity to magnify the antitumor effect, potentially converting a local therapy into a regional or systemic therapy. None of the published animal experiments has provided evidence that the HS-tk/GCV system damages surrounding normal tissues. Therefore, the bystander factor(s) may be useful as infusible anti-tumor drugs themselves, since the toxicity of the bystander effect is apparently limited to tumor cells.

The CD/5-FC system has also been shown to manifest a bystander effect, with a few as 2% CD positive cells being able to destroy all of the cells in culture (55). However, the degree of the bystander effect varies, since not all cells have the same sensitivity to 5-FU (2). The E. coli PNP gene and the cytochrome P450 gene produce active drug products that are released from the genetically-altered cells (29, 30). This feature may have distinct advantages over the HS-tk/GCV mechanism, since a soluble product released within the extracellular spaces of the tumor mass is not limited by gap junction formation. However, these soluble, active drugs will likely affect both dividing and non-dividing cells and may induce substantial toxicity to normal tissues.

22.5 Clinical Applications

22.5.1 In Vivo Gene Transfer

22.5.1.1 Brain Tumors

The in vivo application of murine retroviral vector-mediated gene therapy was first applied to the treatment of human brain tumors in December, 1992 at the National Institutes of Health (NIH; Bethesda, MD). In this protocol, 0.5-1.0x 10^9 HS-tk VPC were stereotactically implanted into multiple areas of the gadolinium-enhancing portion of recurrent brain tumors (56). Seven days later, they received a 14 day course of GCV at 5mg/kg/dose twice daily. All patients had to have a documented recurrence of their tumor after radiotherapy to be eligible. Most patients also had at least one previous surgical resection and chemotherapy prior to receiving gene therapy treatments.

Fifteen patients with recurrent primary brain tumors (12) or metastatic tumors (3) were treated in the phase I NIH trial. No evidence of toxicity related to the implantation of the xenogeneic murine cells or GCV treatment was noted. All patients received 36mg/day of dexamethasone starting several days before injection of the xenogeneic cells. The steroids were tapered to pretreatment levels after GCV was initiated. Four of thirteen evaluable patients (30%) have shown clear evidence of an antitumor effect (>50% decrease in size on MR scans one month apart). One patient continues without tumor >48 months after treatment for a recurrent glioblastoma. This is encouraging since the mean survival at recurrence is 5.8 months (3.43 to 8.8) without treatment (57).

Gene transfer was confirmed, but the efficiency was estimated to be very low suggesting that the bystander effect must be substantial in the responding patients. Low efficiency gene transfer in recurrent tumors is not unexpected since previous treatments produce necrotic, scarred tumors with significantly fewer proliferating cells compared to the animal model. This study is now closed while the data is being analyzed. Due to the lack of toxicity and the suggestion of an antitumor response, additional clinical trials has been approved using stereotactic injection (Tab. 22.3). Intratumoral HS-tk VPC injections have been approved in the U.S.A. for the treatment of recurrent pediatric astrocytomas at St. Jude Childrens Research Hospital and

Table 22.3 Suicide gene transfer clinical trials for cancer in the U.S.

Type of Cancer	No. of Trials	Tissue Transduced	Gene Transferred
CNS tumors	12*	Tumor cells	HS-tk
Ovarian Cancer	4**	Tumor cells	HS-tk
Head/Neck Cancer	1	Tumor cells	HS-tk
Hepatic metastases	1	Tumor cells	HS-tk
Leukemia	1^	T-cells	HS-tk
Malignant Melanoma	1^	T-cells	HS-tk
Mesothelioma	1	Tumor cells	HS-tk
Multiple Myeloma	1^	T-cells	HS-tk
Prostate cancer	1	Tumor cells	HS-tk
Colorectal metastases to the liver	1	Tumor cells	Cytosine deaminase

* All use in vivo gene transfer
** One ex vivo and two in vivo gene transfer trials
^ Ex vivo gene transfer protocols

in adults at Columbia University, the University of Cincinnati, the M.D. Anderson Cancer Center, the University of Florida, the University of Virginia, and the University of California, San Francisco. Genetic Therapy Inc. (Gaithersburg, MD) is the in the process of establishing additional research centers for the evaluation of the stereotactic injection of HS-tk VPC in children and adults.

Results from the initial phase I trial suggest that the critical feature limiting greater clinical success is the efficiency of gene transfer (58). Necrosis and scarring that results from the previous cancer therapies is one factor that limits gene transfer and the bystander effect. Therefore, a new approach combining the direct injection of HS-tk VPC into the walls of the tumor bed at the time of a gross total resection of the recurrent tumor was initiated. Repeated injections of VPC are subsequently administered through an Ommaya reservoir into the tumor bed at 5 week intervals for up to 6 months after the initial surgery. Two weeks after each injection of VPC into the Ommaya reservoir, the patient receives a 14 day course of GCV as described above. The study sites for the adult trials include the University of Iowa, the University of California, San Francisco, the University of Washington, the University of Cincinnati and the University of Texas-Southwestern Medical School (59). The centers involved in the pediatric trial are the Mayo Clinic, the Childrens Hospital of Los Angeles , the University of Washington and the Children's National Medical Center in Washington D.C. (60).

Since most patients die of local progression of their glioblastoma within a 2 cm margin of the resection, the investigators hope that the injection of the VPC 1 cm deep into the unresectable tumor will allow the bystander tumor killing effect to destroy cancer cells within the 2 cm area around the surgical margin. Since repeated therapies have typically been more successful in the treatment of cancer, repeated injections of VPC are administered through the Ommaya reservoir into the tumor bed in responding patients at five week intervals.

More than fifteen patients have been enrolled in the multicenter adult trial since July, 1994. Toxicities have been noted in four patients within a few hours of the injection of VPC into the Ommaya reservoir. These reactions are characterized by the acute onset of high fever, hypertension, meningismus and headache. The symptoms were self-limited and controlled with medication. These adverse side effects were initially thought to be secondary to the passage of xenogeneic VPC from the tumor bed into the subarachnoid space. Using technetium to confirm that the Ommaya reservoir has sealed into the tumor cavity and does not communicate with the subarachnoid space substantially decreased the degree and frequency of the adverse reactions, but did not eliminate them. Further studies to characterize the etiologies of the reactions are in progress. Some of the patients have reportedly shown antitumor responses, but no published clinical data is available at this time.

Table 22.4 Methods to improve the clinical effectiveness of brain tumor gene therapy with the HS tk suicide gene

Achieve more efficient gene delivery throughout the tumor mass by using:
Vectors that can selectively propagate through tumor tissue
Motile vector producer cells that will migrate through the tumor mass
Delivery of genes before the administration of radiation
Enhance the Cytocidal Effect of the Transferred Gene
Increase the dose of the prodrug
Increase the level of expression of the vector genes
Boost the HS-tk/GCV bystander tumor killing effect
Upregulate gap junction formation
Increase the number of HS-tk gene copies per cell
Delivery of the genes before administration of radiation
Induce a concomitant antitumor immune response
Add immunostimulatory cytokine genes
Block the production of immunosuppressive factors by the tumor (e.g. TGF-β)

A next step in the evolution of this therapy is to further improve gene transfer and the bystander effect by treating non-irradiated tumors. Data collected in adult patients with recurrent tumors has been used to justify a trial in humans using HS-tk as an "up front" treatment. At centers throughout North America and Europe, patients initially diagnosed with a glioblastoma will be randomized to receive standard therapy or standard therapy with gene therapy. In an attempt to learn more about the influences of previous treatments on the response to HS-tk therapy, a trial was approved to apply the technique in children (< 6 years of age), who do not routinely receive radiation to the brain to avoid deleterious effects on brain growth. These approaches to the treatment of CNS tumors will be further improved and used in combination as information is collected in current clinical trials (Tab. 22.4).

Three trials have been approved for the direct injection of HS-tk-containing adenoviral vectors into recurrent glioblastoma multiforme and anaplastic astrocytoma at the University of Pennsylvania and Baylor University. These trials have only recently begun to enroll patients because of concern about the potential adverse inflammatory responses induced by the adenovirus (61). Overall, the goals of the retroviral and adenoviral brain tumor trials are to study safety, efficiency of gene transfer, the induction of antitumor immunity and antitumor efficacy (56, 59, 60).

22.5.1.2 Leptomeningeal Carcinomatosis

Scientists initiated a clinical trial in 1994 that involves the direct injection of HS-tk VPC into the ventricular system through an Ommaya reservoir for the treatment of metastatic leptomeningeal carcinomatosis (Tab. 22.3) (62). This trial was based upon animal data that demonstrated lack of toxicity and evidence of efficacy (63, 64). However, the first and only patient treated at NIH developed significant, transient adverse side effects related to the injection of the VPC into the ventricular space. The signs and symptoms seen in this patient were very similar to those described above for patients in the glioblastoma protocol who received VPC through an Ommaya reservoir into the tumor bed. This reaction was unexpected based upon preclinical studies and no additional enrollment of patients is planned until this adverse response and methods to prevent it are better characterized.

22.5.1.3 Mesothelioma

Mesotheliomas are another tumor type that generally remain localized for long periods. The direct injection of recombinant HS-tk adenoviral vectors into mesothelioma tumors has demonstrated significant antitumor activity in rodents (65). Therefore, a human clinical trial has been approved at the University of Pennsylvania using this technique. No published clinical data are available (Tab. 22.3).

22.5.1.4 Ovarian Cancer

Ovarian cancer tends to remain limited to the surface of the peritoneal cavity for an extended period of time in many women. Therefore, the direct injection of viral vectors into the peritoneal cavity many have the potential for significant rates of gene transfer and a greater antitumor efficacy. Safety studies have shown that there are no significant toxicities in the peritoneal cavity related to transfer of the HS-tk gene or the associated bystander effect in rodent models (43). This is consistent with earlier studies that have shown that the transfer of a HS-tk gene into normal cells does not result in toxicity to adjacent normal cells (66). Efficacy studies in mice have also shown a potential for substantial efficacy, especially with adenoviral vectors. Trials have been approved for the direct injection of HS-tk VPC or adenoviral vectors into the peritoneal cavity in women with stage III ovarian cancer at the Iowa Methodist Medical Center and the University of Alabama (Birmingham) respectively (Tab. 22.3).

22.5.1.5 Liver Cancer

The only approved trial using the cytosine deaminase gene is an attempt to treat colorectal cancer metastases of the liver. This particular type of cancer was selected because 5-FU is currently the treatment of choice for colorectal hepatic metastases. The investigators plan to inject recombinant CD adenoviral vectors into the hepatic artery in an attempt to infect the disseminated tumor cells. The patients will then be treated systemically with 5-FC. Transfer of the CD gene into human colorectal tumors in nude mice has produced significant inhibition of tumor growth without toxicity either secondary to gene transfer or the bystander effect (18, 45). The approved trial will be conducted at Cornell University (Tab. 22.3).

A second trial has recently been approved at Mount Sinai Medical Center in New York to use adenoviral vectors to deliver the HS-tk gene directly into metastatic tumors in the liver.

22.5.1.6 Head/Neck Cancer

There is one approved trial using adenoviral vectors to deliver HS-tk into recurrent/persistent squamous cell carcinoma. The group at Johns Hopkins in Baltimore, Md expect to enroll twenty-five patients.

22.5.1.7 Prostate Cancer

Scientists at the Baylor College of Medicine have received approval to directly inject recombinant adenovirus containing the HS-tk gene into locally recurrent prostate cancer. No published results are available at this time.

22.5.2 Ex Vivo Gene Transfer

22.5.2.1 Ovarian Cancer

There is one approved clinical trial that attempts to utilize the HS-tk bystander effect as a therapeutic without gene transfer (Tab. 22.3) (67). The protocol is designed to use a human allogeneic ovarian cancer cell line that has been transduced with an

HS-tk retroviral vector. The transduced cells are irradiated and injected directly into the peritoneal cavity. The premise of the study is that the HS-tk containing cells will attach adjacent to the patients own tumor cells, producing a potent bystander tumor killing effect in the adjacent tumor cells with GCV administration. Animal studies demonstrated evidence of improved survival with this technique. Patients are being enrolled at Tulane University and the University of Rochester. No significant adverse reactions with some evidence clinical efficacy has been reported. No published results are available.

A modification of this initial trial will attempt to immunize the patient with the HER-2/Nm tumor antigen before administration of the HS-tk transduced cells.

22.5.2.2 T-Lymphocyte Gene Therapy Protocols

Suicide genes may also be used as a "safety" against undesirable side effects following the infusion of allogeneic T-cells (68). At the University of Arkansas and Northwestern University, patients who received a bone marrow transplant for multiple myeloma or leukemia respectively will receive T-cells from the bone marrow donor transduced with HS-tk for relapse or persistent disease. At the Fred Hutchinson Cancer Center and the University of Washington, they will be inserting the HS-tk gene into melanoma-specific T-cells. Since the infused cells may induce GVHD, the HS-tk gene has been approved to study the conditional and selective depletion of the infused T-lymphocyte populations in vivo with GCV. In these studies, the HS-tk gene functions as a marker for the infused cells as well as a safety to prevent adverse effects.

22.6 Combination Therapy of HS-tk and Cytokine Genes

While HS-tk appears to be an effective local therapy in vivo, further improvements to convert it to a systemic anticancer therapy are needed. Animal studies have demonstrated that the destruction of tumors with the HS-tk/GCV system can induce systemic immunity in mice as do many cytokine gene expressing tumors (43, 53). Presumably, HS-tk/GCV tumor destruction exposes new epitopes or blocks tumor-mediated immunosuppression allowing more effective immunologic recognition of tumor antigens and enhanced cytotoxic function of T-cells. In order to magnify this new immunologic response, researchers are studying combinations of suicide and cytokine genes. Preliminary experiments have shown that the addition of an interferon-$\alpha 1$ gene to the HS-tk gene enhances the anti-tumor effect against established metastatic friend leukemia cell tumors in mice (69).

22.7 Conclusions

The selective destruction of localized and disseminated cancers is the goal of oncologists. Suicide genes will continue to be a major part of the development of the first group of gene therapy products for cancer because they induce selective, conditional destruction of malignant cells. While suicide genes appear to primarily local therapeutics, they have the potential for synergistic action with immunostimulatory genes because they do not induce immunosuppression like conventional cancer therapies. The next 5-10 years of preclinical and clinical research will likely provide novel new therapies that include suicide genes for the treatment of cancer. The exact nature of the gene delivery systems and the specific clinical targets will be determined as a result of the information gained from these early cancer gene therapy trials.

References

1. Moolten, F. L. (1994). Drug sensitivity ("suicide") genes for selective cancer chemotherapy. Cancer Gene Ther. *1*, 279-287.
2. Harris, J. D., Gutierrez, A. A., Hurst, H. C., Sikora, K., and Lemoine, N. R. (1994). Gene therapy for cancer using tumour-specific prodrug activation. Gene Ther. *1*, 170-175.
3. Deonarain, M. P., Spooner, R. A., and Epenetos, A. A. (1995). Genetic delivery of enzymes for cancer therapy. Gene Ther. 2, 235-244.
4. Colbere-Garapin, F., Chousterman, S., Horodniceanu, F., Kourilsky, P., and Garapin, A.-C. (1979). Cloning of the active thymidine kinase gene of herpes simplex virus type 1 in Escherichia coli K-12. Proc. Natl. Acad. Sci. (USA). *76*, 3755-3759.
5. Elion, G. B. (1980). The chemotherapeutic exploitation of virus-specified enzymes. Adv. Enz. Regulation. *18*, 53-66.
6. Faulds, D. and Heel, R. C. (1990). Ganciclovir. Drugs. *39*, 597-638.
7. Balzarini, J., Morin, K. W., Knaus, E. E., Wiebe, L. I., and De Clercq, E. (1995). Novel (E)-5-(2-iodovinyl)-2'-deoxyuridine derivatives as potential cytostatic agents against herpes simplex virus thymidine kinase gene transfected tumors. Gene Ther. 2, 317-322.
8. Terry, B. J., Cianci, C. W., and Hagen, M. E. (1991). Inhibition of herpes simplex virus type 1 DNA polymerase by [1R(1a,2b,3a)]-9-[2,3-Bis(hydroxymethyl)cyclobutyl]guanine. Mol. Pharm. *40*, 591-596.
9. Smee, D. F., Martin, J. C., Verheyden, J. P. H., and Matthews, T. R. (1983). Anti-herpesvirus activity of the acyclic nucleoside 9-(1,3-dihydroxy-2-propoxymethyl)guanine. Antimicrob. Agents Chemother. *23*, 676-682.
10. Samejima, Y. and Meruelo, D. (1995). "Bystander killing' induces apoptosis and is inhibited by forskolin: plasmid pXtk construction for thymidine-kinase gene transfer to investigate bystander killing. Gene Ther. 2, 50-58.
11. Field, A. K., Davies, M. E., DeWitt, C., Perry, H. C., Liou, R., Germershausen, J., Karkas, J. D., Ashton, W. T., Johnston, D. B. R., and Tolman, R. L. (1983). 9-{[2-hydroxy-1-(hyroxymethyl)ethoxy]methyl}guanine: a selective inhibitor of herpes group virus replication. Proc. Natl. Acad. Sci. (USA). *80*, 4139-4143.
12. Moolten, F. L. (1986). Tumor chemosensitivity conferred by inserted herpes thymidine kinase genes: Paradigm for a prospective cancer control strategy. Cancer Res. *46*, 5276-5281.
13. Moolten, F. L., Wells, J. M., Heyman, R. A., and Evans, R. M. (1990). Lymphoma regression induced by ganciclovir in mice bearing a herpes thymidine kinase transgene. Hum. Gene Ther. *1*, 125-134.
14. Smee, D. F., Boehme, R., Chernow, M., Binko, B. P., and Matthews, T. R. (1985). Intracellular metabolism and enzymatic phosphorylation of 9-(1,3-dihydroxy-2-propoxymethyl)guanine and acyclovir in herpes simplex virus-infected and uninfected cells. Biochem. Pharmacol. *34*, 1049-1056.

15. Cheng, Y.-C., Grill, S. P., Dutschman, G. E., Nakayama, K., and Bastow, K. F. (1983). Metabolism of 9-(1,3-dihydroxy-2-propoxymethyl)guanine, a new anti-herpes virus compound, in herpes-simplex virus-infected cells. J. Biol. Chem. *258*, 12460-12464.
16. Nishiyama, T., Kawamura, Y., Kawamoto, K., Matsumura, H., Yamamoto, N., Ito, T., Ohyama, A., Katsuragi, T., and Sakai, T. (1985). Antineoplastic effects in rats of 5-fluorocytosine in combination with cytosine deaminase capsules. Cancer Res. *45*, 1753-1761.
17. Mullen, C. A., Kilstrup, M., and Blaese, R. M. (1992). Transfer of the bacterial gene for cytosine deaminase to mammalian cells confers lethal sensitivity to 5-fluorocytosine: A negative selection system. Proc. Natl. Acad. Sci. (USA). *89*, 33-37.
18. Mullen, C. A., Coale, M. M., Lowe, R., and Blaese, R. M. (1994). Tumors expressing the cytosine deaminase suicide gene can be eliminated in vivo with 5-fluorocytosine and induce protective immunity to wild type tumor. Cancer Res. *54*, 1503-1506.
19. Richards, C. A., Austin, E. A., and Huber, B. E. (1995). Transcriptional regulatory sequences of carcinoembryonic antigen: Identification and use with cytosine deaminase for tumor-specific gene therapy. Hum. Gene Ther. *6*, 881-893.
20. Senter, P. D., Schreiber, G. J., Hirschberg, D. L., Ashe, S. A., Hellström, K. E., and Hellström, I. (1989). Enhancement of the in vitro and in vivo antitumor activities of phosphorylated mitomycin C and etoposide derivatives by monoclonal antibody-alkaline phosphatase conjugates. Cancer Res. *49*, 5789-5792.
21. Senter, P. D. (1990). Activation of prodrugs by antibody-enzyme conjugates: a new approach to cancer therapy. FASEB J. *4*, 188-193.
22. Rowlinson-Busza, G., Bamias, A., Kraus, T., and Epenetos, A. A. (1992). Antibody-guided enzyme nitrile therapy (Agent): in vitro cytotoxicity and in vivo tumor localization. In: Monoclonal antibodies. Applications in clinical oncology, A. A. Epenetos, ed. (London, England: Chapman & Hall), pp. 111-118.
23. Haenseler, E., Esswein, A., Vitols, K. S., Montejano, Y., Mueller, B. M., Reisfeld, R. A., and Huennekens, F. M. (1992). Activation of methotrexate a-alanine by carboxypeptidase A-monoclonal antibody conjugate. Biochem. *31*, 891-897.
24. Haisma, H. J., Boven, E., van Muijen, M., de Jong, J., van der Vijgh, W. J., and Pinedo, H. M. (1992). A monoclonal antibody b-glucuronidase conjugate as activator of the prodrug epirubicin-glucronide for specific treatment of cancer. Br. J. Cancer. *66*, 474-478.
25. Douglas, S. P., Whitfield, D. M., Radics, L. R., Moolten, F. L., and Krepinsky, J. J. (1991). 5' galactosylation of nucleosides of 1-beta-D-arabinofuranosylcytosine (araC) and 1-beta-D-deoxyribofuranosylcytosine. Glycoconjugate J. *8*, 197.
26. Svensson, H. P., Vrudhula, V. M., Emswiler, J. E., MacMaster, J. F., Cosand, W. L., Senter, P. D., and Wallace, P. M. (1995). In vitro and in vivo activities of a doxorubicin prodrug in combination with monoclonal antibody-b-lactamase conjugates. Cancer Res. *55*, 2357-2365.
27. Mroz, P. J. and Moolten, F. L. (1993). Retrovirally transduced Escherichia coli gpt genes combine selectivity with chemosensitivity capable of mediating tumor eradication. Hum. Gene Ther. *4*, 589-595.
28. Patterson, A. V., Zhang, H., Moghaddan, A., Bicknell, R., Talbot, D. C., Stratford, I. J., and Harris, A. L. (1995). Increased sensitivity to the prodrug 5'-deoxy-5-fluorouridine and modulation of 5-fluoro-2'-deoxyuridine sensitivity in MCF-7 cells transfected with thymidine phosphorylase. Br. J. Cancer. *72*, 669-675.
29. Sorscher, E. J., Peng, S., Bebok, Z., Allan, P. W., Bennett, L. L. Jr., and Parker, W. B. (1994). Tumor cell bystander killing in colonic carcinoma utilizing the Escherichia coli DeoD gene to generate toxic purines. Gene Ther. *1*, 233-238.
30. Wei, M. X., Tamiya, T., Rhee, R. J., Breakefield, X. O., and Chiocca. (1995). Diffusible cytotoxic metabolites contribute to the in vitro bystander effect associated with the cyclophosphamide/cytochrome P450 2B1 cancer gene therapy paradigm. Clinical Cancer Res. *1*, 1171-1177.
31. Bagshawe, K. D. (1995). Antibody-directed enzyme prodrug therapy for cancer: its theoretical basis aand application. Mol. Med. Today. *1*, 424-431.
32. Beck, C., Cayeux, S., Lupton, S. D., Dörken, B., and Blankenstein, T. (1995). The thymidine kinase/ganciclovir-mediated "suicide" effect is variable in different tumor cells. Hum. Gene Ther. *6*, 1525-1530.

33. Ezzedine, Z. D., Martuza, R. L., Platika, D., Short, M. P., Malick, A., Choi, B., and Breakefield, X. O. (1991). Selective killing of glioma cells in culture and in vivo by retrovirus transfer of the herpes simplex virus thymidine kinase gene. New Biologist. *3*, 608-614.
34. Sacco, M. G., Mangiarini, L., Villa, A., Macchi, P., Barbieri, O., Sacchi, M. C., Monteggia, E., Fasolo, V., Vezzoni, P., and Clerici, L. (1995). Local regression of breast tumors following intra-mammary ganciclovir administration in double transgenic mice expressing neu oncogene and herpes simplex thymidine kinase. Gene Ther. 2, 493-497.
35. Short, M. P., Choi, J. K., Lee, A., Malick, A., Breakefield, X. O., and Martuza, R. L. (1990). Gene delivery to glioma cells in rat brain by grafting of a retrovirus packaging cell line. J. Neuroscience Res. *27*, 427-433.
36. Chen, S.-H., Shine, H. D., Goodman, J. C., Grossman, R. G., and Woo, S. L. C. (1994). Gene therapy for brain tumors: Regression of experimental gliomas by adenovirus-mediated gene transfer in vivo. Proc. Natl. Acad. Sci. (USA). *91*, 3054-3057.
37. Trattnig, S., Schindler, E., Ungersbock, K., Schmidbauer, M. Heimberger, K., Hubsch, P. and Stiglbauer, R. (1990). Extra-CNS metastases of glioblastoma: CT and MR studies. J. Comp. Assisted Tomography. *14*, 294-296.
38. Miller, A. D. (1990). Retrovirus packaging cells. Hum. Gene Ther. *1*, 5-14.
39. Ram, Z., Culver, K. W., Walbridge, S., Blaese, R. M., and Oldfield, E. H. (1993). In situ retroviral-mediated gene transfer for the treatment of brain tumors in rats. Cancer Res. *53*, 83-88.
40. Ram, Z., Culver, K. W., Walbridge, S., Frank, J. A., Blaese, R. M., and Oldfield, E. H. (1993). Toxicity studies of retroviral-mediated gene transfer for the treatment of brain tumors. J. Neurosurg. *79*, 400-407.
41. Culver, K. W., Ram, Z., Walbridge, S., Ishii, H., Oldfield, E. H., and Blaese, R. M. (1992). In vivo gene transfer with retroviral vector producer cells for treatment of experimental brain tumors. Science. *256*, 1550-1552.
42. Izquierdo, M., Cortes, M., de Felipe, P., Martin, V., Diez-Guerra, J., Talavera, A., and Perez-Higueras, A. (1995). Long-term rat survival after malignant brain tumor regression by retroviral gene therapy. Gene Ther. *2*, 66-69.
43. Culver, K. W., Moorman, D. W., Muldoon, R. R., Paulsen, R. M. Jr., Lamsam, J. L., Walling, H. W. and Link, C. J. Jr. (1994). Toxicity and immunologic effects of in vivo retrovirus-mediated gene transfer of the herpes simplex-thymidine kinase gene into solid tumors. Cold Spring Harbor Symp. on Quant. Biol. *59*, 685-690.
44. Barba, D., Hardin, J., Ray, J., and Gage, F. H. (1993). Thymidine kinase-mediated killing of rat brain tumors. J. Neurosurg. *79*, 729-735.
45. Hirschowitz, E. A., Ohwada, A., Pascal, W. R., Russi, T. J., and Crystal, R. G. (1995). In vivo adenovirus-mediated gene transfer of the Escherichia coli cytosine deaminase gene to human colon carcinoma-derived tumors induces chemosenitivity to 5-fluorocytosine. Hum. Gene Ther. *6*, 1055-1063.
46. Shewach, D. S., Zerbe, L. K., Hughes, T. L., Roessler, B. J., Breakefield, X. O., and Davidson, B. L. (1994). Enhanced cytotoxicity of antiviral drugs mediated by adenovirus directed transfer of the herpes simplex virus thymidine kinase gene in rat glioma cells. Cancer Gene Ther. 2, 107-112.
47. Chen, C.-Y., Chang, Y.-N., Ryan, P., Linscott, M., McGarrity, G. J., and Chiang, Y. L. (1995). Effect of herpes simplex virus thymidine kinase expression levels on ganciclovir-mediated cytotoxicity and the "bystander effect". Hum. Gene Ther. *6*, 1467-1476.
48. Bi, W. L., Parysek, L. M., Warnick, R., and Stambrook, P. J. (1993). In vitro evidence that metabolic cooperation is responsible for the bystander effect observed with HSV tk retroviral gene therapy. Hum. Gene Ther. *4*, 725-732.
49. Fick, J., Barker, F. G. II., Dazin, P., Westphale, E. M., Beyer, E. C., and Israel, M. A. (1995). The extent of heterocellular communication mediated by gap junctions is predictive of bystander tumor cytotoxicity in vitro. Proc. Natl. Acad. Sci. (USA). *92*, 11071-11075.
50. Freeman, S. M., Abboud, C. N., Whartenby, K. A., Packman, C. H., Koeplin, D. S., Moolten, F. L., and Abraham, G. N. (1993). The "bystander effect": Tumor regression when a fraction of the tumor mass is genetically modified. Cancer Res. *53*, 5274-5283.
51. Colombo, B. M., Benedetti, S., Ottolenghi, S., Mora, M., Pollo, B., Poli, G., and Finocchiaro, G. (1995). The bystander effect": Association of U-87 cell death with ganciclovir-mediated apoptosis of nearby cells and the lack of effect in athymic nude mice. Hum. Gene Ther. *6*, 763-772.

52. Ram, Z., Walbridge, S., Shawker, T., Culver, K. W., Blaese, R. M., and Oldfield, E. H. (1994). The effect of thymidine kinase transduction and ganciclovir therapy on tumor vasculature and growth of 9L gliomas in rats. J. Neurosurg. *81*, 256-260.
53. Barba, D., Hardin, J., Sadelain, M., and Gage, F. H. (1994). Development of anti-tumor immunity following thymidine kinase-mediated killing of experimental brain tumors. Proc. Natl. Acad. Sci. (USA). *91*, 4348-4352.
54. Wu, J. K., Cano, W. G., Meylaerts, S. A. G., Qi, P., Vrionis, F., and Cherington, V. (1994). Bystander tumoricidal effect in the treatment of experimental brain tumors. Neurosurg. *35*,1094-1103.
55. Huber, B. E., Austin, E. A., Richards, C. A., Davis, S. T., and Good, S. S. (1994). Metabolism of 5-fluorocytosine to 5-fluorouracil in human colorectal tumor cells transduced with the cytosine deaminase gene: Significant antitumor effects when only a small percentage of tumor cells express cytosine deaminase. Proc. Natl. Acad. Sci. (USA). *91*, 8302-8306.
56. Oldfield, E. H., Ram, Z., Culver, K. W., and Blaese, R. M. (1993). A clinical protocol: Gene therapy for the treatment of brain tumors using intra-tumoral transduction with the thymidine kinase gene and intravenous ganciclovir. Hum. Gene Ther. *4*, 39-69.
57. Florell, R. C., MacDonald, D. R., Irish, W. D., Bernstien, M., Leibel, S. A., Gutin, P. H., and Cairncross, J. G. (1992). Selection bais, survival, and brachytherapy for glioma. J. Neurosurg. *76*, 179-183.
58. Culver, K. W. and Blaese, R. M. (1994). Gene therapy for adenosine deaminase deficiency and malignant solid tumors. In: Gene Therapeutics, J.A. Wolff, ed. (Boston, USA: Birkhauser), pp. 256-273.
59. Culver, K. W., Van Gilder, J., Link, C. J., Carlstrom, T., Buroker, T., Yuh, W., Koch, K., Schabold, K., Doornbas, S., and Wetjen, B. (1993). Gene therapy for the treatment of malignant brain tumors with in vivo tumor transduction with the herpes simplex thymidine kinase gene/ganciclovir system. Hum. Gene Ther. *5*, 343-377.
60. Raffel, C., Culver, K. W., Kohn, D., Nelson, M., Siegel, S., Gillis, F., Link, C. J. Jr., and Villablanca, J. G. (1994). Gene therapy for the treatment of recurrent pediatric malignant astrocytomas using in vivo tumor transduction with the herpes simplex thymidine kinase gene/ganciclovir system. Hum. Gene Ther. *5*, 863-890.
61. Byrnes, A. P., Rusby, J. E., Wood, M. J. A., and Charlton, H. M. (1995). Adenovirus gene transfer causes inflammation in the brain. Neuroscience. *66*, 1015-1024.
62. Oldfield, E. H., Ram, Z., Chiang, Y., and Blaese, R. M. (1995). Intrathecal gene therapy for the treatment of leptomeningeal carcinomatosis: A phase I/II study. Hum. Gene Ther. *6*, 55-85.
63. Ram, Z., Walbridge, S., Oshiro, E. M., Viola, J. J., Chiang, Y., Mueller, S. N., Blaese, R. M., and Oldfield, E. H. (1994). Intrathecal gene therapy for malignant leptomeningeal neoplasia. Cancer Res. *54*, 2141-2145.
64. Oshiro, E. M., Viola, J. J, Oldfield, E. H., Walbridge, S., Bacher, J., Frank, J. A., Blaese, R. M., and Ram, Z. (1995). Toxicity studies and distribution dynamics of retroviral vectors following intrathecal administration of retroviral vector-producer cells. Cancer Gene Ther. 2, 87-95.
65. Smythe, W. R., Hwang, H. C., Amin, K. M., Eck, S. L., Davidson, B. L., Wilson, J. M., Kaiser, L. R., and Albelda, S. M. (1994). Use of recombinant adenovirus to transfer the herpes simplex virus thymidine kinase (HSVtk) gene to thoracic neoplasms: An effective in vitro drug sensitization system. Cancer Res. *54*, 2055-2059.
66. Plautz, G., Nabel, E. G., and Nabel, G. J. (1991). Selective elimination of recombinant genes in vivo with a suicide retroviral vector. New Biologist *7*, 709-715.
67. Freeman, S. M., McCune, C., Robinson, W., Abboud, C. N., Abraham, G. N., Angel, C., and Marrogi, A. (1995). The treatment of ovarian cancer with a gene modified cancer vaccine: A phase I study. Hum. Gene Ther. *6*, 927-939.
68. Bordignon, C., Bonini, C., Verzeletti, S., Nobili, N., Maggioni, D., Traversari, C., Giavazzi, R., Servida, P., Zappone, E., Benzzi, E., and Porta, F. (1995). Transfer of the HSV-tk gene into donor peripheral blood lymphocytes for in vivo modulation of donor anti-tumor immunity after allogeneic bone marrow transplantation. Hum. Gene Ther. *6*, 813-819.
69. Santodonato, L., Ferrantini, M., Gabriele, L., Proietti, E., Venditti, M., Musiani, P., Modesti, A., Modica, A., Lupton, S. D., and Belardelli, F. (1996). Cure of mice with established metastatic friend leukemia cell tumors by a combined therapy with tumor cells expressing both interferon-a1 and herpes simplex thymidine kinase followed by ganciclovir. Hum. Gene Ther. *7*, 1-10.

23. Tumor Cell Vaccines Using Genetically Modified Cells Coexpressing Cytokines and the T Cell Costimulatory Molecule B7

Sophie Cayeux, Bernd Dörken, Thomas Blankenstein

23.1 Introduction

Among the various strategies used for the immunotherapeutical treatment of cancer are tumor cell vaccines. In this context, the term vaccine refers to a therapeutic rather than a prophylatic modality. Up to date, only limited success has been observed in a number of clinical trials involving tumor cell vaccines (Hoover and Hanna, 1991; Morton et al, 1992; Oettgen and Old, 1991). However, recent knowledge has raised much hope that immunotherapy will eventually be successful in the treatment of patients. 1. It is possible to raise cytotoxic T lymphocytes (CTL) directed against tumors. 2. The first genes encoding tumor derived antigens recognised by CTL have been cloned, e.g. MAGE-1, MAGE-3, MART-1 and tyrosinase expressed by melanomas (Boon et al., 1994), which are normal cellular genes expressed not exclusively by tumor cells but which can be processed and presented in the form of peptides recognized by cytotoxic T lymphocytes (CTL) in association with different MHC class I molecules. 3. Even apparently non-immunogenic tumors engineered to secrete a single additional cytokine can be completely rejected in a tumor-specific and T cell dependent fashion (Blankenstein, 1994; Forni et al., 1994; Blankenstein et al., 1996). Therefore, it is widely believed that most, if not all, tumors possess antigens against which an immune response can be triggered even though these antigens must not necessarily be tumor specific.

Cytokines play a central role in immunoregulation. They usually have a short serum half life, are expressed by various cell types and have overlapping activities. They form an in vivo regulatory network, are involved in a wide range of functions such as cell activation, proliferation, differentiation and suppression thereby regulating most immune responses, inflammation and hematopoeisis. When applied systemically in unphysiological doses, cytokines often have toxic side effects. In many cases, their anti-tumor activity is not mediated by direct cytotoxic or cytostatic effects on tumor cells but by triggering an anti-tumor immune response using short range communication mechanisms between immune and non-immune cells, a finding that led to the application of cytokines by continuous local release using cytokine-gene transfected tumor cells. Increased local levels of cytokines can achieve an extremely high therapeutic index and minimise systemic toxic side effects. In addition, locally secreted cytokines may trigger a systemic immunostimulatory cascade thereby imitating more closely a physiological anti-tumor response.

A large number of cytokines have been transfected up to date in a large number of different tumors, and in many cases a strong immune response which caused tumor rejection was induced by the locally released cytokines. However, depending on the cytokine used, the amount of secreted cytokine or the tumor model, tumor suppression was either complete or partial or not seen at all (Blankenstein et al., 1996). These experiments showed that locally secreted cytokines from gene modified tumor cells could effectively inhibit in vivo tumor growth without significant systemic toxicity. Upon injection of the cytokine-modified tumor cells into mice it was clear that IL-1, IL-2, IL-3, IL-4, IL-6, IL-7, IL-10, IL-12, IL-13, TNF, LT, IFNα, IFNγ, G-CSF, MCAF and IP-10 suppressed tumor growth in at least one tumor model. Basically, cytokine gene modified tumor cells have been studied in two ways: 1) Analysis of the increase of immunogenicity and concommitant decrease of tumorigenicity of the transfected tumor cells themselves and 2) Study of the protective effect of such cells when used as vaccines. In order to investigate the efficiency of gene-modified tumor cells when used as vaccines, mice are pre-immunized with the cytokine gene modified tumor cells and subsequently challenged contralaterally with the parental tumor cells. Alternatively in a therapeutic model, mice are firstly injected with the parental tumor cells which is followed shortly afterwards with treatment of the mice with gene modified cells. Survival of the animals or inhibition of metastasis after a certain period following immunization serves to evaluate the efficacy of the applied vaccines. There is no perfect correlation between loss of tumorigenicity of the transfected tumor cells and their vaccine effects.

Costimulatory molecules are ligand-receptor pairs found on antigen presenting cells (APC) and T cells, respectively. One well characterized system of ligand-receptor molecules involves the B7 family comprising B7.1 (CD80) and B7.2 (CD86) present on APCs. Their counter-receptors are CD28 and CTLA-4 on T cells. Stimulation of T cells by ligation of CD28 leads to IL-2 gene activation and stabilization of mRNA for a number of cytokines (Chen et al., 1993). Recently, a variety of mouse tumor cells were subjected to B7.1 gene transfer. The concept behind this approach is that antigen presenting cells present tumor antigens to T cells in association with MHC molecules and in the presence of costimulatory molecules such as B7, this interaction leads to the activation of T cells and the proliferation and expansion of anti-tumor T cells. The majority of tumor cells lack the B7 molecules and in the absence of B7 the interaction of tumor cells and T cells instead of leading to T cell activation generates T cell anergy, tolerance or apoptosis (Linsley and Lett-better, 1993; Chen et al., 1993; June et al., 1994). Thus, in several tumor models it was shown that B7.1 gene transfected tumors were rejected in syngeneic animals and induced tumor immunity (Chen et al., 1992; Townsend and Allison, 1993; Baskar et al., 1993). Depending on whether the tumor cells expressed MHC class I or class II molecules this effect was mediated by CD8+ or CD4+ T cells. B7.1 expressed by tumor cells seems to act on T cells at the induction and effector phase (Ramarathinam et al., 1994). In addition, it was shown that B7.1 expression increased immunogenicity only of such tumors which per se were to some degree immunogenic and expressed MHC class I molecules (Chen et al., 1994). Transfection of the B7.1 gene in ICAM-1+ and ICAM- tumor cell lines and of ICAM-1 gene in B7.1+ and B7.1- tumor lines showed that the presence of both costimulatory molecules is necessary to costimulate an efficient tumor specific immune response (Cavallo et al., 1995).

From a number of studies which investigated the vaccine efficacy of either cytokine or B7 gene transfected tumor cells, it appeared that despite successful immunization with the genetically modified cells against challenge with the parental tumor in several models, the reproducibility when analysed in different tumor models and vaccine strength, e.g. when analysed in a therapeutical setting, were not satisfying. In order to improve the efficacy of cytokine modified tumor cell vaccines, the gene for the costimulatory molecule B7.1 was cotransfected into tumor cells previously transfected to produce cytokines, e.g. in J558L (Cayeux et al., 1996) and TS/A (Cayeux et al., 1995). These experiments were done because of the following observation: injection of a variety of cytokine gene (IL-2, IL-4, IL-7, TNF, IFNγ) transfected tumor cells into syngeneic mice showed that tumor growth was strongly suppressed by the transfected cytokine (Hock et al., 1993a). Complete rejection of the tumor was seen in a high percentage, but usually not in all of the mice. Immunization of mice with such cells and subsequent contralateral challenge with the parental tumor revealed that, eventhough some mice rejected the challenge tumor, e.g. had developed systemic tumor immunity, this effect was less obvious when higher challenge doses were applied. Importantly, vaccine strength of cytokine gene transfected tumor vaccines was not better than a vaccine consisting of tumor cells to which an adjuvant (Corynebacterium parvum) was mixed (Hock et al., 1993b). The latter type of vaccine has been extensively tested clinically in cancer patients with poor success (Oettgen and Old, 1991). Similarly, vaccine cells transfected to express B7.1 were not effective when tumors were of low immunogenicity (Chen et al., 1994). The conclusion, therefore was that genetically engineered tumor vaccines required further improvement. Another observation which led us to combine B7.1 with cytokine gene transfer into tumor cells was that we found abundant tumor infiltrating T lymphocytes in some cytokine (e.g. IL-7, TNF) gene transfected tumors which, however, did not correlate with the observed systemic tumor immunity (Hock et al., 1993a, b). We reasoned that the T cells detected in the tumor tissue were not appropriately activated, compatible with the above-mentioned two step activation process of T lymphocytes. These observations together with in vitro data on the synergistic effect of e.g. IL-7 and antibodies binding to the B7 ligand (CD28) on T cell prolifera-

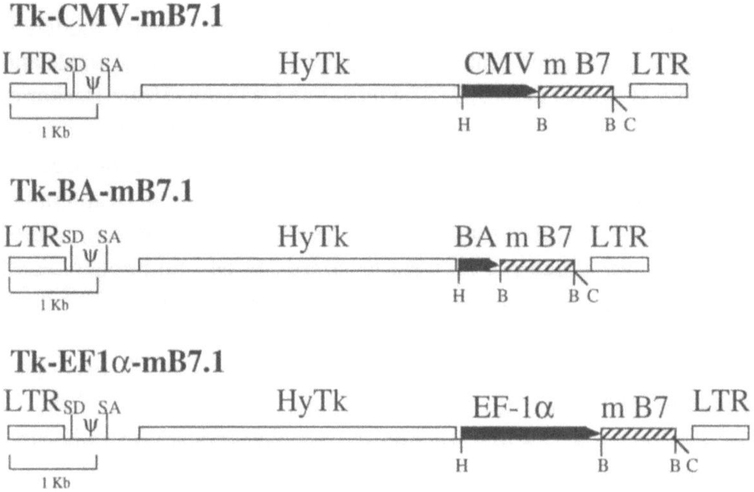

Fig. 23.1: Structure of retroviral vectors used for mouse B7.1 expression. LTR, long terminal repeat; SD, splice donor site; SA, splice acceptor site; HyTk, hygromycin-thymidine kinase fusion gene; CMV, cytomegalus virus promotor; BA, β-actin promotor; EFlα, elongation factor 1α, promotor; mB7.1, mouse B7.1. gene; H, HindIII; B, Bam H1; C, Cla1.

tion, cytokine secretion and IL-2 receptor expression (Costello et al., 1993) provided the rationale for coexpressing B7.1 and cytokines to improve the efficacy of tumor cell vaccines.

23.2 Tumor Cell Vaccines Coexpressing Cytokine and B7.1 Genes

23.2.1 Construction of B7.1 Gene Expression Vectors and Retroviral Gene Transfer Experiments

Because of the general uncertainty whether the gene of interest is expressed in a given retroviral construct, we have constructed three vectors as shown in Figure 23.1. They are based on the vector tgLS(+) HyTK (Lupton et al., 1991). The three retroviral vectors TK-CMV-mB7.1, TK-BA-mB7.1 and TK-EFlα-mB7.1 contain a thymidine kinase/hygromycin fusion gene which can be used for positive and negative selection of infected cells under the long terminal repeat (LTR) promotor control and the mouse B7.1 gene under either the cytomegalovirus (CMV), β-actin (BA) or elongation factor 1α (EFlα) promotor control. All three vectors were converted into recombinant retroviruses by standard procedures using conventional packaging cell lines. Clones Ψ2-TK-CMV-mB7.1 and Ψ2-TK-EF1α-mB7.1 were established which produce a viral titer equivalent to 5×10^4 and 5×10^5 Hygromycin^R colonies/ml, respectively.

The virus-containing supernatant of Ψ2-TK-CMV-mB7.1 cells was used to infect the cell lines J558L, TS/A and their IL-7 and IL-4 producing variants (J558-IL4, J558-IL7 and TS/A-IL7 cells). MHC class I expression of TS/A-B7.1 and J558-B7.1, TS/A-IL7/B7.1 and J558-IL7/B7.1, J558-IL4/B7.1 sublines was the same compared to that of parental cells and all cells were MHC class II negative. The level of B7.1 expression was similar in the B7.1 transduced cells (Cayeux et al., 1995; Cayeux et al., 1996).

23.2.2 Cooperative Effect of B7.1 Expression and IL-7 or IL-4 Secretion on Tumor Rejection

Previous experiments with a variety of different cytokine gene transfected tumors had shown that almost always a certain percentage of the mice developed tumors when the mice were observed over a sufficiently long period of time (Hock et al., 1993a). In the first tumor model, the mammary adenocarcinoma TS/A, neither IL-7 secretion nor B7.1 expression alone by the tumor cells was sufficient to induce tumor rejection in all of the mice whereas IL-7/B7.1 coexpressing cells were reliably rejected in syngeneic animals (Fig. 23.2a) (Cayeux et al., 1995). Similar results were obtained in the second tumor model, the plasmacytoma J558L and variants thereof.

Similarly, the combined immunostimulatory effects of IL-4 and B7.1 conferred an optimal and reliable anti-tumor response. The tumorigenic potential of J558-IL4/B7.1 cells markedly differed from that of single gene transfected cells and J558L cells injected with C. parvum (Fig. 23.2b). While all animals developed a tumor within two weeks when injected with parental J558L cells, none of the mice injected with J558-IL4/B7.1 cells developed a tumor during an observation period of up to five months while J558L cells mixed with 100 μg C. parvum or J558-IL4 or J558-B7.1 cells gave rise to tumors in about 25% of injected mice. Tumors from these different groups differed in growth kinetics and phenotype. J558-B7.1 cell injected mice developed a tumor in most cases without major delay compared to parental cells and expressed undiminished amounts of B7.1 (Cayeux et al., 1996). In contrast, tumors in mice injected with J558-IL4 cells grew only after a long latency period and showed complete loss of IL-4 production (Hock et al., 1993a). This suggested a different mechanism of tumor suppression mediated by IL-4 or B7.1.

23.2.3 Vaccination Efficiency of IL-7/B7.1 or IL-4/B7.1 Coexpressing Cells is Superior to Single Gene Transfectants and to Adjuvant C. parvum

In order to compare the vaccine potency of IL-7/B7.1 cotransfected TS/A cells to that of the clinically extensively tested adjuvant C. parvum or non-proliferating TS/A cells, groups of mice were immunized with TS/A-IL7, -B7.1, -IL7/B7.1 or parental cells mixed with C. parvum (Cayeux et al., 1995). Tumor-free mice were challenged

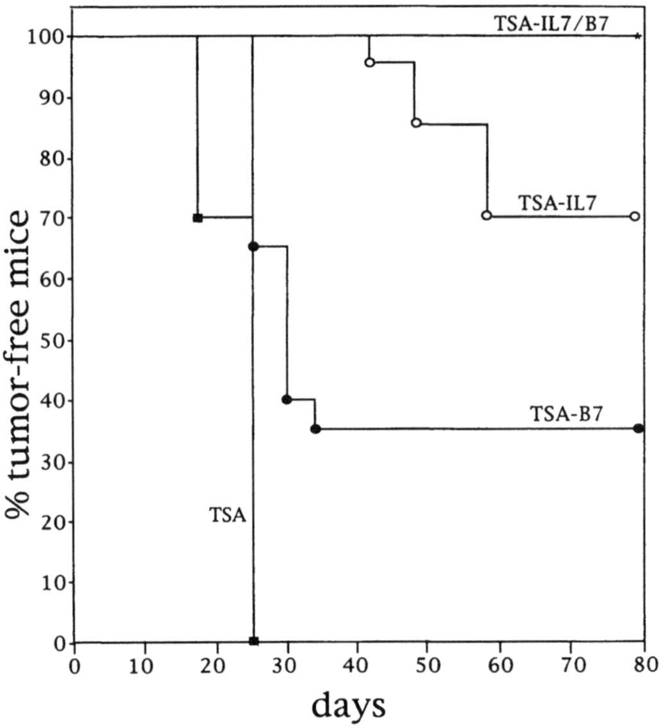

Fig. 23.2a: IL-7 and B7.1 expressed by TS/A cells in a cooperative fashion induce tumor rejection. BALB/c mice were injected subcutaneously with $1x10^6$ TS/A (■), TS/A-TK (□), TS/A-B7.1 (•), TS/A-IL7 (○) or TS/A-IL7/B7.1 (*) cells. 20 cell-injected mice were analysed for each experimental group. (Reproduced with permission from Eur. J. Immunol., Cayeux et al., 1995).

two weeks later contralaterally with non-modified cells. Tumor growth of vaccine cells was only prevented in all mice when cells were irradiated with 100 Gy or when IL7/B7.1 cotransfected cells were used for immunization (Fig. 23.3a). While in the tumor cell/C. parvum group 30% of the mice developed a tumor, mice poorly rejected the TS/A-B7.1 vaccine cells (60%), but those mice which succeeded to reject vaccine cells were well protected and only 5% developed a challenge tumor. In contrast, with IL7, vaccine cells were more effectively rejected (5% developed a vaccine tumor) but mice were less immune to challenge cells (25% developed a challenge tumor). Therefore it appears that B7.1 and IL7 trigger the immune system in different and complementary ways resulting in that TS/A-IL7/B7.1 vaccine cells were completely rejected in all mice and provided protection against TS/A cells in 95% of mice. All of the above mentioned immunization experiments with transfected tumor cells were performed with viable cells.

In a similar way, we analysed whether IL-4/B7.1 cotransfected J558L cells provided a potent vaccination effect and compared their vaccine potency to that of the

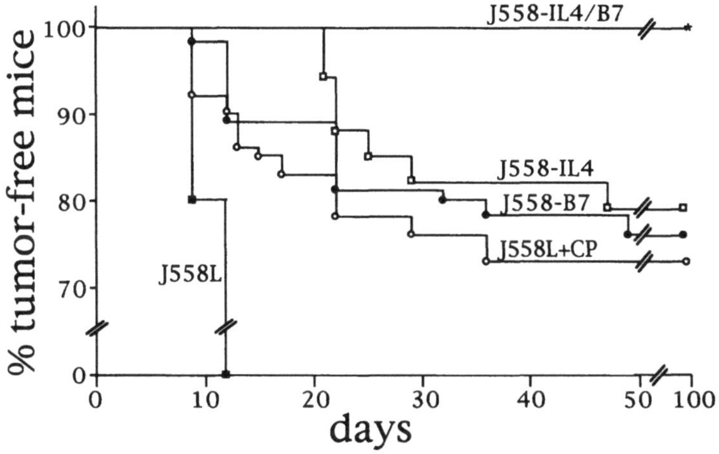

Fig. 23.2b: IL-4 and B7.1 expressed by J558L cells in a cooperative fashion induce tumor rejection. BALB/c mice were injected subcutaneously with 4 x 10^6 J558L (■), J558L + C. parvum (o), J558-B7.1 (•), J558-IL4 (□) or J558-IL4/B7.1 (*) cells. 10 J558L, 37 J558L + C. parvum, 59 J558-B7.1, 34 J558-IL4 and 80 J558-IL4/B7.1 cell-injected mice were analysed. (Reproduced with permission from Human Gene Therapy, Cayeux et al., 1996).

clinically extensively tested adjuvant C. parvum (Cayeux et al., 1996). Large groups of mice were immunized with J558-IL4, -B7.1, -IL4/B7.1 or parental cells mixed with C. parvum and challenged 2 weeks later contralaterally with non-modified cells. Similar as we had previously seen (Hock et al., 1993b), only 7% of the mice immunized with irradiated parental cells and 45% of the mice pretreated with J558-IL4 stayed tumor-free (Fig. 23.3b) (Cayeux et al., 1996). Likewise, from those mice which had rejected the J558-B7.1 vaccine cells, 62% resisted the challenge tumor. Most mice were immune against J558L cells when immunized with J558-IL4/B7.1 cells (78%). From the tumor cell/C. parvum group 45% rejected the challenge tumor. Tumor immunity conferred by transfected J558L cells was specific because all mice (9/9) which had been immunized with various transfected J558L cells and had rejected a subsequent parental J558L tumor challenge developed a tumor when challenged a second time with the unrelated BALB/c tumor TS/A (Tab. 23.1).

23.2.4 Irradiation of Tumor Cells Prior to Immunization Abrogates the Vaccine Effect

The above mentioned immunization experiments with transfected tumor cells were performed with viable cells. We wanted to know how the vaccine effect of viable cells used for immunization compares to that of the same cells irradiated prior to injection. Irradiation of vaccine cells is routinely performed in clinical cancer vaccine trials. Therefore, mice were immunized with J558-IL4/B7.1 cells or a J558L/

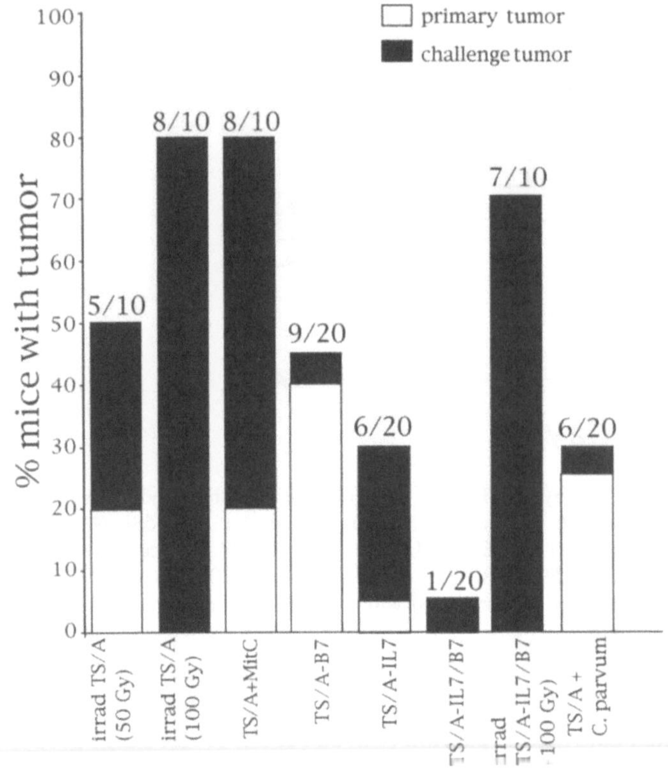

Fig. 23.3a: Vaccine efficacy of IL-7/B7.1 transfected TS/A cells is superior to single gene transfectants and adjuvant C. parvum. Mice were given s.c. injections of $2.5 x 10^5$ of the indicated cells and contralaterally challenged after 2 weeks with $2.5 x 10^5$ TS/A cells. Vaccine cells were either applied as viable cells or irradiated (50 or 100 Gy) or mitomycin C-treated as indicated. The number of tumor-bearing mice/total number of mice in experiment appears at the top of each column. The white part of the columns show the percentage of mice which developed vaccine tumors while the black part of the columns display the percentage of mice that developed challenge tumors. Mice are the same as those shown in Fig. 23.2a. (Reproduced with permission from Eur. J. Immunol., Cayeux et al., 1995).

C. parvum mixture, each either viable or irradiated with 100 Gy, and two weeks later challenged with the parental tumor. All mice (10/10) rejected the challenge tumor when viable J558-IL4/B7.1 cells were used for immunization but 60% (6/10) developed a tumor when the vaccine cells were irradiated. Similarly 95% of mice (19/20) rejected a challenge with T/SA cells when immunized with viable T/SA-IL7/B7.1 cells whereas only 30% (3/10) remained tumor free when immunized with irradiated (100 Gy) T/SA-IL7/B7.1 cells (Fig. 23.4). The loss of vaccine effect was not due to loss of expression of the transfected B7.1 gene because a) irradiated J558-IL4/B7.1 cells after four (85% viable cells) and nine days (10% viable cells) of culture still expressed similar levels of B7.1 compared to non-irradiated cells (data not shown) and

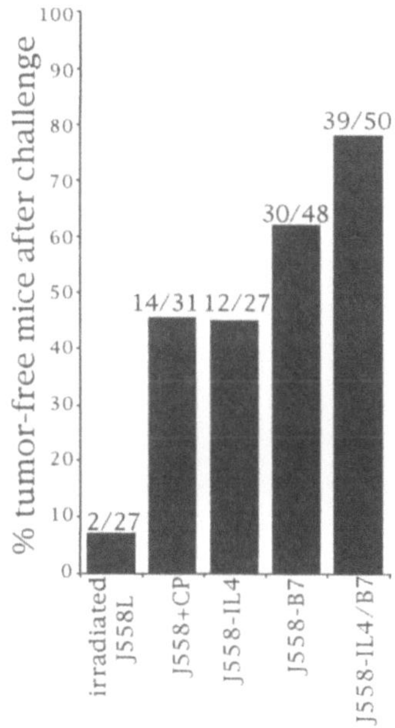

Fig. 23.3b: Vaccine efficacy of IL-4/B7.1 transfected J558L cells is superior to single gene transfectants and adjuvant C. parvum. Mice were given s.c. injections of 4 x 10^6 of the indicated cells and contralaterally challenged after 2 weeks with 4 x 10^6 J558L cells. All vaccine cells were applied as viable cells except in the control group receiving irradiated J558L cells (100 Gy). The number of tumor-free mice/total number of mice in experiment appears at the top of each column. Combined results of several experiments are presented. (Reproduced with permission from Human Gene Therapy, Cayeux et al., 1996).

Table 23.1 Tumor immunity conferred by transfected J558L cells was specific

	Number of mice with tumors
Immunization with gene-modified J558L cells	0/18
1 challenge (after 2 weeks): parental J558L	0/18
2 challenge (after 8 weeks): parental J558L	0/9
2 challenge (after 8 weeks): BALB/c tumor T/SA	9/9

All mice which had been immunized with various gene-modified J558L cells (4 x 10^6 cells) and had rejected a challenge with parental J558L (4 x 10^6) cells, developed tumors when challenged a second time with the unrelated BALB/c tumor T/SA (2.5 x 10^5 cells) but not with J558L cells.

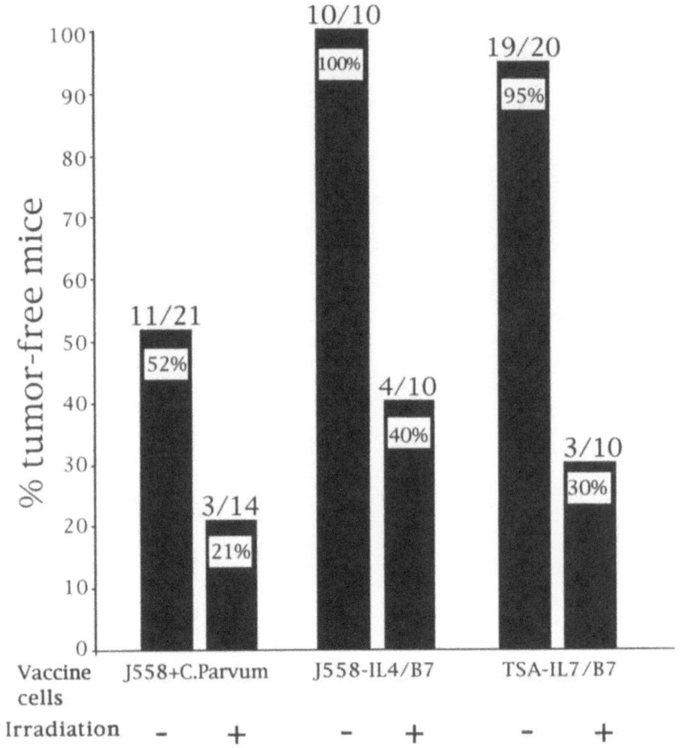

Fig. 23.4: Tumor cell vaccine lose efficiency with irradiation. Mice were immunized either with irradiated or live cells, immunization with parental J558L cells admixed with C. parvum and challenge with parental J558L cells or immunization with J558-IL-4/B7.1 coexpressing cells and challenge with parental J558L cells or immunization with TS/A-IL7/B7.1 coexpressing cells and challenge with parental TS/A cells. The percentage of tumor-free mice after challenge is compared.

b) the tumor cell/C. parvum vaccine similarily lost efficacy upon irradiation. 48% of the mice (10/21) immunized with viable and 79% (11/14) of those immunized with irradiated J558L cells + C. parvum developed a tumor.

The reason for the loss of vaccine effect by irradiation (which is dose-dependent) is currently not known but differences in total antigen concentration, production of immunosuppressive factors or ineffective antigen presentation caused by irradiation could be responsible. Eventhough in several tumor models cytokine gene modified irradiated cells were rather potent vaccines these studies did not compare vaccine efficacy between live and irradiated cells or to a tumor cell/adjuvant mixture and also used lower irradiation doses which can be less detrimental for the vaccine effect but, at least for TS/A cells, does not completely prevent growth of the vaccine cells (Cayeux et al., 1995). In contrast, experiments with GM-CSF gene transfected tumor cells showed that irradiation of the vaccine cells had no effect on generation of tumor immunity (Dranoff et al., 1993). A possible explanation is the mode of T cell activation induced by the different vaccines: GM-CSF seems to activate host antigen presenting cells to reprocess tumor antigens and to induce T cells (Huang et al., 1994), whereas B7 (and/or T cell activating cytokine) transfected tumors may directly present antigens to and activate T cells.

23.3 Mechanisms of Tumor Rejection Induced by Cytokine and/or B7.1 Modified Tumor Cell Vaccines

A comparison between the incidence of tumors and growth rate of cytokine modified tumors in different strains of mice enables the assessment of which types of effector cells are involved in the tumor rejection process. The strains of mice we used for evaluating the mechanisms of cytokine mediated tumor suppression are syngeneic immunocompetent mice, nude mice (nu/nu) which lack functional T cells and SCID mice which are deficient in both, T and B cells. If, for example, a cytokine-modified tumor is rejected in a normal syngeneic host but grows in nude mice, this demonstrates that mature functional T cells are required for tumor rejection.

Furthermore, the role of different T cell subsets can be analysed *in vivo* by depleting different T cell subsets in immunocompetent mice using the *in vivo* application of monoclonal antibodies specific for a particular subset of T cells, e.g. CD4+ or CD8+ cells. Following *in vivo* depletion of defined T cell subsets, the incidence and growth kinetics of cytokine gene modified tumors can be observed. Cytokine modified tumors are often rejected both by T cell independent and a T cell dependent mechanisms (Hock et al., 1993a). The effect of the T cell independent mechanism can be observed in nude mice which lack functional T cells and where tumor growth is retarded. In this case effector cells include inflammatory cells such as granulocytes, macrophages, NK cells and eosinophils. Interestingly, the mechanisms of tumor rejection in the majority of cytokine modified tumors (IL-2, IL-4, TNF, IFN-γ) are redundant, whereby a significant role is played by inflammatory cells and CD8+ T cells, but the relative importance of the different effector cells in a particular tumor model varies. Late outgrowth of cytokine modified tumors is generally observed and reisolation of the tumor in the majority of cases demonstrates loss of expression of the cytokine gene (Hock et al., 1993a). It is probable that a strong *in vivo* selection against cytokine secreting cells influences tumor cell mutation leading to loss of expression of cytokine genes and tumor growth.

A further approach for the characterization of the tumor rejection process is the reisolation of the injected cytokine gene modified tumor cells a few days after tumor cell injection followed by staining of the inflammatory or immune cells infiltrating the tumor tissue by means of immunohistological methods or immunofluorescence. Finally cytotoxic T lymphocytes (CTLs) can be generated *in vitro* from mice immunized with cytokine gene modified tumors. To summarise, using mouse models it is

Table 23.2 Cytokine but not B7.1 transfected J558 cells suppress tumor growth of coinjected parental cell

Coinjection	Tumor incidence	Average + SD appearance of tumors in days
J558-B7.1 + J558L	10/10 (100 %)	12.8 ± 3.3
J558-IL2 + J558L	3/10 (30 %)	32.7 ± 13.3
J558L	10/10 (100 %)	10 ± 0

Syngeneic BALB/c mice were injected with parental 4 x 10^6 J558L cells mixed with either 4 x 10^6 J558-B7.1 or J558-IL2 cells and the effect of the gene-modified cells on the growth of the J558L cells examined. The cytokine transduced cells exert a paracrine effect leading to the inhibition of both the cytokine modified cells and the parental cells whereas the B7.1 transduced J558L have no influence on the growth of the coinjected J558L cells. The mice were observed over a period of 80 days.

clear that the mechanisms of tumor rejection induced by cytokine transduced tumors involve both T cell dependent and T cell independent mechanisms; CD8+ cells are critically important for rejection of MHC Class I+ tumors, the extent of tumor infiltration by T cells does not correlate with the efficiency of tumor regression and the late outgrowth of tumors is correlated with the loss of expression of cytokine genes.

23.3.1 Cytokine but Not B7.1 Transfected Cells Suppress Tumor Growth of Coinjected Parental Cells

Cytokine modified tumor cells have been observed to suppress the growth of coinjected non modified tumor cells (Tepper et al., 1989). We investigated whether B7.1 or cytokine gene transfected J558L cells differed in their ability to suppress the growth of coinjected non-modified cells. Therefore, equal numbers of J558-B7.1 or J558-IL2 cells (which grew more rarely as tumor than J558-IL4 cells) (Hock et al., 1993a) and parental J558L cells were coinjected into BALB/c mice. All mice (10/10) injected with a mixture containing B7.1 transfectants developed tumors which grew as rapidly as parental J558L tumors (Tab. 23.2). On the other hand, only 3/10 mice injected with a J558L/J558-IL2 mixture developed a tumor. In a less dramatic fashion J558-IL4 cells also suppressed the growth of coinjected parental cells (data not shown). The lack of cross protection of B7.1 transfectants further distinguishes the tumor suppressive mechanisms of cytokines and B7.1.

23.3.2 Analysis of Cells Infiltrating Tumor Tissue

Of interest was the analysis of the tumor infiltrating cells found in the different gene transfected tumors. In order to investigate the cellular mechanism of IL-4/B7.1 induced tumor rejection, immunofluorescence analysis of tumor infiltrating T cells was performed. It is important to note, however, that at least for J558-IL4 cells we had

Table 23.3 Analysis of reisolated J558L, J558L cells mixed with C. parvum, J558-IL4 and J558-B7.1 tumors by using immunofluorescence staining

	% CD4+	% CD8+	Ratio CD4:CD8
J558L	11	4	2.8
J558L + C.parvum	15	2	7.5
J558-IL4	23	7	3.3
J558-B7.1	19	21	0.9

Mice were injected subcutaneously with 4 x 10^6 cells and when tumors developed and reached a size of 0.5 cm (on day 21 for J558L + C. parvum, J558-IL4, J558-B7.1 and on day 6 for J558L) the tumor tissue was reisolated and stained for CD8+ and CD4+ T cells. The results are from one representative experiment. J558-IL4/B7.1 cells were rapidly rejected and did not grew as tumor in a single case and could not be examined here.

previously shown that depletion of CD4+ cells did not interfere with tumor suppression whereas CD8+ T cells are mandatory for complete tumor eradication (Hock et al., 1993a). Interestingly, parental tumor cell/C. parvum mixtures gave rise to tumors which were nearly completly devoid of CD8+ T cells (2%), yielding a CD4/ CD8 ratio of 7.5. Compatible with earlier results, IL-4 transfected tumors showed only a moderately increased number of CD8+ T cells (7%, CD4/ CD8 ratio 3.3). A strong increase of CD8+ T cells was found in J558-B7.1 tumors (21%) which meant a CD4/ CD8 ratio of 0.9 (Tab. 23.3). Thus, eventhough tumor growth was similarily suppressed in the different groups, a markedly different CD8+ T cell content was found in the tumors: B7.1 > IL-4 > C. parvum.

A similar analysis was performed to investigate the cellular mechanism of IL-7/ B7.1 induced tumor rejection. Parental TS/A cells, TS/A-IL7, TS/A-B7.1 and TS/A-IL7/B7.1 cells were subcutaneously injected into BALB/c mice and 6, 8 and 10 days later tumors were isolated, single cell suspensions prepared and cells stained using immunofluorescence with mAbs against CD4, CD8, CD25 and CD28 (Cayeux et al., 1995). Both CD4+ and CD8+ T cells were increased in TS/A-B7.1 in comparison to parental TS/A tumors. In TS/A-IL7 tumors an increase of CD4+ T cells was observed. T cells (CD4+ and CD8+) were not further increased in IL7/B7.1 transfected tumors but, if at all, less abundant. However, two colour fluorescence staining for T cell subset markers and activation markers CD28 or CD25 revealed phenotypically different T cells in either IL7 or B7.1 transfected tumors. In TS/A-B7.1 tumors, T cells of both subsets were to a high percentage CD28+ but most T cells were CD25-. In contrast, T cells in TS/A-IL7 tumors were mostly CD25+ and CD28+ cells were essentially absent. For comparison, in parental TS/A tumors only few CD28+ and virtually no CD25+ T cells were detected. Importantly, in TS/A-IL7/B7.1 tumors, most CD4+ and CD8+ cells were CD25+ and/or CD28+. We currently do not know to what extent these cells reflect distinct or overlapping populations. However, to-

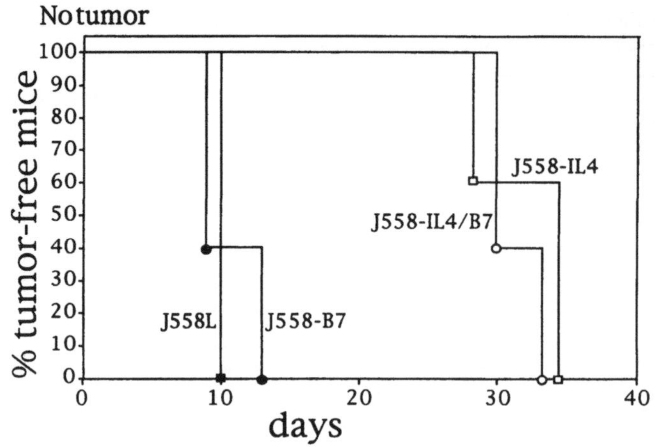

Fig. 23.5: Growth of J558L and its transfectants in SCID mice. Kinetics of tumor progression in SCID mice injected subcutaneously with 4 x 10^6 J558L (■), J558-B7.1 (•), J558-IL4 (□), or J558-1L4/B7.1 (○) cells. In each group 5 mice were analysed.

gether with the fact that only IL7/B7.1 cotransfected cells were reliably rejected and most strongly induced systemic tumor immunity, the local IL-7 secretion and B7.1 expression by the tumor cells seems to improve activation of tumor infiltrating lymphocytes by at least partially distinct pathways.

23.3.3 Growth of Tumor Cell Lines in Nude and SCID Mice

Taking our above mentioned observation that, both, IL-4 and B7.1 single but not double gene transfected cells gave rise to tumors in a substantial number of mice, this could suggest that the cooperative effect of IL-4 and B7.1 to cause tumor rejection results from an improved activation of CD8+ T cells. This assumption could be proved by the demonstration that the concerted tumor suppressive activity of IL-4 and B7.1 would not act in mice lacking T cells. Therefore, IL-4, B7.1 and IL-4/B7.1 transfected cells were injected into nude and SCID mice and tumor growth kinetics were compared. As shown earlier (Hock et al., 1993a; Li et al., 1990) J558-IL4 tumor growth was initially suppressed in both mouse strains but after a latency period all mice developed a tumor (Fig. 23.5). Tumor growth of J558-B7.1 cells was not significantly delayed in immunocompromised mice. The tumor incidence of J558-IL4/B7.1 injected mice was nearly identical compared to that of J558-IL4 injected mice demonstrating the necessity of T cells for the cooperative IL-4/ B7.1 effect.

Similarly we were able to demonstrate that the concerted tumor suppressive activity of IL-7 and B7.1 solely operates via T cells. IL-7, B7.1, IL7/B7.1 and for comparison parental or mock transfected TS/A cells were injected into nude and SCID mice and tumor growth kinetics were compared. Neither IL-7 secretion (Hock et al., 1991) nor B7.1 expression by tumor cells nor both together were able to significantly delay tumor formation in either of the two immunocompromised mouse strains demonstrating the absolute requirement for T cells for the IL-7 and B7.1 induced anti-tumor immune response (Cayeux et al., 1995).

23.4 Conclusion

Previously we have demonstrated that J558L tumor cell vaccines containing single cytokine genes, e.g. IL-2, IL-4, IL-7, TNF or IFN-γ or any possible combination of two cytokine producer cell lines were not more efficient at generating systemic immunity and protecting mice against a tumor cell challenge compared to classical vaccine consisting of tumor cells admixed with an adjuvant (Corynebacterium parvum) (Hock et al., 1993b). Similar observations have also been made in another tumor model (Allione et al., 1994). In contrast, vaccination with IL4/B7.1 or IL7/B7.1 cotransfected cells (as tested both with the plasmacytoma cell line J558L and the mammary adenocarcinoma cell line TS/A) demonstrated that i) gene-modified tumor cell vaccines can be improved by the combined use of cytokine (IL-4 or IL-7) and B7.1, ii) IL-7/B7.1 and IL-4/B7.1 cotransfected cells are superior to a tumor cell/adjuvant vaccine and iii) irradiation of the vaccine cells abrogates the vaccine effect. The reason for the loss of vaccine effect by irradiation is not understood and could be due to variables such as poor antigen presentation, production of immunsuppressive factors or lower antigen concentration resulting from irradiation, but it is not related to loss of expression of the transfected genes by irradiation. Notably, GM-CSF transfected tumors did not lose vaccine potency following irradiation (Dranoff et al., 1993).

Cytokine gene modified tumor cells as vaccines are currently tested in a number of clinical trials (Tepper and Mule, 1994). The combined use of cytokine (IL-7 or IL-4) and B7.1, the use of live vaccine cells, the addition of 'suicide' genes such as the Herpes virus thymidine kinase (TK) gene and/or the cytosine deaminase gene for safety (Culver et al., 1992; Mullen et al., 1992) in a partially HLA matched allogeneic context could become an attractive alternative. An increased number of 'shared' tumor antigens are being described (van der Bruggen, 1991; Schendel et al., 1993) and this legitimises the use of allogeneic vaccine cells although the effect of strong allogeneic antigens on inhibition of development of tumor specific cytotoxic T lymphocytes still remains unclear.

Acknowledgements

This work was supported by a grant from the Deutsche Krebshilfe, Mildred-Scheel-Stiftung, e.V.

References

Allione, A., Consalvo, M., Nanni, P., Lollini, P. L., Cavallo, F., Giovarelli, M., Forni, M., Gulino, A., Colombo, M. P., Dellabona, P., Hock, H., Blankenstein, Th., Rosenthal, F. M., Gansbacher, B., Bosco, M. C., Musso, T., Gusella, L., Forni, G. (1994). Immunizing and curative potential of replicating and nonreplicating murine mammary adenocarcinoma cells engineered with interleukin (IL)-2, IL-4, IL-6, IL-7, IL-10, tumor necrosis factor a, granulocyte-macrophage colony-stimulating factor, and g-Interferon gene or admixed with conventional adjuvants. Cancer Res *54*, 6022-6026.

Baskar, S., Ostrand-Rosenberg, S., Nabavi, N., Nadler, L. M., Freeman, G. J., Glimcher, L. H. (1993). Constitutive expression of B7 restores immunogenicity of tumor cells expressing truncated MHC class II molecules. Proc Natl Acad Sci USA *90*, 5687-5690.

Blankenstein, Th. (1994). Increasing tumour immunogenicity by genetic modification. Eur J Cancer *30A*, 1182-1187.

Blankenstein, Th., Cayeux, S., Qin, Z. (1996). Genetic approaches to cancer immunotherapy. Rev. Physiol. Biochem. Pharmacol. *129*: 1-49.

Boon, T., Cerottini, J. C., Van den Eynde, B., van der Bruggen, P., Van Pel, A. (1994). Tumor antigens recognized by T lymphocytes. Annu Rev Immunol *12*, 337-365.

Cavallo, F., Martin-Fontecha, A., Bellone, M., Heltai, S., Gatti, E., Tornaghi, P., Freschi, M., Forni, G., Dellabona, P., Casorati, G. (1995). Co-expression of B7-1 and ICAM-1 on tumors is required for rejection and the establishment of a memory response. Eur J Immunol *25*, 1154-1162.

Cayeux, S., Beck, C., Aicher, A., Dörken B., Blankenstein, Th. (1995). Tumor cells cotransfected with interleukin 7 and B7.1 genes induce CD25 and CD28 on tumor infiltrating lymphocytes and are strong vaccines. Eur J Immunol *25*, 2325-2331.

Cayeux, S., Beck, C., Dörkenl B., Blankenstein, Th. (1996). Coexpression of interleukin 4 and B7.1 in murine tumor cells leads to improved tumor rejection and vaccine effect compared to single gene transfectants and a classical adjuvant. Hum Gene Therapy *7*, 525-529.

Chen, L., Ashe, S., Brady, W. A., Hellström, I., Hellström, K. E., Ledbetter, J. A., McGowan, P., Linsley, P. S. (1992). Costimulation of antitumor immunity by the B7 counterreceptor for the T lymphocyte molecules CD28 and CTLA-4. Cell *71*, 1093-1102.

Chen, L., Linsley, P. S., Hellström, K. E. (1993). Costimulation of T cells for tumor immunity. Immunol Today *14*, 482-486.

Chen, L., McGowan, P., Ashe, S., Johnston, Y., Li, Y., Hellström, I., Hellström, K. E. (1994). Tumor immunogenicity determines the effect of B7 costimulation on T cell-mediated tumor immunity. J Exp Med *179*, 523-532.

Costello, R. H., Brailly, F., Mallet, C., Mawas, D., Olive (1993). Interleukin-7 is a potent co-stimulus of the adhesion pathway involving CD2 and CD28 molecules. Immunology *80*, 451.

Culver, K. W., Ram, Z., Wallbridge, S., Ishii, H., Oldfield, E. H., Blaese, R. M. (1992). In vivo gene transfer with retroviral vector-producer cells for treatment of experimental brain tumors. Science *256*, 1550.

Dranoff, G., Jaffee, E., Lazenby, A., et al. (1993). Vaccination with irradiated tumor cells engineered to secrete murine granulocyte-macrophage colony-stimulating factor stimulates potent, specific, and long-lasting anti-tumor immunity. Proc Natl Acad Sci USA *90*, 3539-3543.

Forni, G., Parmiani, G., Guarini, A., Foa, R. (1994). Gene transfer in tumor therapy. Annals of Oncology *5*, 789-794.

Hock, H., Dorsch, M., Diamantstein, T., Blankenstein, Th. (1991). Interleukin 7 induces CD4+ T cell-dependent tumor rejection. J Exp Med *174*, 1291-1298.

Hock, H., Dorsch, M., Kunzendorf, U., Qin, Z., Diamantstein, T., Blankenstein, Th. (1993a). Mechanisms of rejection induced by tumor cell targeted gene transfer of interleukin-2, interleukin-4, interleukin-7, tumor necrosis factor or interferon-gamma. Proc Natl Acad Sci USA *90*, 2774-2778.

Hock, H., Dorsch, M., Kunzendorf, U., et al. (1993b). Vaccinations with tumor cells genetically engineered to produce different cytokines, effectivity not superior to a classical adjuvant. Cancer Res *53*, 714-716.

Hoover, H. C., Hanna, M. G. (1991). Immunotherapy by active specific immunization, clinical applications. Colon cancer. In, DeVita, V. T., Helman, S., Rosenberg, S. A., eds. Biologic therapy of cancer, prinicples and pratice. J.B. Lippincott Press, New York, pp. 670-681.

Huang, A. Y. C., Golumbek, P., Ahmadzadeh, M., Jaffee, E., Pardoll, D., Levitzky, H. (1994). Role of bone marrow-derived cells in presenting MHC class I-restricted tumor antigens. Science *264*, 961-965.

June, C. H., Bluestone, J. A., Nadler, L. M., Thompson, C. B. (1994). The B7 and CD28 receptor families. Immunol Today *15*, 321-331.

Li, W., Diamantstein, T., Blankenstein, Th. (1990). Lack of tumorigenicity of interleukin 4 autocrine growing cells seems related to the anti-tumor function of interleukin 4. Mol Immunol 27 1331-1337.

Linsley, P. S., Ledbetter, J. A. (1993). The role of the CD28 receptor during T cell response to antigen. Annu Rev Immunol *11*, 191-212.

Lupton, S. D., Brunton, L. L., Kalberg, V. A., Overell, R. W. (1991). Dominant positive and negative selection using a hygromycin phosphotransferase-thymidine kinase fusion gene. Mol. Cell. Biol. *11*, 3374-3378.

Morton, D. L., Foshag, L. J., Hoon, D. S. B., Nizze, J. A., Wanek, L. A., Chang, C., Davtyan, D. G., Gupta, R. K., Elashoff, R., Irie, R. F. (1992). Prolongation of survival in metastatic melanoma after active specific immunotherapy with a new polyvalent melanoma vaccine. Ann Surgery *216*, 463-482.

Mullen, C. A., Coale, M. M., Levy, A. T., Stetler-Stevenson, W. G., Liotta, L. A., Brandt, S., Blaese, R. M. (1992). Fibrosarcoma cells transduced with the IL-6 gene exhibit reduced tumorigenicity, increased immunogenicity, and decreased metastatic potential. Cancer Res *52*, 6020-6024.

Oettgen, H. F., Old, L. J. (1991). The history of cancer immunotherapy. In, De Vita, V. T., Helman, S., Rosenberg, S. A., eds. Biologic Therapy of Cancer, Principles and Practice. JB Lippincott, 87-119.

Ramarathinam, L., Castle, M., Wu, Y., Liu, Y. (1994). T cell costimulation by B7/BB1 induces CD8 T cell-dependent tumor rejection, An important role of B7/BB1 in the induction, recruitment, and effector function of antitumor T cells. J Exp Med *179*, 1205-1214.

Schendel, D. J., Gansbacher, B., Oberneder, R., Kriegmair, M., Hofstetter, A., Riethmüller G., Segurado, O. G. (1993). Tumor-specific lysis of human renal cell carcinomas by tumor-infiltrating lymphocytes. I. HLA-A2 restricted lysis of autologous and allogeneic tumor lines. J Immunol *151*, 4209-4220.

Tepper, R. I., Pattengale, P. K., Leder, P. (1989). Murine interleukin 4 displays potent anti-tumor activity in vivo. Cell *57*, 503-512.

Tepper, R. I., Mule, J. J. (1994). Experimental and clinical studies of cytokine genemodified tumor cells. Hum Gene Therapy *5*, 153-164.

Townsend, S. E., Allison, J. P. (1993). Tumor rejection after direct costimulation of CD8+ T cells by B7-transfected melanoma cells. Science *259*, 368-370.

Van der Bruggen, P., Traversari, C., Chomez, P., Lurquin, C., De Plaen, E., Van den Eynde, B., Knuth, A., Boon, T. (1991). A gene encoding an antigen recognized by cytolytic T lymphocytes on a human melanoma. Science. *254*, 1643-1647.

24. Dosage Impact on Immunotherapy with Cytokine-Gene Modified Cancer Vaccines

Walter Schmidt

Recent progress in the understanding of the local rather than systemic function of immunostimulatory factors like cytokines and the cloning of the respective cDNAs has given rise to a promising new type of cancer vaccine. These vaccines consist of whole tumor cells, transfected or transduced to express cytokine transgene products. In most cases such modified tumor cells have shown reduced tumorigenicity *in vivo*. Strategies against cancer using cytokine-secreting tumor cells are based on the idea that tumors express tumor-specific antigen(s), which are different from or absent in their normal tissue counterparts. In experimental tumor models some of such cancer vaccines generated potent antitumor immunity sufficient to reject tumor challenge or even cure pre-existing micrometastases. Despite these promising findings, the general requirements for the efficient generation of antitumor immunity are not well understood and conflicting results have been obtained in different models. Recently, with the development of high efficiency gene transfer techniques, technical problems such as low level of transgene expression have been overcome. This allows the adjustment of cytokine expression levels over several orders of magnitude. Using the adenovirus-enhanced transferrinfection (AVET) protocol dosage profile curves have been identified for two cytokines, interleukin-2 (IL-2) and granulocyte-macrophage colony-stimulating factor (GM-CSF). These curves reveal the actual amount of IL-2 or GM-CSF produced by the vaccine as the critical parameter for successful anticancer immunization. Comprehensive investigation of other cytokines should establish further dosage curves. In combination with data about the *in vivo* duration of transgene expression and vaccine cell propagation, the possibilities and limitations of other cytokines for active immunotherapy may be identified.

24.1 Introduction

Cancer is the second leading fatal illness. Although the mechanisms underlying the development of malignancies have been identified, advances in curative treatment for almost all of the most prominent solid malignancies like lung or breast cancer, colon carcinoma, melanoma or prostate cancer, are still missing (1). Metastazising tumor cells, rather than the primary tumor, represent the central problem of neoplastic disease because they cannot be surgically removed and are resistant to chemo- or radiotherapy.

Cancer vaccination, also referred to as active immunotherapy, relies on the immune system for the eradication of malignant single cells spread throughout the body. This approach is based on the idea that tumor cells carry distinct (tumor-specific) antigens (2-5) which are (over-)expressed from mutated or rearranged genes (6), from dysregulated genes (7) or genes shut off in non-malignant tissue counterparts (8). By recognizing these antigens, the immune system should be capable of identifying, attacking and finally eradicating tumor cells. Unfortunately, increasing numbers of cancer cases in humans, e.g. in melanoma (9, 10) demonstrate that tumor cells develop strategies to circumvent their elimination. Neoplastic tissue can actively suppress immunocompetent cells by paracrine or systemic release of immunosuppressive factors like TGF-β (11). Furthermore, malignant cells are derived from normal, self tissues and may be tolerated or ignored by the immune system. Detection of malignant cells may simply not occur because (tumor-specific) antigens are presented at levels too low to bring about immunogenicity.

24.1.1 Adjuvants

The aim of cancer vaccines is to stimulate the immune system to eradicate metastatic disease. The low immunogenicity of whole tumor cells or tumor lysates has been documented in numerous clinical trials where cancer vaccines consisting of tumor cells alone failed to generate effective antitumor immunity (12). The first step in the generation of such immunity is the recognition of the vaccine by cells of the immune system. As a second step, immunological pathways have to be activated to generate tumor-specific immunity to combat metastatic disease. From animal studies, it is becoming increasingly clear that these basic requirements are the key issues for a successful immunization protocol. How can the immune system realize the difference between normal cells, e.g. damaged by injury, which simply have to be removed from the organism and vaccine cells applied for the generation of antitumor immunity? To address the critical issues indicated above, attempts have been made to increase the immunogenicity of vaccine cells or cell debris. Early approaches were based on co-application of immunomodulatory adjuvants like Freunds adjuvants (FA), *Bacillus Calmette-Guerin (BCG)* or *Corynebacterium parvum (C. parvum)* with the vaccine cells. The presence of such adjuvants increased the efficacy of the generation of antitumor immunity in experimental animal models (13). However, in clinical trials such adjuvant strategy has not been a breakthrough for the treatment of cancer patients as yet (12, 14, 15), although there are some survival benefits compared to historical controls (16, 17).

24.1.2 Xenogenization upon Viral Infection

Another endeavour to turn tumor cells into efficacious cancer vaccines has been to xenogenize cells by viral infection, especially with the non-lytic strain of the Newcastle disease virus (NDV). After such modification, tumor cell-based vaccines are recognized efficiently by the immune system and induce antitumor immunity. Mice treated with NDV-infected vaccines were shown to develop protective antimetastatic immunity against murine T cell lymphoma (18) and melanoma (19) without being swamped by a prominent antiviral immune response (20). This concept has been applied in clinical trials (21-24). It will be interesting to see whether the apparently improved survival of colon cancer patients can be reproduced for other cancers.

24.1.3 Transfection with Cytokine Genes

The most recent approach for the generation of effective cancer vaccines utilizes cytokine gene expression systems. Tumor cells have been genetically engineered to secrete cytokine gene products and then applied as cancer vaccines in experimental animal tumor models. The results obtained in these studies have prompted clinicians to investigate such vaccines in clinical trials (25-27).

In general, the paracrine release of a cytokine by such a vaccine places the tumor cells in a highly immunogeneic environment which efficiently attracts cells of the immune system. The strength of the resulting immune response is dependent upon several factors. The cell type used in the vaccine and the conditions under which it is applied (live or inactivated) determines the type of tumor antigens, the amount of these antigens and the length of time for which they are available to the immune system. The nature of the cytokine determines the cells of the immune system which are stimulated, but the amount of cytokine determines the intensity of the stimulus. Further parameters which should also be considered are the transfection procedure, the expression construct, whether stably transfected clones or transiently transfected bulk cultures are used, the vaccination protocol and the tumor burden to be eradicated. All these parameters must be taken into account in order to evaluate the potency of a vaccine. Until recently, this has only been partially investigated. This review will focus on experience with this new approach in animal studies and summarize recent advances, its potential and its limitations.

24.2 Reduced Tumorigenicity upon Cytokine Transgene Expression

An important feature of experimental murine tumor cell lines is their intrinsic tumorigenicity and their ability to form tumors upon transplantation into syngeneic hosts. The minimal tumorigenic cell dose which grows to a tumor in all of the injected animals can differ by several orders of magnitude. For instance, for murine melanoma M-3 the minimal cell dose is 5 x 10^3; for plasmacytoma J558L this dose is 10^6. Cytokine transgene-expressing clones have been generated from such cell lines (see below). Reduced or abolished *in vivo* tumor-forming ability of the clones compared to non-cytokine secreting controls indicated that the cytokine evoked reduction of tumorigenicity. The most intensively studied cytokine in this context has been interleukin-2 (28, 29). IL-2 is likely the most important growth factor for T lymphocytes and a potent stimulator of other effector cells including NK cells (30). Tumor-reactive lymphokine-activated killer cells (LAK) and tumor-infiltrating cells (TILs) could be isolated from head and neck neoplasms after intratumoral injection of recombinant IL-2 protein (31). In animal models interleukin-2 evoked regression of various tumors including sarcoma, adenocarcinoma or mammary neoplasm when applied intra- or peritumorally (32-35). After such treatment, spleen cells could be isolated which specifically killed the parental tumor cells *in vitro* and *in vivo* (36). Similar effects have been obtained with IL-4 (37).

These encouraging results with the recombinant protein stimulated animal studies with tumor cells genetically modified to express the murine or human IL-2 cDNA. It was first published in 1990 that the secretion of IL-2 by stably transfected tumor cells abrogated their tumorigenicity. Growth of tumor cells was abolished in sarcoma and plasmacytoma (38), colon carcinoma and melanoma (39) and fibrosarcoma (40). Other studies completed the picture for interleukin-2 by demonstrating loss of tumorigenicity in other experimental tumor models and high amounts of IL-2 reduced the tumorigenicity more efficiently than low amounts (28, 41-49). Other cytokines were also demonstrated to reduce tumorigenicity. IL-3 (50), IL-4 (51, 52), IL-7 (53-55), IL-12 (56, 57), IFN-γ (58-60) and IFN-α (61, 62) or G-CSF (63, 64) have been demonstrated to reduce tumorigenicity when expressed by transplanted tumor cells. However, other cytokines like IL-5 (65), M-CSF (66) did not affect tumor growth and TGF-β expression actually increased tumorigenicity (67, 68). Studies of IL-6 in different tumors revealed a tumor-suppressive activity for fibrosarcoma and lewis lung carcinoma (69, 70) but there was no impact on the growth of a murine plasmacytoma (71). For TNF-α tumor-suppressive as well as metastases-augmenting effects have been described, dependent on the tumor studied (71-73).

Reduction of tumorigenicity has always been a result of increased immunogenicity, as demonstrated by increased influx of immune-competent cells into the transplanted inoculum. The composition of the infiltrate depends upon the cytokine secreted and cancer type. In the adenocarcinoma TS/A, neutrophilic granulocytes dominated the tumor cell deposits accompanied by CD4+ T cells, macrophages and NK cells if IL-2 was secreted (43) while in plasmacytoma J558L, macrophages and CD8+ cells were identified (13).

24.3 Cancer Vaccines in Experimental Animal Models

24.3.1 Pioneering Studies

The demonstration that cytokine-secreting tumor cells were immunogeneic and infiltrated by host cells raised the question whether these infiltrates only destroy the inoculum or whether in addition systemic immunity is generated. The generation of systemic, long-lasting antitumor immunity is the ultimate goal of a cancer vaccine. To date, it is clear from depletion and cell transfer experiments that such immunity is dependent on the cellular arm of the immune system including CD4+ helper and/or cytotoxic CD8+ T cells (reviewed in 74, 75). However, how this immunity is generated remains to be determined in more detail.

The effects of cancer vaccines genetically modified to release cytokines have been studied intensively in various experimental tumor models. However, until recently, no clear picture had emerged what the requirements for a potent vaccine are, since the same cytokine had different effects in different models. In addition, the experimental systems are difficult to compare. The general experimental parameters, like the choice of tumor model and the cytokine to be expressed are highly variable. Other experimental parameters such as cell number injected, total amount of cytokine released, numbers of vaccinations, choice of injected compartment, vaccination of naive versus metastasis-bearing mice, cytokine/cell number ratio, live versus inactivated vaccine cells, control groups vary even more.

24.4 Vaccination with Viable Tumor Cells

The first investigations of cancer vaccines in experimental models used viable, stably transfected or transduced tumor cells to immunize naive syngeneic mice against subsequent or preexisting tumor challenge with the respective live parental cells. Cytokine transgene products secreted by these vaccines were either IL-2, IFN-γ or IL-4. All three gene-modified vaccines mediated protection against subsequent challenge, as shown for the IL-2 vaccine in the CMS-5 sarcoma (59) and CT-26 colon carcinoma (39), for IFN-γ in CMS-5 (40) and for IL-4 in CT-26 and in the renal tumor model RENCA (52). Protection was tumor-specific since challenge with an unrelated tumor did not lead to rejection of these tumor cells. With IL-4 modified RENCA cells immunization was demonstrated to be effective against a pre-existing micrometastasis. Depletion experiments suggested that CD8+ T cells are essential for the protective immunity generated by cytokine-secreting cancer vaccines. *In vitro* experiments with bulk spleen cell isolates identified tumor cell-specific lysis activity in vaccinated but not in naive mice. From these observations it was concluded that tumor-specific CTLs are responsible for antitumor immunity.

For interleukin-2, a daring model of how this immunity might be generated was proposed. Vaccine cells were seen as artificially antigen-presenting cells which present tumor-specific antigen(s) molecules to prime both CD8+ T cells in the context of MHC class I and to prime CD4+ T cells in the context of MHC class II molecules (39). The production of IL-2 by the vaccine cells was thought to bypass the function of helper T cells (39) for the generation of antitumor immunity.

However as discussed recently, the design of the study described above was not suitable to conclude that the paracrine-release of IL-2 by the vaccine is responsible for the generation of systemic immunity (29, 74, 76). Controls were not included to distinguish between the impact of the cytokine and that of the tumor cells alone. Viable IL-2 secreting tumor cells are rejected from naive mice whereas parental non-secretors form primary tumors. These primary tumors would have to be removed by surgery and animals would have to be readministered with tumor challenge to identify or exclude induction of immunity in the absence of IL-2 production and accurately define the efficacy of that cytokine.

Application of viable tumor vaccines at cell dosages which form solid tumors in the absence of cytokine transgene expression do not allow one to distinguish whether the cytokine is responsible for both rejection of the vaccine and generation of systemic antitumor immunity. It is also conceivable that the cytokine induces the rejection of the vaccine only, and that the tumor cells *per se* are responsible for specific immunity. To discriminate between the impact of the cytokine on the destruction of the vaccine and on the generation of systemic immunity, proper control groups have to be included. Such controls might consist of inactivated, unmodified tumor cells which ideally should be compared to identically treated, cytokine-producing vaccines.

Interestingly, a subsequent study in murine plasmacytoma J558L and melanoma B16 revealed no difference in the vaccination efficiency between a viable IL-4-secreting cancer vaccine and simply mitomycin C- inactivated wild type tumor cell controls (77). IL-4 might therefore be considered as a cytokine with the capability to render tumor cells immunogeneic, leading to their rejection *in vivo*, but which does not necessarily affect the generation of antitumor immunity.

24.4.1 IL-2 in Murine Mammary Carcinomas

For IL-2, a detailed analysis in the murine mammary adenocarcinoma TS/A revealed a different picture. For vaccination studies, stable clones were generated which expressed IL-2 at six different levels, from 30 to 6800 U/10^5 cells/48h. Cells had not been inactivated with mitomycin C. When injected into syngeneic hosts, 100% of them were rejected except for the clone producing the lowest amount of IL-2 (30 units). Here the rejection efficiency was 40% which was shown to arise from an *in vivo* loss of IL-2 expression by this clone. Surviving animals from all IL-2 groups were tested for their ability to reject tumor challenge. Control animals were vaccinated with mitomycin C-inactivated parental cells or cells transfected with a non-IL-2-producing, empty vector (43). Upon challenge with viable TS/A cells all of the animals from the two control groups developed tumors, indicating that no immunity had been generated. The mice which had received one of the above IL-2 clones were protected to a certain degree. However, the percentage of protected animals was dependent upon the level of IL-2 production by the clones. Treatment with the 30 units IL-2 producing clone protected about 20% of the animals. The other five clones showed a protection efficiency of more than 50%. Among these five there were slightly different protection efficiencies. Interestingly, the highest protection efficiency was not mediated by the clone with the highest IL-2 production (6,800 units), but by one which released intermediate amounts (3,600 units).

From the above investigation it seems that the amount of IL-2 produced by viable vaccines plays an important role 1. for the immunogenicity of the vaccine; 2. for the efficiency by which immunized mice can reject tumor challenge. Low levels of IL-2 are clearly not very effective, but high level of IL-2 expression is achievable only by few gene transfer techniques. The utilization of such a protocol might be commendable to ensure the possibility for the generation of the most potent vaccine (see below).

A second study with another mouse mammary tumor, 4TO7, investigated a live IL-2 gene-modified vaccine engineered by retroviral transduction. Here only one IL-2 vaccine with low production was applied, which, however, achieved a protection efficiency of 80%. Vaccination with irradiated, non-transduced parental cells was used as control and protected only 25% of the animals against the subsequent challenge (46).

These two studies with murine mammary tumors compared viable, IL-2 secreting vaccines with inactivated parental cells. Since the viable cells could persist *in vivo* for some time before rejection they could have replicated and significantly increased the amount of tumor antigen as compared to the inactivated controls. Therefore, there was still the possibility that the function of IL-2 was restricted to the rejection process rather than to the generation of immunity.

24.4.2 Comparison with Adjuvants

A comparative study with five viable, cytokine-gene modified cancer vaccines was performed in the poorly tumorigenic plasmacytoma model J558L (minimal cell number to form tumors in all of injected naive animals: 10^6) (13). The impact of single dosage low level IL-2, IL-4, IL-7, TNF-α and IFN-γ secretion (and combinations of two of them) on the protection against tumor challenge was tested. These cytokine-secreting vaccines were compared with irradiated tumor cells and the classical adjuvant C. parvum, admixed with viable tumor cells. In escalating challenge experiments the adjuvant turned out to be superior to all of the cytokine-releasing vaccines. This finding questioned the usefulness of low dosage cytokine-releasing vaccines for active immunotherapy, especially in the view of the fact that the adjuvant was not successful in clinical trials. On the one hand, at least for the imperfect protection by low dosage IL-2, the result from this plasmacytoma study was in agreement with that of the adenocarcinoma TS/A (43). On the other hand, it would be interesting to know what a comparison between a gene-modified J558L vaccine and adjuvant control would have shown if higher levels of cytokine expression could have been achieved.

24.5 Inactivated Cancer Vaccines

The possibility that a tumor might form from live cytokine-secreting cancer vaccines is not acceptable, particularly in a clinical setting. Therefore, an inactivation step performed by irradiation or mitomycin C treatment is also imperative for animal studies. Furthermore, this inactivation step allows one to distinguish precisely between the efficacy of the paracrine cytokine release and the possible contribution of the tumor cells *per se* to the induction of antitumor immunity.

Such irradiation-inactivated vaccines were employed for the treatment of lung metastases in the murine lewis lung carcinoma 3LL-D122 (45, 70, 78). The human IL-2, IL-6 and murine IFN-γ gene were constitutively expressed by the vaccines at least at two different levels. All three cytokines had a beneficial impact on the efficacy of the vaccine since reduction in the growth of orthotopic lung foci was evoked compared to non gene-modified control vaccines. Interestingly, for all three cytokines the highest expression level was the most effective, suggesting that the overall amount of cytokine released by the vaccine is important for the vaccination efficiency.

In the murine bladder cancer MBT-2, a low level human IL-2 releasing irradiated vaccine worked efficiently and cured 60% of the immunized animals from their orthotopically implanted tumor burden. However, when animals were treated with a vaccine which expressed IFN-γ at a single, low level no benefit compared to animals treated with a non cytokine producing control vaccine was achieved (44).

In the B16-F10 melanoma model a comparative study with seven different cytokines including IL-2, IL-4, IL-6 and IFN-γ identified GM-CSF as a molecule for the generation of a potent cancer vaccine (79). In this study irradiated, retrovirally-transduced tumor cells were applied to vaccinate naive mice against later parental tumor cell challenge. Here the vaccines were not stable clones, but transiently transduced B16-F10 cells expressing the respective cytokine at one level. Due to the very high transduction efficiency of the retroviral MFG-vector system used, the transgene expression from most of the vaccines was very high compared to most earlier studies.

Interestingly, using this MFG retrovirus, GM-CSF turned out to be the most potent cytokine. Protection efficiency of up to 100% was achieved. Compared to this, all other cytokines investigated appeared to be ineffective and not significantly better than the controls of untransduced, irradiated cells. The inability of most cytokines to mediate protection against tumor challenge was somewhat surprising, especially for interleukin-2, which had been effective in most other tumor systems investigated.

In the MBT-2 bladder tumor model however, a comparative study with GM-CSF and IL-2 (and other cytokine-) secreting vaccines showed a result different to that obtained with B16-F10. Here, GM-CSF was not superior to IL-2 (80). The IL-2 secreting vaccine achieved a protection efficiency of 60%, whereas the GM-CSF vaccine protected only 40% of the mice against tumor challenge (the others including IFN-γ were even less effective). Only one expression level for both cytokines was investigated which was approximately one order of magnitude lower than that applied to the B16-F10 melanoma.

These results were confirmed with the rat prostate cancer MatLyLu (81). Here, vaccination against a three day subcutaneous micrometastasis with a low level IL-2 expressing vaccine protected 60-100% of the animals against tumor development for the observation time of 45 days. The low level GM-CSF releasing vaccine showed a protection efficiency of 40% and the presence of low level IFN-γ with the vaccine did not have any benefit as compared with the standard control vaccine consisting of non-cytokine producing irradiated cells. Taking these last three studies together, the reasons for the success of IL-2 or GM-CSF in generating potent protection in one tumor model but not in the other and vice versa remained unclear. It is not obvious whether there is a correlation between the respective malignancy and the overall expression of the cytokine (which might be a combination of the expression level and *in vivo* duration).

24.6 Cytokine Dosage as the Critical Parameter for Cancer Vaccination

From these inconsistencies, we decided to systematically and comprehensively investigate IL-2 and GM-CSF secreting cancer vaccines (82). Such a study called for vaccines which express the cytokine transgenes over several orders of magnitudes at a constant cell number. Instead of preparing stable clones with laborious and time consuming selection protocols we decided to generate such vaccines by utilizing the adenovirus-enhanced transferrinfection (AVET). AVET is a gene transfer protocol which allows highly efficient transient gene transfer resulting in very high transgene expression levels (83, 84) of up to 1-2 x 10^5 units IL-2 or 5-10 μg GM-CSF/10^6 cells/day in the tumor cells employed for our studies (unpublished results). Transgene expression can be adjusted to a certain level by variations of the basic transfection protocol. In the AVET transfection system, a cytokine gene expression vector is built into a complex of two polylysine-modified components: chemically inactivated adenovirus particles and receptor ligand molecules, such as transferrin. Applied to tumor cells in culture, the complex binds to the cell surface via ligand- and/or adenovirus receptor interaction to enter via the endosomal pathway. The endosomolytic activity of the virus particle mediates the release of the vector into the cytoplasm thereby preventing degradation and paving the way for transgene expression in the nucleus.

The experimental tumor model we choose was the highly tumorigenic murine melanoma M-3 (minimal tumorigenic cell dose 5 x 10^3). Naive DBA/2 mice were vaccinated subcutaneously twice at a weekly interval with 10^5 irradiated cells. A contralateral challenge with 3 x 10^5 viable, parental cells was applied one week after the second immunization. To identify the impact of the cytokine dosage on the vaccination efficacy, 22 individual IL-2 vaccines were prepared, secreting 22 different levels of the cytokine. The *in vitro* IL-2 expression was adjusted to levels from 0.01 to 7,500 units/10^5 cells/day, a range which covered all values reported for other tumor models so far. For GM-CSF, five vaccines were generated with expression levels from 50 pg to 200 ng/10^5 cells/day. Eight mice per group were administered with the same cytokine-secreting vaccine. Control groups received either irradiated tumor cells (10^5 cells/animal) or tumor cells, transfected with a non-expressing vector (mock-DNA) to distinguish between antitumor effects of the tumor cells alone and the enhancement mediated by the presence of the respective cytokine dosage. For IL-2 we found that the vaccination efficiency was significantly dependent upon the IL-2 dosage produced by the vaccine (Fig. 24.1). The best protection against subsequent challenge was achieved when animals were immunized with vaccines producing "medium" IL-2 levels of 1,000 to 3,000 units/10^5 cells; 88-100% of the animals are protected from tumor take. A stepwise decrease of IL-2 production was accompanied by a corresponding decline in the number of protected animals indicating imperfect anti-tumor immunity. Below the expression of 10 units IL-2/10^5/day, the efficiency approached the background protection of the control groups (approximately 10-15%).

Interestingly, when a vaccine expressed "high" interleukin-2 levels ($\geq$ 5,000 units/10^5 cells/day), no more protection of animals was observed. Application of 5,000 units IL-2/10^5 producing cells allowed the growth of parental M-3 cells in seven out of eight animals challenged and with higher dosages in even eight out of eight animals. This is in agreement with the result from the B16-F10 melanoma, in which the high level IL-2 secreting vaccine failed to mediate protection (79).

With GM-CSF releasing M-3 cells, the dosage impact curve is completely different from the IL-2 based vaccine. Immunity was already high at moderate secretion levels and came to a plateau at maximally obtainable GM-CSF production (Fig. 24.1). This seemed to indicate that GM-CSF is a less critical molecule for the generation of cancer vaccines. Once the dosage is beyond a certain threshold, generation of optimal protection can be expected, whereas for IL-2 expression there is a relatively sharp optimum. This may not only be influenced by the expression level but also by the duration of the IL-2 expression *in vivo*. For the AVET-based M-3 IL-2

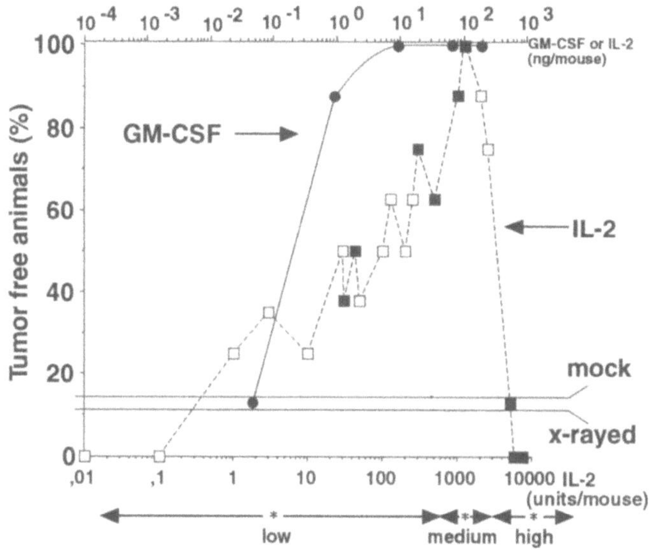

Fig. 24.1: IL-2 and GM-CSF dosage curves.
The broken line shows the protection efficiency of 22 different IL-2 secreting vaccines as tumor free animals in percent (eight animals treated per group), whereas the solid curve depicts the response to GM-CSF secretion. The protection levels (■) and (•) were derived from transfected, irradiated cells alone, values given as (□) from mixtures of irradiated, transfected M-3 cells with irradiated, nontransfected M-3 cells.
X-rayed, control group receiving irradiated M-3 cells as vaccine (46/52 animals tumor positive). mock, control group receiving mock-DNA transfected M-3 cells as vaccines (22/26 animals tumor-positive).

vaccine, the duration of expression was found to be only two to three days for the most effective M-3 IL-2 vaccine (85).

Differences in the persistence of the *in vivo* cytokine expression in a certain experimental tumor may be the reason why different IL-2 expression levels have been described for optimal protection efficacy (86, 87) for colon carcinoma CT-26.

In addition, in these CT-26 studies, human IL-2 was provided in trans by a stably transduced fibroblast line. Irradiated CT-26 tumor cells as the source of CT-26 tumor antigens were admixed with the fibroblasts and injected to immunize naive animals against tumor challenge. With this approach, the optimal IL-2 expression was found to be in the range of 10 to 50 units per vaccine which is about two orders of magnitude lower than the AVET M-3 vaccine. When the fibroblasts released 1,750 units IL-2 per vaccine, a value highly efficacious in the AVET M-3 system, vaccination efficiency was as low as with the non IL-2 producing control vaccine. One possible explanation for this difference is that the M-3 melanoma and CT-26 colon carcinoma require different IL-2 dosage optima. However, it could also be that the *in vivo* duration of IL-2 could be different since transiently transfected M-3 cells were investigated in one case, and stably expressing 3T3 fibroblast clones in the other. It might be speculated that the 3T3 fibroblasts express the IL-2 transgene for a longer period of time than the two to three days determined in the M-3 system. This would mean that different *in vivo* IL-2 expression profiles require adjustment of individual levels, which may make their application for human cancer treatment very difficult.

24.7 Conclusion

The example of IL-2 shows the importance of careful titration of cytokine transgene expression to obtain most effective cancer vaccines. Besides the cytokine itself and its expression level, the protocol by which the vaccines are generated and applied must be taken into account in order to obtain optimal efficacy. These considerations call for comprehensive, comparative animal studies before a certain vaccine formulation can be devised for the treatment of human disease.

Acknowledgements

I would like to acknowledge the support and advice of Max L. Birnstiel, whose contribution to both this review and our work described in it have been invaluable. I would also like to thank Ciaran Morrison and Michael Buschle for critical reading of the manuscript.

References

1. Beardsley, T. A (1994). War not won. Scientific American. January, 118-26.
2. Lurquin, C., Van Pel, A., Mariame, B., De Plaen, E., Szikora, J. P., Janssens, C., Reddehase, M. J., Lejeune, L., and Boon, T. (1989). Structure of the Gene of Tum- Transplantation Antigen P91A: The mutated Exon Encodes a Peptide Recognized with Ld by Cytolytic T Cells. Cell. *58*, 293-303.
3. Boon, T., Cerottini, J. C., Van den Eynde, B., van der Bruggen, P., and Van Pel, A. (1994). Tumor antigens recognized by T lymphocytes. Annu Rev Immunol. *12*, 337-65.
4. Lethe, B., van den Eynde, B., van Pel, A., Corradin, G., and Boon, T. (1992). Mouse tumor rejection antigens P815A and P815B: two epitopes carried by a single peptide. Eur. J. Immunol. *22*, 2283-8.
5. Van den Eynde, B., Lethe, B., Van Pel, A., De Plaen, E., and Boon, T. (1991). The gene coding for a major tumor rejection antigen of tumor P815 is identical to the normal gene of syngeneic DBA/2 mice. J. Exp. Med. *173*, 1373-84.
6. Mandelboim, O., Berke, G., Fridkin, M., Feldman, M., Eisenstein, M., and Eisenbach, L. (1994). CTL induction by a tumour-associated antigen octapeptide derived from a murine lung carcinoma. Nature *369*, 67-71.
7. Roth, C., Rochlitz, C., and Kourilsky, P. (1994). Immune Response against Tumors. Advances in Immunology *57*, 281-351.
8. van der Bruggen, P., Traversari, C., Chomez, P., Lurquin, C., De Plaen, E., Van den Eynde, B., Knuth, A., and Boon, T. A (1991). Gene encoding an antigen recognized by cytolytic T lymphocytes on a human melanoma. Science *254*, 1643-7.
9. Burton, R. C. and Armstrong, B. K. (1994). Recent incidence trends implya non-metastasizing form of invasive melanoma. Melanoma Research *4*, 107-113.
10. Burton, R. C. (1995). Analysis of public education and the implications with regard to nonprogressive thin melanoma. Current Opinion in Oncology *7*, 170-174.

11. Sulitzeanu, D. (1993). Immunosuppressive factors in human cancer. Adv Cancer Res. *60*, 247-67.
12. Oettgen, H. F. and Old, L. J. (1991). The history of cancer immunotherapy. Biologic Therapy of cancer. Principles and Practice.
13. Hock, H., Dorsch, M., Kunzendorf, U., Uberla, K., Qin, Z., Diamantstein, T., and Blankenstein, T. (1993). Vaccinations with tumor cells genetically engineered to produce different cytokines: effectivity not superior to a classical adjuvant. Cancer Res. *53*, 714-6.
14. Hanna, M. G., Jr., Ransom, J. H., Pomato, N., Peters, L., Bloemena, E., Vermorken, J. B., and Hoover, H. C., Jr. (1993). Active specific immunotherapy of human colorectal carcinoma with an autologous tumor cell/bacillus Calmette-Guerin vaccine. Ann N Y Acad Sci. *690*, 135-46.
15. Berd, D., Maguire, H. C., Jr., McCue, P., and Mastrangelo, M. J. (1990). Treatment of metastatic melanoma with an autologous tumor-cell vaccine: clinical and immunologic results in 64 patients. J Clin Oncol. *8*, 1858-67.
16. Morton, D. L., Hoon, D. S., Nizze, J. A., Foshag, L. J., Famatiga, E., Wanek, L. A., Chang, C., Irie, R. F., Gupta, R. K., and Elashoff, R. (1993). Polyvalent melanoma vaccine improves survival of patients with metastatic melanoma. Ann N Y Acad Sci. *690*, 120-34.
17. Barth, A. and Morton, D. L. (1995). The role of adjuvant therapy in melanoma management. Cancer *75*, 726-34.
18. Heicappell, R., Schirrmacher, V., von Hoegen, P., Ahlert, T., and Appelhans, B. (1986). Prevention of Metastatic Spread by Postoperative Immunotherapy with Virally Modified Autologous Tumor Cells. I. Parameters for Optimal Therapeutic Effect. Int. J. Cancer. *37*, 569-577.
19. Plaksin, D., Porgador, A., Vadai, E., Feldman, M., Schirrmacher, V., and Eisenbach, L. (1994). Effective anti-metastatic melanoma vaccination with tumor cells transfected with MHC genes and/or infected with Newcastle disease virus (NDV). Int. J. Cancer *59*, 796-801.
20. von Hoegen, P., Weber, E., and Schirrmacher, V. (1988). Modification of tumor cells by a low dose of Newcastle Disease Virus. Augmentation of the tumor-specific T-cell resonse in the absence of an anti-viral response. Eur. J. Immunol. *18*, 1159-1166.
21. Schlag, P. M., Milleck, J., and Liebrich, W. (1994). Progress with Tumour Vaccines. Clin. Immunother. *2*, 23-31.
22. Schlag, P., Manasterski, M., Gerneth, T., Hohenberger, P., Dueck, M., Herfarth, C., Liebrich, W., and Schirrmacher, V. (1992). Active specific immunotherapy with Newcastle-disease-virus-modified autologous tumor cells following resection of liver metastases in colorectal cancer. First evaluation of clinical response of a phase II-trial. Cancer Immunol. Immunother. *35*, 325-30.
23. Liebrich, W., Schlag, P., Manasterski, M., Lehner, B., Stohr, M., Moller, P., and Schirrmacher, V. (1991). *In vitro* and clinical characterisation of a Newcastle disease virus-modified autologous tumour cell vaccine for treatment of colorectal cancer patients. Eur. J. Cancer *27*, 703-10.
24. Lehner, B., Schlag, P., Liebrich, W., and Schirrmacher, V. (1990). Postoperative active specific immunization in curatively resected colorectal cancer patients with a virus-modified autologous tumor cell vaccine. Cancer Immunol. Immunother. *32*, 173-8.
25. protocols, C. (1996). Human Gene Therapy. *7*, 567-588.
26. protocols, C. (1996). Cancer gene therapy. *3*, 58-68.
27. Stingl, G., Bröcker, E.-B., Mertelsmann, R., Wolff, K., Schreiber, S., Kämpgen, R., Schneeberger, A., Dummer, W., Brennscheid, U., Veelken, H., Birnstiel, M. L., Zatloukal, K., Schmidt, W., Maas, G., Wagner, E., Buschle, M., Giese, M., Kempe, E.-R., Weber, H. A., and Voigt, T. (1996). Phase I Study to the Immunotherapy of Metastatic Malignent Melanoma by a Cancer Vaccine Consisting of Autologous Cancer Cells Transfected with the Human IL-2 Gene. Human gene Therapy *7*, 551-63.
28. Zatloukal, K., Schmidt, W., Cotten, M., Wagner, E., Stingl, G., and Birnstiel, M. L. (1993). Somatic gene therapy for cancer: the utility of transferrinfection in generating 'tumor vaccines'. Gene. *135*, 199-207.
29. Colombo, M. P. and Forni, G. (1994). Cytokine gene transfer in tumor inhibition and tumor therapy: where are we now?. Immunol Today. *15*, 48-51.
30. Trinchiri, G. (1989). Biology of natural killer cells. Adv. Immunol. *47*, 187-201.
31. Rivoltini, L., Gambacorti Passerini, C., Squadrelli Saraceno, M., Grosso, M. I., Cantu, G., Molinari, R., Orazi, A., and Parmiani, G. (1990). *In vivo* interleukin 2-induced activation of lymphokine-activated killer cells and tumor cytotoxic T-cells in cervical lymph nodes of patients with head and neck tumors. Cancer Res. *50*, 5551-7.
32. Pizza, G., Severini, G., Menniti, D., De Vinci, C., and Corado, F. (1984). Tumour regression after intralesional injection of IL-2 in bladder cancer. Int. J. Cancer. *34*, 359.

33. Bubenik, J., Perlmann, P., Indrova, M., Simova, J., Jandlova, T., and Neuwirt, J. (1983). Growth inhibition by interleukin-2 at the tumour/hostinterface. Cancer Immunol. Immunother. *14*, 205.
34. Vaage, J. (1987). Local and systemic effects during Interleukin-2 Therapy of Mouse Mammary tumors. Cancer Research. *47*, 4296-8.
35. Vaage, J. (1988). Local Interleukin 2 Therapy of Mouse Mammery Tumors of Various Immunogenicities. Cancer Research. *48*, 2193-7.
36. Forni, G., Ciovarelli, M., and Santoni, A. (1985). Lymphokine-activated tumor inhibition *in vivo*. J. Immunol. *134*, 1305-1311.
37. Bosco, M., Giovarelli, M., Forni, M., Modesti, A., Scarpa, S., Masuelli, L., and Forni, G. (1990). Low doses of IL-4 injected perilymphatically in tumor-bearing mice inhibit the growth of poorly and apparently nonimmunogenic tumors and induce a tumor-specific immune memory. J. Immunol. *145*, 3136-43.
38. Bubenik, J., Simova, J., and Jandlova, T. (1990). Immunotherapy of cancer using local administration of lymphoid cells transformed by IL-2 cDNA and constitutively producing IL-2. Immunol. Lett. *23*, 287-92.
39. Fearon, E. R., Pardoll, D. M., Itaya, T., Golumbek, P., Levitsky, H. I., Simons, J. W., Karasuyama, H., Vogelstein, B., and Frost, P. (1990). Interleukin-2 production by tumor cells bypasses T helper function in the generation of an antitumor response. Cell *60*, 397-403.
40. Gansbacher, B., Zier, K., Daniels, B., Cronin, K., Bannerji, R., and Gilboa, E. (1990). Interleukin 2 gene transfer into tumor cells abrogates tumorigenicity and induces protective immunity. J. Exp. Med. *172*, 1217-24.
41. Ohe, Y., Podack, E. R., Olsen, K. J., Miyahara, Y., Ohira, T., Miura, K., Nishio, K., and Saijo, N. (1993). Combination effect of vaccination with IL2 and IL4 cDNA transfected cells on the induction of a therapeutic immune response against Lewis lung carcinoma cells. Int J Cancer *53*, 432-7.
42. Ley, V., Langlade Demoyen, P., Kourilsky, P., and Larsson Sciard, E. L. (1991). Interleukin 2-dependent activation of tumor-specific cytotoxic T lymphocytes *in vivo*. Eur. J. Immunol. *21*, 851-4.
43. Cavallo, F., Giovarelli, M., Gulino, A., Vacca, A., Stoppacciaro, A., Modesti, A., and Forni, G. (1992). Role of neutrophils and CD4+ T lymphocytes in the primary and memory response to nonimmunogenic murine mammary adenocarcinoma made immunogenic by IL-2 gene. J. Immunol. *149*, 3627-35.
44. Connor, J., Bannerji, R., Saito, S., Heston, W., Fair, W., and Gilboa, E. (1993). Regression of bladder tumors in mice treated with interleukin 2 gene-modified tumor cells. J. Exp. Med. *177*, 1127-34.
45. Porgador, A., Gansbacher, B., Bannerji, R., Tzehoval, E., Gilboa, E., Feldman, M., and Eisenbach, L. (1993). Anti-metastatic vaccination of tumor-bearing mice with IL-2-gene-inserted tumor cells. Int. J. Cancer *53*, 471-7.
46. Tsai, S. C., Gansbacher, B., Tait, L., Miller, F. R., and Heppner, G. H. (1993). Induction of antitumor immunity by interleukin-2 gene-transduced mouse mammary tumor cells versus transduced mammary stromal fibroblasts. J. Natl. Cancer Inst. *85*, 546-53.
47. Hock, H., Dorsch, M., Kunzendorf, U., Qin, Z., Diamantstein, T., and Blankenstein, T. (1993). Mechanisms of rejection induced by tumor cell-targeted gene transfer of interleukin 2, interleukin 4, interleukin 7, tumor necrosis factor, or interferon gamma. Proc. Natl. Acad. Sci. USA *90*, 2774-8.
48. Haddada, H., Ragot, T., Cordier, L., Duffour, M. T., and Perricaudet, M. (1993). Adenoviral interleukin-2 gene transfer into P815 tumor cells abrogates tumorigenicity and induces antitumoral immunity in mice. Hum. Gene Ther. *4*, 703-11.
49. Russell, S. J., Eccles, S. A., Flemming, C. L., Johnson, C. A., and Collins, M. K. (1991). Decreased tumorigenicity of a transplantable rat sarcoma following transfer and expression of an IL-2 cDNA. Int. J. Cancer *47*, 244-51.
50. Pulaski, B. A., McAdam, A. J., Hutter, E. K., Biggar, S., Lord, E. M., and Frelinger, J. G. (1993). Interleukin 3 enhances development of tumor-reactive cytotoxic cells by a CD4-dependent mechanism. Cancer Res. *53*, 2112-7.
51. Tepper, R. I., Pattengale, P. K., and Leder, P. (1989). Murine interleukin-4 displays potent anti-tumor activity *in vivo*. Cell *57*, 503-12.
52. Golumbek, P. T., Lazenby, A. J., Levitsky, H. I., Jaffee, L. M., Karasuyama, H., Baker, M., and Pardoll, D. M. (1991). Treatment of established renal cancer by tumor cells engineered to secrete interleukin-4. Science *254*, 713-6.
53. Hock, H., Dorsch, M., Diamantstein, T., and Blankenstein, T. (1991). Interleukin 7 induces CD4+ T cell-dependent tumor rejection. J. Exp. Med. *174*, 1291-8.

54. McBride, W. H., Thacker, J. D., Comora, S., Economou, J. S., Kelley, D., Hogge, D., Dubinett, S. M., and Dougherty, G. J. (1992). Genetic modification of a murine fibrosarcoma to produce interleukin 7 stimulates host cell infiltration and tumor immunity. Cancer Res. *52*, 3931-7.
55. Aoki, T., Tashiro, K., Miyatake, S., Kinashi, T., Nakano, T., Oda, Y., Kikuchi, H., and Honjo, T. (1992). Expression of murine interleukin 7 in a murine glioma cell line results in reduced tumorigenicity *in vivo*. Proc. Natl. Acad. Sci. USA *89*, 3850-4.
56. Martinotti, A., Stoppacciaro, A., Vagliani, M., Melani, C., Spreafico, F., Wysocka, M., Parmiani, G., Trinchieri, G., and Colombo, M. P. (1995). CD4 T cells inhibit *in vivo* the CD8-mediated immune response against murine colon carcinoma cells transduced with interleukin-12 genes. Eur. J. Immunol. *25*, 137-46.
57. Tahara, H., Zeh, H. J. r., Storkus, W. J., Pappo, I., Watkins, S. C., Gubler, U., Wolf, S. F., Robbins, P. D., and Lotze, M. T. (1994). Fibroblasts genetically engineered to secrete interleukin 12 can suppress tumor growth and induce antitumor immunity to a murine melanoma *in vivo*. Cancer Res. *54*, 182-9.
58. Watanabe, Y., Kuribayashi, K., Miyatake, S., Nishihara, K., Nakayama, E. I., Taniyama, T., and Sakata, A. A. (1989). Exogenous expression of mouse interferon gamma cDNA in mouse neuroblastoma C1300 cells results in reduced tumorigenicity by augmented anti-tumor immunity. Proc. Natl. Acad. Sci. USA *86*, 9456-9460.
59. Gansbacher, B., Bannerji, R., Daniels, B., Zier, K., Cronin, K., and Gilboa, E. (1990). Retroviral vector-mediated gamma-interferon gene transfer into tumor cells generates potent and long lasting antitumor immunity. Cancer Res. *50*, 7820-5.
60. Esumi, N., Hunt, B., Itaya, T., and Frost, P. (1991). Reduced tumorigenicity of murine tumor cells secreting gamma-interferon is due to nonspecific host responses and is unrelated to class I major histocompatibility complex expression. Cancer Res. *51*, 1185-9.
61. Ferrantini, M., Proietti, E., Santodonato, L., Gabriele, L., Peretti, M., Plavec, I., Meyer, F., Kaido, T., Gresser, I., and Belardelli, F. (1993). Alpha 1-interferon gene transfer into metastatic Friend leukemia cells abrogated tumorigenicity in immunocompetent mice: antitumor therapy by means of interferon-producing cells. Cancer Res. *53*, 1107-12.
62. Ferrantini, M., Giovarelli, M., Modesti, A., Musiani, P., Modica, A., Venditti, M., Peretti, E., Lollini, P. L., Nanni, P., Forni, G., and Bellardelli, F. (1994). IFN-alpha 1 gene expression into a metastatic murine adenocarcinoma (TS/A) results in CD8+ T cell-mediated tumor rejection and development of antitumor immunity. Comparative studies with IFN-gamma-producing TS/A cells. J. Immunol. *153*, 4604-15.
63. Colombo, M. P., Ferrari, G., Stoppacciaro, A., Parenza, M., Rodolfo, M., Mavilio, F., and Parmiani, G. (1991). Granulocyte colony-stimulating factor gene transfer suppresses tumorigenicity of a murine adenocarcinoma *in vivo*. J. Exp. Med. *173*, 889-97.
64. Stoppacciaro, A., Melani, C., Parenza, M., Mastracchio, A., Bassi, C., Baroni, C., Parmiani, G., and Colombo, M. P. (1993). Regression of an established tumor genetically modified to release granulocyte colony-stimulating factor requires granulocyte-T cell cooperation and T cell-produced interferon gamma. J. Exp. Med. *178*, 151-61.
65. Kruger Krasagakes, S., Li, W., Richter, G., Diamantstein, T., and Blankenstein, T. (1993). Eosinophils infiltrating interleukin-5 gene-transfected tumors do not suppress tumor growth. Eur. J. Immunol. *23*, 992-5.
66. Dorsch, M., Hock, H., Kunzendorf, U., Diamantstein, T., and Blankenstein, T. (1993). Macrophage colony-stimulating factor gene transfer into tumor cells induces macrophage infiltration but not tumor suppression. Eur. J. Immunol. *23*, 186-90.
67. Torre Amione, G., Beauchamp, R. D., Koeppen, H., Park, B. H., Schreiber, H., Moses, H. L., and Rowley, D. A. (1990). A highly immunogenic tumor transfected with a murine transforming growth factor type beta 1 cDNA escapes immune surveillance. Proc. Natl. Acad. Sci. USA *87*, 1486-90.
68. Chang, H. L., Gillett, N., Figari, I., Lopez, A. R., Palladino, M. A., and Derynck, R. (1993). Increased transforming growth factor beta expression inhibits cell proliferation *in vitro*, yet increases tumorigenicity and tumor growth of Meth A sarcoma cells. Cancer Res. *53*, 4391-8.
69. Mullen, C. A., Coale, M. M., Levy, A. T., Stetler Stevenson, W. G., Liotta, L. A., Brandt, S., and Blaese, R. M. (1992). Fibrosarcoma cells transduced with the IL-6 gene exhibited reduced tumorigenicity, increased immunogenicity, and decreased metastatic potential. Cancer Res. *52*, 6020-4.
70. Porgador, A., Tzehoval, E., Katz, A., Vadai, E., Revel, M., Feldman, M., and Eisenbach, L. (1992). Interleukin 6 gene transfection into Lewis lung carcinoma tumor cells suppresses the malignant

phenotype and confers immunotherapeutic competence against parental metastatic cells. Cancer Res. *52*, 3679-86.
71. Blankenstein, T., Qin, Z. H., Uberla, K., Muller, W., Rosen, H., Volk, H. D., and Diamantstein, T. (1991). Tumor suppression after tumor cell-targeted tumor necrosis factor alpha gene transfer. J. Exp. Med. *173*, 1047-52.
72. Asher, A. L., Mule, J. J., Kasid, A., Restifo, N. P., Salo, J. C., Reichert, C. M., Jaffe, G., Fendly, B., Kriegler, M., and Rosenberg, S. A. (1991). Murine tumor cells transduced with the gene for tumor necrosis factor-alpha. Evidence for paracrine immune effects of tumor necrosis factor against tumors. J. Immunol. *146*, 3227-34.
73. Qin, Z., Kruger Krasagakes, S., Kunzendorf, U., Hock, H., Diamantstein, T., and Blankenstein, T. (1993). Expression of tumor necrosis factor by different tumor cell lines results either in tumor suppression or augmented metastasis. J. Exp. Med. *178*, 355-60.
74. Dranoff, G. and Mulligan, R. C. Gene Transfer as Cancer Therapy. Advances in Immunology *58*, 417-454.
75. Pardoll, D. M. (1995). Paracrine cytokine adjuvants in cancer immunotherapy. Annu. Rev. Immunol. *13*, 399-415.
76. Tepper, R. I. and Mule, J. J. (1994). Experimental and clinical studies of cytokine gene-modified tumor cells. Hum. Gene Ther. *5*, 153-64.
77. Tepper, R. I., Coffman, R. L., and Leder, P. (1992). An eosinophil-dependent mechanism for the antitumor effect of interleukin-4. Science *257*, 548-51.
78. Porgador, A., Bannerji, R., Watanabe, Y., Feldman, M., Gilboa, E., and Eisenbach, L. (1993). Antimetastatic vaccination of tumor-bearing mice with two types of IFN-gamma gene-inserted tumor cells. J. Immunol. *150*, 1458-70.
79. Dranoff, G., Jaffee, E., Lazenby, A., Golumbek, P., Levitsky, H., Brose, K., Jackson, V., Hamada, H., Pardoll, D., and Mulligan, R. C. (1993). Vaccination with irradiated tumor cells engineered to secrete murine granulocyte-macrophage colony-stimulating factor stimulates potent, specific, and long-lasting anti-tumor immunity. Proc. Natl. Acad. Sci. USA *90*, 3539-43.
80. Saito, S., Bannerji, R., Gansbacher, B., Rosenthal, F. M., Romanenko, P., Heston, W. D., Fair, W. R., and Gilboa, E. (1994). Immunotherapy of bladder cancer with cytokine gene-modified tumor vaccines. Cancer Res. *54*, 3516-20.
81. Vieweg, J., Rosenthal, F. M., Bannerji, R., Heston, W. D., Fair, W. R., Gansbacher, B., and Gilboa, E. (1994). Immunotherapy of prostate cancer in the Dunning rat model: use of cytokine gene modified tumor vaccines. Cancer Res. *54*, 1760-5.
82. Schmidt, W., Schweighoffer, T., Herbst, E., Maass, G., Berger, M., Schilcher, F., Schaffner, G., and Birnstiel, M. L. (1995). Cancer vaccines: the interleukin 2 dosage effect. Proc. Natl. Acad. Sci. USA *92*, 4711-4.
83. Cotten, M., Wagner, E., Zatloukal, K., Phillips, S., Curiel, D. T., and Birnstiel, M. L. (1992). High-efficiency receptor-mediated delivery of small and large (48 kilobase gene constructs using the endosome-disruption activity of defective or chemically inactivated adenovirus particles. Proc. Natl. Acad. Sci. USA *89*, 6094-8.
84. Wagner, E., Zatloukal, K., Cotten, M., Kirlappos, H., Mechtler, K., Curiel, D. T., and Birnstiel, M. L. (1992). Coupling of adenovirus to transferrin-polylysine/DNA complexes greatly enhances receptor-mediated gene delivery and expression of transfected genes. Proc. Natl. Acad. Sci. USA *89*, 6099-103.
85. Maass, G., Schmidt, W., Berger, M., Schilcher, F., Koszik, F., Schneeberger, A., Stingl, G., Birnstiel, M. L., and Schweighoffer, T. (1995). Priming of tumor-specific T cells in the draining lymph nodes after immunization with interleukin 2-secreting tumor cells: three consecutive stages may be required for successful tumor vaccination. Proc. Natl. Acad. Sci. USA *92*, 5540-4.
86. Fakhari, H., Shawler, D. L., Gjerset, R., Naviaux, R. K., Koziol, J., Royston, I., and Sobol, R. E. (1995). Cytokine Gene Therapy with Interleukin-2-Transduced Fibroblasts: Effects of IL-2 Dose on Anti-Tumor Immunity. Human Gene Therapy. *6*, 591-601.
87. Shawler, D. L., Dorigo, O., Gjerset, R. A., Royston, I., Sobol, R. E., and Fakhrai, H. (1995). Comparison of gene therapy with interleukin-2 gene modified fibroblasts and tumor cells in the murine CT-26 model of colorectal carcinoma. J. Immunother. Emphasis Tumor Immunol. *17*, 201-8.

25. Tumor Suppressor Gene Therapy – Growth Arrest and Programmed Cell Death

Michael Strauss, Karsten Brand and Volker Sandig

25.1 Introduction

Gene therapy for malignant diseases is one of the greatest challenges for those scientists working on methods and strategies for gene delivery and on biological mechanisms which can be employed to induce tumor regression. There is only a limited number of basic strategies which show some promise in preclinical models so far. The two major strategies are cytokine-gene aided tumor vaccination (see 23 and 24) and selective prodrug activation (see chapter 22). Whereas the first strategy relies on the strong immunostimulatory effect of a relatively small number of genetically modified cytotoxic T cells or tumor cells, the second one is based on conversion of a nontoxic prodrug into a toxic product by an enzyme-encoding gene where the toxic effect is excerted also on nontransduced dividing tumor cells due to a so-called bystander effect. Alternatively, strategies can be envisaged where the malignant phenotype of a cell is reverted by either inactivating an oncogene or reestablishing an inactivated tumor suppressor gene. The first of these alternatives would require selective inactivation of a mutant transcript by antisense or ribozyme mechanisms and the second would be based on overexpression of a tumor suppressor gene. In both cases, highly efficient gene transfer to almost 100% of the cells in a tumor is required. Although such high efficencies of gene transfer can be obtained *in vitro* and even *in vivo* under certain circumstances, correction of the malignant phenotype by reverting the major oncogenic change in the tumor cells is unlikely to result in normal cells.

It is an established fact that most tumors have lost the ability to undergo apoptosis which is often considered to be the major reason for escape of tumor cells from endogenous surveilance mechanisms. Thus, it would be interesting to find out if growth arrest and/or apoptosis can be reestablished in tumor cells in vivo in order to induce selective cell death of tumor cells under certain circumstances. In this article, we will discuss briefly the molecular basis of growth deregulation involved in tumor development as it is relevant to the new therapy strategies and we further discuss current approaches towards growth inhibition by tumor suppressor gene transfer. Finally, we present first results of combined treatment of tumor cells with two different tumor suppressors mediating growth arrest and subsequent apoptosis.

25.2 Mechanisms of Cell Cycle Regulation

Several genetic changes in different oncogenes or tumor suppressor genes have to occur before a normal cell turns malignant. Most of the oncogenes or tumor suppressors are involved in a certain percentage of tumors but none of them turned out to be involved in the development of every single tumor. There is now growing evidence for the idea that only a very limited number of regulatory pathways exist which are involved in the key decisions about cell's fate including control of the cell cycle. The common feature of multistep carcinogenesis may be the abrogation of at least one of the components of each pathway.

25.2.1 Oncogenes and Tumor Suppressors

A number of extracellular factors have been identified which either stimulate or repress cell division. A cascade of phosphorylation events transmits growth-regulating signals from membrane receptors through the cytoplasm to the nucleus where the signals cause specific changes in gene expression (1-3). Several members of these signal transduction pathways are known or suspected protooncogenes which can be turned into an oncogene by mutations leading to hyperfunction. Tumor suppressors, on the other hand, are genes which actively suppress cell growth and the loss of which leads to tumor development. The existence of tumor suppressors has first been suggested when it was shown that various fusions of normal and malignant cells led to suppression of tumorigenicity (4, 5). The idea of the existence of recessive cancer genes or tumor suppressors was first confirmed by isolation of the retinoblastoma susceptibility gene (*Rb1*). This discovery stimulated the search for tumor suppressors enormously. Since then several other tumor suppressor genes were isolated and investigation of the functions of their gene products is one of the most popular fields in cancer cell biology. The function of the *Rb1* gene product pRb is strongly related to the regulation of cell division (6-9). The protein was originally discovered as one which binds to transforming proteins of DNA tumor viruses like E1a of adenovirus, large T-antigen of SV40 and E7 of oncogenic human papilloma viruses. Later it was shown that pRb is phosphorylated in a cell-cycle dependent manner. It is nonphosphorylated in the early G_1 phase. Nonphosphorylated pRb forms complexes with the transcription factor E2F which is released from this complex by phosphorylation in late G_1 suggesting that its growth inhibiting function is based on binding of E2F which could either result in formation of an active repressor of E2F-dependent genes in G_0/G_1 or in sequestration of normally activating E2F. The current view is that both alternative mechanisms may function with different or even one and the same gene. In addition, stimulation of genes without E2F-binding sites by pRb was shown (9). A larger number of genes encoding proteins of different classes are obviously regulated by pRb which is an indication for its function as a master regulator of genes involved in the transition from a resting differentiated cell to a growing and dividing cell (6-9).

25.2.2 Cell Cycle Regulation by Cyclins and Cyclin-Dependent Kinases

Sequential activation of certain genes is required to promote the progression from the quiescent state (G_0) through G_1 into the phase of DNA replication (S) and subsequently through a second gap phase (G_2)into mitosis (M) (10). It seems obvious that every phase of the cell cycle requires the activity of a defined set of genes which are active only for a certain period of time. One mechanism to ensure this ordered cascade of events is phase-specific phosphorylation and dephosphorylation of transcriptional regulators. These phase-specific phosphorylations are carried out by enzymes called cyclin-dependent kinases (cdks). These cdk catalytic subunits form complexes with regulatory subunits called cyclins (11). The cyclins turn the cdk into an active enzyme and determine the substrate specificity of the kinase by binding specifically to certain target proteins. The cyclins are short-lived proteins whose mRNA and/or protein levels accumulate in a certain phase of the cell cycle.

25.2.3 Inhibitors of Cyclin-Dependent Kinases

Recently, a new class of growth regulators has been identified which are inhibitors of cyclin-dependent kinases (CKI) (12). Whereas one group of CKI including p21, p27, and p57 can inhibit several cdks, members of the second group including p15, p16, p18, and p19 are specific for cdk4/cdk6 (12). p15 was shown to be induced by TGF-ß (13) suggesting that negative growth regulators indeed transmit their signals to pRb by inhibiting its phosphorylation. It seems likely that the other CKIs are induced by so far unknown negative growth-regulating signals. Positive regulators of the kinases (cyclins) and negative regulators (CKIs) seem to compete stoichiometrically for binding to cdks which is at least true for the G_1-specific complexes of cdk4/cdk6 with either D cyclins or p16 (14). This suggests that the balance of growth factors in the environment of the cell would decide about the proportion of cdk4/cdk6 which is in complex with either cyclin D or p16 thereby determining the extent of phosphorylation of pRb. Another tumor suppressor, p53, seems to exert at least one of its functions by inhibiting phosphorylation of pRb via induction of p21 (15-17). This well-characterized response to DNA-damaging agents is characteristic for the so-called check point in late G_1. Thus, it appears that the major target of p53-mediated growth arrest is restriction point control.

25.2.4 G1-Phase Control and the Restriction Point

The term "restriction point" (R) was introduced by Pardee to define the position in the G_1-phase where further progression towards S-phase becomes independent of growth factors (18). A labile R-protein was proposed which must reach a threshold level at the R point. After the restriction point the further progression through the cycle would be entirely dependent on endogenous regulation. D-type cyclins are specific for the G_1 phase and could, therefore, be involved in the regulation of the progression through the restriction point. They are inducible by growth factors (11) and may, therefore, be sensors for positive growth-regulating signals. The only known target for cyclin D is the product of the retinoblastoma susceptibility gene, pRb (19, 20), a well characterized tumor suppressor (5-9). Overexpression of cyclin D1 stimulates progression into the S-phase and shortens the G_1 phase (21, 22). Cyclin D1 is only required for cell-cycle progression in cells which have a functional Rb protein implying that it acts upstream of Rb (23). D1 directs cdk4 or cdk6 to pRb via protein-protein interaction and the cyclin/cdk complex phosphorylates pRb at least in vitro (24). Phosphorylation of pRb begins in mid/late G_1 and, according to the current view, turns the active protein into an inactive one.

pRb functions as the major known repressor of G_1 phase progression (10-12). At least one negatively acting growth factor, TGF-beta, has been shown to function by preventing phosphorylation of pRb probably through inhibition of cdk4 synthesis or activity (25-27). This suggests that pRb serves as both a mediator of exogenous signals and a master regulator of several downstream genes. The schematic drawing of the Rb regulatory pathway is shown in Figure 25.1.

25.2.5 Deregulation of the Restriction Point in Most Tumors

It is particularly striking that the gene coding for p16 was subsequently shown to be a tumor suppressor which is lost at a high rate in certain tumors and even more frequently in tumor-derived cell lines (28, 29). Loss of p16 function can occur by various mechanisms including point mutation, deletion and also by silencing through hypermethylation (30). Most recent studies have demonstrated that the loss of Rb and p16 occurs in a mutually exclusive manner and function of one of the two tumor suppressor genes is lost in almost every tumor cell line (31-34). Interestingly, reintroduction of a functional p16 gene in p16-deficient but Rb-positive tumor cells leads to growth-arrest in G1 whereas a mutant p16 gene derived from a tumor has no growth-inhibiting effect (35-37). Alternatively, considerable overexpression of cyclin D1 has been observed in a significant percentage of tumors of various types due to chromosomal translocation, gene amplification and proviral integration (38, 39). Also, amplification or point mutation of cdk4 have been reported (40-42) with the latter causing loss of complex formation with p16 and therefore resulting in consti-

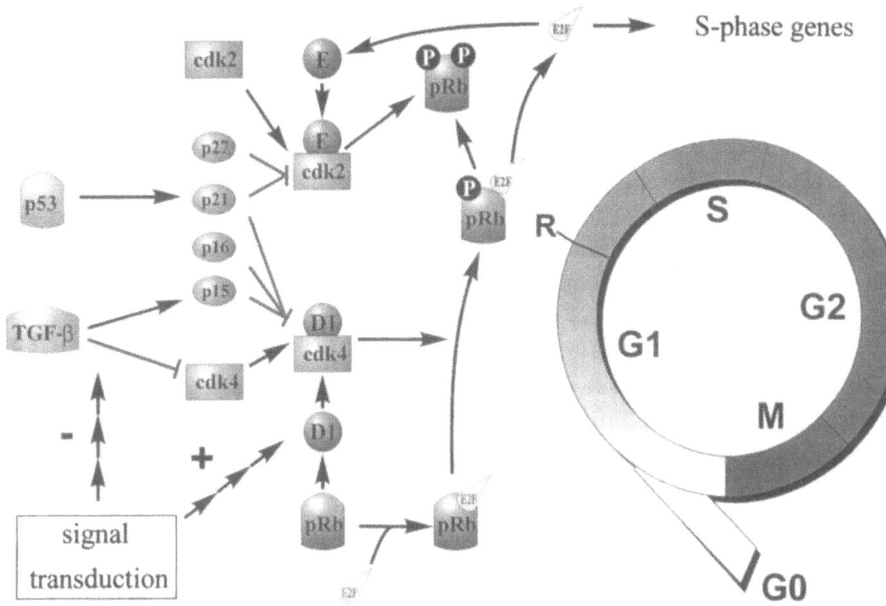

Fig. 25.1: Regulation of G1/S phase progression in the mammalian cell cycle.
According to our current knowledge, pRb is the principal regulator of the G1 restriction point (R). In response to mitogenic stimuli phosphorylation of pRb is primed by cyclin D/cdk4 complexes allowing the release of E2F and transcription of S phase genes. Further phosphorylation by cyclin E/cdk2 is required for S phase progression. A critical step in the activation of cyclin/cdk complexes is the phosphorylation by cdk activating kinase (CAK). Inhibitory signals are transmitted by the action of two protein families of inhibitory peptides (INK4-family and p21-family) that bind to and block the activity of cyclin/cdk comlexes (modified from (9)).

tutive activity of cdk4 (42). Hyperactivity of the cyclin D/cdk4 complex should have the same effect as loss of p16. Overexpression of cyclin D1 may not only be due to chromosomal rearrangements but it could potentially also be caused by oncogene activation further upstream in the signal transduction cascade. Thus, the pathway regulating the restriction point of the cell cycle is composed of pRb as the actual target, cyclin D as the sensor for positive growth-regulating signals and p16 as well as related CKIs as the sensor(s) for negative signals. Cdk4 or cdk6 would serve as a signal mediator. Loss of either Rb or p16 (p15) or, alternatively, overexpression of a D-type cyclin or cdk4 appears to be a common step in tumor development. Whereas the majority of tumors probably carry a mutation or rearrangement in one of the genes of this pathway (31-34, 39) or even in two of the genes (43), in some cases overexpression of D-cyclins may be caused by another oncogene activation in the signal cascade. These results suggest that at least one of the members of the restriction point-controlling Rb pathway has to be deregulated in every tumor (44).

25.3 Regulation of Apoptosis

Most differentiated tissues are continously regenerating with a certain percentage of their cells. In order to keep the number of cells in a solid tissue constant simultaneous death of cells is required. This programmed cell death or apoptosis has first been detected in the liver and was later extensively studied in lymphoid cells. A number of genes have been identified which are involved in the induction or prevention of apoptosis (45). One of the most extensively studied gene products in this context is the tumor suppressor p53. This protein was originally identified as a partner in complexes with SV40 T-antigen which also binds pRb. Simultanous inactivation of pRb and p53 by SV40 and other DNA tumor viruses was suggested to be the sole basis for virus-induced malignant transformation (5). The expression of p53 is strongly induced by irradiation and other DNA-damaging treatments. p53, in turn, induces the expression of a number of other genes including p21 (46) which was shown to be a general inhibitor of cyclin-dependent kinases. Thus, it appears that p53 excerts cell-cycle inhibition in the G_1 phase by inhibiting phosphorylation of pRb through p21-mediated block of cdk4 activity. However, p53 was also shown to induce apoptosis in late G1 which probably occurs when DNA damages cannot be repaired properly. On the other hand, apoptosis in the absence of p53 has been shown suggesting that p53 is an inducer of apoptosis but might not be required for apoptosis and, therefore, not part of the apoptotic machinery (45). There is currently a controversy about the role of p53 in apoptosis because some authors find induction of the major apoptosis-mediating protein *bax* and others do not find it. Thus, we will not go into the details of the apoptotic pathway itself and will rather discuss the role of p53 as an inducer of apoptosis. The apoptotic role of p53 may be its most important contribution to the suppression of tumor cell growth. Where the response of normal cells to p53 seems to be cell cycle arrest, malignant cells are probably more sensitive to p53-induced apoptosis. Thus, p53 positive tumor cells might generally be susceptible to radiation- or chemotherapy-induced apoptosis.

There is a strong tendency to lose p53 function in Rb-deficient cells. To our knowledge, there is no tumor cell line and probably no tumor (besides retinoblastomas) which is Rb-deficient and still p53 positive. Loss of G_1 phase control allows for cell multiplication for a few rounds but would probably be recognized as a serious deregulation by p53, may be due to accumulation of mutations, and would result in apoptosis. Since deregulation of the Rb pathway in cases of cyclin D overexpression or loss of p16 is less pronounced than in the case of loss of Rb, inactivation of p53 is obviously not essential in the first two situations. The presence of a functional Rb has been related to prevention of apoptosis (47, 9). Thus, it appears that cell cycle control and regulation of apoptosis are tightly coordinated by pRb and p53. Loss of both functions probably guarantees unrestricted growth without apoptotic cell death. The question arises whether reconstitution of one or the other tumor suppressor could result in reconstitution of growth control and/or apoptosis.

25.4 Tumor Suppressor Gene Therapy

25.4.1 Rb Gene Transfer

The Rb1 gene is inactivated in all retinoblastomas as well as in various tumors of the adult such as small cell lung carcinoma, prostate, bladder, and mammary carcinomas. Loss of the *Rb1* gene is probably the most efficient means for inactivation of the stringent control of cell cycle progression (43). It occurs at some stage in the course of tumor development and does not need to be the initial event. Regardless of the other genetic changes which have occured in a particular tumor, the loss of Rb function provides the final growth advantage. It should be noted that, apart from retinoblastomas, most or even all Rb-deficient tumors are also p53 negative. The reason for this is that p53 would drive Rb-deficient cells into apoptosis. Thus, such cells would be unable to survive.

Introduction of the normal Rb1 gene has been done not only into retinoblastoma cells but also into osteosarcoma cells. Ectopic Rb1 gene expression affected cell morphology, growth rate, formation of colonies in soft agar, and tumor growth in nude mice (48, 49). However, it is worth mentioning that overexpression of Rb causes efficient growth arrest in most cell types which prevents selection of stable clones to large extent. Thus, inducible Rb gene expression has been suggested as an alternative way to establish Rb-reconstituted cell lines (50). Transfer of the wild-type Rb1 gene has, therefore, not been carried out in a tumor gene therapy setting yet. However, a mutant Rb gene has been used to inhibit hyperproliferation in restenosis (51).

25.4.2 p53 Gene Transfer

The p53 protein is expressed in most normal tissues. It was shown to be involved in control of the cell cycle, transcriptional regulation, DNA replication and induction of apoptosis (52-54). The p53 gene can suppress cell transformation and malignant cell growth (55-58). Mutations in the p53 gene usually lead to an overabundance of the p53 protein caused by increased protein stability and correlates with poor prognosis (59). In addition to its frequent mutation in somatic tissues, p53 germ-line mutations also occur and are associated with the familial cancer called Li-Fraumeni syndrome (60). p53 was shown to cause cell cycle arrest in the G1-phase upon irradiation by inducing the synthesis of a 21kD protein which was found to bind to and inhibit cyclin-dependent kinases. This p21 was dubbed cyclin-interacting protein 1 (Cip1) or wild-type p53-activated fragment 1 (WAF1) by the discoverers (46, 61). As mentioned above, p21 provides the link between p53 and the Rb regulatory pathway. Patients with the Li-Fraumeni syndrome are heterozygous for the mutation, but their tumors are generally homozygous for the mutant allel. Knockout mice with a homozygous deletion of the p53 gene develop a variety of tumors at the age of 6 months whereas heterozygous mice developed less tumors at a slower rate (62). This shows not only that p53 is a real tumor suppressor but also that reestablishment of normal p53 gene function might inhibit tumor growth.

Several earlier studies demonstrated that ectopic expression of the wild-type p53 gene inhibited proliferation *in vitro* of human tumor cell lines either lacking p53 completely or expressing mutant p53 (56-58). p53 gene-transduced tumor cells were less tumorigenic in nude mice. Intratracheal instillation of a retroviral vector carrying the p53 gene prevented the growth of orthotopic human lung cancer cells in nude mice (63). Using a similar retroviral vector in a spheroid tumor model showed that the vector was able to penetrate into the inner tumor mass and also to induce significant apoptotic tumor cell death in lung cancer cells (64). Lung cancers and also the malignant lesions of the epithelium of the esophagus have frequently mutated p53 suggesting that these malignancies could be primary targets for p53 gene transfer by instillation of a suitable vector. Since retroviral vectors have several inherent difficulties in gene delivery *in vivo*, adenoviral vectors with the p53 gene were generated (65). A first-generation E1-deficient vector with the p53 gene under control of the CMV promoter was introduced into human non-small cell lung carcinoma cells deficient in p53 (66). Almost 100% gene transfer was achieved at multiplicities of infection of 50 pfu/cell and high level p53 protein expression was detected. Whereas growth of p53-deficient tumor cells was significantly inhibited almost no effect was found in those cells which had no defect in the p53 gene. If tumor cells were infected at lower doses of the virus (10-15 pfu/cell), significant growth inhibition but no sign of apoptosis was observed in lung cancer lines H358 and H322. However, apoptosis could be induced by subsequent treatment of these cells with radiation or with cisplatin (67). In contrast, direct induction of apoptosis was observed in a lung metastasis line derived from the osteosarcoma line SAOS-2 at the same low titre of the virus (68). Local administration of adenoviral vector with the p53 gene and systemic application of cisplatin generally resulted in a strong tumor-inhibiting effect with massive local apoptosis (67). Tumor-suppressing effects of p53 vectors were demonstrated in various animal models (69, 70). Recently, the first clinical trial for application of p53 gene transfer was published (71). In this study, Roth and colleagues used a retroviral vector to deliver the p53 gene into lung carcinomas of end-stage patients who all died within a few weeks after treatment. In some patients, clear signs of tumor regression correlating with apoptotic cell death were detectable. However, the extent of tumor cell death and regression of the tumor mass is difficult to asses from this study. Since retroviral vectors are not expected to deliver the therapeutic gene very efficiently within the tumor tissue, some apoptosis bystander has to be hypothesised in order to explain the significant effects reported. Lung cancer cells are not only frequently deficient in p53 but also susceptible to the induction of apoptosis by overexpressed p53 making this tumor particularly suitable for gene therapy by p53. The probable reasons for that will be discussed below.

25.4.3 Kinase Inhibitor Gene Transfer

It has been shown that p53 inhibits cell cycle progression via induction of p21 (46, 61, 72-74) and cells overexpressing p21 accumulate in the G_1 phase of the cell cycle (75). Mice lacking p21 have defects in the G_1 checkpoint induced by DNA damage (76) suggesting a role of p21 in mediating p53-induced growth arrest. Thus, it was tempting to test the suitability of p21 as a potential tumor suppressor in gene transfer experiments. Various tumor cell lines were engineered for tetracyclin-regulated expression of p21. Induced expression inhibited tumor cell proliferation *in vitro* and growth of tumors in nude mice (77) indicating the growth-inhibitory potential of p21. An adenoviral vector carrying the p21 gene was generated and delivered to a variety of tumor cell lines. These studies clearly demonstrated that the p21-encoding adenoviral vector had similar growth-inhibiting effects as the p53 vector but failed to induce apoptosis. This result not only indicates that p53-induced apoptosis requires an additional activity of p53 apart from induction of p21 but also suggests that p21 would be less suitable than p53 for tumor therapy where killing of the tumor cells is the ultimate goal. However, data from comparative studies on p53-deficient mouse prostate tumors indicate that growth inhibition of tumors by p21 is more pronounced than the one caused by p53 and, in addition, survival rates of animals who received the p21 vector were much higher than of those who received the p53 vector (78). Thus, the potential side effects of p53 gene transfer in some systems may argue for the use of the gene with a more narrowly targeted function even if it only leads to slowing down of tumor growth.

Another inhibitor of cyclin-dependent kinases, $p16^{INK4}$ (MTS1), is a *bona fide* tumor suppressor. The function of the p16 gene is lost in a variety of tumor types amounting to about 60% of all tumors and reaching 80-90% loss of function in pancreatic tumors and melanomas (28-33, 79-86). As discussed above, the p16 gene product specifically inhibits cdk4 and, thereby, the phosphorylation of pRb (87,88). Thus, it functions directly upstream of pRb (35, 36). Therefore, loss of pRb and loss of p16 are alternative events. Rb-deficient tumors are generally positive for p16 and p16-deficient tumors have normal pRb (44, 45). Thus, reestablishment of p16 function in tumor cells would restore normal growth control (35). Again, Roth and coworkers generated an adenoviral vector with the p16 gene and introduced it into three different non-small cell lung cancer cell lines with homozygous deletions of the p16 gene (89). Cell cycle progression was efficiently inhibited in the G_1 phase in all three tumor cell lines tested *in vitro*. The same vector was also tested in an *in vivo* setting. Human lung cancer cells H460 (5×10^6) were injected into the dorsal flank of nude mice. Tumor nodules (180-220mm^3) were treated 16 days later by direct injection of the Ad-p16 vector. Treatment was repeated at day 18 and day 20. Tumor growth was retarded by about 50% compared to animals injected with a control vector and 34% compared with PBS injected ones (89). This result confirms the tumor suppressing function of p16 but also indicates that no significant regression of tumors can be obtained by this tumor suppressor.

25.4.4 Combined Tumor Suppressor Gene Transfer

As discussed above, cell cycle control and regulation of programmed cell death are interacting mechanisms which partially involve the same players. Since neither p53 nor Rb, p21, and p16 on their own could induce regression of tumors of different types, we decided to try a combinatorial approach which should involve both growth arrest and apoptosis and should function in a variety of tumors. One reason why p53 on its own is not able to induce both events at the same time or subsequently is the failure to induce p21 by overexpression of p53 in several cell lines tested (our unpublished data). The idea was to inhibit cell cycle progression by p16 and to induce apoptosis by p53. It was an open question whether G_1 arrest is really required for efficient apoptosis. However, this question should be answered in the course of the experiments. The first experiments were aiming at the selection of the most efficient vectors. Interestingly, we obtained different results for various AdCMVp16 vectors with regard to growth inhibition in Huh7 hepatocellular carcinoma cells and LOVO colon carcinoma cells. All vector isolates were able to block proliferation in G_1. However, with two virus stocks we did not only see growth arrest but also some degree of apoptosis at day three after infection. It turned out that these two viruses expressed much higher levels of p16 than other stocks which were significantly above the levels expressed in normal cells. Since the apoptotic effect was more pronounced in LOVO cells (normal p53) than in Huh7 cells (mutant p53) we assumed that high level p16 expression cooperates with p53 in induction of apoptosis. We looked next for the effect of p16 overexpression on phosphorylation of pRb by Western blot analysis. This experiment had a surprising outcome. We observed not only a complete block of pRb phosphorylation but also a dramatic reduction of pRb protein levels (90). Since autorepression of pRb expression under certain conditions has been observed before (91) it appeared very likely that the efficient downmodulation of pRb levels was due to the complete inhibition of pRb phosphorylation. We also concluded that downmodulation of pRb levels causes susceptibility towards apoptotic stimuli which is in accordance with the previously demonstrated apoptosis-protective function of pRb (90). Additional overexpression of p53 should aid in driving the cells into apoptotic cell death. Thus, we carried out adenoviral cotransfer of p16 and p53 genes which led to increased apoptosis rates *in vitro* in various tumor cell lines but did not induce apoptosis in normal human fibroblasts. Application of AdCMVp16 and AdCMVp53 to tumors grown subcutanously from the cell line Huh7 in nude mice demonstrated that both vectors on their own induced some growth retardation whereas the combined action of the two tumor suppressors resulted in decrease in tumor size and in complete disappearance of the tumors in some cases within two weeks after application (Fig. 25.2). Although multiple injections (10 to 12) into the

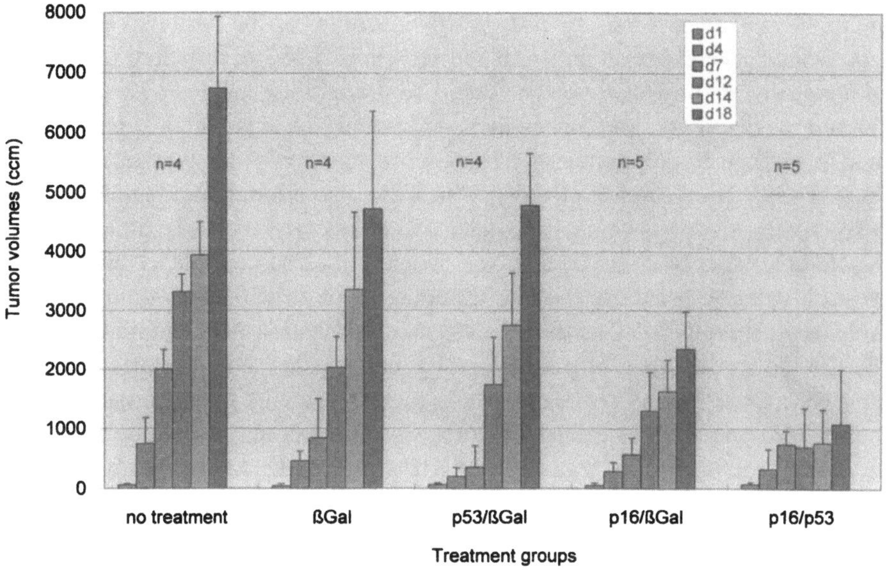

Fig. 25.2: Tumor growth in mice after combinatorial transduction with p16 and p53 genes. Subcutaneous tumors in nude mice were generated by injection of $3x10^6$ human hepatocellular carcinoma cells (HUH7). After 15 days, when tumor nodules (10-100 mm^3) were grown, intratumoral injection of $6x10^9$ pfu of adenoviruses as indicated was performed and repeated 4 days later. Bars represent the means; SDs are indicated.

tumors were performed, 100% infection of tumor cells was very unlikely to occur. Thus, the results suggest that the apoptotic mechanism induced in the transduced tumor cells may have a bystander effect. This requires further exploration. In summary, this new method of combined delivery of two tumor suppressor genes by highly efficient adenoviral vectors offers a strategy for tumor therapy based on two subsequent steps, growth arrest and apoptosis. Whereas growth arrest on its own has already proven to be insufficient for complete inhibition of tumor growth, there is reason to assume that induction of apoptosis on its own might be insufficient without prior release from growth arrest. The combined action of p16 and p53 seems to mimmick events occuring at the G1 check point in p53-positive cells.

25.5 Future Developments

Among the alternative strategies for gene therapy of malignant diseases, tumor suppressor gene transfer stands out due to the natural mechanisms of growth repression which are targeted with the aim to kill tumor cells by programmed cell death as compared to immunologically or suicide-mediated cell death. Since application of single tumor suppressors seems to be restricted to the minority of tumors, the combination of two complementing genes appears to be very efficient in a broad spectrum of tumors. Given the fact that all pRb-positive tumors can be inhibited in the G_1 phase by ectopic overexpression of p16 and presence of p53 is sufficient for induction of apoptosis, p53-positive tumors would be driven into apoptosis by p16 gene transfer and overexpression only. However, additional overexpression of p53 would facilitate this effect. In p53-deficient tumors, a combination of the two genes would be required with the exception of those which are also Rb-deficient. On the other hand, Rb-deficient tumors which are mostly p53 deficient at the same time because they would otherwise be killed by apotosis, would not be susceptible to this kind of combination therapy. For those tumors, introduction of an Rb gene would be most favorable, most suitably a dominant negative phosphorylation-deficient mutant.

Whereas the combination of p16 and p53 genes appears to be the ideal tool for targeting 85% of all tumors at present, various improvements are imaginable. First of all, alternative members of the same regulatory pathways may substitute for them, e.g. the INK4 proteins p15, p18 and p19 or KIP proteins like p27 could replace p16 and members of the bcl-2 family like bax or nbk might replace p53. Further so far unknown members of the two regulatory mechanisms might be possible candidates as well. More work is required to identify alternative representatives of the two complementary groups which are able to serve the same purpose. Combinatorial gene therapy approaches have been tried already with cytokine and suicide genes (92, 93). Combinations of immunostimulatory genes and tumor suppressor genes might have some advantages.

The major hurdle for a successful application of the combined tumor suppressor strategy for growth arrest and apoptosis is the efficiency of gene transfer. Even if there is a significant bystander effect involved in the procedure, fairly efficient gene transfer *in vivo* is required. Adenoviral vectors seem to have the required efficacy. Their well established immune-stimulatory properties might even be advantageous for stimulation of anti-tumor effects. Nonviral as well as retroviral vectors seem to be too inefficient for this purpose.

25.6 Outlook

Tumor gene therapy by tumor suppressor gene transfer is a new field which, in contrast to immunovaccination or suicide gene transfer, requires efficient gene transfer into the majority of tumor cells. Since it appears that apoptotic tumor cells induce a kind of bystander effect on neighboring tumor cells, 100% gene transfer is probably not required to cause regression of tumors. Adenoviral vectors seem to be the most suitable delivery system for tumor suppressors because they are not only the most efficient vectors for gene delivery into tumor tissue *in vivo* but also cause a T-cell response against infected cells which can aid in tumor regression. Further improvements of this strategy should include optimization of the procedure of vector application to the tumor but also improvements of the vector itself. One important modification of the adenoviral vector system was recently suggested by successful selective replication of E1B-deficient adenovirus in p53-deficient tumor cells (94). Whereas such a virus without a therapeutic gene would be able to kill p53-deficient tumor cells by replication, various modifications are imaginable to enable a therapeutic virus to spread out in tumor tissues independend of the p53 status. The final result should be the induction of apoptotic tumor cell death without harming a significant percentage of normal cells. The advanced knowledge about regulation of cell cycle and apoptosis and about deregulation of these processes in tumor cells will allow for the design of improved strategies for targeting of the deregulated pathways by tumor suppressor gene transfer.

Acknowledgement

Work described here was supported by the Danish Cancer Society, Max Planck Society and Fonds der Chemischen Industrie.

References

1. Marshall, C. J. (1995). Specificity of receptor tyrosine kinase signaling: transient versus sustained extracellular signal-regulated kinase activation. Cell *80*, 179-185.
2. Hill, C. S., Wynne, J., and Treisman, R. (1995). The Rho family GTPases RhoA, Rac1, and CDC42Hs regulate transcriptional activation by SRF. Cell *81*, 1159-1170.
3. Hunter, T. (1995). When is a lipid kinase not a lipid kinase? When it is a protein kinase. Cell *83*, 1-4.
4. Harris, H. (1993). How tumour suppressor genes were discovered. FASEB J. *7*, 978-979.
5. Hinds, P. W. and Weinberg, R. A. (1994). Tumor suppressor genes. Curr. Opin. Genet. Dev. *4*, 135-141.
6. Goodrich, D. W. and Lee, W. H. (1993). Molecular characterization of the retinoblastoma susceptibility gene. Biochim. Biophys. Acta *1155*, 43-61.
7. Weinberg, R. A. (1995). The retinoblastom protein and cell cycle control. Cell *81*, 323-330.
8. Bartek, J., Bartkova, J., and Lukas, J. (1996). The retinoblastoma protein pathway and the restriction point. Curr. Opin. Cell. Biol. *8*, 805-814.
9. Herwig, S. and Strauss, M. (1997). The retinoblastoma protein- a master regulator of cell cycle, differentiation and apoptosis. Eur. J. Biochem *246*, 581-601.
10. Pardee, A. B. (1989). G1 events and regulation of cell proliferation. Science *246*, 603-608.
11. Sherr, C. J. (1993). Mammalian G1 cyclins. Cell *73*, 1059-1065.
12. Sherr, C. J. and Roberts, J. M. (1996). Inhibitors of mammalian G1 cyclin-dependent kinases. Genes Dev. *9*, 1149-1163.
13. Hannon, G. J. and Beach, D. (1994). p15INK4B is a potential effector of TGF-beta-induced cell cycle arrest [see comments]. Nature *371*, 257-261.
14. Parry, D., Bates, S., Mann, D. J., and Peters, G. (1995). Lack of cyclin D-Cdk complexes in Rb-negative cells correlates with high levels of p16INK4/MTS1 tumour suppressor gene product. EMBO J. *14*, 503-511.
15. el Deiry, W. S., Tokino, T., Velculescu, V. E., Levy, D. B., Parsons, R., Trent, J. M., Lin, D., Mercer, W. E., Kinzler, K. W., and Vogelstein, B. (1993). WAF1, a potential mediator of p53 tumor suppression. Cell *75*, 817-825.
16. Harper, J. W., Adami, G. R., Wei, N., Keyomarsi, K., and Elledge, S. J. (1993). The p21 Cdk-interacting protein Cip1 is a potent inhibitor of G1 cyclin-dependent kinases. Cell *75*, 805-816.

17. Dulic, V., Kaufmann, W. K., Wilson, S. J., Tlsty, T. D., Lees, E., Harper, J. W., Elledge, S. J., and Reed, S. I. (1994). p53-dependent inhibition of cyclin-dependent kinase activities in human fibroblasts during radiation-induced G1 arrest. Cell *76*, 1013-1023.
18. Pardee, A. B. (1974). A restriction point for control of normal animal cell proliferation. Proc. Natl. Acad. Sci. U. S. A. *71*, 1286-1290.
19. Dowdy, S. F., Hinds, P. W., Louie, K., Reed, S. I., Arnold, A., and Weinberg, R. A. (1993). Physical interaction of the retinoblastoma protein with human D cyclins. Cell *73*, 499-511.
20. Ewen, M. E., Sluss, H. K., Sherr, C. J., Matsushime, H., Kato, J., and Livingston, D. M. (1993a). Functional interactions of the retinoblastoma protein with mammalian D-type cyclins. Cell *73*, 487-497.
21. Quelle, D. E., Ashmun, R. A., Shurtleff, S. A., Kato, J. Y., Bar Sagi, D., Roussel, M. F., and Sherr, C. J. (1993). Overexpression of mouse D-type cyclins accelerates G1 phase in rodent fibroblasts. Genes Dev. *7*, 1559-1571.
22. Resnitzky, D., Gossen, M., Bujard, H., and Reed, S. I. (1994). Acceleration of the G1/S phase transition by expression of cyclins D1 and E with an inducible system. Mol. Cell Biol. *14*, 1669-1679.
23. Lukas, J., Muller, H., Bartkova, J., Spitkovsky, D., Kjerulff, A. A., Jansen Durr, P., Strauss, M., and Bartek, J. (1994). DNA tumor virus oncoproteins and retinoblastoma gene mutations share the ability to relieve the cell's requirement for cyclin D1 function in G1. J. Cell Biol. *125*, 625-638.
24. Kato, J., Matsushime, H., Hiebert, S. W., Ewen, M. E., and Sherr, C. J. (1993). Direct binding of cyclin D to the retinoblastoma gene product (pRb) and pRb phosphorylation by the cyclin D-dependent kinase CDK4. Genes Dev. *7*, 331-342.
25. Ewen, M. E., Sluss, H. K., Whitehouse, L. L., and Livingston, D. M. (1993b). TGF beta inhibition of Cdk4 synthesis is linked to cell cycle arrest. Cell *74*, 1009-1020.
26. Polyak, K., Lee, M. H., Erdjument Bromage, H., Koff, A., Roberts, J. M., Tempst, P., and Massague, J. (1994). Cloning of p27Kip1, a cyclin-dependent kinase inhibitor and a potential mediator of extracellular antimitogenic signals. Cell *78*, 59-66.
27. Ewen, M. E., Oliver, C. J., Sluss, H. K., Miller, S. J., and Peeper, D. S. (1995). p53-dependent repression of CDK4 translation in TGF-beta-induced G1 cell-cycle arrest. Genes Dev. *9*, 204-217.
28. Kamb, A., Gruis, N. A., Weaver-Feldhaus, J., Liu, Q., Harshman, K., Tavtigian, S. V., Stockert, E., Day, R. S., Hohnson, B. E., and Skolnick, M. H. (1994). A cell cycle regulator potentially involved in genesis of many tumor types. Science *264*, 436-440.
29. Nobori, T., Miura, K., Wu, D. J., Lois, A., Takabayashi, K., and Carcon, D. A. (1994). Deletions of the cyclin-dependent kinase-4 inhibitor gene in multiple human cancers. Nature *368*, 753-756.
30. Merlo, A., Herman, J. G., Mao, L., Lee, D. J., Gabrielson, E., Burger, P. C., Baylin, S. B., and Sindransky, D. (1995). 5'CpG island methylation is associated with transcriptional silencing of the tumor suppressor p16/CDKN/MTS1 in human cancer. Nature Med. *1*, 686-692.
31. Otterson, G. A., Kratzke, R. A., Coxon, A., Kin, Y. W., and Kaye, F. J. (1994). Absence of p16^{INK4} protein is restricted to the subset of lung cancer lines that retains wiltype RB. Oncogene *9*, 3375-3378.
32. Okamoto, A., Demetrick, D. J., Spillare, E. A., Hagiwara, K., Hussain, S. P., Bennet, W. P., Forrester, K., Gerwin, B., Serrano, M., Beach, D. H., and Harris, C. C. (1994). Mutations and altered expression of p16^{INK4} in human cancer. Proc. Natl. Acad. Sci. USA *91*, 11045-11049.
33. Tam, S. W., Shay, J. W., and Pagano, M. (1994). Differential expression and cell cycle regulation of the cyclin-dependent kinase 4 inhibitor p16Ink4. Cancer Res. *54*, 5816-5820.
34. Aagaard, L., Lukas, J., Bartkova, J., Kjerulff, A. A., Strauss, M., and Bartek, J. (1995). Aberrations of p16Ink4 and retinoblastoma tumour-suppressor genes occur in distinct sub-sets of human cancer cell lines. Int. J. Cancer *61*, 115-120.
35. Lukas, J., Parry, D., Aagaard, L., Mann, D. J., Bartkova, J., Strauss, M., Peters, G., and Bartek, J. (1995). Retinoblastoma-protein-dependent cell-cycle inhibition by the tumour suppressor p16. Nature *375*, 503-506.
36. Koh, J., Enders, G. H., Dynlacht, B. D., and Harlow, E. (1995). Tumour-derived p16 alleles encoding proteins defective in cell-cycle inhibition. Nature *375*, 506-510.
37. Medema, R. H., Herrera, R. E., Lam, F., and Weinberg, R. A. (1995). Growth suppression by p16ink4 requires functional retinoblastoma protein. Proc. Natl. Acad. Sci. USA *92*, 6289-6293.
38. Motokura, T. and Arnold, A. (1993). Cyclins and oncogenesis. Biochim. Biophys. Acta *1155*, 63-78.
39. Bates, S. and Peters, G. (1995). Cyclin D1 as a cellular proto-oncogene. Semin. Cancer Biol. *6*, 73-82.

40. He, J., Allen, J. R., Collins, V. P., Allalunis Turner, M. J., Godbout, R., Day, R. S., and James, C. D. (1994). CDK4 amplification is an alternative mechanism to p16 gene homozygous deletion in glioma cell lines. Cancer Res. *54*, 5804-5807.
41. Schmidt, E. E., Ichimura, K., Reifenberger, G., and Collins, V. P. (1994). CDKN2 (p16/MTS1) gene deletion or CDK4 amplification occurs in the majority of glioblastomas. Cancer Res. *54*, 6321-6324.
42. Wölfel, T., Hauer, M., Schneider, J., Serrano, M., Wolfel, C., Klehmann Hieb, E., De Plaen, E., Hankeln, T., Meyer zum Buschenfelde, K. H., and Beach, D. (1995). A p16INK4a-insensitive CDK4 mutant targeted by cytolytic T lymphocytes in a human melanoma. Science *269*, 1281-1284.
43. Lukas, J., Aagaatd, L., Strauss, M., Bartek, J. (1995). Oncogenic aberrations of $p16^{INK4/CDKN2}$ and cyclin D1 cooperate to deregulate G1 control. Cancer Res. *55*, 4818-4823.
44. Strauss, M., Lukas, J., Bartek, J. (1995). Unrestricted cell cycling and cancer. Nature Med. *1*, 1245-1246.
45. White, E., (1996). Life, death, and the pursuit of apoptosis. Genes Dev. *10*, 1-15.
46. el Deiry, W. S., Tokino, T., Velculescu, V. E., Levy, D. B., Parsons, R., Trent, J. M., Lin, D., Mercer, W. E., Kinzler, K. W., and Vogelstein, B. (1993). WAF1, a potential mediator of p53 tumor suppression. Cell *75*, 817-825.
47. Haas-Kogan, D. A., Kogan, S. C., Levi, D., Dazin, P., T'Ang, A., Fung, Y.-K., T., Israel, M. A. (1995). Inhibition of apoptosis by the retinoblastoma gene product. EMBO J. *14*, 461-472.
48. Huang, H. J., Yee, J. K., Shew, J. Y., Chen, P. L., Bookstein, R., Friedmann, T., Lee, E. Y., and Lee, W. H. (1988). Suppression of the neoplastic phenotype by replacement of the RB gene in human cancer cells. Science *242*, 1563-1566.
49. Bookstein, R., Shew, J. Y., Chen, P. L., Scully, P., and Lee, W. H. (1990). Suppression of tumorigenicity of human prostate carcinoma cells by replacing a mutated RB gene. Science *247*, 712-715.
50. Gjetting, T., Lukas, J., Bartek, J., Strauss, M. (1995). Regulated expression of the retinoblastoma susceptibility gene in mammary carcinom cells restores cyclin D1 expression and G_1-Phase control. Biol.Chem. Hoppe-Seyler *376*, 441-447.
51. Chang, M. W., Barr, E., Seltzer, J., Jiang, Y. Q., Nabel, J. G., Nabel, E. G., Parmacek, M. S., Leiden, J. M. (1995). Cytostatic gene therapy for vascular proliferative disorders with a constitutively active form of the retinoblastoma gene product. Science *267*, 518-522.
52. Oren, M., Haffner, R. (1995). Biochemical properties and biological effects of p53. Curr. Opin. Gen. Dev. *5*, 84-90.
53. Bates, S. and Vousden, K. H. (1996). p53 in signaling checkpoint arrest or apoptosis. Curr. Opin. Gen. Dev. *6*, 12-19.
54. Ko, L. J., Prives, C. (XXXX). p53: puzzle and paradigm. Genes & Dev. *10*, 1054-1072.
55. Finlay, C. A., Hinds, P. W., and Levine, A. J. (1989). The p53 proto-oncogene can act as a suppressor of transformation. Cell *57*, 1083-1093.
56. Baker, S. J., Markowitz, S., Fearon, E. R., Willson, J. K., and Vogelstein, B. (1990). Suppression of human colorectal carcinoma cell growth by wild-type p53. Science *249*, 912-915.
57. Cheng, J., Yee, J. K., Yeargin, J., Friedmann, T., and Haas, M. (1992). Suppression of acute lymphoblastic leukemia by the human wild-type p53 gene. Cancer Res. *52*, 222-226.
58. Takahashi, T., Carbone, D., Nau, M. M., Hida, T., Linnoila, I., Ueda, R., and Minna, J. D. (1992). Wild-type but not mutant p53 suppresses the growth of human lung cancer cells bearing multiple genetic lesions. Cancer Res. *52*, 2340-2343.
59. Quinlan, D. C., Davidson, A. G., Summers, C. L., Warden, H. E., and Doshi, H. M. (1992). Accumulation of p53 protein correlates with a poor prognosis in human lung cancer. Cancer Res. *52*, 4828-4831.
60. Srivastava, S., Zou, Z. Q., Pirollo, K., Blattner, W., and Chang, E. H. (1990). Germ-line transmission of a mutated p53 gene in a cancer-prone family with Li-Fraumeni syndrome [see comments]. Nature *348*, 747-749.
61. Harper, J. W., Adami, G. R., Wei, N., Keyomarsi, K., and Elledge, S. J. (1993). The p21 Cdk-interacting protein Cip1 is a potent inhibitor of G1 cyclin-dependent kinases. Cell *75*, 805-816.
62. Donehower, L. A., Harvey, M., Slagle, B. L., McArthur, M. J., Montgomery, C. A., Jr., Butel, J. S., and Bradley, A. (1992). Mice deficient for p53 are developmentally normal but susceptible to spontaneous tumours. Nature *356*, 215-221.
63. Cai, D. W., Mukhopadhyay, T., Liu, Y., Fujiwara, T., and Roth, J. A. (1993). Stable expression of the wild-type p53 gene in human lung cancer cells after retrovirus-mediated gene transfer. Hum. Gene Ther. *4*, 617-624.

64. Fujiwara, T., Grimm, E. A., Mukhopadhyay, T., Cai, D. W., Owen Schaub, L. B., and Roth, J. A. (1993). A retroviral wild-type p53 expression vector penetrates human lung cancer spheroids and inhibits growth by inducing apoptosis. Cancer Res. *53*, 4129-4133.
65. Zhang, W. W., Fang, X., Branch, C. D., Mazur, W., French, B. A., and Roth, J. A. (1993). Generation and identification of recombinant adenovirus by liposome-mediated transfection and PCR analysis. Biotechniques *15*, 868-872.
66. Zhang, W.W., Fang, X., Mazur, W., French, B. A., Georges, R. N., and Roth, J. A. (1994). High-efficiency gene transfer and high-level expression of wild-type p53 in human lung cancer cells mediated by recombinant adenovirus. Cancer Gene Ther. *1*, 5-13.
67. Fujiwara, T., Grimm, E. A., Mukhopadhyay, T., Zhang, W. W., Owen Schaub, L. B., and Roth, J. A. (1994b). Induction of chemosensitivity in human lung cancer cells in vivo by adenovirus-mediated transfer of the wild-type p53 gene. Cancer Res. *54*, 2287-2291.
68. Wang, J., Bucana, C. D., Roth, J. A., and Zhang, W. W. (1995). Apoptosis induced in human osteosarcoma cells is one of the mechanisms for the cytocidal effect of Ad5CMV-p53. Cancer Gene Ther. *2*, 9-17.
69. Wills, K. N., Maneval, D. C., Menzel, P., Harris, M. P., Sutjipto, S., Vaillancourt, M. T., Huang, W. M., Johnson, D. E., Anderson, S. C., Wen, S. F., and et al. (1994). Development and characterization of recombinant adenoviruses encoding human p53 for gene therapy of cancer. Hum. Gene Ther. *5*, 1079-1088.
70. Liu, T. J., Zhang, W. W., Taylor, D. L., Roth, J. A., Goepfert, H., and Clayman, G. L. (1994). Growth suppression of human head and neck cancer cells by the introduction of a wild-type p53 gene via a recombinant adenovirus. Cancer Res. *54*, 3662-3667.
71. Roth, J. A., Nguyen, D., Lawrence, D. D., Kemp, B. L., Carrasco, C. H., Ferson, D. Z., Hong, W. K., Komaki, R., Lee, J. J., Nesbitt, J. C., and et al. (1996). Retrovirus-mediated wild-type p53 gene transfer to tumors of patients with lung cancer. Nature Med. *2*, 985-991.
72. Dulic, V., Kaufmann, W. K., Wilson, S. J., Tlsty, T. D., Lees, E., Harper, J. W., Elledge, S. J., and Reed, S. I. (1994). p53-dependent inhibition of cyclin-dependent kinase activities in human fibroblasts during radiation-induced G1 arrest. Cell *76*, 1013-1023.
73. Sherr, C. J. and Roberts, J. M. (1995). Inhibitors of mammalian G1 cyclin-dependent kinases. Genes Dev. *9*, 1149-1163.
74. el Deiry, W. S., Harper, J. W., O'Connor, P. M., Velculescu, V. E., Canman, C. E., Jackman, J., Pietenpol, J. A., Burrell, M., Hill, D. E., Wang, Y., and et al. (1994). WAF1/CIP1 is induced in p53-mediated G1 arrest and apoptosis. Cancer Res. *54*, 1169-1174.
75. Harper, J. W., Elledge, S. J., Keyomarsi, K., Dynlacht, B., Tsai, L. H., Zhang, P., Dobrowolski, S., Bai, C., Connell Crowley, L., Swindell, E., and et al. (1995). Inhibition of cyclin-dependent kinases by p21. Mol. Biol. Cell *6*, 387-400.
76. Deng, C., Zhang, P., Harper, J. W., Elledge, S. J., and Leder, P. (1995). Mice lacking p21CIP1/WAF1 undergo normal development, but are defective in G1 checkpoint control. Cell *82*, 675-684.
77. Chen, Y. Q., Cipriano, S. C., Arenkiel, J. M., and Miller, F. R. (1995). Tumor suppression by p21WAF1. Cancer Res. *55*, 4536-4539.
78. Eastham, J. A., Hall, S. J., Sehgal, I., Wang, J., Timme, T. L., Yang, G., Connell Crowley, L., Elledge, S. J., Zhang, W. W., Harper, J. W., and et al. (1995). In vivo gene therapy with p53 or p21 adenovirus for prostate cancer. Cancer Res. *55*, 5151-5155.
79. Kamb, A., Gruis, N. A., Weaver Feldhaus, J., Liu, Q., Harshman, K., Tavtigian, S. V., Stockert, E., Day, R. S., Johnson, B. E., and Skolnick, M. H. (1994). A cell cycle regulator potentially involved in genesis of many tumor types [see comments]. Science *264*, 436-440.
80. Nobori, T., Miura, K., Wu, D. J., Lois, A., Takabayashi, K., and Carson, D.A. (1994). Deletions of the cyclin-dependent kinase-4 inhibitor gene in multiple human cancers. Nature *368*, 753-756.
81. Hannon, G. J. and Beach, D. (1994). p15INK4B is a potential effector of TGF-beta-induced cell cycle arrest. Nature *371*, 257-261.
82. Marx, J. (1994). A challenge to p16 gene as a major tumor suppressor [news]. Science *264*, 1846.
83. Zhang, S. Y., Klein Szanto, A. J., Sauter, E. R., Shafarenko, M., Mitsunaga, S., Nobori, T., Carson, D. A., Ridge, J. A., and Goodrow, T. L. (1994). Higher frequency of alterations in the p16/CDKN2 gene in squamous cell carcinoma cell lines than in primary tumors of the head and neck. Cancer Res. *54*, 5050-5053.

84. Jen, J., Harper, J. W., Bigner, S. H., Bigner, D. D., Papadopoulos, N., Markowitz, S., Willson, J. K., Kinzler, K. W., and Vogelstein, B. (1994). Deletion of p16 and p15 genes in brain tumors. Cancer Res. *54*, 6353-6358.
85. Washimi, O., Nagatake, M., Osada, H., Ueda, R., Koshikawa, T., Seki, T., and Takahashi, T. (1995). In vivo occurrence of p16 (MTS1) and p15 (MTS2) alterations preferentially in non-small cell lung cancers. Cancer Res. *55*, 514-517.
86. Hussussian, C. J., Struewing, J. P., Goldstein, A. M., Higgins, P. A., Ally, D. S., Sheahan, M. D., Clark, W. H., Jr., Tucker, M. A., and Dracopoli, N. C. (1994). Germline p16 mutations in familial melanoma. Nat. Genet. *8*, 15-21.
87. Serrano, M., Hannon, G. J., and Beach, D. (1993). A new regulatory motif in cell-cycle control causing specific inhibition of cyclin D/CDK4. Nature *366*, 704-707.
88. Serrano, M., Gomez Lahoz, E., DePinho, R. A., Beach, D., and Bar Sagi, D. (1995). Inhibition of ras-induced proliferation and cellular transformation by p16INK4. Science *267*, 249-252.
89. Jin, X., Nguyen, D., Zhang, W. W., Kyritsis, A. P., and Roth, J. A. (1995). Cell cycle arrest and inhibition of tumor cell proliferation by the p16INK4 gene mediated by an adenovirus vector. Cancer Res. *55*, 3250-3253.
90. Sandig, V., Brand, K., Herwig, S., Lukas, J., Bartek, J., Strauss, M. (1997). Adenovirally transferred $p16^{INK/CDKN2}$ and p53 genes cooperate to induce apoptotic tumor cell death. Nature Med. *3*, 313-319.
91. Gill, R. M., Hamel, P. A., Zhe, J., Zacksenhaus, E., Gallie, B. L., and Phillips, P. A. (1994). Characterization of the human Rb1-promotor and of elements involved in transcriptional regulation. Cell Growth Differ. *5*, 467-474.
92. Castleden, S., Helson, J. A., Chong, H., Hart, I., and Vile, G. (1995). The use of combination gene therapies for the treatment of cancer. J. Cell. Biochem. *21(Suppl. A)*, 418.
93. Chen, S. H., Chen, X. H., Wang, Y., Kosai, K., Finegold, M. J., Rich, S. S., and Woo, S. L. (1995). Combination gene therapy for liver metastasis of colon carcinoma in vivo. Proc. Natl. Acad. Sci. USA *92*, 2577-2581.
94. Bischoff, J. R., Kirn, D. H., Williams, A., Heise, C., Horn, S., Muna, M., Ng, L., Nye, J. A., Sampson-Johannes, A., Fattaey, A., McCormick, F. (1996) An adenovirus mutant that replicates selectively in p53-deficient human tumor cells. Science *274*, 373-376.

Contributors

Dr. *Alfred Bahnson*
Department of Human Genetics
University of Pittsburgh
Pittsburgh, PA 15261
USA

Dr. *Edward Ball*
Department of Human Genetics
University of Pittsburgh
Pittsburgh, PA 15261
USA

Prof. Dr. *John A. Barranger*
Department of Human Genetics
University of Pittsburgh
Pittsburgh, PA 15261
USA

Dr. *Christopher Baum*
Department of Cell and Virus Genetics
Heinrich-Pette-Institute
Martinistr. 52
D-20246 Hamburg
Germany

Dr. *Margeret Beeler*
Department of Human Genetics
University of Pittsburgh
Pittsburgh, PA 15261
USA

Prof. Dr. *Thomas Blankenstein*
Max-Delbrück-Center for Molecular
Medicine
Robert-Rössle-Str. 10
D-13122 Berlin
Germany

Dr. *Delphine Bohl*
Laboratoire Rétroviras et Transfert
Génétique
Institut Pasteur
28, rue du Dr. Roux
F-75728 Paris
France

Dr. *Ernst Böhnlein*
SyStemix Inc.
3155 Porter Drive
Palo Alto, CA 94304
USA

Dr. *Karsten Brand*
Humboldt University
Department of Cell Biology at
Max-Delbrück-Center for Molecular
Medicine
Robert-Rössle-Str. 10
D-13122 Berlin
Germany

Dr. *Sophie Cayeux*
Max-Delbrück-Center for Molecular
Medicine
Robert-Rössle-Str. 10
D-13122 Berlin
Germany

Dr. *Günter Cichon*
Humboldt-University at
Max-Delbrück-Center for Molecular
Medicine
D-13122 Berlin
Germany

Prof. Dr. *Charles Coutelle*
Department of Biochemistry and
Molecular Genetics
Imperial College School of Medicine at
St Mary's
GB-London, Paddington W2 1PG
Great Britain

Dr. *Kenneth W. Culver*
OncorPharm, Inc.
200 Perry Parkway
Gaithersburg, MD 20877
USA

Dr. *Jean-François Dedieu*
CNRS URA 1301
Rhône-Poulenc Rorer Gencell
Laboratoire de Génétique des Virus
Oncogènes
Institut Gustave Roussy
39, rue Camille Desmoulins
F-94805 Villejuif Cedex
France

Prof. Dr. *George Dickson*
School of Biological Sciences
Division of Biochemistry
Royal Holloway College
University of London
GB-Egham, Surrey TW20 0EX
Great Britain

Prof. Dr. *Bernd Dörken*
Max-Delbrück-Center for Molecular
Medicine
Robert-Rössle-Str. 10
D-13122 Berlin
Germany

Dr. *James Dunigan*
Department of Human Genetics
University of Pittsburgh
Pittsburgh, PA 15261
USA

Dr. *Hans-Georg Eckert*
Department of Cell and Virus Genetics
Heinrich-Pette-Institute
Martinistr. 52
D-20246 Hamburg
Germany

Prof. Dr. *Fritz Eckstein*
Max-Planck-Institute for Experimental
Medicine
Hermann-Rein-Str. 3
D-37075 Göttingen
Germany

Dr. *Christopher H. Evans*
Department of Molecular Genetics and
Biochemistry
University of Pittsburgh
School of Medicine
Pittsburgh, PA 15261
USA

Dr. *D. J. Fink*
Department of Molecular Genetics and
Biochemistry
University of Pittsburgh
School of Medicine and VAMC
E 1246 Biomedical Science Tower
Pittsburgh, PA 15261
USA

Dr. *Steven C. Ghivizzani*
Department of Molecular Genetics and
Biochemistry

University of Pittsburgh
School of Medicine
Pittsburgh, PA 15261
USA

Dr. *Dalia Ginzberg*
Department of Biological Chemistry
The Life Sciences Institute
The Hebrew University
Jerusalem 91904
Israel

Dr. *Joseph C. Glorioso*
Department of Molecular Genetics and Biochemistry
University of Pittsburgh
School of Medicine and VAMC
E 1246 Biomedical Science Tower
Pittsburgh, PA 15261
USA

Dr. *W. F. Goins*
Department of Molecular Genetics and Biochemistry
University of Pittsburgh
School of Medicine and VAMC
E 1246 Biomedical Science Tower
Pittsburgh, PA 15261
USA

Dr. *Michael Gotthardt*
Max Delbrück Center for Molecular Medicine
Robert-Rössle-Str. 10
D-13125 Berlin
Germany

Dr. *Mirta Grifman*
Department of Biological Chemistry
The Life Sciences Institute
The Hebrew University
Jerusalem 91904
Israel

Prof. Dr. *Walter M. Günzburg*
Institute for Virology
Veterinary University Vienna
Josef-Baumann-Gasse 1
A-1210 Wien
Austria

Dr. *Jean Michel Heard*
Laboratoire Rétroviras et Transfert Génétique
Institut Pasteur
28, rue du Dr. Roux
F-75728 Paris
France

Dr. *Olaf Heidenreich*
Institute for Molecular Biology
Medizinische Hochschule Hannover
D-30623 Hannover
Germany

Dr. *Joachim Herz*
Department of Molecular Genetics
Southwestern University of Texas Medical Center
5323 Harry Hines Blvd.
Dallas, TX 75235
USA

Dr. *Richard Kang*
Department of Molecular Genetics and Biochemistry
University of Pittsburgh
School of Medicine
Pittsburgh, PA 15261
USA

Dr. *Amy Kemp*
Department of Human Genetics
University of Pittsburgh
Pittsburgh, PA 15261
USA

Dr. *D. Krisky*
Department of Molecular Genetics and Biochemistry
University of Pittsburgh
School of Medicine and VAMC
E 1246 Biomedical Science Tower
Pittsburgh, PA 15261
USA

Dr. *Jason Lancia*
Department of Human Genetics
University of Pittsburgh
Pittsburgh, PA 15261
USA

Dr. *Randall Learish*
Department of Human Genetics
University of Pittsburgh
Pittsburgh, PA 15261
USA

Dr. *Efrat Lev-Lehman*
Department of Biological Chemistry
The Life Sciences Institute
The Hebrew University
Jerusalem 91904
Israel

Dr. *Chunming Liu*
Department of Human Genetics
University of Pittsburgh
Pittsburgh, PA 15261
USA

Dr. *Jane Mannion-Henderson*
Department of Human Genetics
University of Pittsburgh
Pittsburgh, PA 15261
USA

Dr. *P. Marconi*
Department of Molecular Genetics and Biochemistry
University of Pittsburgh
School of Medicine and VAMC
E 1246 Biomedical Science Tower
Pittsburgh, PA 15261
USA

Dr. *D. M. McCarty*
Gene Therapy Center
G44 Wilson Hill, CB# 7352
University of North Carolina Chapel Hill
Chapel Hill, NC 27599-7352
USA

Dr. *Joseph Mierski*
Department of Human Genetics
University of Pittsburgh
Pittsburgh, PA 15261
USA

Dr. *Lynn Milich*
Department of Experimental Surgery
Box 2601
Duke University Medical Center
Durham, NC 27710
USA

Dr. *Trina Mohney*
Department of Human Genetics
University of Pittsburgh
Pittsburgh, PA 15261
USA

Dr. *Maya Nimgaonkar*
Department of Human Genetics
University of Pittsburgh
Pittsburgh, PA 15261
USA

Dr. *T. Oligino*
Department of Molecular Genetics and Biochemistry
University of Pittsburgh

School of Medicine and VAMC
E 1246 Biomedical Science Tower
Pittsburgh, PA 15261
USA

Dr. *Cécile Orsini*
CNRS URA 1301
Rhône-Poulenc Rorer Gencell
Laboratoire de Génétique des Virus Oncogènes
Institut Gustave Roussy
39, rue Camille Desmoulins
F-94805 Villejuif Cedex
France

Prof. Dr. *Wolfram Ostertag*
Department of Cell and Virus Genetics
Heinrich-Pette-Institute
Martinistr. 52
D-20246 Hamburg
Germany

Dr. *Michel Perricaudet*
Institut Gustave Roussy
Centre National de la Recherche Scientifique Unité Associée 1301
F-94805 Villejuif
France

Dr. *P. L. Poliani*
Department of Neurology
University of Pittsburgh
School of Medicine and VAMC
E 1246 Biomedical Science Tower
Pittsburgh, PA 15261
USA

Dr. *Alexander L. Rakhmilevich*
Agracetus and Adjunct Pro.
Dept. of Pathology
University of Wisconsin, Medical School
8520 University Green
Middleton, WI 53562
USA

Dr. *R. Ramakrishnan*
Department of Neurology
University of Pittsburgh
School of Medicine and VAMC
E 1246 Biomedical Science Tower
Pittsburgh, PA 15261
USA

Dr. *Erin Rice*
Department of Human Genetics
University of Pittsburgh
Pittsburgh, PA 15261
USA

Prof. Dr. *Paul D. Robbins*
Depts. of Molecular Genetics and Biochemistry
University of Pittsburgh
School of Medicine
Pittsburgh, PA 15261
USA

Dr. *Brian Salmons*
Bavarian Nordic Research Institute and GSF – Forschungszentrum für Umwelt und Gesundheit
Institute for Molecular Virology
D-85758 Oberschleissheim
Germany

Dr. *Richard Jude Samulski*
Gene Therapy Center
Department of Pharmacology
G44 Wilson Hill, CB# 7352
University of North Carolina at Chapel Hill
Chapel Hill, NC 27599-7352
USA

Dr. *Volker Sandig*
Hepa Vec AG at Max-Delbrück-Center for Molecular Medicine
Robert-Rössle-Str. 10
D-13122 Berlin
Germany

Dr. *Walter Schmidt*
Research Institute of Molecular Pathology
Dr.-Bohr-Gasse 7
A-1030 Vienna
Austria

Dr. *Herbert Schuster*
Universitätsklinikum Rudolf Virchow
Franz-Volhard-Klinik
Wiltbergstraße 50
D-13125 Berlin
Germany

Prof. Dr. *Hermona Soreq*
Department of Biological Chemistry
The Life Sciences Institute
The Hebrew University
Jerusalem 91904
Israel

Dr. *Carol Stocking*
Department of Cell and Virus Genetics
Heinrich-Pette-Institute
Martinistr. 52
D-20246 Hamburg
Germany

Prof. Dr. *Michael Strauss*
Humboldt-University
Department of Cell Biology at Max-Delbrück-Center for Molecular Medicine
Robert-Rössle-Str. 10
D-13125 Berlin
Germany

Danish Cancer Society
Division of Cancer Biology
Strandboulevarden 49
DK-2100 Copenhagen
Denmark

Dr. *Bruce A. Sullenger*
Department of Experimental Surgery
Box 2606
Duke University Medical Center
Durham, NC 27710
USA

Dr. *Michael J. Vallor*
Department of Human Genetics
University of Pittsburgh
Pittsburgh, PA 15261
USA

Dr. *Emmanuelle Vigne*
CNRS URA 1301
Rhône-Poulenc Rorer Gencell
Laboratoire de Génétique des Virus Oncogènes
Institut Gustave Roussy
39, rue Camille Desmoulins
F-94805 Villejuif Cedex
France

Dr. *Thomas Wagener*
Department of Cell and Virus Genetics
Heinrich-Pette-Institute
Martinistr. 52
D-20246 Hamburg
Germany

Dr. *Simon Watkins*
Department of Cell Biology and Physiology
University of Pittsburgh
Pittsburgh, PA 15261
USA

Dr. *Thomas E. Willnow*
Max-Delbrück-Center for Molecular Medicine
Robert-Rössle-Str. 10
D-13122 Berlin
Germany

Dr. *Catherine H. Wu*
Department of Medicine
Division of Gastroenterology-Hepatology
University of Connecticut
School of Medicine
263 Farmington Avenue
Farmington, CT 06030
USA

Dr. *George Y. Wu*
Department of Medicine
Division of Gastroenterology-Hepatology
University of Connecticut
School of Medicine
263 Farmington Avenue
Farmington, CT 06030
USA

Dr. *Ning-Sun Yang*
Agracetus and Adjunct Pro., Inc.
Dept. of Pathology
University of Wisconsin, Medical School
8520 University Green
Middleton, WI 53562
USA

Dr. *Patrice Yeh*
CNRS URA 1301
Rhône-Poulenc Rorer Gencell
Laboratoire de Génétique des Virus Oncogènes
Institut Gustave Roussy
39, rue Camille Desmoulins
F-94805 Villejuif Cedex
France

Dr. *Haim Zakut*
The Sackler Faculty of Medicine
Tel Aviv University
The Edith Wolfson Medical Center
Holon 58100
Israel

Index

Bei Fragen zur Produktsicherheit wenden Sie sich bitte an:
If you have any questions regarding product safety,
please contact:

Walter de Gruyter GmbH
Genthiner Straße 13
10785 Berlin
productsafety@degruyterbrill.com